AA001009

2007 International Conference on Power Electronics and Drive Systems

Bangkok, Thailand
27-30 November 2007

Pages 1-479

IEEE Catalog Number:	**CFP07PEL-PRT**
ISBN 10:	1-4244-0644-7
ISBN 13:	978-1-4244-0644-9

Copyright © 2007 by The Institute of Electrical and Electronics Engineers, Inc.
All Rights Reserved

Copyright and Reprint Permissions: Abstracting is permitted with credit to the source. Libraries are permitted to photocopy beyond the limit of U.S. copyright law for private use of patrons those articles in this volume that carry a code at the bottom of the first page, provided the per-copy fee indicated in the code is paid through Copyright Clearance Center, 222 Rosewood Drive, Danvers, MA 01923.

For other copying, reprint or republications permission, write to IEEE Copyrights Manager, IEEE Operations Center, 445 Hoes Lane, Piscataway, New Jersey USA 08854. All rights reserved.

IEEE Catalog Number:	CFP07PEL-PRT
ISBN 10:	1-4244-0644-7
ISBN 13:	978-1-4244-0644-9
LOC:	2006933010

Additional Copies of This Publication Are Available from:

IEEE Service Center
445 Hoes Lane
Piscataway, NJ 08854

Phone:	(800) 678-IEEE
	(732) 981-1393
Fax:	(732) 981-9667
E-mail:	customer-service@ieee.org

Organizers/Committees

Organizers:

Chulalongkorn Univ.
Center of Excellence in Electrical Power Tech., Chulalongkorn Univ.
King Mongkut's Univ. of Tech. Thonburi
King Mongkut's Inst. of Tech. Ladkrabang
King Mongkut's Inst. of Tech. North Bangkok
IEEE Thailand Section
IEEE IAS/PELS Joint Chapter, Singapore Section

Technical Co-Sponsors:

IEEE Power Electronics Society
IEEE Industry Applications Society
IEEE Industrial Electronics Society
IEEE IAS/PELS/IES Joint Chapter, Thailand Section

Organizing Committees

General Chairman	Doncker, R. D.	RWTH-Aachen Univ.
Advisory Board	Lavansiri, D.	Chulalongkorn Univ.
	Jaovisidha, V.	Chulalongkorn Univ.
	Tandhavatana, S.	IEEE Thailand Section
	Pungprasert, V.	IEEE Thailand Section
	Leelarasmee, E.	Chulalongkorn Univ.
	Pichetchumroen, V.	King Mongkut's Inst. of Tech. Ladkrabang
	Yingwatana, A.	King Mongkut's Inst. of Tech. North Bangkok
	Liang, Y. C.	National Univ. of Singapore
	Panda, S. K.	National Univ. of Singapore
General Co-Chairmen	Karnasuta, K.	IEEE Thailand Section
	Vilathgamuwa, D. M.	Nanyang Tech. Univ.
Organizing Committee		
Chairman	Phoomvuthisarn, S.	Center of Excellence in Electrical Power Tech. Chulalongkorn Univ.
Co-Chairman	Kulvitit, Y.	Chulalongkorn Univ.
	Yungyuen, U.	King Mongkut's Univ. of Tech. Thonburi
	Khan-ngern, W.	King Mongkut's Inst. of Tech. Ladkrabang
	Chunkag, V.	King Mongkut's Inst. of Tech. North Bangkok
Technical Program Committee	Khan-ngern, W.	King Mongkut's Inst. of Tech. Ladkrabang
	King Jet, T.	Nanyang Tech. Univ.
	Sangwongwanich, S.	Chulalongkorn Univ.
	Chunkag, V.	King Mongkut's Inst. of Tech. North Bangkok
	Sirisukprasert, S.	Kasetsart Univ.
	Boonyaroonate, I.	King Mongkut's Univ. of Tech. Thonburi
Treasurers	Bunnagulrote, B.	Center of Excellence in Electrical Power Tech. Chulalongkorn Univ.
	Battul, D.	School of Electrical & Electronic Engineering Singapore Polytechnic
Publications	Tarateeraseth, V.	Srinakharinwirot Univ.
	Jangwanitlert, A.	King Mongkut's Inst. of Tech. Ladkrabang
Tutorials	Chunkag, V.	King Mongkut's Inst. of Tech. North Bangkok
	Liutanakul, P.	King Mongkut's Inst. of Tech. North Bangkok
Local Arrangements	Kinnares, V.	King Mongkut's Inst. of Tech. Ladkrabang
	Polmai, S.	King Mongkut's Inst. of Tech. Ladkrabang
	Yutthagowith, P.	King Mongkut's Inst. of Tech. Ladkrabang
	Kittiratsatcha, S.	King Mongkut's Inst. of Tech. Ladkrabang
	Fuengwarodsakul, N.	King Mongkut's Inst. of Tech. North Bangkok
Publicity	Suwankawin, S.	Chulalongkorn Univ
Exhibition	Jangwanitlert, A.	King Mongkut's Inst. of Tech. Ladkrabang
Secretariat	Suwankawin, S.	Chulalongkorn Univ.

International Steering Committee

Acarnley, P.	Univ. of Newcastle	Kennel, R.	Univ. of Wuppertal
Akagi, H.	Tokyo Inst. of Tech.	Kolar, J. W.	Swiss Federal Inst. of Tech. (ETH) Zürich
Alex, Q. H.	North Carolina State Univ.	Lai, J. S.	Virginia Polytechnic Inst. and State Univ.
Amaratunga, G. A. J.	Univ. of Cambridge	Longya, X.	Ohio State Univ.
Bhat, A. K. S.	Univ. of Victoria	Lorenz, R. D.	Univ. of Wisconsin-Madison
Boroyevich, D.	Virginia Polytechnic Inst. and State Univ.	Matsuse, K.	Meiji Univ.
Bose, B. K.	Univ. of Tennessee	Mohan, N.	Univ. of Minnesota
Chan, C.C.	Univ. of Hong Kong	Nakaoka, M.	Yamaguchi Univ.
Clare, J. C.	Univ. of Nottingham	Ninomiya, T.	Kyushu Univ.
Dehong, X.	Zhejiang Univ.	Okuma, S.	Nagoya Univ.
Divan, D.	Georgia Inst. of Tech.	Qian, Z.	Zhejiang Univ.
Elbuluk, M. E.	Univ. of Akron	Rahman, M. A.	Memorial Univ. of Newfoundland
Enjeti, P.	Texas A&M Univ.	Schroeder, D.	Univ. of Munich
Ertan, B.	Middle East Technical Univ.	Sekiya, H.	Chiba Univ.
Forsyth, A.J.	Univ. of Manchester	Sen, P.C.	Queen's Univ.
Green, T. C.	Imperial College	Shoyama, M.	Kyushu Univ.
Guo, X. D.	Harbin Inst. of Tech.	Suetsugu, T.	Fukuoka Univ.
Holtz, J.	Univ. of Wuppertal	Teck, O. B.	McGill Univ.
Hui, R. S. Y.	City Univ. of Hong Kong	Tenti, P.	Univ. of Padova
Husain, I.	Univ. of Akron	Tsai, P. C.	National Tsing Hua Univ.
Jahns, T. M.	Univ. of Wisconsin–Madison	Undeland, T. M.	Norwegian Univ. of Science and Tech.
Jain, P.	Queen's Univ.	Wu, B.	Ryerson Univ.
Jezernik, K.	Univ. of Maribor	Wyk, J. D. V.	Virginia Polytechnic Inst. and State Univ.
Kazimierczuk, M. K.	Wright State Univ.	Zhengming, Z.	Tsing Hua Univ.

History of PEDS Conference

International Conference on Power Electronics and Drive Systems, PEDS Conference, originated in Singapore and the first PEDS conference was held in Singapore in 1995. The aim of the PEDS Conference is to provide a forum for participants from the industry and academia in the area of power electronics and drives to exchange ideas and have interactions. The conference is biennial and since 1995 the PEDS Central Committee, Singapore in collaboration with overseas organizing committees, have organized PEDS Conference series held in various Asia Pacific (IEEE Region 10) countries. All the PEDS conferences are being held in technical co-sponsorship with the IEEE Power Electronics Society and IEEE Industry Applications Society.

PEDS Conference	Venue
PEDS 1995	Singapore
PEDS 1997	Singapore
PEDS 1999	Hong Kong
PEDS 2001	Bali, Indonesia
PEDS 2003	Singapore
PEDS 2005	Kuala Lumpur, Malaysia
PEDS 2007	Bangkok, Thailand

List of Reviewers

Abbaszadeh, K.
Abe, S.
Acarnley, P. P.
Adnani, M. E.
Afjei, E.
Ahmad, G.
Ahmed, M.
Al-Haddad, K.
Amirifar, R.
Ang, S.
Apte, A. A.
Attaviriyanupap, P.
Awad, H.
Azli, N. A.
Baiju, M. R.
Bakan, A. F.
Beig, A. R.
Bhadra, S. N.
Bharanikumar, R.
Bhat, A. K. S.
Bina, M. T.
Biswas, S. K.
Boonyaroonate, I.
Bunlaksananusorn, C.
Chang, K.-T.
Chen, H.
Chen, J.-J.
Cheng, M.-Y.
Chengfeng, Y.
Cheung, N. C.
Chiba, A.
Chien, F. T.
Chiu, H.-J.
Choi, B.
Chou, J.-H.
Chunkag, V.
Clare, J. C.
Colli, V. D.
Corzine, K. A.
Covic, G. A.
Cruden, A.
Dahono, P. A.
Daming, Z.
Dianguo, X.
Doki, S.
Dong-Hee, L.
Duffy, M.
Dzung, P. Q.
Elbuluk, M. E.
Ertugrul, N.
Eskander, M.
Farhangi, S.
Filho, E. R.
Forsyth, A.

Fujiwara, O
Fukuda, S.
Garvey, S. D.
Grabner, H.
Griva, G.
Gueldner, H.
Guo, Y.
Hagh, M. T.
Hakimie, H.
Hamzah, M. K.
Hamzah, N.
Hanamoto, T.
Hava, A. M.
Hayashi, Y.
Hennen, M. D.
Higuchi, T.
Ho, S.-T.
Hofmann, W.
Hori, Y.
Howe, D.
Hsieh, G.-C.
Hua, S.
Huang, L.
Huang, S.-J.
Hung, J. Y.
Hur, J.
Hussien, Z. F.
Idris, N. R. N.
Jain, P. K.
Janakiraman, P. A.
Jangwanitlert, A.
Jerome, J.
Jianxin, S.
Khan, P. K. S.
Khan-Ngern, W.
Kim, I.-S.
Kim, Y.-H.
Kinnares, K.
Kobayashi, S.
Kolar, J. W.
Komurcugil, H.
Kubota, H.
Kulvitit, Y.
Kurokawa, F.
Lafoz, M.
Lai, C.-K.
Lai, J.-S.
Lecci, A.
Lee, D.-C.
Lee, E.-W.
Lee, S. C.
Lee, Y.-S.
Li, D. D.
Li, G.

Li, H.
Li, J.
Li, W.
Liang, Y. C.
Liaw, C.-M.
Lin, C.-H.
Lin, R.-L.
Liserre, C.
Lo, Y.-K.
Loh, A.
Lorenz, L.
Low, K.-S.
Manmek, T
Markadeh, G. A.
Marques, G. D.
Martins, J. F. A.
Matsui, M.
Matsuo, K.
Mekhilef, S.
Morales-Castorena, A.
Morimoto, M.
Morimoto, S.
Mukerjee, R.
Muni, B. P.
Murthy, S. S.
Mutoh, A.
Muyeen, S. M.
Nagaraju, J.
Narayanan, G.
Nho,N. V.
Noguchi, T.
Nussbaumer, T.
Okou, A. F.
Omar, A. M.
Pai, F.-S.
Palandurkar, M. V.
Pan, C.-T.
Panda, S. K.
Patel, H. K.
Phuong, L. M.
Pichetjamroen, V.
Ping, H. W.
Pires, A.
Pires, V.
Polmai, S.
Ponce, M.
Qian, Z.
Rafael, S.
Rahman, M. A.
Ramasamy, A. K.
Rashad, E. E. M.
Ratanapanachote, S.
Rizk, J.
Ruan, X.

Saied, B. M.
Sangwongwanich, S.
Saudemont, C.
Saxena, T. K.
See, K. Y.
Senthilkumar, R.
Shaojun, X.
Sharma, V. K.
Shieh, H.-J.
Shimizu, T.
Shin, G.-H.
Shing, C. S.
Shinnaka, S.
Shuhua, F.
Singh, B.
Sirisukprasert, S.
Soltani, J.
Sopavanit, C.
Staines, C. S.

Sumedha
Sun, K.
Suwankawin, S.
Tahami, F.
Takahashi, R.
Tanaka, T.
Tarnekar, S. G.
Tenti, P.
Thounthong, P.
Tomita, H.
Tseng, K.-J.
Tsui, M.
Vaclavek, P.
Vaez-Zadeh, S.
Veszpremi, K.
Vijayarajan, K.
Vilathgamuwa, D. M.
Villasenor, A. G.
Wang, C.-M.

Wang, H.-P.
Wang, L.
Weiming, M.
Wen, F.-L.
Wolbank, T. M.
Wu, L.
Wu, T.-F.
Xu, D. (David)
Xu, D. (Dehong)
Xu, L.
You, K.
Yousfi, D.
Zhang, X.
Zhengyu, L.
Zhong , Q.-C.
Zhu, J.
Zirn, O.
Zolghadri, M. R.

This page intentionally left blank.

Table of Contents

Fuel cell systems and applications 1
Bernard Davat

Recent Trends iin Power Qualliity Improvements Techniiques 58
Bhim Singh

Power Electronics for Future Utility Applications 213
Rik W. De Doncker, Christoph Meyer, Robert U. Lenke, Florian Mura

Digital Control Generations -- Digital Controls for Power Electronics through the Third Generation 221
Philip T. Krein

Power Electronics and Control of Renewable Energy Systems 226
F. Iov, M. Ciobotaru, D. Sera, R. Teodorescu, F. Blaabjerg

Design and Evaluation of a 60 000 rpm Permanent Magnet Bearingless High Speed Motor 249
T. Schneider, A. Binder

Performance Investigation of Two-, Three- and Four-Phase Bearingless Slice Motor Configurations 257
M.T. Bartholet, S. Silber, T. Nussbaumer, J.W. Kolar

Compensation of Pole Position Estimation Error for Sensor-less IPMSM Drives with DC Link Current
Detection 265
Hisao Kubota, Yusuke Shibano, Takayuki Kobayashi

A Novel Dual-Stator Hybrid Excited Synchronous Wind Generator 270
Liu Xiping, Lin Heyun, Yang Chengfeng, Fang Shuhua, Guo Jian

Application of Multi-level Multi-domain Modeling in the Design and Analysis of a PM Transverse Flux
Motor with SMC Core 275
Youguang Guo, Jianguo Zhu, Dikai Liu, Haiyan Lu, Shuhong Wang

New Approximate 2DOF Digital Controller for DC-DC Converter with Second-Order Differential
Characteristics 280
Eiji Takegami, Kohji Higuchi, Kazushi Nakano, Satoshi Tomioka, Kazushi Watanabe, K.K. Densei-Lambda

High Accuracy CMOS Current Sensing Circuit for Current Mode Control Buck Converter 286
Yuang-Shung Lee, Chih-Jen Hsu

Small Signal Analysis of a dual-switch forward Converter with non-ideal transformer in Current-
Programmed Control 291
Weiping Zhang, Yuzhou Lei, Xiaoqiang Zhang, Yuanchao Liu

High Frequency Transformer Designs for Improving Cross Regulation in Multiple-Output Flyback
Converters 295
Kusumal Chalermyanont, Pairote Sangampai, Anuwat Prasertsit, Surapon Theinmontri

Operation of a wye Connected Three- Level Active Power Filter under Nonideal Conditions 299
H.B. Zhang, A.M. Massoud, S.J. Finney, B.W.Williams, T.C. Lim, H. Hotait

Application of GPRS Techniques for Wide-Area Power Quality Monitoring 305
Shun-Yu Chan, Jen-Hao Teng, David Chang, Li-Yuan Chin

Design and Development of Autotransformer Based 24-Pulse AC-DC Converter fed Induction Motor
Drive 310
Bhim Singh, Vipin Garg, G.Bhuvaneswari

Power Quality Monitoring System Using Real-Time Operating System 318
Krisda Yingkayun, Suttichai Premrudeepreechacharn, Kosol Oranpiroj

Technology Performance Comparison of Triacs Subjected to Fast Transient Voltages 322
L. Gonthier, A. Passal

Table of Contents

On-line Junction Temperature Measurement of CoolMOS Devices 327
Andreas Koenig, Thomas Plum, Peter Fidler, Rik W. De Doncker

Analytical Design of High-Power MTO Thyristors 333
Thomas Plum, Rik W. De Doncker

A Novel Gate Driver with Output Voltage Having Double Source Voltage 338
K. I. Hwu, Y. T. Yau

Effects of Internal Feedback and Gate-Drive Signal on the Turn-off Loss of MOSFET ZVS 342
Youthana Kulvitit, Puckapon Opanuruk, Tanvaa Tansatit

A Novel Bridge Type FCL Based on Single Controllable Switch 350
Wanmin Fei, Yanli Zhang, Qi Wang

A Novel Isolation Power Supply for Gating Multiple Devices in FACTS Equipment 354
Yanli Zhang, Wanmin Fei, Zhengyu Lu

Voltage and Frequency Controller for Parallel Operated Isolated Asynchronous Generators 357
Bhim Singh, Gaurav Kumar Kasal

DSP controlled Semiconductor based High-Voltage Source 363
F. Martin, T. Leibfried, O. Kerz, K. Mossner

Open Switch Fault Diagnosis for a Doubly-Fed Induction Generator 368
W. Sae-Kok, D M Grant

Rapid Analysis & Design Methodologies of High- Frequency LCLC Resonant Inverter as Electrodeless Fluorescent Lamp Ballast 376
Yong-Ann Ang, David Stone, Chris Bingham, Martin Foster

Analysis and Control of Dual-Output LCLC Resonant Converters, and the Impact of Leakage Inductance 382
Y. Ang, C. M. Bingham, M. P. Foster, D. A. Stone

A Novel QR ZCS Switched-Capacitor Bidirectional Converter 388
Yuang-Shung Lee, Yi-Pin Ko, Chien-An Chi

Analysis of a Half - Bridge Inverter for a Small- Size Induction Cooker Using Positive-Negative Phase-Shift Control under ZVS and NON-ZVS Operation 394
P. Achara, P. Viriya, K. Matsuse

Adaptive Phase Control Method for Load Variation of Resonant Converter with Piezoelectric Transformer 401
S. T. Yun, J. M. Sim, J. H. Park, S. J. Choi, B. H. Cho

Adaptation of Motor Parameters in Sensorless PMSM Drives 406
Antti Piippo, Marko Hinkkanen, Jorma Luomi

Development of 150000 r/min, 1.5 kW Permanent- Magnet Motor for Automotive Supercharger 414
Toshihiko Noguchi, Masaru Kano

Analysis and Performance Evaluation of Radial Flux Air-Cored Permanent Magnet Machines with Concentrated Coils 420
P.J. Randewijk, M.J. Kamper, R-J. Wang

Analysis and Experimental Investigation for Field-Control Capability of a Novel Hybrid Excitation Claw-Pole Synchronous Machine 427
Yang Chengfeng, Lin Heyun, Liu Xiping, Fang Shuhua, Guo Jian

A single-Capacitor Turn-off Snubber for Interleaved Boost Converter with Coupled Inductor 433
S.-Y. Tseng, J. Z. Shiang, Y.-H. Su

Buck-Boost Converter Associated with Active Clamp Forward Converter for PV Power System 440
S. Y. Tseng, W. C. Chen, Y. J. Li, J. S. Kuo

Table of Contents

Comparison of Three-Phase DC-DC Converters vs. Single-Phase DC-DC Converters 448
Christian P. Dick, Andreas Konig, Rik W. De Doncker

Applying Modified One-Comparator Counter-Based PWM Control Strategy to Flyback Converter 456
K. I. Hwu,, Y. H. Chen

Analysis of Conducted EMI Reduction on a Boost Converter Using Progressive Inductor Winding Technique 460
Kritsada Saritsiri, Werachet Khan-Ngern

Practical Issues Concerned with Zero sequence component and Harmonic Compensation in Four-Wire systems 465
E. Pashajavid, K. Kanzi, M. Tavakoli Bina

Automated Design and Implementation of Resonant Controllers for Current Control of Shunt Active Filters 470
W. Lenwari, M. Sumner, P. Zanchetta

A Modular Structured Multilevel Inverter Active Power Filter with Unified Constant-Frequency Integration Control for Nonlinear AC Loads 475
P. Y. Lim, N. A. Azli

HCC PWM Control of the Single-Phase Bi- Directional Buck Converter giving IEEE 519 Compliance at any Power Factor 480
A. N. Arvindan, V. K. Sharma

Passive EMI Filter Performance Improvements with Common Mode Voltage Cancellation Technique for PWM Inverter 488
C. Khun, W. Khan-Ngern, M. Kando

Novel Auxiliary Diagnosis Method for State-of-Health of Lead-Acid Battery 493
Yu-Hua Sun, Hurng-Liahng Jou, Jinn-Chang Wu

Electromechanical Model of a Longitudinal Mode Piezoelectric Transformer 498
Shine-Tzong Ho

Latest Development of Transformer Parasitic Inductive Components and Lossless Inductive Snubber-Assisted Series Resonant High-Frequency ZCS-PFM DC-DC Converter for RF Generator 504
Hisayuki Sugimura, Manabu Ishitobi, Bishwajit Saha, Sang Pil Mun, Soon Kurl Kwon, Mutsuo Nakaoka

A General Method for Deciding the Input Filter Capacitance of Flyback Switching AC-DC Converter with Peak Current-Controlled Mode 510
Jiaxin Chen, Jianguo Zhu, Youguang Guo

Design of High Performance and Low Cost Line Impedance Stabilization Network for University Power Electronics and EMC Laboratories 515
D. Sakulhirirak, V. Tarateeraseth, W. Khan-Ngern, N. Yoothanom

A Robust Output Current Control Method with Disturbance Observer for Matrix Converter under Unbalanced Input Voltage 521
Kazuo Oka, Kouki Matsuse

FPGA Design of Single-phase Matrix Converter Operating as Cycloconverter 527
Z. Idris, M.K. Hamzah, A. Saparon, N.R.Hamzah, N.Y. Dahlan

Input and Output Ripple Analysis of AC Chopper 534
Arwindra Rizqiawan, Dessy Amirudin, Deni, Pekik Argo Dahono

A Three-level 4 × 3 Conventional Matrix Converter 541
Runjie Rong, Poh Chiang Loh, Peng Wang, Frede Blaabjerg

A novel primary-side controlled contactless battery charger 546
Yi-Hwa Liu, Shun-Chung Wang, Rong Ceng Leou

Table of Contents

Research on Digital Soft-switch Welding/Cutting Inverter Power Source 551
G.R. Zhu, Z. Liu, X. Li, B.Y. Liu, S.X. Duan, Y. Kang

Design of an Adjustable High Output Voltage Asymmetrically Switched Class D Converter 556
M. Rentzsch, H. Guldner, C. Ditmanson

New Direct High Frequency Soft-Switching Inverter-Fed AC-DC Converter with Voltage Doubler for Consumer Magnetron Drive 563
Hisayuki Sugimura, Bishwajit Saha, Hidekazu Muraoka, Sang Pil Mun, Tomokazu Mishima, Hideki Omori, Mutsuo Nakaoka

Complete loading Characteristics Modeling of an Axial Flux Permanent Magnet Synchronous Machine Using Ck Spline Functions 569
Z. Lakhdari, F. Amrane, L. Adélaide, Ph. Makany

The Bearingless 2-Level Motor 574
P. Karutz, T. Nussbaumer, W. Gruber , J.W. Kolar

Analysis and Design of a Sliding Mode Controller for Buck Converters Operating in DCM with Adaptive Hysteresis Band Control Scheme 581
Hung-Chih Lin, Tsin-Yuan Chang

Buck Converter Simulation Technique Based on the Fourier Transform 587
Acacio M. R. Amaral, A. J. Marques Cardoso

ANALYSIS OF HOPF BIFURCATION IN DC-DC LUO CONVERTER USING CONTINUOUS TIME MODEL 595
A.Kavitha, G.Uma

Analysis of a Mixed-Signal Control for DC-DC Converters based on Hysteresis Modulation And Estimated Inductor Current 600
D. Trevisan, S. Saggini, P. Mattavelli, L. Corradini, P. Tenti

Power Quality Study in Macao 607
Sio-Un Tai, Man-Chung Wong, Ming-Chui Dong, Ying-Duo Han

Some Findings on Harmonic Measurement in Macao 614
Sio-Un Tai, Man-Chung Wong, Ming-Chui Dong, Ying-Duo Han

Coordinated design of PSS and TCSC dynamics model for power system network oscillations 620
M. Tarafdar Haque, A. Roshan Milani, A. Lafzi

An Analytic Approach To Harmonic Analysis of 48-Pulse Voltage Source Inverter 626
B. Geethalakshmi, P. Dananjayan

Detailed losses Analysis of High-Frequency Planar Power Transformer 632
Yu Ma, Peipei Meng, Junming Zhang, Zhaoming Qian

Design of a Nuclear Magnetic Resonance Fast Field Cycling Air Cored Magnet 636
Duarte M. Sousa, Gil D. Marques, Pedro J. Sebastiao,, Antonio C. Ribeiro

Using DFT to Obtain the Equivalent Circuit of Aluminum Electrolytic Capacitors 643
Acácio M. R. Amaral, Gustavo M. Buatti, Hugo Ribeiro, A.J. Marques Cardoso

A Mathematical Analysis on Vector Inversion Generators 648
D. J. Thrimawithana, U. K. Madawala

Novel Multi-Level High Voltage Pulsed Power Generator 654
D. J. Thrimawithana, U. K. Madawala

Potential and Electric Field Distribution Analysis of Field Limiting Ring and Field Plate by Device Simulator 660
C.N. Liao, F.T. Chien, Y.T. Tsai

xii

Table of Contents

Wire and Wireless Linked Remote Control for the Group Lighting System Using Induction Lamps 665
Kyu Min Cho, Jae Eul Yeon, Ma Xian Chao, Hee Jun Kim

Induction Heating with Traveling Magnetic Field for Uniform Heating to Flat Metal 671
T. Sekine, H. Tomita, Y. Saito, S. Obata, S. Yoshimura

Three-Phase (LC)(L)-Type Series-Resonant Converter with Capacitive Output Filter 677
M. Almardy, A.K.S. Bhat

Analysis of a Full-Bridge Inverter for Induction Heating Using Asymmetrical Phase-Shift Control under ZVS and NON-ZVS Operation 685
N. Yongyuth, P. Viriya, K. Matsuse

FPGA-Based Phase-Shift ZVS Full-Bridge DC-DC Converter Using One-Comparator Counter-Based PWM Control Strategy 692
K. I. Hwu, Y. T. Yau

A Simplified Power Control Scheme for Resonant Inverter with Purely Resistive Load 697
Pramoch Dorkmai, Youthana Kulvitit, Tanvaa Tansatit

Voltage Injection Based Initial Rotor Position Estimation Method for Three-Phase Star- Connected Switched Reluctance Machines 703
P. Somsiri, P. Champa, P. Wipasuramonton, K. Tungpimonrut, P. Aree

Control Scheme for Switched Reluctance Drives with Minimized DC-Link Capacitance 710
Christoph R. Neuhaus, Rik W. De Doncker, Nisai H. Fuengwarodsakul

Multiphase Torque-Sharing Concepts of Predictive PWM-DITC for SRM 716
Helge J. Brauer, Martin D. Hennen, Rik W. De Doncker

A New Two Phase Configuration for Switched Reluctance Motor with High Starting Torque 722
E. Afjei, K. Navi, S. Ataei

Application of Power Electronics for Damping of Torsional Vibrations 726
T. Zoller, T. Leibfried, A. M. Miri

Application of Battery Energy Operated System to Isolated Power Distribution Systems 731
Bhim Singh, A. Adya, A.P. Mittal, J.R.P Gupta

Pulse Doubling in 18-Pulse AC-DC Converters 738
Bhim Singh, Sanjay Gairola

Magnetic Field Analysis and Control Strategy of Permanent Magnet Actuator for Low Voltage Vacuum Circuit Breaker 745
Fang Shuhua, Lin Heyun, Yang Chenfeng, Liu Xiping, Guo Jian

Analysis of Transformer Inrush Current under Harmonic Source 749
Chien-Lung Cheng, Jim-Chwen Yeh, Shyi-Ching Chern, Yi-Hung Lan

Voltage Sag Compensation Performance by DSTATCOM with Series Inductor and Energy Storage 755
Sumate Naetiladdanon

Cooperative Operation of Active Power Filters by Instantaneous Complex Power Control 760
Elisabetta Tedeschi, Paolo Tenti, Paolo Mattavelli

Impact of Adjustable Speed PWM drives on Operation and Harmonic Losses of Nonlinear Three Phase Transformers 767
M.A.S. Masoum, Paul S. Moses, Amir S. Masoum

Real-Time Implementation of Voltage Dip Mitigation using D-STATCOM with Fast Extraction of Instantaneous Symmetrical Components 773
Thip Manmek, Chathura P. Mudannayake

Table of Contents

Combined System of Static Synchronous Series Compensation and Passive Filter applied to Wind Energy Conversion System..781
A. Singer, W. Hofmann

Control of active injector for multi-pulse rectifiers operating on variable frequency supplies.....................788
Ismael Araujo-Vargas, Andrew J. Forsyth, F. Javier Chivite-Zabalza

36-pulse hybrid ripple injection for high performance aerospace rectifiers 796
F. Javier Chivite-Zabalza, Andrew J. Forsyth, Ismael Araujo-Vargas

A 48-pulse converter using dc-ripple injection..804
F. Javier Chivite-Zabalza, Andrew J. Forsyth

A Study of Different Possible Switched Mode Chopper Circuits for Multi-Magnet Based DC Electromagnetic Levitation System..812
Subrata Banerjee, Dinkar Prasad, Jayanta Pal

Power Supply with Potential Use in Magnetic Stimulation..817
Duarte M. Sousa, Antonio Ferraz

A Novel Maximum Power Point Tracking Method for the Photovoltaic System..................................824
Hurng-Liahng Jou, Wen-Jung Chiang, Jinn-Chang Wu

Maximum Power Point Algorithm in PV Generation: An Overview...829
Hardik P. Desai, H. K. Patel

A DC-Module-Based Power Configuration for Residential Photovoltaic Power Application 836
Bangyin Liu, Shanxu Duan, Yong Kang

Analysis and Improvement of Maximum Power Point Tracking Algorithm Based on Incremental Conductance Method for Photovoltaic Array..842
Bangyin Liu, Shanxu Duan, Fei Liu, Pengwei Xu

Application of Maximum Power Point Tracker with Self-organizing Fuzzy Logic Controller for Solar-powered Traffic Lights..847
Noppadol Khaehintung, Phaophak Sirisuk

Supply-side Current Harmonics Control of Three Phase PWM Boost Rectifiers Under Distorted and Unbalanced Supply Voltage Conditions 852
Xinhui Wu,, Sanjib K. Panda, Jianxin Xu

A Two-stage Converter with a Coupled-Inductor 858
Hirotaka Nakanishi, Yoshihiro Tomihisa, Terukazu Sato, Takashi Nabeshima, Kimihiro Nishijima, Tadao Nakano

Three-Phase AC to DC Converter with Minimized DC Bus Capacitor and Fast Dynamic Response.......863
U. Kamnarn, Y. Kanthaphayao, V. Chunkag

A Simple Effective Duty Cycle Controller for High Power Factor Boost Rectifier 869
Hussain S. Athab, P. K. Shadhu Khan

A Cost Effective Method of Reducing Total Harmonic Distortion (THD) in Single-Phase Boost Rectifier 874
Hussain S. Athab, P. K. Shadhu Khan

Comparison of Different Methods to Detect Static Air Gap Asymmetry in Inverter Fed Induction Machines..880
T.M. Wolbank, P. Macheiner

Analysis of the Synchronous Torques in a Split Phase Induction Motor..886
P. Scavenius Andersen, D. G. Dorrell, N. C. Weihrauch, P. E. Hansen

On-Line Diagnosis of Three-Phase Closed Loop Induction Motor Drives Using an Eigenvalue aß-Vector Approach 894
J. F. Martins, V. Fernao Pires, A. J. Pires

xiv

Table of Contents

Design and Development of a 36-Pulse AC-DC Converter for Vector Controlled Induction Motor Drive..................899
Bhim Singh, Sanjay Gairola

Comparison of Outer- and Inner-Rotor Switched Reluctance Machines........................907
Martin D. Hennen, Rik W. De Doncker

Optimization of Predesign of Switched Reluctance Machines Cross Section Using Genetic Algorithms......................912
Satit Owatchaiphong, Christian Carstensen, Rik W. De Doncker

Shaft Position for an 8/6 Switched Reluctance Machine: Theoretical concept, FEM analysis and Experimental results........................917
Silviano Rafael, P.J. Costa Branco, A.J. Pires

Sensorless Control of Brushless Doubly-Fed Reluctance Machines using an Angular Velocity Observer....................922
Milutin G Jovanovic, David G Dorrell

A Half-Bridge PV System with Bi-direction Power Flow Controlling and Power Quality Improvement......................930
C.L. Shen, S.T. Peng

Response of DSTATCOM under Voltage Flicker In Farm Wind........................937
K. Aodsup, P. N. Boonchiam, A. Sode-Yome, P. Kongsuk, N. Mithulananthan

A Comparative Study of Fixed Speed and Variable Speed Wind Energy Conversion Systems Feeding the Grid........................941
S.S. Murthy, Bhim Sing, P.K. Goel, S.K. Tiwari

Prediction of Wind Power Generation based on Chaotic Phase Space Reconstruction Models........................949
Dong Lei, Wang Lijie, Hu Shi, Gao Shuang, Liao Xiaozhong

Power Flow Control for Efficiency Improvement in a Forward-Flyback Mixed Converter........................954
Yoshito Kusuhara, Asahi Nakayama, Tamotsu Ninomiya, Shin Nakagawa

Hammerstein Model-Based Robust Control of DC/DC Converters........................959
F. Alonge, F. D'ippolito, T. Cangemi

A New Model Control DC-DC Converter to Improve Dynamic Characteristics........................968
F. Kurokawa, S. Sukita

Fuzzy Incremental Controller for the 3rd Order Buck Converter........................973
M. Veerachary, Deepen Sharma

Design of a Single-Stage Single-Switch Power- Factor-Corrected (S4-PFC) AC/DC Converter........................977
P. Kongthawornwattana, C. Bunlaksananusorn, S. Kittiratsatcha

A DSP-Based Unified Three Phase/Switch/Level Unity Power Factor Rectifier Using Feedback Linearization for DC-Bus Voltage Control........................983
Ali Moallem, Hesameddin Mirzaee Teshnizi, Mohammadreza Zolghadri

A Soft-Switched AC-DC Symmetrical Boost Converter with Power Factor Correction........................989
A. Jangwanitlert, J. Songboonkaew

Education Reforming for Power Electronics........................994
Weiping Zhang, Xiaohan Guan, Dongyan Zhang

A Novel Current Control System for PMSM Considering Effects from Inverter in Overmodulation Range.............999
Smith Lerdudomsak, Shinji Doki, Shigeru Okuma

Modelling of the Feeding Network of a Linear Synchronous Machine and Estimation of Model Parameters........................1006
J. Rost, H. Gueldner, R. Hellinger, A. Weller

Analysis of Losses in Inverter Fed Large Scale Synchronous Machines using 2D FEM Software........................1012
Samer Shisha, Chandur Sadarangani

Table of Contents

Position sensorless control of the Reluctance Synchronous Machine considering High Frequency inductances 1017
H.W. De Kock, M.J. Kamper, O.C. Ferreira, R.M. Kennel

Carrier PWM algorithm in overmodulation range for Multileg Multilevel Inverter 1027
Nguyen Van Nho, Hong Hee Lee

Carrier Based Single-state PWM Technique In multilevel Inverter 1033
Nguyen Van Nho, Quach Thanh Hai, Hong Hee Lee

Implementation of a Single-carrier Multilevel PWM Technique Using Field Programmable Gate Array (FPGA) 1041
N. A. Azli, L. Y. Teng, P. Y. Lim

SPACE VECTOR PWM FOR MULTILEVEL INVERTERS - A FRACTAL APPROACH 1047
Anish Gopinath, M.R. Baiju

Elimination of Harmonics in a Five-Level Diode-Clamped Multilevel Inverter Using Fundamental Modulation 1055
Sule Ozdemir, Engin Ozdemir, Leon M. Tolbert, Surin Khomfoi

Compensation of DC-Link Oscillations of Cascaded H-Bridge Converters 1060
M. Tavakoli Bina, B. Eskandari

Combined DC-Filter and optimized Modulation to Absorb DC-Link Oscillations of Cascaded H-Bridge Converters 1065
M. Tavakoli Bina, B. Eskandari

Control Strategies of a Hybrid Multilevel Converter for Expanding Adjustable Output Voltage Range 1070
Shoji Fukuda, Takatsugu Yoshida, Shigeta Ueda

High Efficiency Single Phase Multi-level Inverter by New Controlled Switch Signal 1078
Ruthapong Kumchaiyo, Itsda Boonyaroonate

FPGA Implementation of Quasi-BLDC Drive 1082
C.S. Soh, C. Bi, K.K. Teo

A Practical Method to Eliminate the Conduction Torque Ripple in BLDCM Using Cascade Topology 1088
Xiaofeng Zhang, Zhengyu Lu, Yu Ma, Zhaoming Qian

Program Architecture for Realizing Design Optimization of a BLDC Motor 1092
Dong-Hun Kim, Giwoo Jeung, Heung-Geun Kim, In Dong Kim

Stable Operation of the Brushless Doubly-Fed Machine (BDFM) 1096
Shiyi Shao, Ehsan Abdi, Richard Mcmahon

Sail Generator Feasibility Study 1102
Ha Pham Ngoc, Yasuaki Matsui, Pathom Attaviriyanupap, Osamu Iso

Braking Circuit of Small Wind Turbine Using NTC Thermistor under Natural Wind Condition 1109
Y. Matsui, A. Sugawara, S. Sato, T. Takeda, K.Ogura

Flywheel Energy Storage Drive for Wind Turbines 1115
K. Veszpremi, I. Schmidt

Theory, Simulation and Experimental Verification of a New Integral Cycle Robust Control Strategy for Self Excited Induction Generators 1123
S.S. Murthy, A.J.P. Pinto

Performance Comparison of DC Link Voltage Controllers in Vector Controlled Boost Type PWM Converter for Wind Turbine System 1129
W. Sudmee, B. Neammanee

Analysis and Design of Class DE Amplifier with Nonlinear Shunt Capacitance 1136
Hiroo Sekiya, Takayuki Watanabe, Tadashi Suetsugu, Marian K. Kazimierczuk

xvi

Table of Contents

A Novel Control Strategy of the Class-D Stereo Audio Amplifier...1142
Kyu Min Cho, Won Seok Oh, Hai Xu, Hee Jun Kim

Robust H_infinity Control Design for PFC Rectifiers...1147
F. Tahami, H. Molla Ahmadian, A. Moallem

Parallel Operation of Power Factor Corrected AC-DC Converter Modules With Two Power Stages.....................1152
Aravind Pothana, Krishna Vasudevan

Noise Radiation of Switched Reluctance Drives...1160
K. A. Kasper, M. Bosing, R. W. De Doncker, S. Fingerhuth, M. Vorlander

Iron Losses in Electrical Machines Due to Non Sinusoidal Alternating Fluxes...1167
J. A. Walker, D. G. Dorrell, E. Ritchie

Design Requirements for Doubly-Fed Reluctance Generators...1174
D. G. Dorrell

A Magnetic Gear Box for application with a Contra-rotating Tidal Turbine...1182
Laxman Shah, A. Cruden, Barry W. Williams

Mechatronic . Advanced Computational Intelligence...1187
D. Schroder, H. Schuster, C. Westermaier

New Space Vector Control Approach for Four Switch Three Phase Inverter (FSTPI)...1195
Phan Quoc Dzung, Le Minh Phuong, Pham Quang Vinh, Nguyen Minh Hoang, Tran Cong Binh

The Development of Artificial Neural Network Space Vector PWM for Four-Switch Three- Phase Inverter...1202
Phan Quoc Dzung, Le Minh Phuong, Pham Quang Vinh

Voltage Losses Compensation Using Artificial Neural Network for Estimation Nonlinear Characteristic of Switches...1208
N. Pothi, S. Premrudeepreechacharn, C. Rakpenthai

A Simple Carrier-Based PWM Method For Three-Phase Four-Leg Inverters Considering All Four Pole Voltages Simultaneously...1213
Nakharet Chudoung, Somboon Sangwongwanich

Inverted Sine Carrier Pulse Width Modulation for Fundamental Fortification in DC-AC Converters...1221
R.Nandhakumar, S.Jeevananthan

Fault Detection and Reconfiguration Technique for Cascaded H-bridge 11-level Inverter Drives Operating under Faulty Condition...1228
Surin Khomfoi, Leon M. Tolbert

Investigation into Harmonic Losses in a PWM Multilevel cascaded H-Bridge Inverter Fed Induction Motor...1236
Prasopchok Hothongkham, Vijit Kinnares

Extend the Use of Auxiliary Circuit to Start up, Shut down, and Balance of the Modified Diode Clamped Multilevel Inverter...1242
Ahmed Ali Ashaibi, S.J. Finney, B.W. Williams, Ahmed Massoud

Five-Level Z-Source Neutral-Point-Clamped Inverter...1247
F. Gao, P. C. Loh, F. Blaabjerg, R. Teodorescu, D. M. Vilathgamuwa

Capacitor Voltage Balancing Using Redundant States for Five-Level Multilevel Inverter...1255
Hadi A Hotait, Ahmed M Massoud, Steve J. Finney, Barry W. Williams

Sliding Mode Repetitive Control of PWM Voltage Source Inverter...1262
Sufen Chen, Y. M. Lai, Siew-Chong Tan, Chi K. Tse

Output Current Ripple Analysis of Five-Phase PWM Inverters...1267
Deni, E. G. Supriatna, P. A. Dahono

Table of Contents

An Improved 'DC-DC Type' High Frequency Transformer-Link Inverter by Employing Regenerative Snubber Circuit 1274
Z. Salam, S. M. Ayob, M. Z. Ramli, N. A. Azli

A Novel Dimming Technique for Cold Cathode Fluorescent Lamp 1278
K. I. Hwu, Y. H. Chen

Time Delay Compensation For A DSP-Based Current-Source Converter Using Observer-Predictor Controller 1284
Huu-Phuc To, Muhammed Fazlur Rahman, Colin Grantham

Implementation of Hysteresis Current Control for Single-Phase Grid Connected Inverter 1290
Krismadinata, Nasrudin Abd Rahim, Jeyraj Selvaraj

Use of Air-Cored Axial Flux Permanent Magnet Generator in Direct Battery Charging Wind Energy Systems 1295
F.G. Rossouw, M.J. Kamper

Transverse Flux Machines for Sustainable Development - Road Transportation and Power Generation 1301
D. Svechkarenko, A. Cosic, J. Soulard, C. Sadarangani

Low Voltage Ride-Through Capability for Wind Turbines based on Current Source Inverter Topologies 1308
Pierluigi Tenca, Andrew A. Rockhill, Thomas A. Lipo

Optimal Control of Direct Driven Feed Axes with Flexible Structural Components 1316
Ekkehard Batzies, Tobias Scholler, Volkmar Welker, Oliver Zirn

Leakage Energy Recovered Narrow Pulsed Voltage Generator Associated with Ultrasound Generator for Liquid Food Sterilization 1321
S. Y. Tseng, Y. D. Chang, P. L. Huang, T. F Wu, Y. M. Chen

Energy Harvesting from Exercise Bicycle 1327
Suchart Janjornmanit, Samart Yachiangkam, Aswin Kaewsingha

Modeling and Analysis of Igniter for HID Lamps 1330
Weiping Zhang, Qiang Cheng

Design of a Single Bi-directional DC-DC Converter for Onboard Energy Improving of Zero Emission Electric Vehicles 1335
Werachet Khan-Ngern

Speed Sensorless Control with Neuron MRAS Estimator of an Induction Machine 1340
Dong Lei, Yang Dong, Liao Xiaozhong

Adaptive Flux model for commissioning of signal injection based zero speed sensorless flux control of induction machines 1346
T.M. Wolbank, M.A. Vogelsberger, R.H. Stumberger

Design and Performance of a Single Stator, Dual Rotor Induction Motor 1352
S. Sinha, N. K. Deb, N. Mondal, S. K. Biswas

Investigation of skew effect on the Performance of Self - Excited Induction Generators 1356
B. Sawetsakulanond, V. Kinnares

Analysis of Double Loops Discrete Single Input PI Fuzzy for Single phase Inverter 1363
S.M. Ayob, Z. Salam, N.A. Azli

A new three-phase varying-band hysteresis current controller for voltage-source inverters 1368
Vinciane Chereau, Francois Auger, Luc Loron

Diode-Assisted Buck-Boost Current Source Inverters 1376
F. Gao, C. Liang, P. C. Loh, F. Blaabjerg

Table of Contents

Single-Stage Fluorescent Lamps Electronic Ballast Using Class-DE Low dv/dt Rectifier for Power-Factor Correction...................1383
Chainarin Ekkaravarodome, Adisak Nathakaranakule, Itsda Boonyaroonate

Output Impedance Design Consideration of Three Control Schemes for Bus Converter in On-Board Distributed Power System...................1388
Seiya Abe, Masahiko Hirokawa, Tamotsu Ninomiya

Optimal Generation Rescheduling for Security Operation of Power Systems Using Optimal Control Theory...................1394
J. Q. Sun, K. W. Chan, D. Z. Fang

Improvement of Transient Response of Thermal Power Plant Using VVVF Inverter...................1398
N. Matsui, F. Kurokawa

A Novel Circuit Topology for Three-Phase Four-Wire Distribution Electronic Power Transformer...................1404
H.Mirmousa, M.R.Zolghadri

A Half-Bridge DC/DC Converter for Plasma Cutting Machine...................1412
N. Sanajit, A. Jangwanitlert

Ripple Estimation for Paralleled Converter System with Automatic Interleaving Function...................1417
Teruhiko Kohama, Ryota Tsunesada, Tamotsu Ninomiya

Design of a New Hysteretic PWM Controller for All Types of DC-to-DC Converters...................1423
Min Lin, Takashi Nabeshima, Terukazu Sato, Kimihiro Nishijima

Implementation of Fuzzy Logic Controller with Bifurcation Control of a Current-mode Boost Converter...................1429
Noppadol Khaehintung, Phaophak Sirisuk, Anantawat Kunakorn

Phase Advance Approach to Expand the Speed Range of Brushless DC Motor...................1434
Binhminh Nguyen, Minh C. Ta

Nonlinear Decoupled Control for a Six-Phase Series-Connected Two Induction Motor Drive Using the Sliding-Mode Technique...................1442
J. Soltani, N. R. Abjadi, Gh. R. Arab Markadeh

The Decoupled Stator Flux and Torque Sliding-Mode Control of Induction Motor Drive Taking the Iron Losses into Account...................1449
M.Hajian, J.Soltani, S.Hosein Nia, G.R.Arab

AN EFFICIENT DIRECT TORQUE CONTROL SCHEME FOR SPLIT PHASE INDUCTION MOTOR...........1455
A. Khajeh, J. S. Moghani, M. Shahbazi

A Method of Speed Sensorless Vector Control Parallel -Connected Dual Induction Motors Fed by One Inverter in a Rotor Flux Feedback Control...................1460
Jun Nishimura, Kazuo Oka, Kouki Matsuse

A Combined Model Flux Observer for Vector Control of Traction Asynchronous Motors...................1465
F. Tahami, S. Chini Foroosh

Torque Ripple Elimination for Doubly-Fed Induction Motors under Unbalanced Source Voltage...................1471
Hong-Geuk Park, Ahmed G. Abo-Khalil, Dong-Choon Lee, Kwang-Myoung Son

Online H8 Speed Control of Sensorless Induction Motors with Rotor Resistance Estimation...................1477
Peda V Medagam, Farzad Pourboghrat

Analysis and Comparative Study on the Performance between Standard and High Efficiency Induction Machines operating as Self - Excited Induction Generators...................1483
B. Sawetsakulanond, V. Kinnaraes

A simple Approach to Capacitance Determination of Self - Exited Induction Generators for Terminal Voltage Regulation...................1489
B. Sawetsakulanond, V. Kinnares

Table of Contents

Symmetrical Components-Based Control Technique of Doubly Fed Induction Generators under Unbalanced Voltages for Reduction of Torque and Reactive Power Pulsations 1495
S. Wangsathitwong, S. Sirisumrannukul, S. Chatratana, W. Deleroi

A New Switching Technique for Direct Torque Control of Induction Motor using Four-Switch Three-Phase Inverter 1501
Phan Quoc Dzung, Le Minh Phuong, Pham Quang Vinh, Nguyen Minh Hoang, Nguyen Xuan Bac

Detection of Some Parameters of Induction Motors a Proposal and Its Verification 1507
H. Bulent Ertan, Volkan Sezgin, Baris Colak

Comparison of Basic Direct Torque Control Designs for Permanent Magnet Synchronous Motor 1514
M. N. Abdul Kadir, S. Mekhilef, W.P. Hew

Improved DSVM-DTC Based Current Sensorless Permanent Magnet Synchronous Motor Drive 1520
Bhim Singh, Devendra Goyal

A High Performance Direct Torque Control Scheme of Permanent Magnet Synchronous Motor 1527
Dong-Hee Lee, Young-Joo An, Eui-Chel Nho

Low Cost Position Sensor for Permanent Magnet Linear Drive 1533
Ralf Wegener, Florian Senicar, Christian Junge, Stefan Soter

Design of One Rotary-linear Permanent Magnet Motor with Two Independently Energized Three Phase Windings 1538
L. Chen, W. Hofmann

Position Estimation of Permanent Magnet Synchronous Motor Using Un-known Input Observer 1543
Masaru Hasegawa, Satoshi Yoshioka, Keiju Matsui

Switched Reluctance Motor Drive for Electric Motorcycle Using HFNN Controller 1549
Chih-Hong Lin

STATE - SPACE AVERAGING, SIMULATION, STABILITY STUDIES FOR STEP UP POSITIVE OUTPUT SWITCHED CAPACITOR DC-DC CONVERTER 1555
E. Jayashree, G. Uma, M. Vaigundamoorthi

Active Clamp Interleaved Boost Converter with Coupled Inductor for High Step-up Ratio Application 1560
S. Y. Tseng, J. Z. Shiang, W. S. Jwo, C. M. Yang

Active Clamp Interleaved Flyback Converter with Single-Capacitor Turn-off Snubber for Stunning Poultry Applications 1567
S. Y. Tseng, C. T. Hsieh, H. C. Lin

Novel Current Feedforward Average Current Mode Control Technique to Improve Output Dynamic Performance of DC-DC Converters 1575
P. Chrin, C. Bunlaksananusorn

Stability Analysis of Cascaded DC-DC Power Electronic System 1581
M. Veerachary, S. Bala Sudhakar

Averaged Switch Modeling of DC/DC Converters using New Switch Network 1586
Chien-Min Lee, Yen-Shin Lai

Soft Transition Operation of UPS in High- Power-Factor Mode of Three-Phase Front- End Rectifier 1590
G. A. Dhomane, H. M. Suryawanshi

Specific Harmonic Power Suppression of Direct- Power-Controlled Current-Source PWM Rectifier 1595
Toshihiko Noguchi, Kohji Sano

Frequency-Controlled LCC Resonant Converter with Synchronous Rectifier 1601
Yu Ma, Xiaogao Xie, Zhaoming Qian

Selection of the Filter Capacitor for Power Supplies using 1-Phase Diode Rectifier 1605
N. Mondal, S. K. Biswas, S. Sinha, N. K. Deb

Table of Contents

High Performance Single-Phase Voltage Regulator with a Simple Circuit Topology 1610
Chien-Ming Wang, Ching-Hung Su, Chang-Hua Lin, Maw-Yang Liu, Kuo-Lun Fang

Small-Signal Modeling of Series Resonant Converter 1615
Weiping Zhang, Peng Mao, Yuanchao Liu

Modelling of Three phase Z-Source Boost Buck Rectifiers 1620
D M Vilathgamuwa, P C Loh, K Karunakar

A NEW SINGLE-PHASE CONTROLLED RECTIFIER USING SINGLE-PHASE MATRIX CONVERTER WITH REGENERATIVE CAPABILITIES 1626
R. Baharom, M.K. Hamzah, A. Saparon, S.Z. Mohammad Noor, N.R.Hamzah

Implementation of Space Vector Modulated 3. to 3 . Matrix Converter Fed Induction Motor 1632
S. Ganesh Kumar, S. Siva Sankar, S. Krishna Kumar, G. Uma

A Single-Phase High-Power-Factor Neutral-pointer Clamped Multilevel Rectifier 1636
Yun Xu, Yunping Zou, Chengzhi Wang, Wei Chen, Bangyin Liu

Two Phase Inverter Drive of Three Phase Motor 1641
Saksit Jangjaempradit, Masayuki Morimoto

Predictive Current Controller for Inverter Fed Medium Voltage Drives with LC Filter 1645
T. Laczynski, A. Mertens

Novel Control Strategy of Instantaneous Power Based CVCF Inverter 1651
Akira Sato, Toshihiko Noguchi

An Improved Parallel Processing UPS Using a Voltage-Controlled Voltage Source Inverter 1657
S.W. Lee, H. Dehbonei, S.H. Ko, S.R. Lee, B.H. Jang, Y.H. Moon, T.K. Ko

A PEMFC/Battery Hybrid UPS System for Backup and Emergency Power Applications 1662
Yuedong Zhan, Jianguo Zhu, Youguang Guo, Hua Wang

Design of the Two Parallel Inverter Modules by Circular Chain Control Technique 1667
K. Piboonwattanakit, W. Khan-Ngern

Investigation of Topologies of Low Voltage Multilevel Inverters 1672
Yanli Zhang, Wanmin Fei, Shoufang Wang

Solution for PWM converter switching for Voltage Source Inverter using Non- Traditional Method 1677
V. Jegathesan, Jovitha Jerome

Piecewise Linear Control Surface for Single Input Nonlinear PI-Fuzzy Controller 1682
S. M. Ayob, Z. Salam, N. A. Azli

Open-Loop Control of a Stepping Motor through IP Network 1686
K. Matsuo, T. Miura, T. Taniguchi

Fuzzy Logic Controller for Electric Vehicle Braking Strategy.....**Fig 4. adjusted due to text re-flow 1691
Xixi. Wang, K.W.Eric Cheng, Xiaozhong Liao, Norbert C. Cheung, Lei Dong

Skid Steering in 4-Wheel-Drive Electric Vehicle 1697
Gao Shuang, Norbert C. Cheung, K. W. Eric Cheng, Dong Lei, Liao Xiaozhong

A Flexible Multi-Pulse Control Strategy for Universal Nail Collator 1703
Chien-Lung Cheng, Shyi-Ching Chern, Jim-Chwen Yeh, Ming-Yi Wu

Cycloconverter Based Three Phase Induction Motor to Replace Flywheel of the Process Machine 1708
M.V. Palandurkar, M. A. Chaudhari, J. P. Modak, S. G. Tarnekar

A Novel Zero-Voltage-Switching Single-Stage High-Power-Factor Electronic Ballast 1712
Chien-Ming Wang, Ching-Hung Su, Chang-Hua Lin, Maw-Yang Liu, Kuo-Lun Fang

Opto-Mechatronic System Design of the LED Projector by Using Brushless DC Motor 1717
Jian-Long Kuo, Tzu-Hsuan Fang

xxi

Table of Contents

The Color Measurement System of PWM-Controlled LCD by Using Back-Propagation Neural Network..............1722
Jian-Long Kuo, Xian-Lin Liu

Gapped Air-cored Power Converter for Intelligent Clothing Power Transfer...1727
Y. Lu, K.W.E.Cheng, Y. L. Kwok, K. W. Kwok, K.W. Chan, N.C.Cheung

Simulation Program for Switching Converters Using Numerical Fourier Transform..................1734
Yoshihiro Tomihisa, Hirotaka Nakanishi, Terukazu Sato, Takashi Nabeshima, Kimihiro Nishijima, Tadao Nakano

The Most Suitable Application of SiC Diode...1740
Tomoaki Makino, Atsushi Hirota, Satoshi Nagai

Multi-Domain System Simulation and Rapid Prototyping of Digital Control Algorithms using VHDL-AMS...1744
P.J. Randewijk

Reforming Power Electronics Laboratory...1752
Xiaohan Guan, Weiping Zhang, Xusen Zhao, Yuanchao Liu

Online performance monitoring and testing of electrical equipment using Virtual Instrumentation........1757
S.S. Murthy, Raghu K. Mittal, Avneesh Dwivedi, G. Pavitra, Sonika Choudhary

A Balancing Strategy and Implementation of Current Equalizer for High Power LED Backlighting................1762
Chang-Hua Lin, Tsung-You Hung, Chien-Ming Wang, Kai-Jun Pai

Modeling of the Parasitical Capacitance Effect in LCD Panel and Corresponding Elimination Strategy................1767
Chang-Hua Lin, Tsung-You Hung, Chien-Ming Wang, Kai-Jun Pai

On-line SOC Estimation of Battery for Wireless Tram Car...1773
Hiroyuki Miyamoto, Masayuki Morimoto, Katsuaki Morita

Narrow- control-bandwidth Operation of Piezoelectric-transformer Converter.............................1777
Weiping Zhang, Xiaoqiang Zhang, Yuzhou Lei, Yuanchao Liu

Modified Map of Variable Active Passive Reactance for Stability Evaluation with Consideration of Capacitor Mode...1782
S. Mohammad Shariatmadar, Jalal Nazarzadeh

Design of the Longitudinal Mode Piezoelectric Transformer...1788
Shine-Tzong Ho

The Comparison of Conducted EMI Emission and Electrical Performances of Lamps....................1794
C. Uyaisom, W. Khan-Ngern

Neural Identification of Average Model of STATCOM using DNN and MLP.................................1799
M. Tavakoli Bina, S. Rahimzadeh

Hybrid Simulation of Power Systems with Dynamic Phasor SVC Transient Model........................1804
E. Zhijun, K. W. Chan, D. Z. Fang

CONTROL OF CURRENT- SOURCE ACTIVE POWER FILTER USING UNIT VECTOR TEMPLATE IN THREE PHASE FOUR WIRE UNBALNCED SYSTEM...1810
K. Vadirajacharya, Pramod Agarwal, H.O. Gupta

Improved Control of Three Phase Active Filters Using Genetic Algorithms.................................1816
Bhim Singh, Varun Singhal

A Fuzzy Adaptive Detecting Approach of Harmonic Currents for Active Power Filter.................1822
Yilong Qu, Weipu Tan, Yihan Yang

Comparative Evaluation of Harmonic Extraction Techniques for Three-Phase Three-Wire Active Power Filter..1827
R. Chudamani, Krishna Vasudevan, C.S. Ramalingam

xxii

Table of Contents

Hybrid Passive Filter Design for Distribution Systems with Adjustable Speed Drives 1834
M.A.S. Masoum, A. Ulinuha, S. Islam, K. Tan

A Graphic User Interface-based Program for Voltage Sag Calculation 1840
T. Tayjasanant, K. Yossombut, P. Sawatpipat

Operational Characteristics of Fault Current Limiting Reactor Combined with Multi- Functional Inverter 1846
S. H. Ko, S. H. Lim, S. R. Lee, S. W. Lee, I. C. Kim, S. H. Ko, H. S. Kim

Low Cost AC Solid State Circuit Breaker 1851
W. Pusorn, W. Srisongkram, W. Subsingha, S. Deng-Em, P. N. Boonchiam

A Variable Gain Control Scheme of Digital Automatic Voltage Regulator for AC Generator 1857
Dong-Hee Lee, Jin-Woo Ahn, Tae-Won Chun

A Graphic User Interface-based Program for Harmonic Impedance Calculation 1862
T. Tayjasanant

The analysis and simulation of power circuits for AC high-voltage converters 1868
Y.Y. Skorokhod, S.I. Volskiy

A Single Stage Flyback PFC Converter for Testing Distance Relay Systems 1875
V. Fernao Pires, J. F. Martins, J. Fernando Silva

H-Infinity Control Theory Apply to New Type Arc-suppression Coil System 1880
Yilong Qu, Weipu Tan, Yihan Yang

Characteristics of a novel topology of a DC-AC Converter for Fuel Cells 1885
K. Fukushima, T. Ninomiya, I. Norigoe, Y. Harada, K. Tsukakoshi, Z. Dai

A Comparative Study of PWM Schemes for Grid Connected PV Cell 1891
Vineeta Agarwal, Alok Vishwakarma

xxiii

This page intentionally left blank.

Fuel cell systems and applications

1. Introduction

2. Principle and different technologies

3. Applications

4. Fuelling fuel cells

5. Conclusion and references

Bangkok, Tutorial 27th November 2007

Bernard DAVAT - INPL - Nancy - France

1. Introduction

Bernard DAVAT - INPL - Nancy - France

Bangkok, Tutorial 27th November 2007

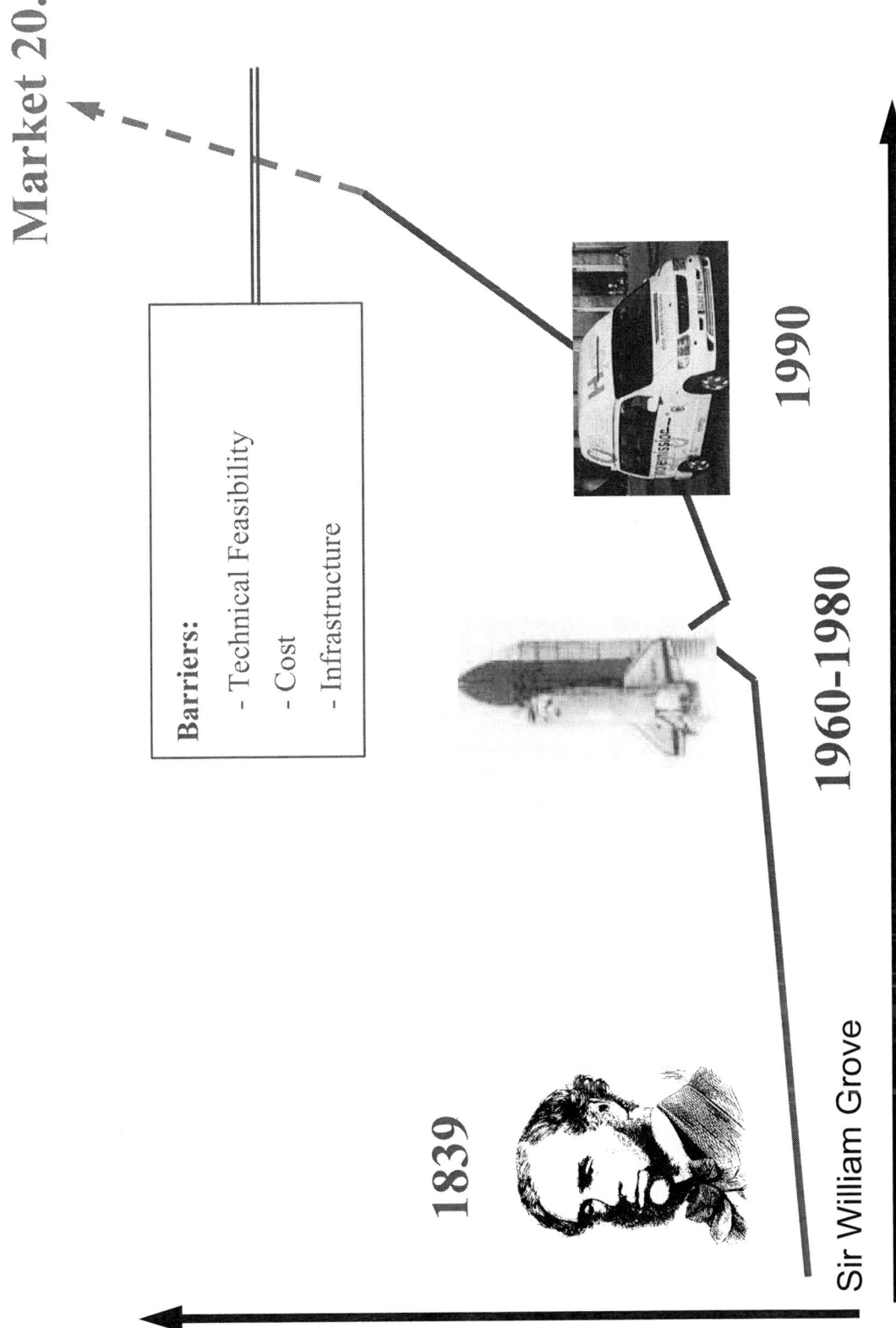

Market 20..

Barriers:
- Technical Feasibility
- Cost
- Infrastructure

1839

Sir William Grove

1960-1980

1990

Bernard DAVAT - INPL - Nancy - France

Bangkok, Tutorial 27th November 2007

Fuel cell principle discovery - W. Grove 1839 Principle

Replace the battery by an amperemeter and a small current is flowing.

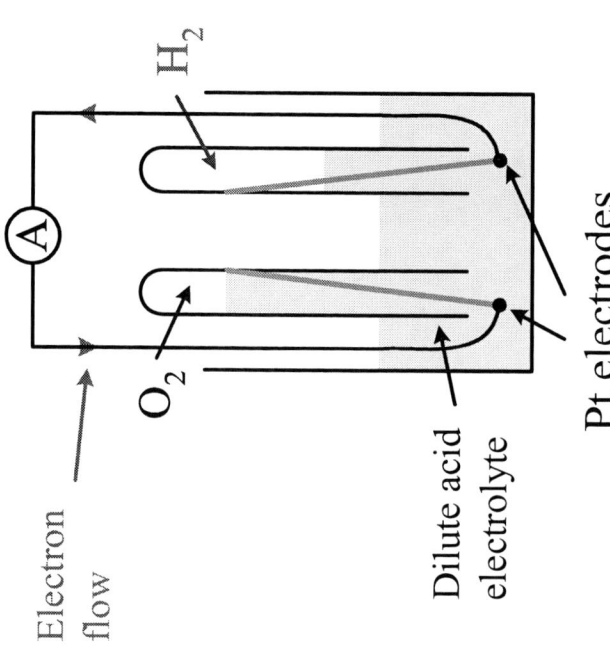

The electrolysis of water is reversible

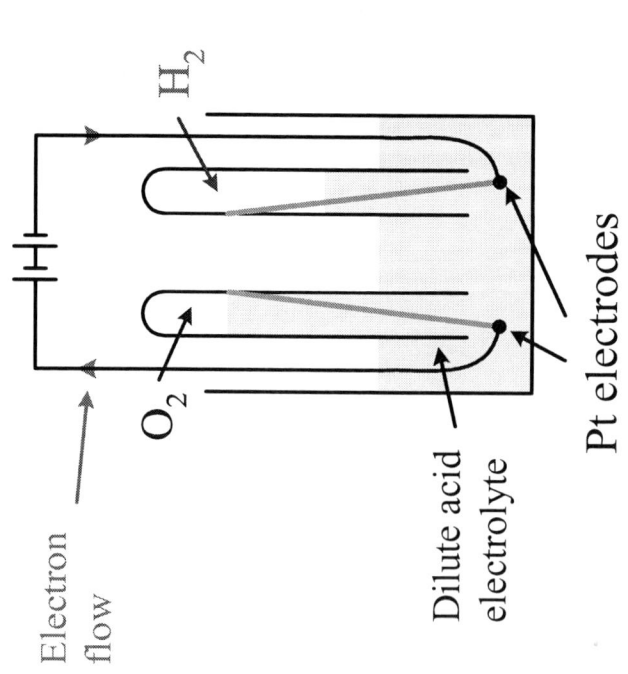

The electrolysis of water

Bernard DAVAT - INPL - Nancy - France   *Bangkok, Tutorial 27th November 2007*

Fuel Cells development

- 1839 Discovery of the effect: R.W. Grove, On the Voltaic Series and the Combination of Gasses by Platinum;

- 1843 Construction of a gas battery;

- 1889 Work by L. Mond and C. Langer result in first alkaline fuel cell. High losses at oxygen **AFC** electrode discovered;

- 1896 W.W. Jacques used molten sodium hydroxide as the electrolyte intending direct conversion of coal;

- 1900 W. Nernst conceptual work in the field of solid oxide fuel cells; **SOFC**

- 1905 F. Haber carried out systematic thermodynamic investigations of hydrogen consuming fuel fuel cells;

- 1932 F.T. Bacon started a long term fuel cell research program;

- 1935 W. Schottky developed the theoretical basics of the solid oxide fuel cells; **SOFC**

- 1938 E. Baur and H. Preis reported on experimental solid oxide fuel cells work; **SOFC**

- 1959 F.T. Bacon built the first working 5 kW alkaline fuel cell stack; **AFC**

- 1964 Polymer electrolyte membrane fuel cell supplied electricity to Gemini spacecraft; **PEMFC**

- 1967 Concept of phosphoric acid fuel cell;

- 1960-80 Alkaline fuel cell were used for Apollo and Space Shuttle; **AFC**

- 1984 Rediscovery of polymer electrolyte membrane fuel cell (Siemens/Ballard) **PEMFC**

Bernard DAVAT - INPL - Nancy - France

Bangkok, Tutorial 27th November 2007

Lessons from fuel cell history - Facts

- Hydrogen is the only fuel offering sufficient reactivity. Even at high temperatures (SOFC) CO-oxidation is by a factor of 10 slower than H_2-oxidation;

- Direct use of coal is not a realistic option;

- Frequently rare or expensive materials are needed for fuel cell construction;

- Construction of efficient electrodes is difficult;

- At low temperatures frequently noble metal catalysts are required;

- The working of a lonely cell does not make a power plant works. Problems of stack and balance.

Bernard DAVAT - INPL - Nancy - France　　　　　*Bangkok, Tutorial 27th November 2007*

2. Principle and different technologies

Bernard DAVAT - INPL - Nancy - France

Bangkok, Tutorial 27th November 2007

Principle

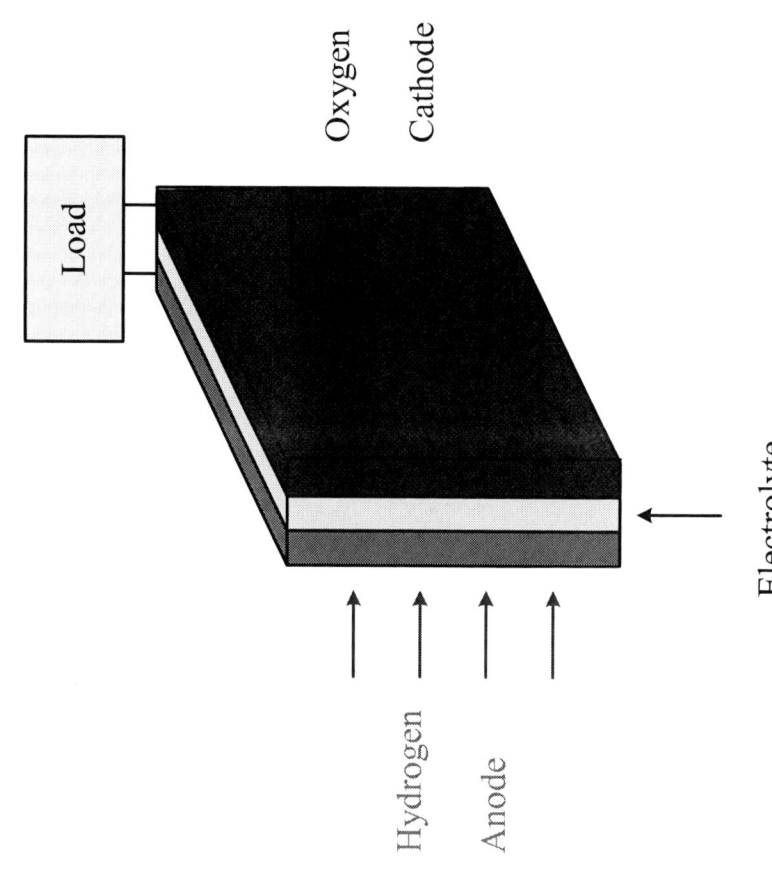

As H_2 and O_2 are never in the same part of the system, this reaction does not produce heat but electricity.

Bernard DAVAT - INPL - Nancy - France

Bangkok, Tutorial 27th November 2007

H_2 is burnt in the simple reaction:

$$H_2 + \frac{1}{2}O_2 \rightarrow H_2O$$

As H_2 and O_2 are separated, two reactions are necessary: one at each electrode. At the anode of an acid electrolyte fuel cell:

$$H_2 \rightarrow 2H^+ + 2e^-$$

This reaction releases energy.

At the cathode, oxygen reacts with electrons and hydrogen ions:

$$\frac{1}{2}O_2 + 2e^- + 2H^+ \rightarrow H_2O$$

Bernard DAVAT - INPL - Nancy - France

Bangkok, Tutorial 27th November 2007

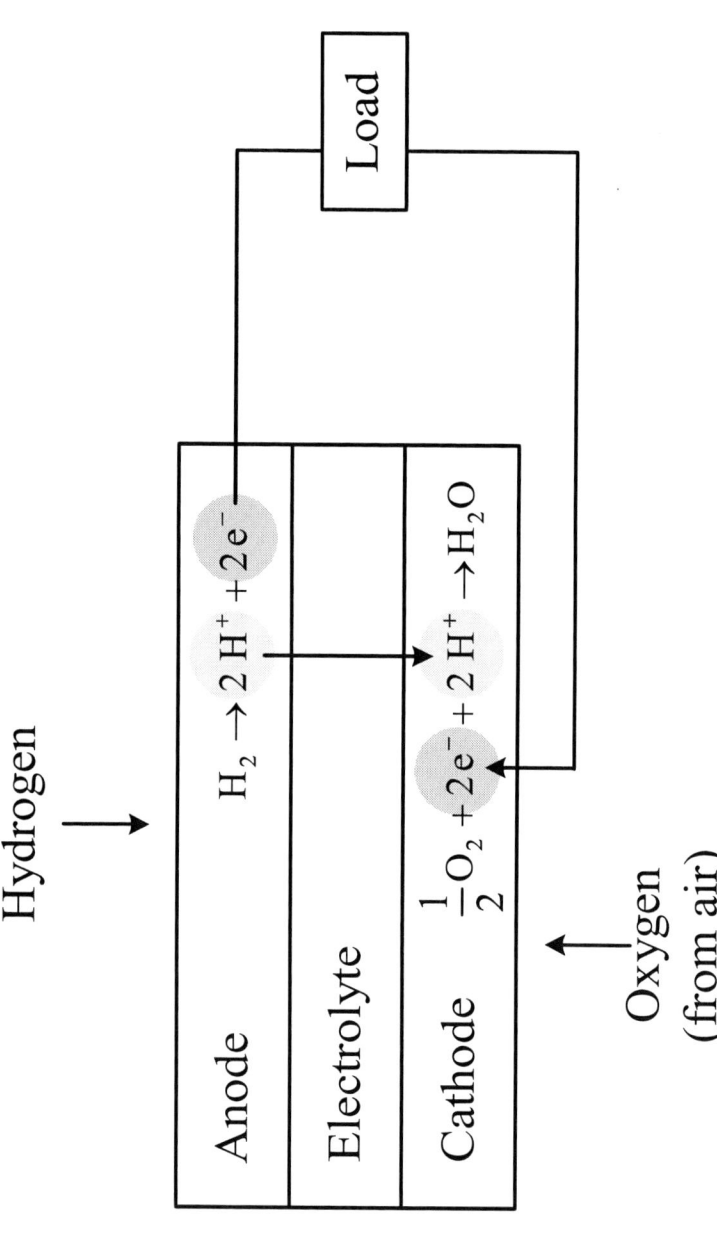

At the anode hydrogen reacts releasing energy. As this energy is limited, the current is limited. To avoid this limitation, there are three main ways which explain the structure of fuel cells: the use of catalysts, raising the temperature and increasing the electrode area.

Bernard DAVAT - INPL - Nancy - France   *Bangkok, Tutorial 27th November 2007*

Elementary cell PEMFC

Electricity is obtained across the cell.

Only H+ can pass through the membrane. Electrons have to go through the external circuit

Bipolar plate distributing on one side air (O_2) and on the other side hydrogen

The results of this reaction is water leaving the cell mainly with the air (vapor)

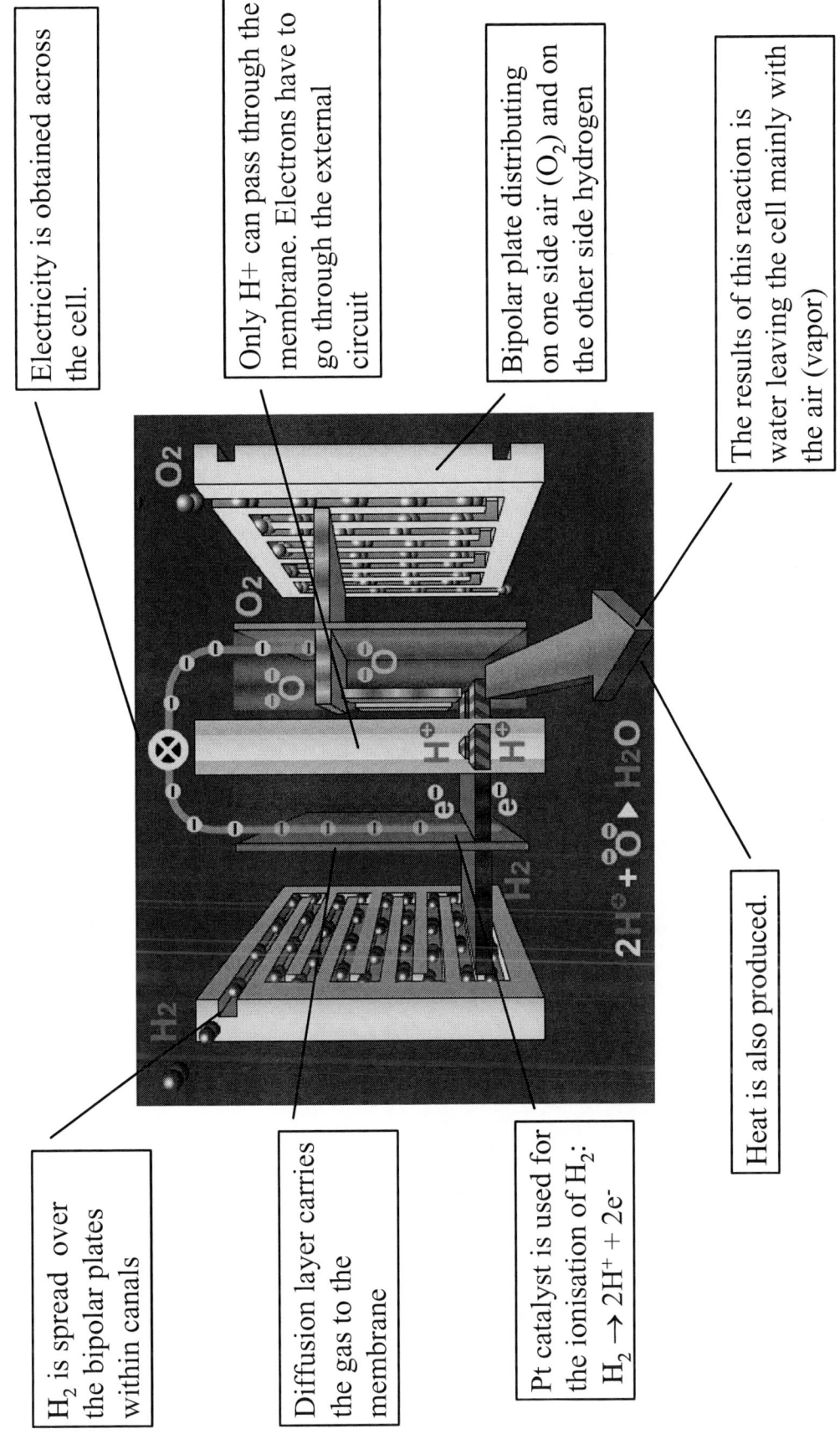

H_2 is spread over the bipolar plates within canals

Diffusion layer carries the gas to the membrane

Pt catalyst is used for the ionisation of H_2:
$H_2 \rightarrow 2H^+ + 2e^-$

Heat is also produced.

Bernard DAVAT - INPL - Nancy - France

Bangkok, Tutorial 27th November 2007

Diffusion layer

Bipolar plate

Bernard DAVAT - INPL - Nancy - France

Bangkok, Tutorial 27th November 2007

Membrane electrode assembly (MEA)

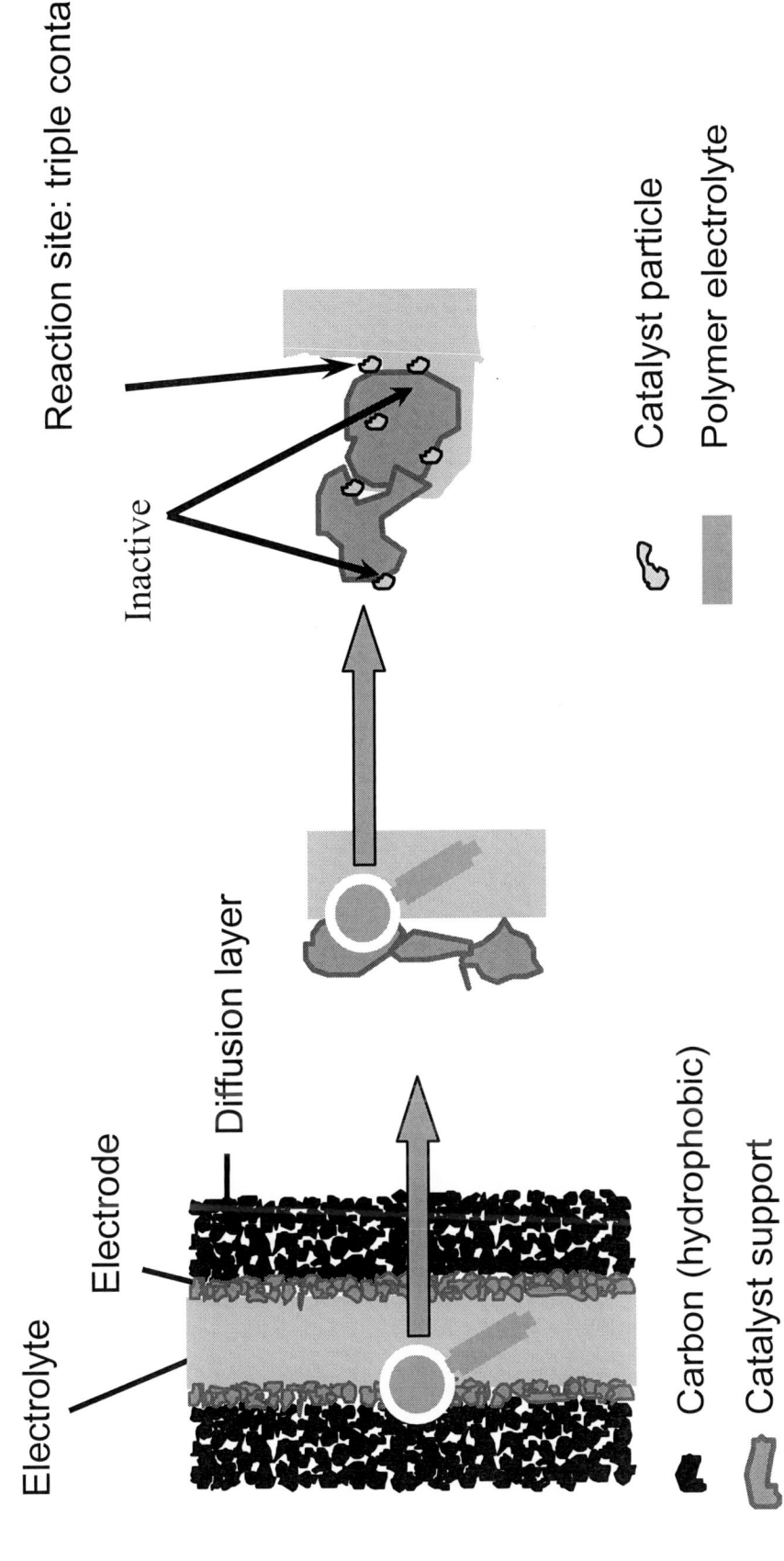

Reaction site: triple contact

Inactive

Electrolyte

Electrode

Diffusion layer

Catalyst particle

Polymer electrolyte

Carbon (hydrophobic)

Catalyst support

Bernard DAVAT - INPL - Nancy - France

Bangkok, Tutorial 27th November 2007

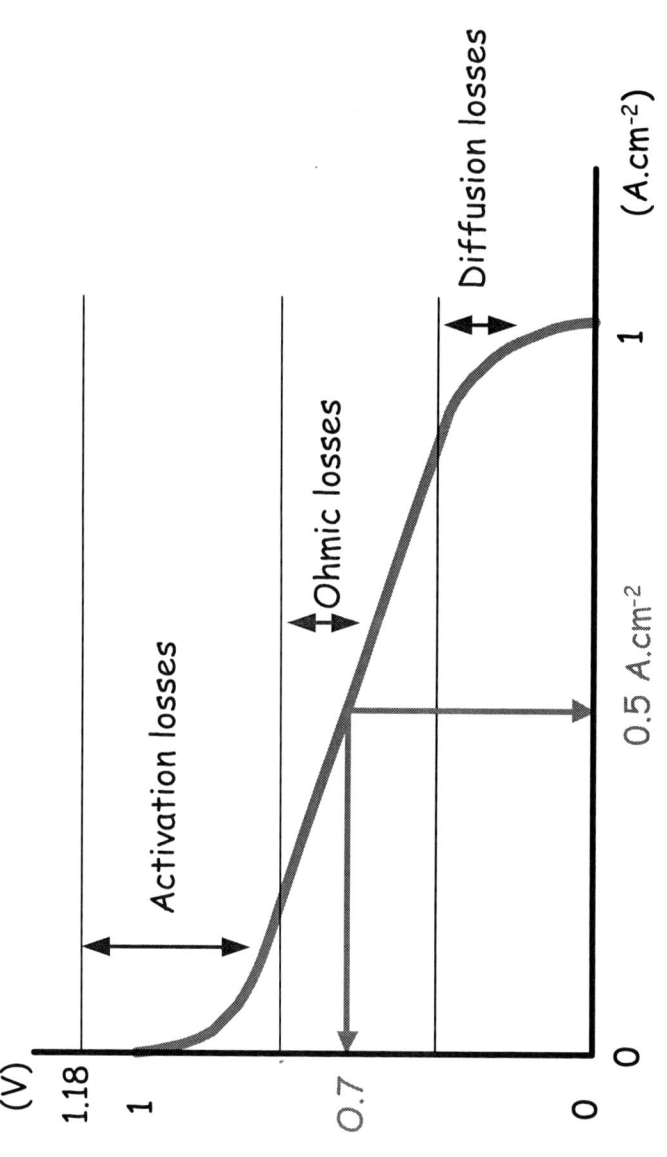

A nominal voltage of about 0.7 V for a current of 0.5 $A.cm^{-2}$. A fuel cell is a low voltage, high current device which necessitates to put in series elementary cells.

Bernard DAVAT - INPL - Nancy - France

Bangkok, Tutorial 27th November 2007

Fuel cell stack

120 cells, 5 kW PEMFC (HPower)

23 cells, 500 W PEMFC (ZSW)

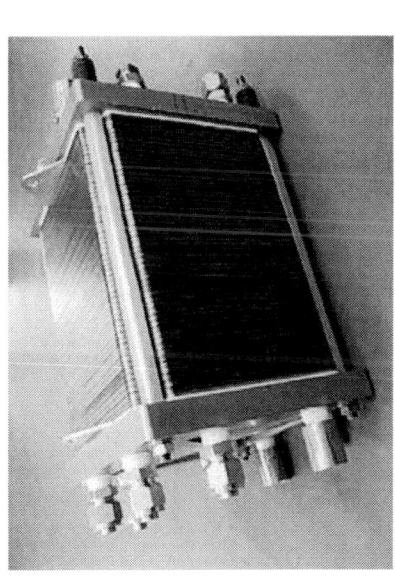

51 cells, 700 W DMFC (IRD)

Stack principle

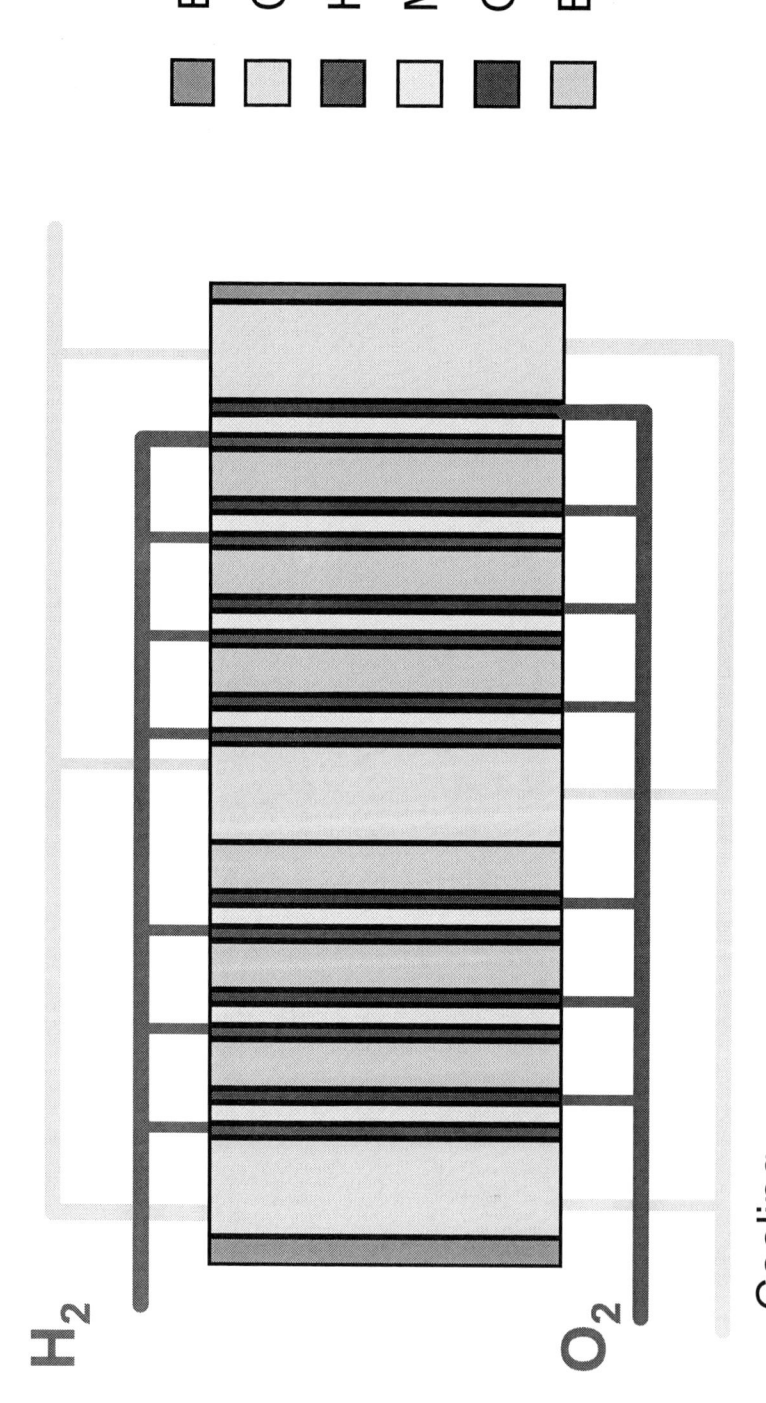

Legend:
- End plate
- Cooling
- Hydrogen
- MEA
- Oxygen/Air
- Bipolar plate

H₂

O₂

Cooling

Bernard DAVAT - INPL - Nancy - France

Bangkok, Tutorial 27th November 2007

Example of fuel cell system

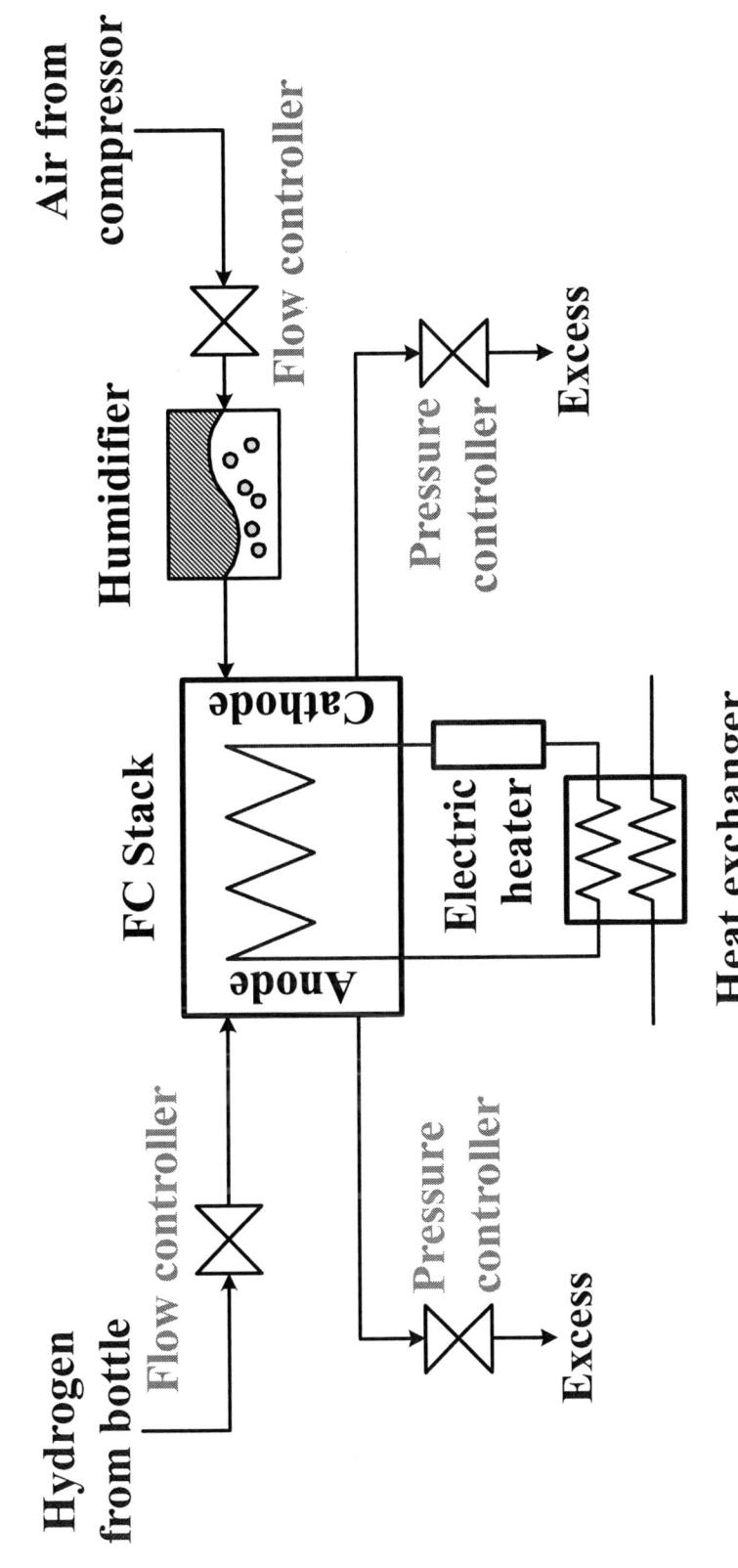

Bernard DAVAT - INPL - Nancy - France

Bangkok, Tutorial 27th November 2007

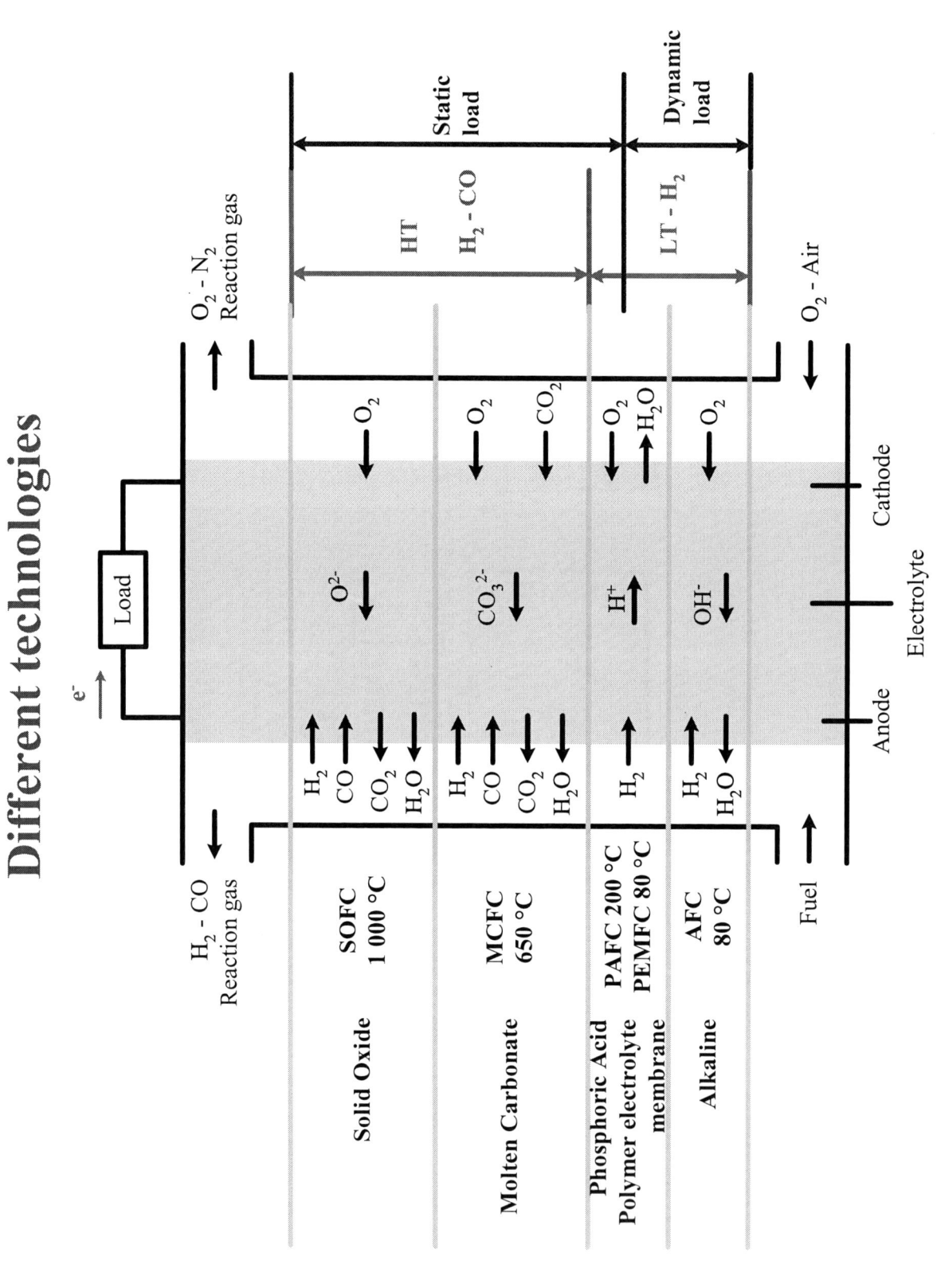

11-MW-PAFC-Power Plant

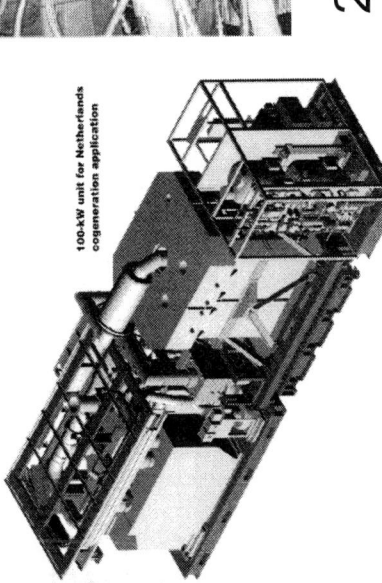
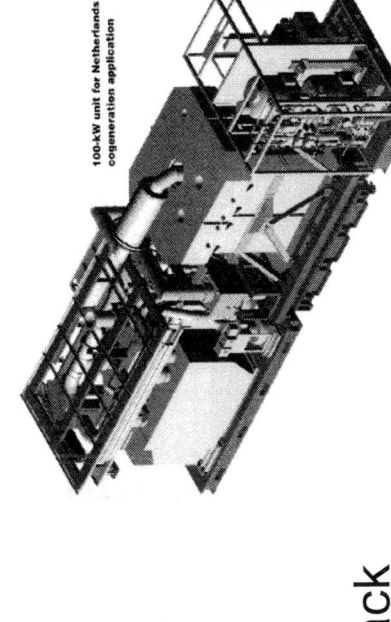

25-kW-PEMFC-Stack (1996)

100-kW-SOFC-System (1998)

48-kW-AFC-System (1990)

250-kW-MCFC-Stack

Bangkok, Tutorial 27th November 2007

Bernard DAVAT - INPL - Nancy - France

3. Applications

Bernard DAVAT - INPL - Nancy - France

Bangkok, Tutorial 27th November 2007

3.1. Stationary applications

Heat and combine power generator (1 - 5 kW)

Electricity generator

Heat and combine power generator (100 - 250 kW)

Bernard DAVAT - INPL - Nancy - France

Bangkok, Tutorial 27th November 2007

Heat and combine power generator (1-5 kW)

Technische Daten:

Durchmesser Zellen: 120 mm
Höhe (Zellenstapel) 518 mm
Anzahl der Zellen: 70
Zellenfläche: 0,7 m^2
Betriebstemperatur: 950 °C
Stackspannung: 39 V
Stackstrom: 27 A
elektrische Leistung: 1053 W

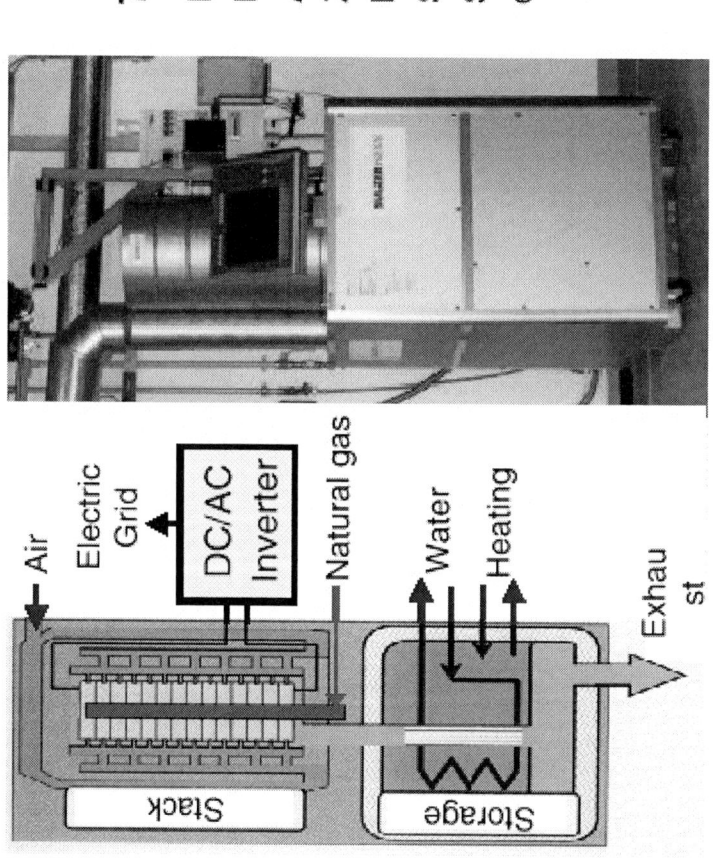

SOFC (Sulzer)

Bernard DAVAT - INPL - Nancy - France

Bangkok, Tutorial 27th November 2007

2.5kW Power Generator, Specifications

Stack type	P01 - 2.5 - 0.3
Stack quantity	1
Electrical Power output, kW	0-2.5
Thermal power output, kW	0-3
Max. electrical efficiency, %	40
Max. heat and power efficiency, %	90
Grid connection V/Hz	230/50
Hydrogen inlet pressure, bar(g)	3-6
Hydrogen consumption, Nm^3/h	2.4
Air supply, Nm^3/h	17
Dimensions, l-w-h, mm	600 x 600 x 1600
Operating temperature range, °C	5-50
Operating humidity, RH %	0-90
Operation and control	PC

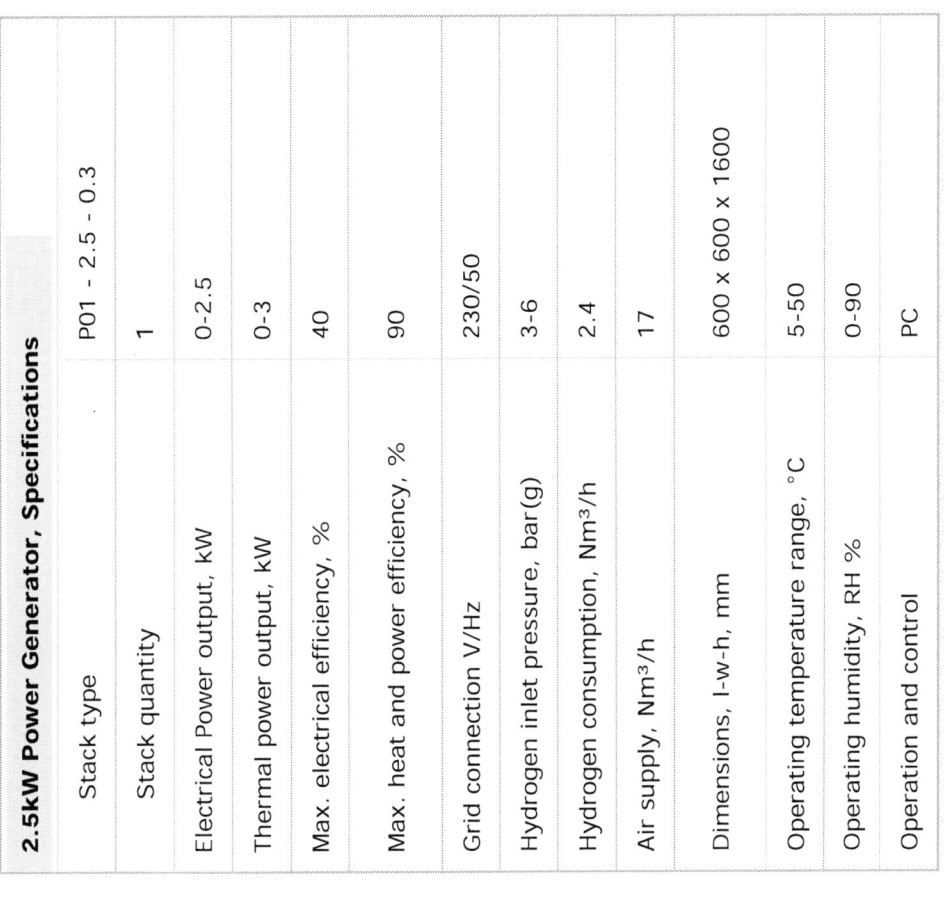

PEMC (IRD)

Bernard DAVAT - INPL - Nancy - France

Bangkok, Tutorial 27th November 2007

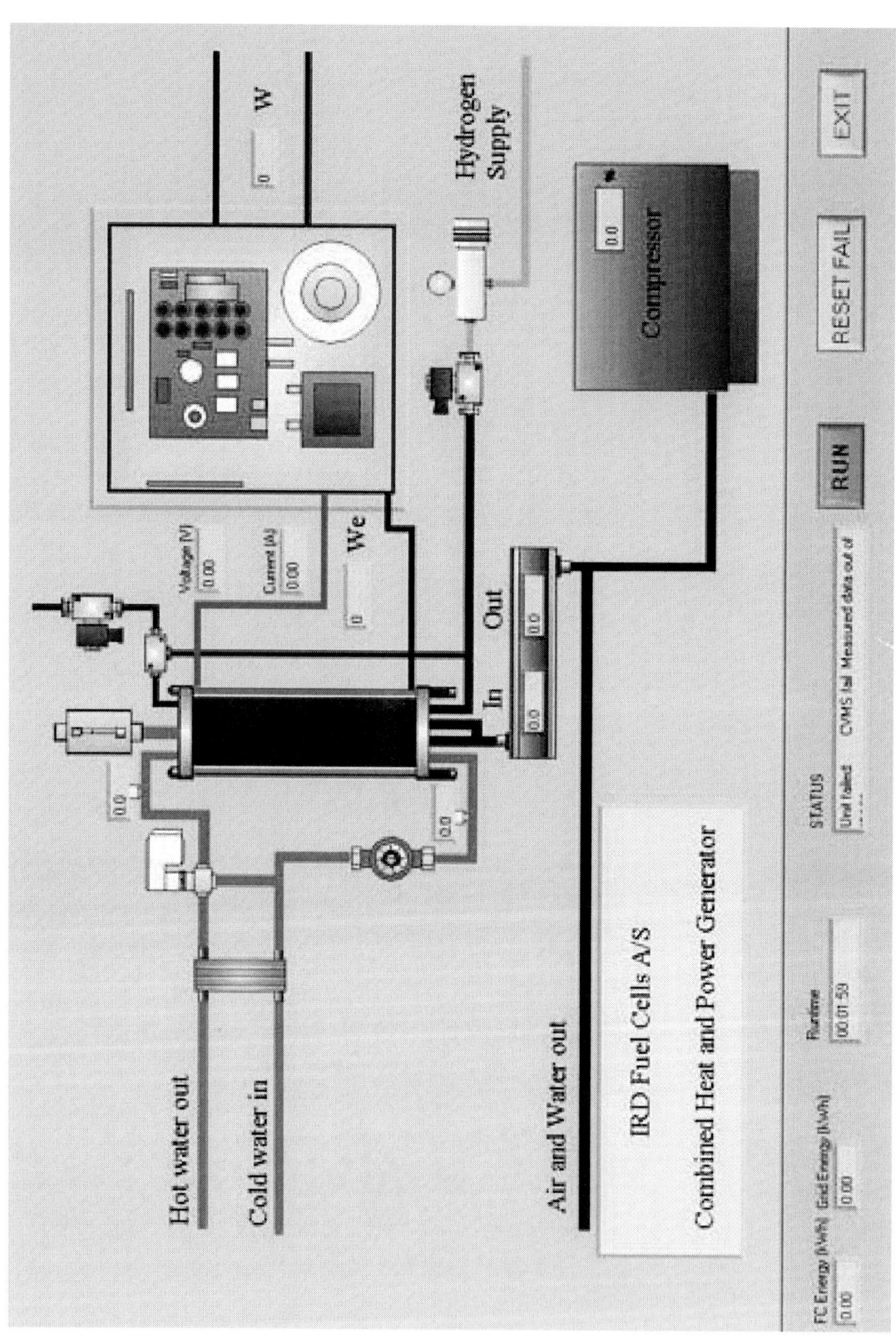

Bernard DAVAT - INPL - Nancy - France

Bangkok, Tutorial 27th November 2007

Many tests in Europe, mainly in Germany with different boiler manufacturers (250 systems from Vaillant, 350 from Sulzer).

For example, 55 Sulzer systems have been proposed to customers in the base of a Global Energy Supplying Program:

- Innovation contribution of 2 000 €;
- Heat subscription of 815 € / year and 4.45 ct / kWh;
- Maintenance: free.

THE EUROPEAN VIRTUAL FUEL CELL POWER PLANT

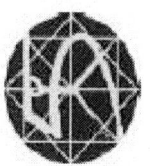

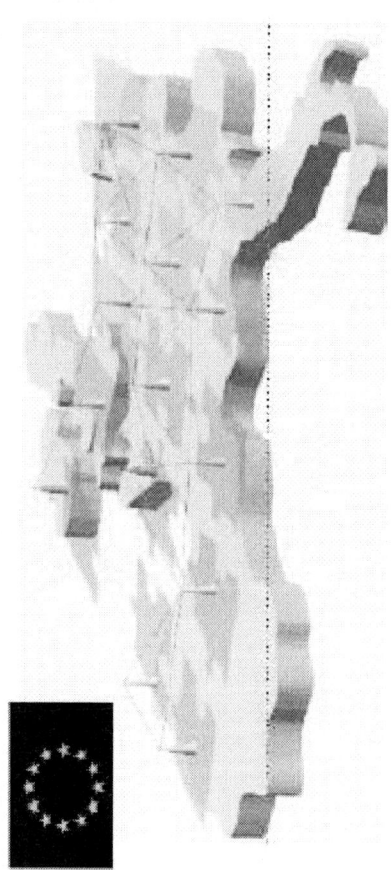

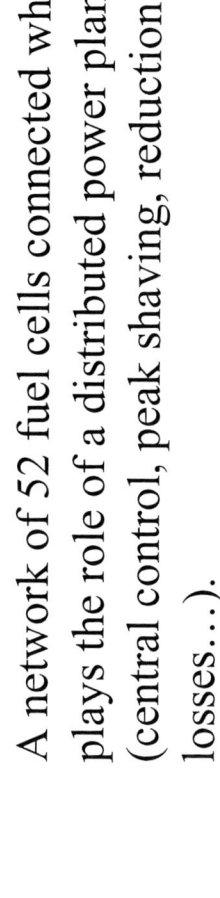

A network of 52 fuel cells connected which plays the role of a distributed power plant (central control, peak shaving, reduction of losses...).

40 months program, 8.6 M€ (36 % from EC).

Bangkok, Tutorial 27th November 2007

Bernard DAVAT - INPL - Nancy - France

Electricity generator

- 20 Cells
- Air supply (internal recycling of damp air)
- Silent
- Sizes 45 x 32 x 40 cm
- Peak power 350 W
- Rated power 250 W
- Losses < 15 W
- Pure H_2 supply

Powerbag ZSW

Bernard DAVAT - INPL - Nancy - France

Bangkok, Tutorial 27th November 2007

0.5 - 5 kW Axane Generator

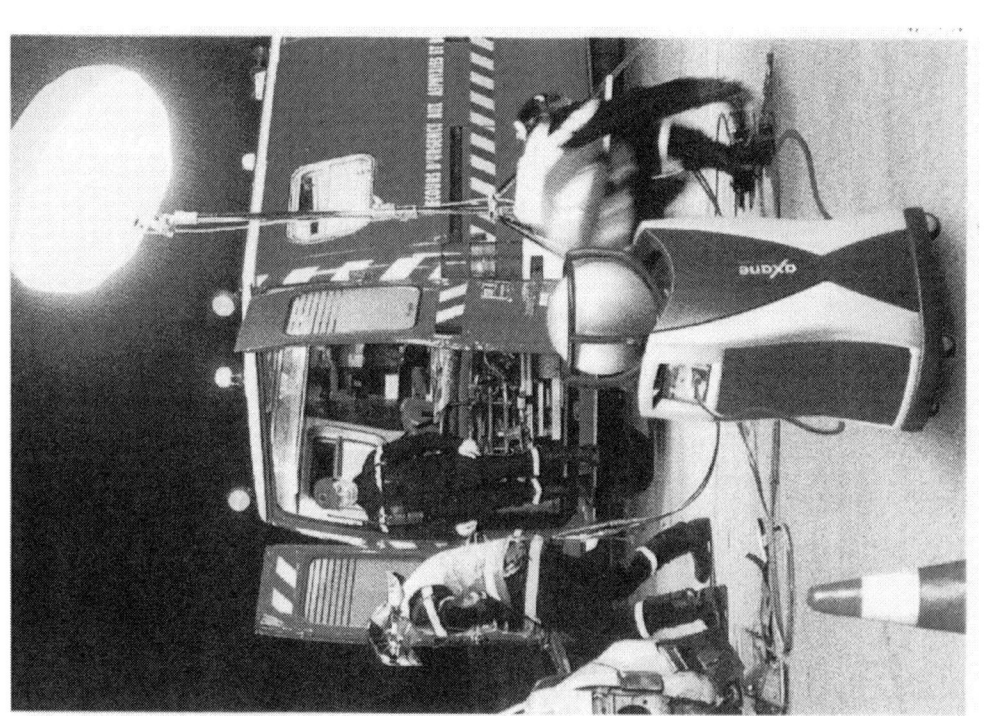

Bernard DAVAT - INPL - Nancy - France

Bangkok, Tutorial 27th November 2007

Heat and combine power generator (100-25 kW)

PEMFC, MCFC, PAFC or SOFC systems.

For example ONSI (now UTC FC) has installed 250 units of PAFC systems all other the world since 1990.

According to the technology of the fuel cell the cogeneration process used warm water (PEMFC, PAFC) or steam (MCFC, SOFC).

Bernard DAVAT - INPL - Nancy - France

Bangkok, Tutorial 27th November 2007

EDF-GDF PAFC

Chelles (Paris) natural gas PAFC
Electrical power 200 kW
Thermal power 220 kW
(for 200 apartments)

Bernard DAVAT - INPL - Nancy - France

Bangkok, Tutorial 27th November 2007

Advantages: Comparison with a gas motor

- a good efficiency even at half load ;

- a reduction of polluting gases.

	ONSI system	Köhler Ziegler motor
CO	$24.8 \ mg/Nm^3$	$325 \ mg/Nm^3$
NOx	$2.7 \ mg/Nm^3$	$250 \ mg/Nm^3$
CO_2	$0.5 \ kg/kWh$ at rate power $0.48 \ kg/kWh$ at half power	$0.68 \ kg/kWh$ at rate power $0.85 \ kg/kWh$ at half power

Bernard DAVAT - INPL - Nancy - France

Bangkok, Tutorial 27th November 2007

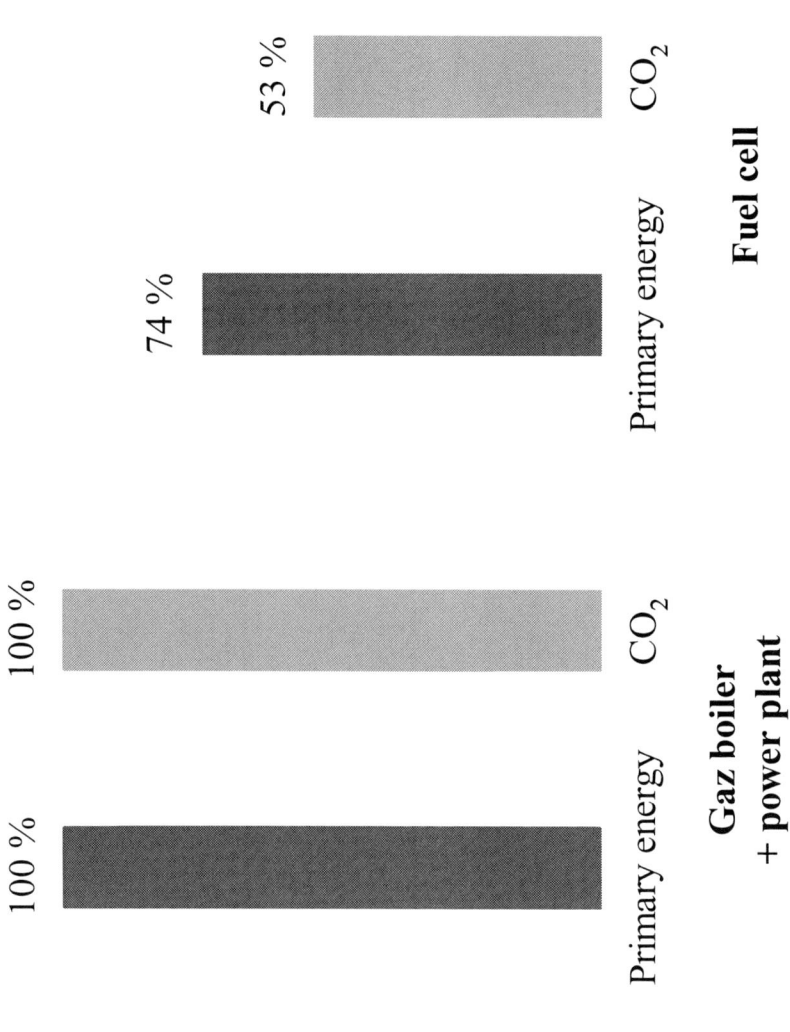

Reduction of energy need and of green house gases (German example)

Bernard DAVAT - INPL - Nancy - France

Bangkok, Tutorial 27th November 2007

3.2. Applications in Transport Cars

Cars

Buses

Bernard DAVAT - INPL - Nancy - France

Bangkok, Tutorial 27th November 2007

Cars

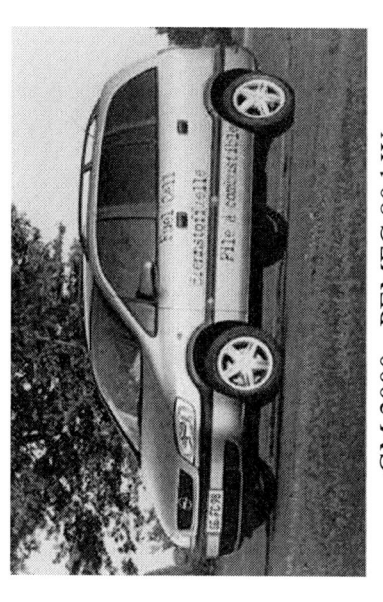

GM 2000 - PEMFC 80 kW

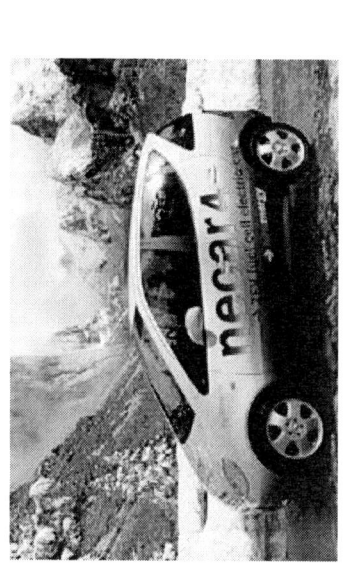

Ballard - Daimler 1999 - PEMFC 70 kW

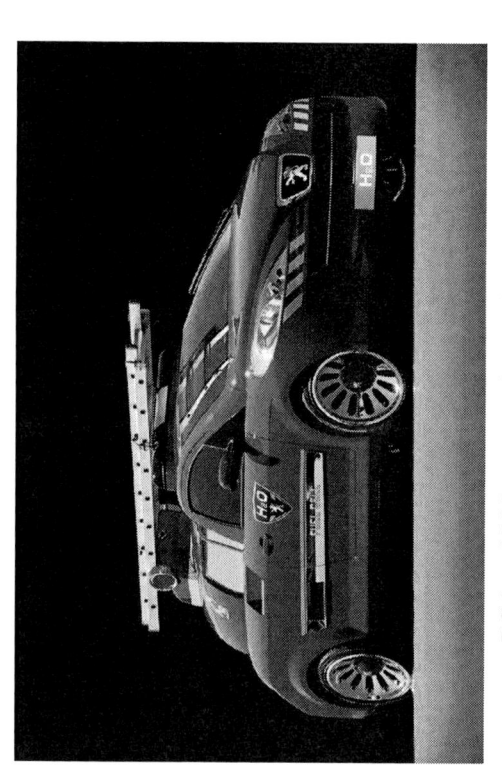

PSA – 2003 – Batteries and 5.5 kW PEMFC

Bangkok, Tutorial 27th November 2007

Bernard DAVAT - INPL - Nancy - France

Daimler Chrysler test project

Daimler Chrysler will test until 2007, 60 Fuel cell A-classes under everyday operating conditions in Asia, North America and Europe.

Refuelling H_2 in Tokyo

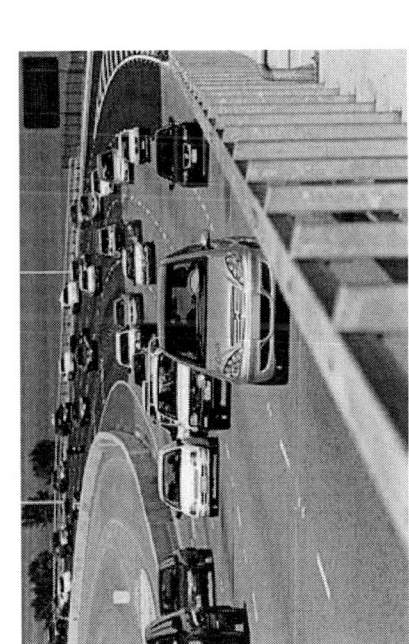

On US highways

The goal is to reach 16 000 km for each vehicle. This means at least a road experience of 1 million km and 60 000 hours.

Bernard DAVAT - INPL - Nancy - France

Bangkok, Tutorial 27th November 2007

Buses

A lot of projects all over the world and a Fuel cell Bus Club in Europe. The CUTE program (Clean Urban Transport in Europe) is actually testing 27 hydrogen fuel cell busses in nine European cities until 2006.

Hydrogen is provided from electrolysis, or steam reforming of oil or from reforming of natural gas to compare the different technologies.

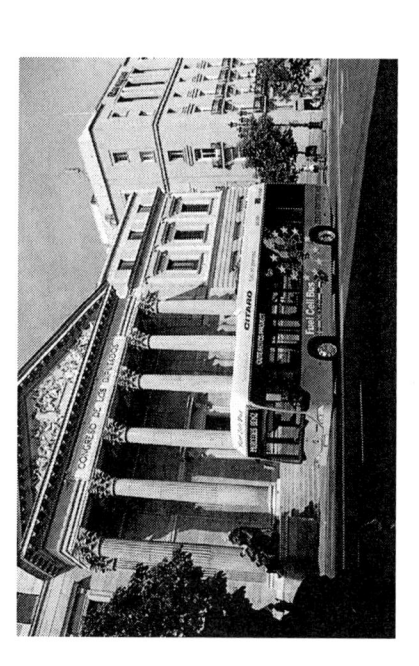

Bernard DAVAT - INPL - Nancy - France

Bangkok, Tutorial 27th November 2007

3.3. Fuel cells for portable applications

Telephone, PC

Battery

Bernard DAVAT - INPL - Nancy - France

Bangkok, Tutorial 27th November 2007

Telephone, PC

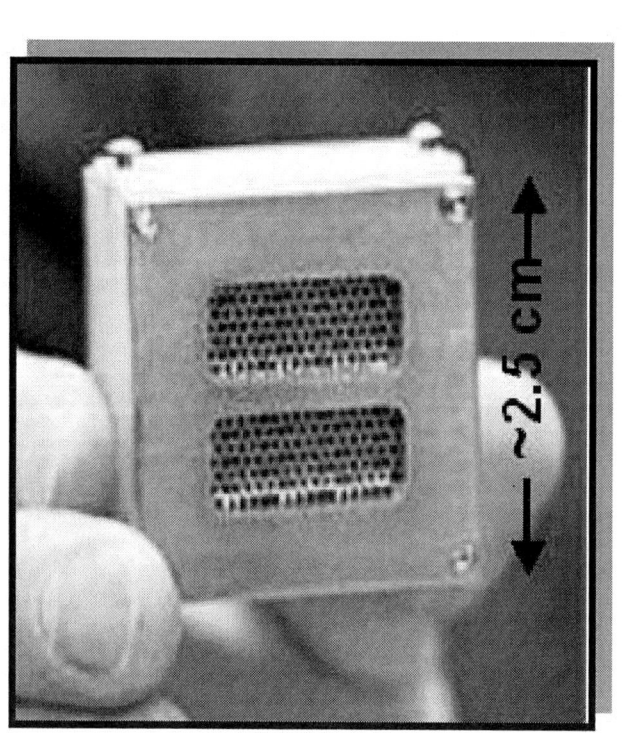

Motorola DEMFC for palmtop
(100 mW, 200 mA, 0,5V)

Bangkok, Tutorial 27th November 2007

Bernard DAVAT - INPL - Nancy - France

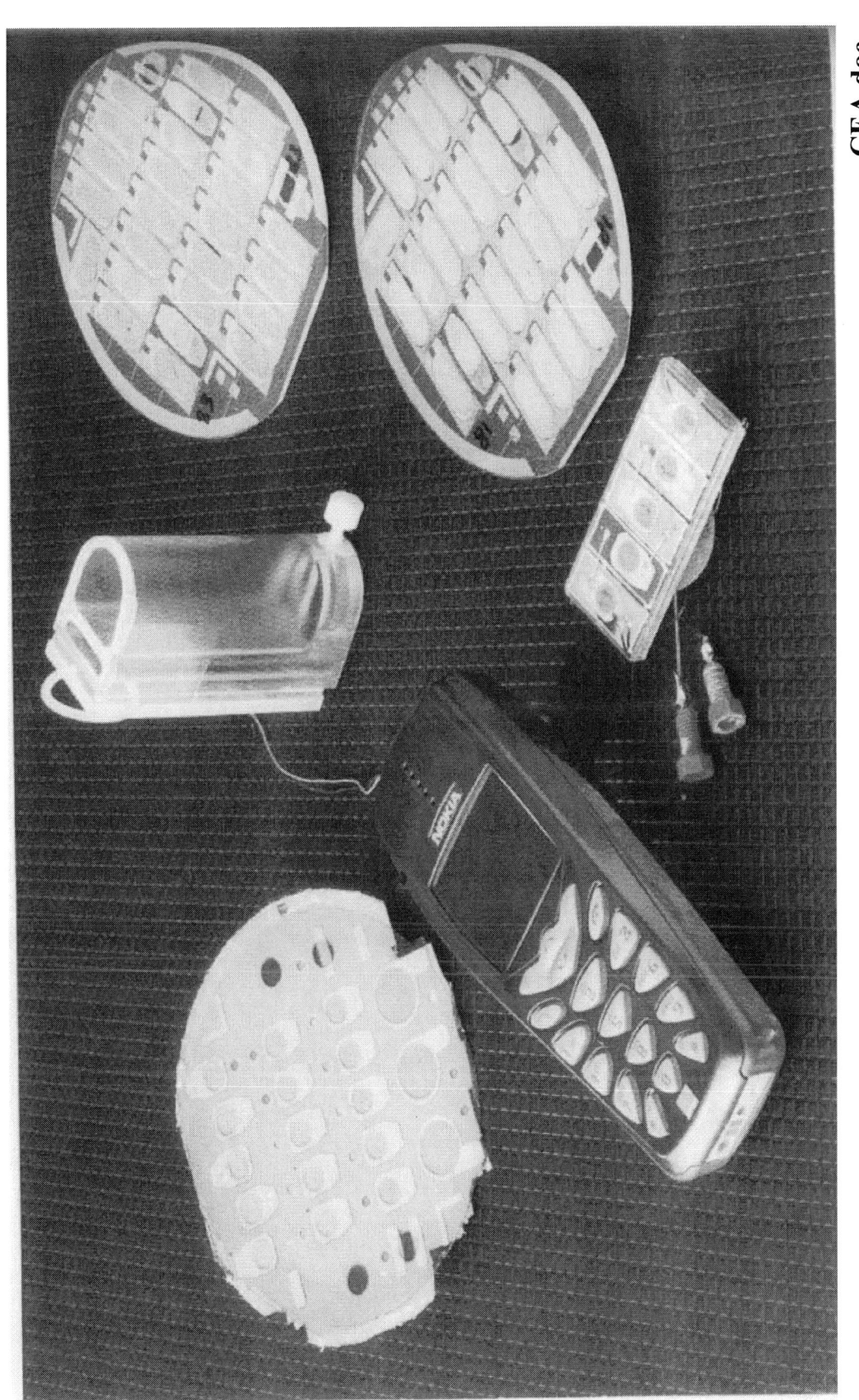

CEA doc.

Bangkok, Tutorial 27th November 2007

Bernard DAVAT - INPL - Nancy - France

Hydrogen battery (Millenium Cell)

A hydrogen battery has two major components: a fuel cell module that provides power and an energy storage module that fuels the fuel cell.

Hydrogen is produced on demand as it is stored on sodium borohydride:

$$NaBH_4 + 2\,H_2O \Longrightarrow 4\,H_2 + NaBO_2 + Heat$$

Fuel is an ambient temperature, non-flammable liquid. Hydrogen produced with this process is of high purity (no carbon or sulfur in the process) and is humidified (heat generates some water vapor)

Bernard DAVAT - INPL - Nancy - France

Bangkok, Tutorial 27th November 2007

	Primary Lithium Sulfur Dioxide	Secondary Lithium-Ion	Primary Zinc-Air	Sodium Borohydride	Methanol
Gravimetric Energy Density Wh/kg	170	110	300	7 100	6 000
Volumic Energy Density Wh/l	190	300	240	7 314	4 800

4. Fuelling fuel cells

One of the main problem: Hydrogen
$(10 \text{ L.kW}^{-1}.\text{mn}^{-1})$

Bernard DAVAT - INPL - Nancy - France

Bangkok, Tutorial 27th November 2007

4.1. Hydrogen as fuel

- Hydrogen is already widely used in different technical applications: food technology, fertilizer and oil refineries;

- Hydrogen can be used in a (solar) hydrogen economy as clean fuel, seasonal energy storage, intercontinal energy vector and chemical raw material (regenerative petrochemistry);

- However: Hydrogen is a „rather bulky fuel".

Bernard DAVAT - INPL - Nancy - France

Bangkok, Tutorial 27th November 2007

4.2. Hydrogen distribution

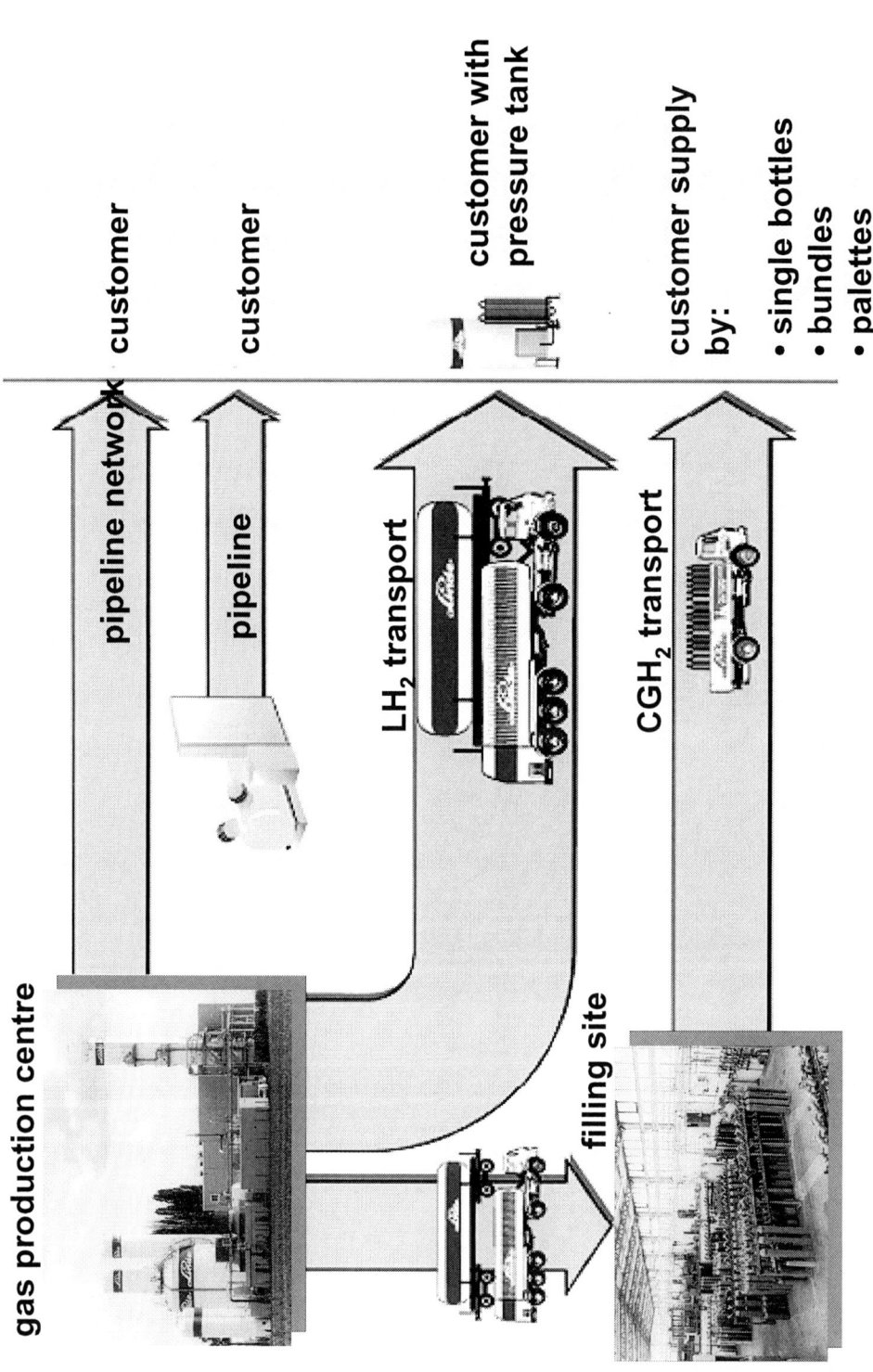

Bernard DAVAT - INPL - Nancy - France

Bangkok, Tutorial 27th November 2007

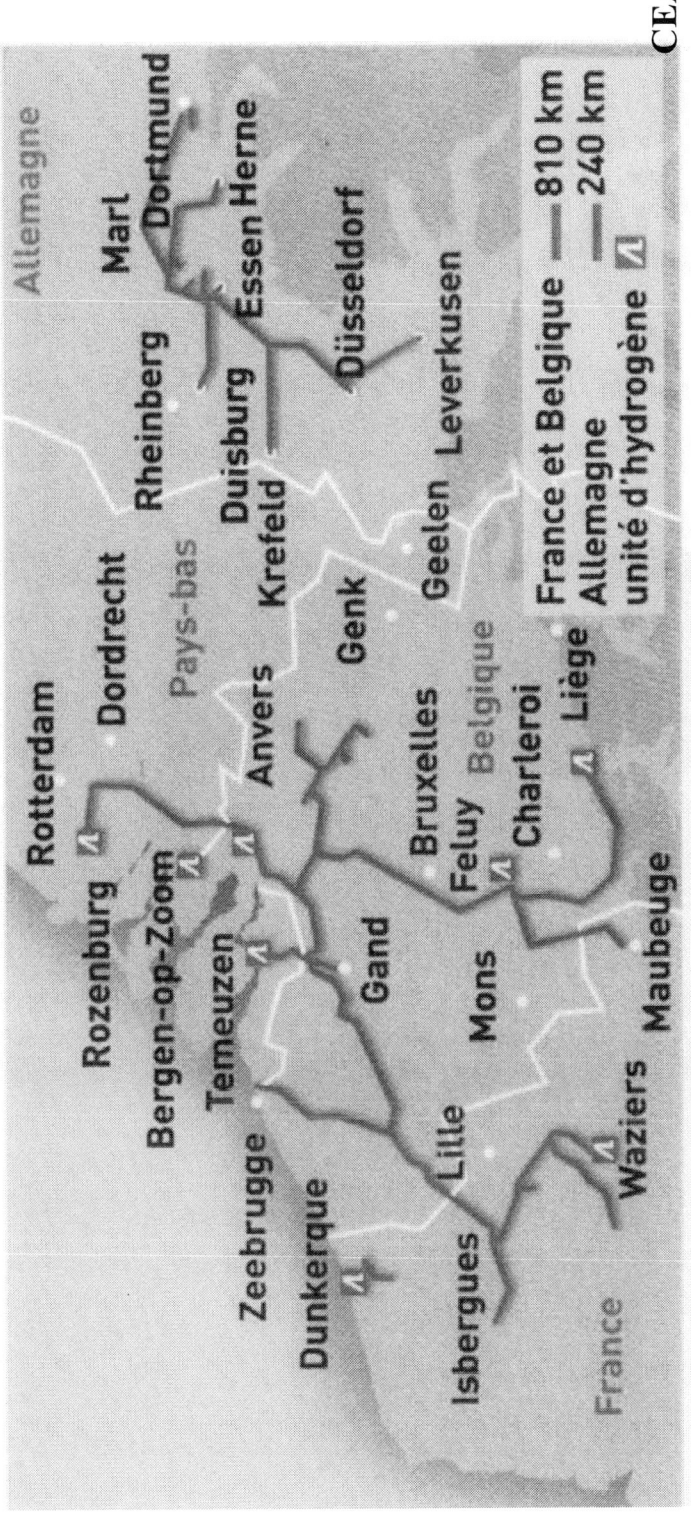

Hydrogen networks between Holland, Belgium and France and in Germany

CEA doc.

Bernard DAVAT - INPL - Nancy - France   *Bangkok, Tutorial 27th November 2007*

4.3. Hydrogen production

Steam reforming of natural gas

Water electrolysis

By-product of chlorine industry

Steam reforming of natural gas

$$CH_4 + 2H_2O \Rightarrow 4H_2 + CO_2 \ (CO)$$

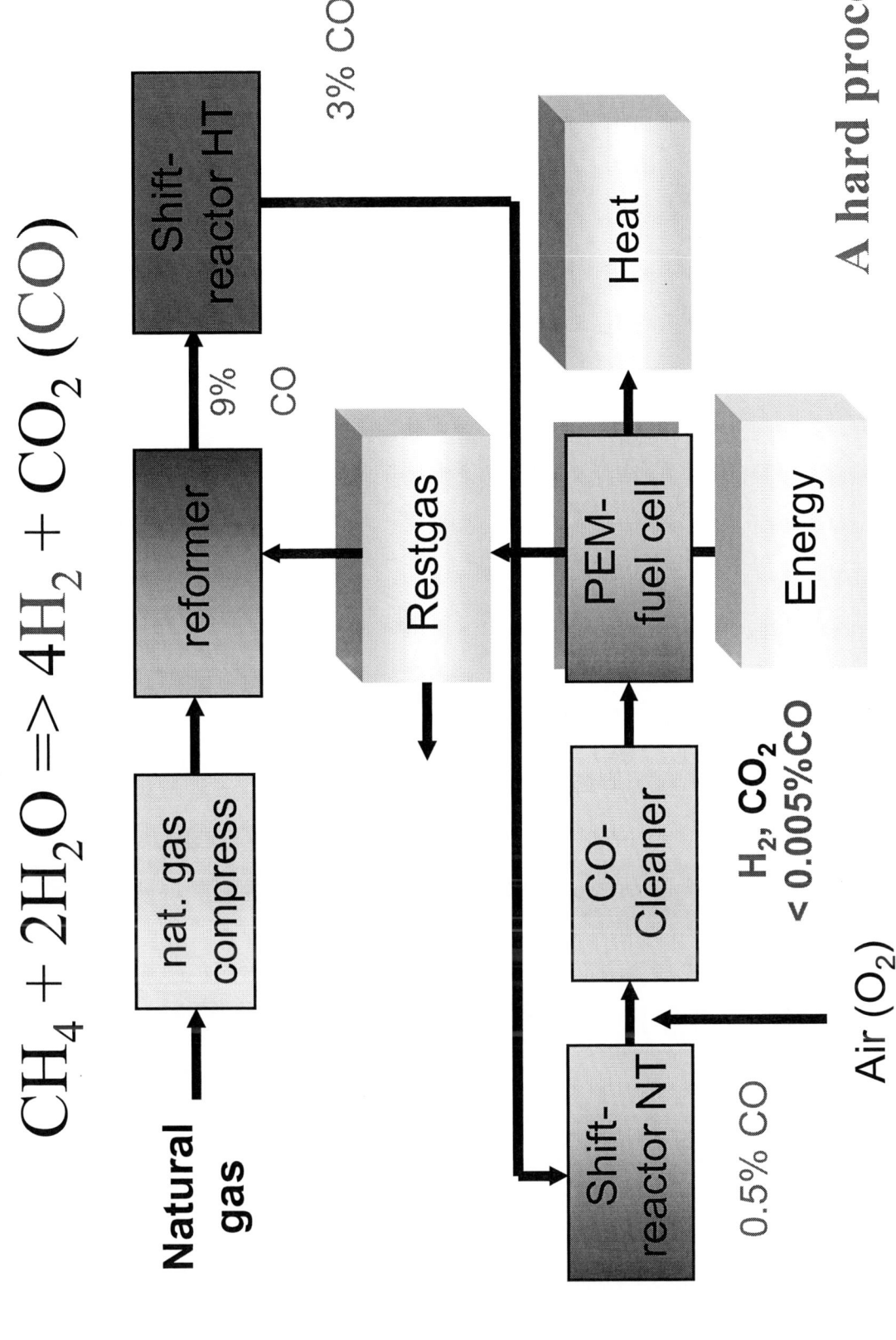

A hard process!

Bernard DAVAT - INPL - Nancy - France

Bangkok, Tutorial 27th November 2007

Water electrolysis

4.5 to 5 kWh/Nm^3 of H$_2$

Efficiency: 50-70 %

Future: High temperature electrolysis

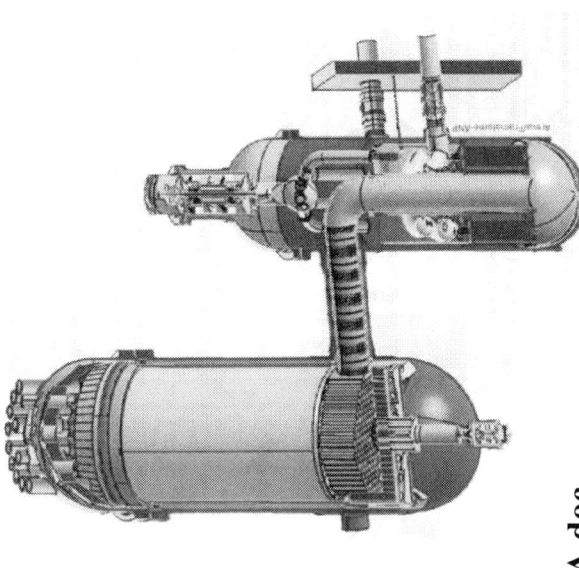

Areva high temperature (850 - 1000 °C) nuclear reactor to produce electricity (300 MWe) and hydrogen.

CEA doc.

Bernard DAVAT - INPL - Nancy - France

Bangkok, Tutorial 27th November 2007

By-product of chlorine industry

$$2\ NaCl\ +\ 2\ H_2O\ \Longrightarrow\ Cl_2\ +\ H_2\ +\ 2\ NaOH$$

Actually, more than 60 % of commercial hydrogen produced in Japan use this process.

Bernard DAVAT - INPL - Nancy - France

Bangkok, Tutorial 27th November 2007

4.4. Hydrogen as an energy vector

Fuel for vehicles: no pollution on the road

Storage of renewable energy

Fuel for vehicles: no pollution on the road

Wang M., Journal of Power Sources, 2002

Bangkok, Tutorial 27th November 2007

Bernard DAVAT - INPL - Nancy - France

Storage of renewable energy

Utsira – demonstration of a hydrogen society

- 2 wind mills
- Hydrogen as energy carrier
- R&D demo-project for 2-3 years
- Ten households completely self-contained

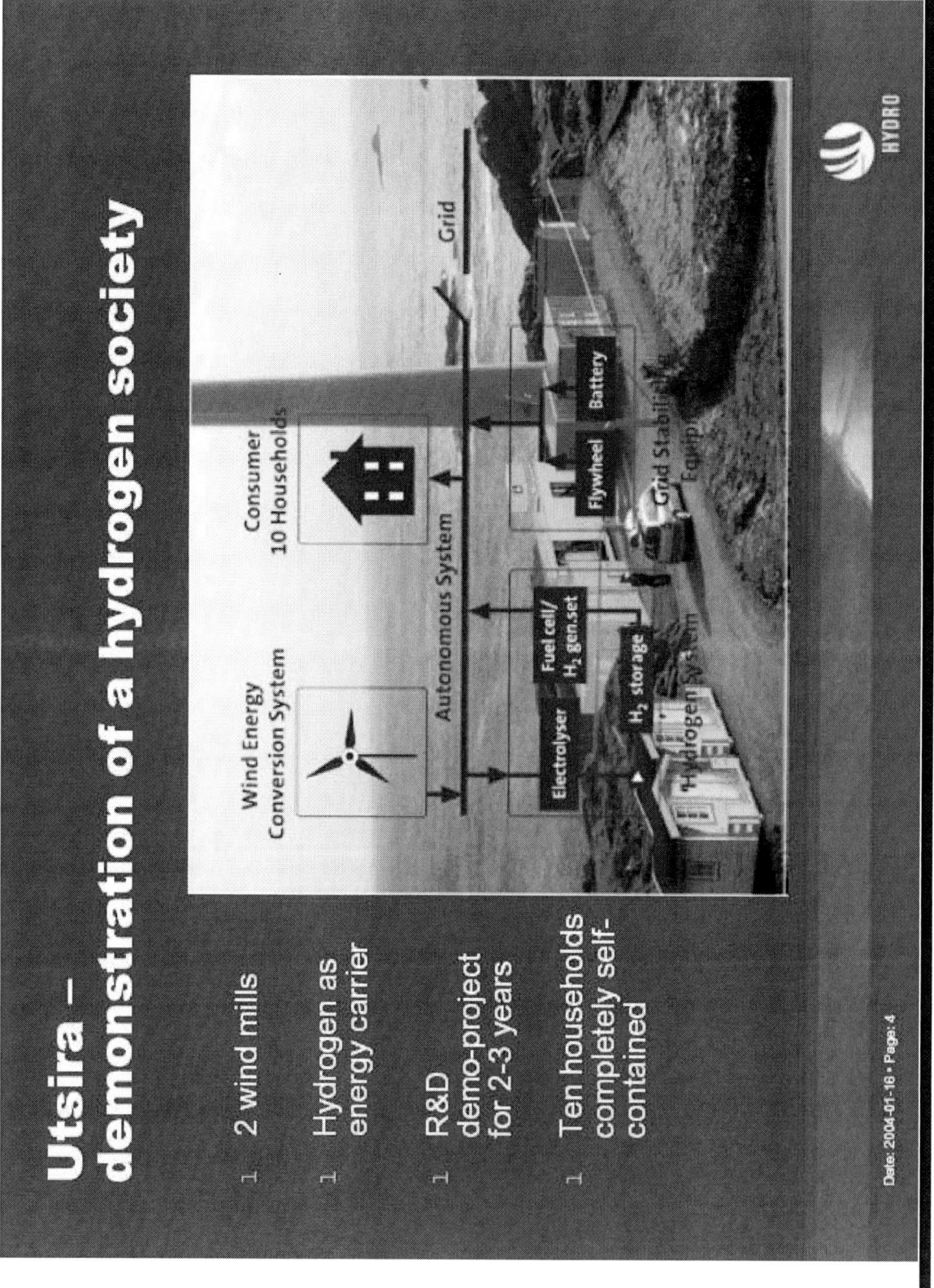

Date: 2004-01-16 • Page: 4

Bernard DAVAT - INPL - Nancy - France

Bangkok, Tutorial 27th November 2007

5. Conclusion and references

Bernard DAVAT - INPL - Nancy - France

Bangkok, Tutorial 27th November 2007

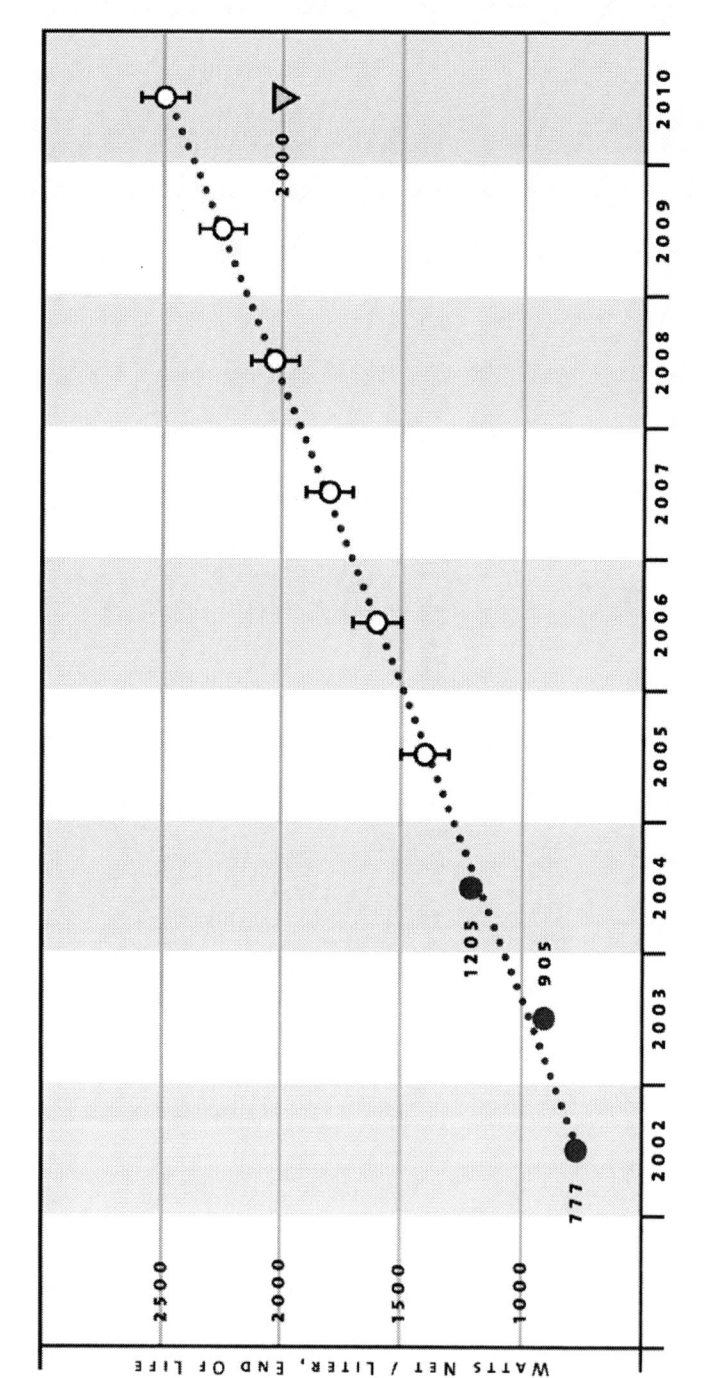

Ballard road map (www.ballard.com)

Bernard DAVAT - INPL - Nancy - France

Bangkok, Tutorial 27th November 2007

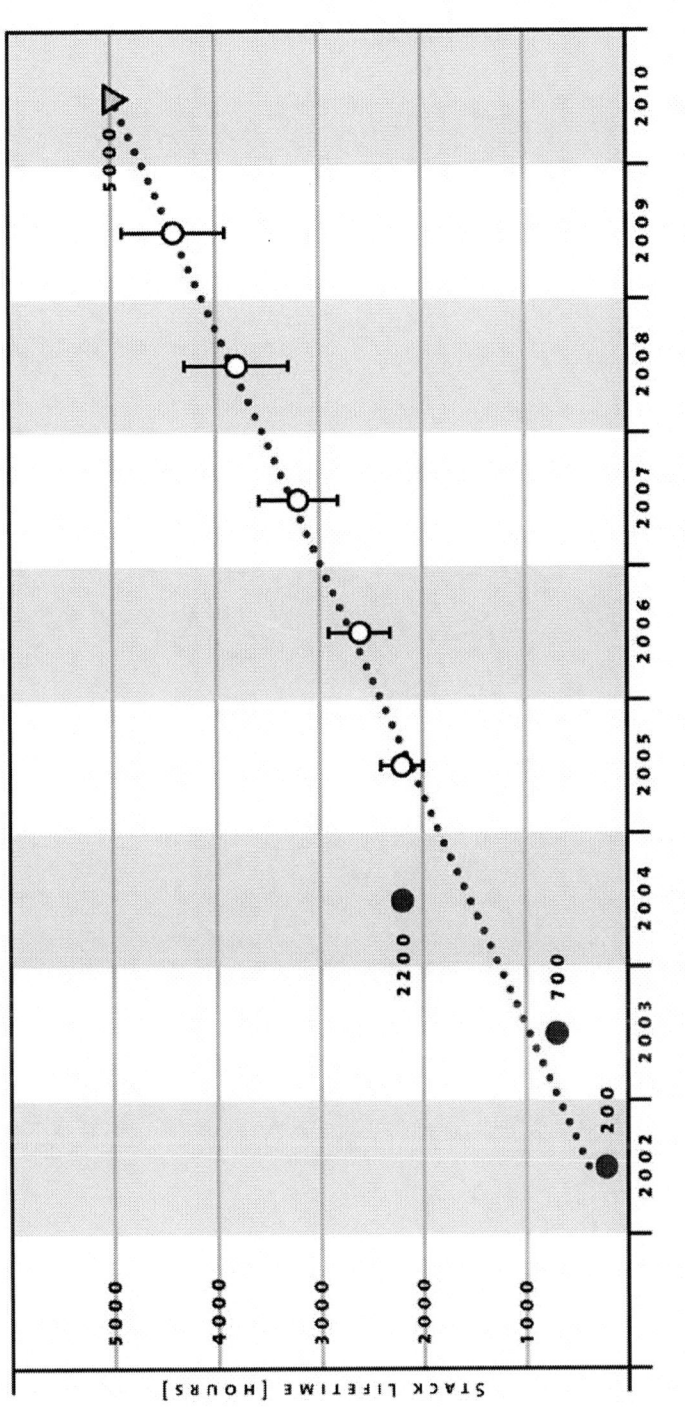

Ballard road map (www.ballard.com)

Bernard DAVAT - INPL - Nancy - France   *Bangkok, Tutorial 27th November 2007*

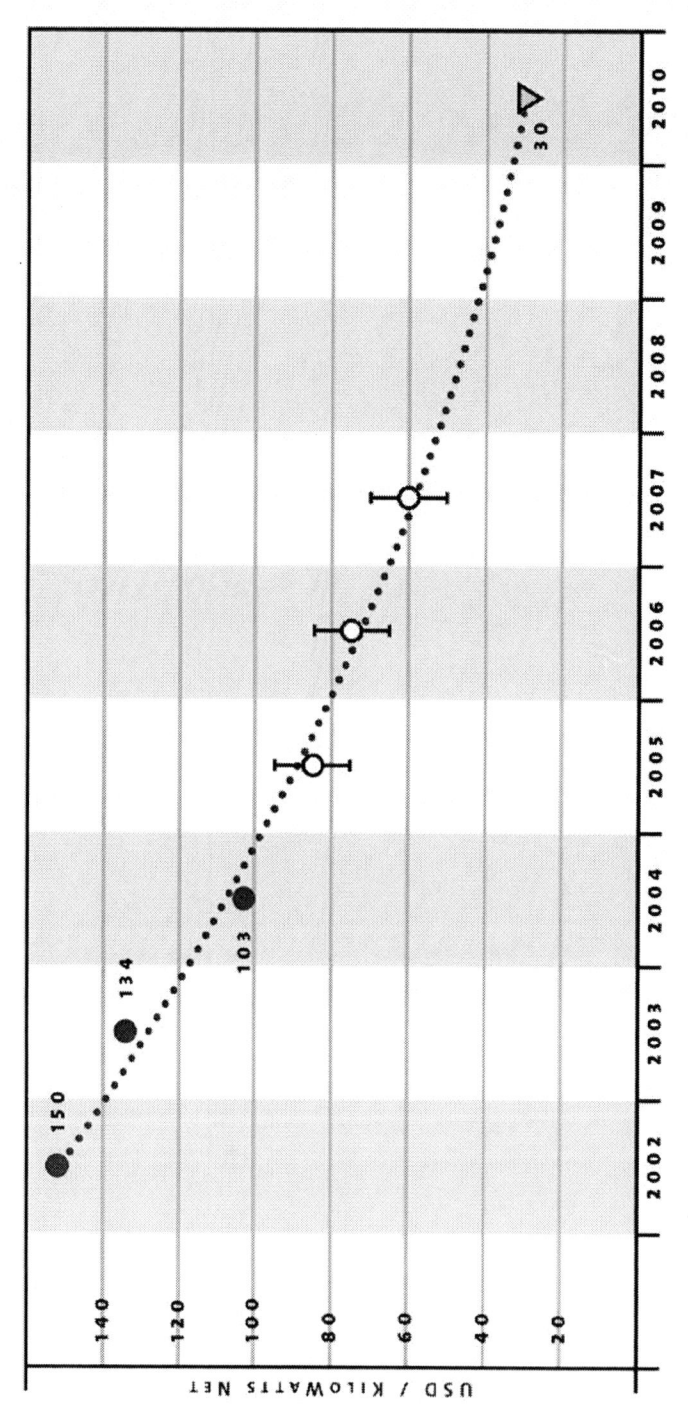

Ballard road map (www.ballard.com)

Bernard DAVAT - INPL - Nancy - France   *Bangkok, Tutorial 27th November 2007*

References

Fuel Cell System Explained

by James Larminie and Andrew Dicks

Wiley

ALPHEA

www.alphea.com

Bernard DAVAT - INPL - Nancy - France

Bangkok, Tutorial 27th November 2007

Tutorial on

Recent Trends in Power Quality Improvements Techniques

By

Prof. Bhim Singh

Deptt of Electrical Engineering

Indian Institute of Technology, Delhi

New Delhi-110016, India

bsingh@ee.iitd.ac.in

bhimsinghr@gmail.com

What are power quality problems ?

- It include all possible situations in which the waveforms of the supply voltage or load current deviate from the sinusoidal waveform at rated frequency with amplitude corresponding to the rated rms value for all three phases of a three-phase system

- Power quality disturbance covers sudden, short duration deviation impulsive and oscillatory transients, voltage dips (or sags), short interruptions, as well as steady- state deviations, such as harmonics and flicker

27/11/07

IIT-Delhi/PEDS-07

Voltage Power Quality Problems

- Voltage Sag
- Voltage Swell
- Voltage Interruption
- Under/ Over Voltage
- Voltage Flicker
- Harmonic Distortion
- Voltage Notching
- Transient Disturbance
- Outage and frequency variation

Voltage Sag

- A voltage sag is a reduction in the RMS voltage in the range of 0.1 to 0.9 p.u. (retained) for duration greater than half a mains cycle and less than 1 minute. Often referred to as a 'sag'. Caused by faults, increased load demand and transitional events such as large motor starting.

Voltage Swell

- A voltage swell is an increase in the RMS voltage in the range of 1.1 to 1.8 p.u. for a duration greater than half a main cycle and less than 1 minute. Caused by system faults, load switching and capacitor switching.

Voltage Interruption

- A *voltage interruption* is the complete loss of electric voltage. Interruptions can be short duration (lasting less than 2 minutes) or long duration. A disconnection of electricity causes an interruption—usually by the opening of a circuit breaker, line recloser, or fuse

27/11/07

IIT-Delhi/PEDS-07

Waveform and RMS voltage during voltage sag

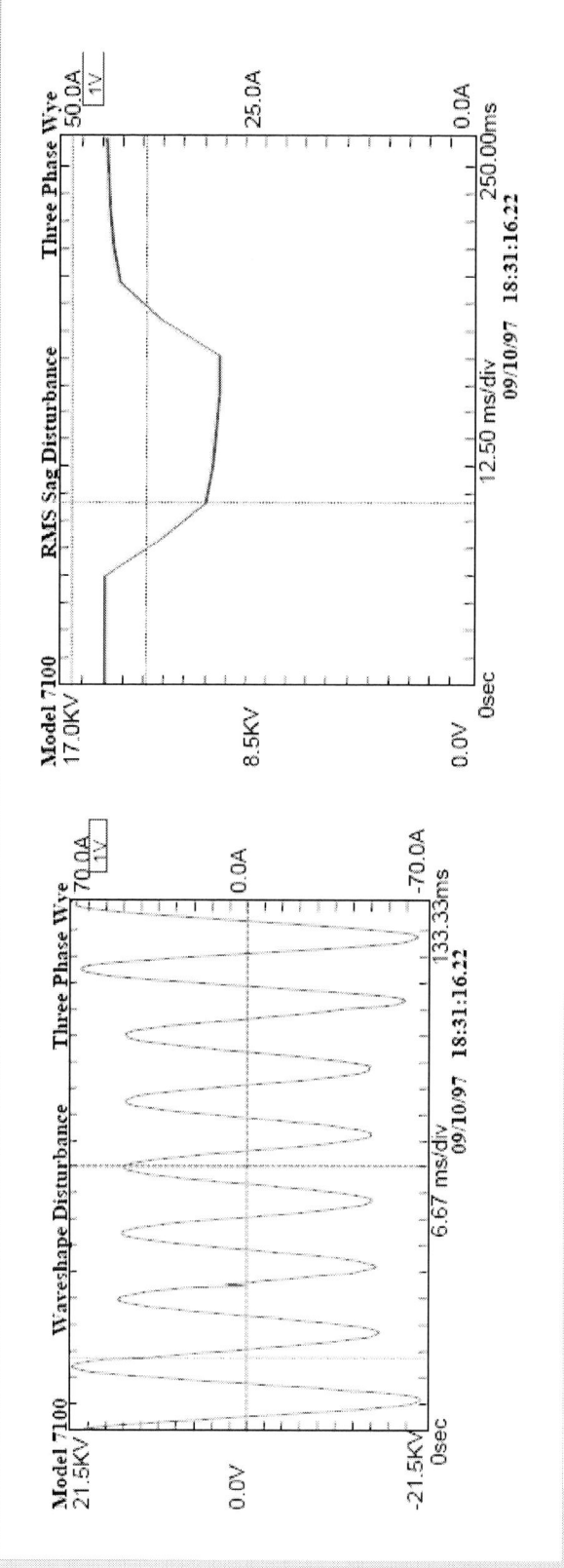

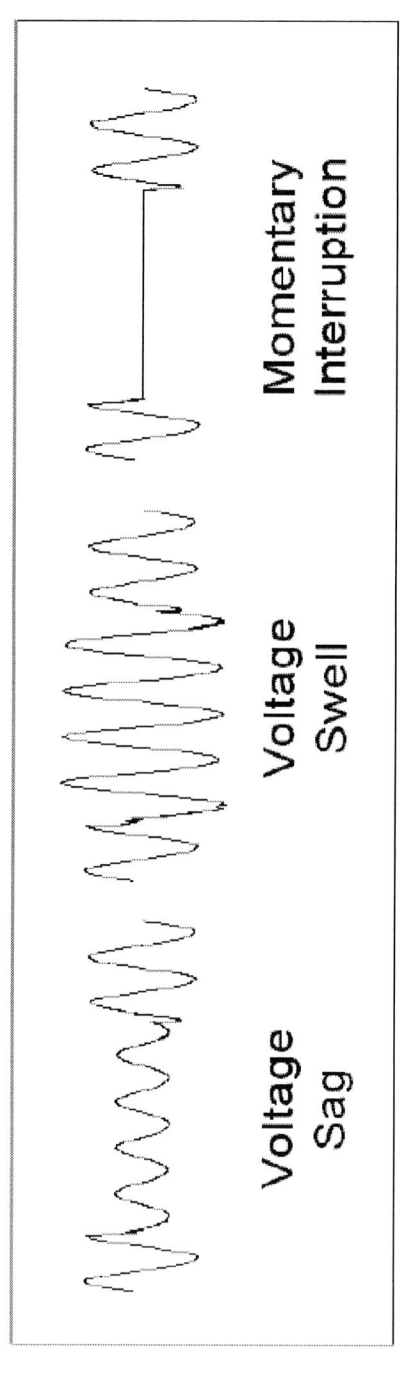

IIT-Delhi/PEDS-07

27/11/07

Over Voltage and Under Voltage

- Long-duration voltage variations that are outside the normal limits (that is, too high or too low) are most often caused by unusual conditions on the power system. For example, out-of-service lines or transformers sometimes cause *under voltage* conditions. These types of root-mean-square (RMS) voltage variations are normally short term, lasting less than one or two days.

- In addition, voltage can be reduced intentionally in response to a shortage of electric supply.

27/11/07

IIT-Delhi/PEDS-07

RMS Measurement of under voltage during one day

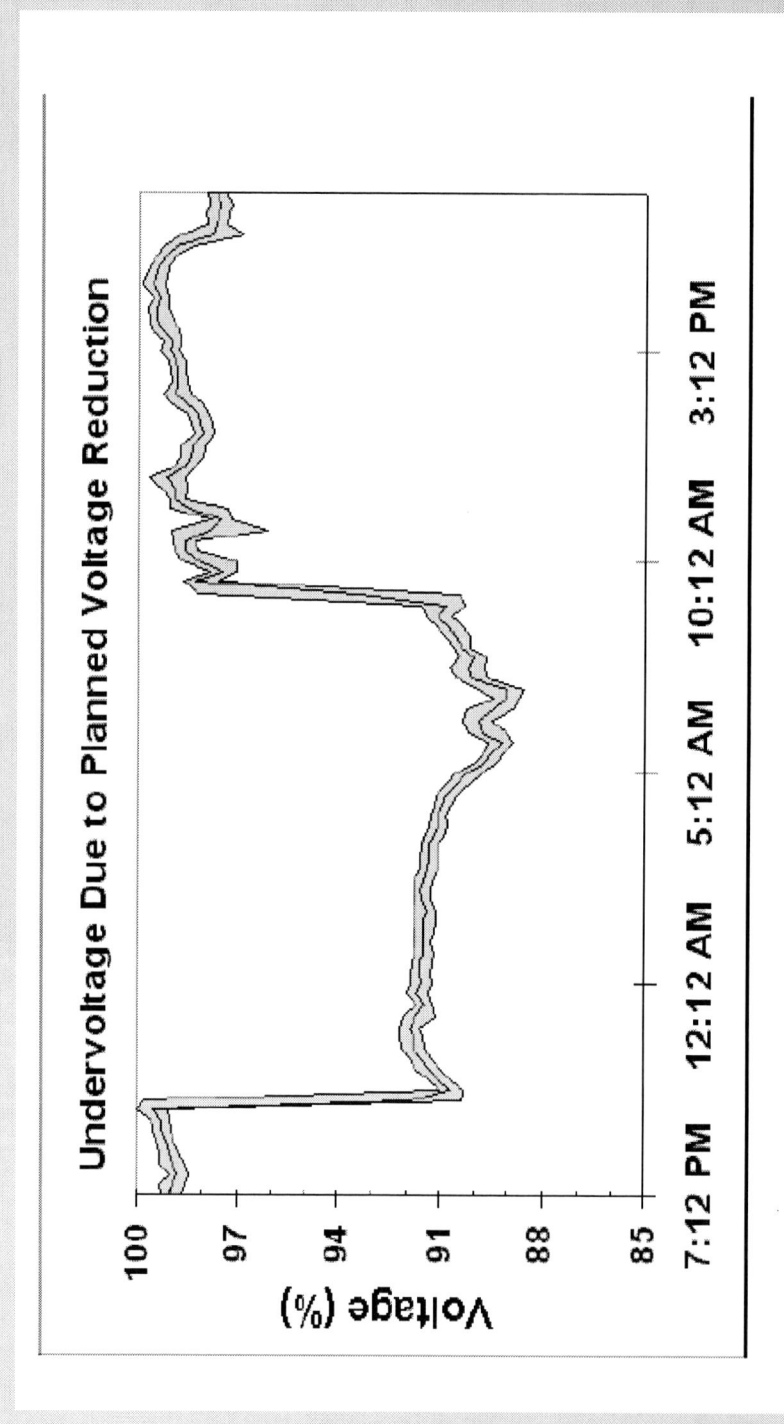

27/11/07 IIT-Delhi/PEDS-07

Voltage Flicker

- A waveform may exhibit *voltage flicker* if its waveform amplitude is modulated at frequencies less than 25 Hz, which the human eye can detect as a variation in the lamp intensity of a standard bulb.

- Voltage flicker is caused by an arcing condition on the power system.

- Flicker problems can be corrected with the installation of filters, static VAR systems, or distribution static compensators

27/11/07

IIT-Delhi/PEDS-07

Example voltage waveforms showing flicker created by an arc furnace

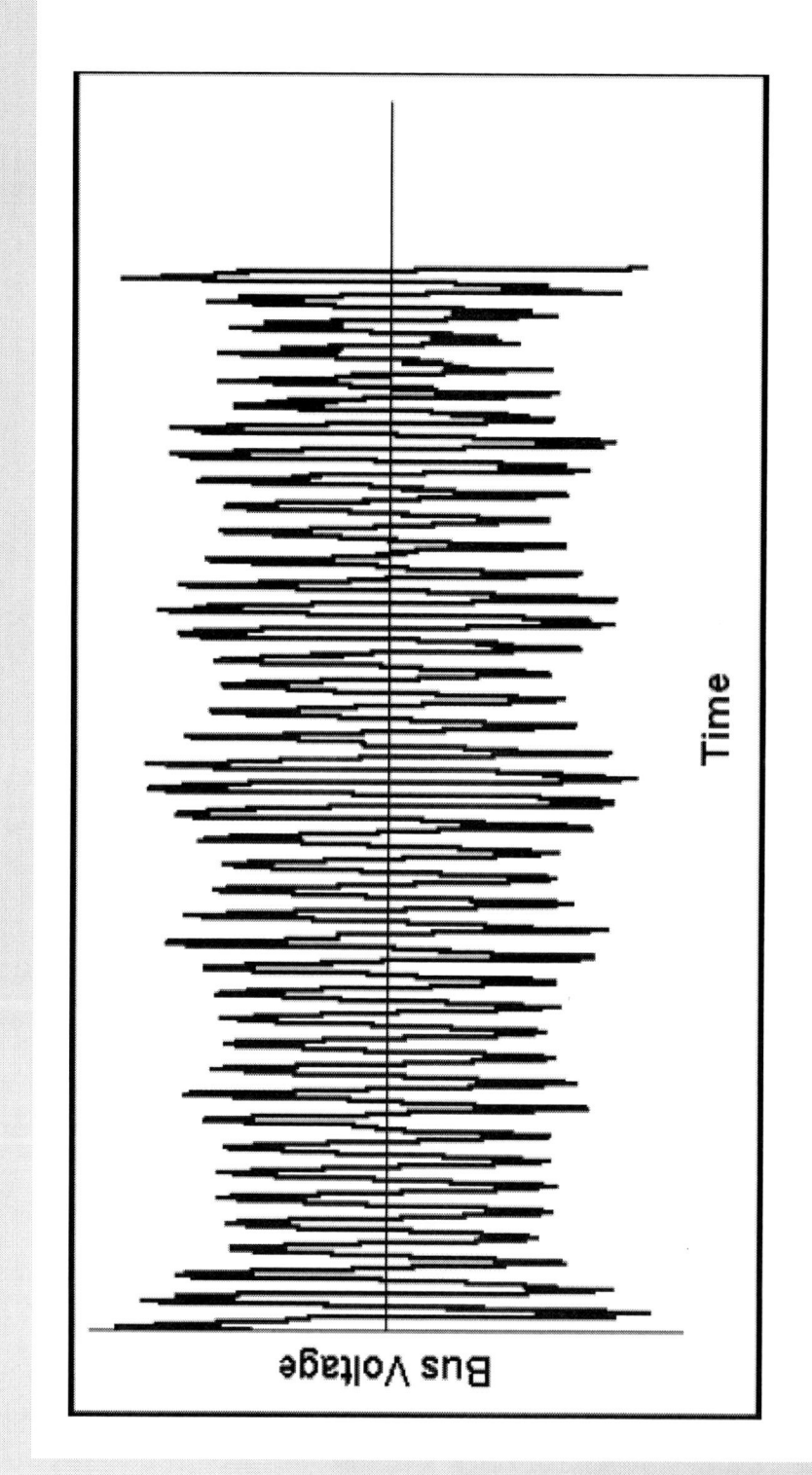

Harmonics Distortion

- Harmonics are periodic sinusoidal distortions of the supply voltage or load current caused by non-linear loads.

- Harmonics are measured in integer multiples of the fundamental supply frequency.

- In commercial facilities, computers, lighting, and electronic office equipment generate harmonic distortion. In industrial facilities, adjustable-speed drives and other power electronic loads can generate significant amounts of harmonics.

- Solutions to problems caused by harmonic distortion include installing active or passive filters at the load or bus, or taking advantage of transformer connections that enable cancellation of zero-sequence components.

27/11/07

IIT-Delhi/PEDS-07

Distorted Voltage Waveforms

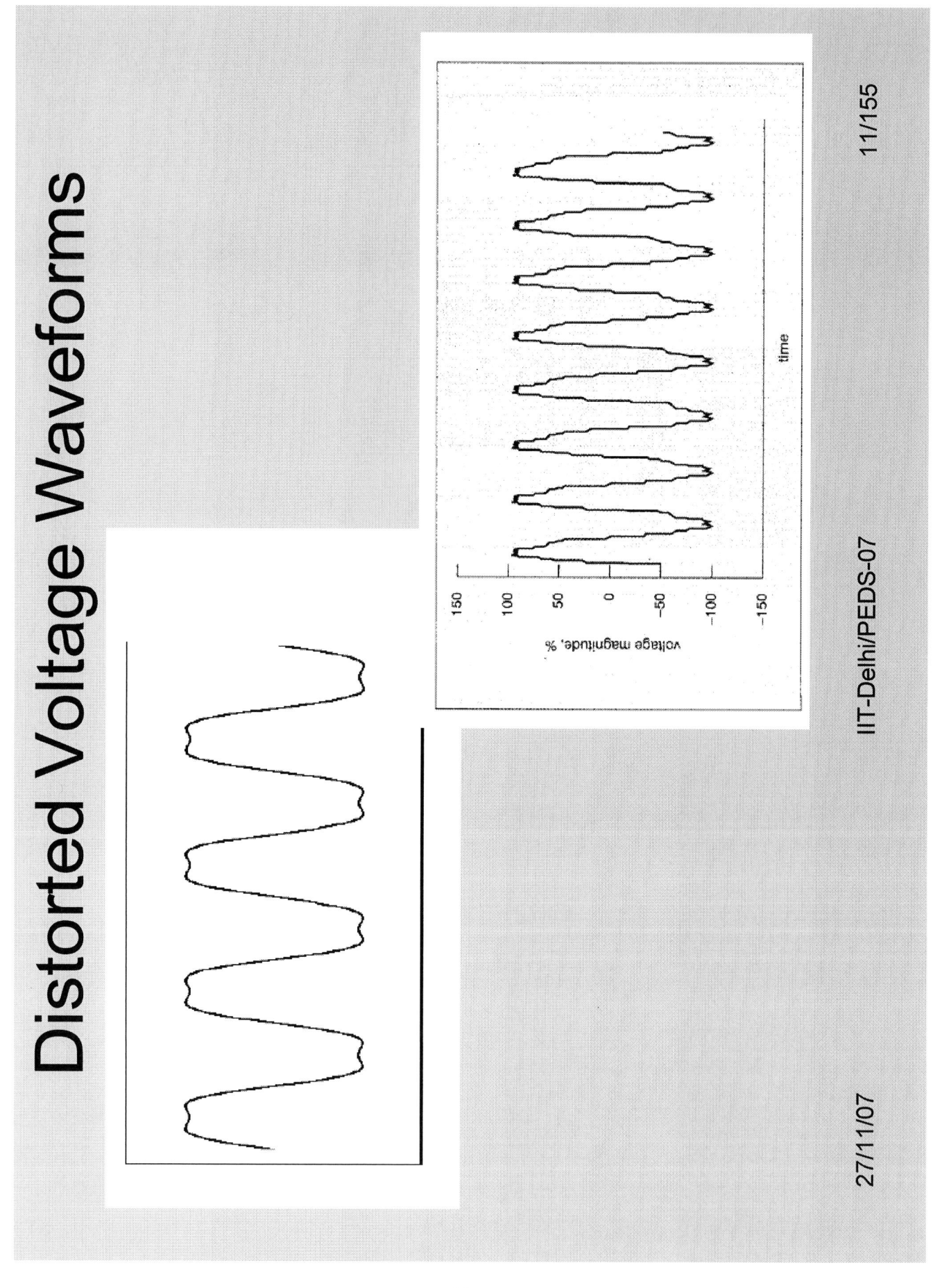

Voltage Notching

- Voltage notching is caused by the commutation of power electronic rectifiers. It is an effect that can raise PQ issues in any facility where solid-state rectifiers (for example, variable-speed drives) are used

- When the drive DC link current is commutated from one rectifier thyristor to the next, an instant exists during which a line-to-line short circuit occurs at the input terminals to the rectifier.

- With this disturbance, any given phase voltage waveform will typically contain four notches per cycle as caused by a six-pulse electronic rectifier

Voltage Notching Waveform

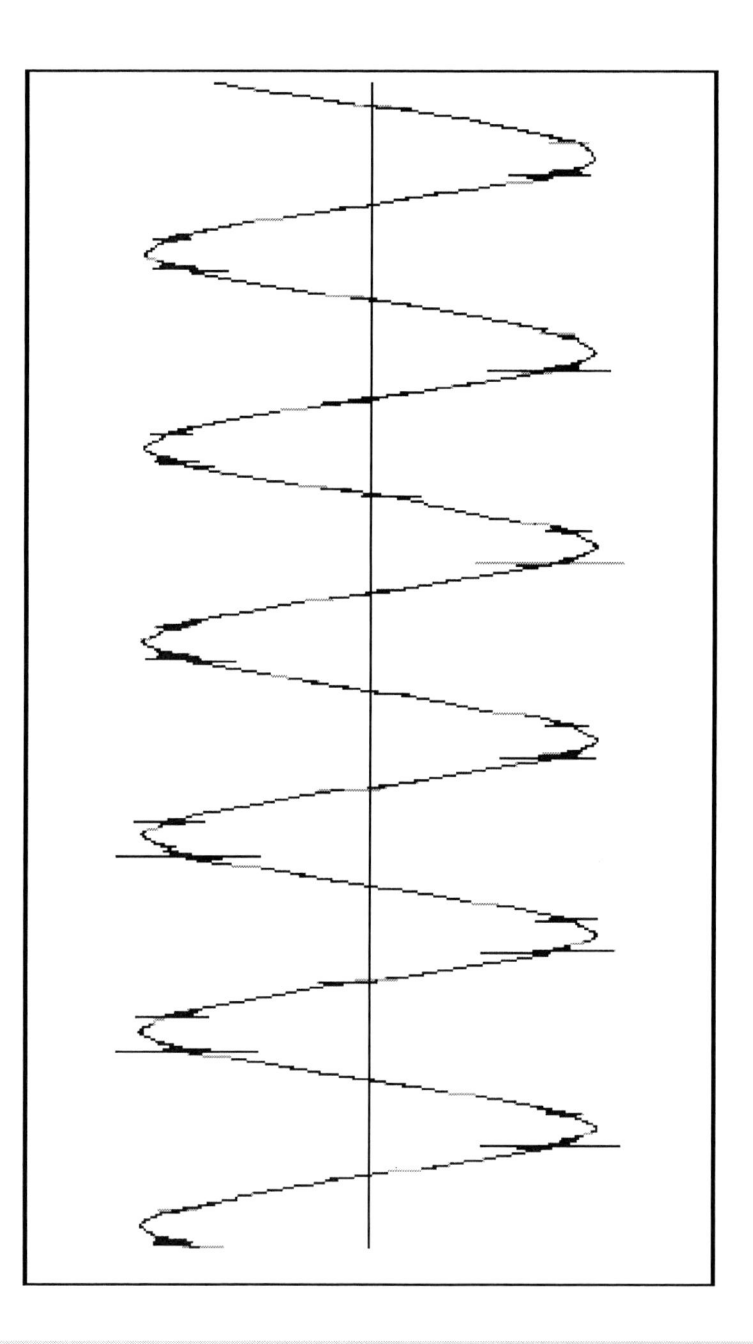

Transient Disturbance

- Transient disturbances are undesirable momentary deviation of the supply voltage or load current and caused by the injection of energy by switching or by lightning.

- Transients are classified in two categories "Impulsive" and "oscillatory"

27/11/07

IIT-Delhi/PEDS-07

14/155

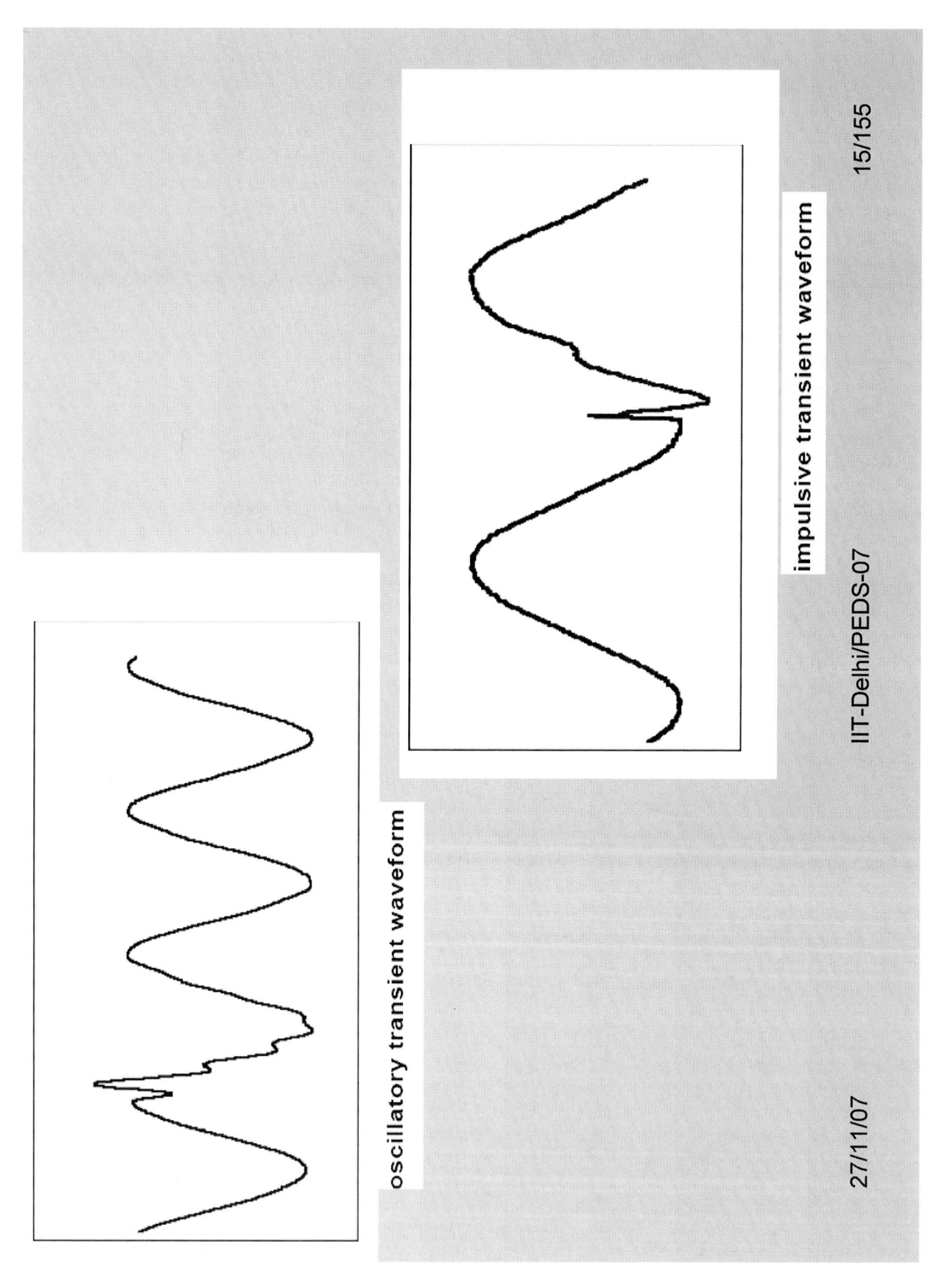

oscillatory transient waveform

impulsive transient waveform

IIT-Delhi/PEDS-07

27/11/07

- Outage

  Outage is defined as an interruption that has duration lasting in excess of one minute.

- Frequency Deviation

  It is a variation in frequency from the nominal supply frequency above/below a predetermined level, normally $\pm$ 0.1%.

Outage

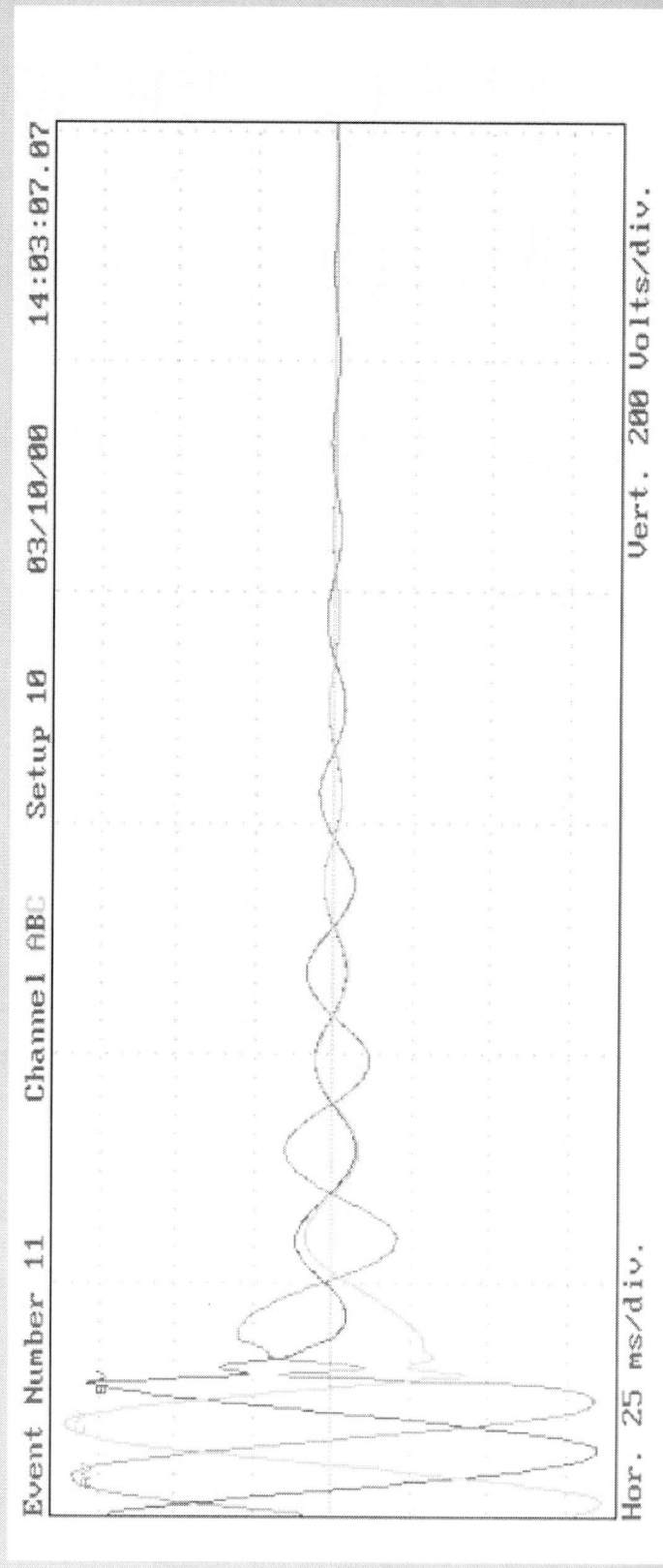

Effects of PQ Quantities

Voltage dips
machine/process downtime, scrap cost, clean up costs, product quality and repair costs all contribute to make these types of problems costly to the end-user

Transients
tripping, component failure, hardware reboot required, software 'glitches', poor product quality

Harmonics
transformer and neutral conductor heating leading to reduced equipment life span; audio hum, video 'flutter', software glitches, power supply failure

Flicker
visual irritation

27/11/07

IIT-Delhi/PEDS-07

Categories of power quality variation

Categories	Spectral Content	Typical Duration	Typical Magnitudes
1.0 Transients			
1.1 Impulsive			
1.1.1 Voltage	> 5 kHz	< 200 μs	
1.1.2 Current	> 5 kHz	< 200 μs	
1.2 Oscillatory			
1.2.1 Low Frequency	< 500 kHz	< 30 cycles	
1.2.2 Medium Frequency	300–2 kHz	< 3 cycles	
1.2.3 High Frequency	> 2 kHz	< 0.5 cycle	
2.0 Short-Duration Variations			
2.1 Sags			
2.1.1 Instantaneous		0.5–30 cycles	0.1–1.0 pu
2.1.2 Momentary		30–120 cycles	0.1–1.0 pu
2.1.3 Temporary		2 sec–2 min	0.1–1.0 pu
2.2 Swells			
2.1.1 Instantaneous		0.5–30 cycles	0.1–1.8 pu
2.1.2 Momentary		30–120 cycles	0.1–1.8 pu
2.1.3 Temporary		2 sec–2 min	0.1–1.8 pu
3.0 Long-Duration Variations			
3.1 Overvoltages		> 2 min	0.1–1.2 pu
3.2 Undervoltages		> 2 min	0.8–1.0 pu
4.0 Interruptions			
4.1 Momentary		< 2 sec	0
4.2 Temporary		2 sec–2 min	0
4.3 Long-Term		> 2 min	0
5.0 Waveform Distortion			
5.2 Voltage	0–100th Harmonic	steady-state	0–20%
5.3 Current	0–100th Harmonic	steady-state	0–100%
6.0 Waveform Notching	0–200 kHz	steady-state	
7.0 Flicker	< 30 Hz	intermittent	0.1–7%
8.0 Noise	0–200 kHz	intermittent	

Current Based Power Quality Problems

- ## Reactive Power Compensation

- ## Voltage Regulation

- ## Current Harmonics Compensation

- ## Load Unbalancing (for 3-phase systems)

- ## Neutral Current Compensation (for 3-phase 4-wire systems)

27/11/07

IIT-Delhi/PEDS-07

Sources of Power Quality Problems

- Power electronic devices
- IT and office equipments
- Arching devices
- Load switching
- Large motor starting
- Embedded generation
- Sensitive equipment
- Storm and environmental related damage

Solution of Power Quality Problems

Flicker Mitigation

- Static Var Compensator
- D-Statcom

Harmonic Mitigation

- Passive Filter
- Active Filter
- Multi-pulse Configuration

27/11/07

IIT-Delhi/PEDS-07

Solution of Power Quality Problems

Mitigation of Voltage Dips and Short Interruption

- Motor-generator set

- Static series compensator

- Dynamic voltage restorer (DVR)

- Static transfer switch

27/11/07

IIT-Delhi/PEDS-07

Other Possible Solutions

- Proper earthing practices

- Online UPS/Hybrid UPS

- Energy storage system

- Ferro- resonant transformer

- Network equipment and design

27/11/07   IIT-Delhi/PEDS-07

Solution of Harmonic Mitigation Using Power Filters

Power Filters

Passive Filters
- Shunt
- Series
- Hybrid

Active Filters
- Shunt
- Series
- Hybrid UPQC

Hybrid Filters
Several Combinations are possible for hybrid of active and passive

27/11/07

IIT-Delhi/PEDS-07

Passive Filters

Passive Filters

✓ Harmonic reduction

✓ Reactive power compensation

✗ Resonance with line impedance

✗ Heavy and bulky

27/11/07

IIT-Delhi/PEDS-07

Passive Filters

1. LC passive filter
2. Tuned passive filters

Tuned Passive Filters

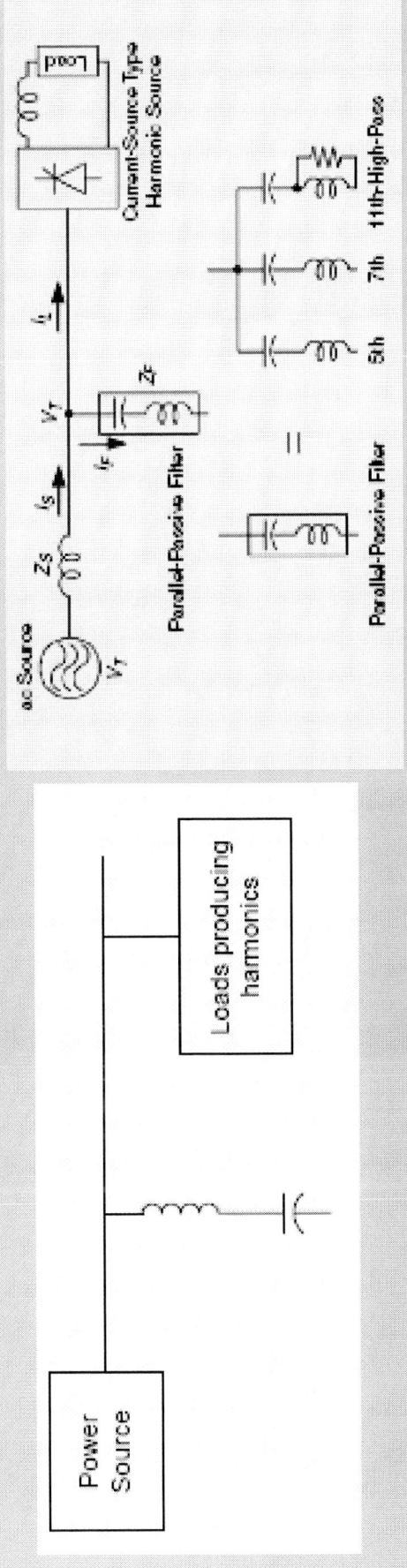

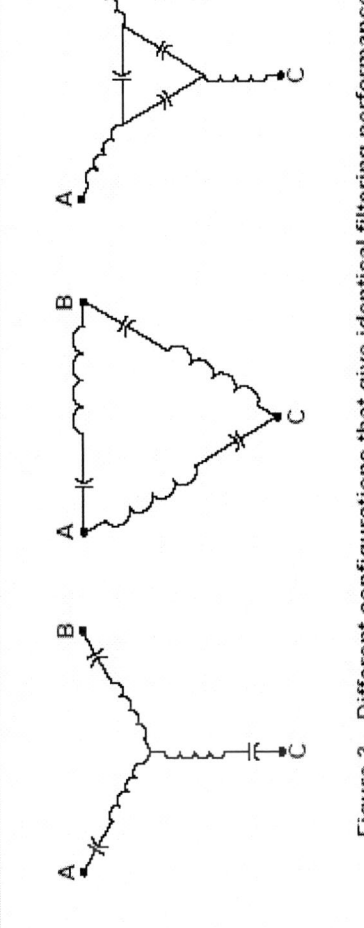

Figure 3—Different configurations that give identical filtering performance

27/11/07

IIT-Delhi/PEDS-07

Tuned Shunt Passive Filters

Three-Phase Three-Wire Nonlinear Loads

PF_{Sha} PF_{Shb} PF_{Shc}

C_5 L_5 C_7 L_7 C_h R_h L_h

Z_{sa} Z_{sb} Z_{sc}

v_{sa} i_{sa}
v_{sb} i_{sb}
v_{sc} i_{sc}

27/11/07

IIT-Delhi/PEDS-07

Tuned Series Passive Filters

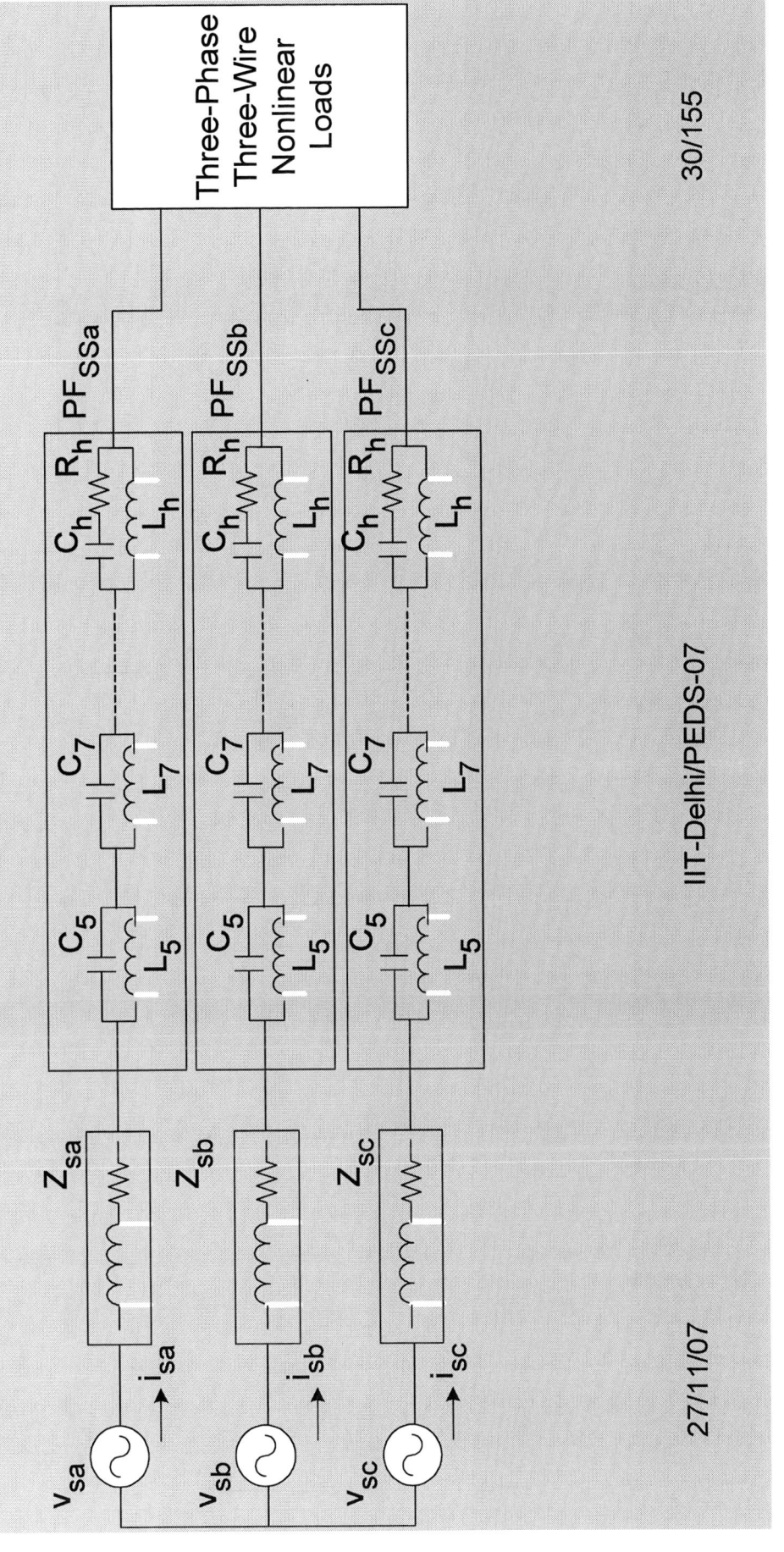

IIT-Delhi/PEDS-07

27/11/07

Tuned Series Passive Filters for 3-Ph 4-Wire System

Tuned Passive Filters

Key filter design considerations include the following

a) Reactive power (kilovar) requirements
b) Harmonic limitations
c) Normal system conditions, including ambient harmonics
d) Normal harmonic filter conditions
e) Contingency system conditions, including ambient harmonics
f) Contingency harmonic filter conditions

27/11/07

Passive Hybrid Filters

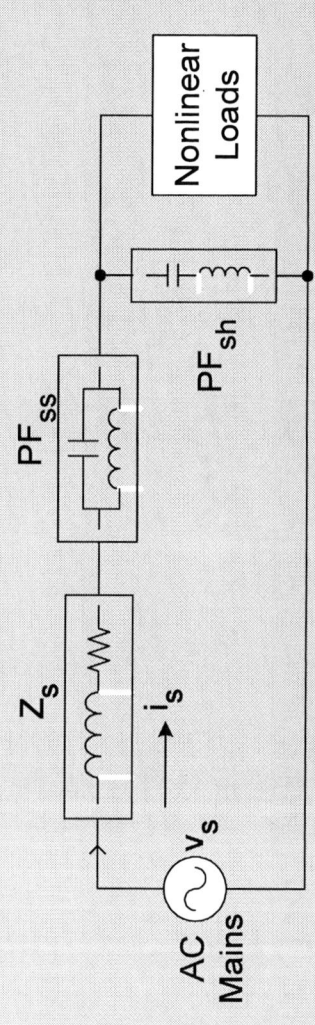

Hybrid Filter as a Combination of Passive-Series (PFss) and Passive-Shunt (PFsh) Filters

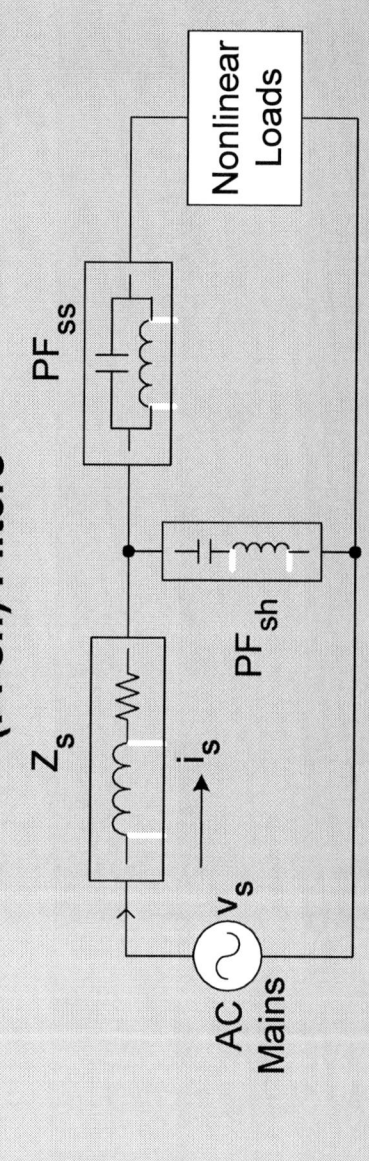

Hybrid Filter as a Combination of Passive-Shunt (PFsh) and Passive-Series (PFss) Filters

27/11/07

IIT-Delhi/PEDS-07

Passive Hybrid Filters

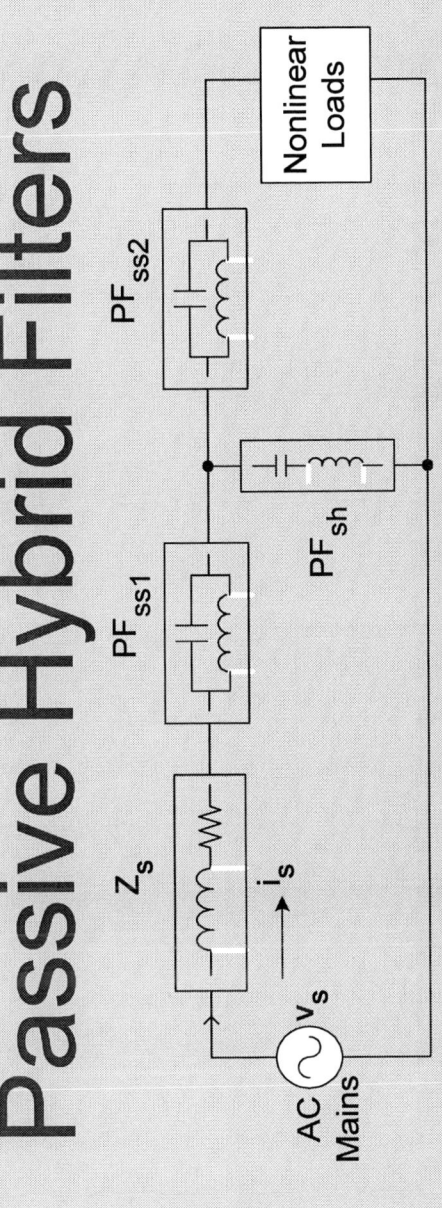

Hybrid Filter as a Combination of Passive-Series (PF_{ss1}), Passive-Shunt (PF_{sh}) and Passive-Series (PF_{ss2}) Filters

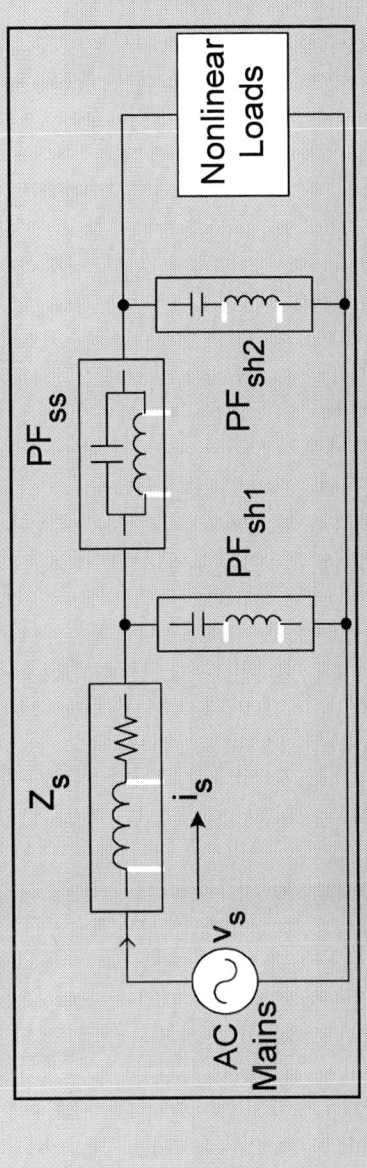

Hybrid Filter as a Combination of Passive-Shunt (PF_{sh1}), Passive-Series (PF_{ss}) and Passive-Shunt (PF_{sh2}) Filters

27/11/07

IIT-Delhi/PEDS-07

Passive Hybrid Filters

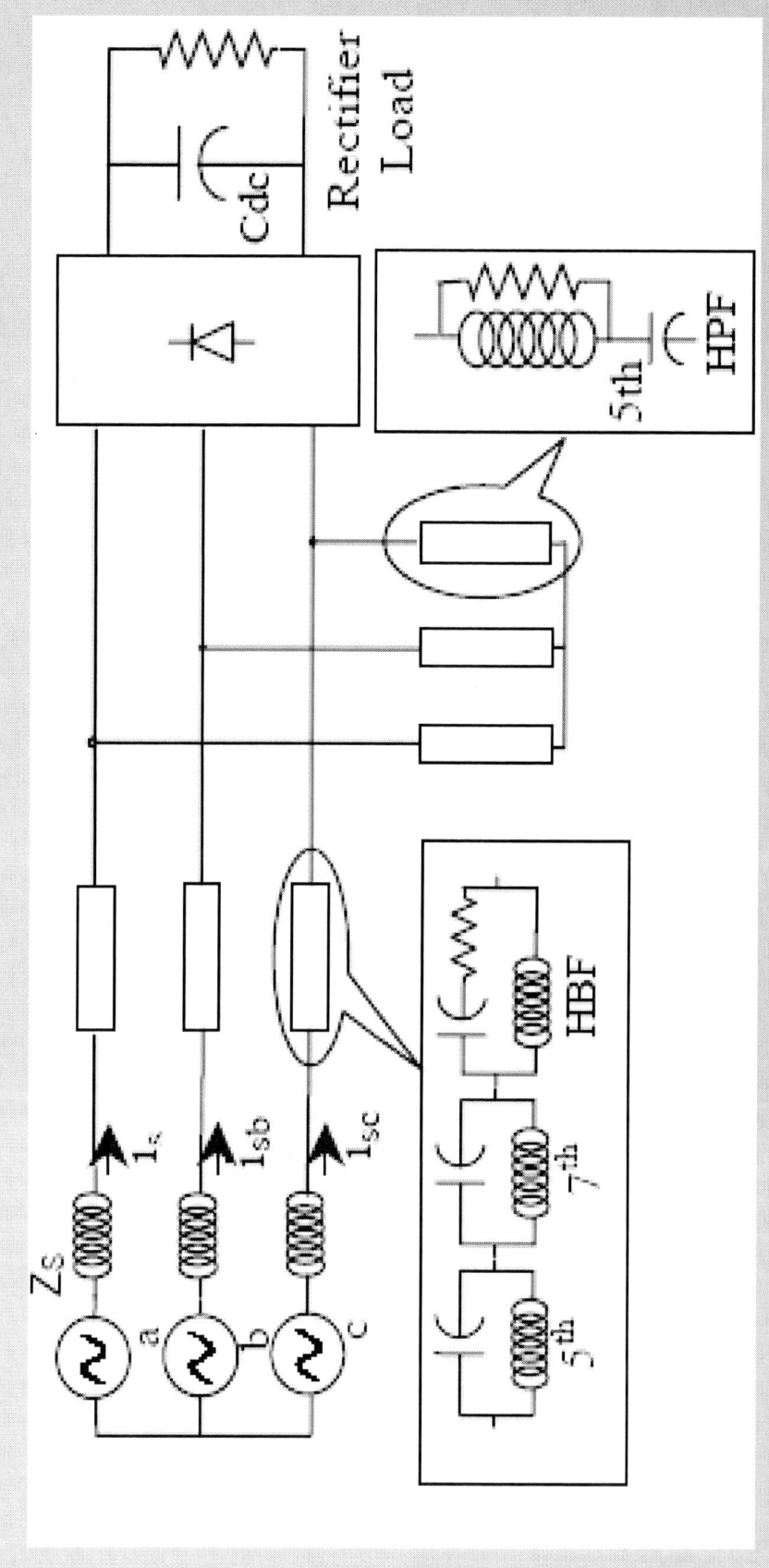

Dynamic response of the system for switching in of shunt and series component of HPF sequentially at t=0.1s and t=0.2s.

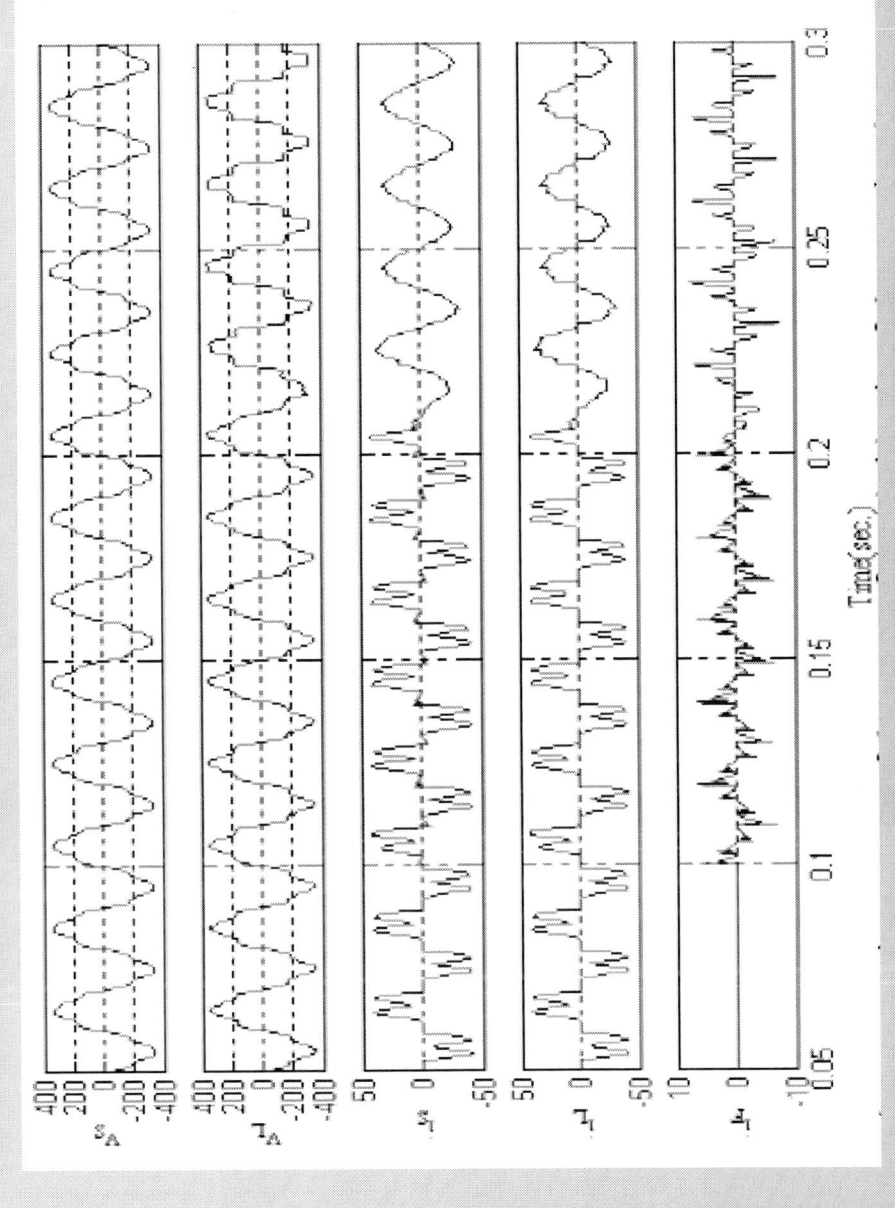

27/11/07

IIT-Delhi/PEDS-07

Harmonic spectrum of load current and source current with S_hPF alone and with Proposed HPF

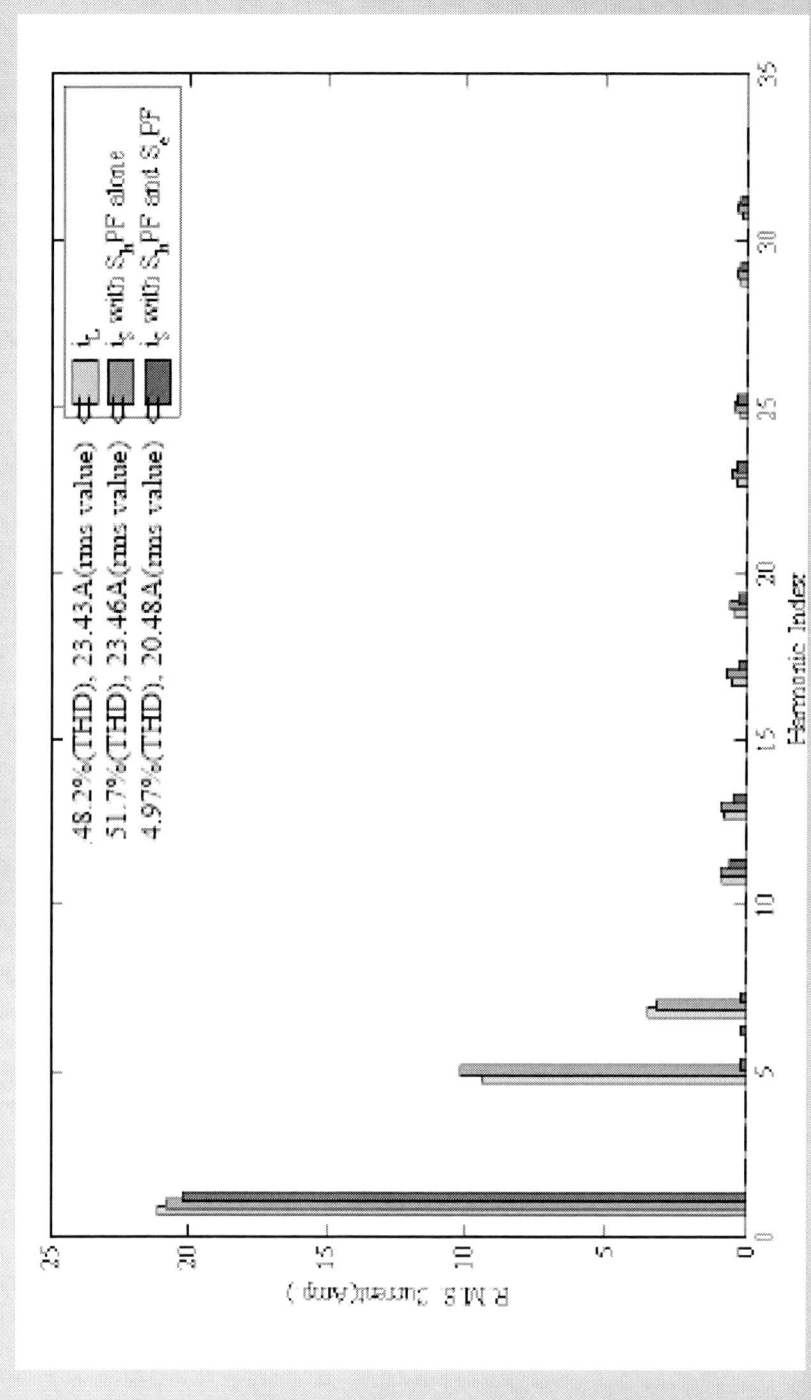

27/11/07

IIT-Delhi/PEDS-07

Dynamic response of proposed HPF under distorted main and varying load conditions.
The Load has been reduced from 15kW to 11.4kW
at t=0.1s, from 11.4kW to 7.67kW at t=0.18s, from 7.67kW to 4.6kW at t=0.26s and at
t=0.34s the load has been reduced to no-load condition

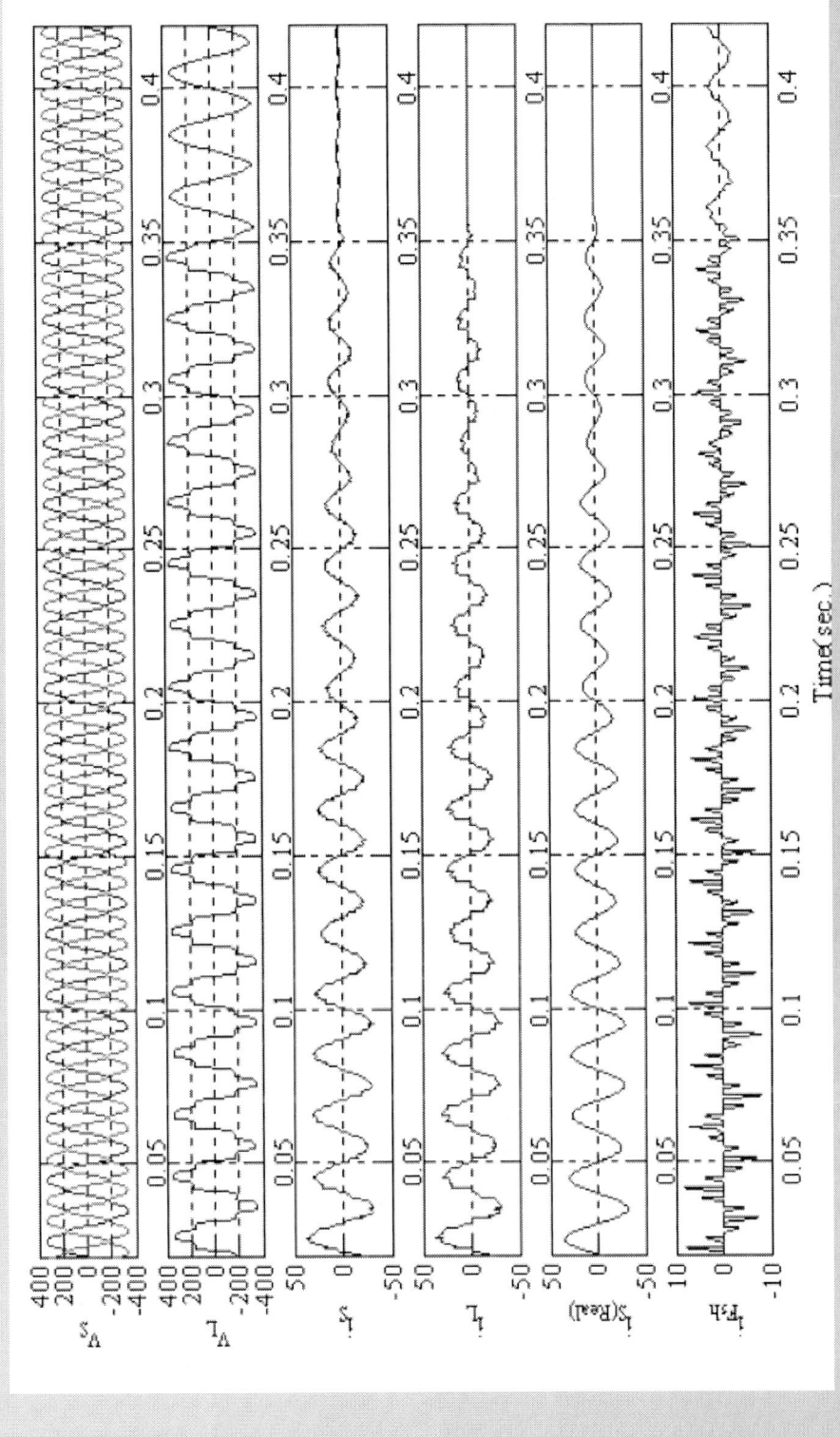

TABLE I.   % THD, RMS VALUE OF SOURCE CURRENT WITH H_yPF_z VOLTAGE AT LOAD TERMINALS AND POWER FACTOR

Load	15kW	11.4kW	7.7kW	4.6kW	No-Load
I_S(RMS)	20.48A	15.9A	10.97A	6.7A	1.4A
% THD	4.97	4.99	5.47	8.26	7.94
pf	0.987	0.997	0.988	0.956	0.068
V_L(RMS)	233.58	239.96	245.5	246.2	239.64

TABLE II.   DESIGNED SET OF PARAMETERS OF THE H_yPF

Series Passive Filter				Shunt Passive Filter		
		R_B	3Ω		R_H	3Ω
C_7	60μF	C_{11}	48μF		C_{HPF}	20μF
L_7	3.4 mH	L_{11}	1.8 mH		L_{HPF}	20 mH
C_5	70μF					
L_5	5.8 mH					

Selection of Power Filters

- Nature of Load (Voltage Fed, Converter Fed or Mixed)

- Type of Supply System (single-phase, three Phase three wire, three phase four wire)

- Compensation required in current (harmonics, reactive power, balancing, neutral current) or voltage (harmonic flicker, unbalance, regulation, sag, swell, spikes, notches)

- Pattern of loads (fixed, variable, fluctuating)

Selection of Power Filters

- Level of compensation required (THD, Individual harmonic reduction meeting specific standard etc.)
- Cost, size, weight
- Efficiency
- Reliability
- Environmental factors (ambient temperature, altitude, pollution, humidity etc)

27/11/07

IIT-Delhi/PEDS-07

Active Filter

Active Filters

✓ Cancel out harmonics

✓ Block resonance

✓ Reactive power management

✗ Costly

✗ Good for retrofit applications

27/11/07

IIT-Delhi/PEDS-07

Active Filters

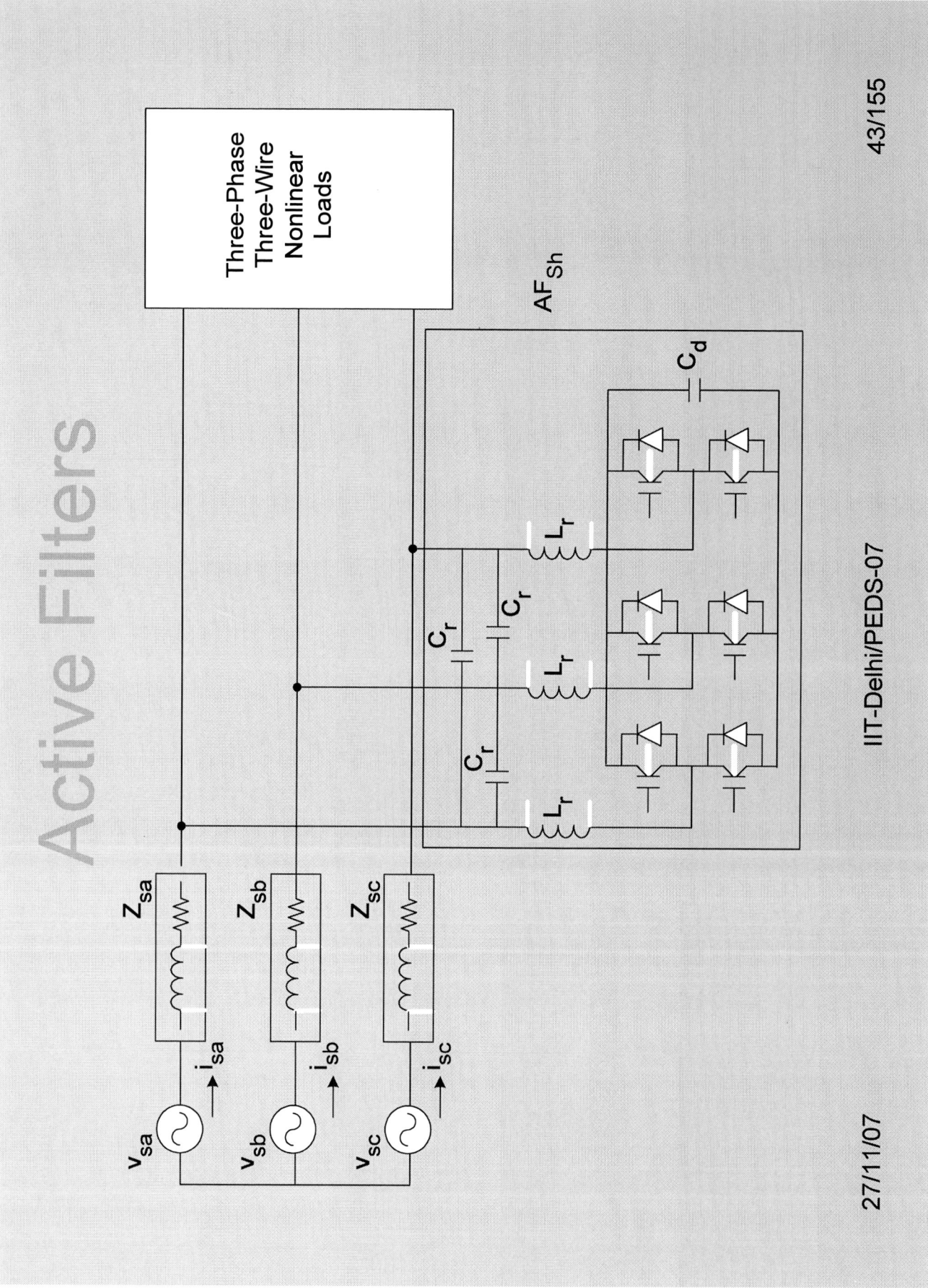

IIT-Delhi/PEDS-07

43/155

27/11/07

Active Filters Basic Principle

NONLINEAR
LOAD

Main Customer
Bus

Interface Filter

IGBT
PWM
Inverter

Controls
and
Gating Signal
Generators

I_L

I_f

I_s

M

27/11/07

IIT-Delhi/PEDS-07

Shunt and Series Active Filters

Current Fed Type AF

Voltage Fed Type AF

Series Type AF

27/11/07

IIT-Delhi/PEDS-07

45/155

Topologies of Active Filters

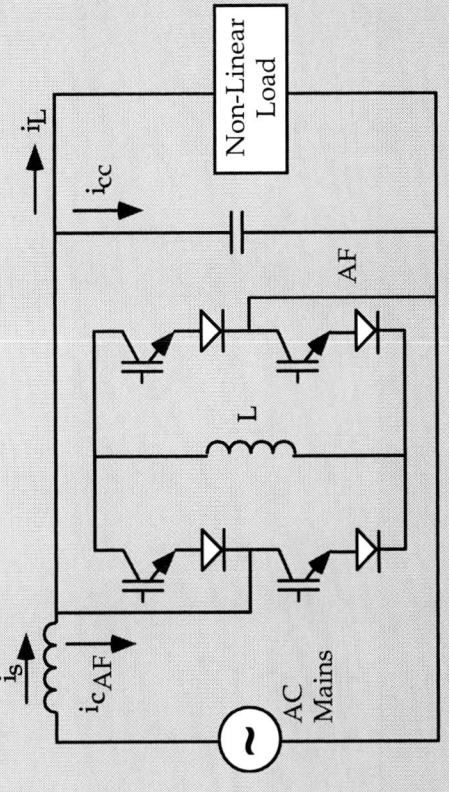

Two Wire Series AF with Current Source Converter

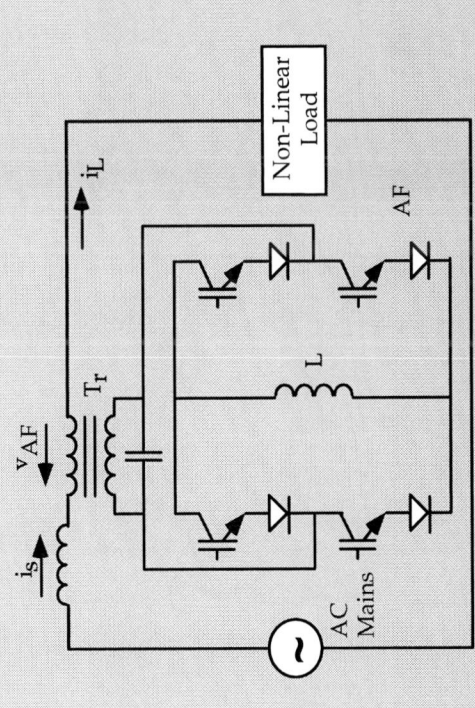

Two Wire Shunt AF with Current Source Converter

27/11/07

IIT-Delhi/PEDS-07

Topologies of Active Filters

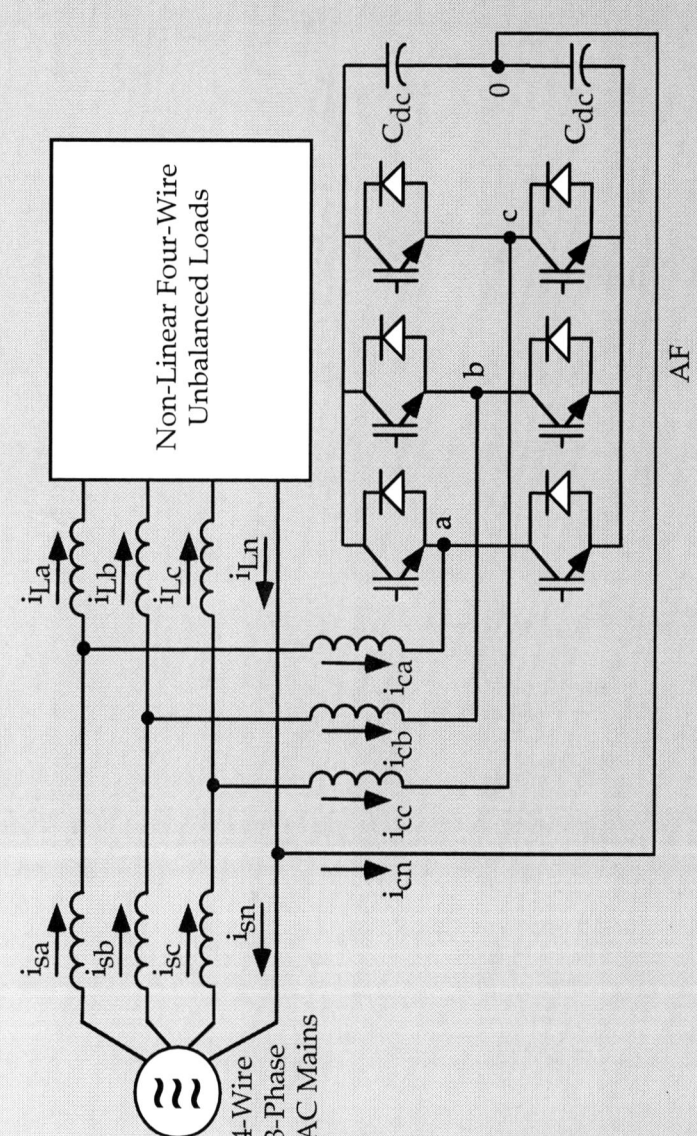

Capacitor Midpoint Four Wire Shunt AF

Topologies of Active Filters

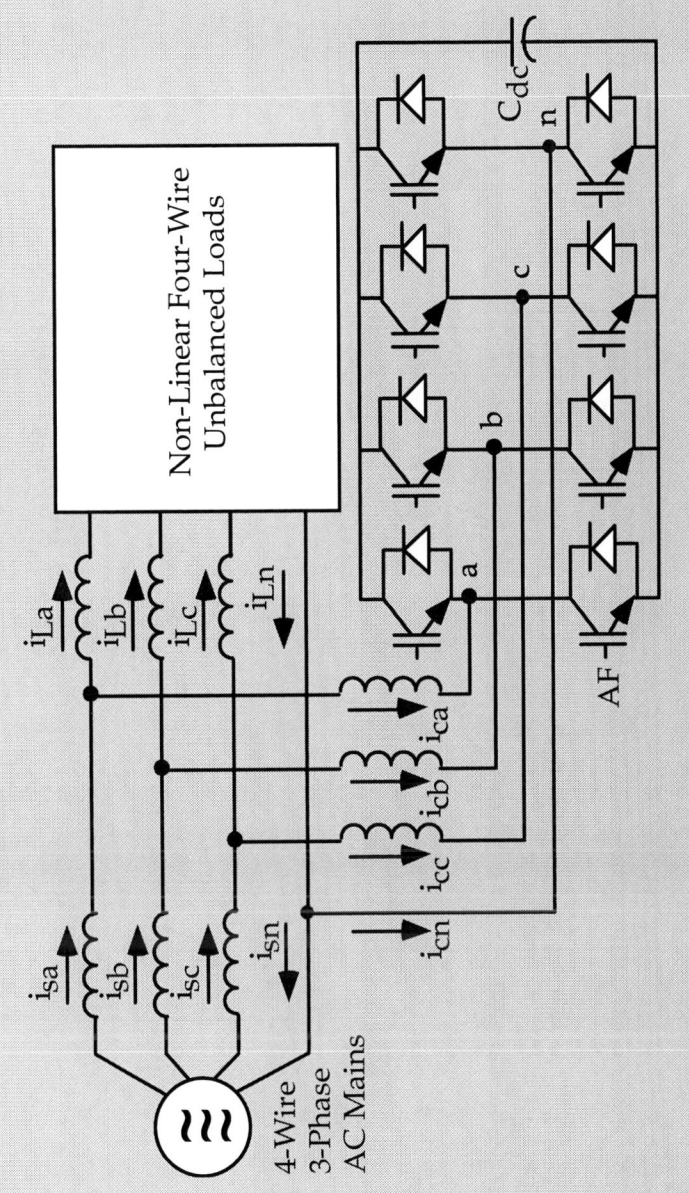

Four Pole, Four Wire Shunt AF

IIT-Delhi/PEDS-07

Topologies of Active Filters

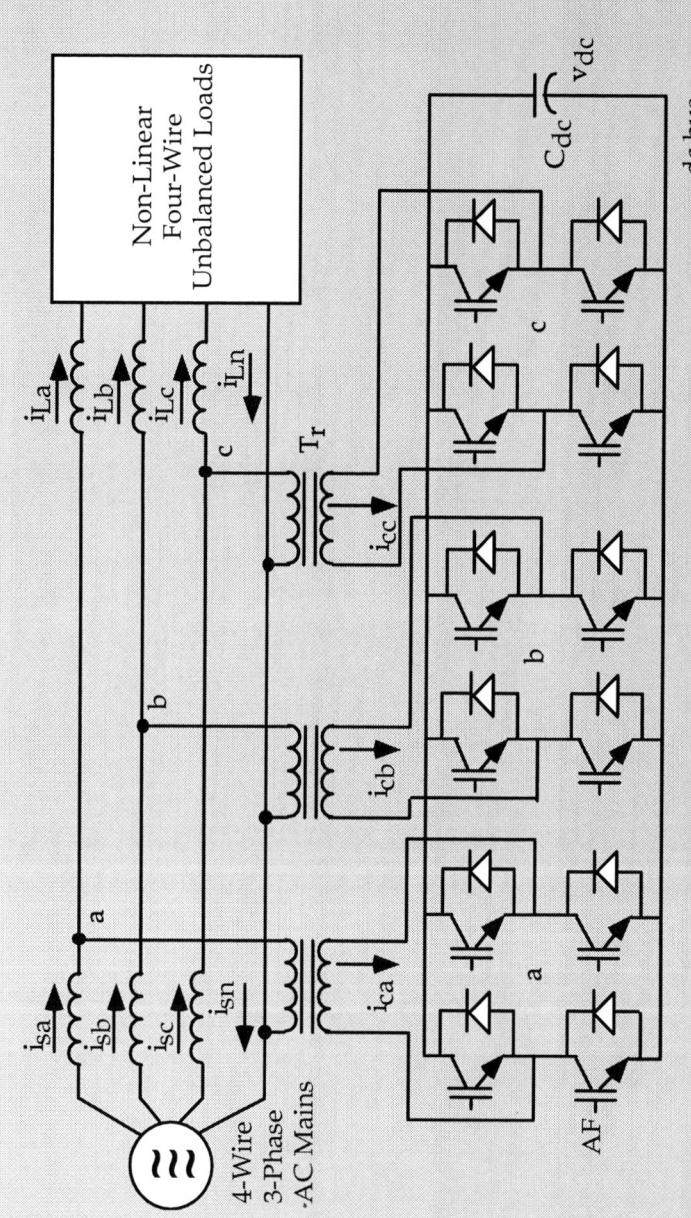

Three Bridge, Four Wire Shunt AF

27/11/07

IIT-Delhi/PEDS-07

Control of Active filter

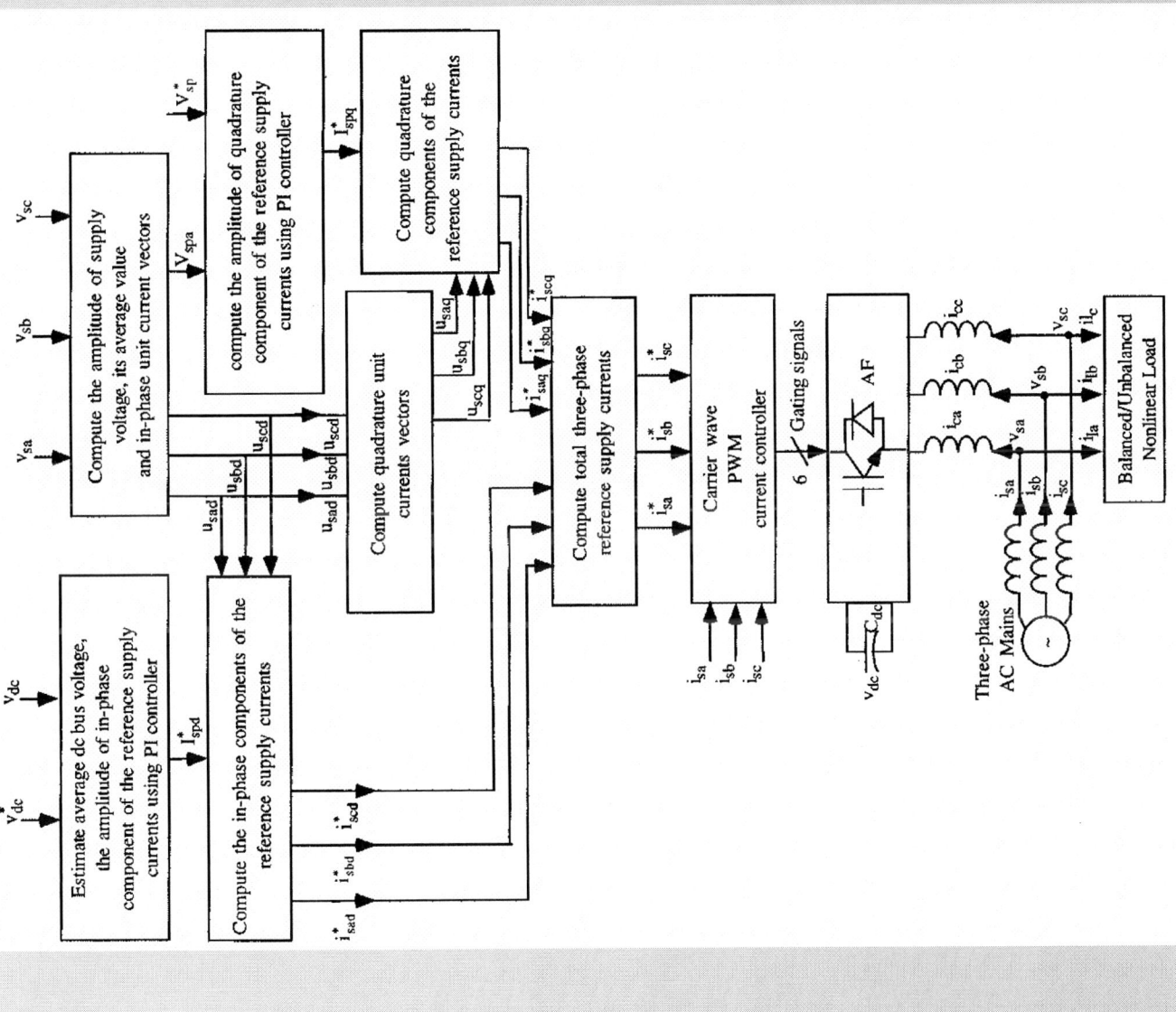

Wave-Forms

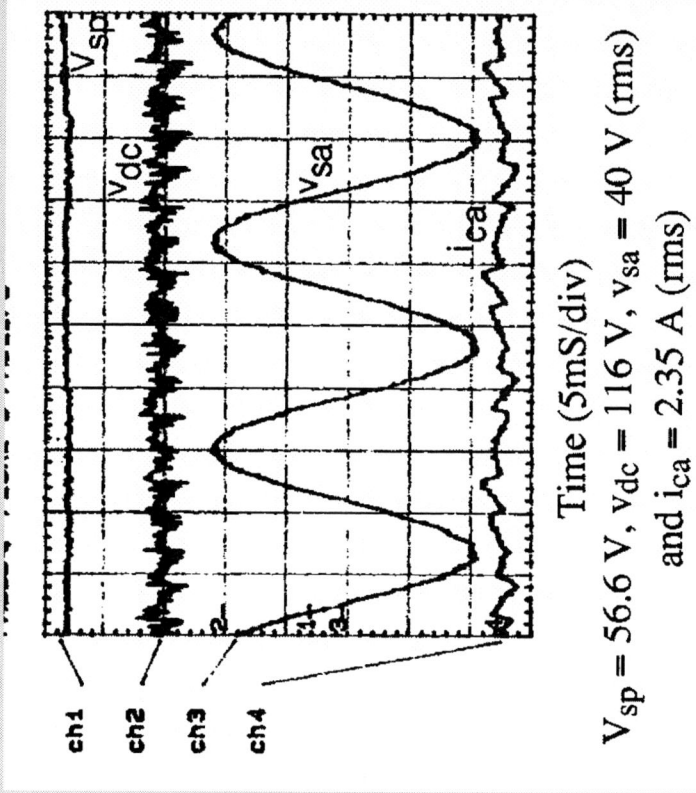

Time (5mS/div)

$V_{sp} = 56.6$ V, $v_{dc} = 116$ V, $v_{sa} = 40$ V (rms)
and $i_{ca} = 2.35$ A (rms)

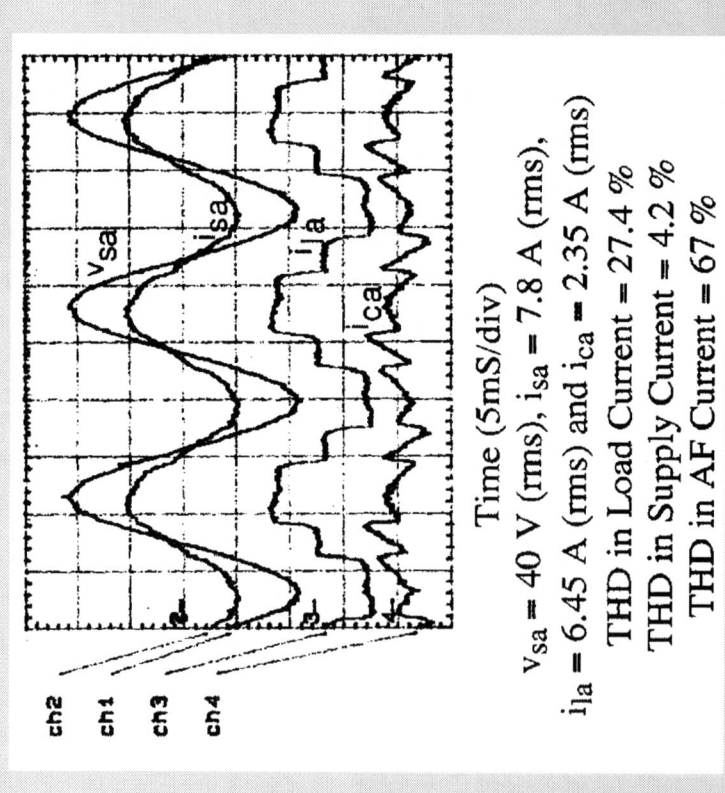

Time (5mS/div)
$v_{sa} = 40$ V (rms), $i_{sa} = 7.8$ A (rms),
$i_{la} = 6.45$ A (rms) and $i_{ca} = 2.35$ A (rms)
THD in Load Current = 27.4 %
THD in Supply Current = 4.2 %
THD in AF Current = 67 %

27/11/07

IIT-Delhi/PEDS-07

Basic Structure: Series Active Filter

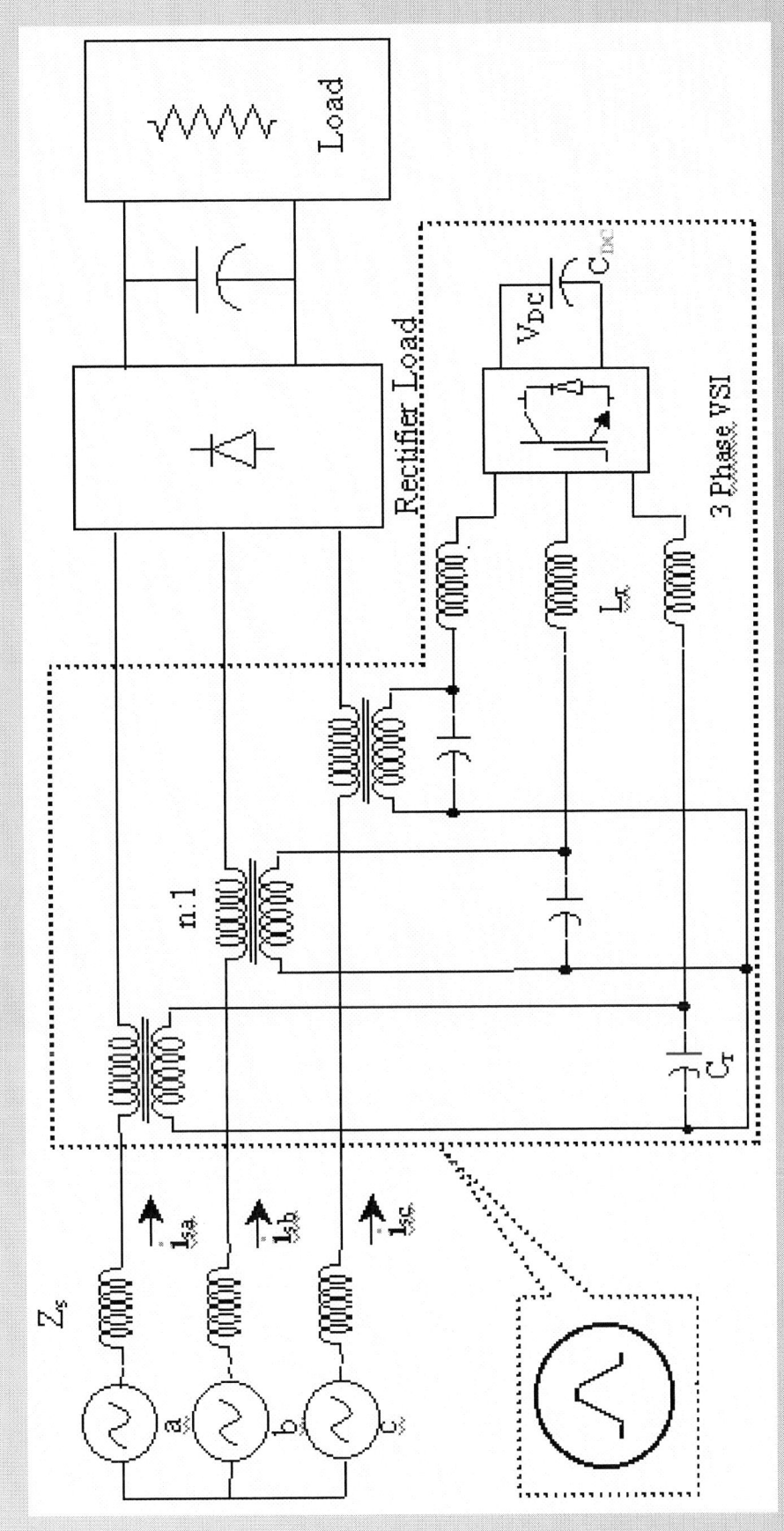

27/11/07

IIT-Delhi/PEDS-07

Simulated Performance SeAF
(Dynamic Response: Load Change)

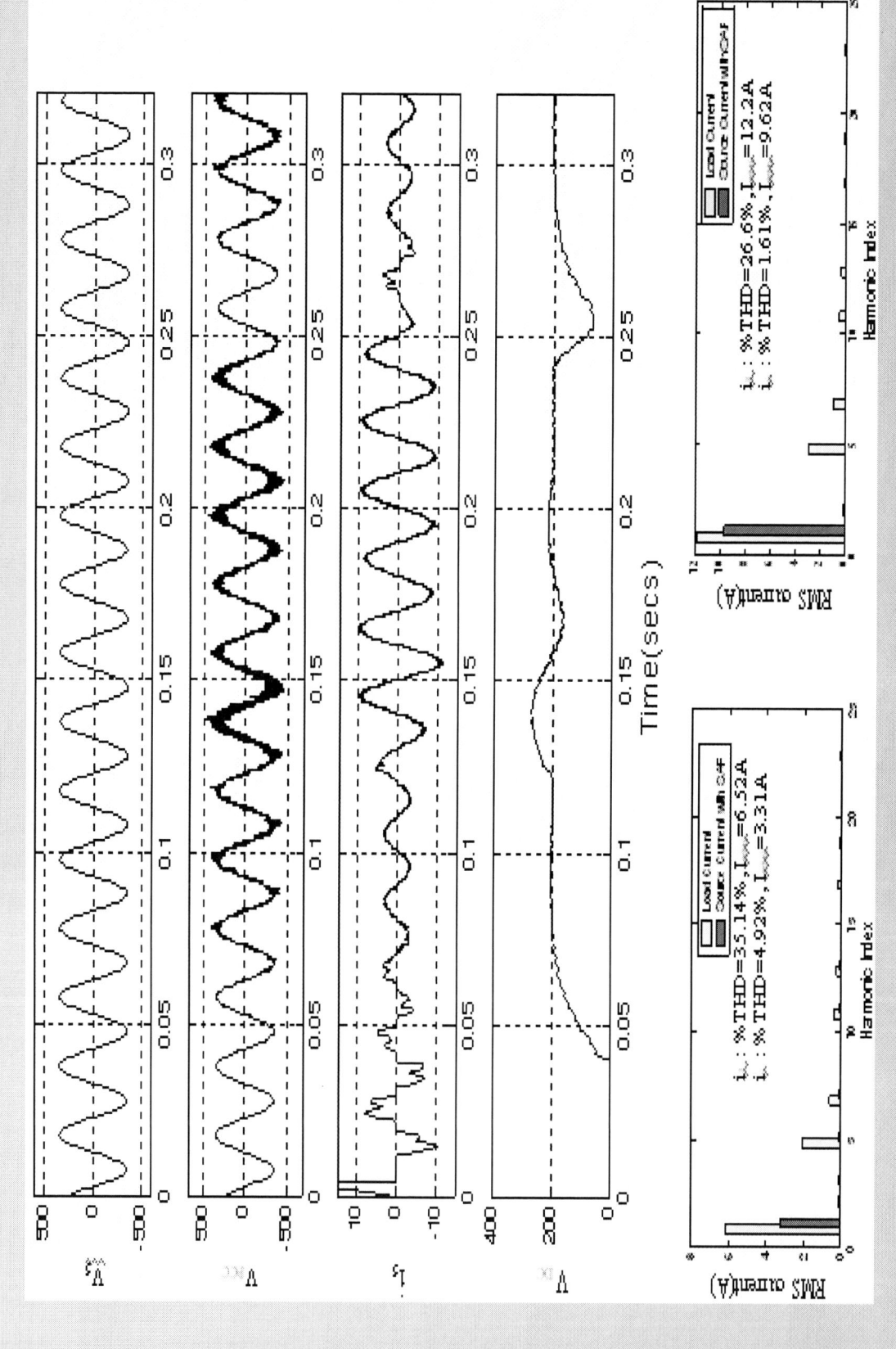

Experimental Performance (CT-5:1)

$V_{THD} = 1.1\%$

$I_{LTHD} = 35.1\%$

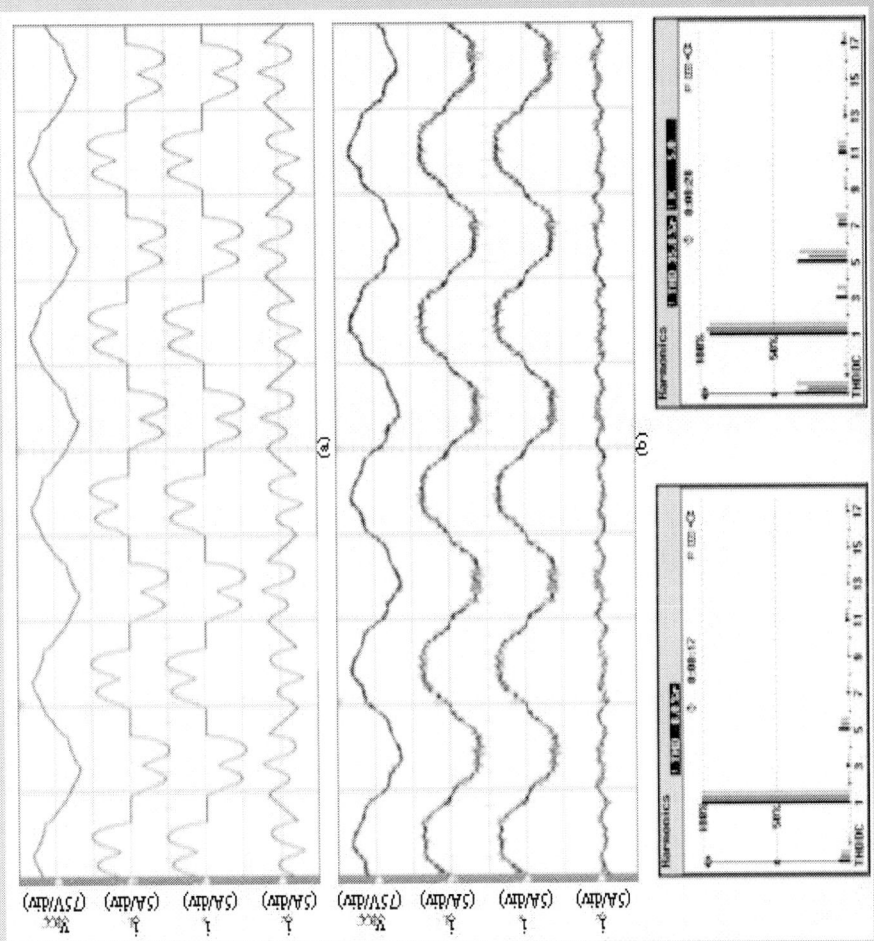

27/11/07

IIT-Delhi/PEDS-07

Experimental Performance (CT-1:1)

$V_{THD} = 1.1\%$

$I_{LTHD} = 35.1\%$

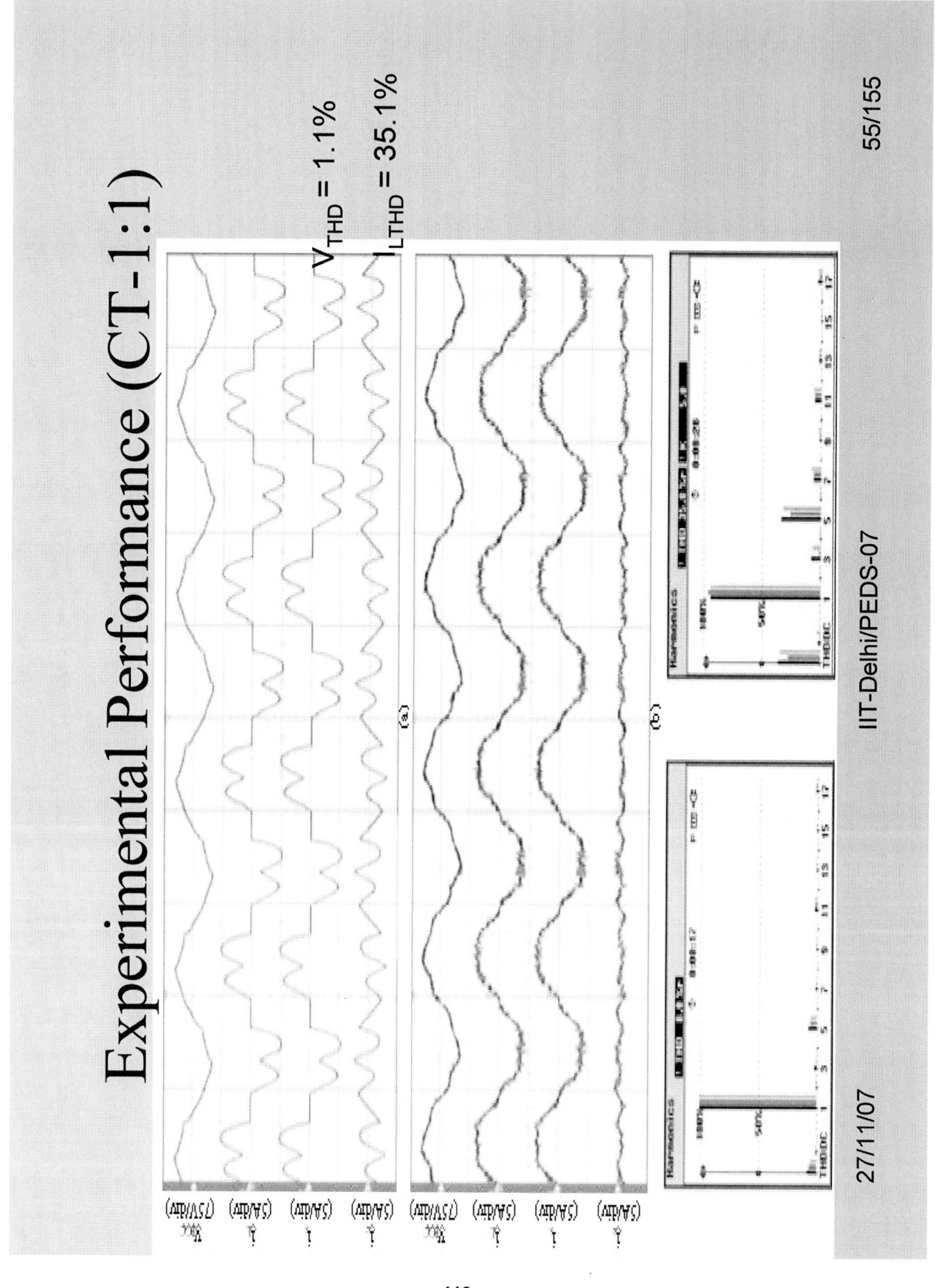

27/11/07

IIT-Delhi/PEDS-07

Hybrid Filters

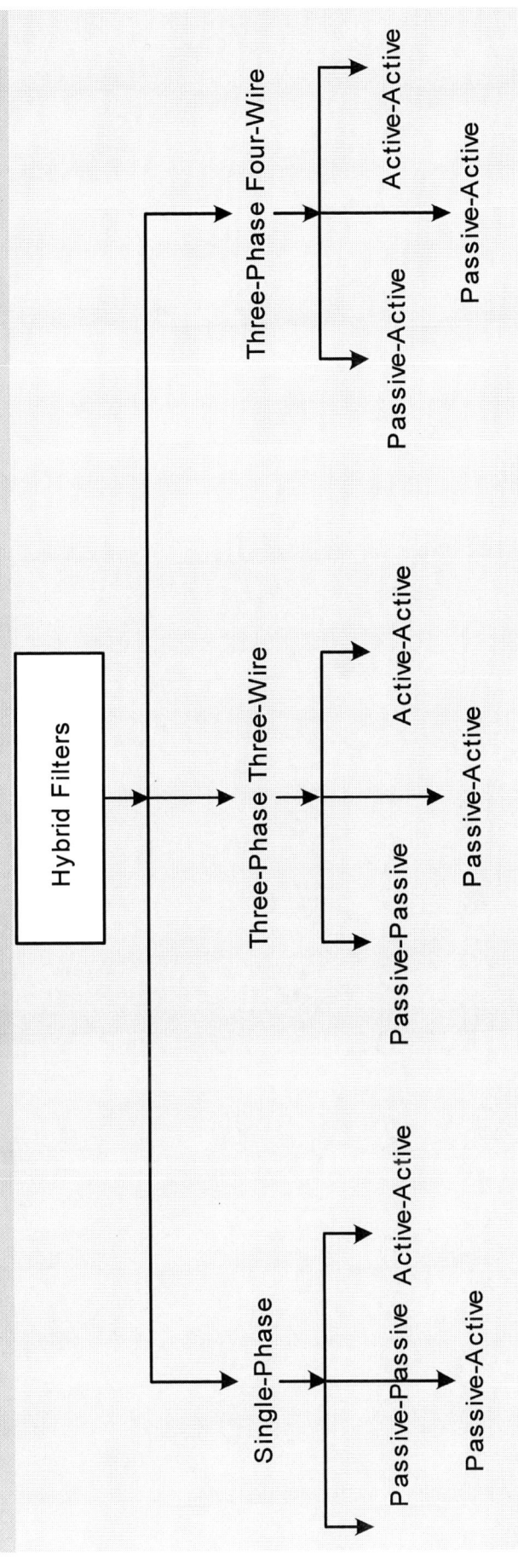

FIG. 1 - Classification of Hybrid Filters for Power Quality Improvements.

Hybrid Filters

✓ Cancel out harmonics

✓ Block resonance

✓ Reactive power management

✓ Less costly

✗ Good for retrofit applications

Hybrid Filters

Hybrid of Active and Passive Filters

Hybrid Filter as a Combination of Series Connected Passive-Series (PF_{ss}) and Active-Series (AF_{ss}) Filters

Hybrid Filter as a Combination of parallel Connected Passive-Series (PFss) and Active-Series (AFss) Filters

Hybrid Filters

Hybrid of Active and Passive Filters

Hybrid Filter as a Combination of Passive-Shunt (PFsh) and Active-Series (AFss) Filters

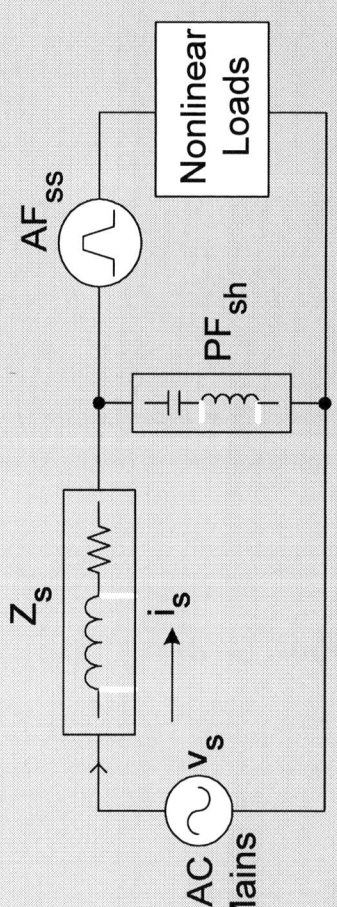

Hybrid Filter as a Combination of Active-Shunt (AFsh) and Passive-Series (PFsh) Filters

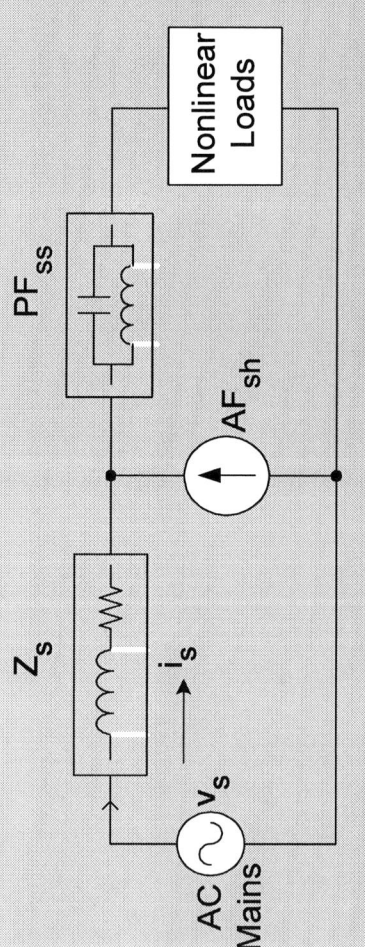

Hybrid Filters

Hybrid of Active and Passive Filters

Hybrid Filter as a Combination of Active-Shunt (AFsh) and Passive-Shunt (PFsh) Filters

Hybrid Filter as a Combination of Series Connected Passive-Shunt (PFsh) and Active-Shunt (AFsh) Filters

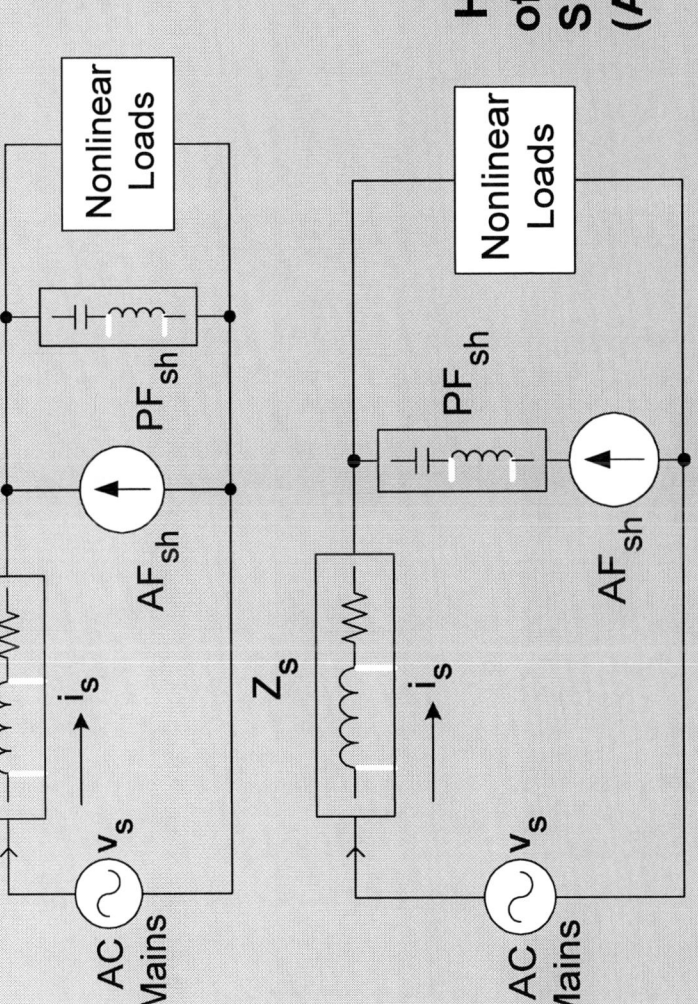

27/11/07

IIT-Delhi/PEDS-07

Hybrid Filters

Hybrid of Active and Passive Filters

Hybrid Filter as a Combination of Passive-Series (PFsh) and Active-Shunt (AFsh) Filters

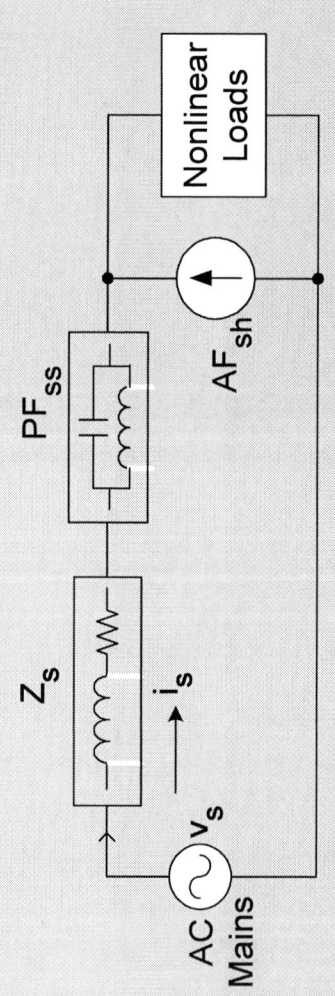

Hybrid Filter as a Combination of Active-Series (AFss) and Passive-Shunt (PFsh) Filters

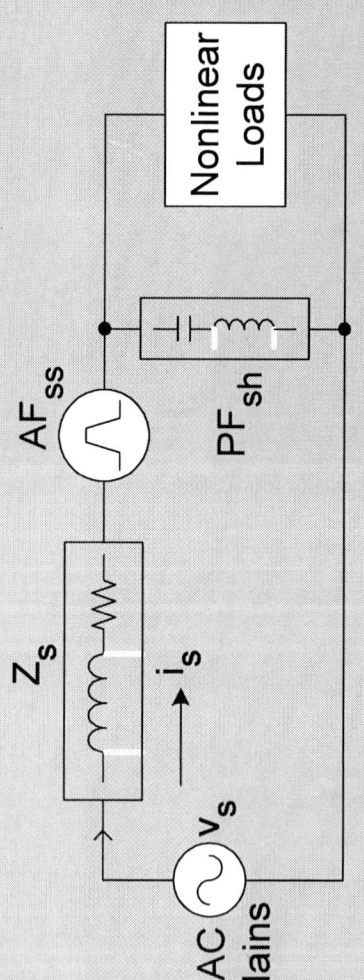

27/11/07

IIT-Delhi/PEDS-07

Parallel Hybrid Power Filter (PHF)

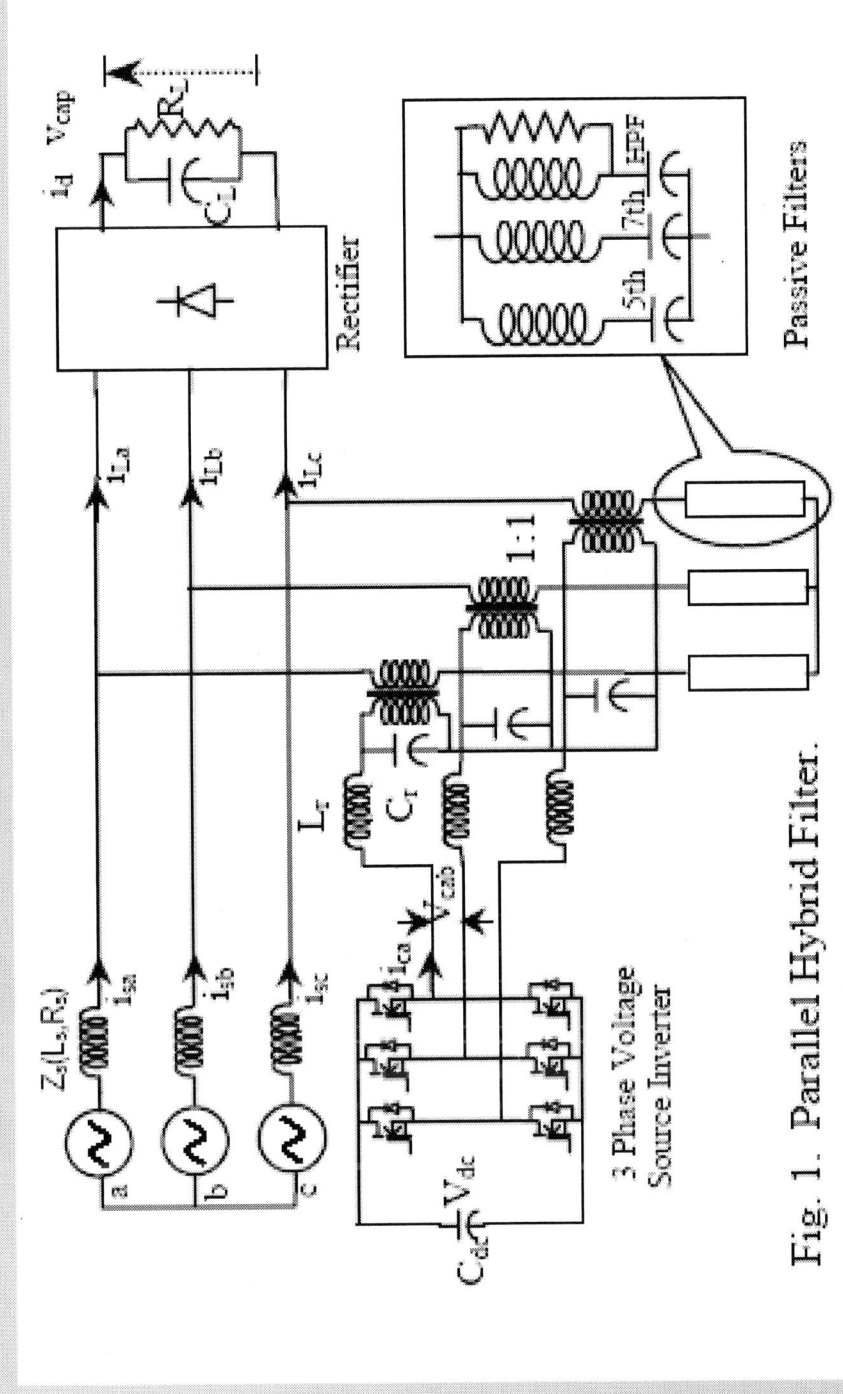

Fig. 1. Parallel Hybrid Filter.

Control Scheme of PHF

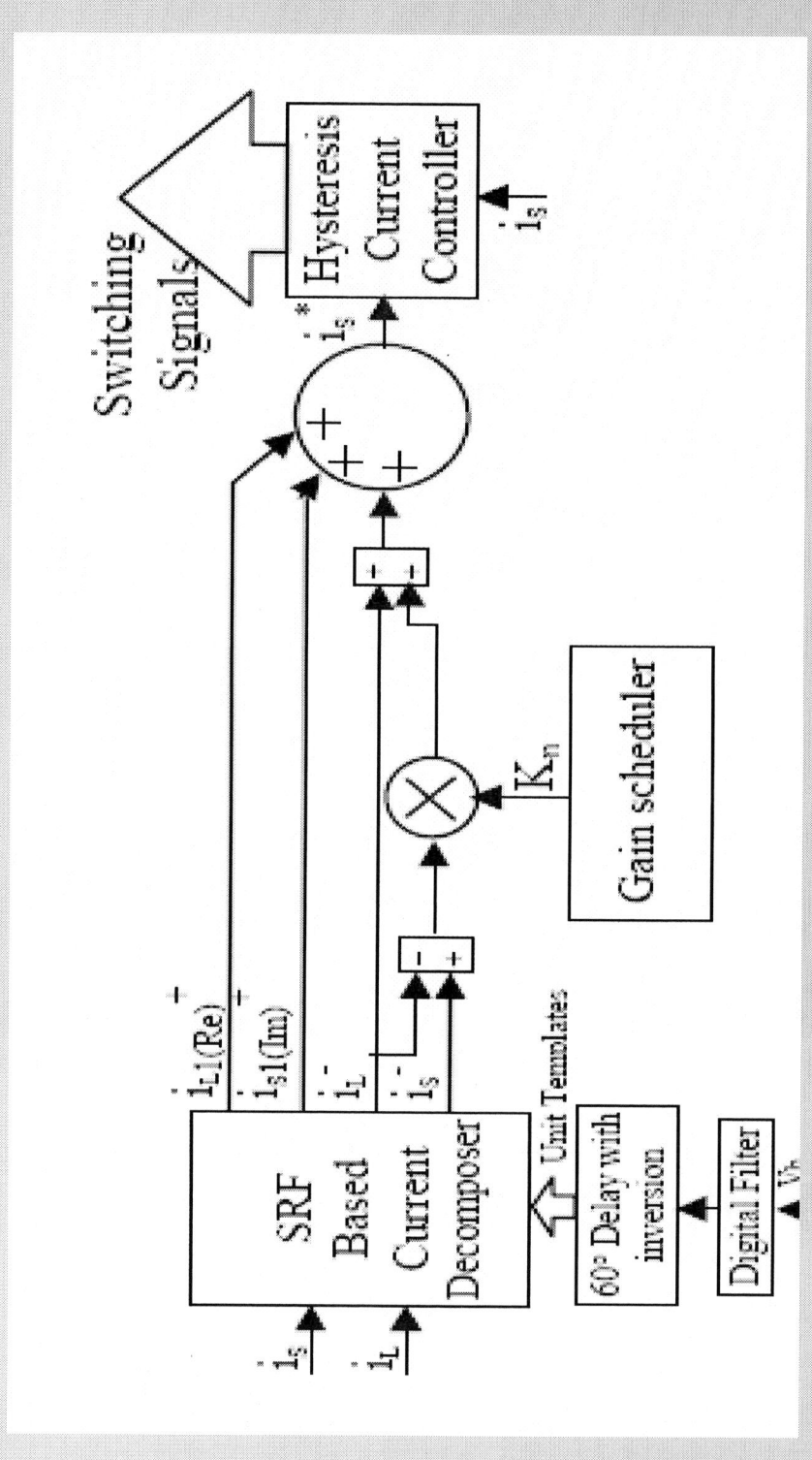

27/11/07

IIT-Delhi/PEDS-07

Load current with Passive filter alone along with harmonic spectrum
before and after the unbalance in passive filters

During balance
$I_{THD} = 44.79\%$

During un-balance
$I_{THD} = 28.28\%$

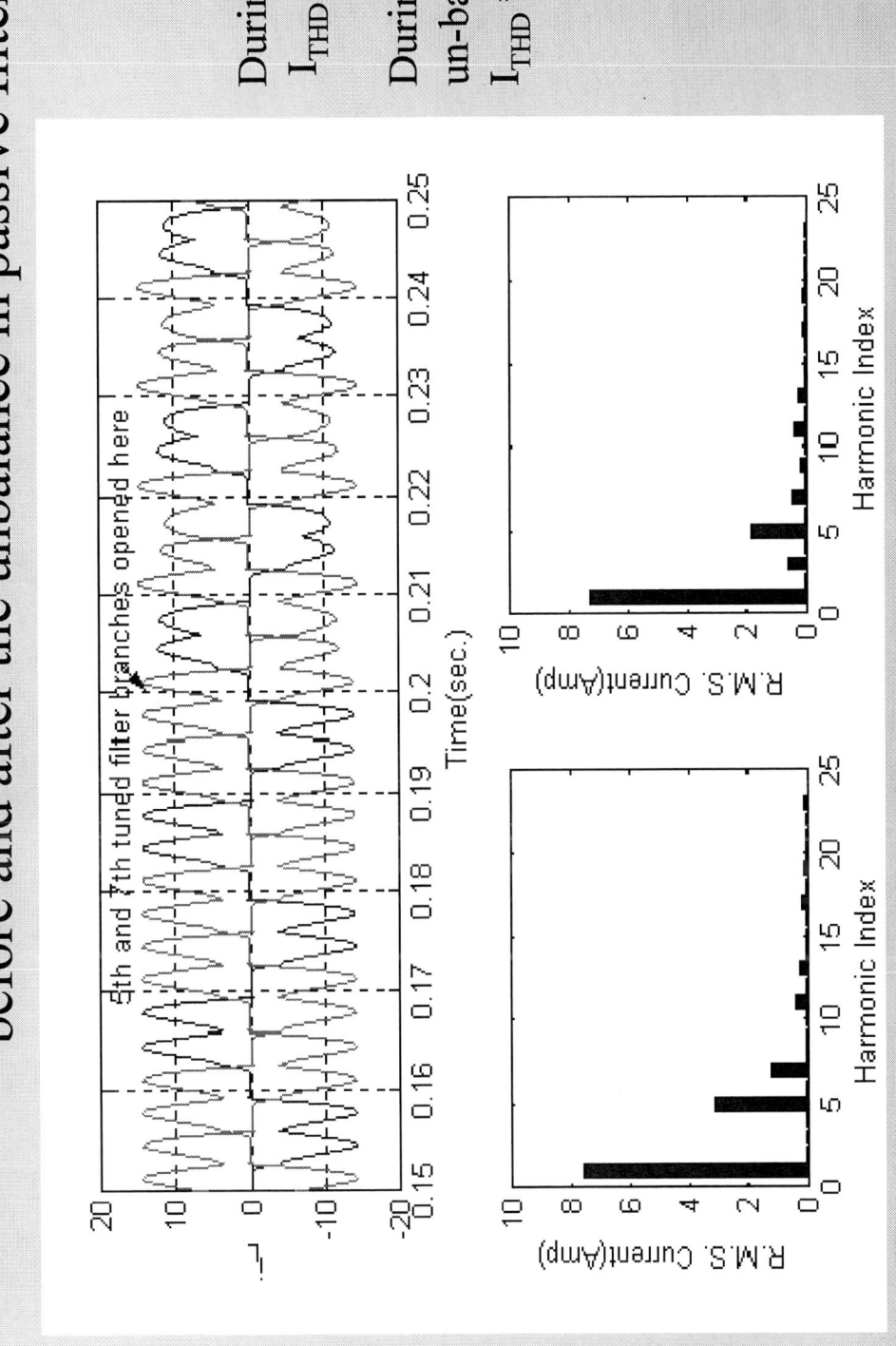

27/11/07

IIT-Delhi/PEDS-07

Source current with Passive filter alone along with harmonic spectrum before and after the unbalance in passive filters

During balance
I_{THD} = 8.36

During un-balance
I_{THD} = 32.88%

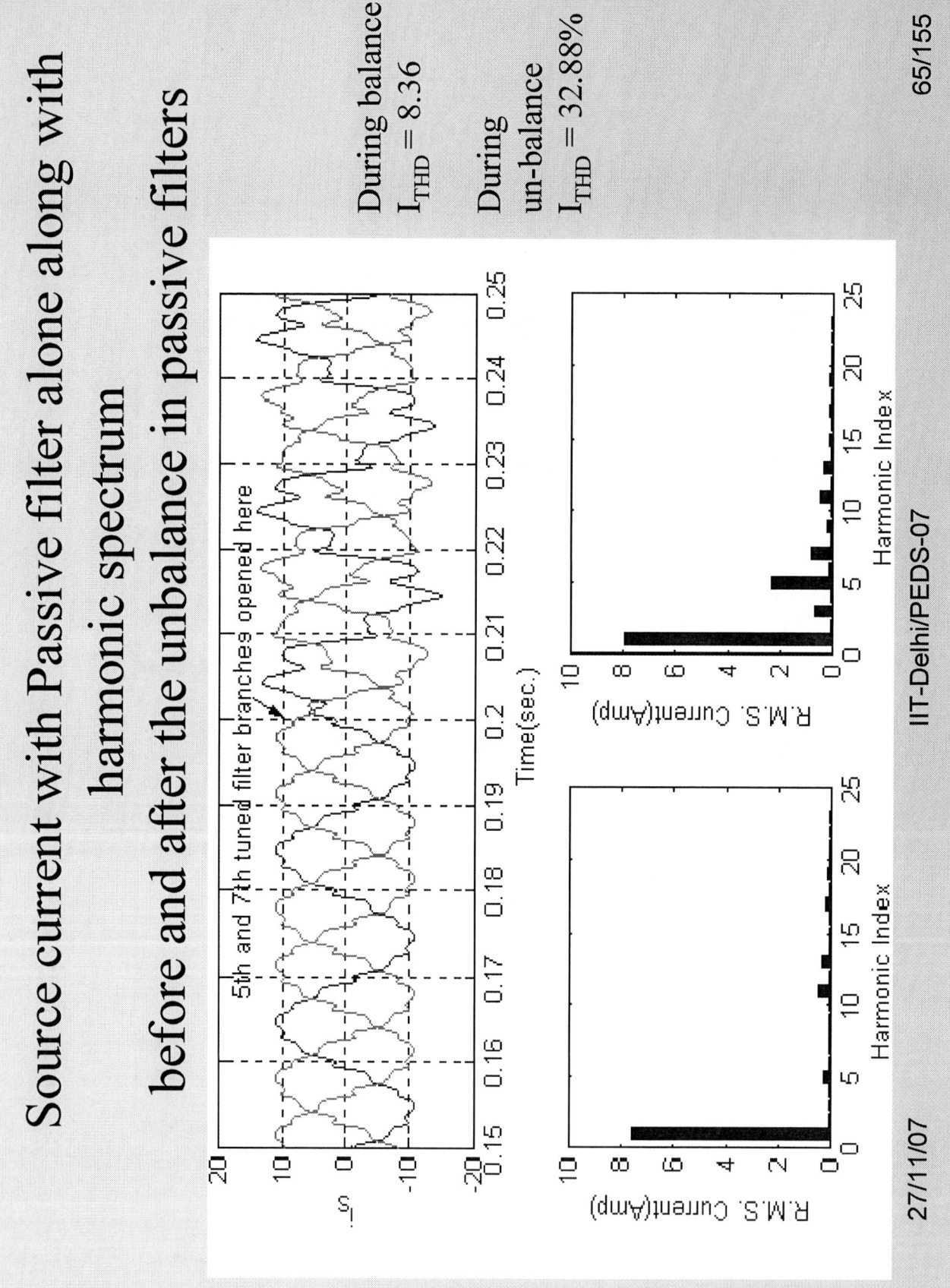

27/11/07

IIT-Delhi/PEDS-07

Passive filter current when used alone along with harmonic spectrum before and after the unbalance in passive filters

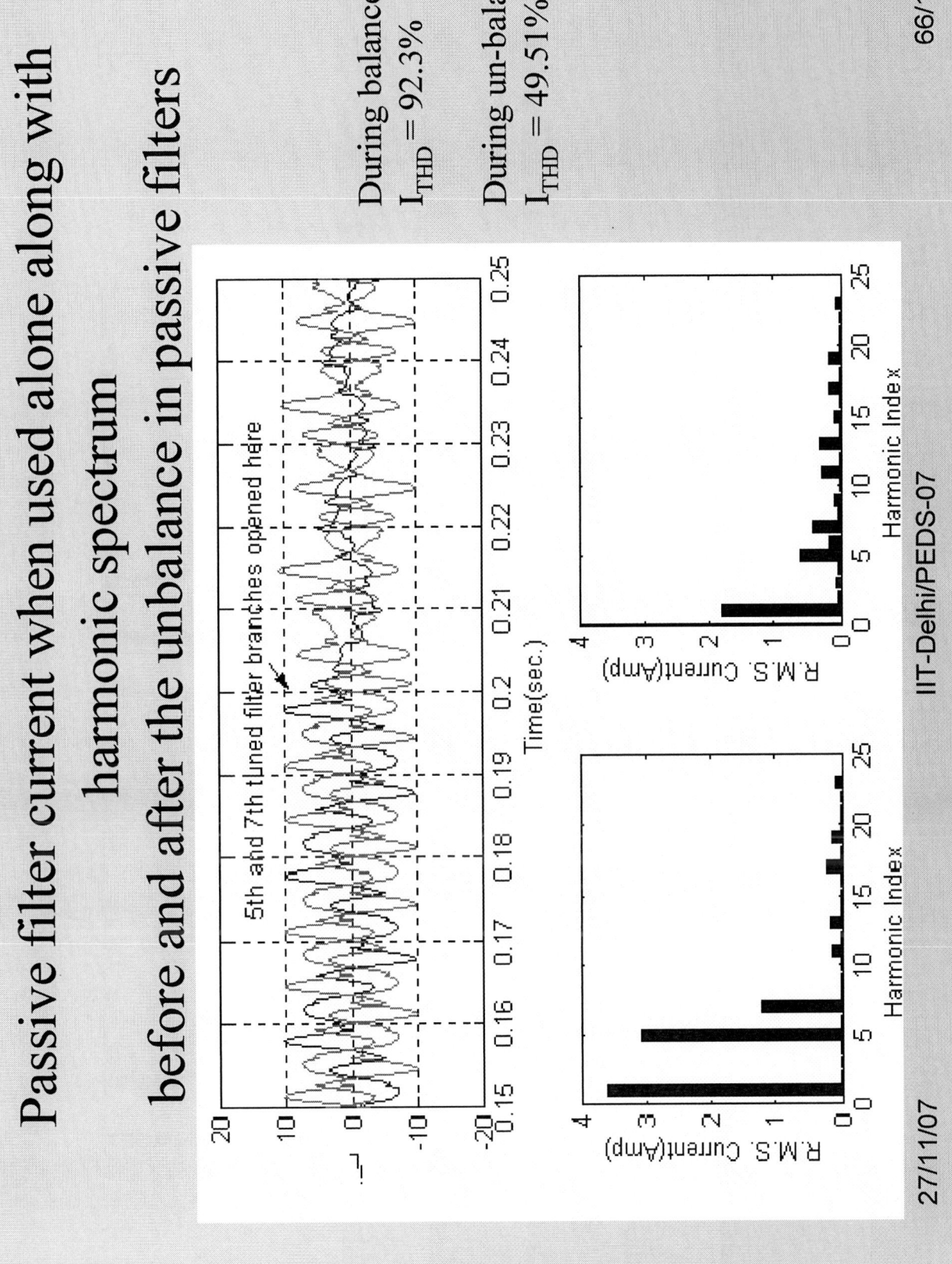

During balance
$I_{THD} = 92.3\%$

During un-balance
$I_{THD} = 49.51\%$

27/11/07

IIT-Delhi/PEDS-07

Load current with PHF along with harmonic spectrum before and after the unbalance in passive filters

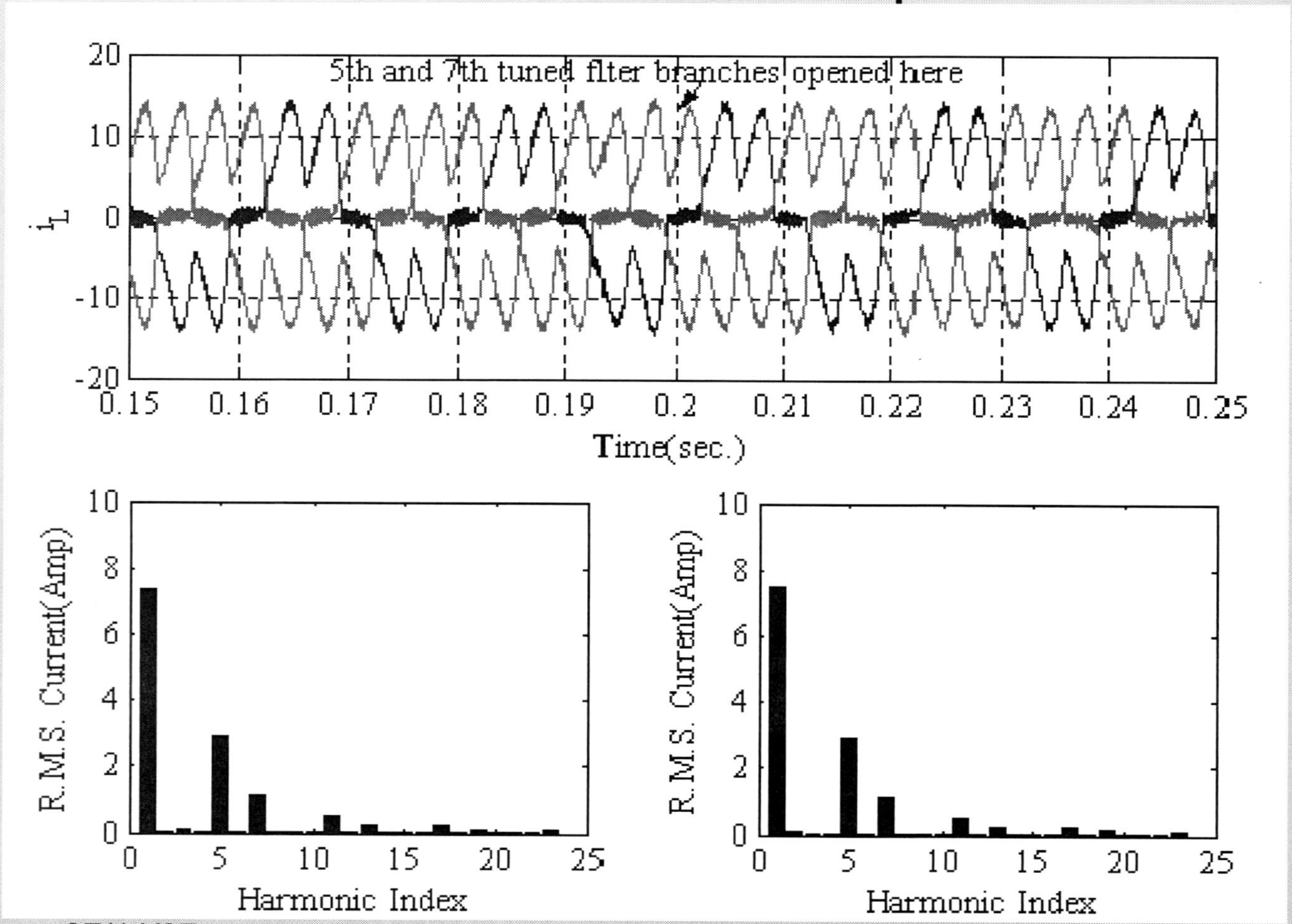

During balance I_{LTHD} = 42.37%

During un Balance I_{LTHD} = 42.82%

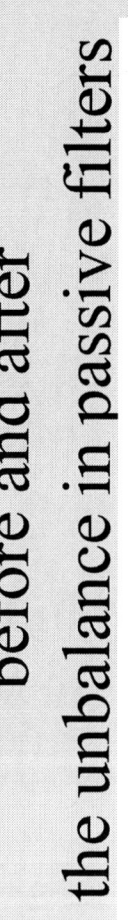

27/11/07

IIT-Delhi/PEDS-07

Source current with PHF along with harmonic spectrum before and after the unbalance in passive filters

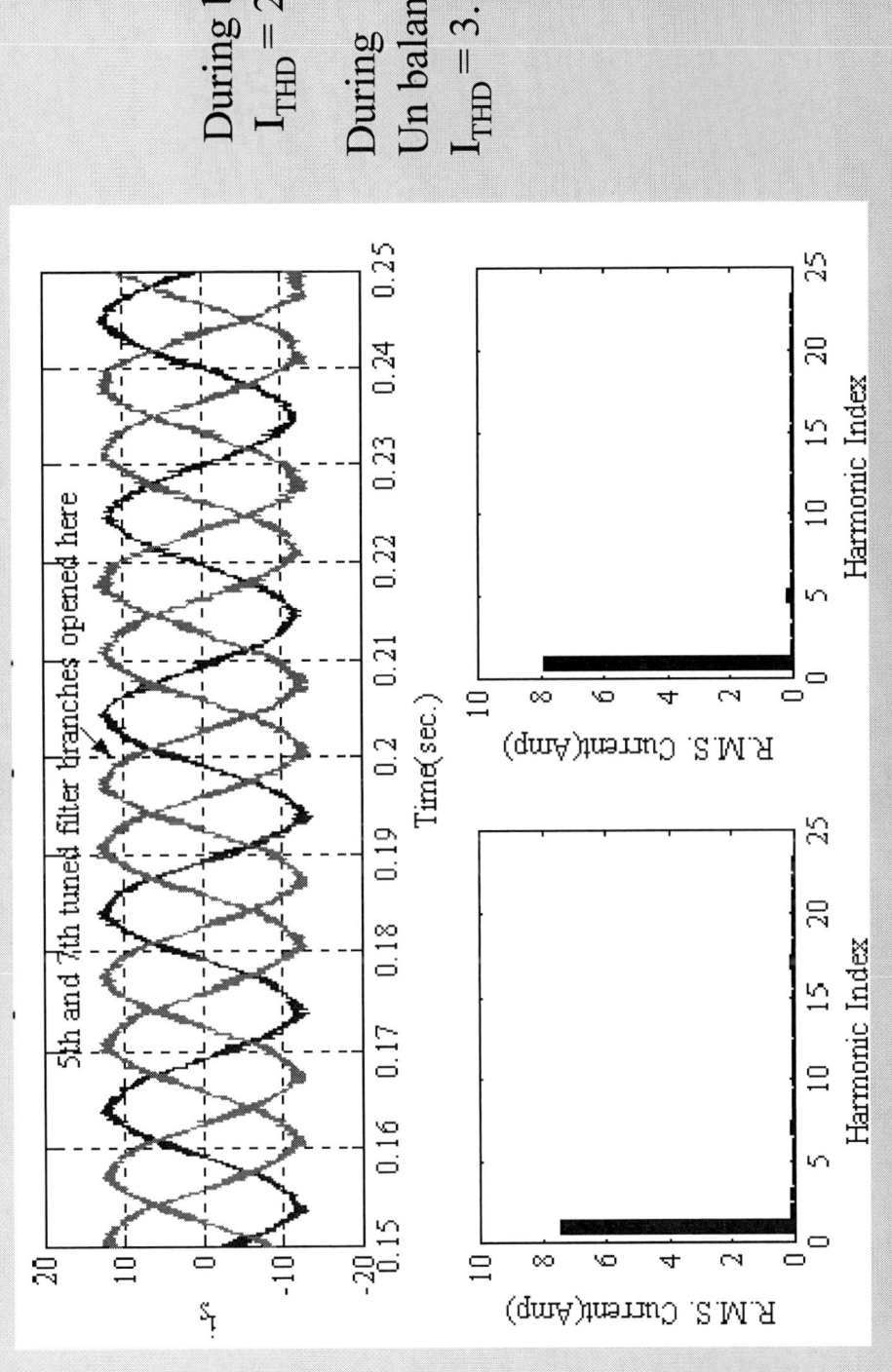

During balance
$I_{THD} = 2.97\%$

During
Un balance
$I_{THD} = 3.13\%$

27/11/07

IIT-Delhi/PEDS-07

Performance of PHF under load dynamics

Initially load is 2.62kW, at t=0.2s the load is increased to 5.23kW and at t=0.36s is again reverted to 2.62kW

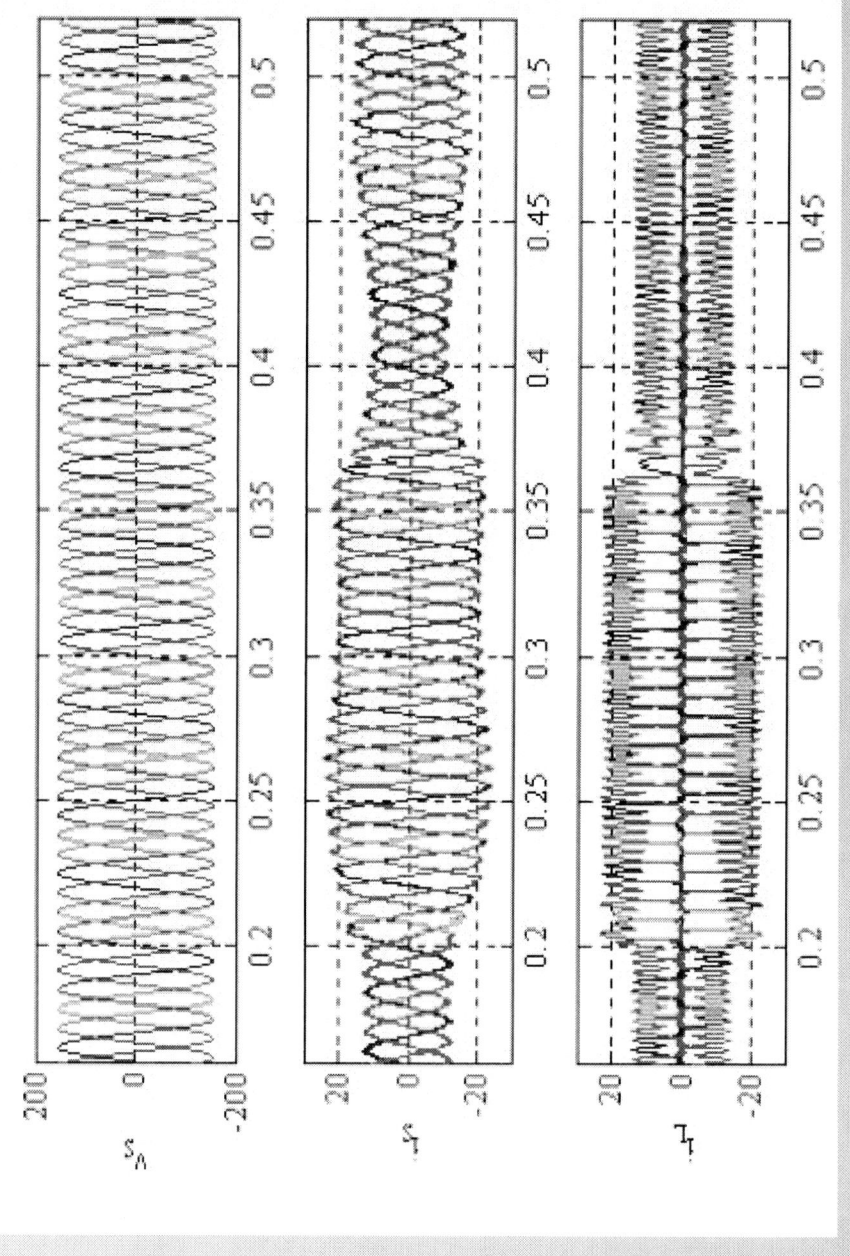

27/11/07   IIT-Delhi/PEDS-07

Steady state performance of passive filters alone and of PHF with diode rectifier for harmonic compensation along with harmonic spectrum of load and source currents (coupling transformer 1:1).

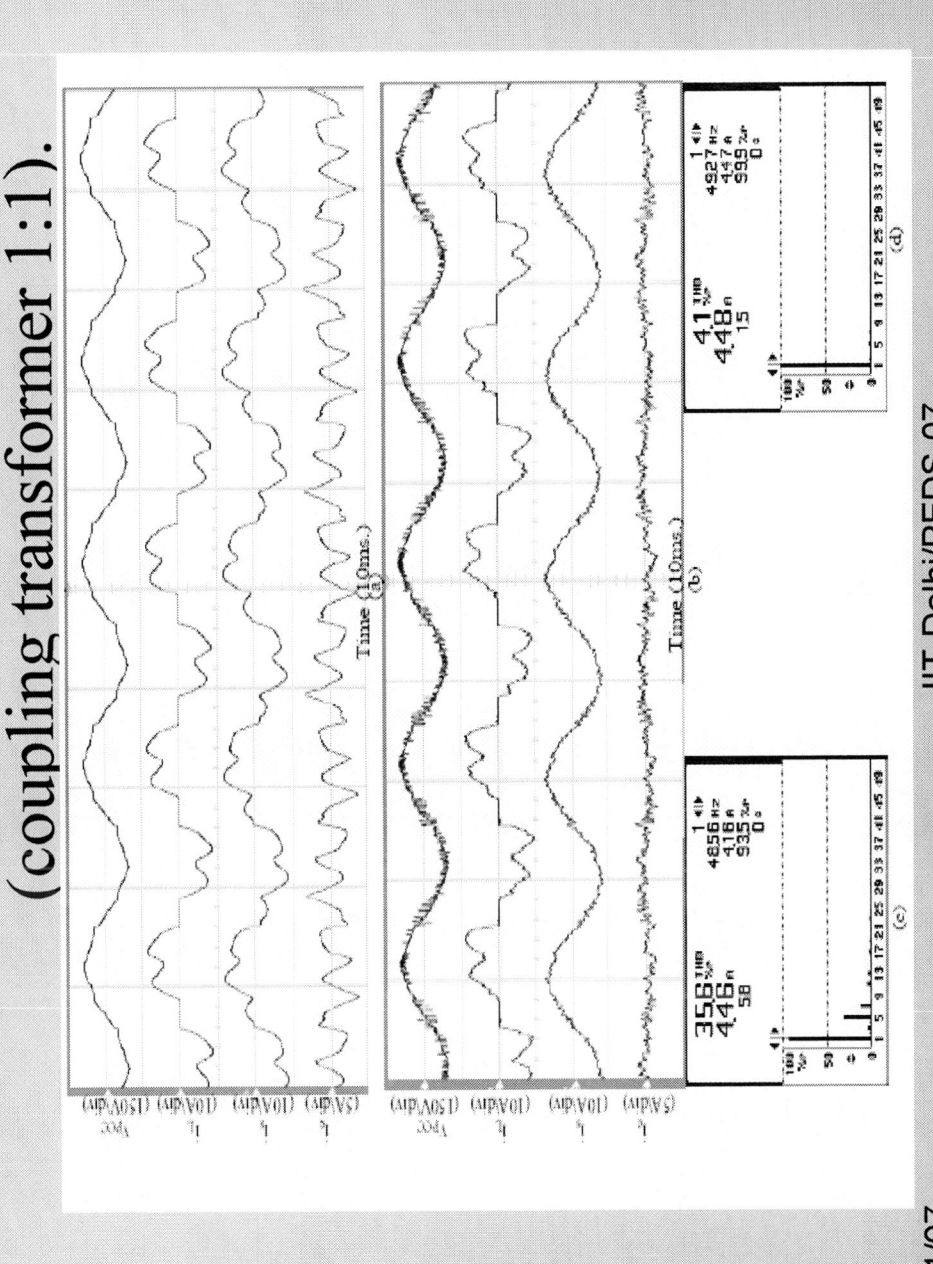

TABLE I
%**THD** AND **RMS** VALUE OF LOAD CURRENT, SOURCE CURRENT WITH PF AND PHF FOR OPERATION OF PHF (SIMULATION STUDY)

Current		RMS Current (A)			% THD		
		Ph. a	Ph. b	Ph. c	Ph. a	Ph. b	Ph. c
Load Current (with PF alone)	B	8.25	8.25	8.25	44.79	44.85	44.86
	U	7.53	8.06	8.23	28.28	41.47	42.07
Source Current with PF	B	7.56	7.56	7.56	8.36	8.36	8.37
	U	8.34	6.94	8.83	32.88	21.5	21.4
Passive Filter current	B	4.89	4.90	4.90	92.3	93	93.06
	U	2.02	3.87	4.74	49.51	96	83.6
Load Current with PHF	B	8.14	8.15	8.14	42.37	43.24	42.27
	U	8.06	8.15	8.09	42.82	41.72	42.82
Source Current with PHF	B	7.90	7.95	7.97	2.97	3.34	3.09
	U	7.41	7.6	7.8	3.13	3.7	4.7

27/11/07   IIT-Delhi/PEDS-07

TABLE II

SET OF PARAMETERS FOR THE PROTOTYPE SHUNT PASSIVE FILTERS

C_5	40µF	C_7	25µF	C_{11}	40µF
L_5	10 mH	L_7	8.2 mH	L_{11}	2 mH
R_5	0.24Ω	R_7	0.2Ω	R_{11}	0.17Ω
				R_h	2.5Ω

TABLE III

%THD AND RMS VALUE OF LOAD CURRENT , SOURCE CURRENT WITH PF AND PHF FOR OPERATION OF PHF (EXPERIMENTAL STUDY)

Current	% THD	RMS Current (A)
	Other Load	Other Load
Load Current	35.1	4.46
Source Current with PF	19.2	4.58
Source Current with SHF	4.0	4.48

Custom Power Devices
(Static Compensators)

Distribution Static Compensator

(DSTATCOM)

Distribution Voltage Restorer

(DVR)

Unified Power Quality Compensator

(UPQC)

Distribution Static Compensator (DSTATCOM)

Functions

- ❖ Reactive Power Compensation
- ❖ Voltage Regulation
- ❖ Unbalance Compensation (for 3-phase systems)
- ❖ Neutral Current Compensation (for 3-phase 4-wire systems)

DSTATCOM 1-Phase 2-Wire

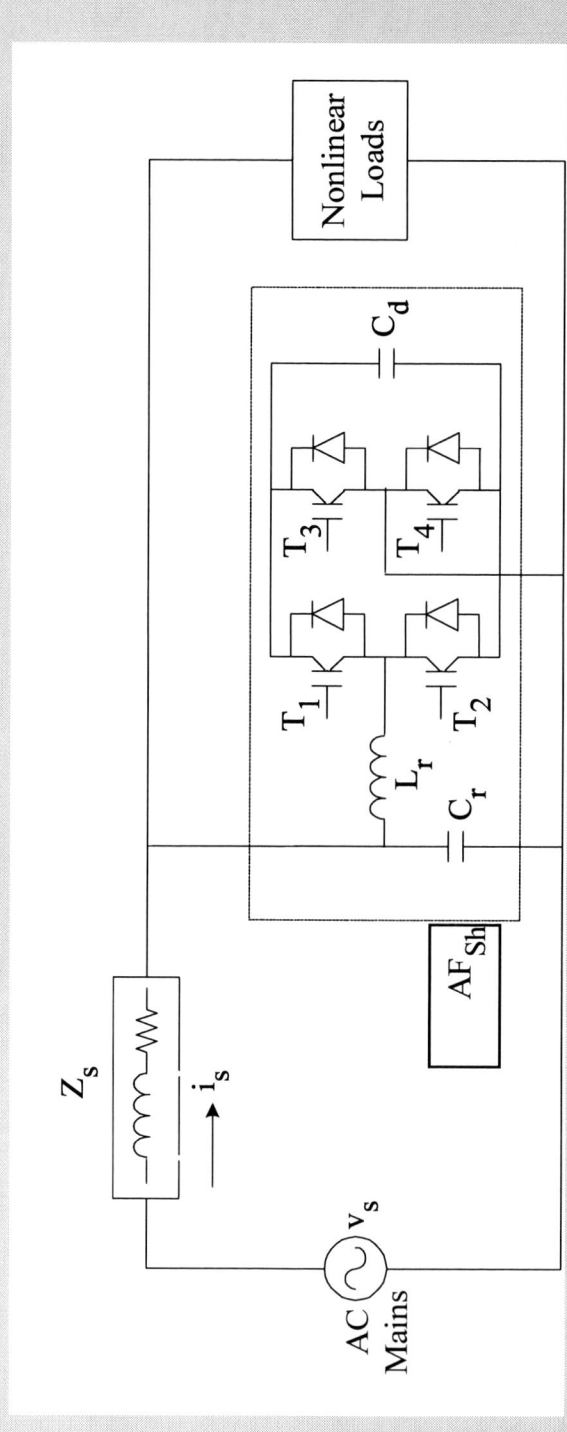

27/11/07

IIT-Delhi/PEDS-07

DSTATCOM 3-Phase 3-Wire

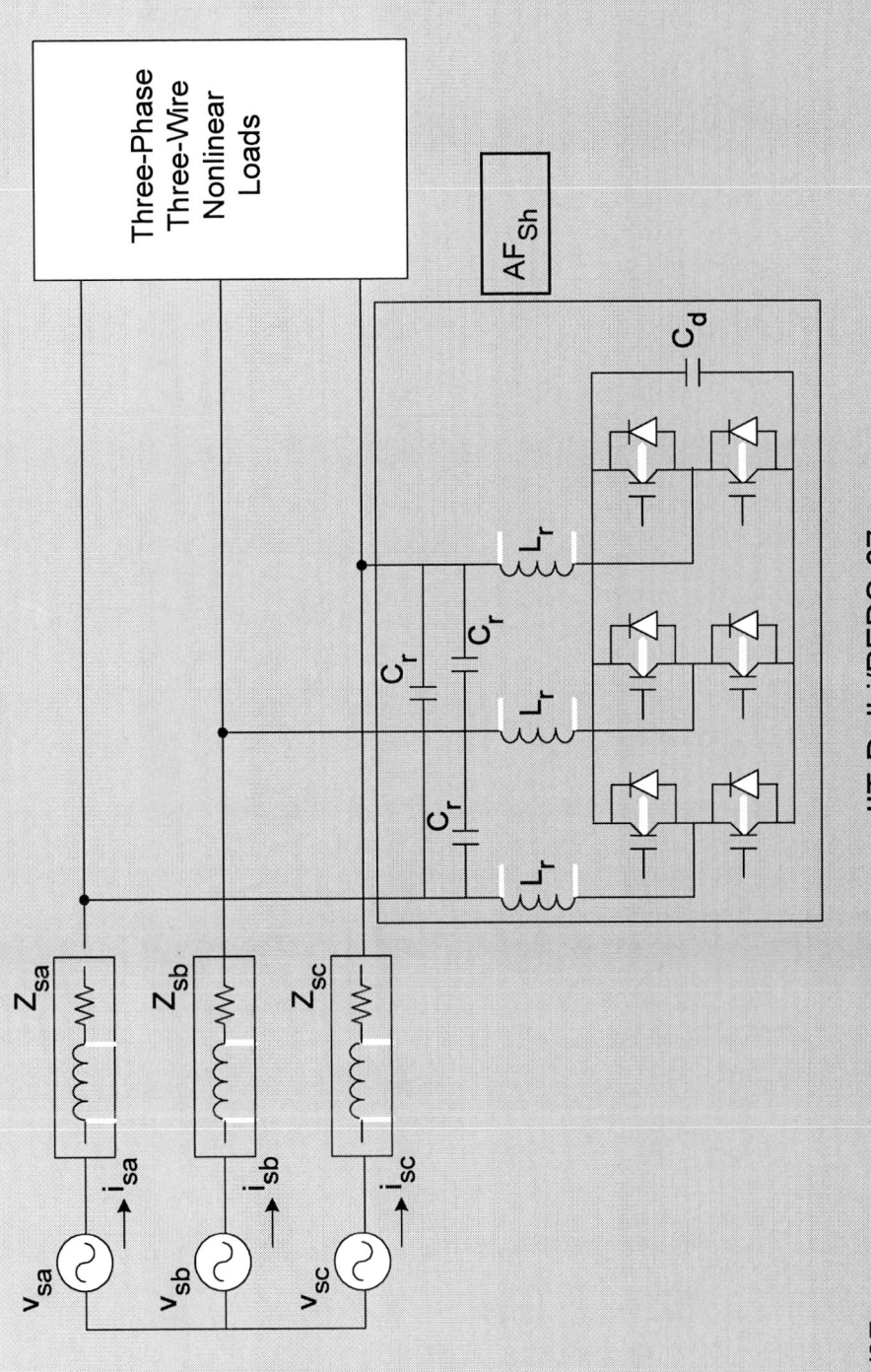

27/11/07

IIT-Delhi/PEDS-07

Simulated Performance of 3-Phase 3-Wire DSTATCOM

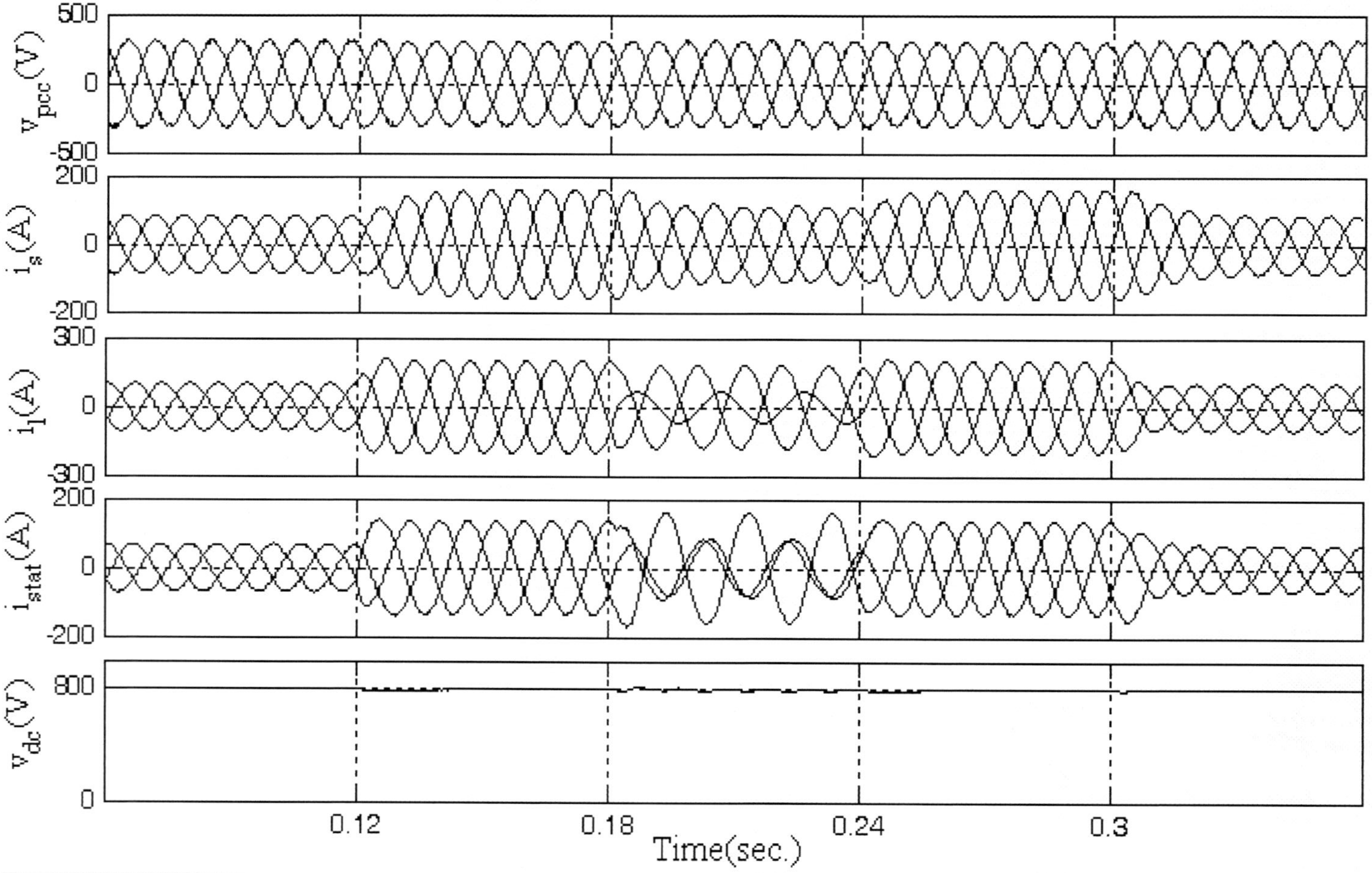

Dynamic performance of DSTATCOM load change (38kW to 71kW) at t=0.12s, for operation under unbalance from t=0.18s to t=0.24s similar dynamics in reverse sequence henceforth from t=0.24s to t=0.36s

27/11/07   IIT-Delhi/PEDS-07

DSTATCOM 3-Phase 4-Wire Mid-Point Capacitor Topology

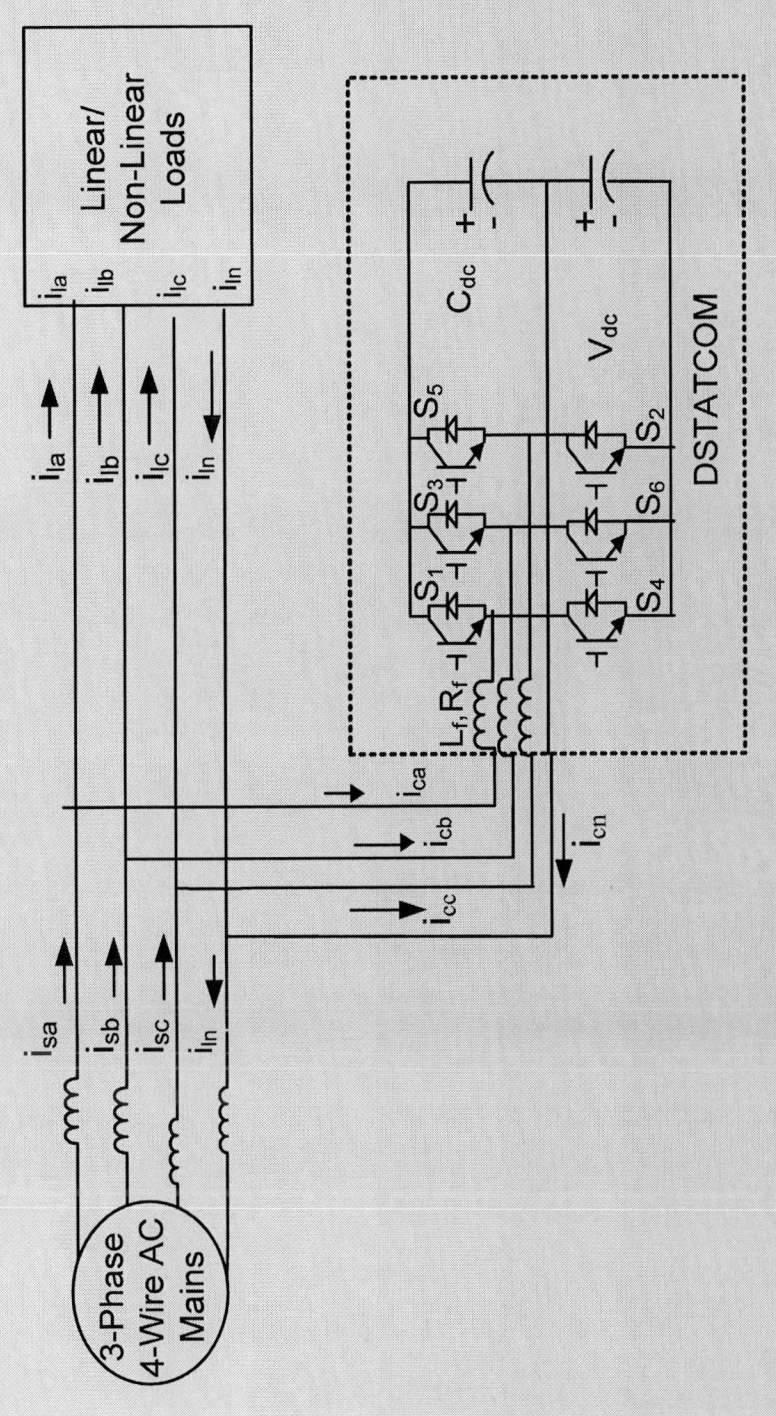

27/11/07

IIT-Delhi/PEDS-07

Four Pole Topology of DSTATCOM

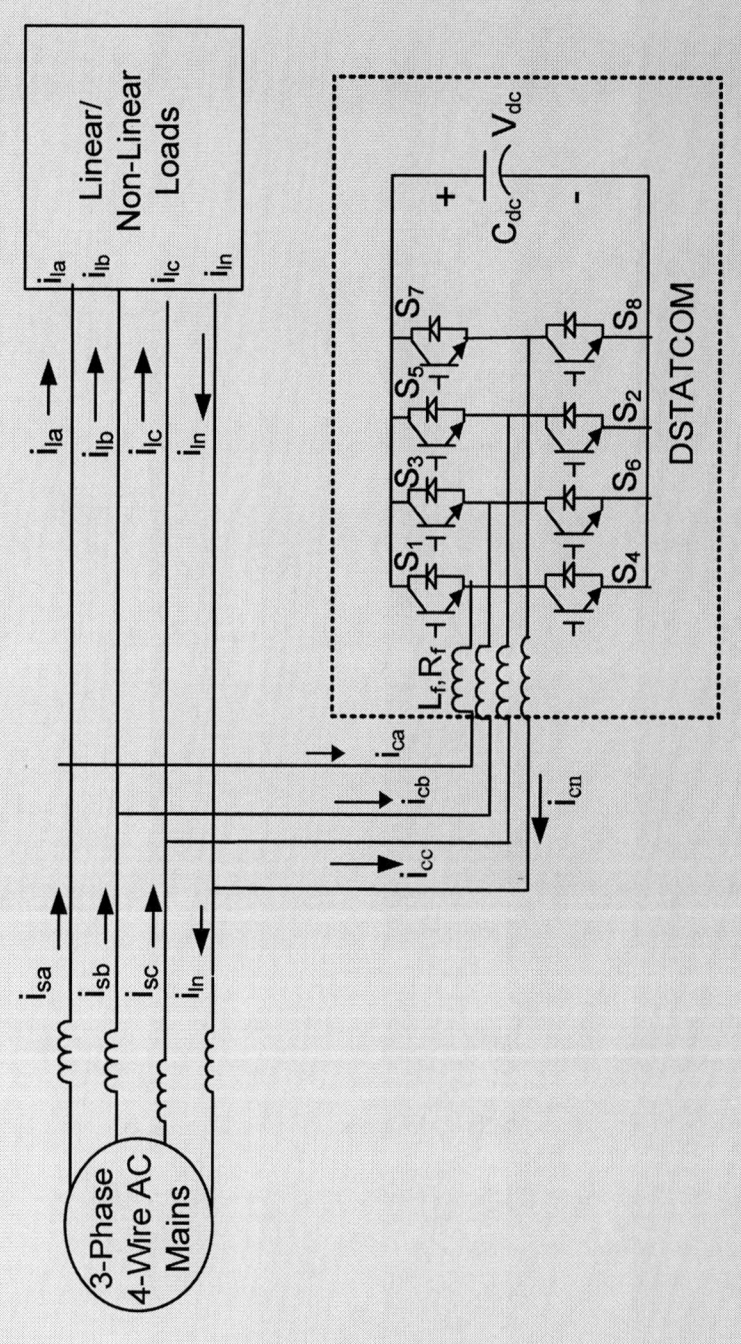

Three Single-Phase VSC Topology of DSTATCOM

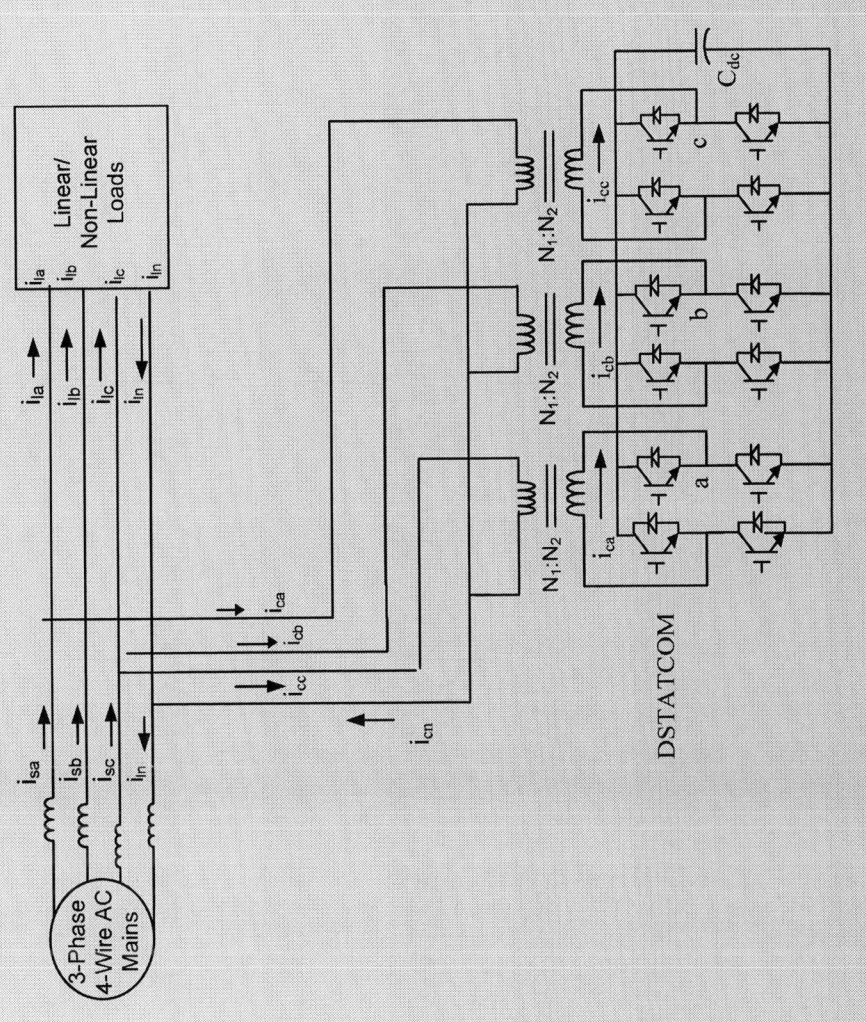

27/11/07

IIT-Delhi/PEDS-07

Simulated Performance of 3-Phase 4-Wire DSTATCOM

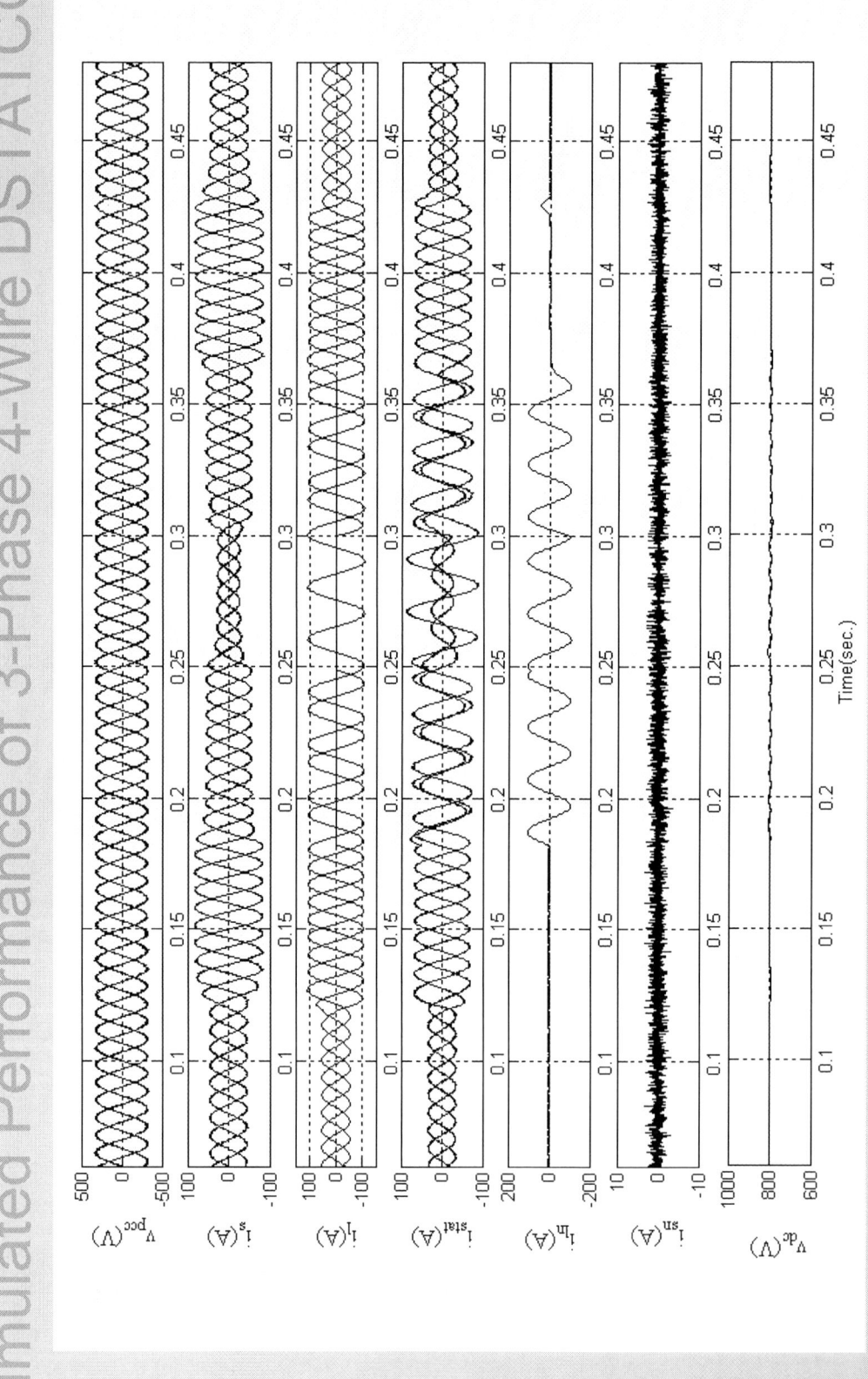

Dynamic performance of DSTATCOM for load change (20kW to 38.5kW) at t=0.12s, for operation under two phase from from t=0.18s to t=0.24s at load (26.8kW) from t=0.24s to t=0.3s at single phase load (13.4kW) IIT-Delhi/PEDS-07

Distribution Voltage Restorer (DVR)

Functions

❖ Reactive Power Compensation

❖ Voltage Regulation

❖ Compensation for Voltage sag and Swell

❖ Unbalance Voltage Compensation (for 3-phase systems)

27/11/07

IIT-Delhi/PEDS-07

DVR 1-Phase 2-Wire

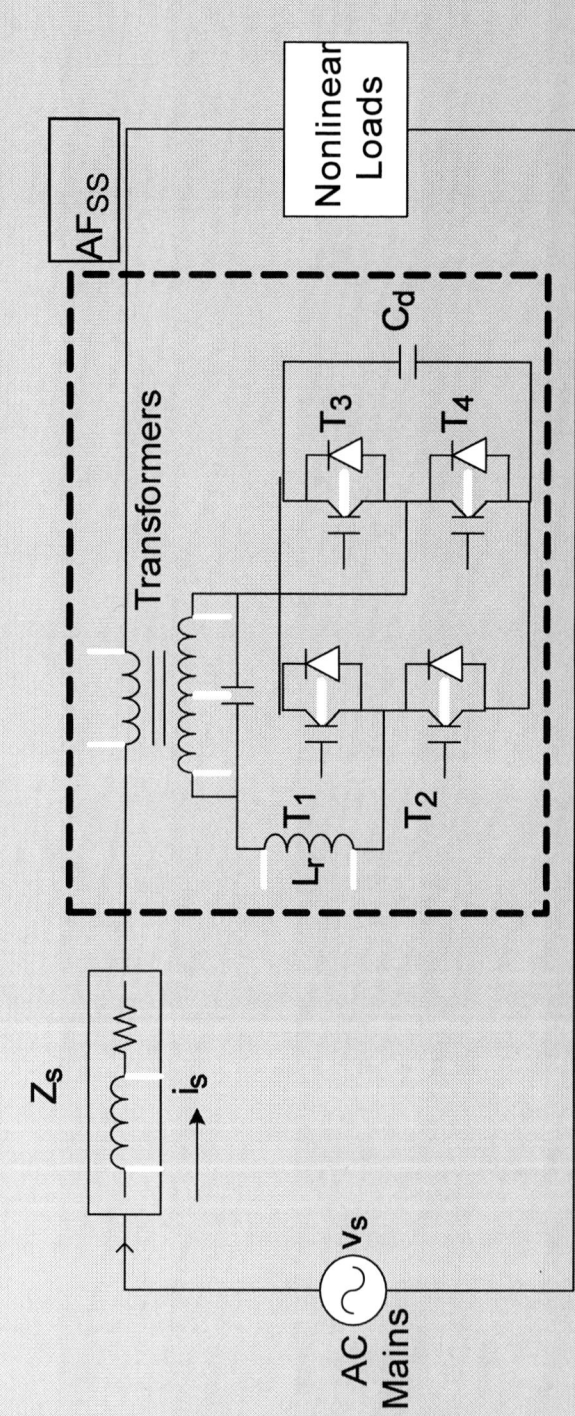

DVR 3-Phase 3-Wire

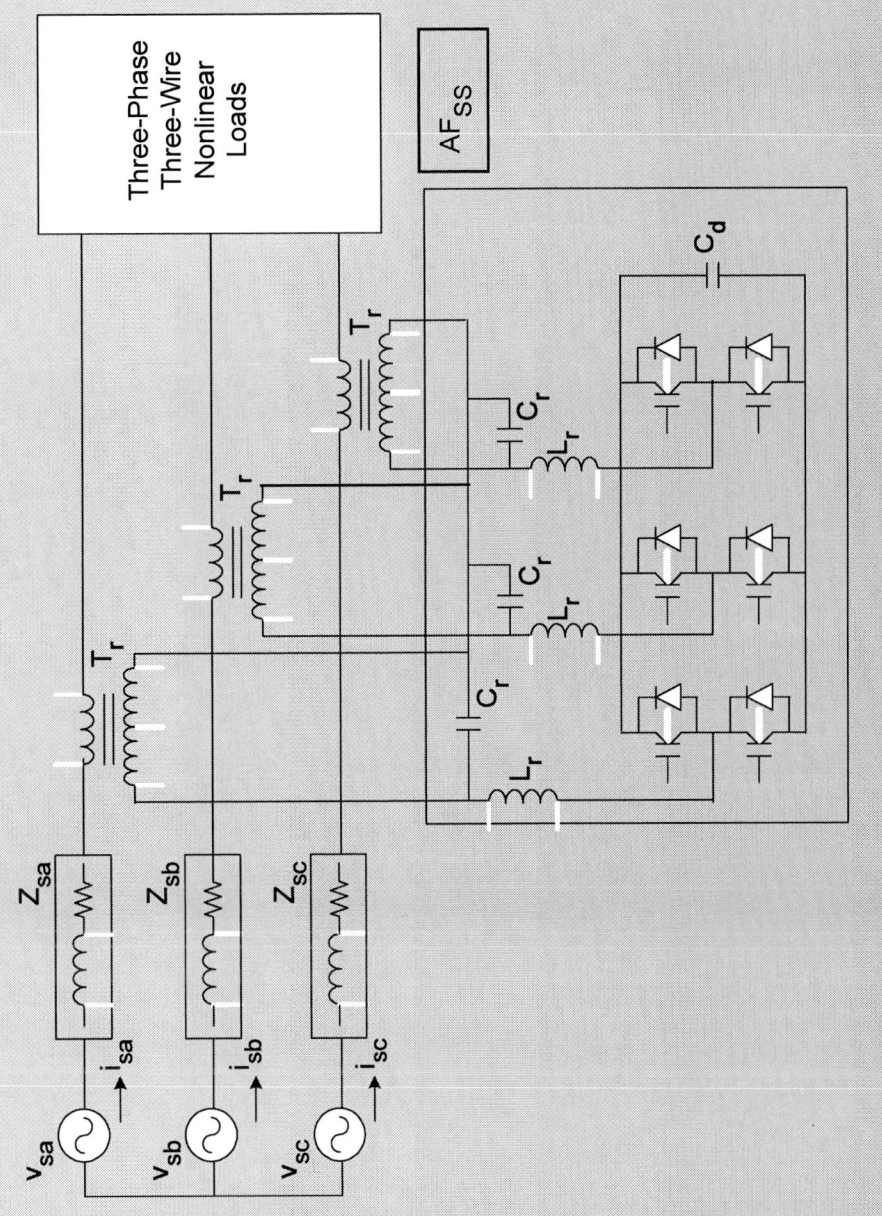

27/11/07

IIT-Delhi/PEDS-07

Performance of DVR During a 20% Sag in Supply Side

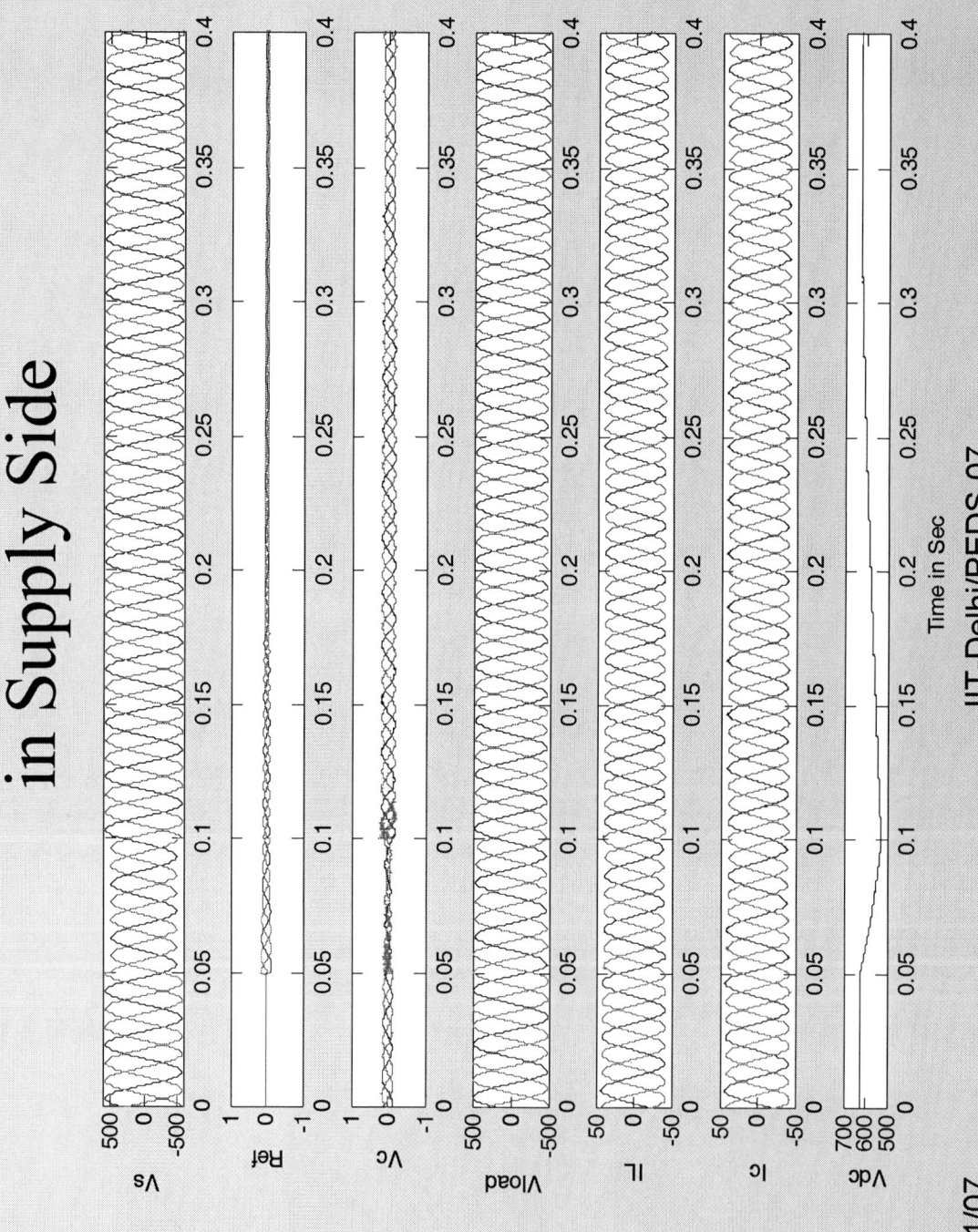

27/11/07

IIT-Delhi/PEDS-07

Performance of DVR During a 20% Swell in Supply Side

27/11/07

IIT-Delhi/PEDS-07

Performance of DVR During Single Phase Sag

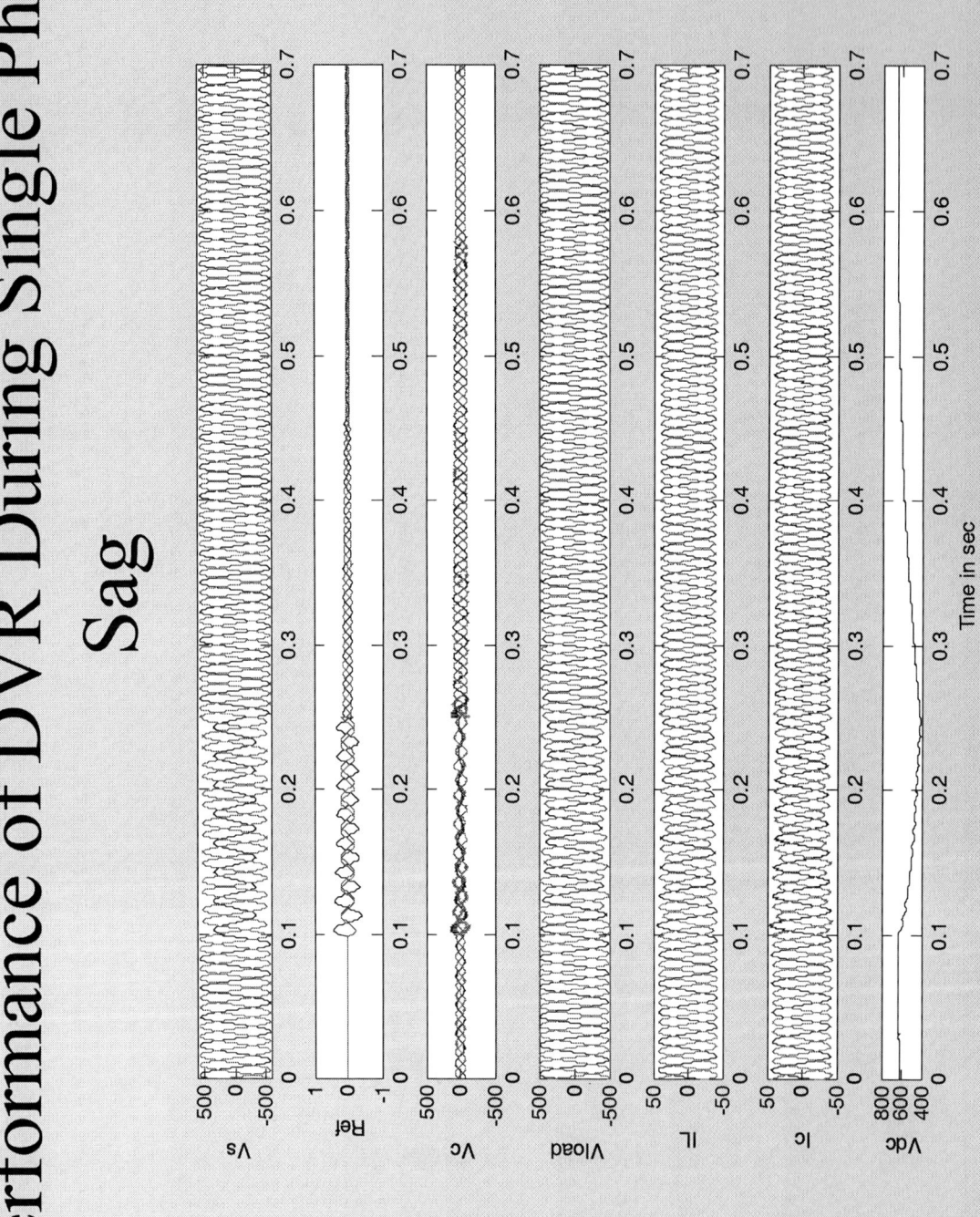

27/11/07

IIT-Delhi/PEDS-07

Performance of DVR During Harmonics in Supply Side

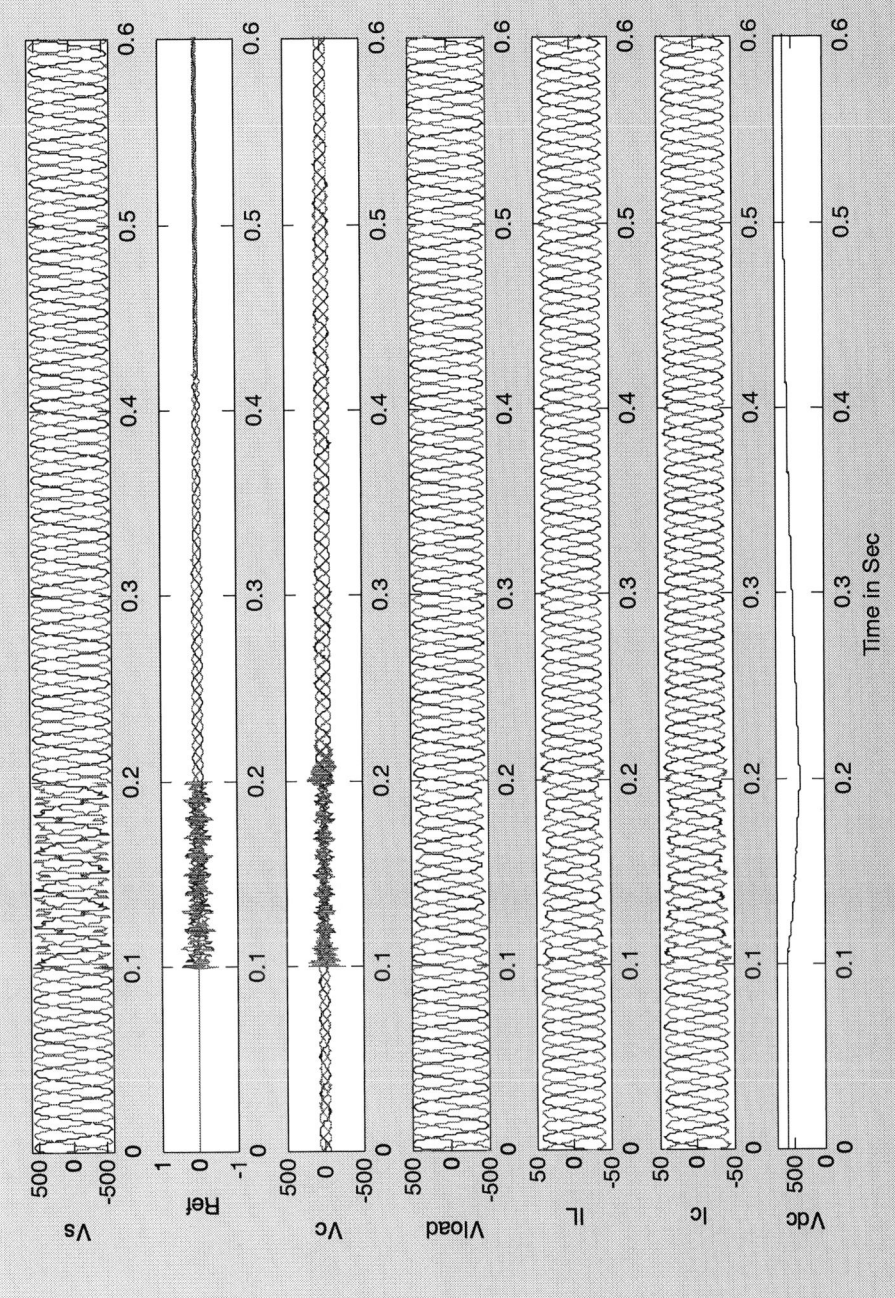

27/11/07

IIT-Delhi/PEDS-07

DVR 3-Phase 4-Wire

Three-Phase Four-Wire Nonlinear Loads

AF_{SS}

IIT-Delhi/PEDS-07

27/11/07

Unified Power Quality Compensator (UPQC)

Functions

- **Reactive Power Compensation**
- **Voltage Regulation**
- **Compensation for Voltage sag and swell**
- **Unbalance Compensation for current and voltage (for 3-phase systems)**
- **Neutral Current Compensation (for 3-phase 4-wire systems)**

UPQC 1-Phase 2-Wire

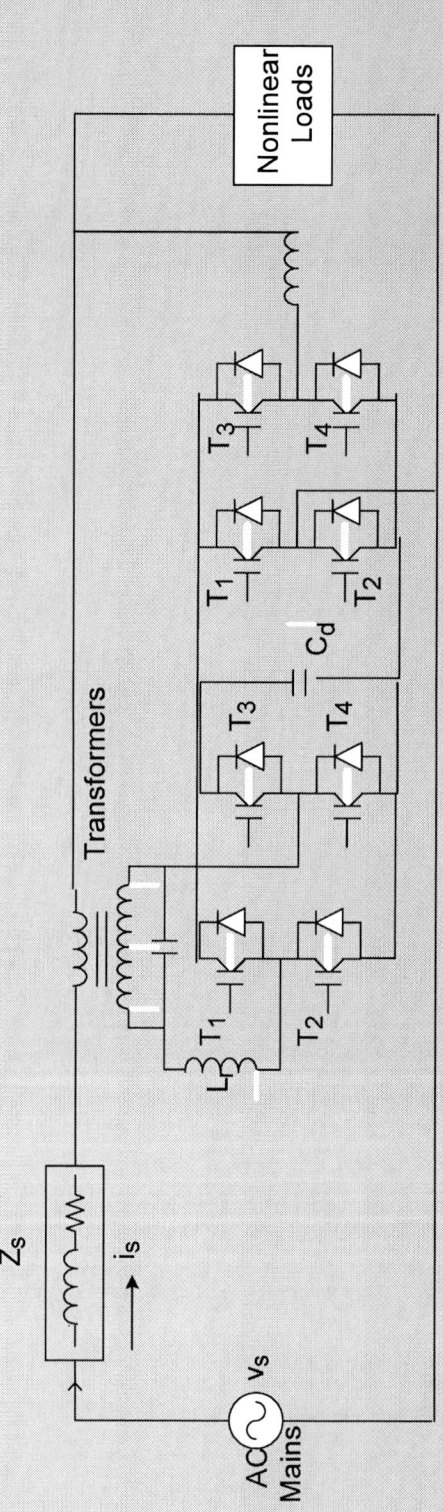

UPQC 3-Phase 3-Wire

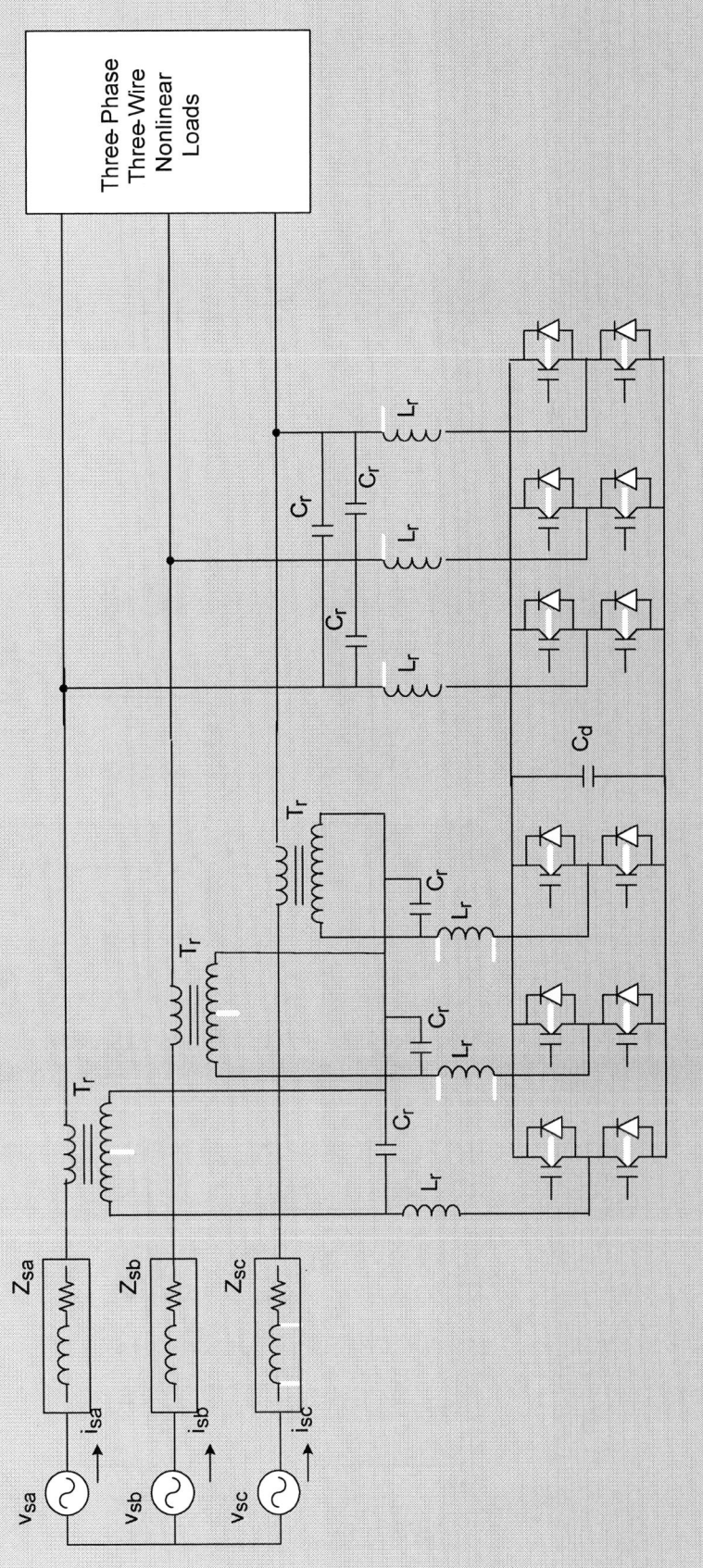

27/11/07   IIT-Delhi/PEDS-07

Right Shunt and Left Shunt UPQC

- Convert the feeder (source) current (i_s) to balanced sinusoids through the shunt compensator.
- Convert the load voltage (v_l) to balanced sinusoids through the series compensator and also regulate it to a desired value.

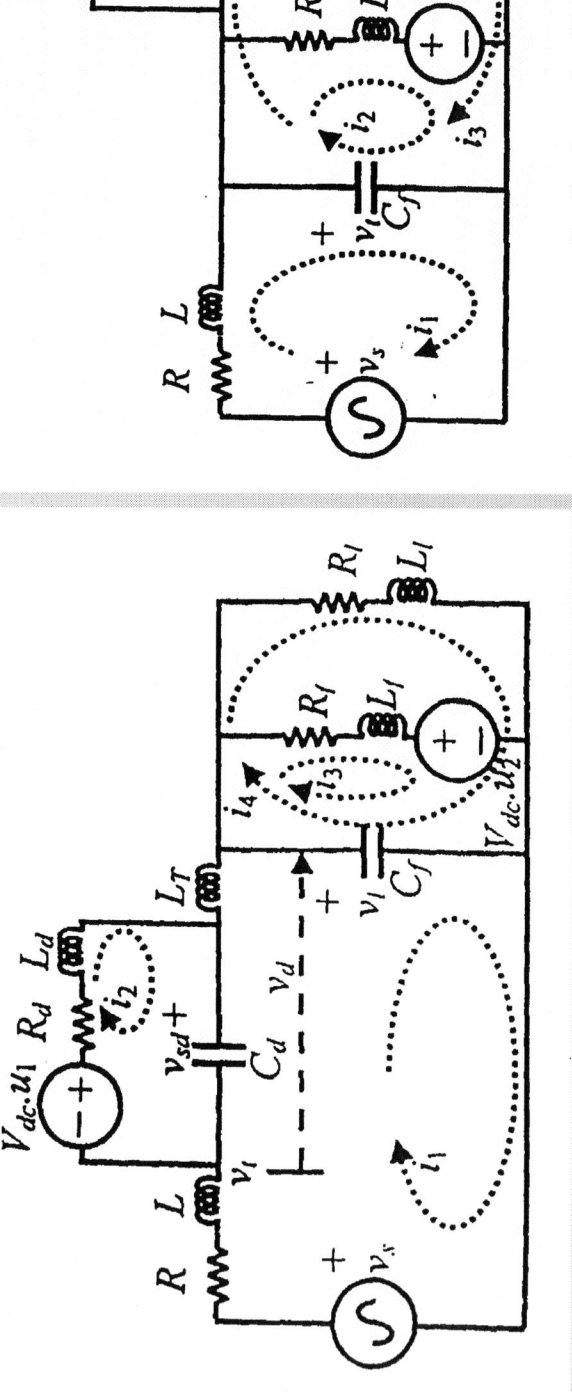

Equivalent circuit of a left-shunt UPQC

Right Shunt UPQC

Improved Power Quality Based Converter (IPQC)

27/11/07

IIT-Delhi/PEDS-07

Functions of IPQC

- Reduced harmonic currents
- High power factor
- Low EMI and RFI at input AC mains
- Well regulated and good quality DC output
- Rating from fraction of Watt to MW power in large number of applications

Applications of IPQC

- DC power supplies
- Telecommunication power supply
- Improved power factor ballast
- Power supplies for equipments like computers, medical equipment, printers, scanners etc
- Electrical welding

Supply Based Classification

Improved Power Quality Converters

Single Phase

Unidirectional

Bi-directional

Three phase

Unidirectional

Bi-directional

27/11/07

IIT-Delhi/PEDS-07

Classification of Single Phase IPQC

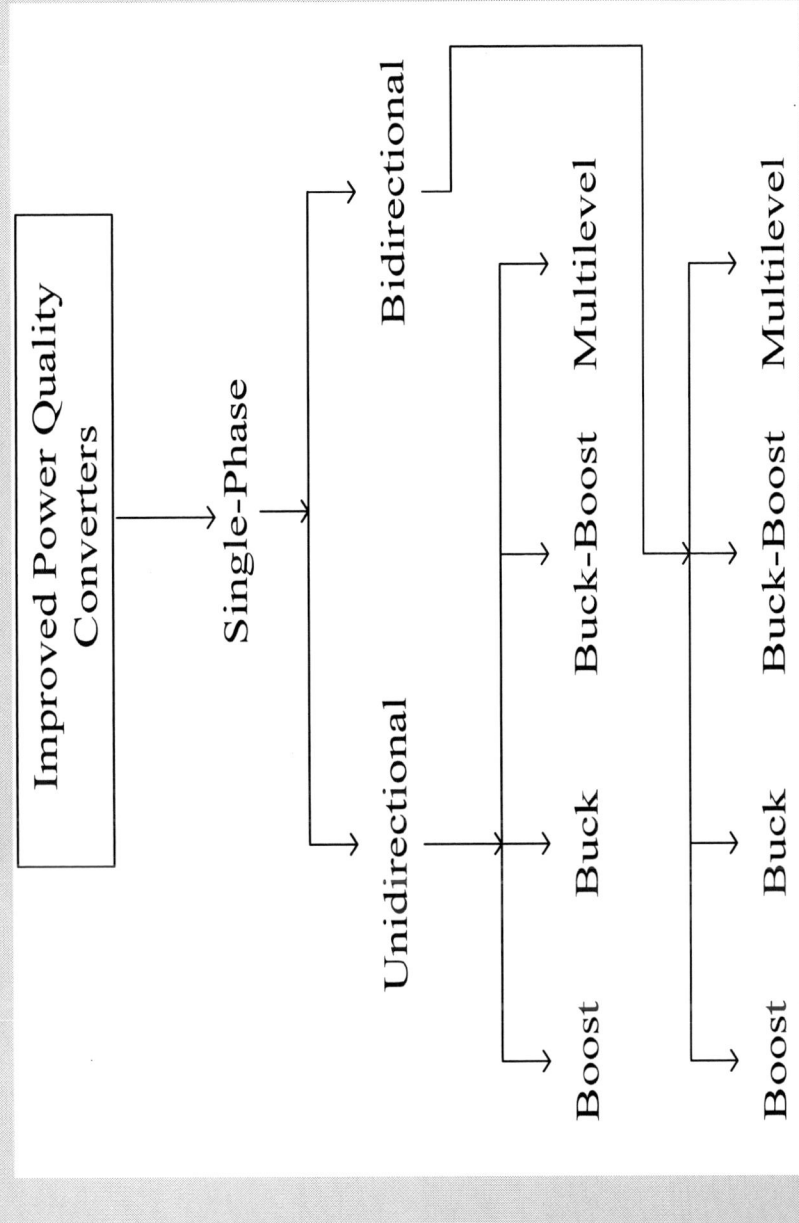

27/11/07

IIT-Delhi/PEDS-07

Classification of Three Phase IPQC

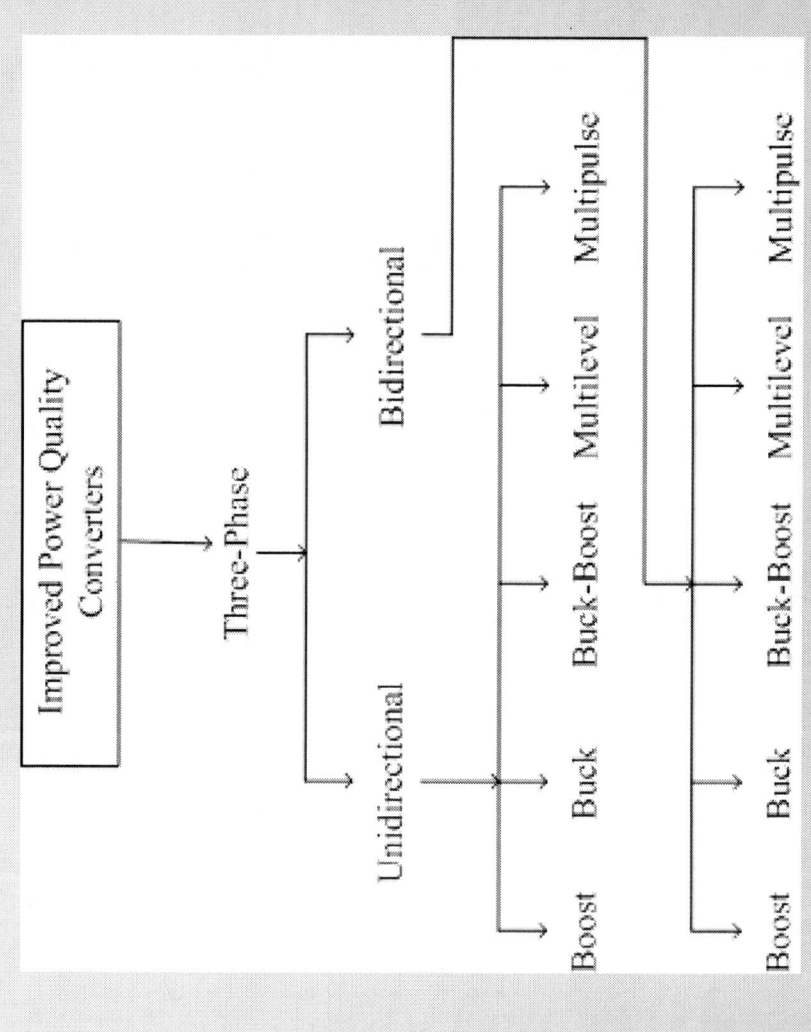

27/11/07

IIT-Delhi/PEDS-07

Topology Based Classification

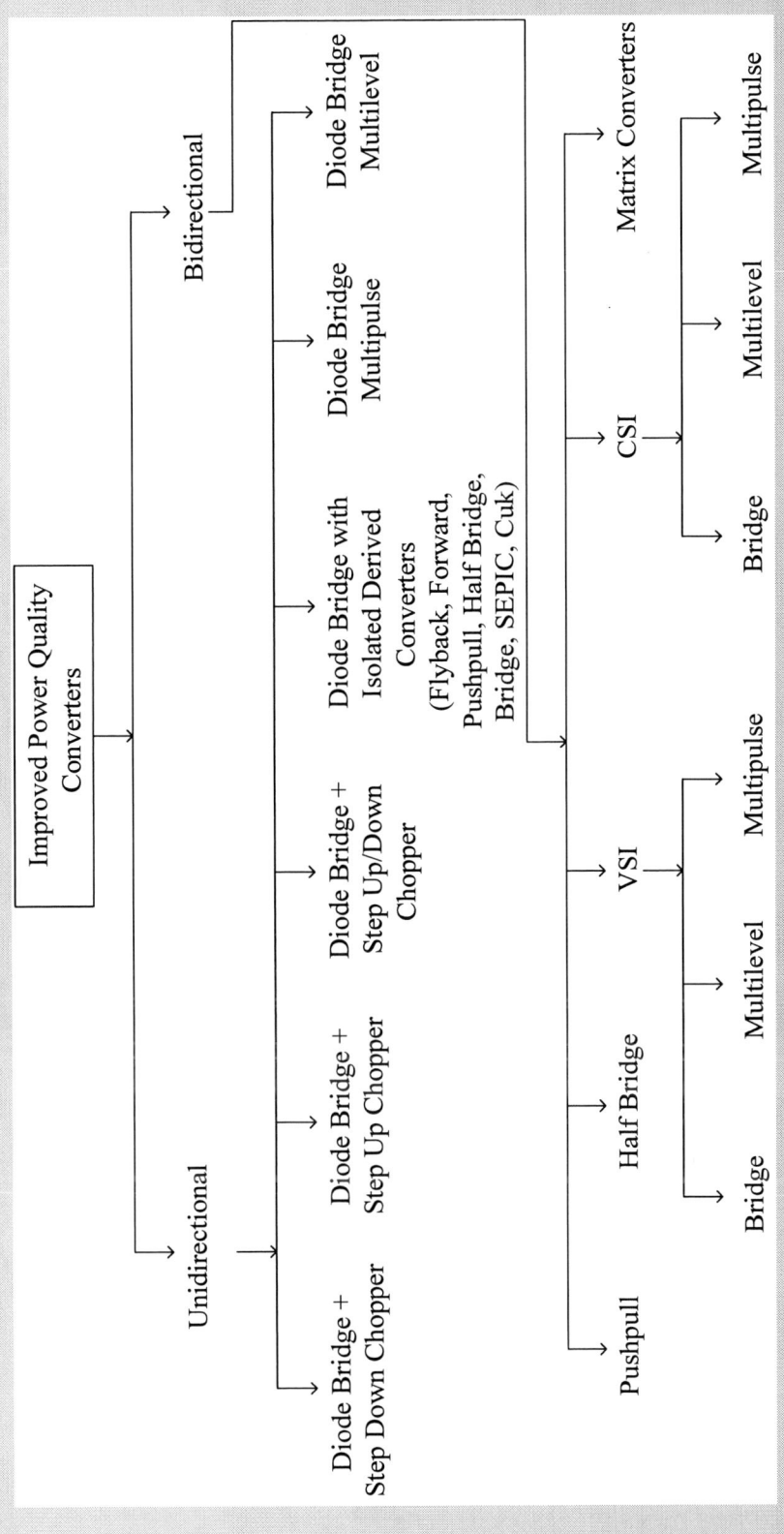

27/11/07

IIT-Delhi/PEDS-07

Single Phase Boost Converter

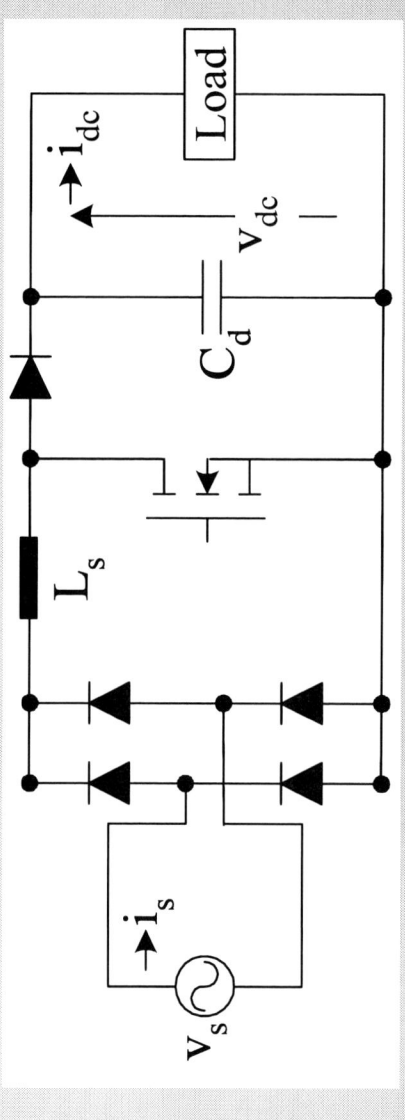

Single-Phase Unidirectional Boost Converter.

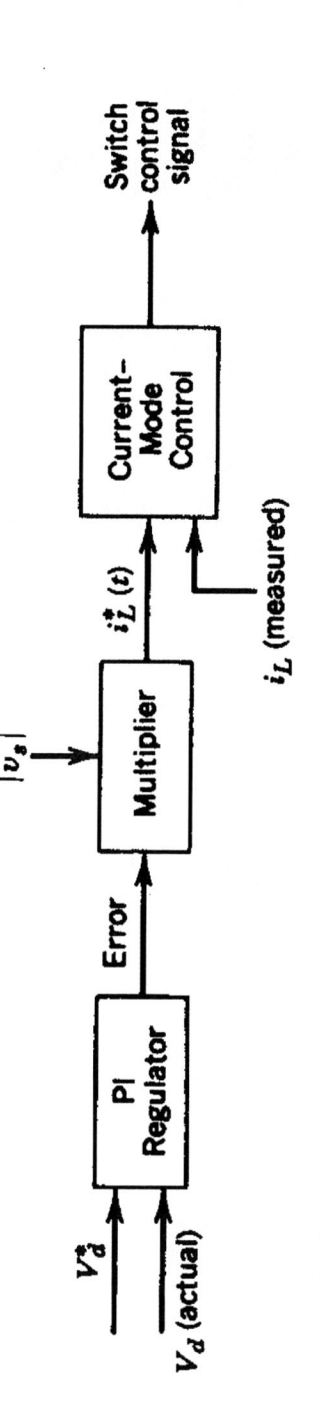

27/11/07

IIT-Delhi/PEDS-07

Single-Phase Unidirectional Boost Converter.

Single Phase Boost Converter Control

Experimental Waveforms

Input Power

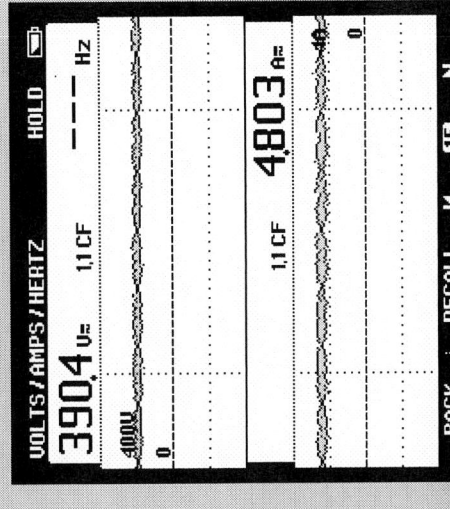

Output DC Voltage and Current

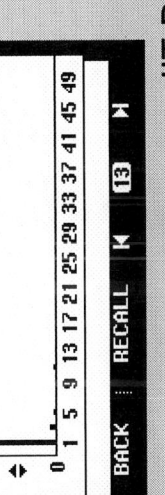

Input AC Voltage and Current

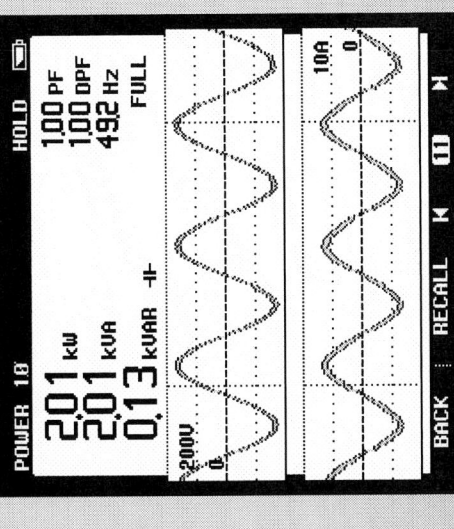

Input Current harmonic Spectrum

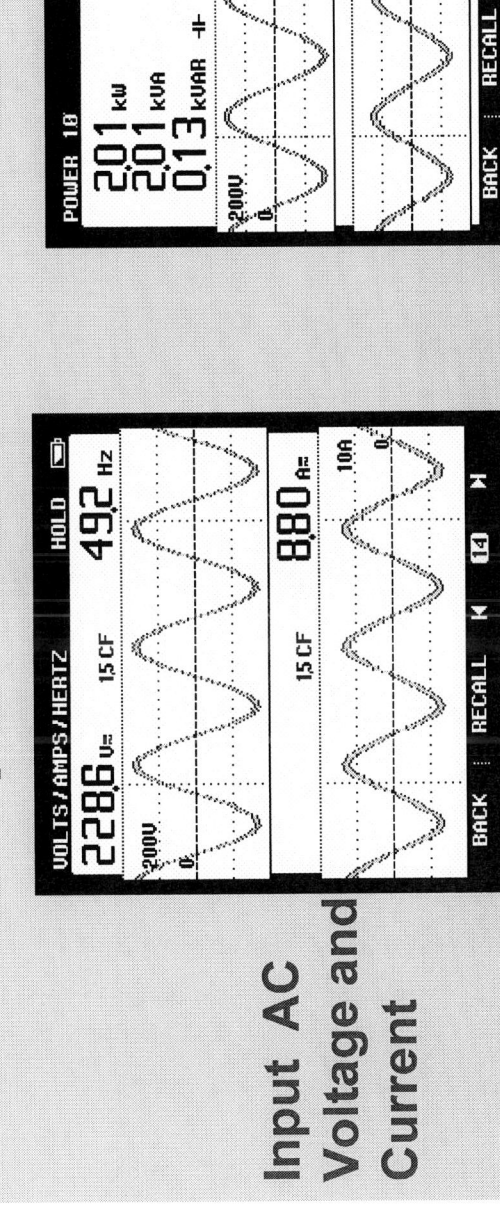

27/11/07

IIT-Delhi/PEDS-07

Single Phase Boost IPQC

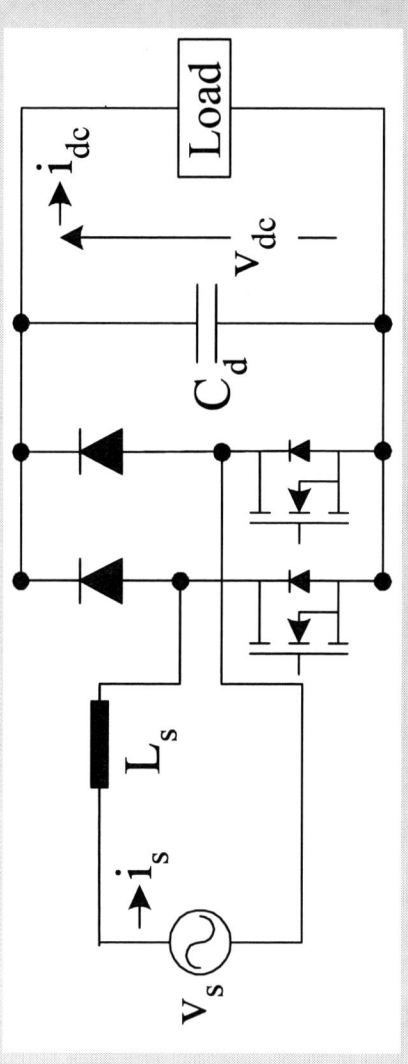

Symmetrical Two Device Single-Phase Unidirectional Boost Converter

Single Phase Boost IPQC

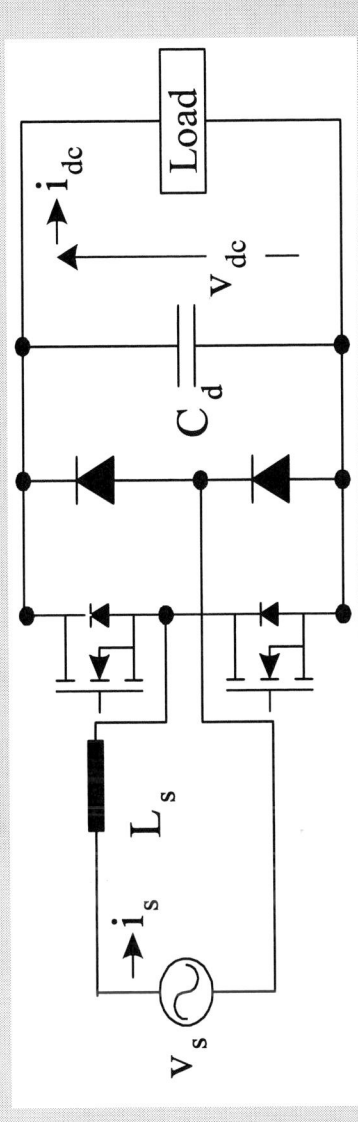

Asymmetrical Two Device Single-Phase Unidirectional Boost Converter

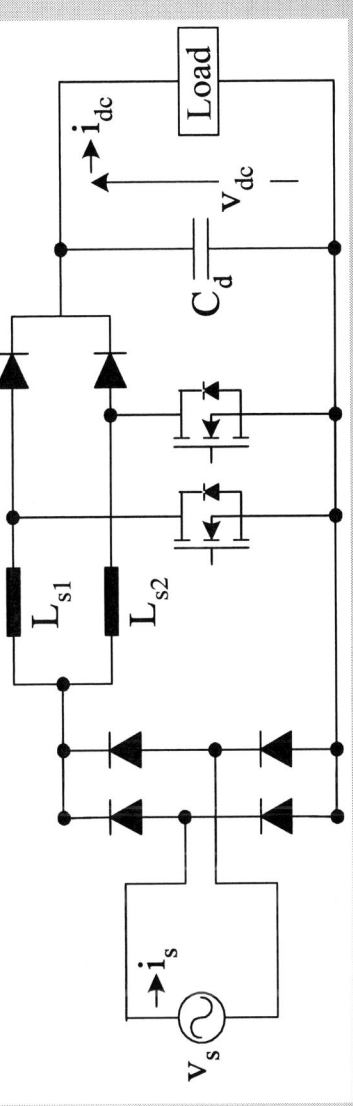

Interleaved Two Cell Single-Phase Unidirectional Boost Converter

27/11/07

IIT-Delhi/PEDS-07

Single Phase Boost IPQC

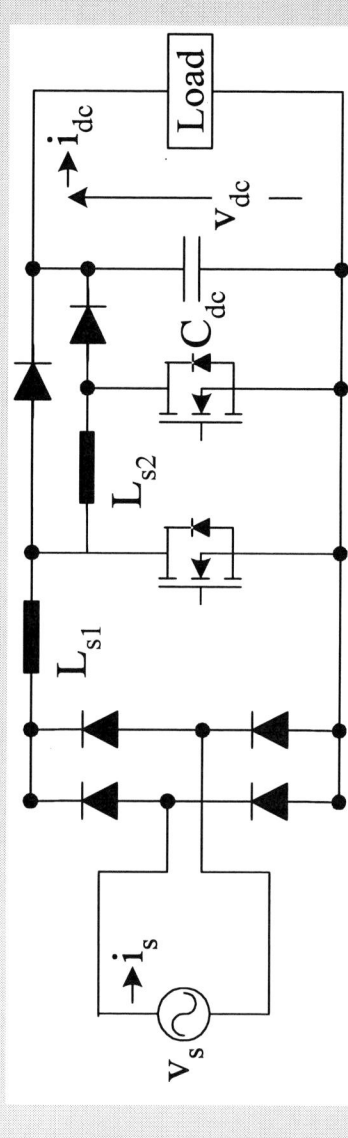

Single-Phase Unidirectional Boost Converter with High Frequency Active EMI Filter.

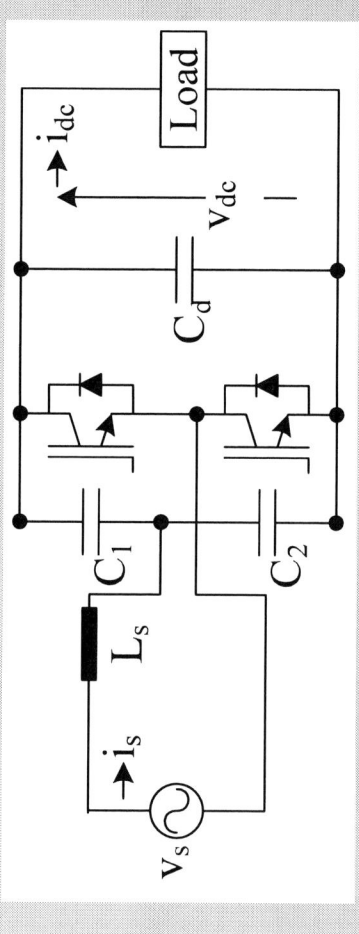

Single-Phase Half Bridge Bidirectional Boost Converter.

27/11/07

IIT-Delhi/PEDS-07

Single Phase Boost IPQC

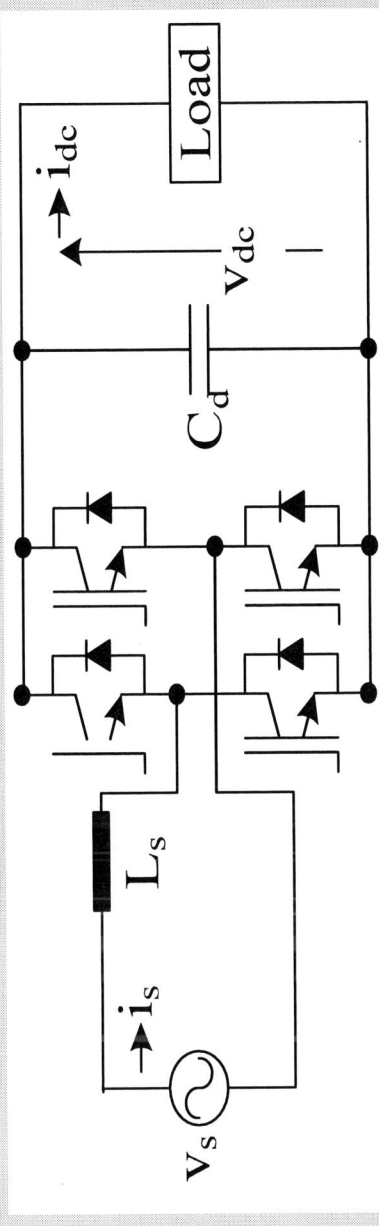

Single-Phase VSI Full Bridge Bidirectional Boost Converter

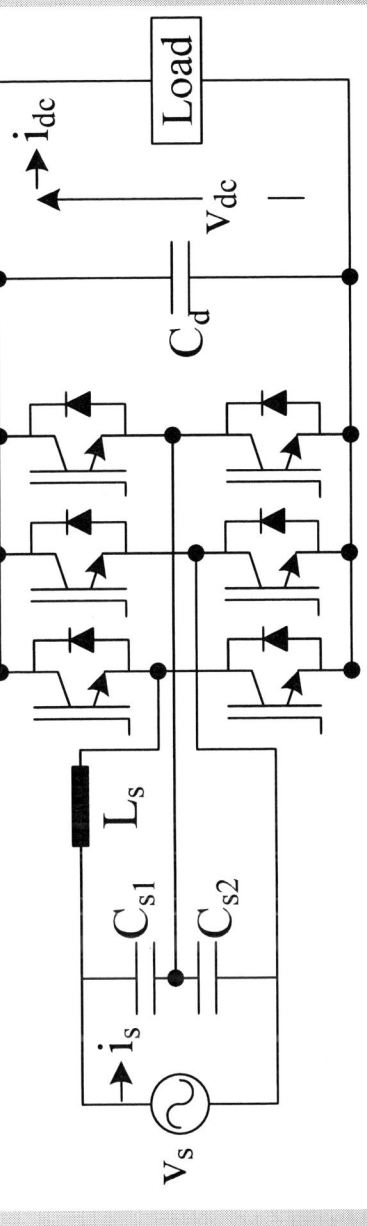

Single-Phase Bridge Bidirectional Boost Converter with DC Ripple Compensation using AC Mid Point Capacitors and Third Leg.

27/11/07

IIT-Delhi/PEDS-07

Single Phase Boost IPQC

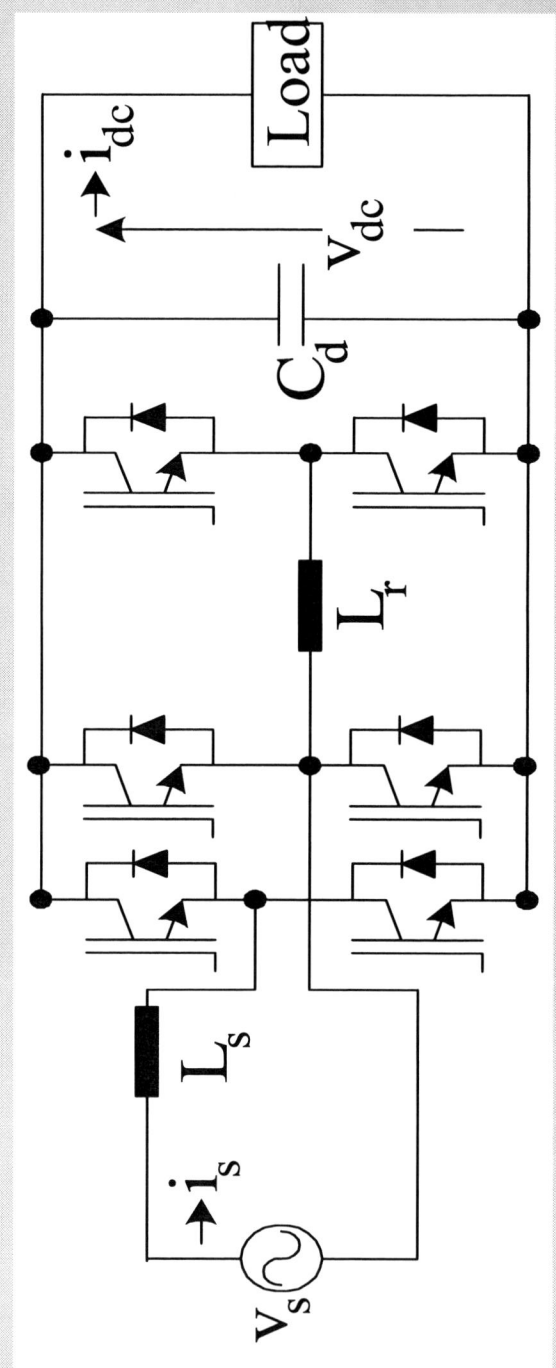

Single-Phase Bridge Bidirectional Boost Converter with DC Ripple Compensation using an Inductor and Third Leg.

27/11/07

IIT-Delhi/PEDS-07

Single Phase Buck IPQC

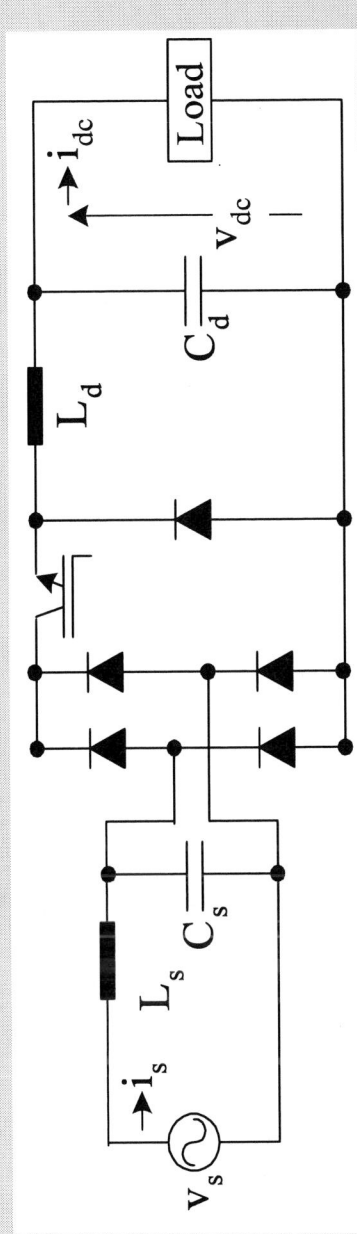

Single-Phase Unidirectional Buck Converter with Input AC Filter.

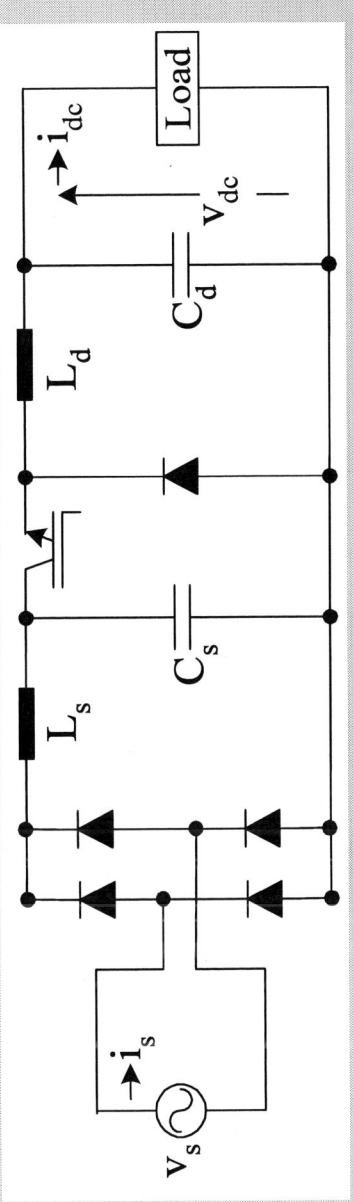

Single-Phase Unidirectional Buck Converter with Input DC Filter.

Single Phase Buck IPQC

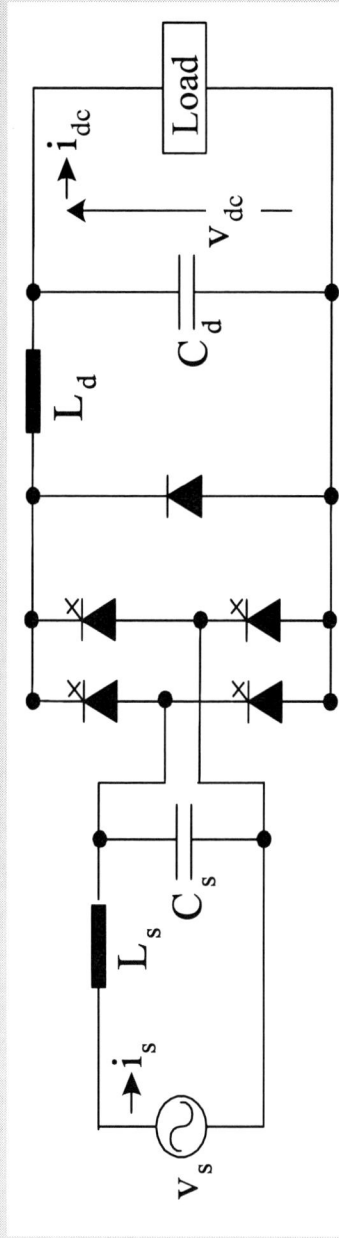

GTO Bridge Based Unidirectional Buck Converter

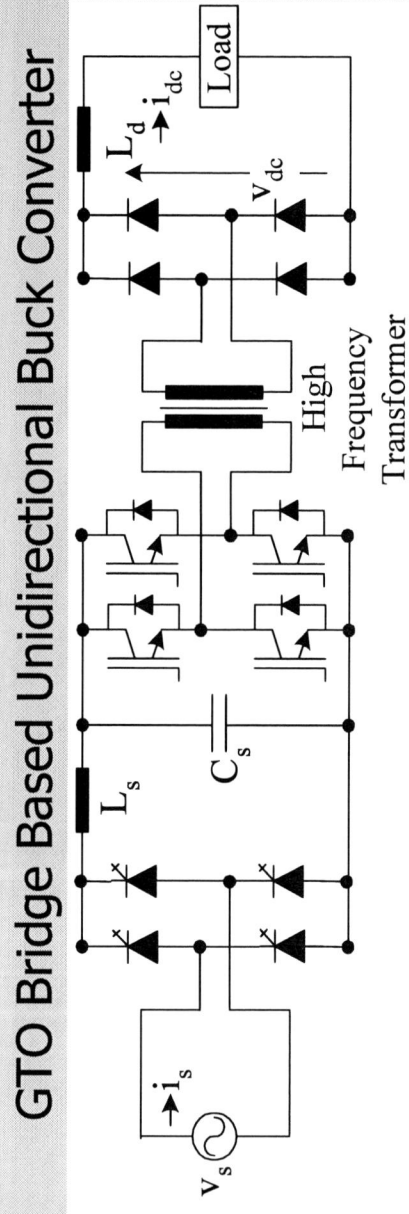

Single-Phase Unidirectional Buck Converter with High Frequency Isolated DC-DC Buck Stage.

27/11/07

IIT-Delhi/PEDS-07

Single Phase Buck IPQC

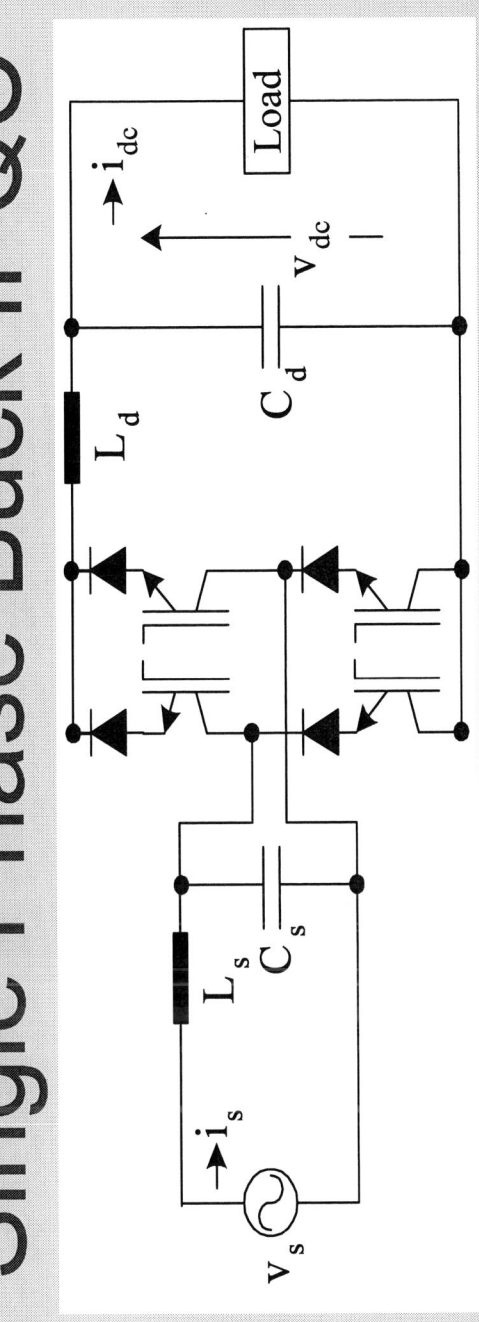

Single-Phase Bidirectional Buck Converter.

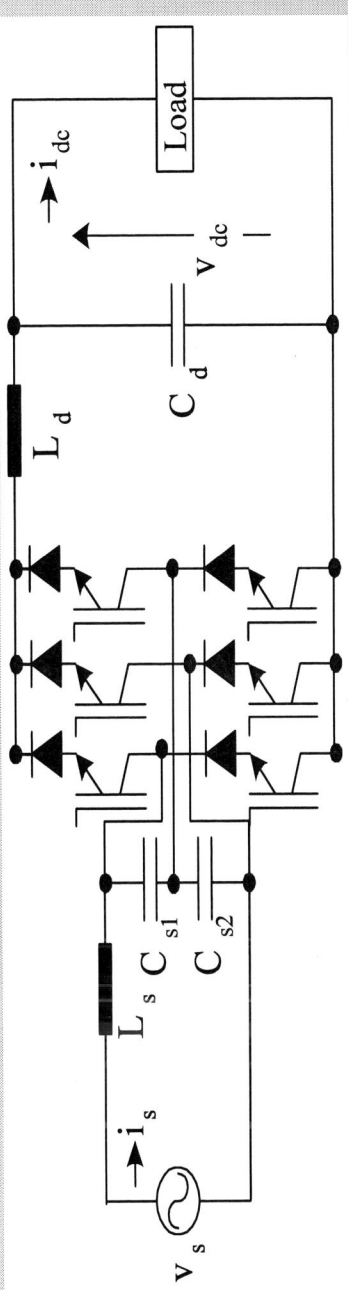

Single-Phase Bidirectional Buck Converter with a Neutral Leg.

Single Phase Buck-Boost IPQC

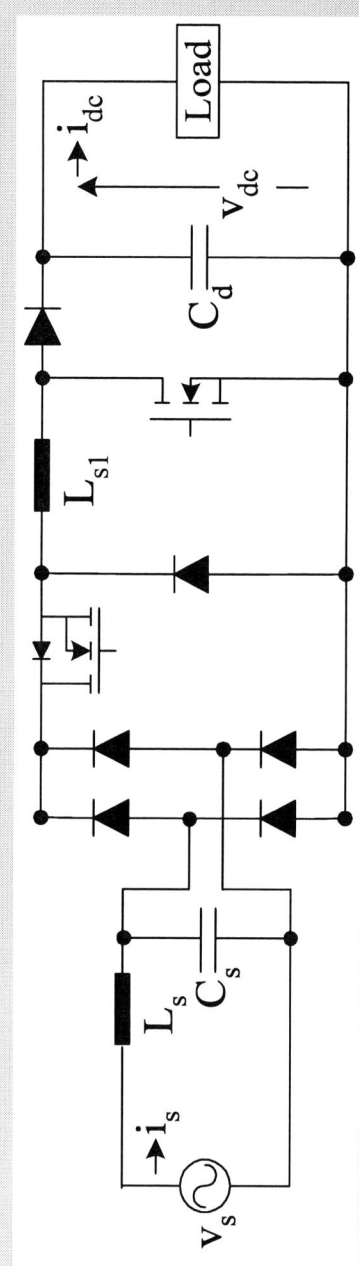

Single-Phase Cascaded Unidirectional Buck-Boost Converter.

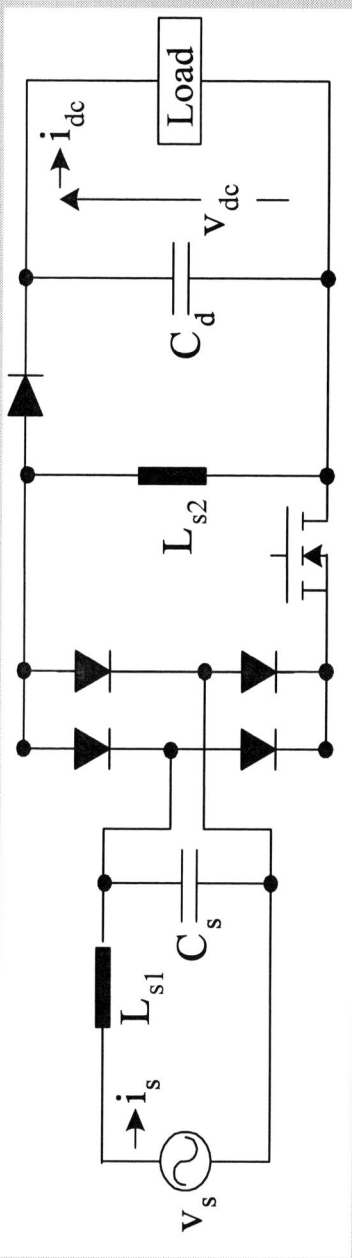

Single Device Single-Phase Unidirectional Buck-Boost Converter

Single Phase Buck-Boost IPQC

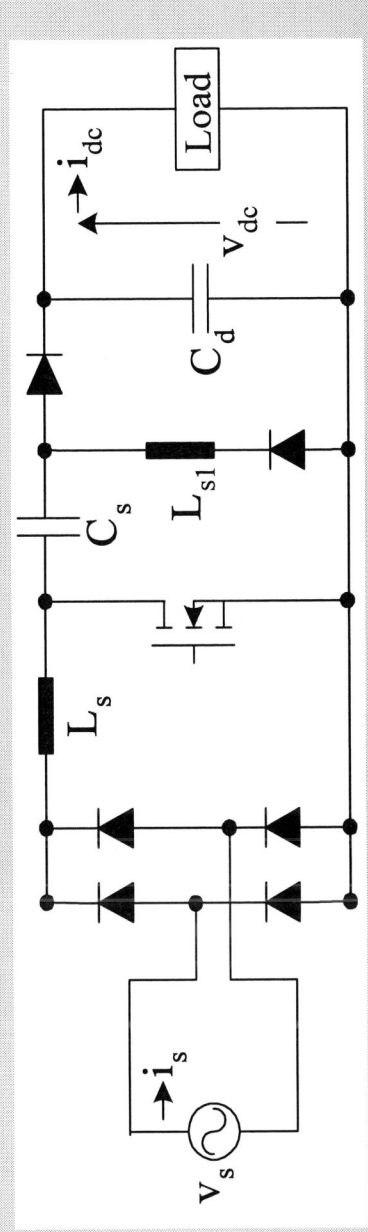

SEPIC-Derived Single-Phase Unidirectional Buck-Boost Converter.

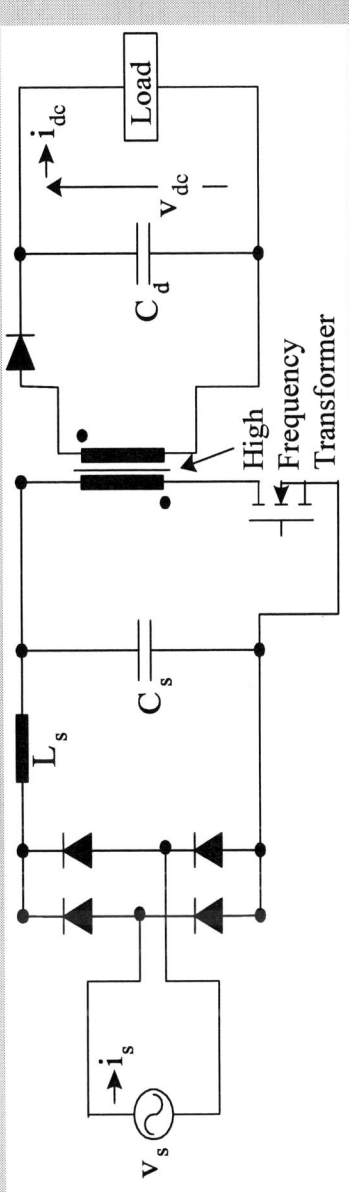

Flyback Based Single-Phase Unidirectional Buck-Boost Converter

Single Phase Buck-Boost IPQC

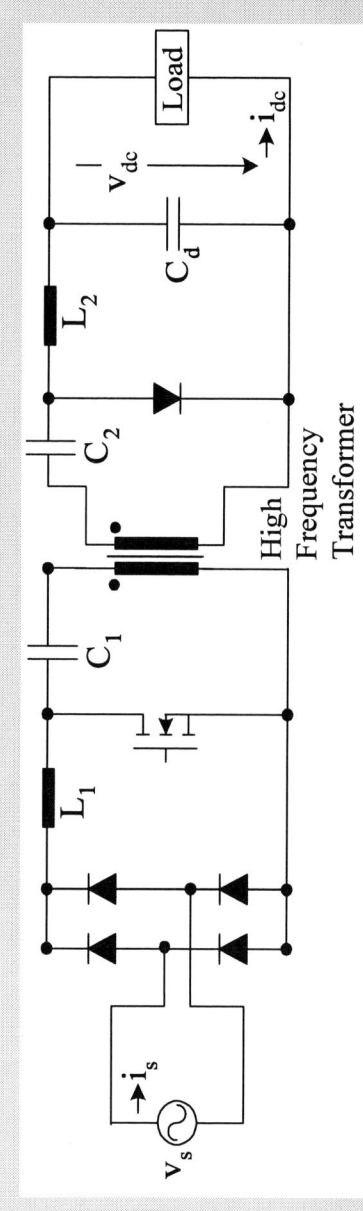

Isolated Cuk Derived Single-Phase Unidirectional Buck-Boost Converter.

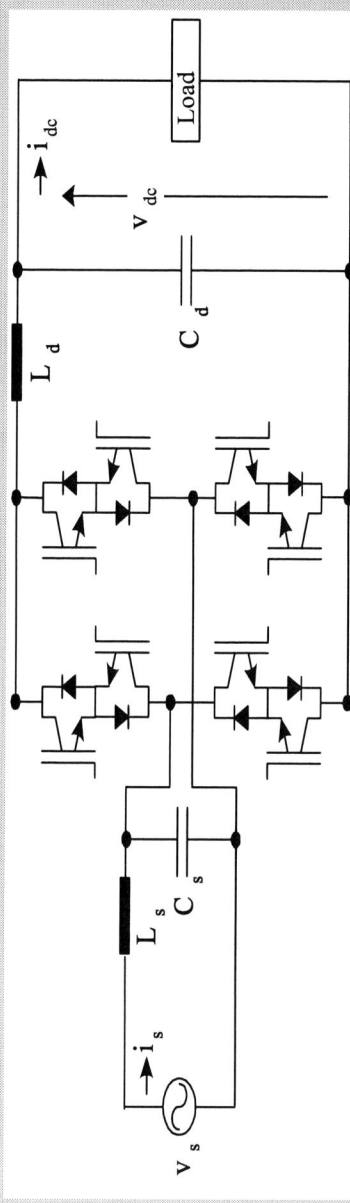

Single Phase Bidirectionnel Buck-Boost Converter

27/1/07

IIT-Delhi/PEDS-07

Single Phase Multi-level IPQC

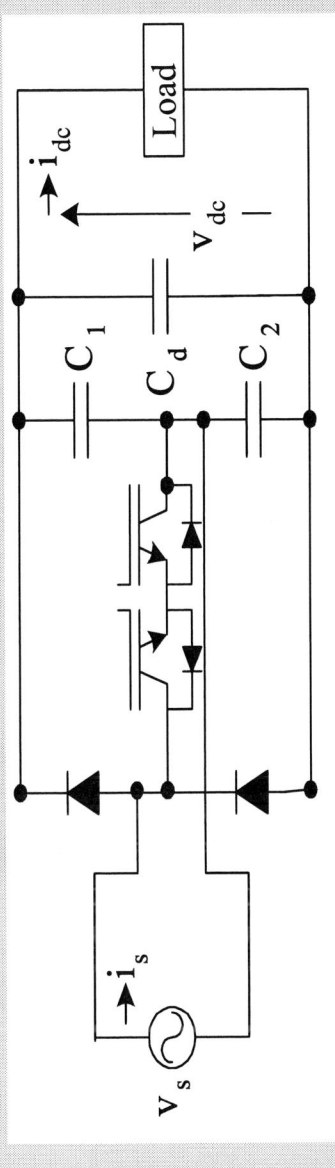

Half Bridge Unidirectionnel Multilevel Converter.

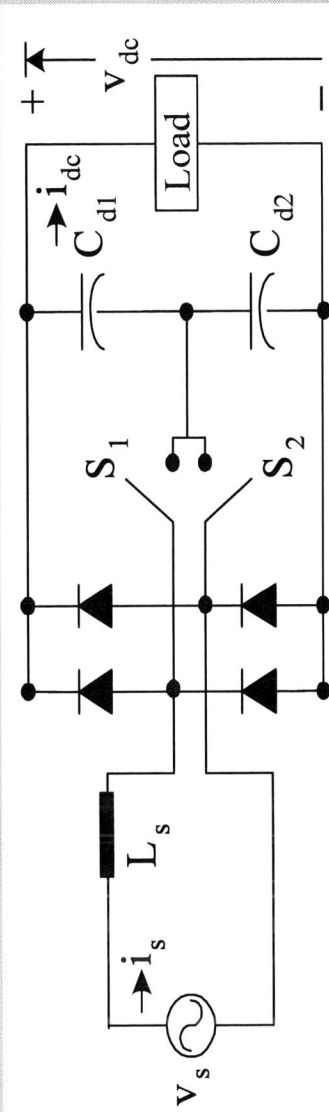

Two Bidirectionnel Switch Unidirectionnel Multilevel Converter.

27/11/07

IIT-Delhi/PEDS-07

Single Phase Multi-level IPQC

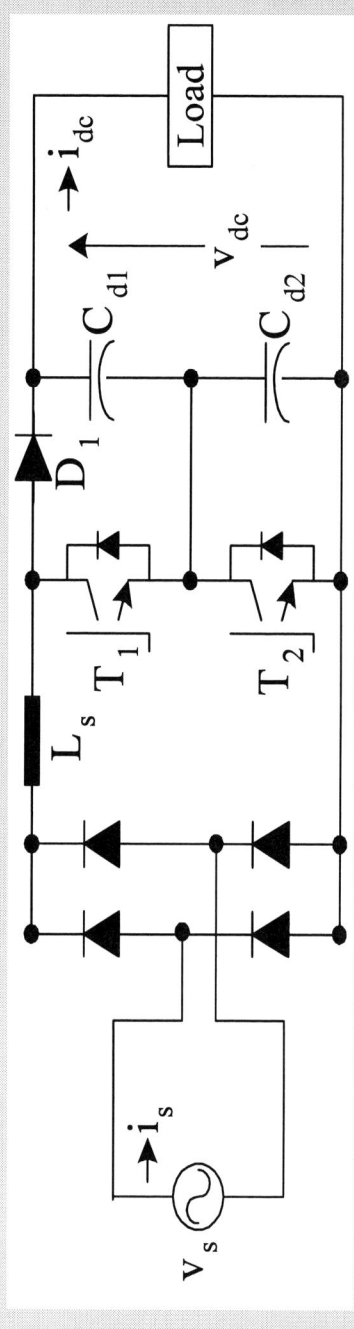

Two Switch Mid Point Unidirectionnel Multilevel Converter.

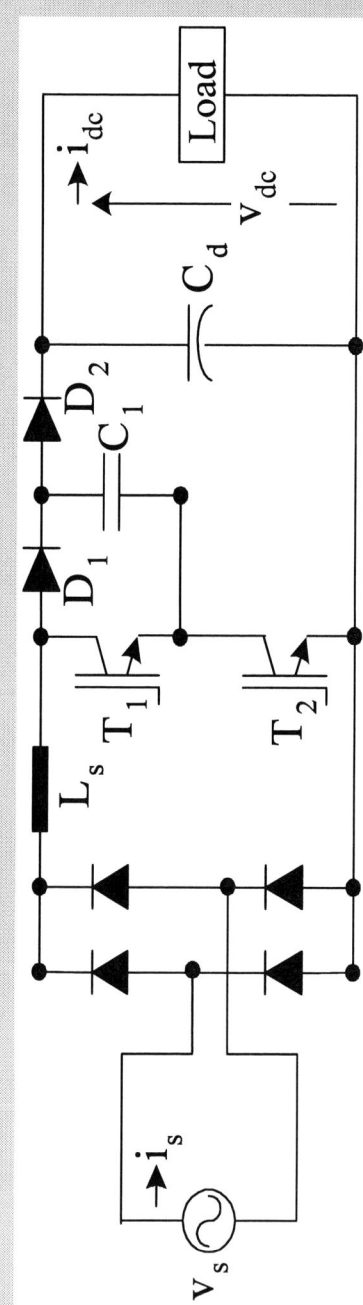

Adapted Unidirectionnel Multilevel Converter

Single Phase Multi-level IPQC

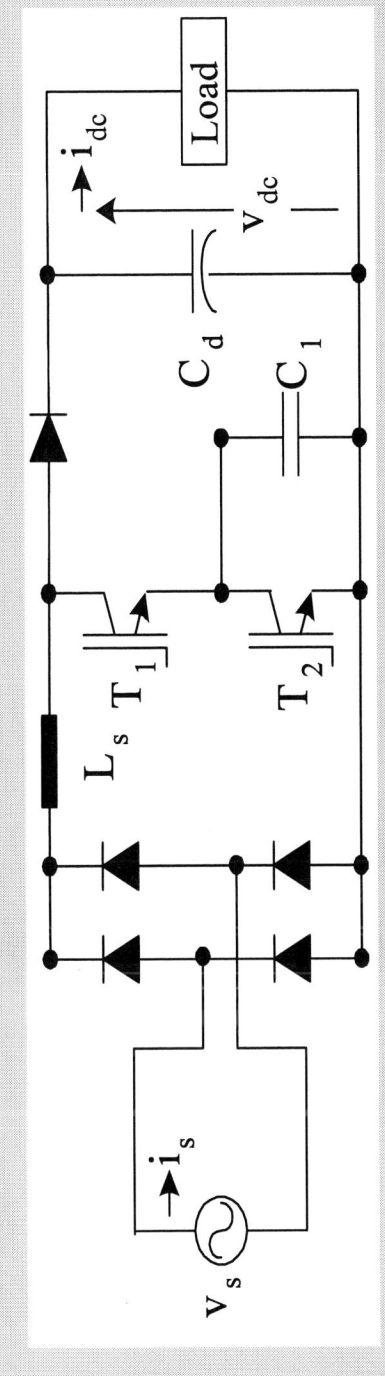

Modified Adapted Unidirectional Multilevel Converter

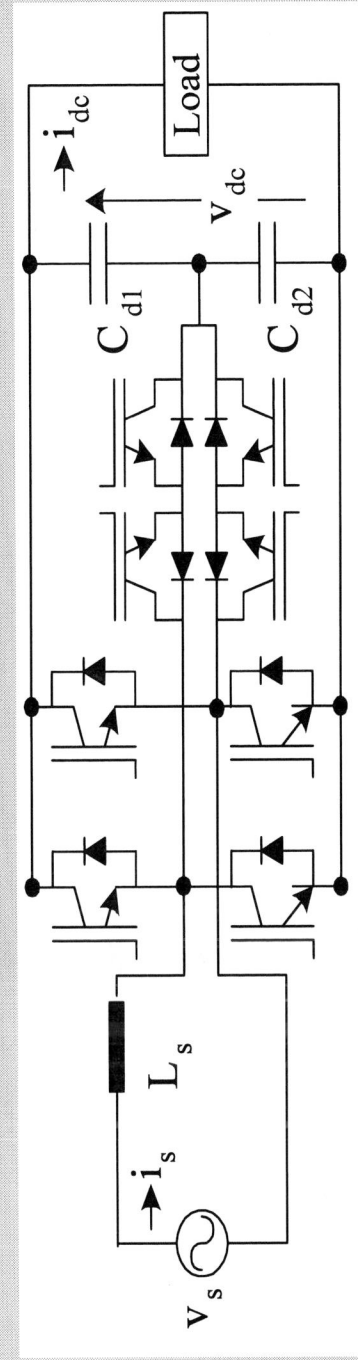

Single Phase Bidirectionnel Three Level Converter using Two Bidirectional Switches.

27/11/07

IIT-Delhi/PEDS-07

Single Phase Multi-level IPQC

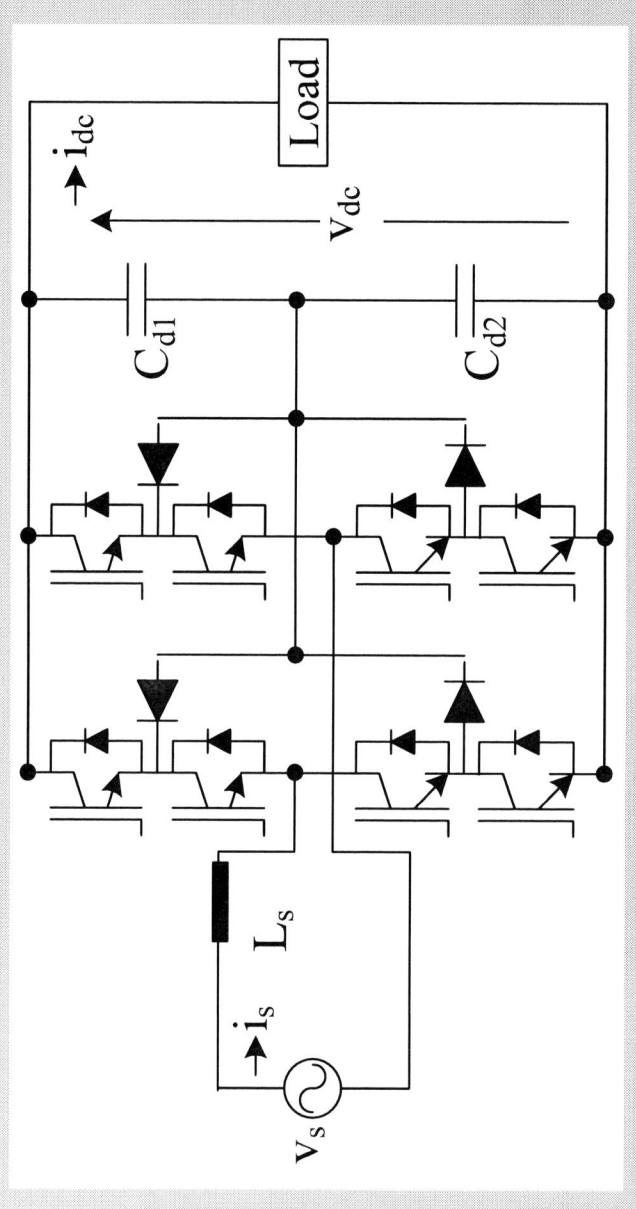

Single-Phase Bidirectional Diode Clamped Three Level Converter.

27/11/07   IIT-Delhi/PEDS-07

Single Phase Multi-level IPQC

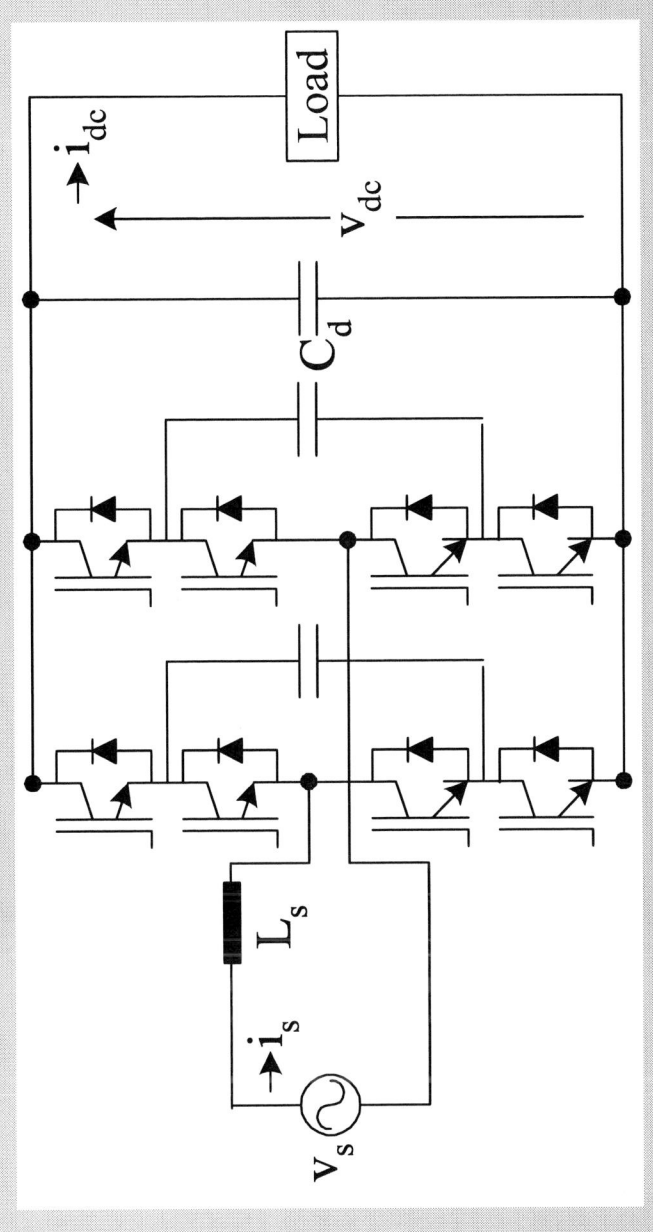

Single-Phase Bidirectional Flying Capacitor Clamped Three Level Converter.

27/11/07

IIT-Delhi/PEDS-07

Single Phase Multi-level IPQC

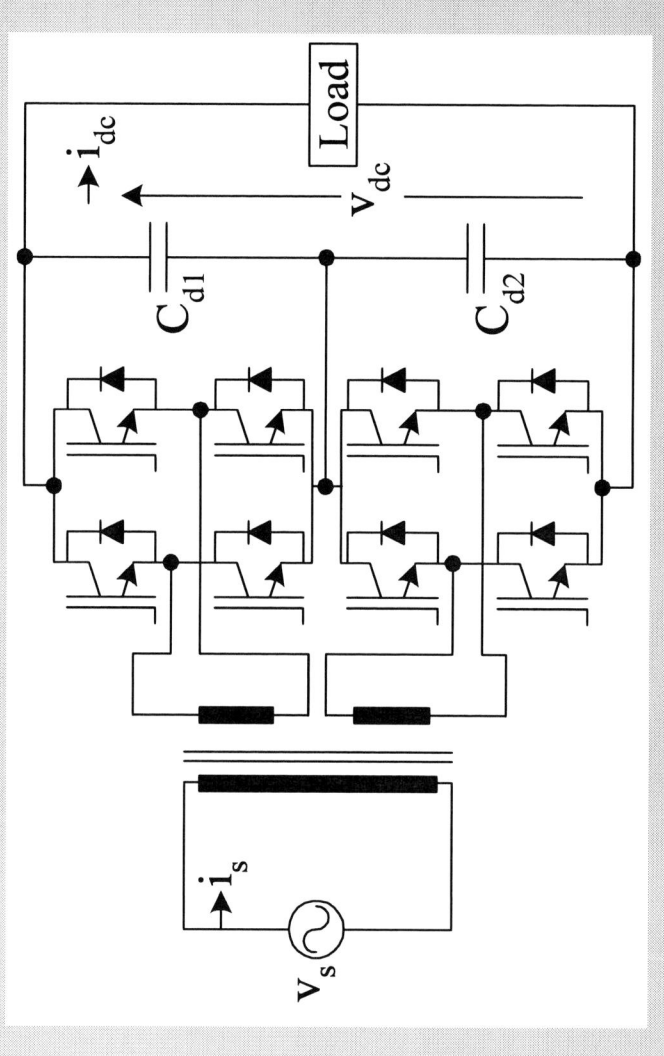

Single-Phase Bidirectional Cascaded Five Level Converter.

IIT-Delhi/PEDS-07

Three Phase
Improved Power Quality Converters

IIT-Delhi/PEDS-07

27/11/07

Classification of Three Phase IPQC

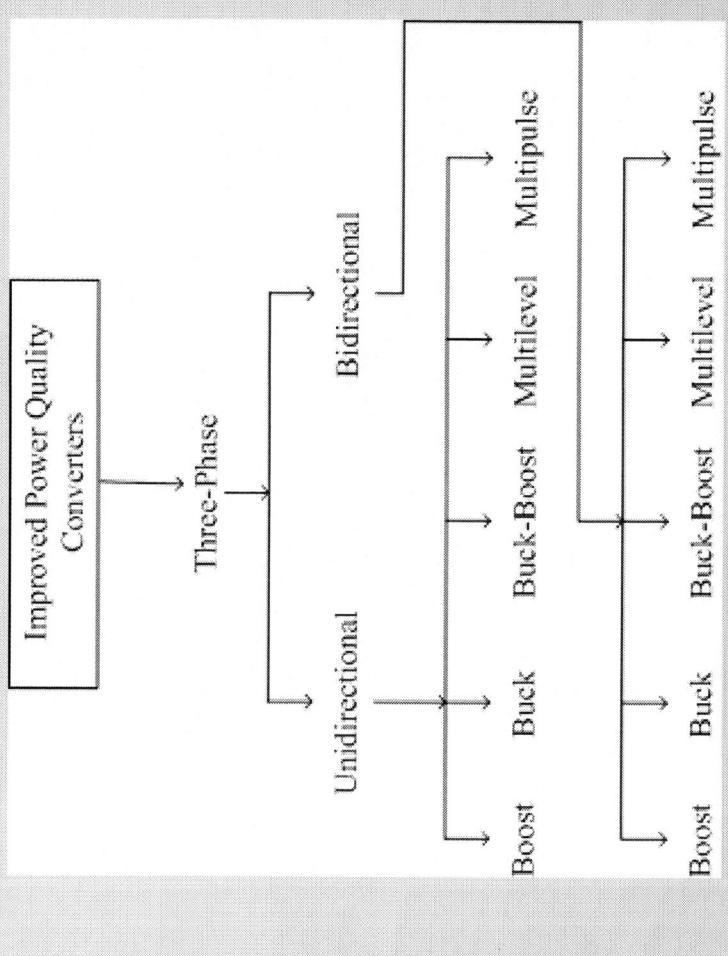

27/11/07

IIT-Delhi/PEDS-07

123/155

Three Phase Boost IPQC

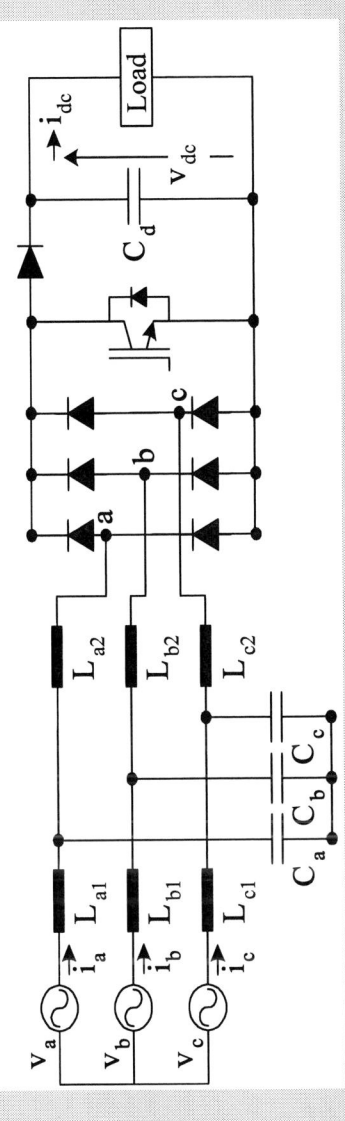

Three-Phase Single Switch Unidirectional Boost Converter.

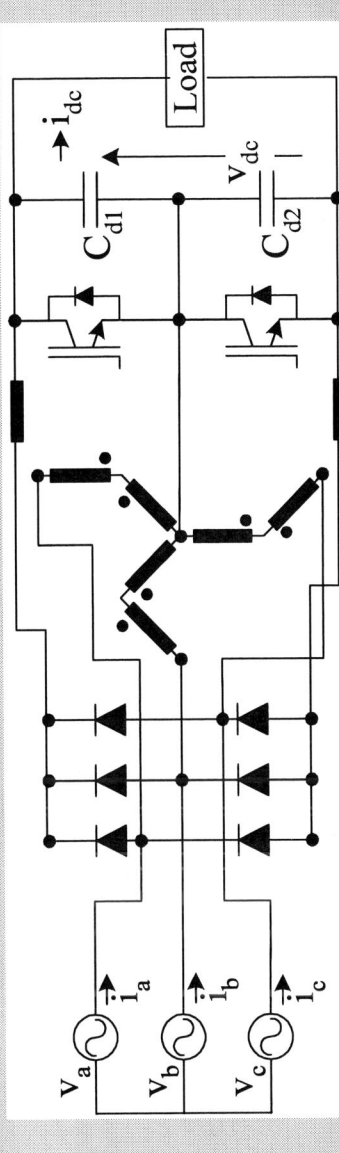

Three-Phase Two Switch Unidirectional Boost Converter using Zigzag Injection Transformer (Minnesota Rectifier)

27/11/07

IIT-Delhi/PEDS-07

Three Phase Boost IPQC

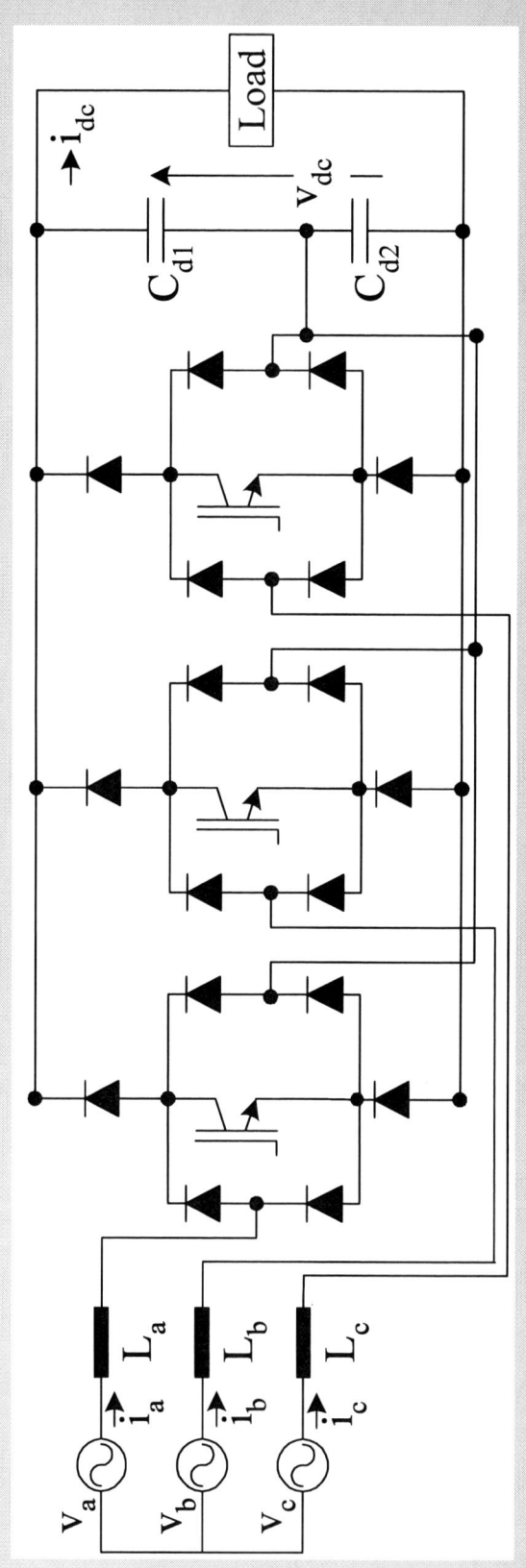

Three-Phase Three Switch Unidirectional Boost Converter (Vienna Rectifier)

Three Phase Boost IPQC

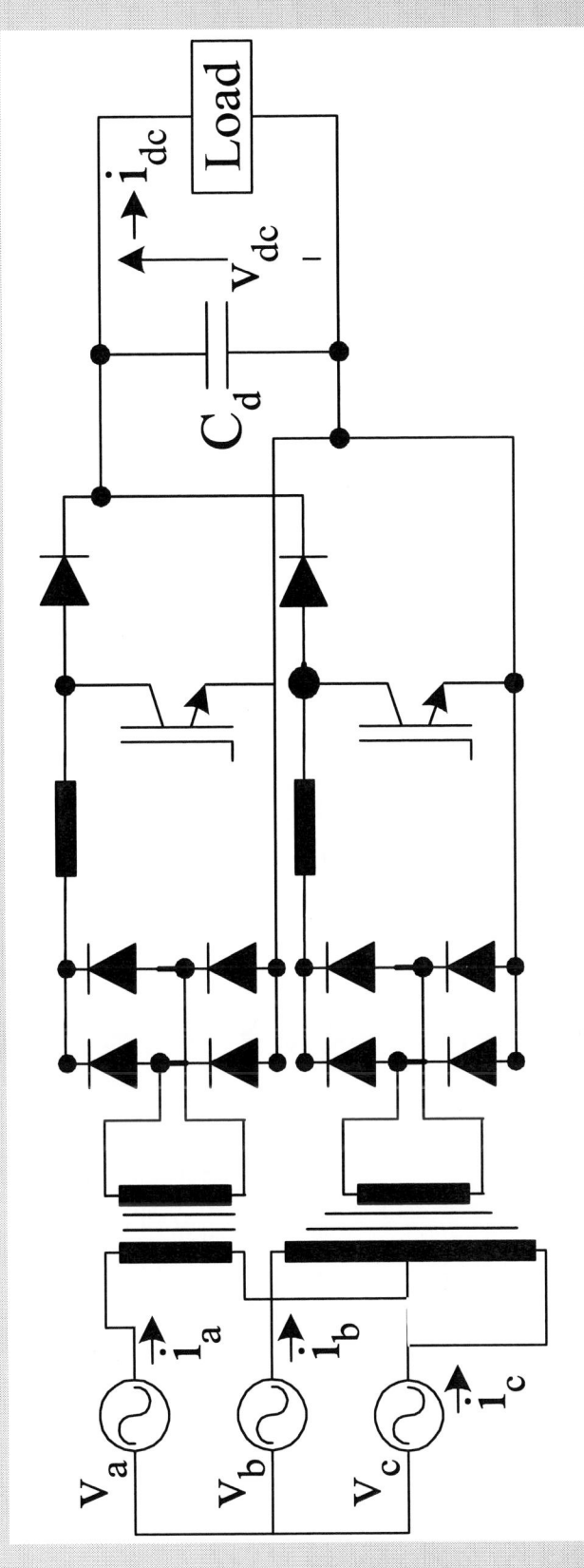

Three-Phase Unidirectional Boost Converter using Isolated Scott Connection Transformers.

Three Phase Boost IPQC

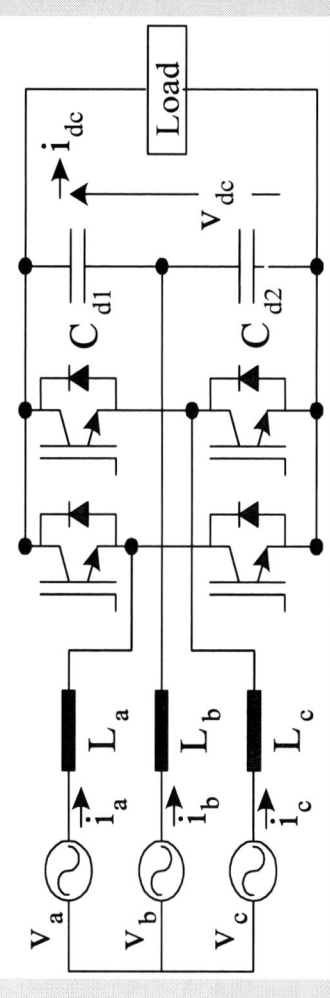

Four Switch Three-Phase Bidirectional Boost Converter.

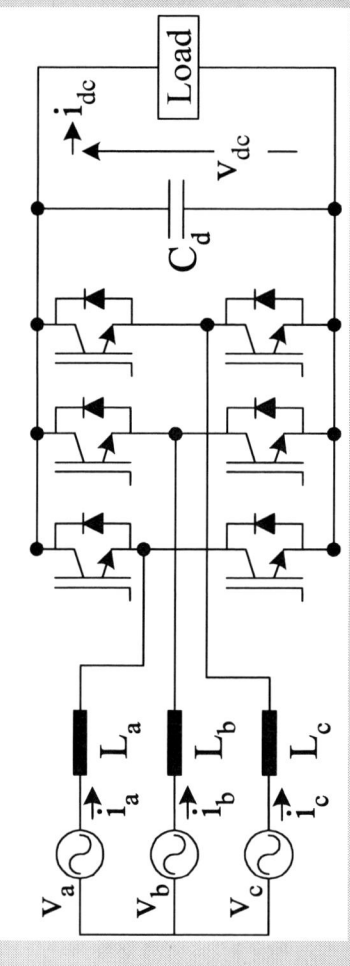

VSI Bridge Based Three-Phase Bidirectional Boost Converter.

27/11/07

IIT-Delhi/PEDS-07

Three Phase Boost IPQC

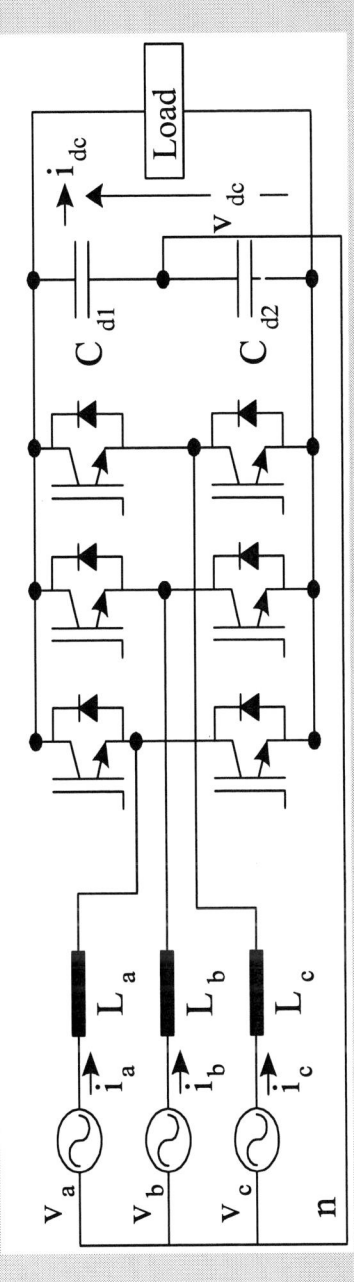

Four Wire Three-Phase Bidirectional Boost Converter.

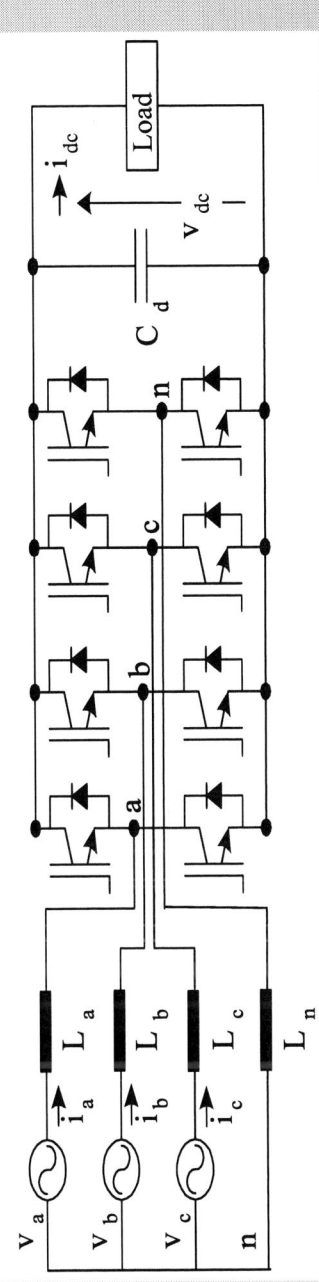

Four-Legged Three-Phase Bidirectional Boost Converter.

27/11/07

IIT-Delhi/PEDS-07

Three Phase Buck IPQC

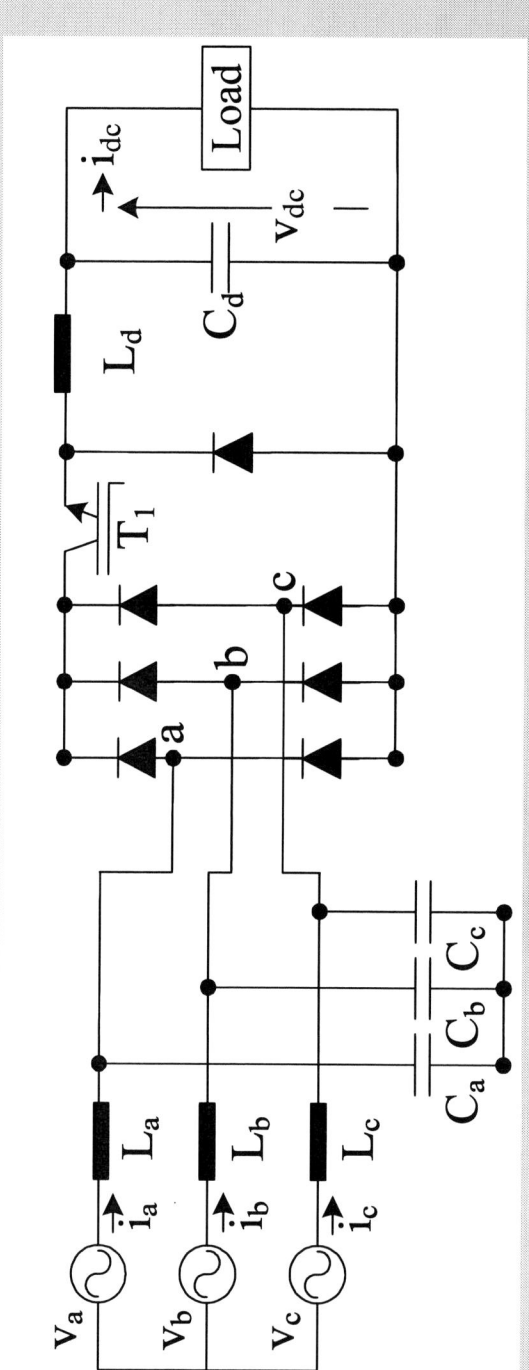

Single Switch Three-Phase Unidirectional Buck Converter.

27/11/07

IIT-Delhi/PEDS-07

Three Phase Buck IPQC

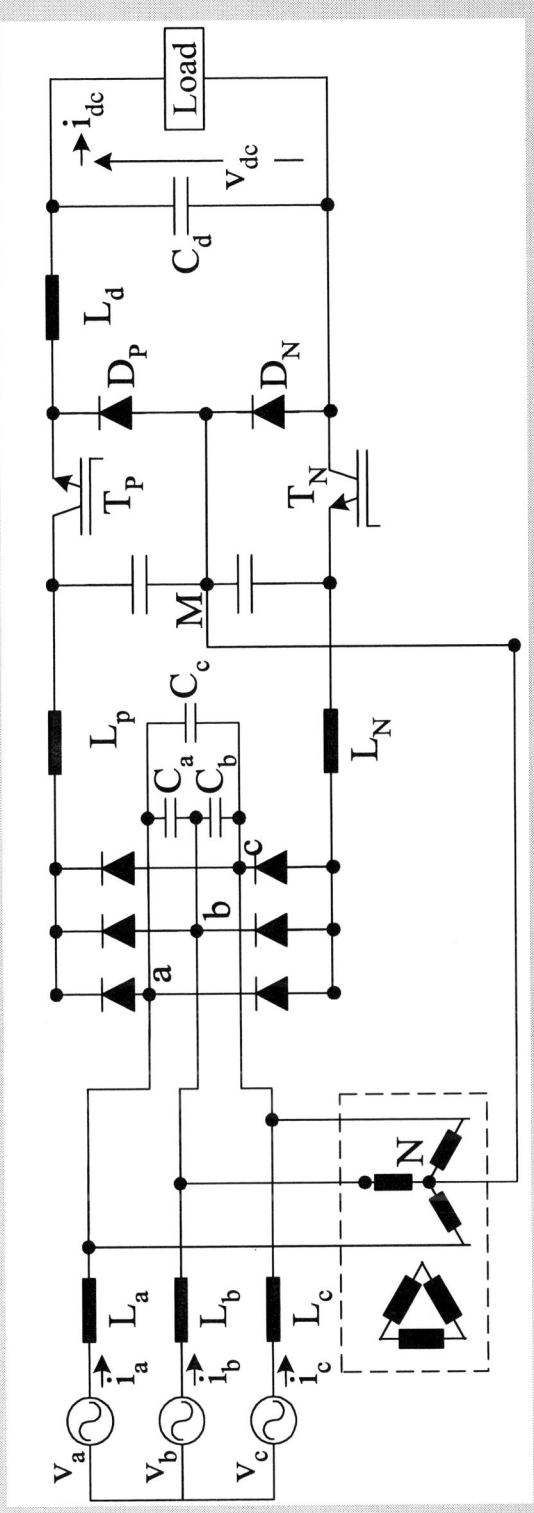

Two Switch Three-Phase Unidirectional Buck Converter.

27/11/07

IIT-Delhi/PEDS-07

Three Phase Buck IPQC

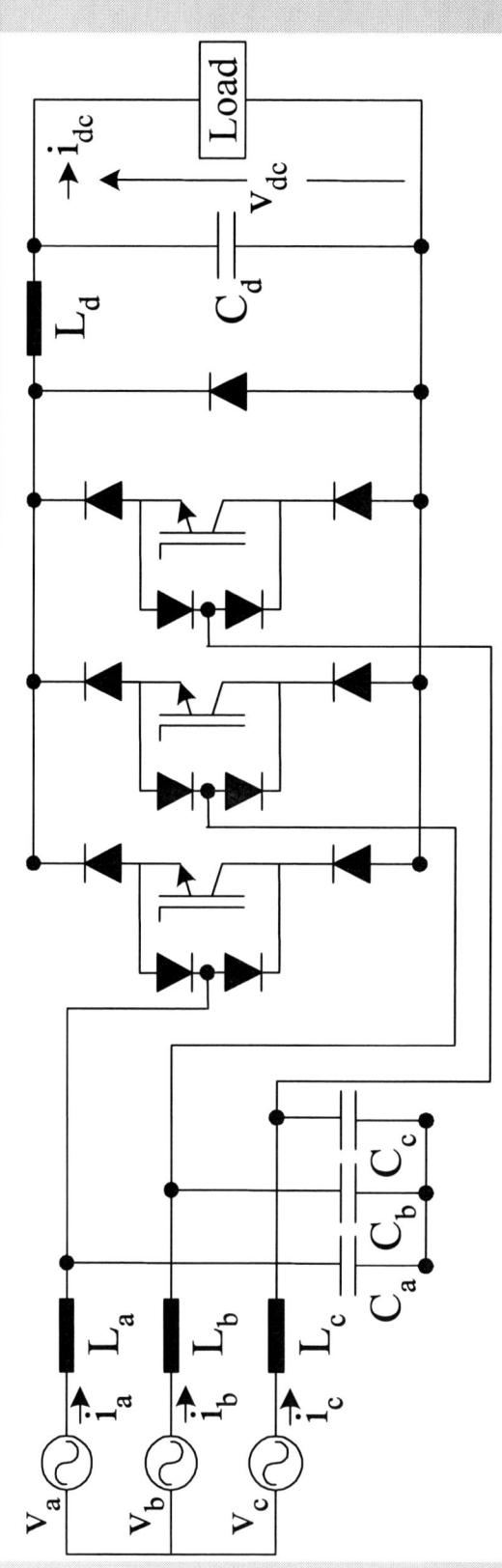

Three Switch Three-Phase Unidirectional Buck Converter.

Three Phase Buck IPQC

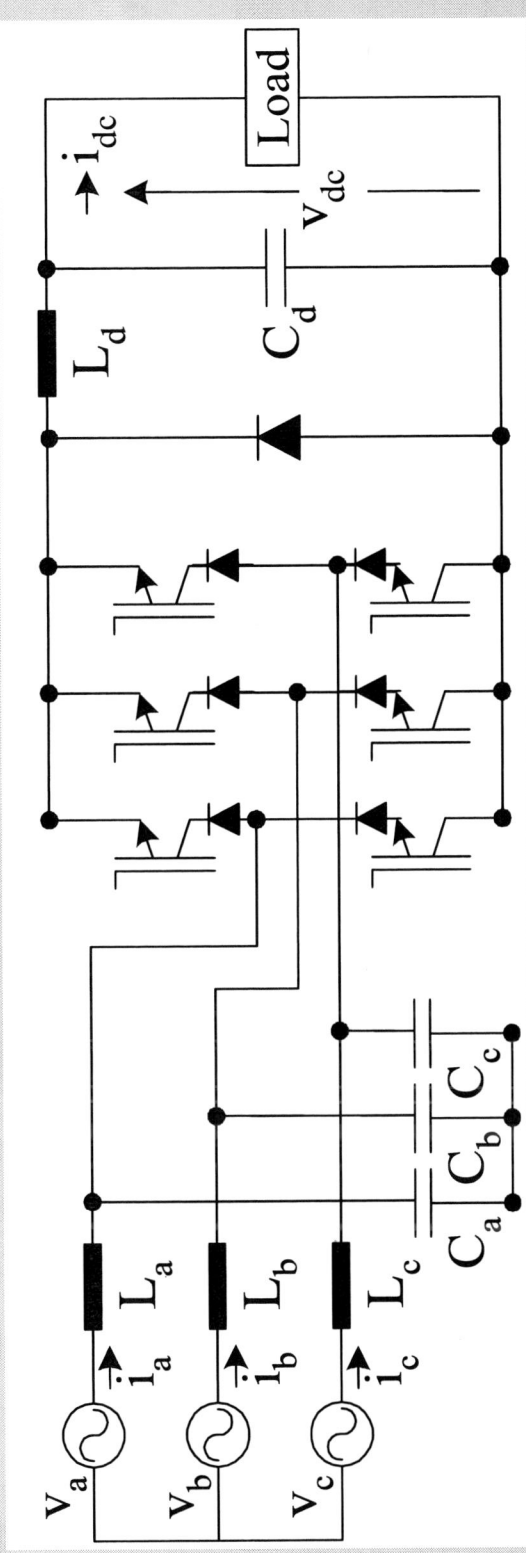

Three-Phase CSI Based Three-Phase Unidirectional Buck Converter.

27/11/07   IIT-Delhi/PEDS-07

Three Phase Buck IPQC

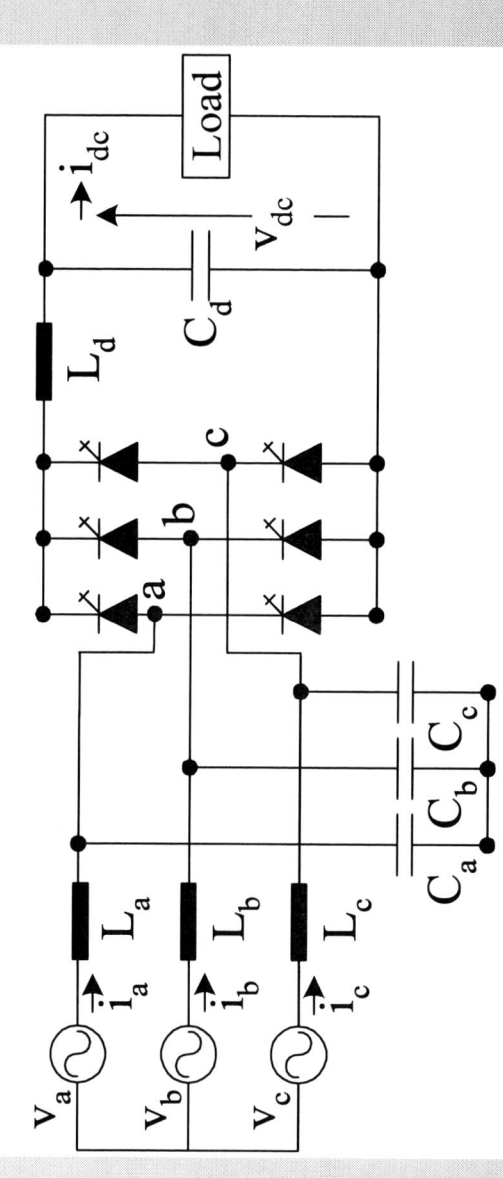

GTO Based Three-Phase Bidirectional Buck Converter.

Three Phase Buck IPQC

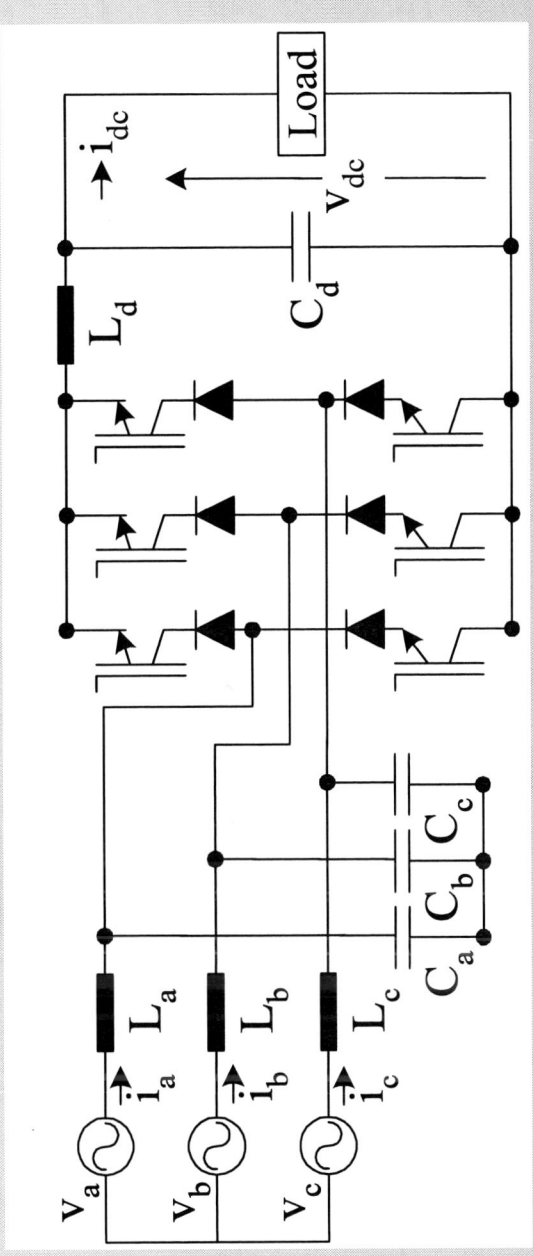

IGBT Based Three-Phase Bidirectional Buck Converter.

Three Phase Buck IPQC

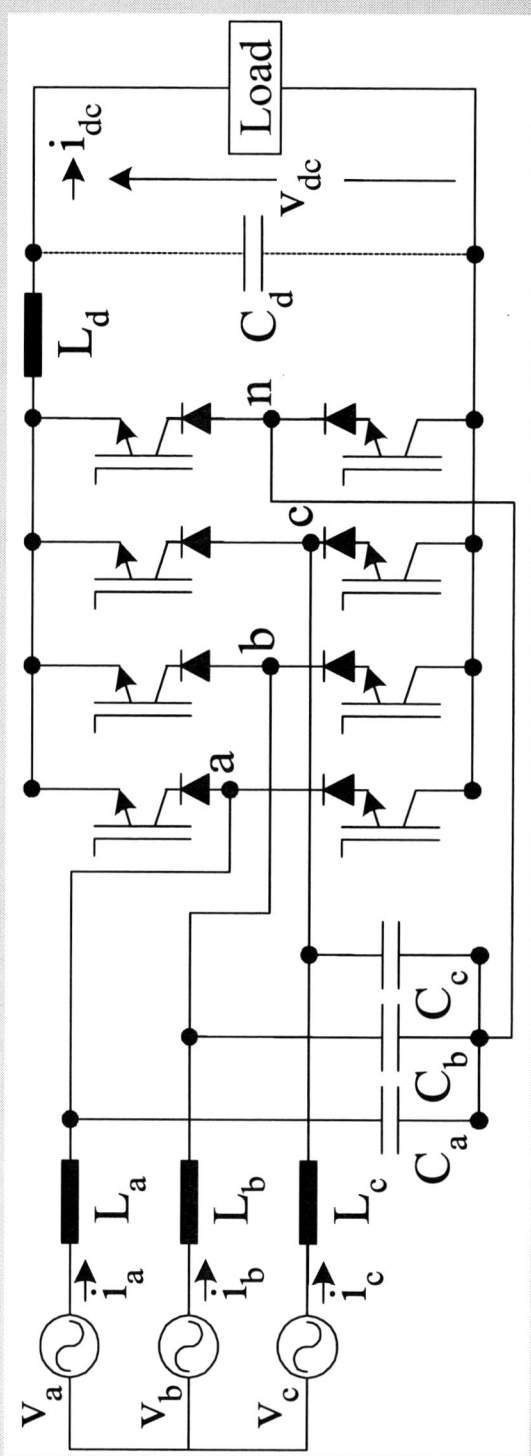

Four Pole Three-Phase Bidirectional Buck Converter.

Three Phase Buck-Boost
IPQC

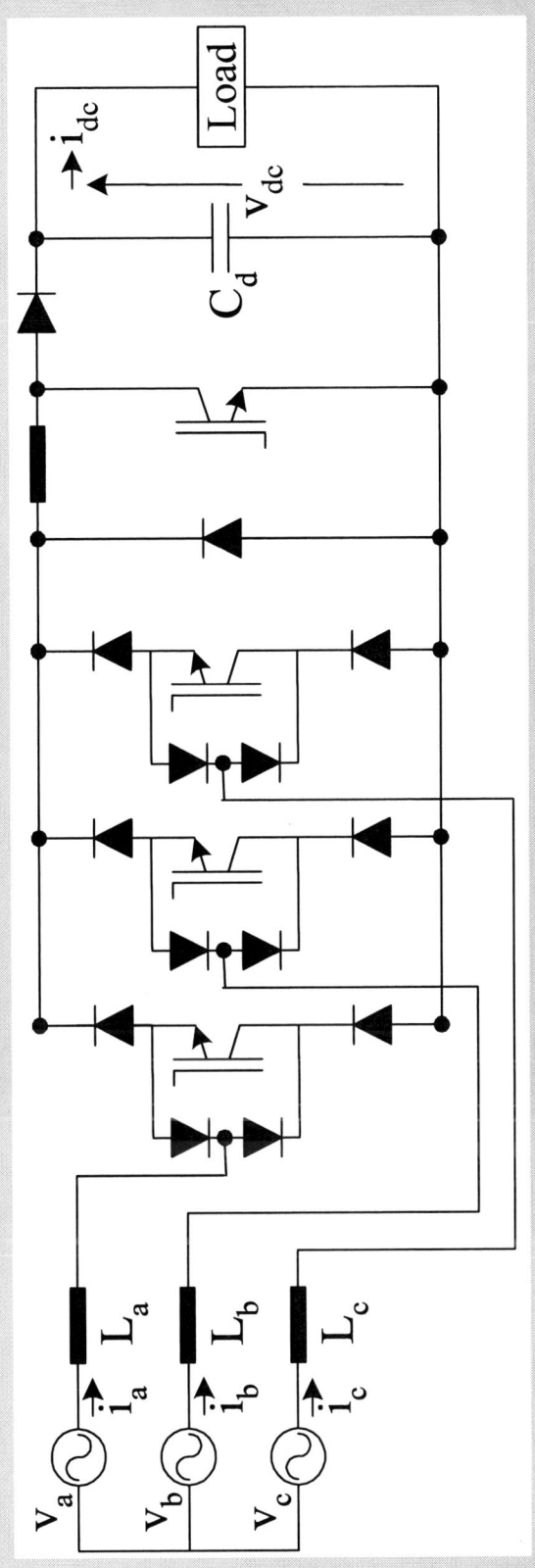

Three-Phase Four Switch Unidirectional Buck-Boost Converter.

Three Phase Buck-Boost IPQC

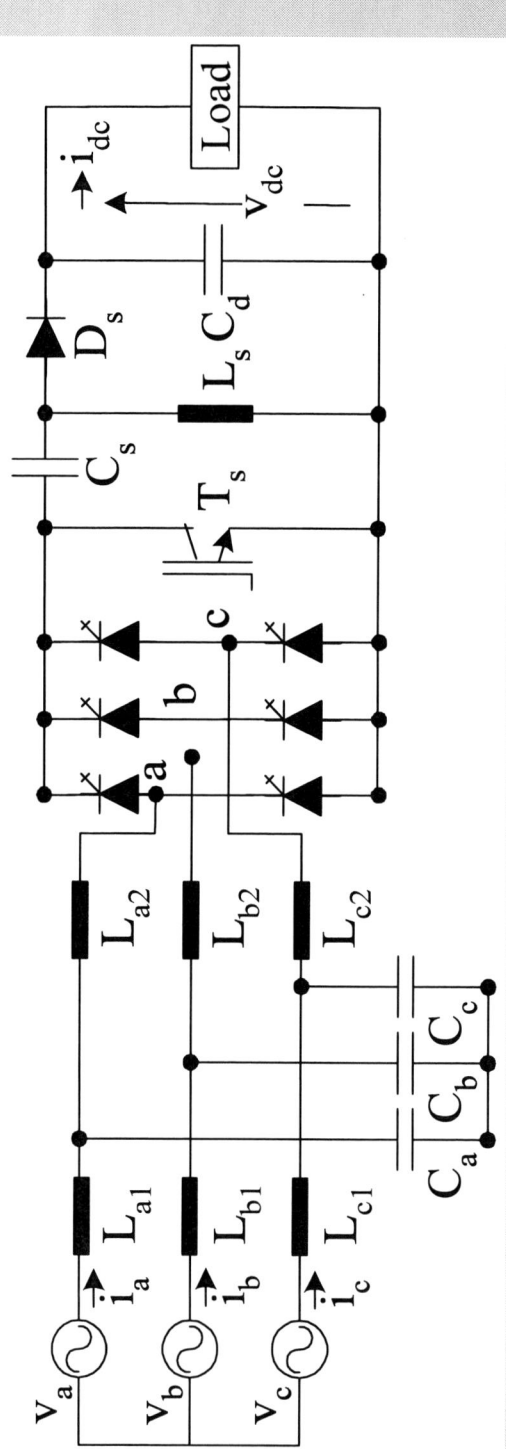

Three-Phase SEPIC Derived Unidirectional Buck-Boost Converter.

Three Phase Buck-Boost IPQC

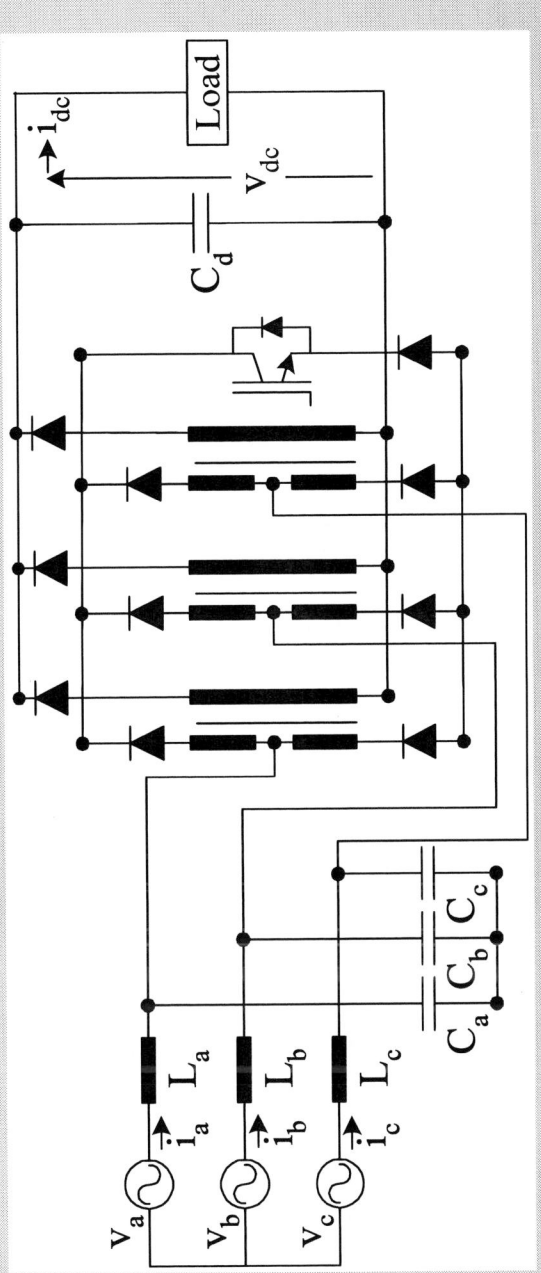

Three-Phase Flyback Derived Unidirectional Buck-Boost Converter

27/11/07

IIT-Delhi/PEDS-07

Three Phase Buck-Boost IPQC

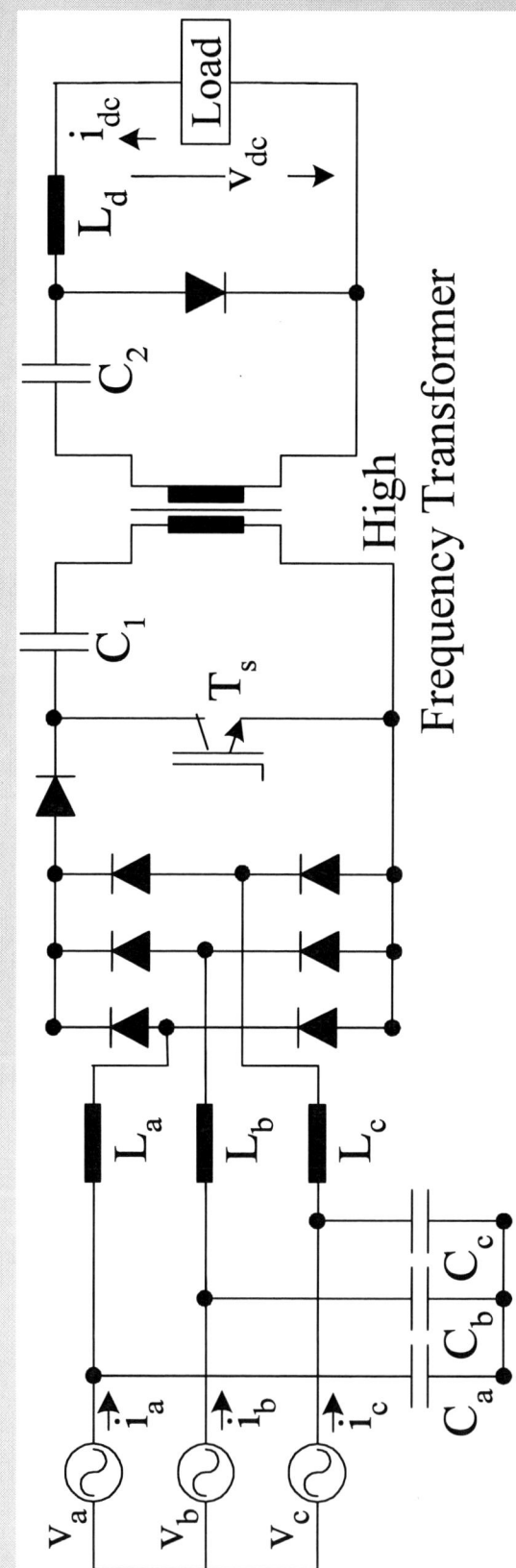

Three-Phase Isolated Cuk Derived Unidirectional Buck-Boost Converter

27/11/07

IIT-Delhi/PEDS-07

Three Phase Buck-Boost IPQC

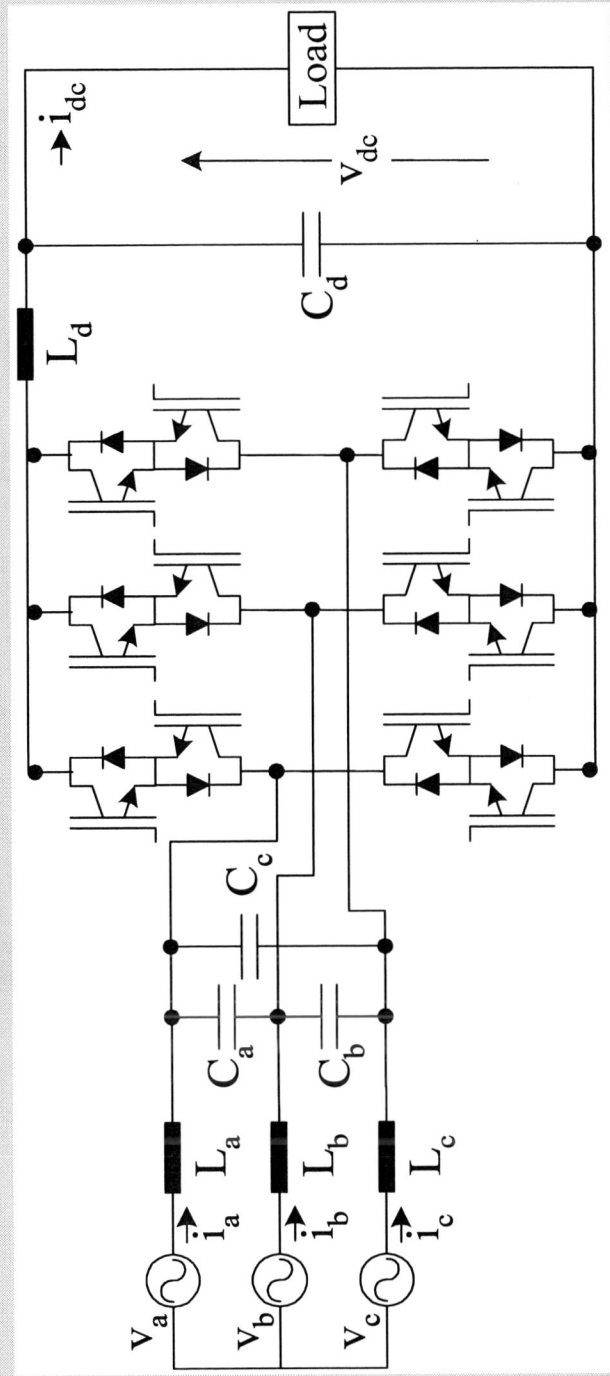

Matrix Converter Based Three-Phase Bidirectional Buck-Boost Converter.

Three Phase Multilevel IPQC

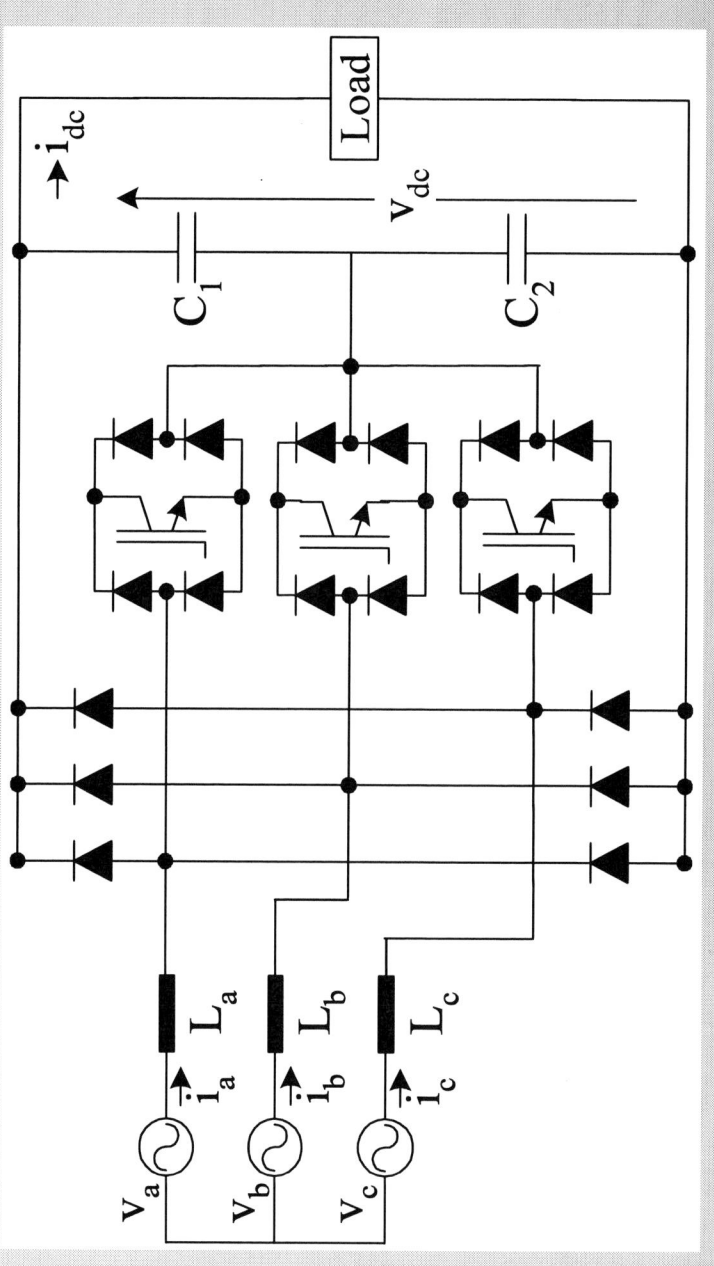

Three Switch Three-Phase Unidirectional Three Level Converter.

27/11/07

IIT-Delhi/PEDS-07

Three Phase Multilevel IPQC

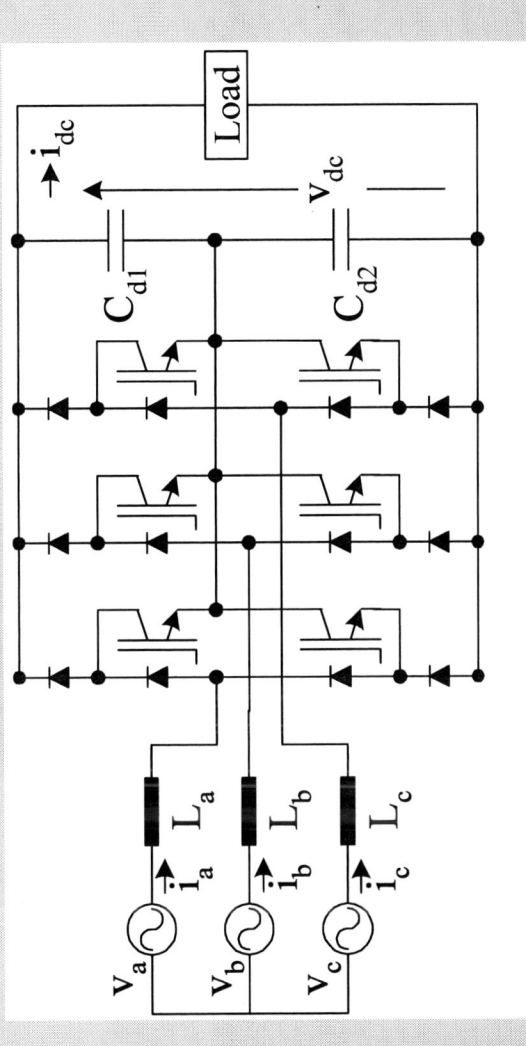

Six Switch Three-Phase Three Level Unidirectional Converter.

Three Phase Multilevel IPQC

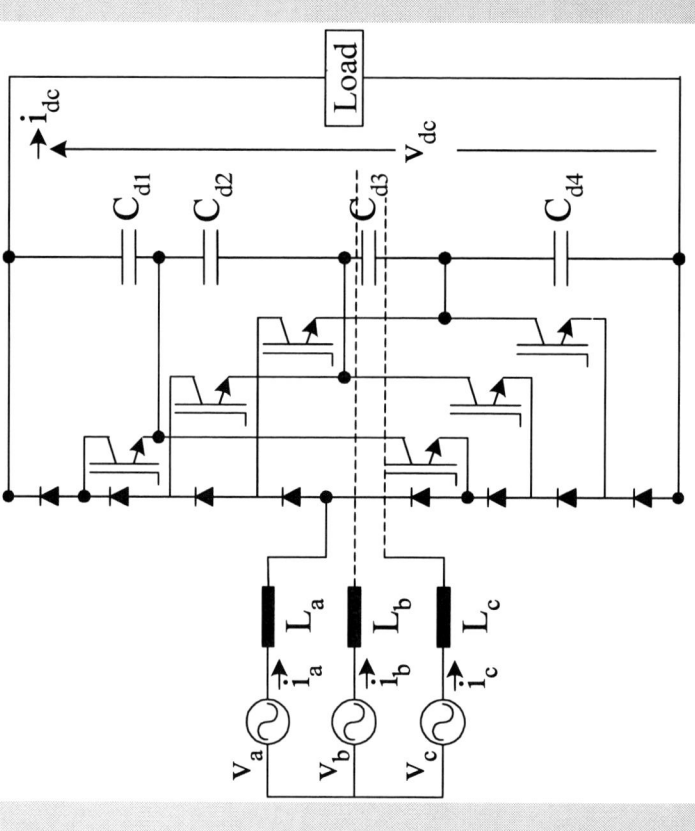

Three-Phase Unidirectional Five Level Converter.

Three Phase Multilevel IPQC

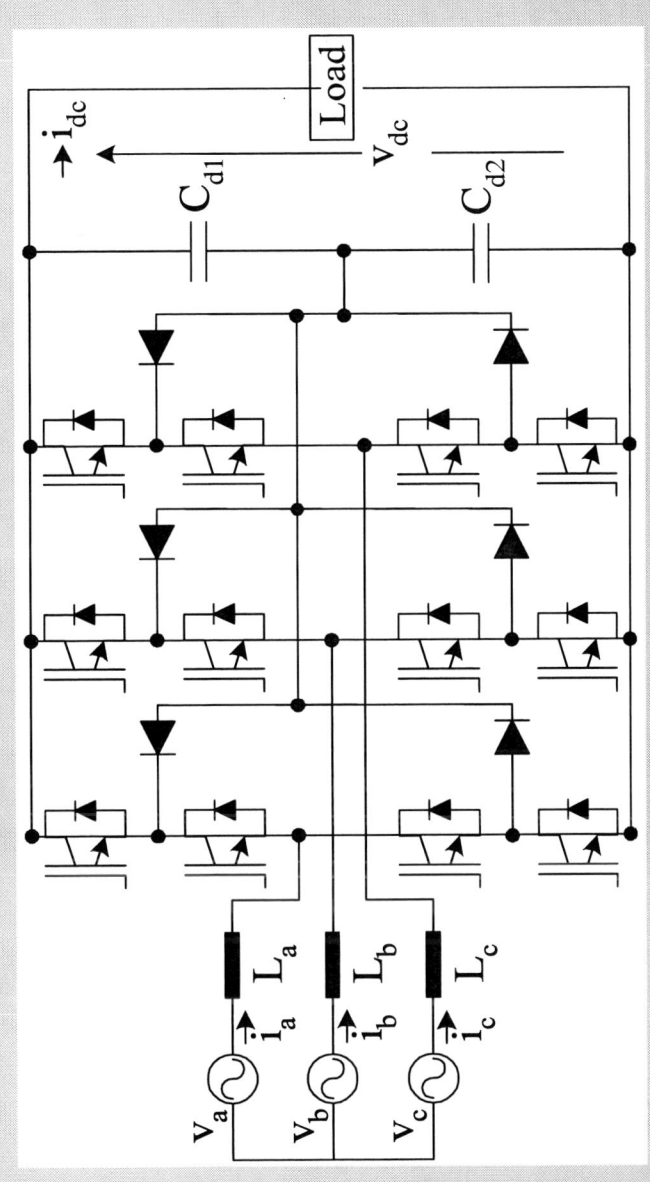

Three-Phase Three Level Diode Clamped Bidirectional Converter.

IIT-Delhi/PEDS-07

27/11/07

Three Phase Multilevel IPQC

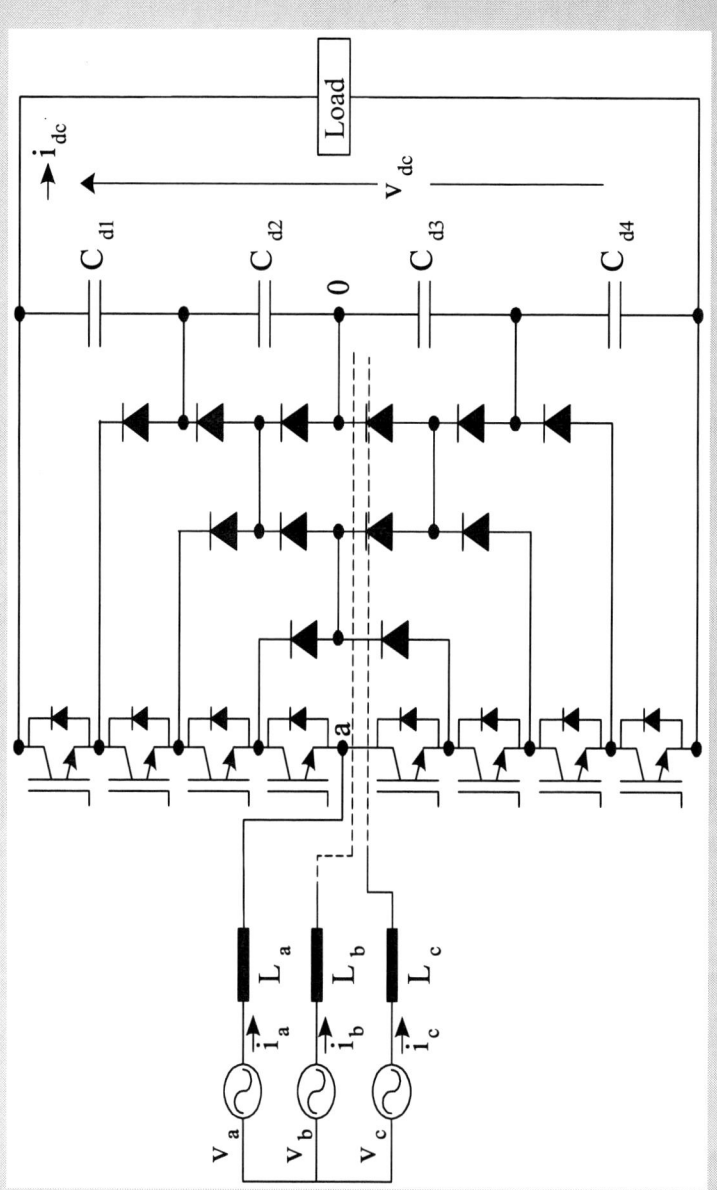

Three-Phase Five Level Diode Clamped Bidirectional Converter.

27/11/07

IIT-Delhi/PEDS-07

Three Phase Multilevel IPQC

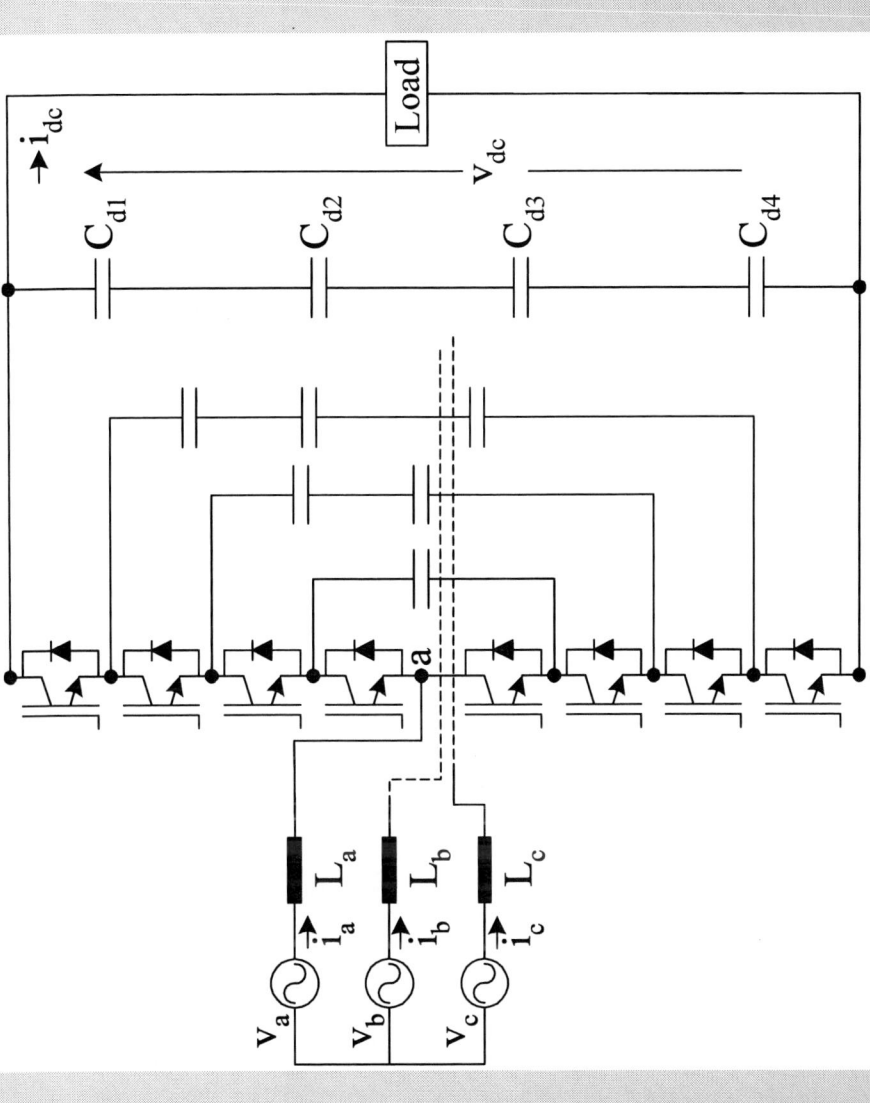

Three-Phase Five Level Flying Capacitor Bidirectional Converter.

27/11/07

IIT-Delhi/PEDS-07

Selection Criterion of IPQC

Number of phases in AC mains (Single-Phase, Three-Phase)

Required level of power quality in input (permitted PF, CF, THD)

Type of output DC voltage (constant, variable, etc.)

Power-flow (unidirectional and bi-directional)

Number of quadrants (one, two or four)

Nature of DC output (isolated, non-isolated)

Requirement of DC output (buck, boost and buck-boost)

Required level of power quality in DC output (voltage ripple, voltage regulation, sag and swell)

Type of DC loads (linear, nonlinear, etc.)

27/11/07

IIT-Delhi/PEDS-07

Selection Criterion of IPQC

Cost

Size

Weight

Efficiency

Noise level (EMI, RFI, etc.)

Rating (W, kW, MW, etc.)

Reliability

Number of DC outputs

Environment (ambient temperature, altitude, pollution level, humidity, types of cooling, etc.)

References

- H. Akagi, Y. Kanazawa, A. Nabae, "Instantaneous Reactive Power Compensators Comprising Switching Devices Without Energy Storage Components", IEEE Trans. on Industry Applications, Vol.IA-20, No.3, May/June 1984, pp.625-630.

- Bhim Singh, Kamal-Al-Haddad and Ambrish Chandra, "A review of active filters for power quality Improvement," *IEEE Trans. on Industrial Electronics*, vol. 46, no. 5, Oct. 1999. pp 960-970.

- Ambrish Chandra, Bhim Singh, B.N. Singh and Kamal-Al-Haddad and "An improved control algorithm of shunt active filter for voltage regulation, harmonic elimination, power-factor correction, and balancing of nonlinear loads" IEEE Trans on Power Electronics Vol. 15, no 3, pp 495 – 507, May 2000.

- B. Singh, V. Verma, A. Chandra and K. Al-Haddad. "Hybrid filters for power quality improvement" IEE Proc.-Gener. Transm. Distrib., Vol. 152, No. 3, pp. 365-378,May 2005

- El-Habrouk, M., Darwish, M.K., and Mehta, P.: 'Active power filters: A review', IEE Proc., Electr. Power Appl. vol. 147, pp. 493–413, 2000.

References

- Arindam Ghosh, and Gerard Ledwich "*Power Quality Enhancement using custom power devices*", Kluwer's Power Electronics and Power System series, U.S.A, 2002.

- A. Ghosh, and G. Ledwich, "Compensation of distribution system voltage using DVR," *IEEE Trans. Power Delivery*, vol. 17, pp. 1030 – 1036, Oct. 2002.

- A. Ghosh, K. Jindal, A. Joshi, "Design of a capacitor-supported dynamic voltage restorer (DVR) for unbalanced and distorted loads," *IEEE Trans. Power Electron.*, vol.19, pp. 405-413, Jan. 2004.

- I. Etxeberria-Otadui, U. Viscarret, S. Bacha, M. Caballero, and R. Reyero, "Evaluation of different strategies for series voltage sag compensation," in *Proc. IEEE PESC'02*, vol. 4, 2002, pp. 1797 – 1802.

- Chi-Jen Huang, Shyh-Jier Huang, and Fu-Sheng Pai, "Design of dynamic voltage restorer with disturbance-filtering enhancement," *IEEE Trans. Power Electron.*, vol. 18, pp. 1202 – 1210, Sept. 2003.

References

- IEEE Guide for harmonic control and reactive compensation of Static Power Converters, IEEE Std. 519-1992.

- G. T. Heydt, "Electric Power Quality", Stars in a Circle Publications, second edition, 1994, Avarua, Rarotonga, Cook Islands.

- R. C. Duagan, M. F. Mcgranaghan and H. W. Beaty, "Electric Power System Quality", McGraw-Hill, 1221 Avenue of the Americas, New York, NY 10020

- M. H. J. Bollen, "Understanding Power Quality Problems", Standard Publishers Distributors, First Indian Edition, 2001, Delhi.

- C. Sankaran, "Power Quality" CRC Press, New York, 2002,

- R. C. Duagan, M. F. Mcgranaghan and H. W. Beaty, "Electric Power System Quality", McGraw-Hill, 1221 Avenue of the Americas, New York, ISBN0-7803-3464-7, 1996.

References

- G. J. Porter and J. A. V. Sciver, "Power Quality Solutions: Case Study for Troubleshooters", Fairmont Press, Inc., 1999.

- J. Arrillaga, N. R. Watson and S. Chen, "Power System Quality Assessment", John Wiley & Sons, Inc., New York, 2000.

- M. H. J. Bollen, "Understanding Power Quality Problems", Standard Publishers Distributors, First Indian Edition, 2001, Delhi.

- J. Schlabbach, D. Blume and T. Stephanblome, "Voltage Quality in Electrical Power Systems, 2001, U. K. ISBN 0-85296-975-9.

- IEEE Guide for Application and Specification of Harmonic Filters, IEEE Standard 1573, 2003.

References

- D. Borojevic, "Analog vs. digital design Three-phase power factor correction-Part 2," in Proc. HFPC'94, 1994, pp. 322-348.

- D. Boroyevich and S. Hiti, *Three-phase PWM converter: Modeling and Control Design.* Seminar 9, IEEE-APEC'96, 1996.

- P. Enjeti and I. Pitel, Design of Three-Phase Rectifier Systems with Clean Power Characteristics, Tutorial, PESC'99, 1999.

- J. W. Kolar and J. Sun, Three-Phase Power Factor Correction Technology, Seminar 1& 4, PESC'01, 2001.

- H. Mao, F. C. Y. Lee, D. Boroyevich, "Review of high-performance three-phase power-factor correction circuits," *IEEE Trans. Ind. Electron.*, vol. 44, pp. 437-446, August 1997.

27/11/07

IIT-Delhi/PEDS-07

References

- J. W. Kolar and H. Ertl, "Status of the techniques of three-phase rectifier systems with low effects on the mains," in *Proc. IEEE INTELEC'99*, 1999.

- B. Singh, B.N. Singh, A. Chandra, K. Al-Haddad, A. Pandey, and D.P. Kothari, "A review of single-phase improved power quality AC-DC converters," *IEEE Transactions on Industrial Electronics*, vol. 50, no. 5, pp. 962 - 981, October 2003.

- B. Singh, B.N. Singh, A. Chandra, K. Al-Haddad, A. Pandey, and D.P. Kothari, "A Review of Three-Phase Improved Power Quality AC–DC Converters," *IEEE Transactions on Industrial Electronics*, vol. 51, no. 3, pp. 641 - 660, June 2004.

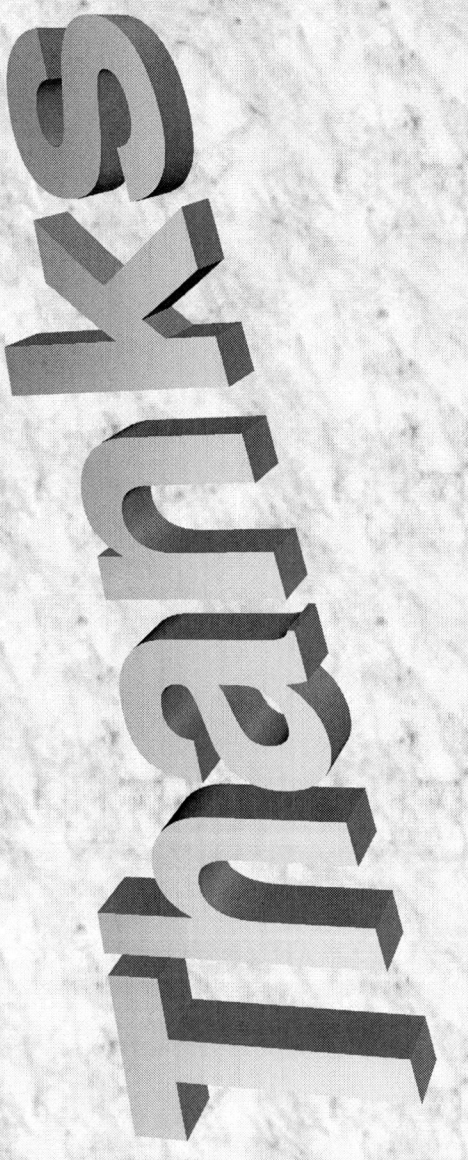

IIT-Delhi/PEDS-07

27/11/07

Power Electronics for Future Utility Applications

Rik W. De Doncker, Christoph Meyer, Robert U. Lenke, Florian Mura
Institute for Power Generation and Storage Systems (PGS)
E.ON Energy Research Center (E.ON ERC)
RWTH Aachen University, Germany
www.eonerc.rwth-aachen.de

Abstract— Medium-voltage converters, originally developed for industrial drives (e.g. in steel and paper mills), have nowadays entered utility applications. For several years, the growing need for power quality in distribution systems in conjunction with the large scale integration of renewable energy sources has boosted the demand for new technologies. Together with communication systems, power electronics are the key enabling technology to meet these challenges. This paper addresses several utility applications for power electronics, some of which are in use already today though market penetration is still low. In the field of high power inverters, conventional 3-level hard-switching converters as presently used e.g. in wind turbines are presented alongside with soft-switching converters and their possible applications such as STATCOMs or mini-turbines. The possibility of using DC instead of AC transmission and distribution systems will be discussed. DC systems have already been used for several decades to transmit bulk power. New opportunities for the use of modern VSC-HVDC will be shown, such as DC distribution systems. For this novel application, new technologies are needed, such as multi-megawatt DC-DC converters and DC circuit breakers; possible concepts for these technologies will be addressed. Cycloconverters, a rather conventional technology, have been applied recently in innovative ways to increase the efficiencies of very high-power pumped-hydro storage systems. Due to the fact that the development of new converter systems is always strongly related to the available device technology, future high-power devices will finally be discussed.

I. INTRODUCTION

Over the past decades, the worldwide interest in renewable energy sources has risen significantly. The limitation of fossil fuels like oil and gas, the increasing cost of these primary energy sources and the impact of the climate change have stimulated the research in the area of alternative electrical energy supplies. Consequently, the share of decentralized power systems in the electricity infrastructure has increased considerably. Most dispersed generation systems require power electronics for the conversion and control of electrical energy. Furthermore, power electronic circuits can provide additional services for grid stability and power quality purposes. Thus, power electronic systems represent a key-enabling technology to cope with the challenges of tomorrow's electricity distribution systems. However, in order to make power electronic technology successful, capital investment and life cycle cost (LCC) of these systems will have to be further reduced. In the following, present and future utility applications for power electronics will be presented.

II. CONVERTERS FOR SMALL AND MEDIUM SIZE GENERATION AND POWER CONTROL SYSTEMS

A. Converters for Wind Turbine Applications

A very up-to-date application for power electronic converters with ratings up to 5 MW is wind energy. In 2006, turbines with a total power of 15 GW were installed worldwide, leading to an accumulated installed power of more than 74 GW [1]. Fast growth of the wind energy market is expected to continue during the next years. This also implies opportunities for suppliers of power electronic equipment, since converters are required in all wind turbine installations in order to decouple the low-speed rotor from the grid frequency. To comply with the grid codes of utilities, the converter has to provide control of active and reactive power, filtering of harmonics and fault ride-through capability. Today, full-scale converters typically used with direct-driven synchronous generators, as well as partial scale converters required for doubly-fed induction generators (DFIG) are usually realized as back-to-back voltage source converters. Here, the three-level neutral-point-clamped (NPC) inverter is the standard solution due to its low harmonic distortion and high efficiency [2]. As in industrial converter applications, alternative topologies such as the Flying Capacitor Inverter (FLC) [3] and matrix converters are a research topic and might emerge in the future. Innovative developments might be supported by the currently observed shift to medium-voltage generators, driven by the ever-increasing power ratings that are expected to reach 10 MW within the next decade. As for semiconductors, this might challenge the dominance of Insulated Gate Bipolar Transistors (IGBTs) in favor of highly efficient and robust Gate Commutated Thyristor (GCT) switches.

B. Converters for Power Quality Enhancement

Flexible AC transmission systems (FACTS) for high and medium-voltage grids have become increasingly popular over the last decade, as they squeeze out remaining transmission capacities in notoriously overloaded power supply systems, provide industrial customers with excellent voltage quality and increase grid stability, which can become important with a rising degree of renewable generation. The scope of applications for these power electronic flow control and compensation systems might be improved in the future by a combination with energy storages such as batteries and flywheels [4].

STATCOMs are FACTS systems capable of reactive power compensation, harmonic filtering and phase load balancing.

978-1-4244-0644-9/07/$25.00 ©2007 IEEE

They are used e.g. in factories and transportation systems, where they keep supply voltages stable under demanding conditions. Therefore, their usage increases with transmission and distribution grids becoming more sensitive to disturbances. Figure 1 shows the topology of a proposed STATCOM system that can be directly connected to an industrial 6.6 kV grid because of the series operation of 6.5 kV IGBTs in a 5-level flying-capacitor-clamped (FLC) inverter [5]. With this number of levels, the flying-capacitor topology is very attractive due to equal loss distribution in the devices and reduced overall cost compared to the diode-clamped multi-level inverter topology found in most 3-level medium-voltage drive inverters. However, the capacitors make for a large part of system size and cost.

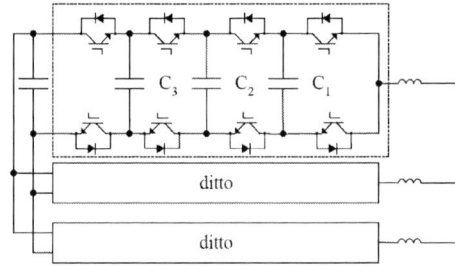

Fig. 1: 5-Level FLC inverter for STATCOM [5]

An increase of the operation frequency will yield a substantial reduction in capacity demand, which is largely determined by the tolerable voltage ripple during switching periods. This can be achieved by soft-switching FLC inverters as shown in Fig. 2. Auxiliary resonant commutated pole (ARCP) inverter topology [6] is used, since current and voltage rating of the hard-switched devices do not have to be adjusted [7]. The cost for the additional resonant elements and auxiliary switches is overcompensated by the reduced capacity demand [8]. This application is a good example about how the intelligent application of soft-switching techniques can reduce system cost and increase the performance of medium-voltage inverters.

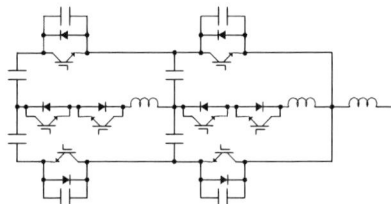

Fig. 2: Soft-switching 3-Level flying capacitor clamped inverter [7]

C. Converters for High-Speed Mini-Turbine Generators

The integration of volatile renewable energy sources like wind power into the grid requires a certain amount of reserve capacity in order to maintain a reliable energy supply. Although biogas-fired micro-turbines (up to some 10 kW) are

capable of delivering power-on-demand, their output power is not sufficient for industrial plants or municipal electricity distribution systems. Instead, fluctuating wind power fed into the grid could be compensated by means of future medium-size turbines which can be powered up very quickly and which are based on energy sources amenable to storage, e.g. biogas. In principle, the basic concept of micro-turbines can be adopted for these mini-turbines in the envisaged power range of several megawatts. The power station basically comprises three components, which are a gas turbine and a generator that is speed controlled by a power electronic converter. With respect to maximum turbine efficiencies, the variable quality of biofuels is leading to different optimal turbine speeds. Thus, instead of using conventional fixed-speed turbines, concepts for speed-variable mini-turbines are currently under investigation.

The turbine is directly connected to a high speed generator, thereby making the bulky mechanical gearbox used in conventional turbine systems obsolete. However, this concept implies the use of a high-power high-speed generator, which does not exist today. Due to the high speed, the rotor of the generator is subjected to extremely high centrifugal forces. Thus, if synchronous or induction generators are used, measures have to be taken to prevent the flaking of coils or permanent magnets in the rotor. Apart from generator concepts based on conventional synchronous or induction machines, the switched reluctance machine is particularly well suited for the envisaged application due to its simple and robust construction of the rotor. Furthermore, to eliminate the medium-voltage transformer, the generator has to be coupled directly to the medium-voltage grid by means of an intermediate power electronic converter. Main advantages of this transformer- and gearless design are the enhanced transportability and the elimination of environmentally hazardous oil. Since the grid side converter is connected directly to a medium-voltage grid, the use of multi-level topologies as well as state-of-the-art series connection of switching devices is necessary. Moreover, due to the high rotational speed of the gas turbine and the generator, the machine converter needs to operate at very high switching frequencies. Therefore, soft-switching topologies have to be taken into consideration [9], since otherwise switching frequencies of several kHz can not be realized with existing devices. In order to further augment the overall efficiency of the system, such a power station could also be used for combined heat and power (CHP) generation.

III. STORAGE SYSTEMS

Due to the growing market share of volatile renewable energy sources in electricity production, energy storage systems that are able to deliver power-on-demand become an indispensable part of modern power systems. By withdrawing electrical energy from the grid and releasing it if a supply shortfall occurs, fast reacting storage systems contribute to the stabilization of the electricity delivery infrastructure. Concerning the classification of storage systems with regard to their application in power systems, one has to distinguish between the utilization for power quality purposes and energy management. Power quality applications, such as flicker compensation, require high power to be delivered within a very short time, typically in the range of a few cycles up to seconds. On the other hand, the aforementioned large-scale integration of renewable energy sources into the grid, load leveling and peak shaving represent energy management applications which are demanding high energy for time spans up to several hours. In the following, the application of power electronics in two very different types of storage systems, pumped-hydro and electrochemical storage systems, will be discussed. Other technologies, e.g. Compressed Air Energy Storage (CAES) [10] and flywheels will not be further addressed in this paper.

A. Pumped-Hydro

Pumped-hydro systems represent a well-established technology for bulk energy storage that is used throughout the world. With regard to both power and stored energy, these systems are currently the largest storage systems for electrical energy. In a pumped-hydro system, water is transferred from one reservoir to another, thereby storing or releasing energy. Thus, the electrical machines connected to the turbines are operating in motor and generator mode, respectively. However, the feasibility of such systems depends strongly on the geographical properties of the terrain, and locations for new installations are difficult to find. Today, the total pumped-storage power in operation attains more than 90 GW [11]. The majority of these units are conventional hydro-power plants employing synchronous machines operating at constant speed in order to attain high efficiencies. On the other hand, new pumped-storage units are equipped with doubly-fed induction machines driven by cyclo-converters, thereby enabling the turbine to operate at variable speed. The main advantages of these adjustable speed pump-storage systems are the improved efficiency at partial loads, increased lifetime of the turbine and the highly dynamic control of the power delivered into the grid. Thus, pumped-hydro systems are able to play an important role for the stabilization of the electricity delivery infrastructure.

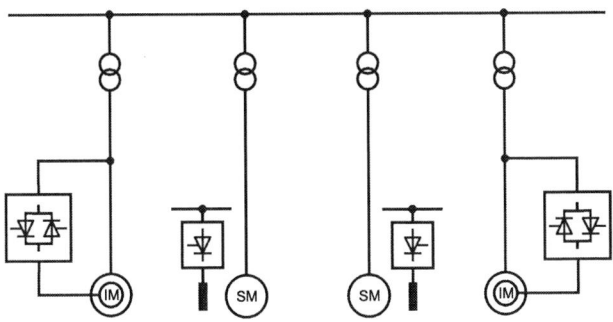

Fig. 3: Principal layout of the Goldisthal pumped-hydro plant with two 100 MVA cyclo-converters

Today, six adjustable speed pumped-storage systems are operating in Japan [12]. The only European unit is located in Goldisthal, Germany. After a planning period of about 30 years and a construction period of seven years, it was finally taken into operation in 2003. The principal layout of the pumped-hydro power station is shown in figure 3. It is equipped with two fixed-speed synchronous machines (rated power 331 MVA each) and two doubly-fed induction machines (rated power 340 MVA each) [13]. The total storage capacity amounts to 8.5 GWh. The induction machines are driven by cyclo-converters with a power rating of 100 MVA, being one of the largest drive inverters in the world. Two anti-parallel 12-pulse thyristor bridges are feeding each rotor phase. In order to provide sufficient redundancy, three independent thyristor bridges are operated in parallel to carry the high current. The converters are designed to vary the speed of the induction machines from -10 % to +4 % of their rated speed.

B. Electrochemical Storages

To connect a Battery Energy Storage System (BESS) to the grid, its DC voltage has to be converted to a three-phase AC voltage by means of power electronic converters. In the case of a coupling to a future medium-voltage DC distribution grid (see section IV), the voltage has to be boosted to the grid level. Thus, power electronics are indispensable for the connection of BESS to both AC and DC grids. The converters ensure the proper charging and discharging of the battery in order to increase its lifetime, as well as the decoupling of the storage device from fluctuations of the grid. Typical applications for electrochemical storage systems are VAR compensation, load leveling and peak shaving.

Since several decades, lead-acid batteries are the most widely used storage system for electrical energy. Due to their low initial cost, they will remain the workhorse for stationary applications. Drawbacks of the lead-acid technology, however, are the low specific gravimetric energy content and their limited cycle life caused by accelerated ageing. Lead-acid BESS have been built up to a rated power and energy of 10 MW/40 MWh [14], [15]. Nickel-cadmium (NiCd) BESS represent another well-established battery technology. They are mainly used for high power applications where energy has to be supplied during several minutes. Although being more expensive per watt-hour than lead-acid batteries, Ni-Cd offers a significantly better life-cycle cost [16]. The largest nickel-cadmium BESS currently in operation is located in Fairbanks, Alaska. It is capable of providing 27 MW for 15 minutes [16], [14]. Sodium-sulfur (NaS) batteries are operated in Japan since the beginning of the 1990s [17]. They feature a high energy density, (factor 3 to 5 compared to lead-acid batteries), low maintenance requirements and potential for low-cost mass production [17]. In 2004, Tokyo Electric Power Company (TEPCO) installed two NaS battery energy storage systems rated at 8 MW and providing 60 MWh of energy [18].

IV. HVDC/MVDC TRANSMISSION AND DISTRIBUTION

A. HVDC Applications

a) Classical HVDC: DC technology could take a leading role in future energy transmission and distribution. While todays 80 GW of worldwide HVDC capacity still seems modest in comparison with classical AC transmission, rapid growth is observed over the last decade. Classical thyristor-based HVDC systems increasingly help emerging economies to supply the booming energy demand of their industrial hubs from remote generation facilities. Examples are two 3 GW, +/-500 kV links connecting inner China's Three Gorges hydro-plants to coastal regions (each being 900 kilometers in length), and a projected connection between water-rich chinese Yunnan province and Thailand with similarly impressive ratings [19]. In the near future, "Ultra" HVDC technology of +/-800 kV voltage rating is about to be introduced for transmission capacities beyond 6 GW and distances exceeding 1000 km [20]. At least 12 of these systems are in planning stages; their realization alone would almost double today's installed HVDC capacity [21]. Further applications for classical HVDC are the coupling of island grids, such as targeted by projects for links between Iceland and Scotland (500 MW) and between mainland Italy and the island of Sardinia (SAPEI, 1 GW). A novel application is the interconnection of electricity markets for energy trading, such as realized by the 700 MW / 580 km NORNED link between Norway and the Netherlands, which will feature world's longest underwater high-voltage cable connection when taken into operation in 2007. Finally, classical HVDC systems continue to be the technology of choice for very high power back-to-back grid couplers, as currently installed at the Al Fadhili facility in Saudi Arabia for an asynchronous grid interconnection with five other gulf states (3 x 600 MW) [22].

b) VSC-HVDC (HVDC Light® and others): Since their first installation 10 years ago, IGBT-based HVDC systems have paved the way for DC technology in novel applications thanks to their distinctive advantages over conventional HVDC systems. Among these are the ability to supply very weak grids (even single consumers such as oil rigs), the feasibility of multi-terminal systems, and a reduced demand for estate at the terminal sites. They are currently available up to 1 GW capacity at +/-300 kV, but can also be economically viable in small systems down to power ratings of some ten MW. Recent milestone installations of such VSC-HVDC links are the 350 MW Estlink connection between Finland and Estonia over 350 km of combined sea and land cable and the 84 MW supply of Troll offshore gas compressor facility in the North Sea. Since VSC-HVDC systems are able to transmit power over longer distances than AC lines, a key role could fall to them in interconnecting remote offshore wind farms to mainland power grids [23]. In Germany, offshore wind farms are intended to produce an ambitious 15 % of total electricity demand by 2030 [24]. Other countries are likely to develop similar targets. Together with offshore generation, the role of VSC-HVDC technology in power systems could hence significantly increase within the next decades.

B. MVDC Collector Grids for Offshore Wind Farms

Within the issues that make the realization of offshore wind farms still challenging - from costly foundations to corrosive air - economic collector grids that connect single turbines to an onshore link are receiving much attention. It has been proposed to use medium voltage DC connections rather than conventional AC links in order to obtain weight savings and higher efficiencies [25], [26]. In Fig. 4, a possible configuration of a large grid connected wind farm is shown. The turbines are connected to a low loss MVDC collector grid, which feeds into the HVDC onshore link through a high power DC/DC conversion stage, possibly located on a central platform in the farm. This central DC/DC converter consists of modular building blocks, intelligently interconnected to provide the necessary high-voltage level. In addition, surplus energy might be stored in future high power battery energy storage systems (BESS) or used for hydrogen generation, each of these facilities requiring electronic MVDC/HVDC transformers. Not shown are additional DC/DC converters at the wind turbines, which could be used to adjust the turbine generator output to an appropriate medium-voltage level.

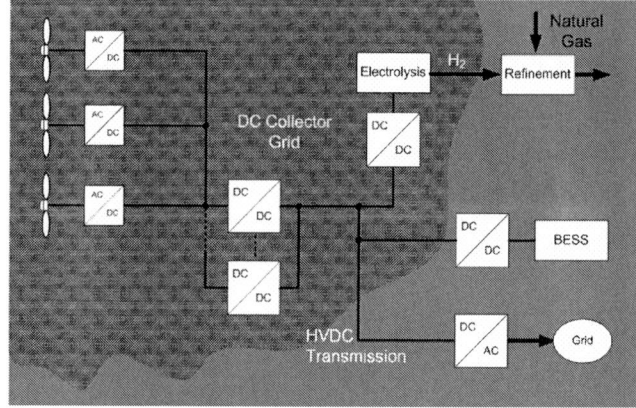

Fig. 4: Offshore wind farms and storage systems connected to the main grid over medium and high voltage DC converters

C. MVDC Distribution Grids

Medium- and low voltage DC technology has the potential to substantially improve future power distribution systems [27], [28], [29], [30], [31]. A far-looking perspective envisioning MVDC rings as power backbones of interconnected urban areas is shown in Fig. 5. The increasing number of small distributed generators and the imminent ubiquity of power electronic front-ends in consumer devices make it more challenging for conventional AC systems to sustain a reliable supply with adequate power quality. As an example, short-circuit power ratings often have to be increased where generation and storage facilities in the megawatt range are connected to existing grids. This can result in larger conductors, modified transformers and more expensive protection devices. Similar measures, supplemented by filters prone to inefficiency, are necessary to mitigate harmonics caused by

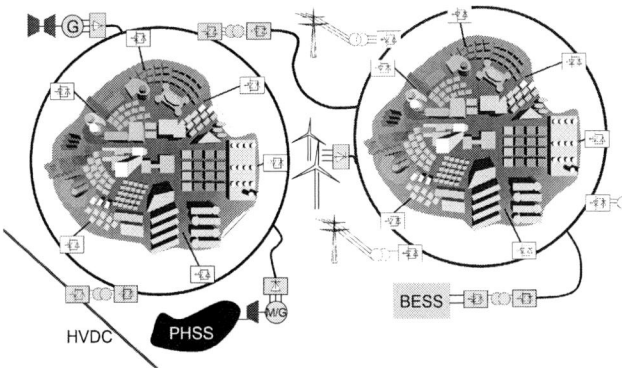

Fig. 5: "City of Tomorrow": Interconnected MVDC rings serve as backbones of the distribution grid. Bulk power is supplied from conventional AC and HVDC transmission grids. Smaller generators and storage systems are directly connected to the MVDC infrastructure. Components are coupled with the help of electronic transformers

power electronic loads. Today, a large part of the growing costs of AC grid enhancements is shifted to those requesting to become connected - by means of ever tighter grid codes. In the future, a paradigm shift to an MVDC infrastructure might become the overall more economical and ecological response to the changing demands of modern power generation, storage and consumption. Thanks to an inherently high short-circuit power and the obsolence of grid inverters, an MVDC distribution would drastically simplify the integration of small and medium generators and storage systems relying on power electronic interfaces, among those small wind farms, (bio-)gas turbines, small pumped-hydro facilities (PHSS) and battery energy storage systems (BESS). Power quality can be maintained excellent with much reduced filtering and compensation effort due to fast electronic voltage control and the inherent filter properties of cable capacitances. High ampacity DC cables might even be the most economical solution in situations where AC overhead lines are not an option due to legal difficulties in obtaining way-of-rights, which already today is a prohibitively difficult endeavor in many places. Superconducting cable technology might become an interesting option for ultra-high efficiency grids [32].

Competitive life cycle cost of large scale MVDC distribution grids will largely depend on affordable and efficient power electronic transformer units for coupling grids of different voltages (HVDC/MVDC, HVAC/MVDC, MVDC/MVDC, MVDC/Low voltage distribution), and on the availability of protection equipment against the high fault currents. Cost restrictions for DC usage will be less tight in areas without grid access, e.g. in undeveloped world regions but also places like ski resorts, where small DC grids, combined with storage systems, can foster the self-supply from renewable sources and help to reduce the dependence on diesel generators or alike [33], [34]. Pioneering applications like these and the

abovementioned wind farm collector grids are expected to contribute to cost reductions in DC system technology.

D. Key Components for DC Applications

While future MVDC solutions promise better economics for wind farms and distribution grids, two key components are vital to their success: High power DC/DC converters and fast DC circuit breakers.

c) High Power DC/DC Converters: High power galvanically isolated DC/DC converters, or electronic transformers, are vital for the interconnection of DC systems. They typically consist of a medium-frequency AC transformer connected to the two DC systems by an input inverter and an output rectifier. For illustration, Fig. 6 shows the SRC3 topology, which has distinctive advantages in very high-power applications. Capacitors in series with the transformer reactance enable soft-switching of all devices when operated close to resonant frequency. Moreover, the three-phase layout offers improved performance compared to single-phase versions [35]. The topology can be easily adapted for bi-directional operation by using active inverters on both sides of the AC link.

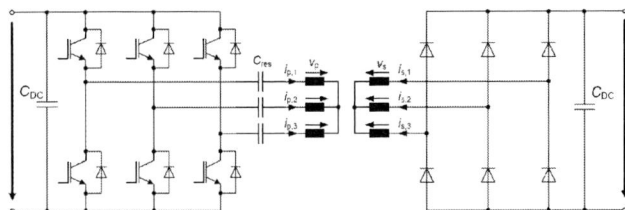

Fig. 6: Three-phase series-resonant DC/DC converter (SRC3)

Operating the transformer at some kilohertz results in much smaller and lighter core and winding sizes than used for 50/60 Hz transformers, which is a major reason for choosing this frequency range. However, a sharp increase of specific losses at elevated frequencies prohibits the use of conventional core laminations in this case. High saturation amorphous iron alloys, used already today in efficient 60 Hz distribution transformers, are well suited materials for high power medium-frequency designs. Calculations suggest that 25 MW medium-voltage DC/DC converter modules could be built using the SRC3 topology, with efficiencies exceeding those of conventional transformers [26]. Since switching losses are significantly reduced, the design can fully profit from the application of GCT devices with low on-state voltages. Very high power centralized DC/DC conversion stages, such as those suggested above for the use in wind farms, can be built by clustering converter modules in a parallel/series connection as shown in Fig. 7.

d) DC Circuit Breakers: Fast switching action is essential for circuit breakers in DC grids, since the absence of in-line transformer and generator reactances cause very high di/dt values during faults. Snubbered mechanical circuit breakers, such as shown in Fig. 8, achieve turn-off times below 100 µs and therefore fit the demands of MVDC applications well. When the mechanical breaker opens, arcing

217

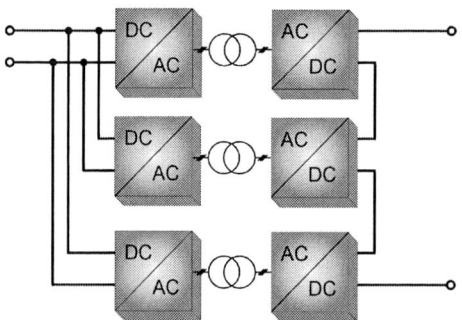

Fig. 7: Parallel/series connection of DC/DC converter modules to form

is prevented thanks to a slowed voltage increase that results from the load current flowing into a snubber capacitor. At sufficiently high capacitor voltages (e.g. 30 kV), the varistor conducts and provides a path for the safe demagnetization of any line inductance, forcing the current finally to zero. In order to prevent the current from commutating back to the capacitor if voltage sags during line demagnetization, two series thyristors are turned off after the varistor has ignited. Other fast hybrid breaker concepts have been presented, which however have the disadvantage of requiring expensive active turn-off devices [36].

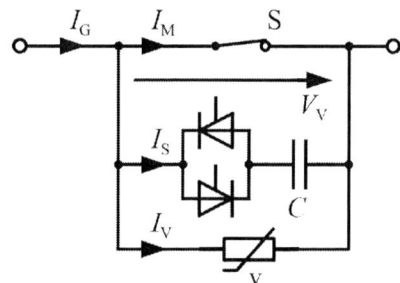

Fig. 8: Fast snubbered mechanical circuit breaker for DC application

V. ENABLING TECHNOLOGIES

In order to increase the power and voltage levels of power electronic converters, the development of novel devices is required. In the following, two high-power devices, the Internally Commutated Thyristor and the MOS Turn-Off Thyristor will be briefly addressed.

e) Internally Commutated Thyristor (ICT): In high power voltage source inverters, GCTs have advantages over IGBTs due to their superior on-state behavior. Since GCTs are latching devices, they offer significantly lower conduction losses than IGBTs. However, the robust operation of GCTs requires a very low inductive path between the gate drive unit (GDU) and the semiconductor device. Consequently, IGCT (Integrated Gate Commutated Thyristor) [37] devices featuring a close connection of GCT housing and GDU were developed. The integration of critical parts of the gate drive into the

housing of the GCT, resulting in the Internally Commutated Thyristor (ICT), can further improve GCT performances [38]. In the ICT, the turn-off unit consisting of MOSFETs and capacitors is placed inside the press-pack housing (figure 9a). The remaining parts of the gate drive unit can be connected via cable to the ICT.

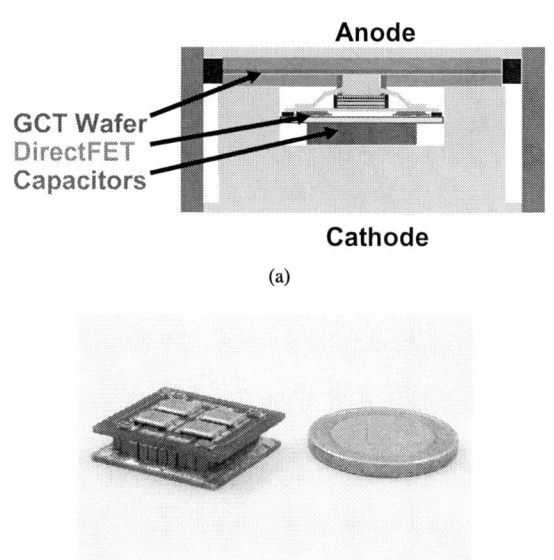

Fig. 9: Cross section (a)) and internal turn-off unit (b)) of the ICT [39]

Since the turn-off unit has to withstand temperatures of up to 125 °C and the space inside the press pack is limited, a careful selection of the components is required. Consequently, a complete redesign of the internal turn-off unit has been performed, thereby reducing its size and volume drastically (figure 9b). For the capacitors, Multi Layer Ceramic Capacitors (MLCCs) exhibiting lower ESR and ESL values than electrolytic capacitors are chosen. By employing high temperature materials, e.g. X7R with a continous operating temperature up to 125 °C, operation of the capacitors in vicinity to the wafer becomes feasible. Furthermore, the low inductive and ohmic resistance of the integrated turn-off unit enables the reduction of the capacitor voltage. As for the MOSFETs, DirectFETs® offering an improved volumetric ratio between silicon and packaging, are used [40].

In addition to the standard functionality of a gate driver, two supplementary features, a zero-voltage and a short-circuit detection, were implemented into the gate drive unit. The former is particularly useful for the envisaged application of ICTs in soft-switching inverters, since these converters often require the turn-on command to be delayed until the voltage across the semiconductor switch has become sufficiently low.

f) MOS Turn-Off Thyristor (MTO): Like the Emitter Turn-Off Thyristor (ETO) [41], the MOS-turn-off thyristor (MTO) [42] belongs to the family of combined bipolar-MOSFET semiconductor devices, combining a thyristor structure with a power MOSFET. In the future, the turn-off MOSFETs of the MTO will be integrated at semiconductor level, yielding extremely low commutation inductances and hence superior switching behavior. Compared to IGCTs, the complexity of the gate driver is significantly reduced, while maintaining similar switching characteristics. Todays MTO

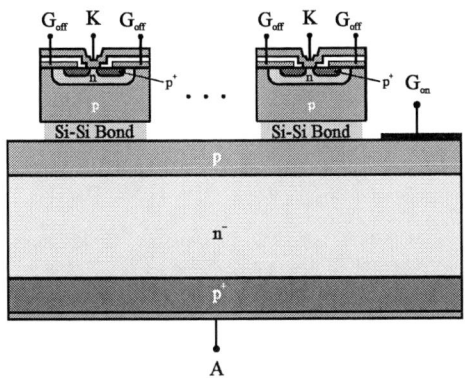

Fig. 10: Cathode gated MTO structure

devices are using bond wires in order to connect the MOSFET to the bipolar structure. Since these bond wires cause critical stray-inductances in the commutation path and are susceptible to damages due to thermal cycling, they significantly lower the reliability of the devices. Applying the direct wafer-bonding technology [43], these problems can be circumvented. The structure of the novel MOS-Turn-Off Thyristor based on silicon-silicon bonding is shown in figure 10. The drain sides of several p-channel MOSFETs are bonded directly to the gate layer of the bipolar structure. For the turn-on of the MTO, a current pulse is injected at the gate contact. By applying a negative voltage at the MOSFET's gate the device is switched off.

VI. CONCLUSION

Increasing need for availability and power quality in combination with the fast growth of distributed generation have promoted the use of power electronic systems in transmission and distribution grids. Growing pressure for innovative solutions will demand power electronics to take an even larger role in future grid applications. Medium-voltage inverters will become key interfaces for grid components with moderate power ratings. Direct connection to medium voltage levels requires the intelligent use of multi-level topologies. Soft-switching techniques can improve today's inverter concepts by means of increased switching frequencies, essential for the operation of high-speed mini-turbines and flywheel storages, and reduced system cost, as has been demonstrated for STATCOMs. The performance of large pumped-hydro and compressed air energy storage systems will be enhanced with the help of cyclo-converter drives larger than ever built before. On the transmission level, HVDC systems are projected to grow in capacity and contribute to bridge large distances between generation and load centers as in China. They will

further be used increasingly to couple grids for energy trade and supply reliability. IGBT-based VSC-HVDC technology has opened new opportunities for offshore power transmission. Wind farms will become connected to onshore grid access points over VSC-HVDC links. Medium voltage collector grids might be a cost-effective option for the interconnection of turbines within farms. MVDC grids as future backbones of urban and rural electrification are a long-term vision for high capacity, high performance power systems with extensive distributed generation. New technologies such as multi-megawatt DC/DC converters and fast DC circuit breakers will become necessary for these novel applications. Since developments in power electronic conversion systems go hand in hand with advances in device technology, future high-power devices like ICT and MTO will contribute to further reduce life-cycle-costs and improve converter performance.

REFERENCES

[1] *Global Wind 2006 Report.* Global Wind Energy Council, 2007. [Online]. Available: http://www.gwec.net

[2] D. Krug, S. Bernet, and S. Dieckerhoff, "Comparison of state-of-the-art voltage source converter topologies for medium voltage applications," in *Industry Applications Conference, 2003. 38th IAS Annual Meeting. Conference Record of the*, vol. 1, 2003, pp. 168–175 vol.1.

[3] T. Meynard and H. Foch, "Multi-level conversion: high voltage choppers and voltage-source inverters," in *Power Electronics Specialists Conference, 1992. PESC '92 Record., 23rd Annual IEEE*, 1992, pp. 397–403 vol.1.

[4] P. Lundberg, "Presentation: Power electronics and energy storage," in *ECPE Seminar on Energy Storage Technologies*, Aachen, 2007.

[5] K. Fujii, U. Schwarzer, and R. W. De Doncker, "Comparison of hard-switched multi-level inverter topologies for STATCOM by loss-implemented simulation and cost estimation," in *Power Electronics Specialists Conference, 2005. PESC '05. IEEE 36th*, Recife, 2005, pp. 340–346.

[6] R. De Doncker and J. Lyons, "The auxiliary resonant commutated pole converter," in *Industry Applications Society Annual Meeting, 1990., Conference Record of the 1990 IEEE*, 7-12 Oct. 1990, pp. 1228–1235vol.2.

[7] C. Turpin, L. Deprez, F. Forest, F. Richardeau, and T. A. Meynard, "A ZVS imbricated cell multilevel inverter with auxiliary resonant commutated poles," *IEEE Transactions on Power Electronics*, vol. 17, no. 6, pp. 874–882, Nov. 2002.

[8] K. Fujii, "Characterization and optimization of soft-switched multi-level converters for STATCOMs," Ph.D. dissertation, RWTH Aachen University, Aachen, Germany, 2007.

[9] P. Koellensperger, C. Meyer, U. Schwarzer, S. Schroder, and R. De Doncker, "High-power converter systems for high-speed generators," in *Power Electronics and Applications, 2005 European Conference on*, 2005, pp. 9 pp.–.

[10] D. J. Swider, "Compressed Air Energy Storage in an Electricity System With Significant Wind Power Generation," *Energy Conversion, IEEE Transaction on*, vol. 22, no. 1, pp. 95–102, 2007.

[11] Electricity Storage Association (ESA). [Online]. Available: http://www.electricitystorage.org

[12] J.-K. Lung, Y. Lu, W.-L. Hung, and W.-S. Kao, "Modeling and dynamic simulations of doubly fed adjustable-speed pumped storage units," *Energy Conversion, IEEE Transaction on*, vol. 22, no. 2, pp. 250–258, 2007.

[13] A. Bocquel and J. Janning, "Analysis of a 300 MW variable speed drive for pump-storage plant applications," in *Power Electronics and Applications, 2005 European Conference on*, 2005, pp. 10 pp.–.

[14] J. McDowall, "Integrating energy storage with wind power in weak electricity grids," *Journal of Power Sources*, vol. 162, no. 2, pp. 959–964, Nov. 2006. [Online]. Available: http://www.sciencedirect.com/science/article/B6TH1-4GYNY5P-4/2/0c46c9de4bc17119dc88da48f9f022d2

[15] L. Walker, "10-MW GTO converter for battery peaking service," *Industry Applications, IEEE Transactions on*, vol. 26, no. 1, pp. 63–72, 1990.

[16] J. McDowall, "High power batteries for utilities - the world's most powerful battery and other developments," in *Power Engineering Society General Meeting, 2004. IEEE*, 2004, pp. 2034–2037 Vol.2.

[17] M. Kamibayashi and K. Tanaka, "Recent sodium sulfur battery applications," in *Transmission and Distribution Conference and Exposition, 2001 IEEE/PES*, vol. 2, 2001, pp. 1169–1173 vol.2.

[18] A. Bito, "Overview of the sodium-sulfur battery for the IEEE Stationary Battery Committee," in *Power Engineering Society General Meeting, 2005. IEEE*, 2005, pp. 1232–1235 Vol. 2.

[19] S. Chusanapiputt, "The studies of transmission system interconnection between PR China and Thailand," in *Power System Technology, 2000. Proceedings. PowerCon 2000. International Conference on*, vol. 1, Perth, WA, Dec. 2000, pp. 169–172.

[20] (2007) ABB ultra high voltage dc systems. [Online]. Available: http://www.abb.com/cawp/gad02181/f665be70ddd7edb3c1256fe2002cdf0d.aspx

[21] (2006) HVDC projects listing, working group on HVDC and FACTS bibliography and records, IEEE transmission and distribution committee.

[22] A. Majeed, H. Karim, N. AlMaskati, and S. Sud. Presentation: The status of the GCC electricity grid system interconnection, IEEE power engineering society, panel session, Denver, june 2004. [Online]. Available: http://www.abb.com/cawp/gad02181/f665be70ddd7edb3c1256fe2002cdf0d.aspx

[23] W. Lu and B.-T. Ooi, "Optimal acquisition and aggregation of offshore wind power by multiterminal voltage-source HVDC," *IEEE Transactions on Power Delivery*, vol. 18, no. 1, pp. 201–206, Jan. 2003.

[24] (2007) German Federal Ministry for the Environment, Nature Conservation and Nuclear Safety: Offshore wind power deployment in Germany, brochure. [Online]. Available: http://www.bmu.de/english/renewable_energy/downloads/doc/print/38731.php

[25] S. Lundberg, "Wind farm configuration and energy efficiency studies - series DC versus AC layouts," Ph.D. dissertation, Chalmers University of Tehcnology, Göteborg, Sweden, 2006.

[26] C. Meyer, "Key components for future offshore DC grids," Ph.D. dissertation, RWTH Aachen University, Aachen, Germany, 2007.

[27] M. E. Baran and N. R. Mahajan, "DC distribution for industrial systems: opportunities and challenges," *IEEE Transactions on Industry Applications*, vol. 39, no. 6, pp. 1596–1601, Nov./Dec. 2003.

[28] D. J. Hammerstrom, "AC versus DC distribution systems: Did we get it right?" in *Power Engineering Society General Meeting, 2007. IEEE*, Tampa, FL, USA, June 2007, pp. 1–5.

[29] D. Nilsson and A. Sannino, "Efficiency analysis of low- and medium-voltage DC distribution systems," in *Power Engineering Society General Meeting, 2004. IEEE*, June 2004, pp. 2315–2321.

[30] G. S. Thandi, R. Zhang, K. Xing, F. C. Lee, and D. Boroyevich, "Modeling, control and stability analysis of a PEBB based DC DPS," *IEEE Transactions on Power Delivery*, vol. 14, no. 2, pp. 497–505, Apr. 1999.

[31] D. Salomonsson and A. Sannino, "Low-voltage DC distribution system for commercial power systems with sensitive electronic loads," *IEEE Transactions on Power Delivery*, vol. 22, no. 3, pp. 1620–1627, July 2007.

[32] B. W. McConnell, "Applications of high temperature superconductors to direct current electric power transmission and distribution," *IEEE Transactions on Applied Superconductivity*, vol. 15, pp. 2142–2145, June 2005.

[33] Y. Ito, Y. Zhongqing, and H. Akagi, "DC microgrid based distribution power generation system," in *Power Electronics and Motion Control Conference, 2004. IPEMC 2004. The 4th International*, vol. 3, Aug. 2004, pp. 1740–1745.

[34] B. Han, G. Ledwich, and G. Karady, "Study on resonant fly-back converter for DC distribution system," *IEEE Transactions on Power Delivery*, vol. 14, no. 3, pp. 1069–1074, July 1999.

[35] J. Jacobs, A. Averberg, S. Schroder, and R. De Doncker, "Multi-phase series resonant DC-to-DC converters: Transient investigations," in *Power Electronics Specialists Conference, 2005. PESC '05. IEEE 36th*, Recife, 2005, pp. 1972–1978.

[36] C. Meyer, M. Kowal, and R. W. De Doncker, "Circuit breaker concepts for future high-power DC-applications," *2005. Fourtieth IAS Annual Meeting. Conference Record of the 2005 Industry Applications Conference*, vol. 2, pp. 860–866, Oct. 2005.

[37] P. Steimer, H. Gruning, J. Werninger, E. Carroll, S. Klaka, and S. Linder, "IGCT-a new emerging technology for high power, low cost inverters," in *Industry Applications Conference, 1997. Thirty-Second IAS Annual Meeting, IAS '97., Conference Record of the 1997 IEEE*, vol. 2, 1997, pp. 1592–1599 vol.2.

[38] P. Koellensperger and R. De Doncker, "Optimized Gate Drivers for Internally Commutated Thyristors (ICTs)," in *Industry Applications Conference, 2006. 41st IAS Annual Meeting. Conference Record of the 2006 IEEE*, vol. 5, 2006, pp. 2269–2275.

[39] P. Koellensperger and R. W. De Doncker, "The Internally Commutated Thyristor - A new GCT with integrated turn-off unit," in *4th International Conference on Integrated Power Systems (CIPS06), Proceedings of the*, 2006.

[40] A. Sawle, C. Blake, and D. Maric, "Novel power mosfet packaging technology doubles power density in synchronous buck converters for next generation microprocessors," in *Applied Power Electronics Conference and Exposition, 2002. APEC 2002. Seventeenth Annual IEEE*, vol. 1, 2002, pp. 106–111 vol.1.

[41] Y. Li, A. Huang, and F. Lee, "Introducing the emitter turn-off thyristor (ETO)," in *Industry Applications Conference, 1998. Thirty-Third IAS Annual Meeting. The 1998 IEEE*, vol. 2, 1998, pp. 860–864 vol.2.

[42] D. Piccone, R. De Doncker, J. Barrow, and W. Tobin, "The MTO thyristor – a new high power bipolar MOS thyristor," in *Industry Applications Conference, 1996. Thirty-First IAS Annual Meeting, IAS '96., Conference Record of the 1996 IEEE*, vol. 3, 1996, pp. 1472–1473 vol.3.

[43] D. Detjen, S. Schroder, T. Plum, and R. De Doncker, "Novel MTO-design based on silicon-silicon bonding," in *Power Electronics Congress, 2002. Technical Proceedings. CIEP 2002. VIII IEEE International*, 2002, pp. 27–32.

Digital Control Generations -- Digital Controls for Power Electronics through the Third Generation

Philip T. Krein

Grainger Center for Electric Machinery and Electromechanics
Department of Electrical and Computer Engineering
University of Illinois at Urbana-Champaign
Urbana, Illinois   61801   USA

Abstract – **Digital control in power electronics can be divided into three "generations." First-generation digital controls use digital "outside the loop" in communications, setup, and supervisory roles. Second generation digital controls use digital processes "inside the loop," including discrete-time feedback loops and sometimes even digital signal processing. Today, first-generation digital methods are expanding quickly, as new communication protocols and adjustable analog loops become common. Even companies that continue to design analog controls for power electronics often include these types of digital processes. Second-generation digital controls are a hot topic right now, as real-time digital controllers become feasible. In third-generation digital controls, the digital process functions directly with individual switches to push performance up to the physical limits of power electronics. A digital switch decides when it must turn on or off. The control is on direct switch timing rather than a converter duty ratio or a setting. Extreme performance is possible with this approach, such as converters that do not exhibit output disturbances when confronted with load or line step changes. The talk compares these different arenas, all of which are current active topics in power electronics, and shows what can become possible as the third generation develops[1].**

I. INTRODUCTION

Digital control in power electronics, treated now as a hot topic area and debated in many forums, has a long history. The general approaches and contexts are perhaps better understood by dividing the developments into three "generations." In this paper, the concept of digital control generations is introduced. The emphasis is on third-generation techniques that are beginning to capture interest in research laboratories. Although the general value of digital control continues to be a matter of debate, the broad question of how control adds value in a power electronic system, and whether digital techniques offer special value, is the underlying issue that motivates this work. The concepts in this paper may be of value in discussions of digital control.

As digital control is discussed, it is important to keep in mind the fundamental analog processing to be accomplished. We are not free to create arbitrary digital representations of energy, in contrast to most communications and information

processing applications. Power electronics in the end is characterized by large-signal nonlinear systems with analog functions. No matter how much digital processing is involved, its merit is always determined by the ability to better perform these analog functions.

The generations are defined as follows:

- First-generation digital control: digital processing *outside* a control loop, in a management or supervisory role.
- Second-generation digital control: digital processing *inside* a control loop.   The ultimate formulation includes digital loop designs and real-time control processes.
- Third-generation digital control: digital processing is responsible for the moment-by-moment direct action of active switching devices in a converter. The ultimate formulation is a *digital switch* with built-in computational capability that functions in real time as the device operates.

Although the digital control generations defined in this paper have a certain time evolution sequence, they are not meant to imply obsolescence. First-generation digital controls, which have the longest history, are quickly becoming dominant and are not likely to leave the stage in the foreseeable future. Second-generation digital controls seem to be at the heart of present debates. Third-generation controls open the way to unique performance improvements, but are rare.

II. FIRST-GENERATION DIGITAL CONTROL IN POWER ELECTRONICS

The earliest power electronics controllers, dating to the TL494 and similar chips, were some of the first mixed-mode integrated circuits. These ICs include simple logic along with oscillators and amplifiers, and thus combine digital and analog functions. In this sense, digital control has been a fundamental aspect of power electronics for 40 years or more.

In this paper, first-generation digital controls are assumed to be more than just mixed-mode circuits. In first-generation digital controls, a digital process *manages* a power electronic process. The objectives typically include communication, programming, or protection. Motor drives were an early example of first-generation controls. When electronic

---

[1] This paper is provided under the Distinguished Lecturer Program of the IEEE Power Electronics Society.   This material is based in part upon work supported by the U.S. National Science Foundation under Grant No. ECS-0621643.

978-1-4244-0644-9/07/$25.00 ©2007 IEEE

adjustable-speed drives emerged in the 1970s, many already had displays and internal interactions governed by digital logic. Modern drives are designed with dedicated digital signal processors [1, 2], which often manage nearly all the power electronics through computer control. A more recent example is the PMBus™ architecture for power supply communications and interaction [3].

Today, first-generation digital controls for power electronics are widespread. In addition to the PMBus architecture, various smart battery charging interfaces and other communication configurations are becoming common. Dc-dc converters for processors often have external digital settings to support adaptive output voltage. Even those manufacturers "dedicated" to analog power management have embraced digital communication and control interfaces. As a result, first-generation techniques are not really part of the present debate about digital control, and should be taken as a routine extension of other control methods in power electronics.

First-generation controls provide a wide array of advantages. Two crucial advantages are the ability to managing event-driven actions and the ability to provide numerical settings. Event-driven actions, such as responses to overloads, transitions among various modes, or even the ability to control different converter topologies bring fundamental performance advantages to this class of control. Since real-time performance demands are avoided in digital part of the system, these capabilities are possible without compromise in dynamic performance. Numerical settings, including gains, output reference values, or operating frequency add software-like flexibility to hardware devices. Other potential advantages include communications interfaces and control buses, memory for various programming functions, and the ability for IC designers to add new features as blocks. The latter allows a vendor to create comprehensive product families from a single base design.

Many present first-generation implementations emphasize communication and basic settings, but a range of opportunities remain. Potential innovations include variable-gain tuning, in which gain settings depend on actual line or load conditions, frequency tuning to or from resonances, and various types of control tuning. On-line calibration and active digital trimming, common in many drive applications, can be extended to most power converters. The use of various frequency-domain techniques, such as Fourier Transforms for compensation [4], nonlinear filters [5], and more sophisticated signal processing for fault detection, has been a topic of previous study that is well worth a closer look.

III. SECOND-GENERATION DIGITAL CONTROLS IN POWER ELECTRONICS

In second-generation control, the digital process moves inside the control loop and operates a power converter in real time. Like first-generation approaches, the basic technique is not new. Once motor drives moved to digital PWM processes more than a decade ago, many of them used complete digital loops for operation and control. At the research level, complete digital controls were presented almost twenty years ago [6]. In motor drive applications, computation time is usually ample and computation cost is a modest fraction of total system cost. The net result has been early adoption of all-digital implementations in that industry.

In power supplies and dc-dc converters, real-time operation tends to work against second-generation designs, which are often characterized by intensive analog-digital (A-D) conversion requirements and short computation time windows. The development of second-generation digital controls for these applications is perhaps the most active topic in digital control for power electronics and is the subject of controversy. Many designers still question the value of digital implementations compared to conventional analog hardware.

To see the rate challenge, consider a counter-based digital PWM generator intended to support 250 kHz switching for a dc-dc converter. If this device provides 0.1% pulse-width resolution, its clock must run at 250 MHz or more. A PWM generator to support 500 kHz switching with 16-bit pulse-width resolution demands 31 ps time resolution. This requires a 33 GHz clock. Resolution and operating requirements such as these, which have little meaning in the context of analog controls, quickly become unwieldy in a digital application. Digital controls of this type can chatter and operate in limit cycles [7], although known methods such as integral controls can help avoid the problems.

Many second-generation digital controls involve a direct mapping from analog implementations to discrete-time implementations. This practice supports the numerical setting advantages of digital controls, but does not fundamentally alter performance compared to analog implementations. Much of the controversy about digital control in power electronics today is concerned with whether a discrete-time implementation offers special advantages over an analog version given a conventional average-model controller. In this paper, the controversy is not entirely germane: there are valid analog controls that use external digital management, as in first-generation digital control, and evolution towards digital control need not place real-time digital processing inside a loop.

A. Discussion of sampling issues

Given the linkage between second-generation digital control and A-D/D-A conversion, sampling challenges become a significant aspect. One relatively misunderstood aspect is the Nyquist rate, which reflects the results of sampling theory. As is well known [8], a bandlimited signal can be reconstructed perfectly from properly selected samples taken at higher than the Nyquist rate. This rate is normally taken as half the period associated with the signal band limit. It is tempting to infer that any periodic signal can be reconstructed from samples taken at half the period, but this is not correct.

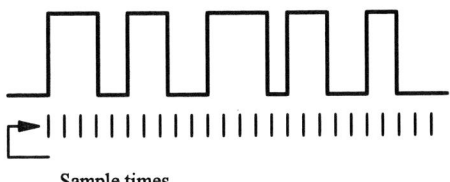

Sample times

Fig. 1. A square wave with arbitrary duty ratio cannot be reconstructed from uniform samples. Here samples taken five times per switching cycle are not adequate.

Consider a square wave of unknown (but constant) duty ratio, as might be measured as the voltage drop across a switch in a power converter or as the ESR jump on a capacitor. As Fig. 1 shows, *no* set of uniform samples, no matter how often they are taken, will permit the waveform to be sampled and reconstructed perfectly. This is because for arbitrary duty ratio the probability of sampling exactly at the switching instant is zero. What does the Nyquist rate not apply here? A square wave is not a bandlimited signal, so conventional sampling theory does not apply.

Curiously, the fundamental Nyquist rate problem associated with the square wave in Fig. 1 can be circumvented. The integral of the square wave yields a triangle wave, as in Fig. 2. This waveform, although also not bandlimited, is easy to reconstruct if the sampling rate is sufficient to ensure two samples during each rising portion and two during each falling portion. This means that for any duty ratio between 0 and 1, a sampling frequency suitable to permit perfect reconstruction can always be found. Indeed, the underlying square wave can be reconstructed from the same samples by taking the derivative of the computed triangle. The sampling frequency is not really a Nyquist rate in the conventional sense, but waveform reconstruction is possible. It is also clear that non-uniform samples can be used to advantage: if samples are taken just before and just after each switch operation, the information needed to reconstruct the waveform will be available.

The possibility of reconstructing square waves from limited samples gives rise to the notion of integral sampling [9], but in contrast to the hold process in [9], samples are to be taken during the integration process. The addition of a single analog block – the integrator – adds considerable signal processing capability to a power converter since it support signal reconstruction from a small number of well-place samples. Notice that the process uses general knowledge about the shape of the waveform (a square wave or triangle) instead of knowledge about its frequency limits. In effect, a time-domain sampling theorem has been identified in place of a more conventional frequency-domain theorem.

B. Real-time limits for second-generation controls

In second-generation designs, real-time digital control must push the limits. Concerns include the conversion speed of A-D and D-A converters, the time needed for computation, and time needed to obtain low-noise samples. Precision, both in terms of time resolution and quantization, becomes an

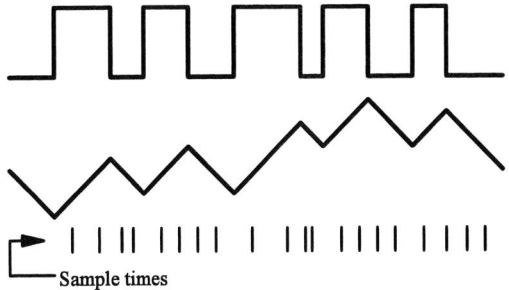

Sample times

Fig. 2. The integral of the square wave (a triangle) can be reconstructed if two samples are available during each rising and falling portion.

important issue. A fundamental question is how a control can determine whether the output has reached the desired value and that the converter should enter steady state.

A important way to manage extreme resolution requirements is to employ dithering or noise shaping methods [10, 11]. In both approaches, a large number of switching periods is used as a group to deliver the desired pulse width. For example, a PWM process with only 10% resolution can deliver effective 1% output resolution if a group of ten cycles is employed. In dithering, the local duty ratio variation needed to deliver higher resolution is randomized. Thus, a desired 54% duty ratio in a process that can deliver only 50% and 60% values would be obtained by random combinations of 50% and 60% in the right proportions. In noise shaping, the process is not random, and instead is characterized by a high-pass filter that shifts the output quantization noise away from the baseband duty ratio modulation.

As microprocessors improve, real-time computational limits become less important in second-generation controls. Today's DSPs, for instance, perform multiple complicated arithmetic steps in a single clock cycle. At clock frequencies above 100 MHz, there is time for several hundred computations per switching period, even for dc-dc converters operating at up to 500 kHz. One commercial product [12] processes six inverter channels for high-fidelity audio output based on a complete second-generation implementation. Other vendors provide sophisticated adaptive controls in second-generation devices [13]. These examples suggest that second-generation digital controls will continue to be an area of active growth for years to come.

IV. THIRD-GENERATION DIGITAL CONTROLS IN POWER ELECTRONICS

In any switching power converter, the true control actuation is the time at which switches operate. At the most basic level, the control question is to determine when to operate each switching device in the network to achieve a set of performance objectives. Beyond the implementation of real-time digital control in a closed-loop power converter is the challenge of direct switch control to address this question. The issue can be considered in a manner analogous to averaging: in second-generation digital controls, the control generally computes a desired duty ratio. A counter implements the final step of a PWM process. The control is altering pulse width, rather than direct timing.

Third-generation digital controls act on information to determine specific time-domain action of each switching device. The ultimate objective is the *digital switch*, an intelligent switching device that operates at just the right times to achieve objectives. The objectives could represent any performance aspect needed by the user. Many of them do not lend themselves to analog controls. For example, in a given dc-dc converter, the detailed performance objectives might include the following:

- Deliver an output voltage that is within 0.5% of a specified reference.
- Do not allow the current to exceed a given dynamic limit.
- Deliver the voltage while minimizing internal converter losses.
- Avoid certain frequency bands to prevent noise problems.
- Respond to load changes as rapidly as possible while continuing to meet output tolerance requirements.

Objectives like these mix steady-state, dynamic, and protection requirements. They imply computation challenges such as those associated with loss minimization and electromagnetic interference (EMI). At the most basic level, performance objectives translate into difficult control requirements: determine when to operate the next switching device in a sequence, such that loss is minimized, EMI is avoided, and steady-state requirements are met. In general, it might be possible to formulate an optimal control problem for a set of objectives:

Optimal control problem formulation

Given a set of n switches and a time interval T, find times $t_{i,n}$ for these switches to minimize a performance objective function $J(x,t)$ that is a function of states x and time.

With enough constraints and well-defined objective functions, this problem can be solved. For example, in a dc-dc converter in which there is one active switch and the switching period is constrained to be fixed, there is a unique time that delivers the correct output in steady state. This is just the well-known average duty ratio. The general problem deals with dynamics rather than steady state, and a suitable problem formulation should have the switching frequency as a dependent variable rather than a constraint, but at least the steady-state operation is relatively well defined. The general problem, in which there are many performance objectives representing both static and dynamic requirements, may not be tractable, however.

Third-generation controls are the subject of present research in a few groups. An early example that follows the general approach is given in [14], although the geometric controls introduced much earlier by Burns [15] are straightforward to represent in terms of third-generation digital methods. Dead-time optimization [16, 17] is a partial

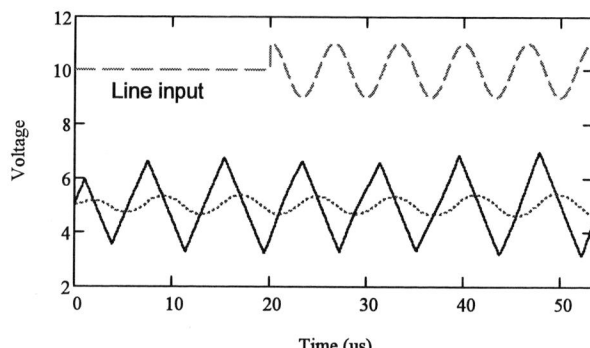

Fig. 3. Hysteresis controlled buck converter responding when a 150 kHz line disturbance is imposed at 20 µs. Top trace: input voltage. Triangle: inductor current. Dotted trace: output voltage across capacitor. Output capacitor is small to show ripple.

third-generation example, in which detailed switch timing is controlled inside a loop to minimize loss.

Fig. 3 provides a hint – based on an analog control – of what might be possible with third-generation controls. The waveforms shown are the input voltage, output voltage, and output current for a buck converter. The converter operates with a hysteresis control to maintain the output at 5 V. The steady-state switching frequency is about 120 kHz. At a time of 20 µs, a 150 kHz sinusoidal disturbance is imposed at the input. The output voltage is virtually unaffected by this: the hysteresis control is "choosing" switch operation times to null out this disturbance, and succeeds even though the disturbance is faster than the effective switching frequency.

Fig. 4, from [18], shows the disturbance response for a buck converter operating from a high-performance second-generation digital control. In this case, a 25% step input disturbance is detected and the duty ratio is immediately adjusted to the new correct value. Even so, the output voltage shows a second-order response to the step, and peaks at about 5% above its nominal output. Fig. 5, also from [18], shows response for the same converter to the same disturbance with a third-generation digital control. Now the switching times are being computed to cancel the disturbance in the output voltage, and the transient is almost impossible to see in the traces.

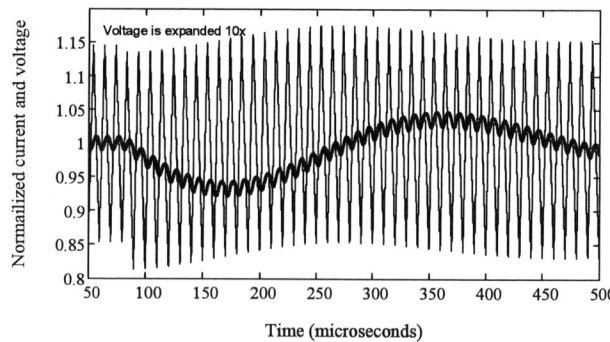

Fig. 4. Line step response of buck converter given instant change to the new duty ratio. From [18].

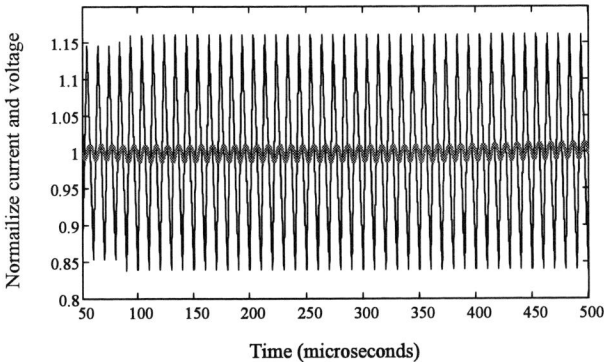

Fig. 5. Line step response of buck converter based on computed null condition. Light trace: inductor current. Bold trace: capacitor voltage, expanded ten times. From [18].

Ref. [18] provides additional examples of third-generation methods, including circuits that maintain undisturbed output during extreme line and load transient conditions. Hardware implementation is possible, and an early discussion appeared in [19].

V. CONCLUSION

Although there is continuing debate about digital control in power electronics, the reality is more a question of degree than a distinction between analog and digital controls. First-generation digital controls for power electronics, with their supervisory and management functions outside the control loop, are rapidly becoming ubiquitous. In applications in which simple low-cost analog blocks can provide the necessary performance, digital augmentation adds features and capabilities at minimal cost. Second-generation digital controls perform complete closed-loop processes in real time, and this is where challenges of digital controls come into focus. It seems unlikely that first-generation methods will become obsolete any time soon, and the present debate is more about the pace of migration to the second generation than it is about digital control in general.

Third-generation digital controls, in contrast, offer performance possibilities that are extremely difficult to achieve without fast computation. Since they determine actual moment-by-moment switch action in real time, third-generation controls can eliminate disturbances and compensate for fast variations. They can also optimize losses and take advantage of nonlinear control methods for power converters.

REFERENCES

[1] Texas Instruments, "C2000 Controllers Overview," [Online] http://focus.ti.com/dsp/docs/dspplatformscontenttp.tsp?sectionId=2&familyId=110&tabId=510.

[2] Analog Devices, "Analog Devices Part Number – ADSP-21991," [Online] http://www.analog.com/en/prod/0,2877,ADSP%252D21991,00.html.

[3] R. V. White, D. Durant, "Understanding and Using PMBus™ Data Formats," *Proc. IEEE Applied Power Electronics Conf.*, 2006, pp. 834-840.

[4] S. M. Williams, R. G. Hoft, "Adaptive frequency domain control of PWM switched power line conditioner," *IEEE Trans. Power Electronics*, vol. 6, no. 4, pp. 665-670, 1991.

[5] W. C. Karl, S. B. Leeb, L. A. Jones, J. L. Kirtley, G. C. Verghese, "Applications of a class of nonlinear filters to problems in power electronics," *Rec., IEEE Power Electronics Specialists Conference*, 1990, pp. 35-62.

[6] H. Matsuo, F. Kurokawa, K. Higashi, "Dynamic characteristics of the digitally controlled DC-DC converter," *IEEE Trans. Power Electronics*, vol. 4, no. 4, pp. 419-426, 1989.

[7] S.R. Sanders, "On limit cycles and the describing function method in periodically switched circuits," *IEEE Trans. Circuits and Systems I: Fundamental Theory and Appl.*, vol. 40, pp. 564-572, Sept. 1993.

[8] P. Z. Peebles, *Communication System Principles*. Reading, MA: Addison-Wesley, 1976.

[9] A. G. J. Holt, J. J. Hill, R. Linggard, "Integral sampling," *Proc. IEEE*, vol. 61, no. 5, pp. 679-680, May 1973.

[10] S. R. Sander, A. V. Peterchev, "Quantization resolution and limit cycling in digitally controlled PWM converters," *IEEE Trans. Power Electronics*, vol. 18, no. 1, pp. 301-308, 2003.

[11] S. R. Norsworthy, R. Schreier, G. C. Temes, *Delta-Sigma Data Converters*. New York: IEEE Press, 1996.

[12] "Freescale's Symphony™ Class D Digital Amplifier Solution. [Online] http://www.freescale.com/files/wireless_comm/doc/white_paper/SYMPHONYCLSDWP.pdf

[13] "Integrated Step-down DC-DC Converter with Power Management," Zilker Labs. [Online] available, http://www.zilkerlabs.com/121/ZL2105_Overview.htm.

[14] I. Celanovic, I. Milosavljevic, D. Boroyevich, and J. Guo, "A new distributed digital controller for the next generation of power electronics building blocks," *IEEE Applied Power Electronics Conference*, 2000, pp. 889-894.

[15] W. W. Burns, III and T. G. Wilson, "Analytic Derivation and Evaluation of a State-Trajectory Control Law for DC-to-DC Converters," in *IEEE Power Electronics Specialists Conf. Rec.*, 1977, pp. 70-85.

[16] J. Kimball, P. T. Krein, "Real-time optimization of dead time for motor control inverters," in *Rec., IEEE Power Electronics Specialists Conf.*, 1997, pp. 597-600.

[17] V. Yousefzadeh, D. Maksimović, "Sensorless optimization of dead times in dc-dc converters with synchronous rectifiers," in *Proc. IEEE Applied Power Electronics Conference*, 2005, pp. 911-917.

[18] P. T. Krein, "Feasibility of geometric digital controls and augmentation for ultrafast dc-dc converter response," *Proc. IEEE Workshop on Computers in Power Electronics (COMPEL)*, 2006, pp. 48-56.

[19] G. E. Pitel, P. T. Krein, " Transient reduction of dc-dc converters via augmentation and geometric control," *Proc. IEEE Power Electronics Specialists Conf.*, 2007, pp. 1652-1657.

Power Electronics and Control of Renewable Energy Systems

F. Iov, M. Ciobotaru, D. Sera, R. Teodorescu, F. Blaabjerg
Aalborg University, Institute of Energy Technology
Pontoppidanstraede 101, DK-9220 Aalborg East, Denmark
fi@iet.aau.dk, mpc@iet.aau.dk, des@iet.aau.dk, ret@iet.aau.dk, fbl@iet.aau.dk

Abstract – **The global electrical energy consumption is still rising and there is a demand to double the power capacity within 20 years. The production, distribution and use of energy should be as technological efficient as possible and incentives to save energy at the end-user should also be set up. Deregulation of energy has in the past lowered the investment in larger power plants, which means the need for new electrical power sources may be very high in the near future. Two major technologies will play important roles to solve the future problems. One is to change the electrical power production sources from the conventional, fossil (and short term) based energy sources to renewable energy resources. Another is to use high efficient power electronics in power generation, power transmission/distribution and end-user application. This paper discuss some of the most emerging renewable energy sources, wind energy and photovoltaics, which by means of power electronics are changing from being minor energy sources to be acting as important power sources in the energy system.**

I. Introduction

In classical power systems, large power generation plants located at adequate geographical places produce most of the power, which is then transferred towards large consumption centers over long distance transmission lines. The system control centers monitor and regulate the power system continuously to ensure the quality of the power, namely frequency and voltage. However, now the overall power system is changing, a large number of dispersed generation (DG) units, including both renewable and non-renewable sources such as wind turbines, wave generators, photovoltaic (PV) generators, small hydro, fuel cells and gas/steam powered Combined Heat and Power (CHP) stations, are being developed [1], [2] and installed. A wide-spread use of renewable energy sources in distribution networks and a high penetration level will be seen in the near future many places. E.g. Denmark has a high power capacity penetration (> 20%) of wind energy in major areas of the country and today 18% of the whole electrical energy consumption is covered by wind energy. The main advantages of using renewable energy sources are the elimination of harmful emissions and inexhaustible resources of the primary energy. However, the main disadvantage, apart from the higher costs, e.g. photovoltaic, is the uncontrollability. The availability of renewable energy sources has strong daily and seasonal patterns and the power demand by the consumers could have a very different characteristic. Therefore, it is difficult to operate a power system installed with only renewable generation units due to the characteristic differences and the high uncertainty in the availability of the renewable energy sources. This is further strengthened as no real large energy storage systems exist.

The wind turbine technology is one of the most emerging renewable energy technologies. It started in the 1980'es with a few tens of kW production power to today with multi-MW size wind turbines that are being installed. It also means that wind power production in the beginning did not have any impact on the power system control but now due to their size they have to play an active part in the grid. The technology used in wind turbines was in the beginning based on a squirrel-cage induction generator connected directly to the grid. By that power pulsations in the wind are almost directly transferred to the electrical grid. Furthermore there is no control of the active and reactive power, which typically are important control parameters to regulate the frequency and the voltage. As the power range of the turbines increases those control parameters become more important and it is necessary to introduce power electronics [3] as an interface between the wind turbine and the grid. The power electronics is changing the basic characteristic of the wind turbine from being an energy source to be an active power source. The electrical technology used in wind turbine is not new. It has been discussed for several years [6]-[50] but now the price pr. produced kWh is so low, that solutions with power electronics are very attractive.

This paper will first discuss the basic development in power electronics and power electronic conversion. Then different wind turbine configurations will be explained both aerodynamically and electrically. Also different control methods will be shown for a wind turbine. They are now also installed in remote areas with good wind conditions (off-shore, on-shore) and different possible configurations are shown and compared. Next the PV-technology is discussed including the necessary basic power electronic conversion. Power converters are given and more advanced control features described. Finally, a general technology status of the wind power and the PV technology is presented demonstrating still more efficient and attractive power sources for the future.

978-1-4244-0644-9/07/$25.00 ©2007 IEEE

II. MODERN POWER ELECTRONICS

Power electronics has changed rapidly during the last thirty years and the number of applications has been increasing, mainly due to the developments of the semiconductor devices and the microprocessor technology. For both cases higher performance is steadily given for the same area of silicon, and at the same time they are continuously reducing in price. A typical power electronic system, consisting of a power converter, a load/source and a control unit, is shown in Fig. 1.

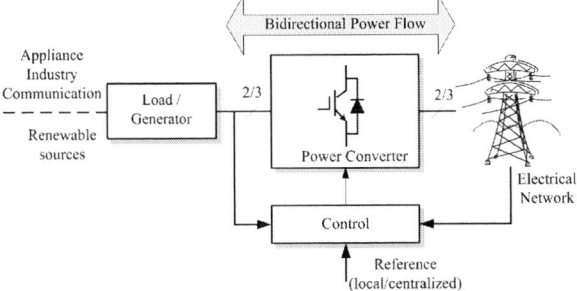

Fig. 1. Power electronic system with the grid, load/source, power converter and control.

The power converter is the interface between the load/generator and the grid. The power may flow in both directions, of course, dependent on topology and applications.

Three important issues are of concern using such a system. The first one is reliability; the second is efficiency and the third one is cost. For the moment the cost of power semiconductor devices is decreasing 1÷5 % every year for the same output performance and the price pr. kW for a power electronic system is also decreasing. An example of a mass-produced and high competitive power electronic system is an adjustable speed drive (ASD). The trend of weight, size, number of components and functions in a standard Danfoss Drives A/S frequency converter can be seen in Fig. 2. It clearly shows that power electronic conversion is shrinking in volume and weight. It also shows that more integration is an important key to be competitive as well as more functions become available in such a product.

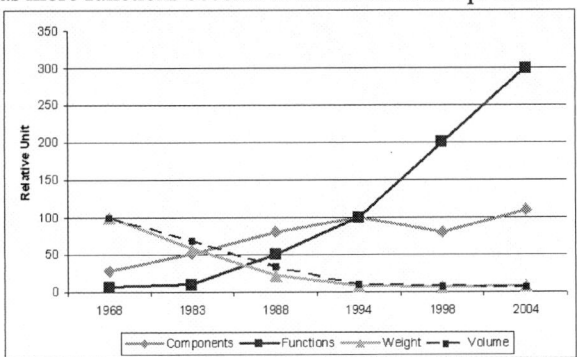

Fig. 2. Development of standard adjustable speed drives for the last four decades.

The key driver of this development is that the power electronic device technology is still undergoing important progress.

Fig. 3 shows different power devices and the areas where the development is still going on.

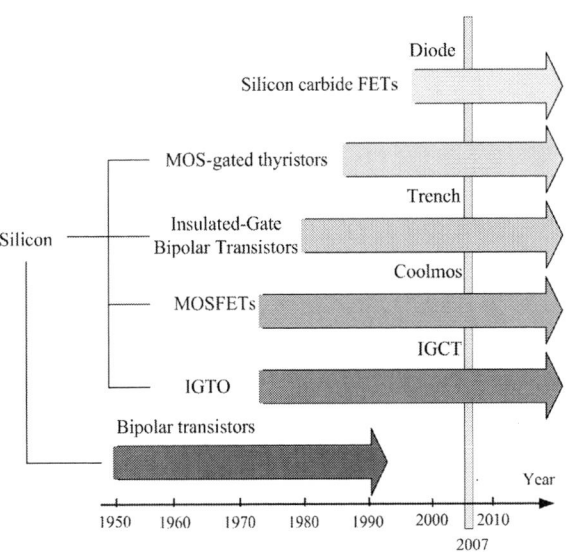

Fig. 3. Development of power semiconductor devices in the past and in the future [36].

The only power device which is not under development any more is the silicon-based power bipolar transistor because MOS-gated devices are preferable in the sense of easy control. The breakdown voltage and/or current carrying capability of the components are also continuously increasing. Important research is going on to change the material from silicon to silicon carbide, which may dramatically increase the power density of power converters.

III. WIND ENERGY CONVERSION

Wind turbines capture power from the wind by means of aerodynamically designed blades and convert it to rotating mechanical power. The number of blades is normally three. As the blade tip-speed should be lower than half the speed of sound the rotational speed will decrease as the radius of the blade increases. For multi-MW wind turbines the rotational speed will be 10-15 rpm. The most weight efficient way to convert the low-speed, high-torque power to electrical power is to use a gear-box and a standard fixed speed generator as illustrated in Fig. 4.

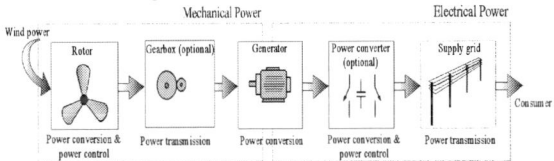

Fig. 4. Converting wind power to electrical power in a wind turbine [19].

The gear-box is optional as multi-pole generator systems are possible solutions. Between the grid and the generator a power converter can be inserted.

The possible technical solutions are many and a technological roadmap starting with wind energy/power and converting the mechanical power into electrical power is shown in Fig. 5. The electrical output can either be ac or dc. In the last case a power converter will be used as interface to the grid.

227

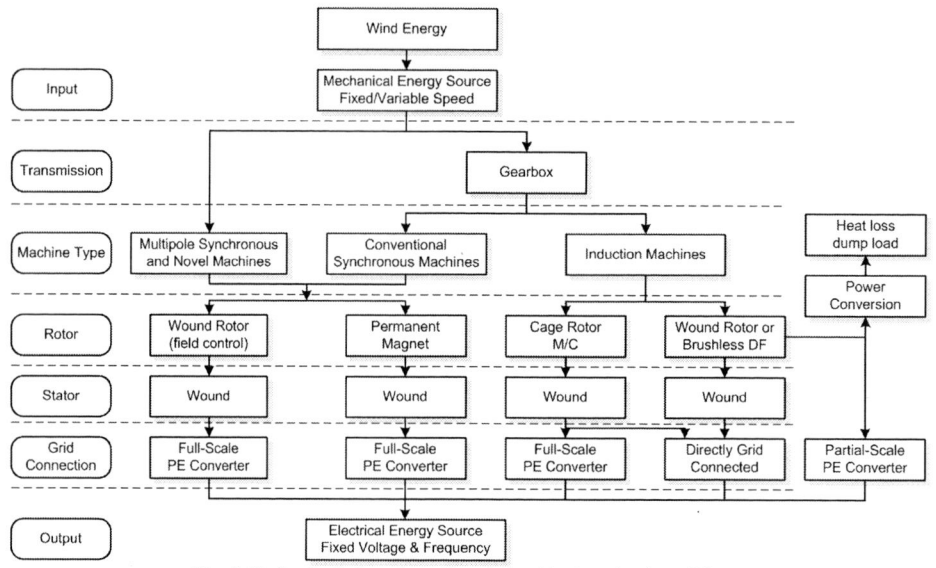

Fig. 5. Technological roadmap for wind turbine's technology [3].

A. Control methods for wind turbines

The development in wind turbine systems has been steady for the last 25 years and four to five generations of wind turbines exist and it is now proven technology. It is important to be able to control and limit the converted mechanical power at higher wind speed, as the power in the wind is a cube of the wind speed. The power limitation may be done either by stall control (the blade position is fixed but stall of the wind appears along the blade at higher wind speed), active stall (the blade angle is adjusted in order to create stall along the blades) or pitch control (the blades are turned out of the wind at higher wind speed) [6], [7]. The basic output characteristics of these three methods of controlling the power are summarized in Fig. 6.

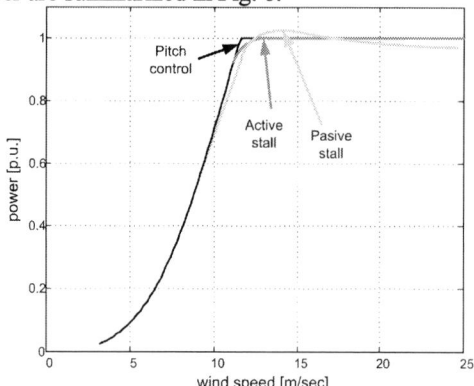

Fig. 6. Power characteristics of different fixed speed wind turbine systems.

Another control variable in wind turbine system is the speed. Based on this criterion the wind turbines are classified into two main categories [6], [7]; namely fixed speed and variable speed wind turbines respectively.

A fixed speed wind turbine has the advantages of being simple, robust, reliable, well proven and with low cost of the electrical parts. Its direct drawbacks are the uncontrollable reactive power consumption, mechanical stress and limited power quality control. Due to its fixed speed operation, wind speed fluctuations are converted to mechanical torque fluctuations,

beneficially reduced slightly by small changes in generator slip, and transmitted as fluctuations into electrical power to the grid. The power fluctuations can also yield large voltage fluctuations in the case of a weak grid and thus, significant line losses [6], [7].

The variable speed wind turbines are designed to achieve maximum aerodynamic efficiency over a wide range of wind speed. By introducing the variable speed operation, it is possible to continuously adapt (accelerate or decelerate) the rotational speed of the wind turbine to the wind speed v, in such a way that tip speed ratio is kept constant to a predefined value corresponding to the maximum power coefficient. Contrary to a fixed speed system, a variable speed system keeps the generator torque nearly constant, the variations in wind being absorbed by the generator speed changes.

Seen from the wind turbine point of view, the most important advantages of the variable speed operation compared to the conventional fixed speed operation are: reduced mechanical stress on the mechanical components such as shaft and gearbox, increased power capture and reduced acoustical.

Additionally, the presence of power converters in wind turbines also provides high potential control capabilities for both large modern wind turbines and wind farms to fulfill the high technical demands imposed by the grid operators [6], [7], [8] and [23], such as: controllable active and reactive power (frequency and voltage control); quick response under transient and dynamic power system situations, influence on network stability and improved power quality.

B. Wind Turbine Concepts

The most commonly applied wind turbine designs can be categorized into four wind turbine concepts. The main differences between these concepts concern the generating system and the way in which the aerodynamic efficiency of the rotor is limited during above the rated value in order to prevent overloading. These concepts are presented in detail in the following paragraphs.

1) Fixed Speed Wind Turbines (WT Type A)

This configuration corresponds to the so called Danish concept that was very popular in 80's. This wind turbine is fixed speed controlled machine, with asynchronous squirrel cage induction generator (SCIG) directly connected to the grid via a transformer as shown in Fig. 7.

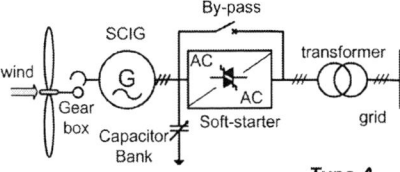

Fig. 7. Fixed speed wind turbine with directly grid connected squirrel-cage induction generator.

This concept needs a reactive power compensator to reduce (almost eliminate) the reactive power demand from the turbine generators to the grid. It is usually done by continuously switching capacitor banks following the production variation (5-25 steps) Smoother grid connection occurs by incorporating a soft-starter. Regardless the power control principle in a fixed speed wind turbine, the wind fluctuations are converted into mechanical fluctuations and further into electrical power fluctuations. These can yield to voltage fluctuations at the point of connection in the case of a weak grid. Because of these voltage fluctuations, the fixed speed wind turbine draws varying amounts of reactive power from the utility grid (in the case of no capacitor bank), which increases both the voltage fluctuations and the line losses.

Thus, the main drawbacks of this concept are: does not support any speed control, requires a stiff grid and its mechanical construction must be able to support high mechanical stress caused by wind gusts.

2) Partial Variable Speed Wind Turbine with Variable Rotor Resistance (WT Type B)

This configuration corresponds to the limited variable speed controlled wind turbine with variable rotor resistance, known as OptiSlip (Vestas™) as presented in Fig. 8.

It uses a wound rotor induction generator (WRIG) and it has been used by the Danish manufacturer Vestas Wind Systems since the mid 1990's.

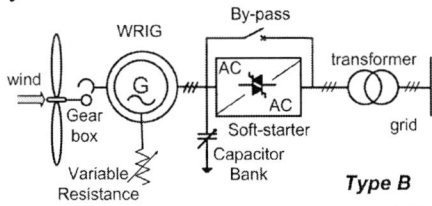

Fig. 8. Partial variable speed wind turbine with variable rotor resistance.

The generator is directly connected to the grid. The rotor winding of the generator is connected in series with a controlled resistance, whose size defines the range of the variable speed (typically 0-10% above synchronous speed). A capacitor bank performs the reactive power compensation and smooth grid connection occurs by means of a soft-starter. An extra resistance is added in the rotor circuit, which can be controlled by power electronics Thus, the total rotor

resistance is controllable and the slip and thus the power output in the system are controlled. The dynamic speed control range depends on the size of the variable rotor resistance. Typically the speed range is 0-10% above synchronous speed. The energy coming from the external power conversion unit is dumped as heat loss. In [24] an alternative concept using passive component instead of a power electronic converter is described. This concept achieves 10% slip, but it does not support controllable slip.

3) Variable Speed WT with partial-scale frequency converter (WT Type C)

This configuration, known as the doubly-fed induction generator (DFIG) concept, corresponds to the variable speed controlled wind turbine with a wound rotor induction generator (WRIG) and partial-scale frequency converter (rated to approx. 30% of nominal generator power) on the rotor circuit as shown in Fig. 9.

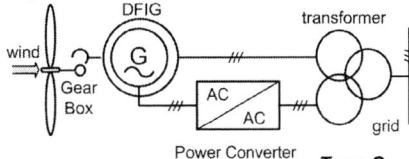

Fig. 9. Variable speed wind turbine with partial scale power converter.

The stator is directly connected to the grid, while a partial-scale power converter controls the rotor frequency and thus the rotor speed. The power rating of this partial-scale frequency converter defines the speed range (typically ±30% around synchronous speed). Moreover, this converter performs the reactive power compensation and a smooth grid connection. The control range of the rotor speed is wide compared to that of OptiSlip. Moreover, it captures the energy, which in the OptiSlip concept is burned off in the controllable rotor resistance. The smaller frequency converter makes this concept attractive from an economical point of view. Moreover, the power electronics is enabling the wind turbine to act as a more dynamic power source to the grid. However, its main drawbacks are the use of slip-rings and the protection schemes in the case of grid faults.

4) Variable Speed Wind Turbine with Full-scale Power Converter (WT Type D)

This configuration corresponds to the full variable speed controlled wind turbine, with the generator connected to the grid through a full-scale frequency converter as shown in Fig. 10.

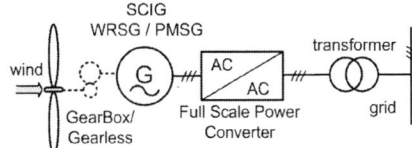

Fig. 10. Variable speed wind turbine with full-scale power converter.

The frequency converter performs the reactive power compensation and a smooth grid connection for the entire speed range. The generator can be electrically excited (wound rotor synchronous generator WRSG) or

229

permanent magnet excited type (permanent magnet synchronous generator PMSG). The stator windings are connected to the grid through a full-scale power converter.

Some variable speed wind turbines systems are gearless – see dotted gearbox in Fig. 10. In these cases, a bulky direct driven multi-pole generator is used. The wind turbine companies Enercon, Siemens Wind Power, Made and Lagerwey are examples of manufacturers using this configuration.

C. System comparison of wind turbines.

Comparing the different wind turbine topologies in respect to their performances will reveal a contradiction between cost and performance to the grid [5], [7]. A technical comparison of the main wind turbine concepts, where issues on grid control, cost, maintenance, internal turbine performance is given in Table 1.

Table 1. System comparison of wind turbine configurations.

System	Type A	Type B	Type C	Type D
Variable speed	No	No	Yes	Yes
Control active power	Limited	Limited	Yes	Yes
Control reactive power	No	No	Yes	Yes
Short circuit (fault-active)	No	No	No/Yes	Yes
Short circuit power	contribute	contribute	contribute	limit
Control bandwidth	1-10 s	100 ms	1 ms	0.5-1 ms
Standby function	No	No	Yes +	Yes ++
Flicker (sensitive)	Yes	Yes	No	No
Softstarter needed	Yes	Yes	No	No
Rolling capacity on grid	Yes, partly	Yes, partly	Yes	Yes
Reactive compensator (C)	Yes	Yes	No	No
Island operation	No	No	Yes/No	Yes
Investment	++	++	+	0
Maintenance	++	++	0	+

D. Control of Wind Turbines

Controlling a wind turbine involves both fast and slow control dynamics. Overall the power has to be controlled by means of the aerodynamic system and has to react based on a set-point given by a dispatched center or locally with the goal to maximize the power production based on the available wind power. The power controller should also be able to limit the power. An example of an overall control scheme of a wind turbine with a doubly-fed generator system is shown in Fig. 11 [5], [37].

Below maximum power production the wind turbine will typically vary the speed proportional with the wind speed and keep the pitch angle θ fixed. At very low wind the speed of the turbine will be fixed at the maximum allowable slip in order not to have over voltage. A pitch angle controller limits the power when the turbine reaches nominal power. The generated electrical power is done by controlling the doubly-fed generator through the rotor-side converter. The control of the grid-side converter is simply just keeping the dc-link voltage fixed. Internal current loops in both converters are used which typically are linear PI-controllers, as it is illustrated in Fig. 11. The power converters to the grid-side and the rotor-side are voltage source converters.

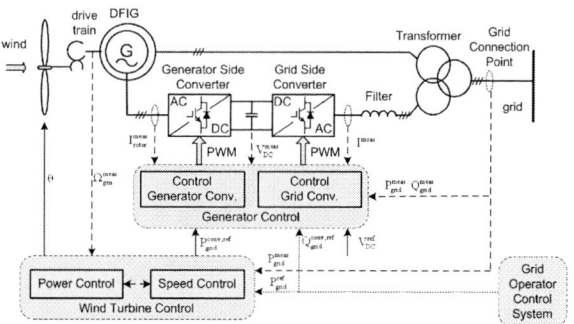

Fig. 11. Control of a wind turbine with doubly-fed induction generator (WT Type C).

Another solution for the electrical power control is to use the multi-pole synchronous generator. A passive rectifier and a boost converter are used in order to boost the voltage at low speed. The system is industrially used today and it is shown in Fig. 12.

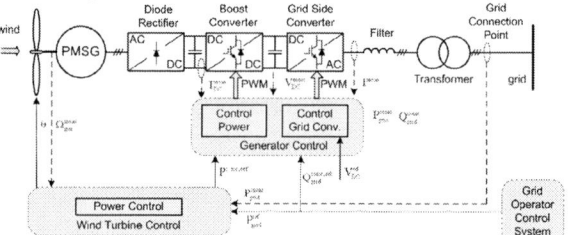

Fig. 12. Control of active and reactive power in a wind turbine with multi-pole synchronous generator (WT Type D).

A grid-side inverter is interfacing the dc-link to the grid. Common for both systems are they are able to control active and reactive power to the grid with high dynamics

E. Wind Farm Configurations

In many countries energy planning is going on with a high penetration of wind energy, which will be covered by large offshore wind farms. These wind farms may in the future present a significant power contribution to the national grid, and therefore, play an important role on the power quality and the control of complex power systems. Consequently, very high technical demands are expected to be met by these generation units, such as to perform frequency and voltage control, regulation of active and reactive power, quick responses under power system transient and dynamic situations, for example, to reduce the power from the nominal power to 20 % power within 2 seconds. The power electronic technology is again an important part in both the system configurations and the control of the offshore wind farms in order to fulfill the future demands.

One off-shore wind farm equipped with power electronic converters can perform both active and reactive power control and also operate the wind turbines in variable speed to maximize the energy captured and reduce the mechanical stress and acoustical noise. This solution is shown in Fig. 13 and it is in operation in Denmark as a 160 MW off-shore wind power station.

The active stall wind farms based on wind turbine Type A (see Fig. 7) are directly connected to the grid. A reactive power compensation unit is used in the connection point as shown in Fig. 14.

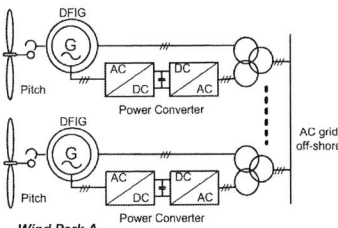

Fig. 13. DFIG based wind farm with an AC grid connection.

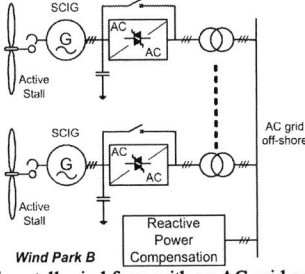

Fig. 14. Active stall wind farm with an AC grid connection.

For long distance power transmission from off-shore wind farm, HVDC may be an interesting option. In an HVDC transmission system, the low or medium AC voltage at the wind farm is converted into a high dc voltage on the transmission side and the dc power is transferred to the on-shore system where the DC voltage is converted back into AC voltage as shown in Fig. 15. The topology may even be able to vary the speed on the wind turbines in the complete wind farm [47], [48].

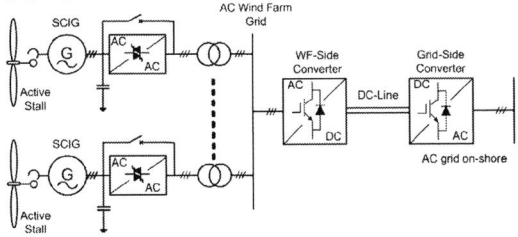

Fig. 15. Active stall wind farm with a DC-link grid connection.

Another possible DC transmission system configuration is shown in Fig. 16, where each wind turbine has its own power electronic converter, so it is possible to operate each wind turbine at an individual optimal speed. A common DC grid is present on the wind farm while a full scale power converter is used for the on-shore grid connection.

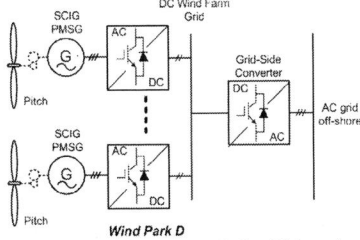

Fig. 16. Wind farm with common DC grid based on variable speed wind turbines with full scale power converter.

A comparison of these possible wind farm topologies is given in Table 2.

As it can be seen the wind farms have interesting features in order to act as a power source to the grid. Some have better abilities than others. Bottom-line will always be a total cost scenario including production,

investment, maintenance and reliability. This may be different depending on the planned site.

Table 2. Comparison of wind farm topologies.

	Wind Park A	Wind Park B	Wind Park C	Wind Park D
Individual speed control	Yes	No	Yes	No
Control active power electronically	Yes	No	Yes	Yes
Control reactive power	Yes	Centralized	Yes	Yes
Short circuit (active)	Partly	Partly	Yes	Yes
Short circuit power	Contribute	Contribute	No	No
Control bandwidth	10-100 ms	200ms - 2s	10 -100 ms	10 ms – 10 s
Stand by-function	Yes	No	Yes	Yes
Soft-starter needed	No	Yes	No	No
Rolling capacity on grid	Yes	Partly	Yes	Yes
Redundancy	Yes	Yes	No	No
Investment	+	++	+	+
Maintenance	+	++	+	+

F. Grid connection requirements

Some European countries have at this moment dedicated grid codes for wind power. These requirements reflect, in most of the cases, the penetration of wind power into the electrical network or a future development is prepared.

The requirements for wind power cover a wide range of voltage levels from medium voltage to very high voltage. The grid codes for wind power address issues that make the wind farms to act as a conventional power plant into the electrical network. These requirements have focus on power controllability, power quality, fault ride-through capability and grid support during network disturbances. According to several references [6] and [8] in some of the cases these requirements are very stringent.

1) Active power control

According to this demand the wind turbines must be able to control the active in the Point-of-Common-Coupling (PCC) in a given power range. The active power is typically controlled based on the system frequency e.g. Denmark, Ireland, Germany [51]-[57] so that the power delivered to the grid is decreased when the grid frequency rise above 50 Hz. A typical characteristic for the frequency control in the Danish grid code is shown in Fig. 17.

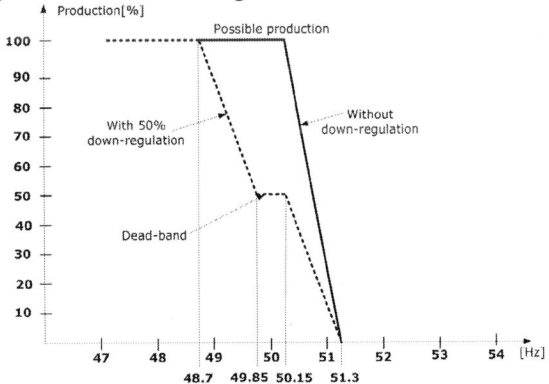

Fig. 17. Frequency control characteristic for the wind turbines connected to the Danish grid [52].

On the contrary other grid codes, e.g. Great Britain [58] specifies that the active power output must be kept constant for the frequency range 49.5 to 50.5 Hz, and a drop of maximum 5% in the delivered power is allowed when frequency drops to 47 Hz.

Curtailment of produced power based on system operator demands is required in Denmark, Ireland, Germany and Great Britain.

Currently, Denmark has the most demanding requirements regarding the controllability of the produced power. Wind farms connected at the transmission level shall act as a conventional power plant providing a wide range of controlling the output power based on Transmission System Operator's (TSO) demands and also participation in primary and secondary control [52]. Seven regulation functions are required in the wind farm control. Among these control functions, each one prioritized, the following must be mentioned: delta control, balance control, absolute production and system protection as shown in Fig. 18.

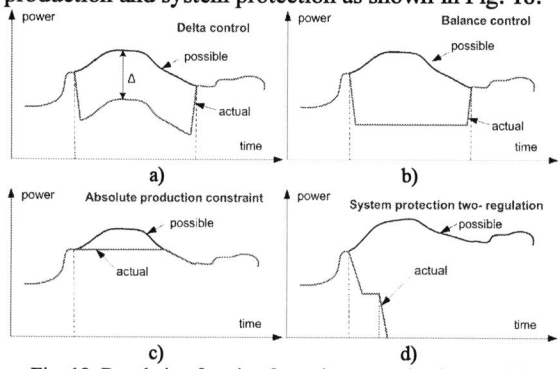

Fig. 18. Regulation function for active power implemented in wind farm controller required by the Danish grid codes: a) delta control, b) balance control, c) absolute production constraint and d) system protection.

2) Reactive power control and voltage stability

Reactive power is typically controlled in a given range. The grid codes specify in different ways this control capability. The Danish grid code gives a band for controlling the reactive power based on the active power output as shown in Fig. 19.

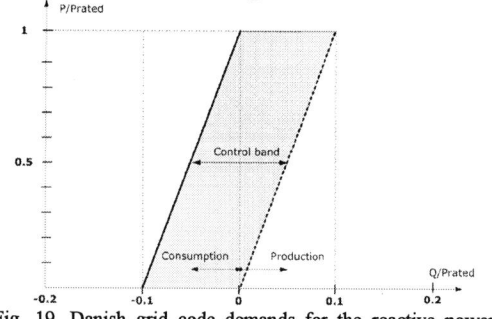

Fig. 19. Danish grid code demands for the reactive power exchange in the PCC [51], [52].

The Irish grid code specifies e.g. the reactive power capability in terms of power factor as shown in Fig. 20.

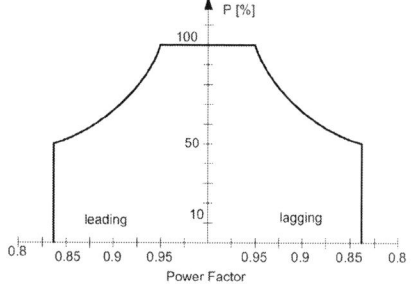

Fig. 20. Requirements for reactive power control in the Irish grid code for wind turbines [54].

The German transmission grid code for wind power specifies that the wind power units must provide a reactive power provision in the connection point without limiting the active power output as shown in Fig. 21.

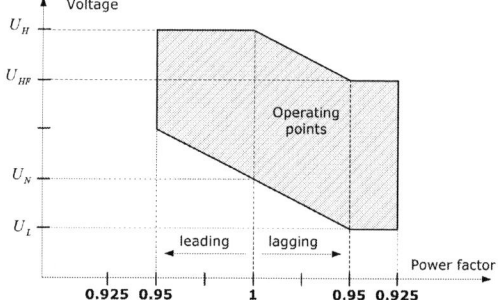

Fig. 21. Requirements for reactive power provision of generating units without limiting the active power output in the German transmission grid code [55], [56].

3) Power Quality

Power quality issues are addressed especially for wind turbines connected to the medium voltage networks. However, some grid codes, e.g. in Denmark and Ireland have also requirements at the transmission level.

Mainly two standards are used for defining the power quality parameters namely: IEC 61000-x-x and EN 50160. Specific values are given for fast variations in voltage, short term flicker severity, long term flicker severity and the total harmonic distortion. A schedule of individual harmonics distortion limits for voltage are also given based on standards or in some cases e.g. Denmark custom harmonic compatibility levels are defined. Interharmonics may also be considered [51].

4) Ride through capability

All considered grid codes requires fault ride-through capabilities for wind turbines. Voltage profiles are given specifying the depth of the voltage dip and the clearance time as well. One of the problems is that the calculation of the voltage during all types of unsymmetrical faults is not very well defined in some grid codes. The voltage profile for ride-through capability can be summarized as shown in Fig. 22.

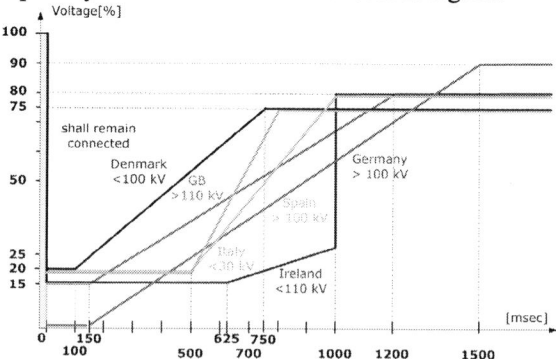

Fig. 22. Voltage profile for fault ride-through capability in European grid codes for wind power [7].

Ireland's grid code is very demanding in respect with the fault duration while Denmark has the lowest short circuit time duration with only 100 msec. However, Denmark's grid code requires that the wind turbine shall remain connected to the electrical network during successive faults which is a technical challenge.

On the other hand Germany and Spain requires grid support during faults by reactive current injection up to 100% from the rated current [55], [56] and [59] as shown in Fig. 23.

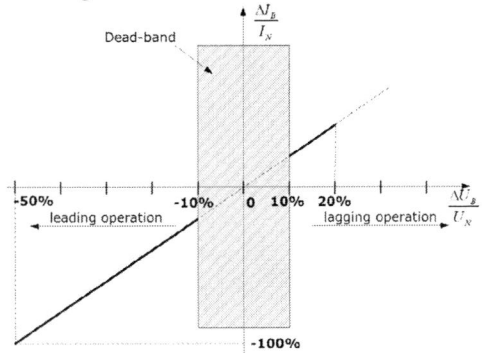

Fig. 23. Reactive current support during faults as specified in the German grid code [55].

This demand is relative difficult to meet by some of the wind turbine concepts e.g. active stall wind turbine with directly grid connected squirrel cage induction generator (WT Type A).

A summary regarding the interconnection requirements for wind power in Europe is given in detail in Appendix I.

IV. SOLAR ENERGY POWER CONVERSION

Photovoltaic (PV) power supplied to the utility grid is gaining more and more visibility due to many national incentives [65]. With a continuous reduction in system cost (PV modules, DC/AC inverters, cables, fittings and man-power), the PV technology has the potential to become one of the main renewable energy sources for the future electricity supply.

The PV cell is an all-electrical device, which produces electrical power when exposed to sunlight and connected to a suitable load. Without any moving parts inside the PV module, the tear-and-wear is very low. Thus, lifetimes of more than 25 years for modules are easily reached. However, the power generation capability may be reduced to 75% ~ 80% of nominal value due to ageing. A typical PV module is made up of around 36 or 72 cells connected in series, encapsulated in a structure made of e.g. aluminum and tedlar. An electrical model of PV cell is shown in Fig. 24.

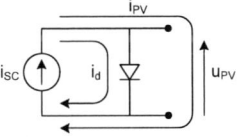

Fig. 24. Electrical model and characteristics of a PV cell.

Several types of proven PV technologies exist, where the crystalline (PV module light-to-electricity efficiency: η = 10% - 15%) and multi-crystalline (η = 9% - 12%) silicon cells are based on standard microelectronic manufacturing processes.

Other types are: thin-film amorphous silicon (η = 10%), thin-film copper indium diselenide (η = 12%), and thin-film cadmium telluride (η = 9%). Novel technologies such as the thin-layer silicon (η = 8%) and the dye-sensitised nano-structured materials (η = 9%) are in their early development. The reason to maintain a high level of research and development within these technologies is to decrease the cost of the PV-cells, perhaps on the expense of a somewhat lower efficiency. This is mainly due to the fact that cells based on today's microelectronic processes are rather costly, when compared to other renewable energy sources.

The series connection of the cells benefit from a high voltage (around 25 V ~ 45 V) across the terminals, but the weakest cell determines the current seen at the terminals.

This causes reduction in the available power, which to some extent can be mitigated by the use of bypass diodes, in parallel with the cells. The parallel connection of the cells solves the 'weakest-link' problem, but the voltage seen at the terminals is rather low.

Typical curves of a PV cell current-voltage and power-voltage characteristics are plotted in Fig. 25a and Fig. 25b respectively, with insolation and cell temperature as parameters.

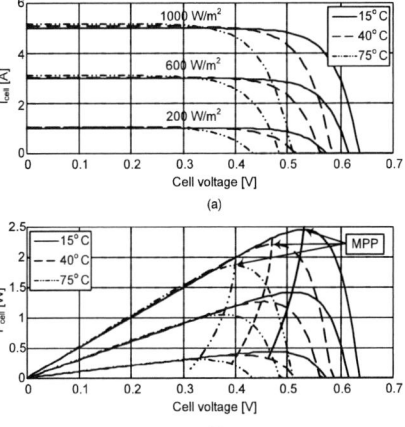

Fig. 25. Characteristics of a PV cell. Model based on the British Petroleum BP5170 crystalline silicon PV module. Power at standard test condition (1000 W/m2 irradiation, and a cell temperature of 25°C): 170 W @ 36.0 V [4].

The graph reveals that the captured power is determined by the loading conditions (terminal voltage and current). This leads to a few basic requirements for the power electronics used to interface the PV module(s) to the utility grid.

An overview of the power converter topologies for PV systems including their control techniques is given in the following sections. Next grid monitoring methods including grid voltage monitoring, grid impedance estimation and islanding detection are presented.

A. Structures for PV systems

The general block diagram of a grid connected photovoltaic system is similar with the one shown in Fig. 1. It consists of a PV array, a power converter with a filter, a controller and the grid utility.

The PV array can be a single panel, a string of PV panels or a multitude of parallel strings of PV panels. Centralized or decentralized PV systems can be used as depicted in Fig. 26.

1) Central inverters

In this topology the PV plant (typical > 10 kW) is arranged in many parallel strings that are connected to a single central inverter on the DC-side (Fig. 26a). These inverters are characterized by high efficiency and low cost pr. kW. However, the energy yield of the PV plant decreases due to module mismatching and potential partial shading conditions. Also, the reliability of the plant may be limited due to the dependence of power generation on a single component: a failure of the central inverter results in that the whole PV plant is out of operation.

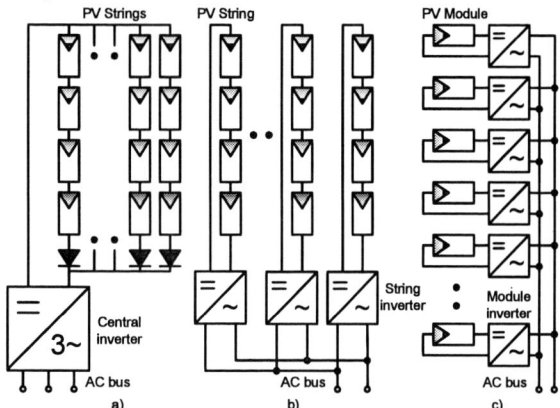

Fig. 26 Structures for PV systems: a) Central inverter, b) String inverter and c) Module integrated inverter [71].

2) String inverters

Similar to the central inverter, the PV plant is divided into several parallel strings. Each of the PV strings is assigned to a designated inverter, the so-called "string inverter" (see Fig. 26b). String inverters have the capability of separate Maximum Power Point (MPP) tracking of each PV string. This increases the energy yield by the reduction of mismatching and partial shading losses. These superior technical characteristics increase the energy yield and enhance the supply reliability. String inverters have evolved as a standard in PV system technology for grid connected PV plants.

An evolution of the string technology applicable for higher power levels is the multi-string inverter. It allows the connection of several strings with separate MPP tracking systems (via DC-DC converter) to a common DC-AC inverter. Accordingly, a compact and cost-effective solution, which combines the advantages of central and string technologies, is achieved. This multi-string topology allows the integration of PV strings of different technologies and of various orientations (south, north, west and east). These characteristics allow time-shifted solar power, which optimizes the operation efficiencies of each string separately. The application area of the multi-string inverter covers PV plants of 3-10 kW.

3) Module integrated inverter

This system uses one inverter for each module (Fig. 26c). This topology optimizes the adaptability of the inverter to the PV characteristics, since each module has its own Maximum Power Point (MPP) tracker. Although the module-integrated inverter optimizes the energy yield, it has a lower efficiency than the string inverter. Module integrated inverters are characterized by a more extended AC-side cabling, since each module of the PV plant has to be connected to the available AC grid (e.g. 230 V/ 50 Hz). Also, the maintenance processes are quite complicated, especially for facade-

integrated PV systems. This concept can be implemented for PV plants of about 50- 400 W peak.

B. Topologies for PV inverters

The PV inverter technology has evolved quite a lot during the last years towards maturity [66]. Still there are different power configurations possible as shown in Fig. 27.

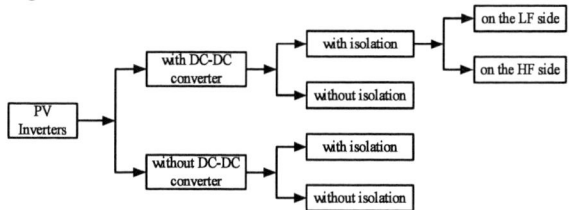

Fig. 27. Power configurations for PV inverters.

The question of having a dc-dc converter or not is first of all related to the PV string configuration. Having more panels in series and lower grid voltage, like in US and Japan, it is possible to avoid the boost function with a dc-dc converter. Thus a single stage PV inverter can be used leading to higher efficiencies.

The issue of isolation is mainly related to safety standards and is for the moment only required in US. The drawback of having so many panels in series is that MPPT is harder to achieve especially during partial shading, as demonstrated in [67]. In the following, the different PV inverter power configurations are described in more details.

1) PV inverters with DC-DC converter and isolation

The isolation is typically acquired using a transformer that can be placed on either the grid frequency side (LF) as shown in Fig. 28a or on the high-frequency (HF) side in the dc-dc converter as shown in Fig. 28b. The HF transformer leads to more compact solutions but high care should be taken in the transformer design in order to keep the losses low.

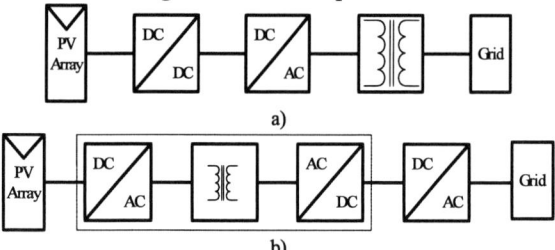

Fig. 28. PV inverter system with DC-DC converter and isolation transformer: a) on the Low Frequency (LF) side and b) on the High Frequency (HF) side.

In Fig. 29 a PV inverter with an HF transformer using an isolated push-pull boost converter is presented [68].

In this solution the dc-ac inverter is a low cost inverter switched at the line frequency. The new solutions on the market are using PWM dc-ac inverters with IGBT's switched typically at 10-20 kHz leading to a better power quality performance.

Other solutions for high frequency dc-dc converters with isolation include: full-bridge isolated converter, Single-Inductor push-pull Converter (SIC) and Double-Inductor Converter (DIC) [69].

234

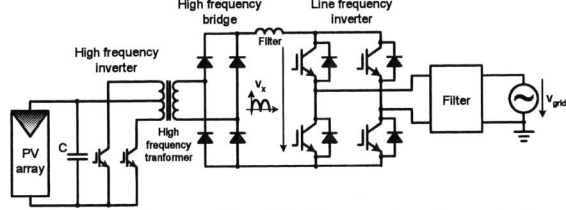

Fig. 29. PV inverter with a high frequency transformer in the dc-dc converter.

In order to keep the magnetic components small high switching frequencies in the range of 20 – 100 kHz are typically employed. The full-bridge converter is usually utilized at power levels above 750 W. The advantages of this topology are: good transformer utilization – bipolar magnetization of the core, good performance with current programmed control – reduced DC magnetization of transformer. The main disadvantages in comparison with push-pull topology are the higher active part count and the higher transformer ratio needed for boosting the dc voltage to the grid level.

The single inductor push-pull converter can provide boosting function on both the boosting inductor and transformer, reducing the transformer ratio. Thus higher efficiency can be achieved together with smoother input current. On the negative side higher blocking voltage switches are required and the transformer with tap point puts some construction and reliability problems.

Those shortcomings can be alleviated using the double inductor push-pull converter (DIC) where the boost inductor has been split into two. Actually this topology is equivalent with two inter-leaved boost converters leading to lower ripple in the input current. The transformer construction is simpler not requiring a tap point. The single disadvantage of this topology remains the need for an extra inductor.

2) PV inverters with DC-DC converter without isolation

In some countries as the grid-isolation is not mandatory, more simplified PV inverter design can be used, like shown in Fig. 30a.

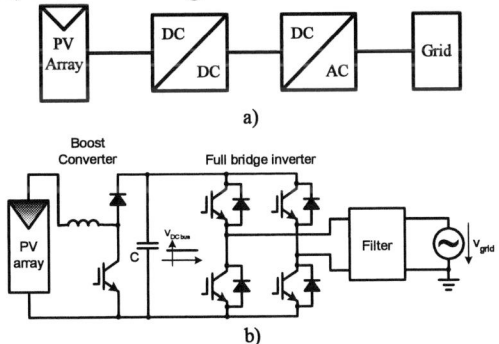

Fig. 30. PV inverter system with DC-DC converter without isolation transformer a) General diagram and b) Practical example with boost converter and full-bridge inverter.

In Fig. 30b a practical example [70] using a simple boost converter is shown.

3) PV inverters without DC-DC converter and with isolation

The block diagram of this topology is shown in Fig. 31.

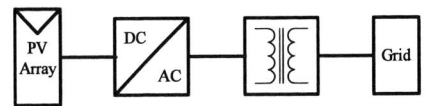

Fig. 31. General diagram of a PV system without DC-DC converter and with isolation transformer.

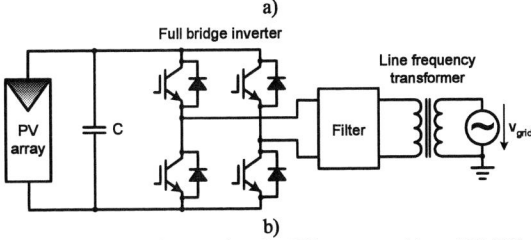

Fig. 32. Practical example of a PV system without DC-DC converter and with a full-bridge converter and isolation grid side transformer.

A PV inverter topology is presented in Fig. 32, in which a line frequency transformer is used. For higher power levels, self-commutated inverters using thyristors may be used [70].

4) PV inverters without DC-DC converter and without isolation

This topology is shown in Fig. 33a.

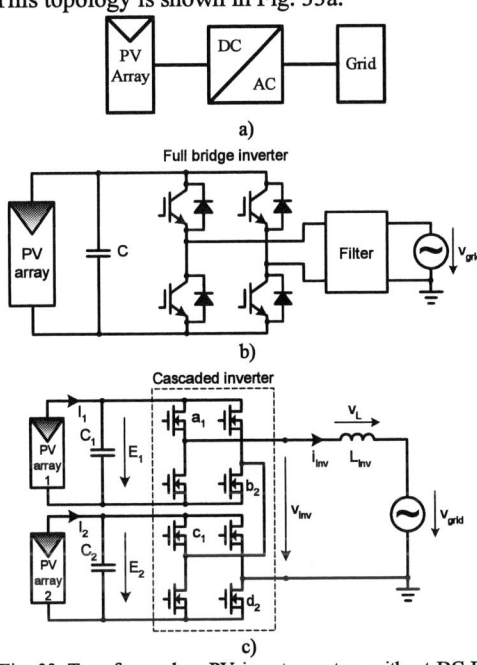

Fig. 33. Transformer-less PV inverter system without DC-DC converter: a) general diagram, b) typical example with full-bridge inverter and c) multilevel inverter.

In Fig. 33b, a typical transformer-less topology is shown using PWM IGBT inverters. This topology can be used when a large number of PV panels are available connected in series producing in excess of the grid voltage peak at all times.

Another interesting PV inverter topology without boost and isolation can be achieved using a multilevel concept. Grid connected photovoltaic systems with a five level cascaded inverter is presented in Fig. 33c [68]. The redundant inverter states of the five level cascaded inverter allow for a cyclic switching scheme which minimizes the switching frequency, equalizes stress evenly on all switches and minimizes the voltage ripple on the DC capacitors.

C. Control of PV inverters

Based on the above presented power converter topologies it can be concluded that two main structures are used in PV applications namely the double-stage conversion (DC to DC plus DC to AC) and the single stage conversion (DC to AC only). Therefore, the next sections present the control techniques used for these topologies.

1) Control of DC-DC boost converter

In order to control the output dc-voltage to a desired value, a control system is needed which automatically can adjust the duty cycle, regardless of the load current or input changes. There are at least two types of control for the dc-dc converters: the direct duty-cycle control and the current control [71].

Direct duty cycle - The output voltage is measured and then compared to the reference. The error signal is used as input in the compensator, which will calculate it from the duty-cycle reference for the pulse-width modulator as shown in Fig. 34a.

Current control - The converter output is controlled by the choice of the transistor peak current. The control signal is a current and a simple control network switches on and off the transistor such its peak current follows the control input. The current control (Fig. 34b), in the case of an isolated boost push-pull converter has some advantages against the duty-cycle control e.g. simpler dynamics (removes one pole from the control to output transfer function). Also as it uses a current sensor it can provide a better protection of the switch by limiting the current to acceptable levels.

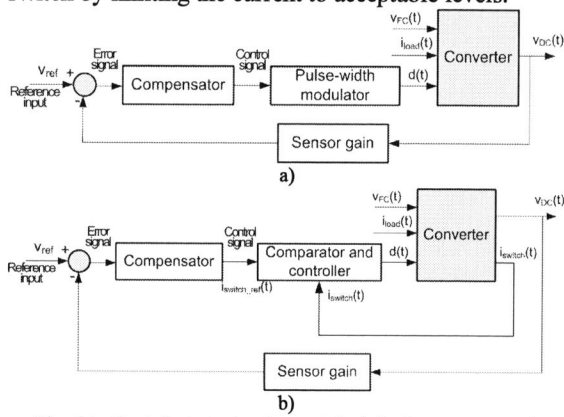

Fig. 34. Control strategies for switched dc-dc converters a) direct duty-cycle control and b) current control.

Among the drawbacks of the current control it can be mentioned that it requires an extra current sensor and it has a susceptibility to noise and thus light filtering of the feedback signals is required.

2) Control of DC-AC converter

For the grid-connected PV inverters in the range of 1-5 kW, the most common control structure for the DC-AC grid converter is using a current-controlled H-bridge PWM inverter which has a low-pass output filter. Typically L-filters are used but the new trend is to use LCL filters that have a higher order filter (3rd) which leads to a more compact design. The drawback is that due to its own resonance frequency it can produce stability problems and special control design is required

[72]. A typical single-stage PV grid-connected converter with an LCL filter is shown in Fig. 35.

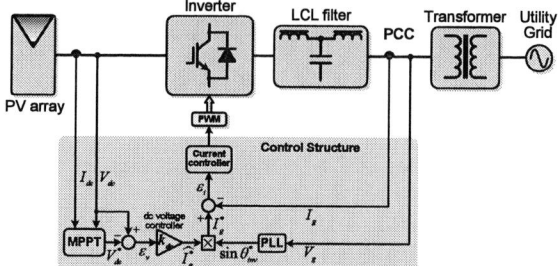

Fig. 35. Single-stage PV grid-connected system.

The main elements of the control structure are the synchronization algorithm based on PLL, the MPPT, the input power control, the grid current controller including the PWM generator.

The harmonics level in the grid current is still a controversial issue for PV inverters. The IEEE 929 standard from year 2000 allows a limit of 5% for the current Total Harmonic Distortion (THD) factor with individual limits of 4% for each odd harmonic from 3rd to 9th and 2% for 11th to 15th while a recent draft of European IEC61727 suggests something similar. These levels are far more stringent than other domestic appliances such as IEC61000-3-2 as PV systems are viewed as generation sources and so they are subject to higher standards than load systems.

Classical PI control with grid voltage feed-forward (v_{ff}) [13], as depicted in Fig. 36a, is commonly used for current-controlled PV inverters, but this solution exhibits two well known drawbacks: inability of the PI controller to track a sinusoidal reference without steady-state error and poor disturbance rejection capability. This is due to the poor performance of the integral action.

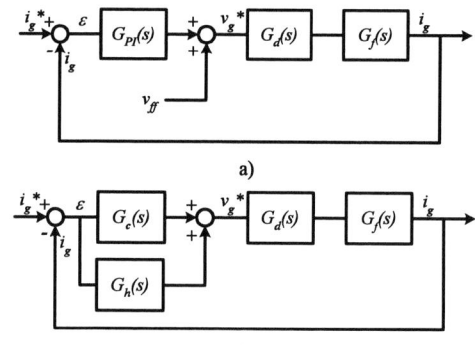

Fig. 36. The current loop of a PV inverter: a) with PI controller and b) with P+Resonant (PR) controller.

In order to get a good dynamic response, a grid voltage feed-forward (v_{ff}) is used, as depicted in Fig. 26a. This leads in turn to stability problems related to the delay introduced in the system by the voltage feedback filter.

In order to alleviate these problems, a second order generalized integrator (GI) as reported in [72], [73] and [74] can be used. The GI is a double integrator that achieves an infinite gain at a certain frequency, also called resonance frequency, and almost no gain exists outside this frequency. Thus, it can be used as a notch filter in order to compensate the harmonics in a very selective way. This technique has been primarily used in three-phase active filter applications as reported in [73]. Another approach reported in [72] where a new type of stationary-frame regulators called P+Resonant (PR) is introduced and applied to three-phase PWM inverter control. In this approach the PI dc-compensator is transformed into an equivalent ac-compensator, so that it has the same frequency response characteristics in the bandwidth of concern. The current loop of the PV inverter with PR controller is depicted in Fig. 36b.

The harmonic compensator (HC) $G_h(s)$ as defined in [75] is designed to compensate the selected harmonics 3rd, 5th and 7th as they are the most prominent harmonics in the current spectrum. A processing delay typical equal to sampling time for the PWM inverters is introduced in [72].

Thus it is demonstrated the superiority of the PR controller in respect to the PI controller in terms of harmonic current rejection.

The issue of stability when several PV inverters run in parallel on the same grid becomes more and more important, especially when LCL filters are used. Thus, special attention is required when designing the current control.

3) Maximum Power Point Tracking (MPPT)

In order to capture the maximum power available from the PV array, a Maximum Power Point Tracker (MPPT) is required. The maximum power point of PV panels is a function of solar irradiance and temperature as depicted in Fig. 25. This function can be implemented either in the dc-dc converter or in the DC-AC converter. Several algorithms can be used in order to implement the MPPT like:

a) Perturb and Observe method

The most commonly used MPPT algorithm is the Perturb and Observe (P&O), due to its ease of implementation in its basic form. Fig. 25 shows the characteristic of a PV array, which has a global maximum at the MPP. Thus, if the operating voltage of the PV array is perturbed in a given direction and dP/dV > 0, it is known that the perturbation is moving the operating point towards the MPP. The P&O algorithm would then continue to perturb the PV array voltage in the same direction. If dP/dV < 0, then the change in operating point moved the PV array away from the MPP, and the P&O algorithm reverses the direction of the perturbation. [76] A problem with P&O is that it oscillates around the MPP in steady state operation. It can also track into the wrong direction, away from the MPP, under rapidly increasing or decreasing irradiance levels [77]-[79]. There are several variations of the basic P&O that have been proposed to minimize these drawbacks. These include using an average of several samples of the array power and dynamically adjusting the magnitude of the perturbation of the PV operating point.

b) Improved P&O method for rapidly changing irradiance

The method performs an additional measurement of power in the middle of the MPPT sampling period without any perturbation, and based on these measurements, it calculates the change of power due to the varying irradiation, [80] according to Fig. 37.

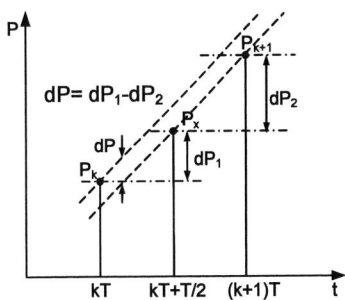

Fig. 37. Measurement of the power between two MPPT sampling instances.

Assuming that the rate of change in the irradiation is constant over one sampling period of the MPPT, the dP caused purely by the MPPT command can be calculated as:

$$dP = dP_1 - dP_2 = (P_x - P_k) - (P_{k+1} - P_x) = 2P_x - P_{k+1} - P_k \quad (1)$$

The resulting 'dP' reflects the changes due to the perturbation of the MPPT method.

Using the above calculation in the flowchart of the dp-P&O method, (see Fig. 38) can be avoided the confusion of the MPPT due to the rapidly changing irradiation.

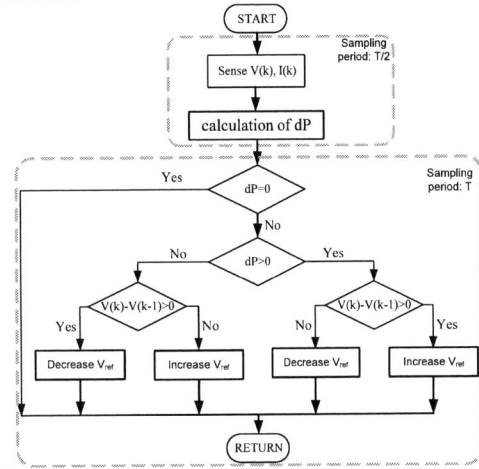

Fig. 38. The flowchart of the dp-P&O method.

The experimental results show that the dP-P&O method performs superior to the traditional P&O during rapidly changing irradiance, resulting in higher dynamic efficiency, see Fig. 39.

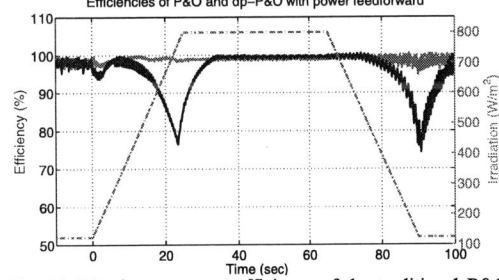

Fig. 39. The instantaneous efficiency of the traditional P&O method can decrease to below 80% during rapidly increasing and decreasing irradiation, while the efficiency of dP-P&O is not affected.

c) Incremental conductance method

The incremental conductance algorithm seeks to overcome the limitations of the P&O algorithm by using the PV array's incremental conductance to compute the sign of dP/dV without a perturbation. It does this using an expression derived from the condition that, at the MPP, dP/dV = 0. Beginning with this condition, it is possible to show that, at the MPP dI/dV = -I/V [76] and [81]. Thus, incremental conductance can determine that the MPPT has reached the MPP and stop perturbing the operating point. If this condition is not met, the direction in which the MPPT operating point must be perturbed can be calculated using the relationship between dI/dV and -I/V. This relationship is derived from the fact that dP/dV is negative when the MPPT is to the right of the MPP and positive when it is to the left of the MPP. This algorithm has advantages over perturb and observe in that it can determine when the MPPT has reached the MPP, where perturb and observe oscillates around the MPP. Also, incremental conductance can track rapidly increasing and decreasing irradiance conditions with higher accuracy than perturb and observe [76]. However, because of noise and errors due to measurement and quantization, this method can also produce oscillations around the MPP; and it can also be confused in rapidly changing atmospheric conditions [77]. One disadvantage of this algorithm is the increased complexity when compared to perturb and observe. This increases real-time computational time, and slows down the sampling frequency of the array voltage and current.

d) Parasitic capacitance method

The parasitic capacitance method is a refinement of the incremental conductance method that takes into account the parasitic capacitances of the solar cells in the PV array. Parasitic capacitance uses the switching ripple of the MPPT to perturb the array. To account for the parasitic capacitance, the average ripple in the array power and voltage, generated by the switching frequency, are measured using a series of filters and multipliers and then used to calculate the array conductance. The incremental conductance algorithm is then used to determine the direction to move the operating point of the MPPT. One disadvantage of this algorithm is that the parasitic capacitance in each module is very small, and will only come into play in large PV arrays where several module strings are connected in parallel. Also, the DC-DC converter has a sizable input capacitor used to filter out the small ripple in the array power. This capacitor may mask the overall effects of the parasitic capacitance of the PV array.

e) Constant voltage method

This algorithm makes use of the fact that the MPP voltage changes only slightly with varying irradiances, as depicted in Fig. 25. The ratio of VMP/VOC depends on the solar cell parameters, but a commonly used value is 76% [76] and [82]. In this algorithm, the MPPT momentarily sets the PV array current to zero to allow a measurement of the array's open circuit voltage. The array's operating voltage is then set to 76% of this measured value. This operating point is maintained for a set amount of time, and then the cycle is repeated. A problem with this algorithm is that the available energy is wasted when the load is disconnected from the PV array; also the MPP is not always located at 76% of the array's open circuit voltage [76].

4) Input power control for PV applications

For PV applications, the input power control can be realized through the use of either DC-DC converter or DC-AC converters. The control strategies of the input power in the case of a power configuration of PV system without DC-DC converter (single-stage PV converter) are presented in the following. The implementation of the MPPT could be realized in two different ways in this case:

– the output of the MPPT is the AC current amplitude reference;

– the output of the MPPT is the DC voltage reference.

In the first case the MPPT block has I_{pv} and V_{pv} as inputs and the output variable is the AC current amplitude reference ($\hat{I}_{ref}$) as depicted in Fig. 40a [83].

In the second case the MPPT block has the same inputs (I_{pv} and V_{pv}) but the output variable of the algorithm is the dc voltage reference (V^*_{pv}). The dc voltage controller (P or PI controller) is used to control the DC voltage loop to produce the AC current amplitude reference ($\hat{I}_{ref}$). Then the AC current amplitude reference is multiplied by $sin(\theta)$, which is captured from a phase-looked-loop (PLL) circuit to produce the output current reference command I_{ref} of the inverter. This topology is described in Fig. 40b [84] and [85].

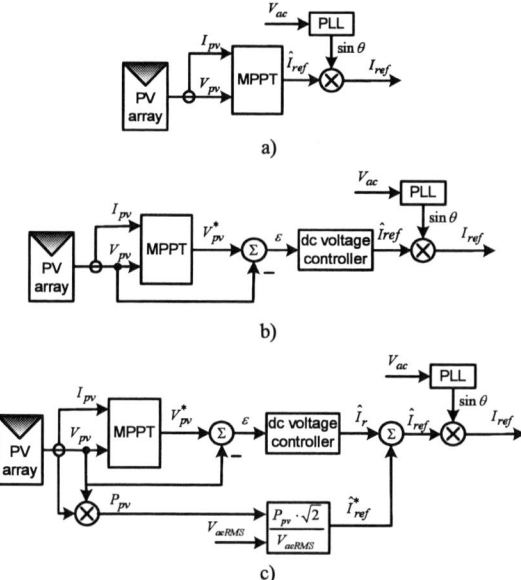

Fig. 40. Control structures of the input power. a) the output of MPPT is the ac current amplitude reference ($\hat{I}_{ref}$), b) the output of the MPPT is the dc voltage reference (V^*_{pv}) and a dc voltage controller is used, c) new control structure where a feed-forward of the input power is used.

In Fig. 40c a new control strategy of input power is proposed. The new element introduced is a power feed-forward. The computed value of the current amplitude reference using the PV power (P_{pv}) and the RMS value of the ac voltage (V_{acRMS}) is added to the output value of the dc voltage controller ($\hat{I}_r$) resulting in an ac-current amplitude reference ($\hat{I}_{ref}$). Using the input power feed-forward the dynamic of the PV system is improved being known the fact that the MPPT is rather slow.

D. PV systems - Grid monitoring

1) Grid voltage monitoring

The increased penetration of DPGS connected to the electrical grid based on sources such as PV necessitates better grid condition detection in order to meet standard specifications in terms of power quality and safety.

Grid-connected converter systems rely on accurate and fast detection of the phase angle, amplitude and frequency of the utility voltage to guarantee the correct generation of the reference signals. This is also required by the relevant grid codes which are country specific and can vary also in respect to the generation system (e.g. PV systems, wind turbines, fuel cell, etc). The grid codes may refer to different standards for distributed generation systems. These standards impose the operation conditions of the grid-connected converter systems in terms of grid voltage amplitude and frequency. Considering grid voltage monitoring requirements for interconnection of PV systems to the grid, the standard IEC61727 [86] and IEEE 929 [87] are given as examples. These standards apply to utility-interconnected PV power systems operating in parallel with the utility and utilizing static (solid-state) non-islanding inverters for the conversation of DC to AC.

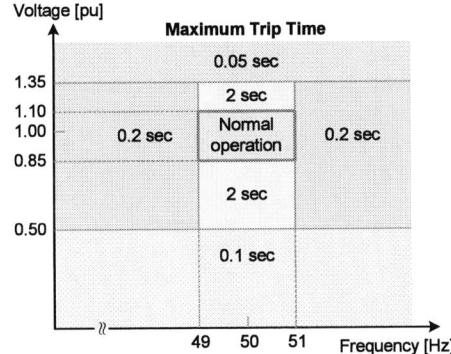

Fig. 41. Maximum trip times for both voltage amplitude and frequency according to the standard IEC61727 [86].

Fig. 41 shows the boundaries of operation in respect to grid voltage amplitude and frequency. A continuous operation area between 0.85 and 1.10 pu and ± 1 Hz around the nominal frequency is defined. Abnormal conditions can arise on the utility system that requires a response from the grid-connected PV system. This response is to ensure the safety of utility maintenance personnel and the general public, as well as to avoid damage to connected equipment, including the PV system. The abnormal utility conditions of concern are the grid voltage amplitude and frequency excursions above or below the values stated in Fig. 41. If the voltage amplitude or frequency exceeds the predefined limits, the grid-connected PV system has to cease to energize the utility line within the specified time interval. As it can be noticed from Fig. 41, the most restrictive requirement is when the maximum trip time is 0.05 seconds for a grid voltage amplitude excursion above 1.35 pu. An accurate and fast grid voltage monitoring algorithm is required in order to comply with these requirements.

Fig. 42 presents the principle of the grid voltage monitoring which consists in obtaining the parameters of the grid voltage as presented in (2).

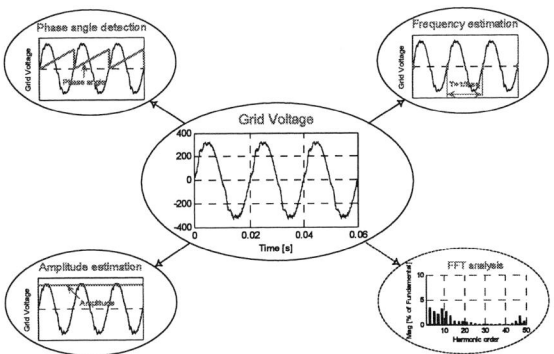

Fig. 42. Grid voltage monitoring principles.

$$v(t) = \underbrace{\widehat{V} \cdot \sin(\omega \cdot t)}_{Fundamental} + \underbrace{\sum \widehat{V}_h \cdot \sin(\omega_h \cdot t + \theta_h)}_{Harmonics} \quad (2)$$

The voltage equation is divided in two main parts: the fundamental and the harmonics. The grid phase angle ($\omega \cdot t$) is mostly used for synchronization. Moreover, the detection of the grid phase angle can also be used for anti-islanding detection algorithms [88]. The frequency of the grid voltage (ω) is used for over/under frequency detection algorithms but also to provide information to the control system (such as resonant controllers or filters which need to adjust the resonance frequency). The amplitude of the grid voltage ($\widehat{V}$) is required for over/under voltage and to provide information to the control system (such as power feed forward loop). Additional information such as harmonic content of the grid voltage can be required for some algorithms (e.g. harmonics monitoring for the passive anti-islanding methods [88] or active power filters applications.

a) Grid voltage monitoring techniques – Overview

Different algorithms are used in order to monitor the grid voltage. In the technical literature numerous methods using different techniques are presented. These methods can be organized in three main categories:

- methods based on Zero-Crossing Detection (ZCD),
- methods based on Phase-Locked Loop (PLL)
- methods based on arctangent function ($\tan^{-1}$).

A simple method of obtaining the phase and frequency information is to detect the zero-crossing point of the grid voltage [89]-[91]. This method has two major drawbacks as described in the following.

Since the zero crossing point can be detected only at every half cycle of the utility frequency, the phase tracking action is impossible between the detecting points and thus the fast dynamic performance can not be obtained [92]. Some work has been done in order to alleviate this problem using multiple level crossing detection as presented in [93].

Significant line voltage distortion due to notches caused by power device switching and/or low frequency harmonic content can easily corrupt the output of a conventional zero-crossing detector [94]. Therefore, the zero-crossing detection of the grid voltage needs to obtain its fundamental component at the line frequency. This task is usually made by a digital filter. In order to avoid the delay introduced by this filter numerous techniques are used in the technical literature. Methods based on advanced filtering techniques are presented in

239

[94]-[98]. Other methods use Neural Networks for detection of the true zero-crossing of the grid voltage waveform [99]-[101]. An improved accuracy in the integrity of the zero-crossing can also be obtained by reconstructing a voltage representing the grid voltage [102]-[105].

However, starting from its simplicity, when the two major drawbacks are alleviated by using advanced techniques, the zero-crossing method proves to be rather complex and unsuitable for applications which require accurate and fast tracking of the grid voltage.

The arctangent function technique is another solution for detecting the phase angle and frequency of the grid voltage. An orthogonal voltage system is required in order to implement this technique. This method is used in adjustable speed drives applications in order to transform the feedback signals to a reference frame suitable for control purposes [19]. However, this method has the drawback that requires additional filtering in order to obtain an accurate detection of the phase angle and frequency in the case of a distorted grid voltage. Therefore, this technique is not suitable for grid-connected converter applications.

Recently, there has been an increasing interest in PLL techniques for grid-connected converter systems [106]. Usually, the PLL technique is mainly applied in communication technologies. Though, it has been proven that its application in the grid-connected converter systems was a success [91], [92], [106]-[126]. Used for such systems, the PLL is a grid voltage phase detection algorithm. The main task of the PLL algorithm is to provide a unitary power factor operation of a grid-connected converter system. This task involves synchronization of the converter output current with the grid voltage, and to provide a clean sinusoidal current reference to the current controller. Moreover, using the PLL, the grid voltage parameters such as amplitude and frequency, can be easily monitored.

Like in the case of the arctangent function technique, an orthogonal voltage system is required for the PLL algorithm. In a three-phase system, the grid voltage information can easily be obtained through the Clarke Transformation. However, for a single-phase system, the grid voltage is much more difficult to acquire [91]. Therefore, more attention should be paid for the generation of the orthogonal voltage system.

The general structures of a single-phase and three-phase PLL including the grid voltage monitoring are presented in Fig. 43a and Fig. 43b respectively. Usually, the main difference among different single-phase PLL methods is the orthogonal voltage system generation structure.

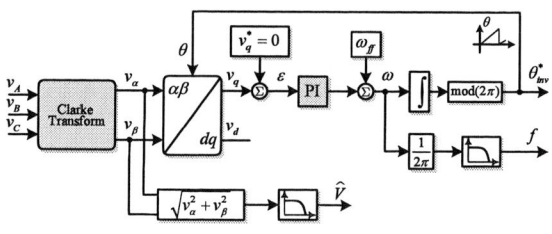

b)

Fig. 43. General structure of a: a) single-phase PLL and b) three-phase PLL.

Next paragraph discusses about techniques used for generating the orthogonal voltage systems. The structure responsible for generating the orthogonal voltage system is called orthogonal signal generator.

b) Orthogonal signal generators

In the technical literature, some techniques for generating the orthogonal voltage components from a single-phase input signal are described, some of which are compared in [106] and [127]. An easy technique of generating the orthogonal voltage system in a single-phase system incorporates a transport delay function, which is responsible for introducing a phase shift of 90 degrees with respect to the fundamental frequency of the input signal [115]. A related method, but more complex of creating a phase shift of 90 degrees, uses the Hilbert Transformation [106] and [110]. Other methods of generating the orthogonal voltage system are based on inverse Park Transformation [106], [115], [122] and [126], using resonant structures such as Second Order Generalized Integrator (SOGI) [117] or Kalman estimator-based filter [112].

2) Grid impedance estimation

In order to comply with certain stringent standard requirements for islanding detection such as the German standard VDE 0126-1-1 [128] for grid-connected PV systems, it is important to estimate the impedance of the distribution line (grid). The standard requirement is to isolate the supply within 5 s after an impedance change of 1 ohm. Therefore, the PV inverters should make use of an online estimation technique in order to meet these regulation requirements. Moreover, the estimation of the grid impedance can also be used in order to increase the stability of the current controller by adjusting its parameters online (see Fig. 46). If the variation is mainly resistive then the damping of the line filter is significant and makes the PV inverter control more stable. As it can be noticed from Fig. 45, if the variation is mainly inductive, then the bandwidth of the controller decreases [129]. Also, in this case, due to the additional inductance of the grid, the tuning order of the line filter becomes lower and the filter will not fulfill the initial design purpose. In order to alleviate this problem, the gain scheduling method can be used for adjusting online the current controller parameters, as presented in Fig. 46. Therefore, besides the standard requirements the knowledge about the grid impedance value is an added feature for the PV inverter [130].

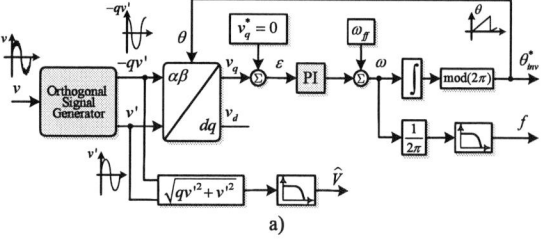

a)

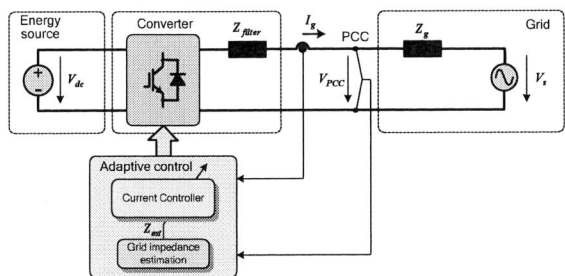

Fig. 44. Adaptive control of the grid-connected inverter [138].

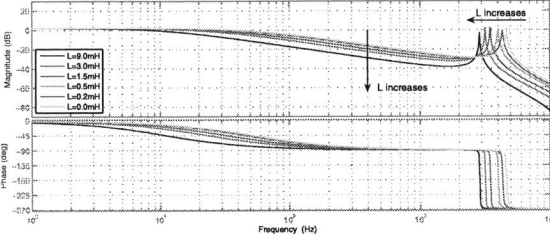

Fig. 45. Bode plot of plant for different values of the grid inductance L in case of using an LCL filter.

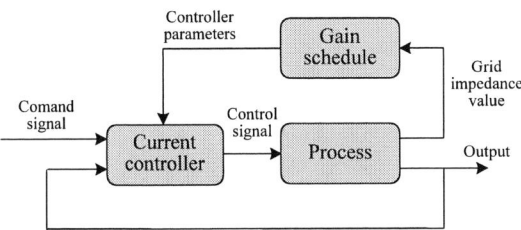

Fig. 46. Gain scheduling method [138].

According to [130] different techniques, as presented in [131]-[136] can be used for line impedance measurements. It is noticeable that, usually, these methods use special hardware devices. Once the inputs are acquired by voltage and current measurement, the processing part follows, typically involving large mathematical calculations in order to obtain the impedance value.

The state of the art divides the measuring solutions into two major categories: the passive and the active methods.

The passive method uses the non characteristic signals (line voltages and currents) that are already present in the system. This method depends on the existing background distortion of the voltage [137] and, in numerous cases, the distortion has neither the amplitude nor the repetition rate to be properly measured. This will not be interesting for implementing it in a PV inverter.

Active methods make use of deliberately "disturbing" the power supply network followed by acquisition and signal processing [131], [132], [133] and [135]. The way of "disturbing" the network can vary, therefore, active methods are also divided into two major categories: transient methods and steady-state methods.

Other two new active methods for estimating the grid impedance are presented in [138] and [139]. The method presented in [138] is based on producing a small perturbation on the output of the power converter that is in the form of periodical variations of active and reactive power (PQ variations). The control diagram for

the implementation of this technique is shown in Fig. 47.

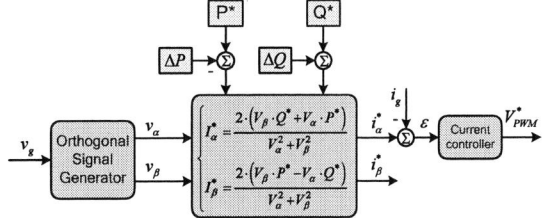

Fig. 47. Control diagram of the PQ control technique [138].

The main idea is to make the power converter working in two operation points (see Fig. 48) in order to solve the equation of the equivalent grid impedance.

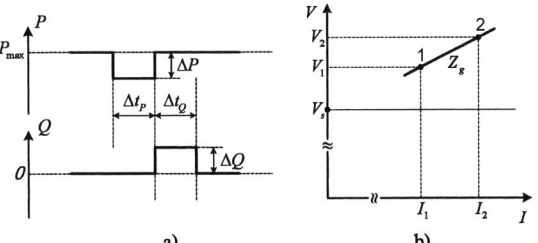

Fig. 48. a) Principle for the variation of active (P) and reactive (Q) power; b) Power converter working in two operation points [138].

During the perturbation, measurements of voltage and current are performed and signal processing algorithms are used in order to estimate the value of the grid impedance.

The method proposed in [139] is based on producing a perturbation on the output of the power converter that is in the form of periodical injection of one or two voltage harmonic signals (see Fig. 49). The single harmonic injection uses a 600 Hz signal and the double harmonic injection uses a 400 Hz and 600 Hz signals, respectively. During the perturbation, the current response(s) at the same frequency as the injected signal(s) is/are measured. The value of the grid impedance is estimated using two different signal processing algorithms. The DFT technique is used for the single harmonic injection and the statistic technique is used for the double harmonic injection (see Fig. 50).

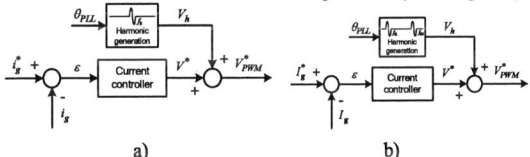

Fig. 49. Harmonic injection methods [139]: a) single harmonic injection; b) double harmonic injection.

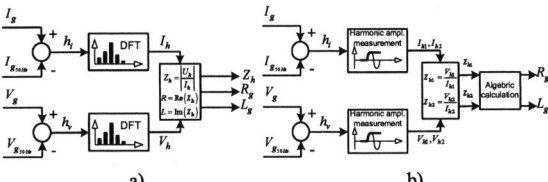

Fig. 50. Grid impedance estimation algorithms [139]: a) single harmonic injection; b) double harmonic injection.

3) Islanding detection

A grid-connected PV system shall cease to energize the utility line from a de-energized distribution line irrespective of connected loads or other generators within specified time limits. This is to prevent back-feeding to the line, also called islanding, which could create hazardous situation for utility maintenance personnel and the general public. Although the probability of islanding occurrence is extremely low [158], standards dealing with the interconnection of inverter based photovoltaic system with the grid require that an effective anti-islanding method is incorporated into the operation of the inverter [87], [140], [141].

The German standard VDE 0126-1-1 [128] for grid-connected PV systems requires isolating the supply within 5 s after an impedance change of 1 ohm. The test setup proposed by this standard is shown in Fig. 51.

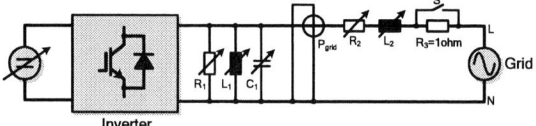

Fig. 51. Test setup for the German standard VDE 0126-1-1 [128].

According to IEEE 929-2000 standard, a PV inverter shall cease to energize the utility line in ten cycles or less when subjected to a typical islanded load in which either of the following is true:

• There is at least a 50% mismatch in real power load to inverter output (that is, real power load is <50% or >150% of inverter power output).

• The islanded-load power factor is <0.95 (lead or lag).

If the real-power-generation-to-load match is within 50% and the islanded-load power factor is >0.95, then a PV inverter will cease to energize the utility line within 2 seconds whenever the connected line has a quality factor of 2.5 or less.

The test setup for the IEEE 929-2000 is depicted in Fig. 52.

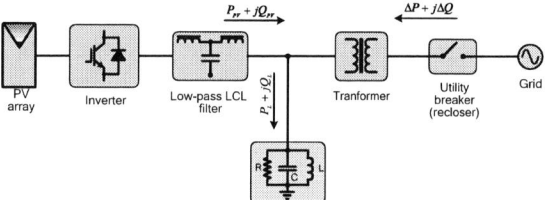

Fig. 52. Islanding operation test setup for IEEE 929-2000 standard [87].

There are numerous islanding detection methods for grid-connected PV systems reported in the technical literature [142]-[157] and their development has been summarized in a number of recent technical papers [147] and reports [142], [143]. They can be classified into two broad categories, namely, passive and active which can be inverter built or utility supported. The passive methods are based on the detection of the following:

• Over-voltage/under-voltage protection (OVP/UVP) [142], [144].

• Over-frequency/under-frequency protection (OFP /UFP) [142], [144].

• Voltage phase jump [142], [144], [147].

• Voltage harmonic monitoring [144], [147].

• Current harmonic monitoring.

However, passive methods have a number of weaknesses and inability to detect islanding. The use of

non-detection zones (NDZs) is used as a measure of performance for both these techniques as well as the active ones in a number of papers [152], [154]. An evaluation of different but most widely-used passive anti-islanding methods is offered for passive methods in [136] and an excellent overview report for both passive and active methods is available in [142].

Active methods have been developed in order to overcome the limitations of the passive methods. In simple terms, active methods introduce perturbations in the inverter output power for a number of parameters as follows:

• Output power variation either real or reactive [144], [155].

• Active frequency drift or frequency shift up/down [142], [147]-[151].

• Sliding mode or slip-mode frequency shift [142], [147], [151].

• Sandia frequency shift or accelerated frequency drift or active frequency drift with positive feedback [147], [150].

• Impedance estimation [138], [139].

• Detection of impedance at a specific frequency or monitoring of harmonic distortion [142], [157].

• Sandia voltage shift [142].

• Frequency jump [142].

In a recent paper, it has been shown that although the effectiveness of passive methods can be established by non-detection zones [146] as represented by the power mismatch space (ΔP vs. ΔQ), in active frequency drifting methods their performance can be evaluated by using load parameter space based on the values of the quality factor and resonant frequency of the local load [154].

Although most of the papers have been concentrated on PV inverters, islanding detection is also needed for all other inverter based systems using different sources such as fuel cells [140], [155]. The algorithm proposed in [155] is an active method and continuously perturbs the reactive power supplied by the inverter by as much as ±5% while monitoring the utility voltage and frequency simultaneously. When islanding occurs, the deviation of the frequency taking place results in a real power reduced to 80%. A drop in voltage positively confirms islanding which in turn results in the inverter being successfully disconnected.

Many papers have concentrated on single-phase inverters and others also address three-phase technology [143], using DQ implementation [156]. Recently, the power mismatch for the 3rd and 5th harmonics and the implementation of an active anti-islanding method using resonant controllers was reported in [157].

Although numerous techniques exist and their implementation varies as it has been discussed so far, it is important to note that a recommendation for robust software based algorithms would simplify matters for the easier adoption of the most robust and simplest technique of all, and this should be kept as a guide for the further development of the anti-islanding technology [158].

V. STATUS AND TRENDS

A. Wind power

The wind turbine market was dominated in the last years by ten major companies [6], [48] and [50]. At the end of 2005 the wind turbine market share by manufacturer was as shown in Fig. 53.

The Danish company VESTAS Wind Systems A/S was still on the top position among the largest manufacturers of wind turbines in the world, followed by GE Wind, as the second largest in the world. German manufacturers ENERCON, Gamesa and Suzlon are in third, fourth and fifth positions, respectively. Notice that, the first four largest suppliers (Vestas, Gamesa, Enercon, GE Wind) had much larger markets with the first leading positions, compared to the others.

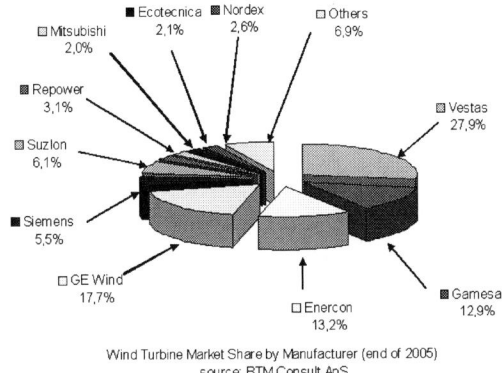

Fig. 53. Wind turbine market share by manufacturer (end of 2005).

Nowadays, the most attractive concept seemed to be the variable speed wind turbine with pitch control. Out of the Top Five-suppliers, only Siemens Wind Power (ex Bonus) used the 'traditional' active stall fixed speed concept, while the other manufacturers had at least one of their two largest wind turbines with the variable speed concept.

However, recently Siemens Wind Power has released the multi-megawatt class variable speed full-scale power converter wind turbine based on the squirrel-cage induction generator. The most used generator type was the induction generator (WRIG and SCIG). Only ENERCON and GE wind used the synchronous generator (WRSG). Only one manufacturer, ENERCON, offered a gearless variable speed wind turbine. All wind turbines manufacturers used a step-up transformer for connection of the generator to the grid.

A trend towards the configuration using a doubly-fed induction generator concept (Type C) with variable speed and variable pitch control, can be identified. In order to illustrate this trend, a dedicated investigation of the market penetration for the different wind turbine concepts is presented in [6]. The analysis cover approximately 75% of the accumulated world power installed at the end of 2004 as shown in Fig. 54.

Full-scale power converter based wind turbines have a relative constant market share over the years, while the interest for the variable-rotor resistance wind turbines (Type B) have fall down in the considered period.

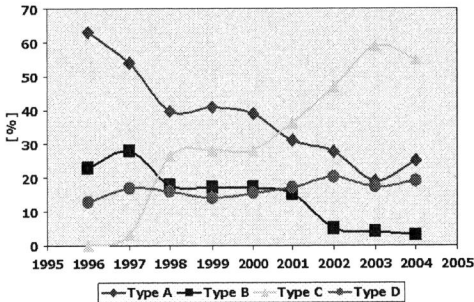

Fig. 54. Wind World share of yearly installed power for the considered wind turbine concepts (see Fig. 7 to Fig. 10).

B. Solar power

PV solar electricity is also a booming industry; since 1980, when terrestrial applications began, annual installation of photovoltaic power has increased to above 750 MWp, the cumulative installed PV power in 2004 reaching approximately 2.6 GWp [159] and [160].

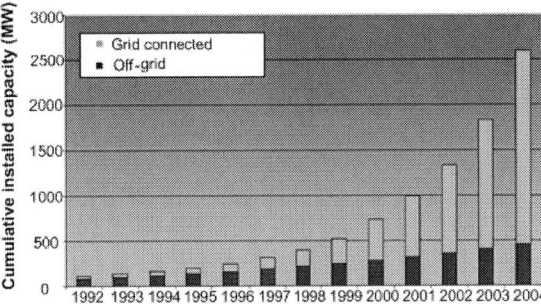

Fig. 55. Cumulative installed capacity from 1992 to 2004 in the IEA-PVPS reporting countries (source: IEA-PVPS, http://www.iea-pvps.org).

The annual rate of growth has varied between 20% in 1994 to over 40% in 2000, but the growth between 2002 and 2003 of 36% has been similar to the latest three years. As in the previous years the vast majority of new capacity was installed in Japan, Germany, and USA, with these three countries accounting for about 88% of the total installed in the year [160].

Historically the main market segments for PV were the remote industrial and developing country applications where PV power over long term is often more cost-effective than alternative power options such as diesel generator or mains grid extension. According to the IEA-PVPS, since 1997, the proportion of new grid-connected PV installed in the reporting countries rose from 42% to more than 93% in 2004 [160] (see Fig. 55).

According to [161], the prices for PV modules are around €5.7/Wp in Europe, with the lowest prices of: €3.10/Wp for monocrystalline modules, €3.02/Wp for polycrystalline modules and €2.96/Wp for thin film modules.

The prices for PV modules in the recent years are shown in Fig. 56.

In addition to the PV module cost, the cost and reliability of PV inverters are basic issues if market competitive PV supply systems are the aim. The inverter cost share represents about 10-15% of the total investment cost of a grid connected system.

243

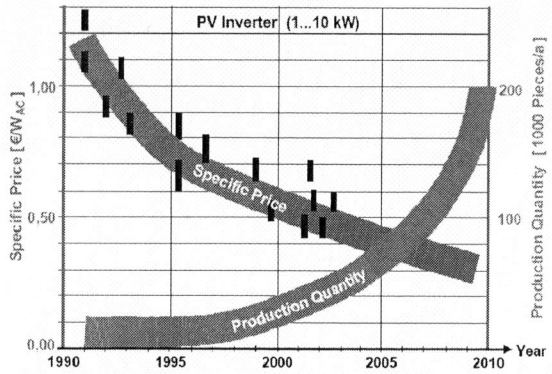

Fig. 56. Development and prognoses of specific cost and production quantity for the PV inverter of nominal powers between 1 and 10 kW during two decades (¦ indicates specific prices of products on the market [162].

The development of PV inverter specific cost (€/WAC) in small to medium power range (1-10 kW) is illustrated in Fig. 56. It can be seen that the inverter cost of this power class has decreased by more than 50% during the last decade. The main reasons for this reduction are the increase of the production quantities and the implementation of new system technologies (e.g. string-inverters). A further 50 % reduction of the specific cost is anticipated during the coming decade. The corresponding specific cost is expected to achieve about 0.3 €/WAC by the year 2010, which requires the implementation of specific measures for the development and the manufacturing processes [162].

VI. CONCLUSION

The paper discusses the applications of power electronic for both wind turbine and photovoltaic technologies. The development of modern power electronics has been briefly reviewed. The applications of power electronics in various kinds of wind turbine generation systems and offshore wind farms are also illustrated, showing that the wind turbine behavior/performance is significantly improved by using power electronics. They are able to act as a contributor to the frequency and voltage control by means of active and reactive power control.

Furthermore, PV systems are discussed including technology, inverters and their control methods.

Finally, a status of the wind turbine and PV market is given and some future trends are highlighted. Both wind and PV will be important power sources for the future energy system.

VII. REFERENCES

[1] S. Heier, "Grid integration of wind energy conversion systems", translated by Rachel Waddington, John Wiley, 1998. ISBN-10: 0-47-197143X.

[2] E. Bossanyi, "Wind Energy Handbook", John Wiley, 2000.

[3] L.H. Hansen, L. Helle, F. Blaabjerg, E. Ritchie, S. Munk-Nielsen, H. Bindner, P. Sørensen and B. Bak-Jensen, "Conceptual survey of Generators and Power Electronics for Wind Turbines", Risø-R-1205(EN), 2001.

[4] F. Blaabjerg, and Z. Chen, "Power electronics as an enabling technology for renewable energy integration", Journal of Power Electronics, vol. 3, no.2, 2003, pp. 81-89.

[5] F. Blaabjerg, F. Iov, R. Teodorescu, Z. Chen, "Power Electronics in Renewable Energy Systems", keynote paper presented at EPE-PEMC Conference, 2006, Portoroz, Slovenia, pp. 1-17.

[6] A.D. Hansen, F. Iov, F. Blaabjerg, L.H. Hansen, "Review of contemporary wind turbine concepts and their market penetration", Journal of Wind Engineering, 28(3), 2004, pp. 247-263.

[7] F. Iov, F. Blaabjerg, "UNIFLEX-PM. Advanced power converters for universal and flexible power management in future electricity network – Converter applications in future European electricity network". Deliverable D2.1, EC Contract no. 019794(SES6), February 2007, p. 171, (available on line www.eee.nott.ac.uk/uniflex/Deliverables.htm).

[8] D. Milborrow, "Going mainstream at the grid face. Examining grid codes for wind", Windpower Monthly, September 2005, ISSN 109-7318.

[9] F. Iov, A.D. Hansen, P. Sørensen, N.A. Cutululis, "Mapping of grid faults and grid codes". Risø-R-1617(EN) (2007) 41 p. (available online at www.risoe.dk).

[10] Z. Chen, E. Spooner, "Grid Power Quality with Variable-Speed Wind Turbines", IEEE Trans. on Energy Conversion, Vol. 16, No.2, June 2001, pp. 148-154.

[11] F. Iov, Z. Chen, F. Blaabjerg, A. Hansen, P. Sorensen, "A New Simulation Platform to Model, Optimize and Design Wind Turbine", Proc. of IECON, 2002, Vol. 1, pp. 561-566.

[12] S. Bolik, "Grid Requirements Challenges for Wind Turbines", Proc. of Fourth International Workshop on Large-Scale Integration of Wind Power and Transmission Networks for Offshore Windfarms, 2003.

[13] E. Bogalecka, "Power control of a doubly fed induction generator without speed or position sensor", Proc. of EPE, 1993, Vol.8, pp. 224-228.

[14] O. Carlson, J. Hylander, K. Thorborg, "Survey of variable speed operation of wind turbines", Proc. of European Union Wind Energy Conference, Sweden, 1996, pp. 406-409.

[15] M. Dahlgren, H. Frank, M. Leijon, F. Owman, L. Walfridsson, "Wind power goes large scale", ABB Review, 2000, Vol.3, pp. 31-37.

[16] M.R. Dubois, H. Polinder, J.A. Ferreira, "Comparison of Generator Topologies for Direct-Drive Wind Turbines", IEEE Nordic Workshop on Power and Industrial Electronics (Norpie 2000), Aalborg-Denmark, pp. 22-26.

[17] L.H. Hansen, P.H. Madsen, F. Blaabjerg, H.C. Christensen, U. Lindhard, K. Eskildsen, "Generators and power electronics technology for wind turbines", Proc. of IECON '01, Vol. 3, 2001, pp. 2000-2005.

[18] Z. Chen, E. Spooner, "Wind turbine power converters: a comparative study", Proc. of PEVD, 1998, pp. 471-476.

[19] M.P. Kazmierkowski, R. Krishnan, F. Blaabjerg,"Control in Power Electronics-Selected problems", Academic Press, 2002. ISBN 0-12-402772-5.

[20] Å. Larsson, "The Power quality of Wind Turbines", Ph.D. report, Chalmers University of Technology, Göteborg, Sweden, 2000.

[21] R. Pena, J.C. Clare, G.M. Asher, "Doubly fed induction generator using back-to-back PWM converters and its application to variable speed wind-energy generation". IEE proceedings on Electronic Power application, 1996, pp. 231-241.

[22] J. Rodriguez, L. Moran, A. Gonzalez, C. Silva, "High voltage multilevel converter with regeneration capability", Proc. of PESC, 1999, Vol.2, pp.1077-1082.

[23] P. Sørensen, B. Bak-Jensen, J. Kristian, A.D. Hansen, L. Janosi, J. Bech, " Power Plant Characteristics of Wind Farms", Proc. of the Int. Conf. in Wind Power for the 21st Century, 2000.

[24] K. Wallace, J.A. Oliver, "Variable-Speed Generation Controlled by Passive Elements", Proc. of ICEM, 1998, pp. 1554-1559.

[25] S. Bhowmik, R. Spee, J.H.R. Enslin, "Performance optimization for doubly fed wind power generation systems", IEEE Trans. on Industry Applications, Vol. 35, No. 4 , July-Aug. 1999, pp. 949-958.

[26] Z. Saad-Saoud, N. Jenkins, "The application of advanced static VAr compensators to wind farms", IEE Colloquium on Power Electronics for Renewable Energy, 1997, pp. 6/1 - 6/5.

[27] J.B. Ekanayake, L. Holdsworth, W. XueGuang, N. Jenkins, "Dynamic modelling of doubly fed induction generator wind turbines", IEEE Trans. on Power Systems, Vol. 18 , No. 2 , May 2003 , pp.803-809.

[28] D. Arsudis, "Doppeltgespeister Drehstromgenerator mit Spannungszwischenkreis Umrichter in Rotorkreis für Wind Kraftanlagen, Ph.D. Thesis, 1998, T.U. Braunschweig, Germany.

[29] D. Arsudis, "Sensorlose Regelung einer doppelt-gespeisten Asynchronmaschine mit geringen Netzrückwirkungen", Archiv für Elektrotechnik, Vol. 74, 1990, pp. 89-97.

[30] T. Matsuzaka, K. Trusliga, S. Yamada, H. Kitahara, "A variable speed wind generating system and its test results". Proc. of EWEC '89, Part Two, 1989, pp. 608-612.

[31] R.S. Barton, T.J. Horp, G.P. Schanzenback, "Control System Design for the MOD-5A 7.3 MW wind turbine generator". Proc. of DOE/NASA workshop on Horizontal-Axis Wind Turbine Technology Workshop, 1984, pp. 157-174.

[32] O. Warneke, "Einsatz einer doppeltgespeisten Asynchronmaschine in der Großen Windenergie-anlage Growian", Siemens-Energietechnik 5, Heft 6, 1983, pp. 364-367.

[33] L. Gertmar, "Power Electronics and Wind Power", Proc. of EPE 2003, paper 1205.

[34] F. Blaabjerg, Z. Chen, S.B. Kjær, "Power Electronics as Efficient Interface in Dispersed Power Generation Systems", IEEE Trans. on PE, Vol. 19, No. 4, 2004, pp. 1184-1194.

[35] E.N. Hinrichsen, "Controls for variable pitch wind turbine generators", IEEE Trans. on Power Apparatus and Systems, Vol. 103, No. 4, 1984, pp. 886-892.

[36] B.J. Baliga, "Power IC's in the saddle", IEEE Spectrum, July 1995, pp. 34-49.

[37] A.D. Hansen, C. Jauch, P. Soerensen, F. Iov, F. Blaabjerg. "Dynamic Wind Turbine Models in Power System Simulation Tool DigSilent", Report Risoe-R-1400 (EN), Dec. 2003, ISBN 87-550-3198-6 (80 pages).

[38] T. A. Lipo, "Variable Speed Generator Technology Options for Wind Turbine Generators", NASA Workshop on HAWTT Technology, May 1984, pp. 214-220.

[39] K. Thorborg, "Asynchronous Machine with Variable Speed", Appendix G, Power Electronics, 1988, ISBN 0-13-686593-3, pp. G1.

[40] D. Arsudis, W. Vollstedt, "Sensorless Power control of a Double-Fed AC-Machine with nearly Sinusoidal Line Currents", Proc. of EPE, 1989, pp. 899-904.

[41] M. Yamamoto, O. Motoyoshi, "Active and Reactive Power control for Doubly-Fed Wound Rotor Induction Generator", Proc. of PESC, 1990, Vol. 1, pp. 455-460.

[42] O. Carlson, J. Hylander, S. Tsiolis, "Variable Speed AC-Generators Applied in WECS", European Wind Energy Association Conference and Exhibition, October 1986, pp. 685-690.

[43] J.D. van Wyk, J.H.R. Enslin, "A Study of Wind Power Converter with Microcomputer Based Maximal Power Control Utilising an Oversynchronous Electronic Schertives Cascade", Proc. of IPEC, 1983, Vol. I, pp. 766-777.

[44] T. Sun, Z. Chen, F. Blaabjerg, "Flicker Study on Variable Speed Wind Turbines With Doubly Fed Induction Generators". IEEE Trans. on Energy Conversion, Vol. 20, No. 4, 2005, pp. 896-905.

[45] T. Sun, Z. Chen, F. Blaabjerg, "Transient Stability of DFIG Wind Turbines at an External Short-circuit-Fault". Wind Energy, 2005, Vol. 8, pp. 345-360.

[46] L. Mihet-Popa, F. Blaabjerg, I. Boldea, "Wind Turbine Generator Modeling and Simulation Where Rotational Speed is the Controlled Variable". IEEE Transactions on Industry Applications, 2004, Vol. 40, No. 1. pp. 3-10.

[47] F. Iov, P. Soerensen, A. Hansen, F. Blaabjerg, "Modelling, Analysis and Control of DC-connected Wind Farms to Grid", International Review of Electrical Engineering, Praise Worthy Prize, February 2006, pp.10, ISSN 1827-6600.

[48] F. Iov, P. Soerensen, A. Hansen, F. Blaabjerg, "Modelling and Control of VSC based DC Connection for Active Stall Wind Farms to Grid", IEE Japan Trans. on Industry Applications, April 2006, Vol. 126-D, No. 5.

[49] A. Cameron, E. de Vries, "Top of the list", Renewable Energy World, James & James, January-February 2006, Vol. 9, No. 1, pp. 56-66, ISSN 1462-6381.

[50] A. Cameron, "Changing winds", BTM's world market update, Renewable Energy World, Pennwell Co., July-August 2006, Vol. 9, No. 4, pp. 28-41, ISSN 1462-6381.

[51] EnergiNet – Grid connection of wind turbines to networks with voltages below 100 kV, Regulation TF 3.2.6, May 2004, p. 29.

[52] Energinet - Grid connection of wind turbines to networks with voltages above 100 kV, Regulation TF 3.2.5, December 2004, p. 25.

[53] ESB Networks – Distribution Code, version 1.4, February 2005.

[54] CER – Wind Farm Transmission Grid Code Provisions, July 2004.

[55] E.ON-Netz – Grid Code. High and extra high voltage, April 2006.

[56] VDN – Transmission Code 2003. Network and System Rules of the German Transmission System Operators, August 2003.

[57] VDN – Distribution Code 2003. Rules on access to distribution networks, August 2003.

[58] National Grid Electricity Transmission plc – The grid code, Issue 3, Revision 17, September 2006.

[59] Gambica Technical Guide - Managing Harmonics. A guide to ENA Engineering Recommendation G5/4-1, 4th Edition, 2006, The Energy Networks Association.

[60] REE – Requisitos de respuesta frente a huecos de tension de las instalaciones de produccion de regimen especial, PO 12.3, November 2005.

[61] ENEL – DK 5400 - Criteri di allacciamento di clienti alla rete AT della distribuzione, October 2004.

[62] ENEL - – DK 5740 - Criteri di allacciamento di impianti di produzione alla rete MT di ENEL distribuzione, February 2005.

[63] TERNA - Codice di trasmissione, dispacciamento, sviluppo e sicurezza della rete, 2006.

[64] CEI 11/32, Appendice N.6 – Normativa impianti di produzione eolica, February 2006 (draft).

[65] IEA International Energy Agency: Trends in Photovoltaic Applications. Survey report of selected IEA countries between 1992 and 2003. Source: http://www.oja-services.nl/iea-pvps/products/download/rep1_13.pdf.

[66] H. Haeberlin, "Evolution of Inverters for Grid connected PV systems from 1989 to 2000", Proc. of Photovoltaic Solar Energy Conference, 2001.

[67] T. Shimizu, M. Hirakata, T. Kamezawa, H. Watanabe, "Generation Control Circuit for Photovoltaic Modules", IEEE Trans. On Power Electronics, Vol. 16, No. 3, May, 2001, pp. 293-300.

[68] M. Calais, V.G. Agelidis, L.J. Borle, M.S. Dymond, "A transformerless five level cascaded inverter based single phase photovoltaic system", Proc. of PESC, 2000, Vol. 3, pp. 1173-1178.

[69] R.W. Erickson, D. Maksimovic, "Fundamentals of Power Electronics", Kluwer Academic Pub; March 1, 1997, ISBN: 0-412-08541-0, 773 pages.

[70] M. Calais, J. Myrzik, T. Spooner, V.G. Agelidis, "Inverters for single-phase grid connected photovoltaic systems - An overview", Proc. of PESC '02, 2002, Vol. 4, pp. 1995 – 2000.

[71] F. Blaabjerg, R. Teodorescu, Z. Chen, M. Liserre, "Power Converters and Control of Renewable Energy Systems", Proc. of ICPE, 2004, pp. 1-19.

[72] R. Teodorescu, F. Blaabjerg, M. Liserre, U. Borup, " A New Control Structure for Grid-Connected PV Inverters with Zero Steady-State Error and Selective Harmonic Compensation", Proc. of APEC, 2004, Vol. 1, pp. 580-586.

[73] S. Fukuda and T. Yoda, "A novel current-tracking method for active filters based on a sinusoidal internal mode", IEEE Trans. on Ind. App., 2001, Vol.37, No. 3, pp. 888-895.

[74] X. Yuan, W. Merk, H. Stemmler and J. Allmeling, "Stationary-Frame Generalized Integrators for Current Control of Active Power Filters with Zero Steady-State Error for Current Harmonics of Concern Under Unbalanced and Distorted Operating Conditions", IEEE Trans. on Ind. App., Vol. 38, No. 2, 2002, pp. 523-532.

[75] M. Ciobotaru, R. Teodorescu, F. Blaabjerg, "Control of single-stage single-phase PV inverter", Proc. of EPE'05, 10 pages, ISBN : 90-75815-08-5.

[76] D.P. Hohm, M.E. Ropp, "Comparative Study of Maximum Power Point Tracking Algorithms Using an Experimental, Programmable, Maximum Power Point Tracking Test Bed". IEEE Proc. of Photovoltaic Specialists Conference, 2000, Pages:1699-1702.

[77] N. Femia, G. Petrone, G. Spagnuolo, M. Vitelli, "Optimizing sampling rate of P&O MPPT technique", Proc. of PESC, 2004, Vol. 3, pp. 1945-1949.

[78] A. Brambilla, M. Gambarara, A. Garutti, F. Ronchi, "New approach to photovoltaic arrays maximum power point tracking", Proc. of PESC, 1999, Vol. 2, pp. 632-637.

[79] X. Liu, L.A.C. Lopes, "An improved perturbation and observation maximum power point tracking algorithm for PV arrays", Proc. of PESC, 2004, Vol. 3, Pages: 2005 - 2010.

[80] D. Sera, T. Kerekes, R. Teodorescu, and F. Blaabjerg, "Improved MPPT method for rapidly changing environmental conditions," in Industrial Electronics, 2006 IEEE International Symposium on, Vol. 2, 2006, pp.1420-1425.

[81] K.H. Hussein, I. Muta, T. Hoshino, M. Osakada, "Maximum photovoltaic power tracking: an algorithm for rapidly changing atmospheric conditions". IEE Trans. on Generation, Transmission and Distribution, Jan. 1995, Vol. 142, No. 1, pp. 59-64.

[82] W. Swiegers, Enslin J.H.R.: "An integrated maximum power point tracker for photovoltaic panels", Proc. of ISIE, 1998, Vol. 1, pp. 40-44.

[83] T.J. Liang, Y.C. Kuo and J.F. Chen, "Single-stage photovoltaic energy conversion system", IEE Proceedings Electric Power Applications, 2001, Vol. 148, No. 4, pp. 339-344.

[84] Y.C. Kuo and T.J. Liang, "Novel Maximum-Power-Point-Tracking Controller for Photovoltaic Energy Conversion System", IEEE Trans. on Industrial Electronics, 2001, Vol. 48, No. 3, pp. 594-601.

[85] M. Nikraz, H. Dehbonei, C.V.N. Curtin, "Digital control of a voltage source inverter in photovoltaic applications", Proc. of PESC, 2004, Vol. 5, 2004, pp. 3266-3271.

[86] "Characteristics of the utility interface for photovoltaic (PV) systems," IEC 61727-2002, 2002.

[87] IEEE Standard 929-2000: IEEE Recommended practice for utility interface of photovoltaic (PV) systems.

[88] M. Francesco De, L. Marco, D.A. Antonio, and P. Alberto, "Overview of Anti-Islanding Algorithms for PV Systems. Part I: Passive Methods," Proc. of EPE-PEMC, 2006, pp. 1878-1883.

[89] F. M. Gardner, "Phaselock Techniques", Publisher: Wiley-Interscience, 1979, Vol. 2nd edition, ISBN-10: 0471042943, 304 pages.

[90] F. Mur, V. Cardenas, J. Vaquero, and S. Martinez, "Phase synchronization and measurement digital systems of AC mains for power converters", Proc. of CIEP, 1998, pp. 188-194.

[91] J. W. Choi, Y.K. Kim, and H.G. Kim, "Digital PLL control for single-phase photovoltaic system", IEE Trans. on Electric Power Applications, 2006, Vol. 153, pp. 40-46.

[92] S.K. Chung, "A phase tracking system for three phase utility interface inverters", IEEE Trans. on Power Electronics, 2000, Vol. 15, pp. 431-438.

[93] C. T. Nguyen and K. Srinivasan, "A New Technique for Rapid Tracking of Frequency Deviations Based on Level Crossings," IEEE Trans. on Power Apparatus and Systems, 1984, Vol. PAS-103, pp. 2230-2236.

[94] B.P. McGrath, D.G. Holmes, J.J.H. Galloway, "Power converter line synchronization using a discrete Fourier transform (DFT) based on a variable sample rate", IEEE Trans. on Power Electronics, 2005, Vol. 20, pp. 877-884.

[95] O. Vainio, S. J. Ovaska, and M. Polla, "Adaptive filtering using multiplicative general parameters for zero-crossing detection", IEEE Trans. on Industrial Electronics, 2003, vol. 50, pp. 1340-1342.

[96] S. Valiviita, S. J. Ovaska, and J. Kyyra, "Adaptive signal processing system for accurate zero-crossing detection of cycloconverter phase currents", Proc. of PCC, 1997, Vol.1, pp. 467-472

[97] O. Vainio and S. J. Ovaska, "Noise reduction in zero crossing detection by predictive digital filtering," IEEE Trans. on Industrial Electronics, 1995, vol. 42, pp. 58-62.

[98] R.W. Wall, "Simple methods for detecting zero crossing", Proc. of IECON, 2003, Vol.3, pp. 2477-2481.

[99] S. Valiviita, "Neural network for zero-crossing detection of distorted line voltages in weak AC-systems", Proc. of IMTC, 1998, Vol.1, pp. 280-285.

[100] S. Das, P. Syam, G. Bandyopadhyay, and A.K. Chattopadhyay, "Wavelet transform application for zero-crossing detection of distorted line voltages in weak AC-systems", Proc. of INDICON, 2004, pp. 464-467.

[101] S. Valiviita, "Zero-crossing detection of distorted line voltages using 1-b measurements", IEEE Trans. on Industrial Electronics, 1999, Vol. 46, pp. 917-922.

[102] R. Weidenbrug, F. P. Dawson, and R. Bonert, "New synchronization method for thyristor power converters to weak", IEEE Trans. on Industrial Electronics, 1993, Vol. 40, pp. 505-511.

[103] D.M. Baker and V.G. Agelidis, "Phase-locked loop for microprocessor with reduced complexity voltage controlled oscillator suitable for inverters," Proc. of PEDES, 1998, pp. 464-469 Vol.1.

[104] D. Nedeljkovic, J. Nastran, D. Voncina, and V. Ambrozic, "Synchronization of active power filter current reference to the network", IEEE Trans. on Industrial Electronics, 1999, vol. 46, pp. 333-339.

[105] D. Nedeljkovic, V. Ambrozic, J. Nastran, and D. Hudnik, "Synchronization to the network without voltage zero-cross detection", Proc. of MELECON, 1998, Vol. 2, pp. 1228-1232.

[106] S. M. Silva, B. M. Lopes, B. J. C. Filho, R. P. Campana, and W. C. Bosventura, "Performance evaluation of PLL algorithms for single-phase grid-connected systems," Proc. of IAS, 2004, Vol.4, pp. 2259-2263.

[107] W. Tsai-Fu, S. Chih-Lung, N. Hung-Shou, and L. Guang-Feng, "A 1phi-3W inverter with grid connection and active power filtering based on nonlinear programming and fast-zero-phase detection algorithm", IEEE Trans. on Power Electronics, 2005, Vol. 20, pp. 218-226.

[108] P. Rodriguez, A. Luna, M. Ciobotaru, R. Teodorescu, F. Blaabjerg, "Advanced Grid Synchronization System for Power Converters under Unbalanced and Distorted Operating Conditions", Proc. of IECON, 2006, pp. 5173-5178.

[109] L. R. Limongi, R. Bojoi, C. Pica, F. Profumo, and A. Tenconi, "Analysis and Comparison of Phase Locked Loop Techniques for Grid Utility Applications", Proc. of PCC, 2007, pp. 674-681.

[110] M. Saitou, N. Matsui, and T. Shimizu, "A control strategy of single-phase active filter using a novel d-q transformation", Proc. of IAS, 2003, Vol. 2, pp. 1222-1227.

[111] P. Rodriguez, J. Pou, J. Bergas, J. I. Candela, R. P. Burgos, and D. Boroyevich, "Decoupled Double Synchronous Reference Frame PLL for Power Converters Control", IEEE Trans. on Power Electronics, 2007, vol. 22, pp. 584-592.

[112] K. De Brabandere, T. Loix, K. Engelen, B. Bolsens, J. Van den Keybus, J. Driesen, and R. Belmans, "Design and Operation of a Phase-Locked Loop with Kalman Estimator-Based Filter for Single-Phase Applications", Proc. of IECON, 2006, pp. 525-530.

[113] P. Rodriguez, J. Pou, J. Bergas, I. Candela, R. Burgos, and D. Boroyevic, "Double Synchronous Reference Frame PLL for Power Converters Control", Proc. of PESC, 2005, pp. 1415-1421.

[114] T. Ostrem, W. Sulkowski, L. E. Norum, and C. Wang, "Grid Connected Photovoltaic (PV) Inverter with Robust Phase-Locked Loop (PLL)", Proc. of TDC, 2006, pp. 1-7.

[115] M. Ciobotaru, R. Teodorescu, and F. Blaabjerg, "Improved PLL structures for single-phase grid inverters", Proc. of PELINCEC, 2005, pp. 1-6.

[116] S. Shinnaka, "A New Frequency-Adaptive Phase-Estimation Method Based on a New PLL Structure for Single-Phase Signals", Proc. of PCC, 2007, pp. 191-198.

[117] M. Ciobotaru, R. Teodorescu, and F. Blaabjerg, "A New Single-Phase PLL Structure Based on Second Order Generalized Integrator", Proc. of PESC, 2006, pp. 1-6.

[118] V. Kaura and V. Blasko, "Operation of a phase locked loop system under distorted utility conditions", Proc. of APEC, 1996, Vol.2, pp. 703-708.

[119] S. K. Chung, "Phase-locked loop for grid-connected three-phase power conversion systems", IEE Trans. on Electric Power Applications, 2000, vol. 147, pp. 213-219.

[120] A. W. Krieger and J. C. Salmon, "Phase-locked loop synchronization with gated control," Proc. of CCECE, 2005, pp. 523-526.

[121] A. V. Timbus, R. Teodorescu, F. Blaabjerg, M. Liserre, and P. Rodriguez, "PLL Algorithm for Power Generation Systems Robust to Grid Voltage Faults", Proc. of PESC, 2006. pp. 1-7.

[122] L. N. Arruda, S. M. Silva, and B. J. C. Filho, "PLL structures for utility connected systems", Proc. of IAS, 2001, Vol. 4, pp. 2655-2660.

[123] E. S. Sreeraj and K. Chatterjee, "Power Factor Improvement in One Cycle Controlled Converter", Proc. of ISIE, 2006, pp. 1454-1460.

[124] S. K. Chung, H. B. Shin, and H. W. Lee, "Precision control of single-phase PWM inverter using PLL compensation", IEE Trans. on Electric Power Applications, 2005, Vol. 152, pp. 429-436.

[125] A. Timbus, M. Liserre, R. Teodorescu, and F. Blaabjerg, "Synchronization Methods for Three Phase Distributed Power Generation Systems. An Overview and Evaluation", Proc. of PESC, 2005, pp. 2474-2481.

[126] L. N. Arruda, B. J. Cardoso Filho, S. M. Silva, S. R. Silva, and A. S. A. C. Diniz, "Wide bandwidth single and three-phase PLL structures for grid-tied", Proc. of Photovoltaic Specialists Conference, 2000, pp. 1660-1663.

[127] J. Salaet, S. Alepuz, A. Gilabert, and J. Bordonau, "Comparison between two methods of DQ transformation for single phase converters control. Application to a 3-level boost rectifier", Proc. of PESC, 2004, Vol.1, pp. 214-220.

[128] DIN VDE 0126-1-1, "Automatic disconnection device between a generator and the public low-voltage grid", June 2005.

[129] M. Liserre, R. Teodorescu, and F. Blaabjerg, "Stability of grid-connected PV inverters with large grid impedance variation", in Proc. of PESC, 2004, pp. 4773–4779.

[130] L. Asiminoaei, R. Teodorescu, F. Blaabjerg, U. Borup, "Implementation and Test of an Online Embedded Grid Impedance Estimation Technique for PV Inverters", IEEE Trans. on Industrial Electronics, 2005, vol.52, no.4, pp. 1136-1144.

[131] M. Sumner, B. Palethorpe, D. Thomas, P. Zanchetta, M.C. Di Piazza, "Estimation of power supply harmonic impedance using a controlled voltage disturbance", Proc. of PESC, 2001, vol.2, pp. 522-527.

[132] M.C. Di Piazza, P. Zanchetta, M. Sumner, D.W.P. Thomas, "Estimation of load impedance in a power system", Proc. of Harmonics and Quality of Power Conference, 2000, vol.2, pp. 520-525.

[133] M. Sumner, B. Palethorpe, D.W.P. Thomas, P. Zanchetta, M.C. Di Piazza, "A technique for power supply harmonic impedance estimation using a controlled voltage disturbance", IEEE Trans. on Power Electronics, 2002, vol.17, no.2, pp. 207-215.

[134] J.P. Rhode, A.W. Kelley, M.E. Baran, "Line impedance measurement: a nondisruptive wideband technique", Proc. of IAS, 1995, vol.3, pp. 2233-2240.

[135] N. Ishigure, K. Matsui, F. Ueda, "Development of an on-line impedance meter to measure the impedance of a distribution line", Proc. of ISIE, 2001, vol.1, pp. 549-554.

[136] Tsukamoto, M.; Ogawa, S.; Natsuda, Y.; Minowa, Y.; Nishimura, S., "Advanced technology to identify harmonics characteristics and results of measuring," Harmonics and Quality of Power, 2000. Proceedings. Ninth International Conference on, vol.1, pp., 341-346.

[137] K.O.H. Pedersen, A.H. Nielsen, and N.K. Poulsen, "Short-circuit impedance measurement," IEE Trans. on Generation, Transmission and Distribution, 2003, vol. 150, no. 2, pp. 169–174.

[138] M. Ciobotaru, R. Teodorescu, P. Rodriguez, A. Timbus and F. Blaabjerg, "Online grid impedance estimation for single-phase grid-connected systems using PQ variations", Proc. of PESC, 2007, pp. 2306-2312.

[139] M. Ciobotaru, R. Teodorescu and F. Blaabjerg, "On-line grid impedance estimation based on harmonic injection for grid-connected PV inverter", Proc. of ISIE, 2007, pp. 2437-2442.

[140] IEEE Standard 1547-2003: IEEE Standard for interconnecting distributed resources with electric power systems.

[141] A. Woyte, K. De Brabandere, D.V. Dommelen, R. Belmans, and J. Nijs, "International harmonization of grid connection guidelines: adequate requirements for the prevention of unintentional islanding", Progress in Photovoltaics: Research and Applications, 2003, Vol. 11, pp. 407-424.

[142] W. Bower and M Ropp, "Evaluation of islanding detection methods for photovoltaic utility-interactive power systems", IEA Task V Report IEA-PVPS T5-09, March 2002.

[143] Z. Ye, R. Walling, L. Garces, R. Zhou, L. Li and T. Wang, "Study and development of anti-islanding control for grid-connected inverters", National Renewable Energy Laboratory, NREL/SR-560-36243, May 2004.

[144] H. Kobayashi, K. Takigawa and E. Hashimoto, "Method for preventing islanding phenomenon on utility grid with a number of small scale PV systems", Proc. of. Photovoltaic Specialists Conference, 1991, pp. 695-700.

[145] A. Kitamura, M. Okamoto, F. Yamamoto, K. Nakaji, H. Matsuda, K. Hotta, "Islanding phenomenon elimination study at Rokko test center", Proc. of Photovoltaic Specialists Conference, 1994, Vol. 1, p. 759-762.

[146] Z. Ye, A. Kolwalkar, Y. Zhang, P. Du and R. Walling, "Evaluation of anti-islanding schemes based on non-detection zone concept", IEEE Trans. on Power Electronics, 2004, Vol. 19, No. 5, pp. 1171-1176.

[147] M.E. Ropp, M. Begovic and A. Rohatgi, "Prevention of islanding in grid-connected photovoltaic systems", Progress in Photovoltaics: Research and Applications, 1999, Vol. 7, pp. 39-59.

[148] M.E. Ropp, M. Begovic and A. Rohatgi, "Analysis and performance assessment of the active frequency drift method of islanding prevention", IEEE Trans. on Energy Conversion, 1999, Vol. 14, No. 3, pp. 810-816.

[149] S. Yuyama, T. Ichinose, K. Kimoto, T. Itami, T. Ambo, C. Okado, K. Nakajima. S. Hojo, H. Shinohara, S. Ioka and M. Kuniyoshi, "A high-speed frequency shift method as a protection for islanding phenomena of utility interactive PV systems", Solar Energy Materials and Solar Cells, 1994, Vol. 35, pp. 477-486.

[150] P. Sanchis, L. Marroyo and J. Coloma, "Design methodology for the frequency shift method of islanding prevention and analysis of its detection capability", Progress in Photovoltaics: Research and Applications, 2005, Vol. 13, pp. 409-428.

[151] G.A. Smith, P.A. Onions and D.G. Infield, "Predicting islanding operation of grid connected PV inverters", IEE Trans. Electrical Power Applications, 2000, Vol. 147, No 1, pp. 1-5.

[152] M.E. Ropp, M. Begovic, A. Rohatgi, G.A. Kern, H. Bonn and S. Gonzalez, "Determining the relative effectiveness of islanding detection methods using phase criteria and non-detection zones", IEEE Trans. on Energy Conversion, 2000, Vol. 15, No. 3, pp. 290-296.

[153] G.K. Hung, C.C. Chang and C.L. Chen, "Automatic phase-shift method for islanding detection of grid-connected photovoltaic inverters", IEEE Trans. on Energy Conversion, 2003, Vol. 18, No. 1, pp. 169-173.

[154] V. John, Z. Ye and A. Kolwalkar, "Investigation of anti-islanding protection of power converter based distributed generators using frequency domain analysis", IEEE Trans. on Power Electronics, 2004, Vol. 19, No. 5, pp. 1177-1183.

[155] L.A.C. Lopes and H. Sun, "Performance assessment of active frequency drifting islanding detection methods", IEEE Trans. on Energy Conversion, 2006, Vol. 21, No. 1, pp. 171-180.

[156] C. Jeraputra and P.N. Enjeti, "Development of a robust anti-islanding algorithm for utility interconnection of distributed fuel cell powered generation", IEEE Trans. on Power Electronics, 2004, Vol. 19, No. 5, pp. 1163-1170.

[157] Z. Ye, L. Li, L. Garces, C. Wang, R. Zhang, M. Dame, R. Walling and N. Miller, "A new family of active anti-islanding schemes based on DQ implementation for grid-connected inverters", Proc. of PESC, 2004, pp. 235-241.

[158] N. Cullen, J. Thornycroft and A. Collinson, "Risk analysis of islanding of photovoltaic power systems within low voltage distribution networks", IEA Report PVPS T5-08, March 2002.

[159] European Photovoltaic Industry Association: EPIA Roadmap. Source: http://www.epia.org/04events/docs/EPIAroadmap.pdf.

[160] IEA International Energy Agency: Trends in Photovoltaic Applications. Survey report of selected IEA countries between 1992 and 2003. Source: http://www.oja-services.nl/iea-pvps/products/download/rep1_13.pdf.

[161] G. Cramer , M. Ibrahim and W. Kleinkauf, "PV System Technologies: State-of-the-art and Trends in Decentralized Electrification." Science Direct-Refocus, Vol. 5, pp. 38-42. source: www.sciencedirect.com, www.re-focus.net.

[162] Mohammad Shahidehpour, Fred Schwartz, "Don't Let the Sun Go Down on PV". IEEE Power and Energy Magazine, 2004, Vol. 2, No. 3, pp. 40-48.

Appendix I. Review of connection requirements for wind power in European grid codes [7].

		Denmark		Ireland	Germany	Great Britain	Spain	Italy (draft)
Voltage Level		DS	TS	DS(TS)	TS(DS)	TS(DS)	TS	> 35 kV
Power Level		all	all	≥5MW	all	all	all	> 10 MW
Tolerance over frequency range		yes	yes	yes	yes	yes	-	yes
Frequency	Frequency control	all	all	all	all	all	-	> 25 MW
	MW Curtailment	20-100% P_r	20-100% P_r	yes	yes	-	-	-
	Maximum Ramp Rates	10-100% P_r/min	10-100% P_r/min	1-30 MW/min	yes	-	-	<20% P_r/min
Voltage	Voltage Control	no	no	yes	no	no	-	no
	Reactive Power Control	yes	yes	yes	yes	yes	-	yes
Voltage quality	Fast voltage variations	≤ 3%	≤ 3%	-	≤ 2%	≤ 3%-	-	EN 50160
	Short Term Flicker Severity	-	≤ 0.3	≤ 0.35	-	≤0.8	-	EN 50160
	Long Term Flicker Severity	≤ 0.35	≤ 0.2	≤ 0.35	≤ 0.46	≤0.6	-	EN 50160
	Harmonic Compatibility Levels	Specific levels	-	Specific Levels[1]	EN 50160	IEC 61000-3-2	-	EN 50160
	THD	-	≤ 1.5%	≤ 1.5%	≤ 8%	N/A	-	EN 50160
Fault ride-through	Fault duration	100 msec	100 msec	625 msec	150 msec	140 msec	500 msec	500 msec
	Min voltage	25%U_r	25%U_r	15%U_r	0%U_r	15%U_r	20%U_r	20% U_r
	Recovery time	1 sec	1 sec	3 sec	1.5 sec	1.2 sec	1 sec	0.3 sec
	Voltage profile	2, 3-ph	1, 2, 3- ph	1, 2, 3- ph	generic	generic	generic	generic
	Reactive current injection	no	no	no	Up to 100%	no	Up to 100%	no
Island operation		not required	not required	not required	not required	not required	not required	not required
Black start capability		not required	not required	may	if required	not required	not required	not required
Signals, Communication and Control	Availability	yes	yes	yes	yes	yes	-	yes
	Active power output	yes	yes	yes	yes	yes	-	yes
	Reactive power output	yes	yes	yes	yes	yes	-	yes
	MW Curtailment	yes	yes	yes	yes	yes	-	-
	Frequency control	yes	yes	yes	yes	yes	-	-
	Circuit breaker status	yes	yes	yes	yes	yes	-	yes
	Meteorological data: Wind speed, wind direction, air pressure and temperature	yes	yes	yes	-	-	-	yes

1) Harmonic compatibility levels are given in general for loads and installations. DSO shall provide a schedule of individual limits where appropriate.

Design and Evaluation of a 60 000 rpm Permanent Magnet Bearingless High Speed Motor

T. Schneider, A. Binder

Institute for Electrical Energy Conversion, Technische Universität Darmstadt, Darmstadt, Germany
Tel.: +49 6151 65263, Fax: +49 6151 166033, Email: tschneider@ew.tu-darmstadt.de

Abstract— Bearingless motors are an alternative to motors with active magnetic bearings and have been investigated in low and medium speed range. The aim of this project is to introduce bearingless motor technology into high speed applications, e.g. compressors or special pumps. The design process and test results of a 60 000 rpm, 500 W prototype high speed bearingless PM motor are presented. Motor design is done using analytical equations accompanied by finite element (FE) calculations. Measurements on the existing two-pole prototype verify the different design steps. The successful operation at 60 000 rpm indicates that the bearingless motor technology is well suited for high speed applications.

Keywords: bearingless motors, magnetic levitation, high-speed machines

I. INTRODUCTION

High speed motors in direct drive applications like compressors, spindles or flywheel energy storage are often suspended with active magnetic bearings (Fig. 1a). Besides many advantages, e.g. low maintenance and low friction, magnetic bearings have some disadvantages, such as increased shaft length, reduced shaft resonance frequencies, need for special iron laminations and dc choppers, which increase system complexity and cost. Bearingless motors can be a good alternative as they include levitation force generation into the active motor parts, reducing the number of additional components [1]. In Fig. 2 the torque and force generation inside a two-pole bearingless motor are shown schematically. The interaction of the two-pole rotor field with a two-pole stator field will generate torque (Fig. 2a). Adding an additional stator field with four poles will generate a lateral force (Fig. 2b). Rotor north and stator south pole in the upper part of the machine attract each other, while the two opposing south poles in the lower half will repel each other. As a result, a vertical lateral force F_y is generated. In general, lateral forces can be generated inside a p_1 pole pair motor, if a second winding system with $p_2 = p_1 \pm 1$ pole pairs is added to the stator [2]. Radial forces generated by this additional winding can be used to provide a non-contact suspension of the shaft. Different arrangements of bearingless motors are conceivable and reported in literature [3]–[10]. Fig. 1a) shows the setup of a conventional high speed drive using two active magnetic bearings. In Fig. 1b) one active magnetic bearing is replaced

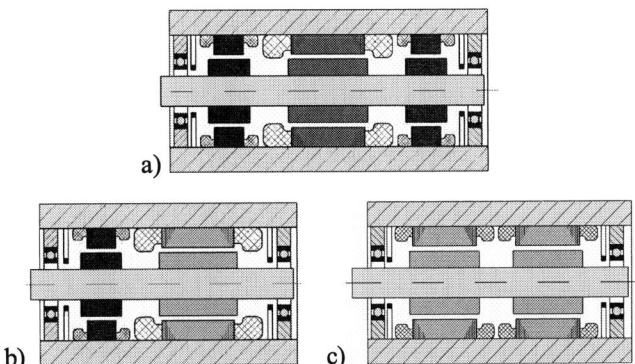

Fig. 1: (a) Radial active magnetic suspension of a high-speed rotor, alternative arrangements with one (b) or two (c) bearingless motors

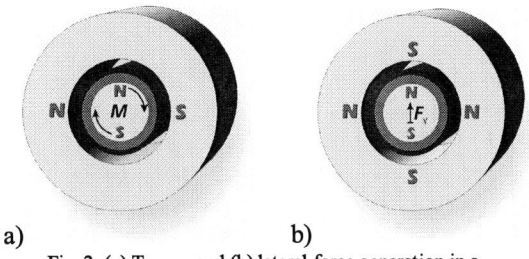

Fig. 2: (a) Torque and (b) lateral force generation in a bearingless motor with a two-pole PM rotor

by a bearingless motor. Together with the remaining active magnetic bearing two radial suspension points are present in this arrangement. In Fig. 1c) the total motor power is split into two bearingless half-motors, which together provide the total torque and two radial suspension points. In case of big axial forces or strict requirements to axial positioning, an axial magnetic bearing needs to be added. Very short bearingless disc motors operate with only one single suspension point. The radial displacement is controlled by the levitation winding, whereas the axial displacement and tilting can be adjusted by the axial magnetic pull of the motor [3]. While bearingless disc-type low-speed motors have already reached market maturity in applications such as medical pumps or clean room drive systems [4], bearingless technology is not yet established in high speed applications [5]. The idea of force generation inside the motor itself as shown in Fig. 1b) is not limited to a certain motor topology. In the past, different motor types, such as induction, synchronous and reluctance motors have been investigated as bearingless motors [6]–[8].

978-1-4244-0644-9/07/$25.00 ©2007 IEEE

II. THEORY OF BEARINGLESS MOTORS

The force vector $\vec{F}$ generated by a bearingless motor as shown in Fig. 2b) is evaluated analytically using two of the nine components of *Maxwell*'s stress tensor $\mathbf{F}_m$ inside the air gap [9].

$$\vec{F} = \begin{pmatrix} F_n \\ F_t \end{pmatrix} = \oint_A \frac{1}{2\mu_0} \begin{pmatrix} B_n^2 - B_t^2 \\ 2B_n B_t \end{pmatrix} dA \tag{1}$$

$$dA = r_{si} d\alpha \cdot l_{Fe}, \qquad \alpha = 0 \dots 2\pi$$

with B_n and B_t being the normal and tangential component of the air gap flux density, μ_0 the permeability of free space, and dA representing the inner stator surface area, calculated from the radius of the stator bore r_{si}, the iron length l_{Fe} and the circumferential angle α. Considering only fundamental components, the p_1 pole pair motor field $B_{\delta,1}$, e.g. generated by permanent magnets mounted onto the rotor surface, is represented by equ. (2). The p_2 pole pair electric loading A_2 (3) excites the additional p_2 pole pair levitation field $B_{\delta,2}$, which is described by equ. (4).

$$B_{\delta,1}(t,\alpha) = \hat{B}_{\delta,1} \cos(p_1\alpha - \omega_1 t - \gamma_1) \tag{2}$$

$$A_2(t,\alpha) = \hat{A}_2 \cos(p_2\alpha - \omega_2 t - \gamma_2) \tag{3}$$

$$B_{\delta,2}(t,\alpha) = \hat{B}_{\delta,2} \sin(p_2\alpha - \omega_2 t - \gamma_2) \tag{4}$$

To distinguish between drive and levitation parameters, subscript "1" is used to label drive winding related parameters, whereas "2" refers to the levitation system. Both magnetic fields and electric loading depend on circumferential angle α, angular frequency ω, time t and phase angle γ. The superposition of these two magnetic fields in the air gap of an electric machine leads to a normal component $B_n(t,\alpha) = B_{\delta,1}(t,\alpha) + B_{\delta,2}(t,\alpha)$ and a tangential flux density component of $B_t(t,\alpha) = \mu_0 A_2(t,\alpha)$. Application of (1) and presuming $p_2 = p_1 \pm 1$ and $\omega_1 = \omega_2$ results in

$$\begin{pmatrix} F_x \\ F_y \end{pmatrix} = \pi \cdot l_{Fe} \cdot r_{si} \left(\frac{\hat{B}_{\delta,1}\hat{B}_{\delta,2}}{2\mu_0} \pm \frac{\hat{B}_{\delta,1}\hat{A}_2}{2} \right) \cdot \begin{pmatrix} \sin(\gamma_1 - \gamma_2) \\ \pm\cos(\gamma_1 - \gamma_2) \end{pmatrix}. \tag{5}$$

The first summand can be interpreted as *Maxwell* pull forces generated by the interaction of two different magnetic fields $B_{\delta,1}$ and $B_{\delta,2}$, whereas the second summand represents *Lorentz* forces generated by the interaction of the p_1 pole pair air gap flux density $B_{\delta,1}$ with the p_2 pole pair current loading A_2. The direction of the force depends only on the relative position of the levitation field $B_{\delta,2}$ with respect to motor field $B_{\delta,1}$, given by $\gamma_1 - \gamma_2$. Replacing the flux density $B_{\delta,2}$ in (5) by the current loading A_2 according to

$$\hat{B}_{\delta,2}(r = r_{si}) = \mu_0 \cdot \sqrt{2} \cdot k_{w,2} \cdot A_2 \frac{r_{si}}{p_2 \cdot \delta} \tag{6}$$

leads to

$$\begin{pmatrix} F_x \\ F_y \end{pmatrix} = \pi \cdot l_{Fe} \cdot r_{si} \frac{\hat{B}_{\delta,1}}{\sqrt{2}} A_2 \left(\frac{r_{si}k_{w,2}}{p_2\delta} \pm 1 \right) \cdot \begin{pmatrix} \sin(\gamma_1 - \gamma_2) \\ \pm\cos(\gamma_1 - \gamma_2) \end{pmatrix}, \tag{7}$$

with $k_{w,2}$ representing the winding factor of the levitation winding. If only the maximum force F in x- or y-direction is considered, i.e. $(\gamma_1 - \gamma_2) = 0°$ or $90°$, equ. (7) can be used to calculate the required electric loading A_2 for a desired lateral force F and given air gap flux density $B_{\delta,1}$:

$$A_2 = \frac{\sqrt{2} \cdot F}{\pi \cdot l_{Fe} \cdot r_{si} \cdot \hat{B}_{\delta,1}} \left[\frac{r_{si} \cdot k_{w,2}}{p_2 \cdot \delta} \pm 1 \right]^{-1}. \tag{8}$$

According to (7), the field orientation of A_2 with respect to $B_{\delta,1}$, given by $\gamma_1 - \gamma_2$, leads to a convenient method for controlling the direction of the generated lateral force $\vec{F}$. For that, we replace the current loading A_2 in (7) by the levitation current I_2 according to

$$A_2 = \frac{2mN_{s,2}I_2}{2r_{si}\pi}, \tag{9}$$

with m as the number of stator phases and $N_{s,2}$ being the number of turns per phase of the three-phase levitation winding. If the instantaneous orientation of the rotor field γ_1 is measured by a rotor position sensor, the orientation of the levitation field γ_2 can be used to adjust the direction of the lateral force via the feeding inverter. Hence, if the phase angle difference $(\gamma_1 - \gamma_2)$ is $90°$, levitation and drive field are aligned (see (2),(4)), and a force in x-direction is generated. If the phase angle difference is $0°$, drive and levitation field are perpendicular and a force in y-direction is obtained. The three-phase levitation current system is transformed into a component I_d that generates a field in line with the rotor field, and one that generates a field perpendicular to it, which is referred to as I_q. Therefore, combining (7) and (9) and defining $I_{2d} = I_2 \sin(\gamma_1 - \gamma_2)$ and $I_{2q} = I_2 \cos(\gamma_1 - \gamma_2)$, the lateral forces F_x and F_y can be expressed as

$$\begin{pmatrix} F_x \\ F_y \end{pmatrix} = l_{Fe} m N_{s,2} \frac{\hat{B}_{\delta,1}}{\sqrt{2}} \left(\frac{r_{si}k_{w,2}}{p_2\delta} \pm 1 \right) \cdot \begin{pmatrix} I_{2,d} \\ \pm I_{2,q} \end{pmatrix} = k_i \cdot \begin{pmatrix} I_{2,d} \\ \pm I_{2,q} \end{pmatrix}. \tag{11}$$

Established mathematical procedures, known from the field oriented control of drives, are therefore adapted to the operation of the levitation system. While in drive systems the q-current is used for torque generation and the d-current for field weakening, in the $2p_2$-pole levitation system these two components are used to generate independent forces in two perpendicular directions.

Apart from currents in the levitation winding, the radial displacement x of the rotor will also cause a lateral force,

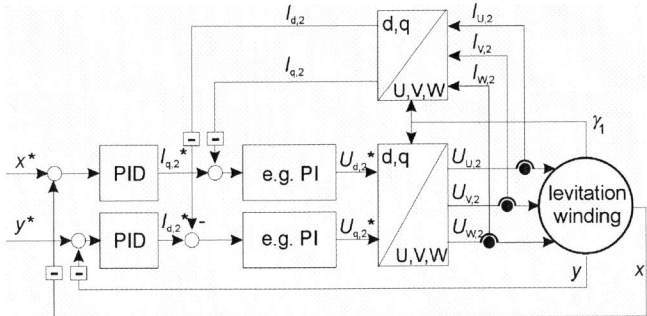

Fig. 3: Cascaded position control with nested current control loop and U-V-W to *d-q* transformations

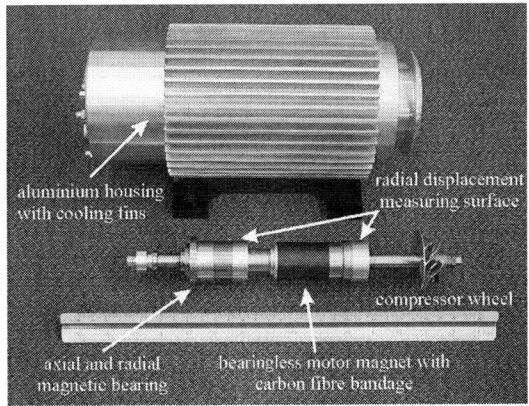

Fig. 4: Prototype components of the bearingless high-speed motor

which is well known as single-sided magnetic pull in electric machines with rotor eccentricity [10]. This force is oriented into the direction of the minimum air gap $\delta_{\min} = \delta - x$.

$$p_1 = 1: \qquad F = \frac{\tau_{p,1} l_{Fe}}{4\mu_0} \hat{B}_\delta^2 \frac{x}{\delta} \sim x \qquad (12a)$$

$$p_1 > 1: \qquad F = \frac{\tau_{p,1} l_{Fe}}{2\mu_0} \hat{B}_\delta^2 p_1 \frac{x}{\delta} \sim x \qquad (12b)$$

The superposition of the lateral force vector generated by the three-phase levitation current system (11) and due to the single-sided radial pull, oriented in the direction of the eccentricity vector $\vec{e} = (x, y)$ (12), leads to the total force on the rotor:

$$\vec{F} = \begin{pmatrix} F_x \\ F_y \end{pmatrix} = k_i \begin{pmatrix} I_d \\ \pm I_q \end{pmatrix} + k_x \begin{pmatrix} x \\ y \end{pmatrix} \qquad (13)$$

This equation shows, that from a control point of view, a bearingless motor operated with field oriented control behaves identically to an active magnetic bearing, with two characteristic parameters, the force-current factor k_i and the force-displacement factor k_x. In active magnetic bearings two different DC control currents I_x and I_y, flowing in separate coils, generate forces in two perpendicular directions [11]. These two currents resemble the d- and q-components of the stator current in the bearingless motor case. Active magnetic bearing control algorithms developed in the past can directly be adapted to the control of bearingless motors. Cascaded position and current control is shown in Fig. 3. An inner control loop (e.g. PI-controller) controls the current components I_d and I_q, whose setpoint values are generated by a cascaded outer position control loop (e.g. PID-controller).

III. DESIGN OF A BEARINGLESS HIGH-SPEED SYNCHRONOUS MOTOR FOR 60 000 RPM

To introduce bearingless motor technology into the high-speed range a prototype bearingless synchronous motor with a rated power $P_N = 500$ W at a rated speed of $n_N = 60\,000$ rpm is investigated (Table 2). The prototype machine can be loaded

in the test bench with an eddy-current brake for low and medium speed and with a turbo charger compressor wheel attached to the shaft for high speed. As the turbo charger compressor wheel generates an axial force, a magnetic axial bearing is required along with two radial suspension points. These are realized with a combination of one bearingless motor (motor and radial bearing) and one combined radial-axial magnetic bearing with permanent magnet bias excitation. The integration of the radial and axial levitation force generation into one unit leads to a rather short shaft length. The total shaft design is based on the arrangement depicted in Fig. 1b. Fig. 4 shows the stator and rotor components of the PM synchronous prototype motor.

1) Electromagnetic design

The electromagnetic design of a permanent magnet bearingless motor is done in two steps. First, the drive system is designed following the design rules for high speed permanent magnet machines. However, only a certain share of the total slot area may be occupied by the drive winding to ensure enough space for the levitation winding, which will be added to the stator in a second step.

a) Design of the drive system

The rotor consists of a massive shaft made of magnetic stainless steel. A diametrically magnetized two-pole ($p_1 = 1$) permanent magnet ring made of bonded neodymium iron boron material is glued onto the shaft and fixed by a carbon fiber bandage. The bandage has a small diameter undersize Δd and is axially pressed onto the magnet. Stretching of the bandage generates the required pressure between magnet and bandage that ensures structural integrity of the rotor at high-speed operation. Despite the high rotational speed of 60 000 rpm, the small rotor outer diameter $d_{ro} = d_{si} - 2\delta = 29.2$ mm leads to a moderate surface velocity of $v = \pi n d_{ro} = 92$ m/s $= 330$ km/h, so a small bandage thickness h_b of about 1 mm is used. Guidelines for the design of carbon fiber bandages for high-speed machines can be found in [12]. Stator iron losses $P_{d,Fe}$ are proportional to $(\sigma_{Hy}(f/50) + \sigma_{Ft}(f/50)^{1.5...2})B^2$, with σ_{Hy} and σ_{Ft} representing hysteresis

TABLE I
GEOMETRY OF BEARINGLESS PM MOTOR PROTOTYPE

Parameter	Symbol	
Number of stator slots	Q	12
Iron stack length	l_{Fe}	36 mm
Stator outer diameter	d_{sa}	60 mm
Stator inner diameter	d_{si}	32 mm
Mechanical air gap width	δ_m	1.4 mm
Magnet height	h_m	3.5 mm
Magnet remanence (GPM-12D), 20°C	B_R	0.7…0.8 T

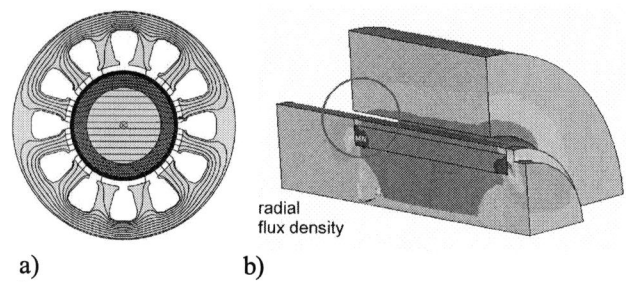

a)   b)

Fig. 5: a) No-load 2D-FE calculation, b) no-load 3D-FE calculation (slotting neglected) showing flux reduction on one stator end due to axial fringing, caused by close proximity of the magnet to the magnetic shaft

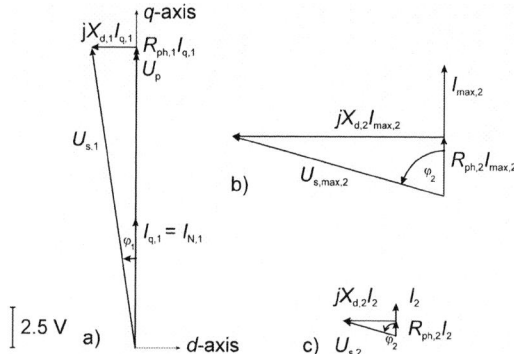

Fig. 6: Phasor diagrams of bearingless prototype motor at rated speed (drawn to scale): a) drive winding at rated torque, b) levitation winding at maximum levitation force $F = 18$ N, c) levitation winding at weight force $F_{BM,0} = 4.6$ N

and eddy current losses per kg of the iron sheets at 50 Hz, 1 T. Due to the high fundamental stator frequency $f_N = n_N \cdot p_1 = 1000$ Hz, iron losses must be kept low by using special high frequency iron sheets and by reducing the air gap flux density to $B_{\delta,1} = 0.3 \ldots 0.4$ T (see III.1.d). The small air gap flux density is achieved by appropriate design of magnet height, bandage size and air gap width (Table 1). While 2D-FE calculations, using a magnet remanence of $B_R = 0.7$ T at 20 °C (worst case, see Table 1), yield a no-load air gap flux density of $B_{\delta,1} = 0.332$ T (Fig. 5a), 3D calculations reveal that axial fringing will reduce the flux linkage with the stator winding by 6 % (Fig. 5b). On one end, the rotor magnet edge is aligned with the stator end and pushed against a shaft rim, which is made of magnetic steel, so some flux lines will not penetrate the stator surface. Therefore, the iron stack length of $l_{Fe} = 36$ mm is reduced by an axial fringing factor of $k_{fr} = 0.94$ to an active iron length of $l'_{Fe} = 34$ mm. On the other end of the rotor, the magnet is slightly longer than the stator in order to be used as speed sensor in combination with two *Hall*-switches. This magnet overhang avoids fringing on the second stator side.

The bearingless motor is supplied by an inverter, which operates at a DC link voltage of $U_{DC} = 65$ V and provides a maximum r.m.s. phase voltage of $U_{ph,max} = 21$ V, with a maximum r.m.s. phase current of $I_{ph,max} = 9$ A. The back e.m.f. U_p of the drive winding should be about 90% of $U_{ph,max}$, to avoid the necessity of field weakening. This voltage is used to determine the number of turns of the drive winding $N_{s,1}$. According to

$$U_p = \sqrt{2}\pi f_N \cdot k_{w,1} N_{s,1} \frac{2}{\pi} \tau_p l_{Fe} k_{fr} B_{\delta,1} \quad , \qquad (14)$$

$N_{s,1} = 12$ yields a back e.m.f. of $U_p = 18.5$ V. The stator winding is realized with five strands in hand ($a_i = 5$, $q_{Cu,1} = 1.76$ mm²) as a single layer winding with $q_1 = 2$ slots per pole and phase, giving a winding factor of $k_{w,1} = 0.966$. The pole pitch is $\tau_{p,1} = \pi d_{si}/(2p_1) = 50.3$ mm. The fundamental stator frequency for nominal speed is $f_N = 1000$ Hz. The phase resistance at maximum winding temperature 155°C (Thermal Class F, acc. to IEC 60034-1), including current displacement of first order (unequal current sharing between parallel strands per turn) and second order (non-uniform current distribution in each conductor cross section) is $R_{ph,155°C,1} = 0,041\ \Omega$ [10]. The

synchronous inductance, comprising of the magnetizing inductance $L_{h,1}$ and the sum of slot, winding overhang and tooth tip leakage inductances $L_{\sigma,1}$ is $L_{d,1} = 51\ \mu H$. Assuming q-current operation ($I_q = I_N$, $I_d = 0$), the air gap power is calculated as

$$P_{\delta,N} = 3 \cdot U_p \cdot I_q = 3 \cdot 18.5V \cdot 9A = 500\ W \ , \qquad (15)$$

giving an air gap torque of

$$M_{\delta,N} = \frac{P_\delta}{2\pi n_N} = 0.08\ \text{Nm} \qquad . \qquad (16)$$

The output power is lower due to losses (see section III.1.d). Table 2 summarizes no-load and rated load data of the drive winding. The r.m.s.-value of the fundamental of the rated phase voltage $U_{N,1}$ and the fundamental power factor $\cos\varphi$ are calculated from the phasor diagram shown in Fig. 6a).

$$U_{N,1} = \sqrt{(U_p + R_{ph,155°C,1}I_{N,1})^2 + (\omega_s L_{d,1}I_{N,1})^2} = 19.1\ V \quad (17)$$

$$\cos\varphi_1 = \frac{U_p + R_{ph,155°C,1}I_{N,1}}{U_{N,1}} = 0.988 \qquad (18)$$

b) Design of the levitation winding

To design the levitation winding, the required lateral force of the bearingless motor $F_{BM,0}$ must be determined. Fig. 7 shows the shaft cross section with the location of the centre of

TABLE II
NO-LOAD, RATED MOTOR AND LEVITATION DATA

Parameter	Symbol	
No-load air gap flux density	$B_{\delta,1}$	0.332 T
Motor data		
Number of poles	$2p_1$	2
Rated motor phase voltage (r.m.s.)	$U_{N,1}$	19.1 V
Rated motor phase current (r.m.s.)	$I_{N,1}$	9 A
Rated motor speed	n_N	60 000 rpm
Rated stator frequency	f_N	1000 Hz
Rated power factor (fundamental)	$\cos\varphi_1$	0.988
Rated air gap power	$P_{\delta,N}$	500 W
Levitation data		
Number of poles	$2p_2$	4
Rated lateral force	F_N	10 N
Rated levitation current (r.m.s.)	$I_{N,2}$	5 A
Rated phase voltage (r.m.s.)	$U_{N,2}$	7.75 V
Weight lateral force	$F_{BM,0}$	4.6 N
Weight levitation current (r.m.s.)	$I_{W,2}$	2.3 A
Force-current factor	k_i	2 N/A
Force-displacement factor	k_x	6.2 N/mm
Rated power factor (fundamental)	$\cos\varphi_2$	0.26

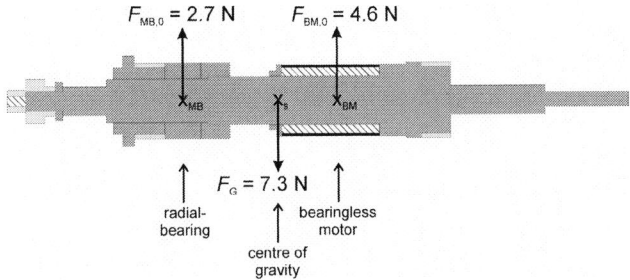

Fig. 7: Prototype shaft with centre of gravity and weight force sharing between radial bearing and bearingless motor

gravity x_S and the points of force application of the radial bearing x_{MB} and of the bearingless motor x_{BM}. The sharing of the shaft weight force $F_G = 7.3$ N between the two radial bearing points is calculated with the law of force and torque equilibrium (Fig. 7). Without any external radial force applied to the shaft, the bearingless motor must provide $F_{BM,0} = 4.6$ N to lift the rotor. The rated force of the bearingless motor is chosen to be 2.2 times the weight force $F_N = 10$ N, generated with a rated levitation current of $I_{N,2} = 5$ A. At maximum inverter output current $I_{max,2} = 9$ A, a maximum lateral force of $F_{max} = 18$ N can be generated, as the iron is unsaturated. However, this force is only available for short term transient disturbances to avoid overheating of the motor winding. As the levitation winding pole count must be $p_2 = p_1 \pm 1$, a four pole levitation winding must be added to the two pole motor winding. With a total of $Q = 12$ stator slots, we get $q_2 = Q/(2mp_2) = 1$ slot per pole and phase for the levitation winding. This winding is designed as a single layer winding, so no short-pitching is possible. Therefore, the winding factor of the levitation winding is $k_{w,2} = 1$. Equ. 8 yields a required rated current loading of $A_{N,2} = 107$ A/cm for a rated levitation force of $F_N = 10$ N. If the rated levitation current loading shall be provided with a rated levitation current of $I_{N,2} = 5$ A, the following number of turns per phase must be chosen:

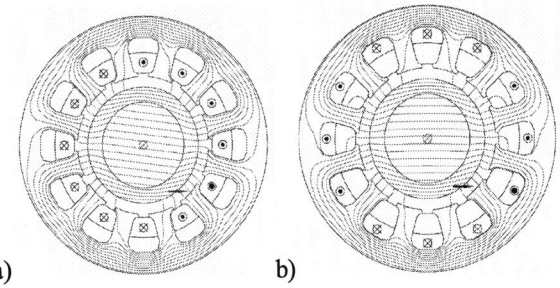

a)   b)

Fig. 8: FE-verification of a) torque (two-pole current system) and b) lateral force (four-pole current system) (program: FEMAG)

$$N_{s,2} = \frac{A_{N,2}d_{si}\pi}{2mI_{N,2}} = 36 \,. \tag{19}$$

The two coil groups of the four pole single layer winding are connected in series (no parallel branches, $a_a = 1$), so the number of turns per coil is

$$N_{c,2} = \frac{N_{s,2}a_a}{p_2q_2} = 18 \,. \tag{20}$$

The copper cross section of the levitation winding is $q_{Cu,2} = 0.4$ mm², resulting in a stator resistance at maximum temperature 155°C (Thermal Class F), including current displacement effects at 1000 Hz, of $R_{ph,155°C,2} = 0,41$ Ω. The synchronous inductance, comprising of main and leakage inductance, is $L_{d,2} = 238$ µH. As the rotating two-pole magnetic field of the drive system cannot induce any voltage into the four-pole winding, no back e.m.f. exists in the levitation winding. The phasor diagram consists only of the phase voltage $U_{s,2}$ and the voltage drop across $R_{ph,2}$ and $X_{d,2} = 2\pi f \cdot L_{d,2}$ (Fig. 6). The main data of the levitation winding is stated in Table 2. The electromagnetic motor utilization C is given by apparent air gap power S_δ versus rotor active volume and speed, hence for q-current operation $P_\delta = S_\delta = P_{\delta N}$. So *ESSON's* number is $C = P_{\delta,N}/(d_{si}^2 l_{Fe} n) = 0.226$ kWmin/m³. As 41% of the slot copper cross section is required for the levitation winding, C is reduced to 100% − 41% = 59% in comparison to a PM motor of the same torque without levitation winding. The thermal loading for rated torque and weight force compensation is $A \cdot J = A_{N,1} \cdot J_{N,1} + A_{W,2} \cdot J_{W,2} = 613$ A/cm·A/mm². For a totally enclosed motor with natural air cooling this requires an increased cooling surface of the aluminum housing with cooling fins.

c) Design verification with FE calculations

In order to verify torque and force generation of the bearingless prototype motor, FE-calculations are carried out and the results are compared to the machine performance as expected from analytical equations. Fig. 8a) shows the results of the magnetostatic torque calculation. The two-pole drive winding, located at the slot openings, is fed with rated q-current $I_N = I_q = 9$ A. The air gap torque is found to be

$M_{\delta,\text{FE}} = 0.078$ Nm, which is in good accordance with the calculated torque of 0.08 Nm from equ. (16). In Fig. 8b), the four-pole levitation winding is supplied with rated levitation current $I_{N,2} = 5$A. The vertical lateral force generated under these circumstances is $F_N = F_y = 10.56$ N, which is only 5.6% bigger than expected from equ. (11). Due to the low iron saturation the analytical results gained for $\mu_{\text{Fe}} \to \infty$ are well fitting to the FE results. For higher magnetic utilization the FE results would yield lower forces. When changing the orientation of the levitation current by 90°, the force in y-direction gets zero, and rated force $F_N = 10.56$ N is generated in a horizontal direction. The results of analytical and FE-calculations are compared to measurements in section 5.

d) Losses inside the prototype motor

In high-speed synchronous machines, the total losses comprise mainly of the following components: resistive losses with consideration of current displacement effects $P_{d,\text{Cu}}$, iron losses in the stator iron P_{Fe}, air friction losses P_{fr} and additional eddy current losses due to harmonic field components and inverter supply. The I^2R loss component is calculated according to (21).

$$P_{d,\text{Cu},(1,2)} = 3 \cdot R_{\text{ph},155°C,(1,2)} \cdot I^2_{N,(1,2)} \tag{21}$$

The no-load iron losses $P_{d,\text{Fe},0}$ due to the rotating permanent magnet field are calculated separately for teeth and yoke, due to different local peak flux densities $\hat{B}_t$ and $\hat{B}_y$:

$$P_{d,\text{Fe},0,(t,y)} = m_{(t,y)}k_{(t,y)} \cdot \left(\sigma_{\text{hy}}\frac{f}{50\,\text{Hz}} + \sigma_{\text{Ft}}\left(\frac{f}{50\,\text{Hz}}\right)^x\right)\left(\frac{\hat{B}_{(t,y)}}{1\,\text{T}}\right)^2 \tag{22}$$

(typical high speed sheets: $\sigma_{\text{hy}} = 0.4$ W/kg, $\sigma_{\text{Ft}} = 0.5$ W/kg, $x = 1.5 \ldots 1.8$)

with $m_{(t,y)}$ being teeth or yoke iron mass and $k_{(t,y)}$ representing deterioration factors due to manufacturing, e.g. $k_t = 1.8$, $k_y = 1.3$ [10]. The frequency f describes the magnetic pulsation with the tooth or yoke peak flux density $\hat{B}_{(t,y)}$. Iron losses at rated load can be calculated from the no-load iron losses with the square of the total voltage:

$$P_{d,\text{Fe},N} = \left(\frac{U_{s,1}}{U_p}\right)^2 \cdot P_{d,\text{Fe},0} \tag{23}$$

For the calculation of air friction losses, the shaft shown in Fig. 7 is separated into 17 sections. Some of these sections are rotating inside an air gap, while others can be considered as rotating in free space. The losses of each section are

$$P_{d,\text{Fr},N} = \sum_{i=1}^{17} c_{w,i}\pi\rho_{\text{air}}(2\pi n_N)^2 r_i^4 l_i \tag{24}$$

with $c_{w,i}$ depending on both the *Reynolds* number Re and the *Taylor* number Ta according to [13]. The motor losses are summarized in Table 3. Additional rotor losses due to field

TABLE III
LOSSES AND EFFICIENCY OF THE PROTOTYPE MOTOR

Parameter	Symbol	
Electrical input power	$P_{\text{el,in}}$	510 W
Resistive losses in drive winding	$P_{d,\text{Cu},1}$	10 W
No-load iron losses	$P_{\text{Fe},0}$	23.5 W
Rated iron losses	$P_{\text{Fe},N}$	25.0 W
Rated air friction losses	$P_{\text{fr},N}$	51.6 W
Total losses in drive system	P_d	86.6 W
Mechanical output power	P_m	423.4 W
Rated motor efficiency	η_N	83 %
Resistive losses in lev. winding		
a) at $I_2 = I_{N,2}$	$P_{d,\text{Cu},N,2}$	30.75 W
b) at $I_2 = I_{W,2} = 2.3$ A	$P_{d,\text{Cu},2}$	6.5 W

harmonics, e.g. slotting effects, are negligible due to the large magnetic air gap. The contribution of current harmonics, caused by the inverter supply is negligible, as a PWM frequency of 40 kHz causes a nearly sinusoidal phase current.

2) Mechanical Design

The mechanical design of high speed machines needs special attention to ensure structural integrity at high speed, including flexural bending resonance effects of the shaft and rotor balancing. As it can be seen in Fig. 4 and 7, the rotating parts of the axial and radial magnetic bearing, the rotor magnet, the carbon fiber bandage and the high-strength aluminum distance sensor ring are mounted on the shaft, most of them by shrink-fitting. These fittings must be carefully designed to ensure that under all thermal and speed values (e.g. overspeed 1.2 n_N) the residual pressure between shaft and attached part is larger than zero, and that the mechanical stress inside the component does not exceed its yield strength $R_{p0.2}$. These calculations are done using analytical formulas for rotating rings under inner and outer pressure [14]. As an example, the maximum tangential stress σ_t at 20% overspeed $n_{\text{max}} = 72\,000$ rpm and at 150 °C, which represents worst case conditions, inside the aluminum ring, used to detect the radial displacement, is $\sigma_{t,\text{max}} = 358$ N/mm². This value does not exceed the yield strength of high-strength aluminum $R_{p0.2} = 500$ N/mm². Fig. 9 shows the calculated natural bending frequencies and bending modes of the shaft. The calculations are done without consideration of gyro effects. The rigid body oscillations are strongly influenced by the stiffness of the suspension, whereas the natural bending modes are nearly independent from bearing stiffness. The calculations in Fig. 9 were performed assuming a bearing stiffness of $c = 200$ N/mm. The rated speed of 1000 Hz (60 000 rpm) is with 43 % of the first natural bending frequency $f_{\text{nb},1} = 2323$ Hz well below the 70% margin, so the rotor can be treated as a rigid body. As such, the rotor was balanced in two planes. Negative balancing was chosen. Mass was removed from the rotor by drilling in two planes to achieve a balance quality of $Q = 1$ mm/s. In case of a power or controller failure in the levitation system, the rotor needs to decelerate safely from maximum speed to standstill. Therefore, backup ball bearings are installed on each shaft end to catch the rotor. On one side, a standard ball bearing is used,

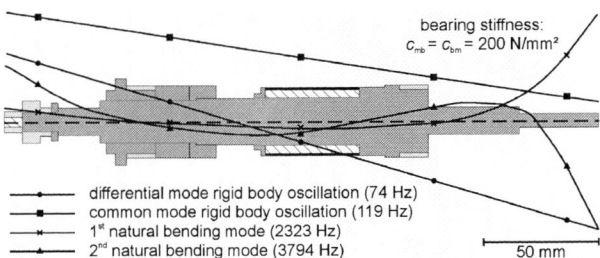

Fig. 9: Calculated natural bending frequencies and modes of prototype shaft.

whereas on the second side, a combination of two angular-contact ball bearings is used to operate as an axial backup bearing, too. The clearance of all backup bearings in radial and axial direction is 0.15 mm.

IV. PROTOTYPE SYSTEM

The bearingless high-speed prototype motor is fed by two commercially available digital magnetic bearing controllers. One unit supplies five axes of a magnetic bearing system with DC current and is prepared to read the signals of displacement sensors to detect the radial and axial shaft position. This device is used to control the combined axial/radial magnetic bearing. In the second unit two independent three-phase current systems can be generated, which are used to feed drive and levitation winding of the bearingless motor. Two separate three-phase inverters could be used for that purpose, too. The core of the controller is a digital signal processor, which contains the control software. The control of the levitation winding corresponds to the control scheme shown in Fig. 3. In an inner control loop, the levitation current is decomposed into d- and q-components, whose setpoint values are generated by a superimposed position controller. As radial bearing and bearingless motor are supplied by two separate inverters, both bearing planes are controlled independently. The motor is operated with a PWM frequency of 40 kHz. The shaft position is measured with displacement sensors and supplied to the controllers. As the field oriented control of both drive and levitation winding requires the instantaneous angular position of the shaft, *Hall*-switches are used to determine the shaft rotation. Furthermore, the bearingless motor and the radial/axial bearing are equipped with temperature sensors in the windings for thermal protection.

V. EXPERIMENTAL INVESTIGATION OF PROTOTYPE MOTOR

The prototype motor has been built and successfully tested in its full speed range up to 60 000 rpm. Fig. 10 shows the prototype motor on the test bed with the turbo charger compressor wheel attached to the shaft. As the air gap flux density is one of the key parameters for both torque and lateral force generation, this quantity is measured via the induced back e.m.f. U_p. The compressor wheel is mounted on the shaft and propelled with compressed air to operate the machine as a PM synchronous generator at no load. The induced line-to-line voltage of the drive winding is measured in a speed range up to 13 500 rpm. Fig. 11 shows the measured and calculated

Fig. 10: Prototype motor with compressor wheel on the test bed

values of the induced no-load voltage (back e.m.f.) U_p. The average air gap flux density calculated from these measurements with equ. (14) is $B_{\delta1,\text{meas}} = 0.3335$ T, which is only 0.5 % deviation from the value obtained from FE-calculations (Table 2). The lateral force of the bearingless motor is measured by attaching calibrated weights from 100 g to 400 g to the shaft drive end. If this additional weight is considered at the shaft end in Fig. 7, the new lateral force of the bearingless motor can be calculated. In Fig. 12, the measured lateral force and torque are compared to expected values obtained from analytical and FE calculations. The measurement was done for each additional weight at different speeds from 3 000 rpm to 24 000 rpm, showing that the levitation current is nearly independent from rotational speed. The torque measurement in dependence of the drive currents I_1

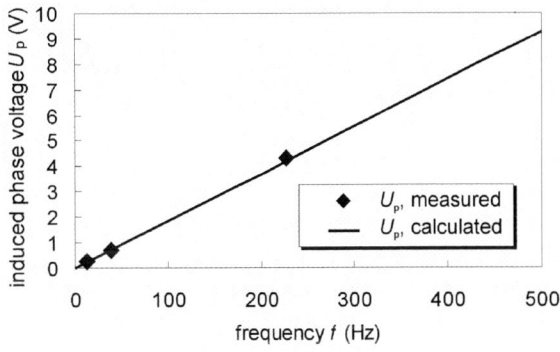

Fig. 11: Calculated and measured back e.m.f. U_p at different frequencies

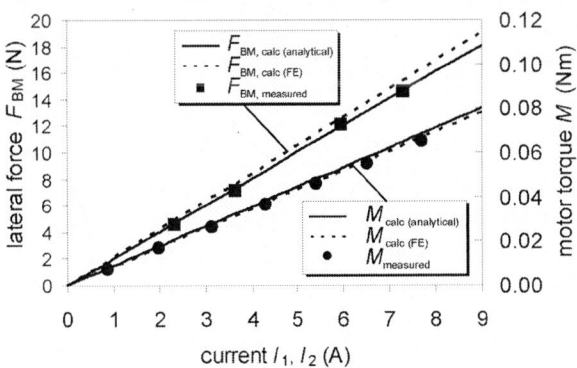

Fig. 12: Comparison of measured lateral force generation (average values from different speeds 3 000 rpm ... 24 000 rpm) and torque generation with analytical and FE values

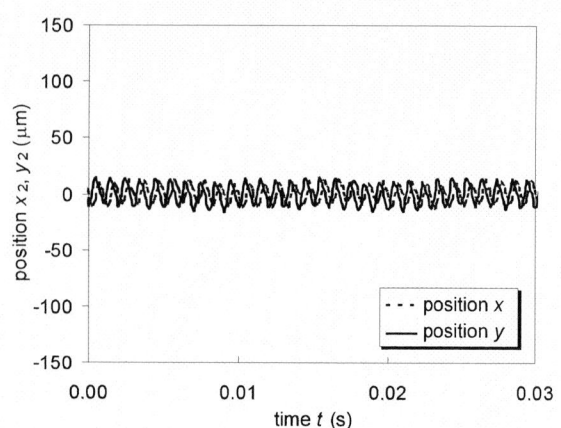

Fig. 13: Measured x,y-shaft position at n = 60 000 rpm. The PM machine is running at no-load without compressor wheel.

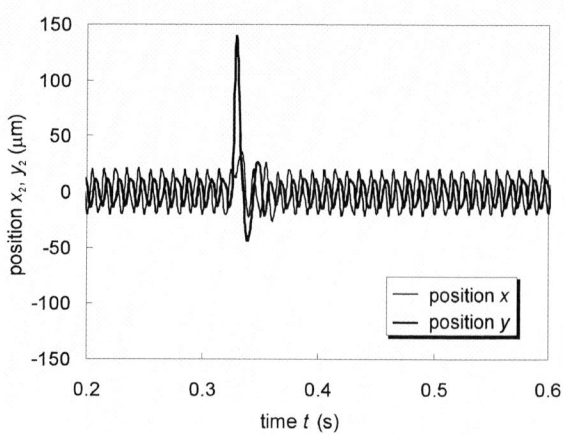

Fig. 14: Measured transient response of the levitation control to an impulse disturbance force at n = 6 000 rpm, no-load operation

TABLE IV
COMPARISON OF ANALYTICAL AND FE-CALCULATIONS WITH MEASUREMENTS

Parameter	Calc.	FE	Measured
Rated back e.m.f.	18.5 V	-	18.6 V
Rated torque M_N	0.08 Nm	0.078 Nm	0.0765 Nm*
Rated lateral force F_N	10 N	10.56 N	10.02 N
Lev. voltage for F = 4.6 N	3.50 V	-	3.35 V
Lev. current for F = 4.6 N	2.29 A	2.18 N	2.32 A

*measured at n = 12 000 rpm

was done using an eddy-current brake at a speed of 12 000 rpm. An aluminum disc with a diameter of 90 mm was attached to the shaft and placed in the air gap of an eddy-current brake. The braking torque generated from air friction was calculated and added to the measured braking torque of the eddy-current brake. As the comparison to analytical and FE-calculations shows, both torque and lateral force generation show good accordance with the expected values (Table 4). Fig. 13 shows the measured shaft position x and y at nominal speed n = 60 000 rpm and no-load operation. The maximum shaft displacement is ±17 μm. The shaft is reliably kept within the backup bearing clearance of ±150 μm. In Fig. 14, the same signals at 6 000 rpm with an impulse disturbance force acting on the shaft end are displayed. This force was caused by a vertical stroke on the shaft drive end. Again, the shaft is safely kept within the backup bearing clearance, and the disturbance decays within less than 5 periods ($\approx$ 50 ms).

VI. CONCLUSION

The performance of the high speed bearingless prototype motor meets the expectations and shows that bearingless motor technology has the capacity to be successfully implemented in high speed applications. The accuracy of the analytical design equations is verified both by FE calculations and measurements. Experience gathered during the construction of this small power bearingless high speed motor will be incorporated in a future design of a bearingless motor with a similar speed range but considerably higher power.

VII. ACKNOWLEDGMENT

This research project is carried out in close collaboration with *Levitec GmbH, Lahnau*, Germany. The authors would like to thank *Levitec* for their constant support and useful assistance during the course of this project.

REFERENCES

[1] Schoeb, R.: *Beiträge zur lagerlosen Asynchronmaschine*, PhD Thesis, ETH Zürich, Switzerland, 1993
[2] Sequenz, H.: *Die Wicklungen elektrischer Maschinen*, Vol. 1, Springer, Vienna, 1950
[3] Silber, S.; Amrhein, W.; Bösch, P.; Schoeb, R.; Barletta, N.: *Design aspects of bearingless slice motors*, IEEE/ASME Transactions on Mechatronics, Vol. 6, No. 10, Dec. 2005, p. 611-617
[4] Schoeb, R., et. al.: *A bearingless motor for a left ventricular assist device (LVAD)*, 7th Int. Symp. on Magn. Bearings, Zürich, Switzerland, Aug. 23-25, 2000, p. 383-388
[5] Salazar, A.; Chiba, A.; Fukao, T.: *A review of development in bearingless motors*, 7th Int. Symp. on Magn. Bearings, Zürich, Switzerland, Aug. 23-25, 2000, p. 335-340
[6] Ooshima, M.; Chiba, A.; Fukao, T.; Rahman, M.: *Design and analysis of permanent magnet-type bearingless motors*, IEEE Transaction on Industrial Electronics, Vol. 43, No. 2, April 1996, p. 292-299
[7] Redemann, Chr.; Meuter, P.; Ramella, A.; Gempp, T.: *30 kW bearingless canned motor pump on the test bed*, 7th Int. Symp. on Magn. Bearings, Zürich, Switzerland, Aug. 23-25, 2000, p. 189-194
[8] Hertel, L.; Hofmann, W.: *Design and test results of a high speed bearingless reluctance motor*, 8th Europ. Conf. on Power Electr. and Appl., EPE, Lausanne, Switzerland, Sept. 7.-9., 1999, p. 1464-1470
[9] Bikle, U.: *Die Auslegung lagerloser Asynchronmaschinen*, PhD Thesis, ETH Zürich, Switzerland, No. 13180, 1999
[10] Vogt, K.: *Berechnung elektrischer Maschinen*, VCH-Verlag, Weinheim, 1996
[11] Schweitzer, G.; Traxler, A.: *Active magnetic bearings: basics, properties and applications of magnetic bearings*, VFH-Verlag, ETH Zürich, Switzerland, 1994
[12] Schneider, T.; Binder, A.; Klohr, M.: *Fixation of Buried and Surface mounted Magnets in High-Speed Permanent Magnet Synchronous Motors*, IEEE Transactions on Industry Applications, Vol. 42, No. 4, July/August 2006, p. 1031-1037
[13] Mack, M.: *Luftreibungsverluste bei elektrischen Maschinen kleiner Baugröße*, PhD Thesis, Stuttgart University, Germany, 1967
[14] Beitz, W.; Grothe, K.-H.: *Dubbel – Taschenbuch für den Maschinenbau*, Ed. 19, Springer Verlag, Berlin, 1997

Performance Investigation of Two-, Three- and Four-Phase Bearingless Slice Motor Configurations

M.T. Bartholet*, S. Silber**, T. Nussbaumer***, J.W. Kolar*

* ETH Zurich, Power Electronic Systems Laboratory, 8092 Zurich, Switzerland, bartholet@lem.ee.ethz.ch
** LCM, Linz Center of Mechatronics GmbH, 4040 Linz, Austria
*** Levitronix GmbH, Technoparkstrass 1, 8005 Zurich, Switzerland

Abstract—The fact that bearingless slice motors (BSM) are widely used in pump systems in the semiconductor industry and for medical applications has caused the attention of other industries for this emerging technology. Here, costs, power consumption and pump volume play an important role. Since the mechanical setup of the motor has a strong impact on these issues five different motor and converter setups are comparatively evaluated and discussed in this paper. The comparison will be carried out for two-, three- and four-phase BSM concepts based on performance indices such as power losses, power electronics requirements and cost-related realisation issues.

I. INTRODUCTION

In today's bearingless pump systems, which have successfully been launched on the market, a two-phase bearingless motor with a symmetrical configuration of the drive and bearing windings is used to actively control the impeller of the pump. This pump setup offers several advantages compared to conventional pumps currently employed in semiconductor and medical applications. The various benefits for the handling of ultra-pure and aggressive fluids in these markets are described in literature [1], [2].

Pharmaceutical, biotechnology and food processing applications are similar to the processes in the semiconductor industry in terms of purity requirements. They demand a high degree of sterility and precision to ensure the quality of the end product, eg. drugs, enzymes in biochemical processes or dairy, cereal and beverages in food processing applications. Standard centrifugal pumps cannot be used in these applications, since their ceramic housing does not sustain the hot steam, which is commonly used to sterilise the process plant. Therefore, mainly tube pumps are used in these applications. Although their maintenance costs are significantly higher compared to those of bearingless pumps their overall price is still lower than that of today's commercially available bearingless pump systems. Thus, if one manages to reduce the overall cost of a bearingless pump they can become very attractive for these markets.

Furthermore, a big potential for next generation bearingless pump systems is located in applications, where magnetically coupled pumps are currently used to deliver hazardous materials, e.g. chromic acid, sodium hypochlorite or sulphur dioxide. The biggest advantage of a bearingless pump compared to magnetically coupled pumps is the fact, that they can run dry without a destruction of their bearings and thus offer an extended lifetime in comparison to standard magnetically coupled pumps. Another very attractive area for more cost effective bearingless pump systems lies in the plating market. The problem with magnetically coupled pumps in that area is the fact that the plating material tends to be deposited in the narrow bearing gap which can result in a locking of the bearing. This problem especially arises in copper, gold and nickel plating processes and often results in large down times of the production plant and maintenance costs of the pumps. Again, both factors can be significantly reduced with the use of a bearingless pump. Additional potential markets for BSM pumps with a strong demand for long durability without maintenance are heating and cooling pumps. An overview about the future application areas for bearingless pump systems is given in **Fig. 1**.

Recent research has mainly focused on the power electronics part of the system in order to decrease the complexity of the system and hence the manufacturing cost. As a result, new converter concepts featuring higher power density, higher efficiency and a lower number of power switches have been developed [3]. However, in order to attract the before described application areas the bearingless slice motor itself must be taken into account as well.

Due to the fact that the design of a BSM offers a lot of constructive freedom several different topologies have been developed over the last years [4]-[9]. They often highly differ in the way how the bearing forces and the motor torque are generated. Looking at the winding perspective, they can broadly be categorized into two groups. Namely, into those which comprise of a dual set of winding configurations [4]-[6] and a second group which only has a single set of winding configurations that carries both the torque and the levitation currents [7]-[9].

However, only one type has reached readiness for marketing in pump applications so far. Its temple motor design is depicted in **Fig. 2**. There it can be seen that the stabilisation of the impeller of the pump is realised with contactless magnetic bearings that are placed around the claws which are carrying the flux of the bearing and drive system. The six spatial degrees of freedom of the rotor are stabilised magnetically through the housing wall. This is done passively for

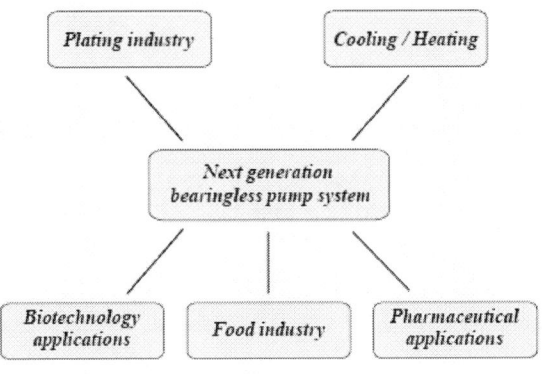

Fig. 1: Future application areas of next generation bearingless pump systems.

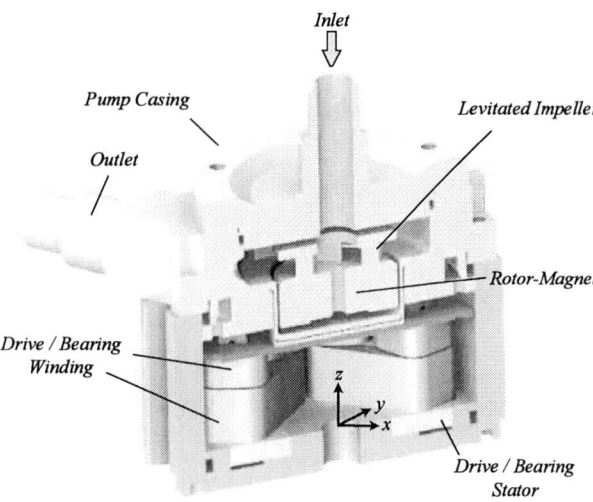

Fig. 2: Schematic of the basic principle of the bearingless centrifugal pump.

three of them, i.e. the axial displacement (in *z*-direction) and the angular displacement (tilting in *x*- and *y*-direction). The three remaining degrees of freedom are controlled actively, i.e. the radial displacement (in *x*- and *y*-direction) and the rotation of the rotor. Therefore, the active motor part generates the driving torque as well as the radial magnetic bearing forces. With this, an extremely compact design can be accomplished.

In order to achieve the before-mentioned complexity and cost reduction in this paper two-, three- and four-phase BSM configurations consisting of the motor and the power electronics are comparatively evaluated based on performance indices regarding volume, losses and cost with a strong focus on the suitability for future application. First, the force and torque model utilized to calculate the performance parameters is briefly presented in **section II**. Starting with the characteristics of the two-phase BSM with separated bearing and drive systems for torque and force generation, three-and four-phase motor configurations are then presented in **section III**. The comparison of these concepts is afterwards carried out in **section IV** where a detailed comparison of the copper, the iron and the power electronics losses is presented that occur in the different embodiments. Furthermore, in **section V** the necessary VA requirement needed to guarantee a safe operation of each motor is calculated. Finally, in **section VI**, cost-related realisation issues are discussed for the different concepts and the suitability of the presented assemblies for future applications of bearingless pump systems is evaluated.

II. FORCE AND TORQUE CALCULATION

The analytical force and torque model applied for the performance analysis of the hereafter discussed BSM configurations is explained in detail in [10]. Thus, only a brief summary of the theoretical fundamentals will be given here. The underlying simulations which provide the basis for the mathematical calculations of the losses and power requirements have been carried out with the electromagnetic field simulation program Maxwell [11] for all of the discussed motor embodiments.

With the precise knowledge of the electromagnetic field variables in the air gap a general force and torque model can be derived with the use of the Maxwell stress tensor $\boldsymbol{T}_M$ [12]

$$\boldsymbol{T}_M = \mu \begin{bmatrix} H_t^2 - \frac{1}{2}H^2 & H_t H_n & H_t H_z \\ H_n H_t & H_n^2 - \frac{1}{2}H^2 & H_n H_z \\ H_z H_t & H_z H_n & H_z^2 - \frac{1}{2}H^2 \end{bmatrix}. \quad (1)$$

Here, $\boldsymbol{H}$ is the magnetic field intensity, which is composed of

$$\boldsymbol{H} = \begin{bmatrix} H_t & H_n & H_z \end{bmatrix}^T, \quad H = |\boldsymbol{H}|, \quad (2)$$

and μ is the permeability. The mechanical stress σ acting on a surface element can then be calculated with

$$\boldsymbol{\sigma} = \boldsymbol{T}_M \boldsymbol{e}_n, \quad (3)$$

where $\boldsymbol{e}_n$ represents the vector perpendicular to the stator surface (cf. **Fig. 3**). Furthermore, it is assumed that the permeability of the ferromagnetic stator is much higher than that of air and thus the tangential component of the flux density H_{1t} in the air gap can be neglected for the following calculation of the torque and the forces responsible for the levitation of the impeller.

With this, the mechanical tension σ_{12} on the interface between air (medium 1) and stator iron (medium 2) can be approximated by

$$\boldsymbol{\sigma}_{12} = \begin{bmatrix} \dfrac{B_{1n}^2}{2\mu_0} \\ B_{1n} J_s \\ 0 \end{bmatrix}, \quad (4)$$

where J_s is the current density distribution on the stator surface, which is assumed of cylindrical shape, and $\boldsymbol{B}_{1n}$ is the normal component of the flux density in air.

The force and torque acting on the rotor of the BSM are then determined by the surface integral

$$F = \oint_A \sigma_{12} dA, \quad (5)$$

where A represents the area of the surface.

The currents in the drive and bearing systems which provide the basis for the comparison of the hereafter presented motor configurations are obtained with this mathematical model that is described in more detail in [13].

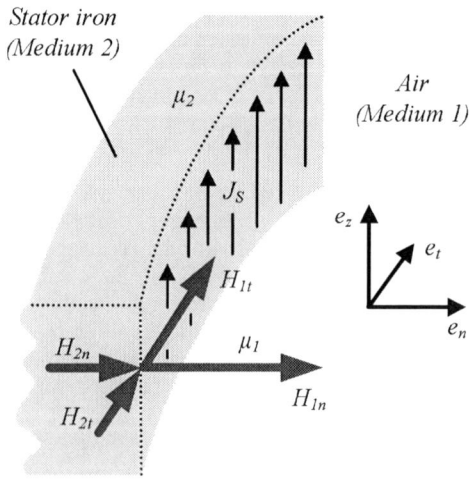

Fig. 3: Stator surface with current density distribution and field strength in the air gap.

258

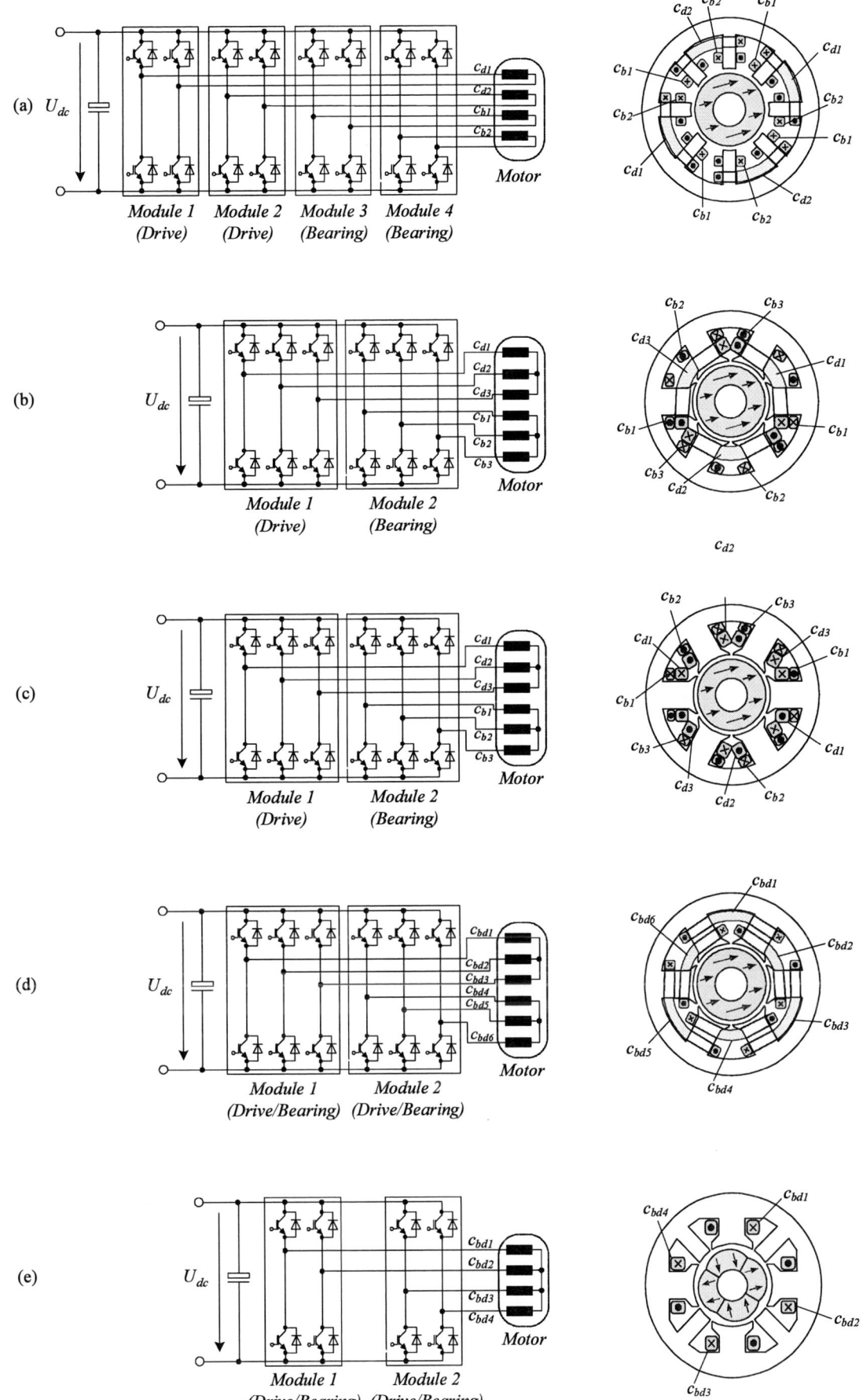

Fig. 4: Investigated BSM systems (a)-(e) consisting of the converter and the corresponding motor configuration.

III. INVESTIGATED BSM CONFIGURATIONS

With regard to the previously discussed performance indices different BSM configurations suitable for pump applications in general are assessed in the following. The targeted pump applications are specified by the maximum rotational speed, the rated torque and the maximum bearing force. As a base for the comparison the following assumptions are considered for the design of the BSM:

- The required rated drive power at the chosen operating point is $P_{DR} = 1200\text{W}$.
- The bearing system is designed in order to ensure the sufficient levitation forces for the whole operating range.
- The rotor diameter and the stator bore, respectively, are the same for all the chosen configurations.
- At rated torque, the current density in the windings is the same for all the chosen configurations.
- For all the presented configurations, the maximum allowable copper volume is chosen in order to reduce copper losses to the minimum.
- The flux density in the iron circuit is assumed the same for motor configurations (a)-(d). Furthermore, these configurations are realised as a temple motor whereas configuration (e) is from disc shape.

With these assumptions the assessment of the bearingless motors depicted in **Fig. 4** has been carried out. The configurations are the following:

- Fig. 4(a) shows a two-phase BSM with 12 separate windings for the drive and bearing system and a two-pole ($p = 1$) permanent magnet (PM) rotor. This symmetrical winding configuration represents the standard setup in today's bearingless pump systems [14] and possesses a total of eight claws. In order to independently generate the levitation forces and the motor torque eight full bridges are needed consisting of totally 16 power transistors. However, as has been shown in recent research [3], this motor embodiment can also be operated with only six half bridges. In order to ensure full control flexibility special modulation schemes must be employed [15]. For the here presented comparison the standard full bridge configuration will be used for sake of better comparability with the other concepts.

- Fig. 4(b) shows a three-phase BSM with nine separate windings and six claws for the drive and bearing systems and a two-pole ($p = 1$) PM rotor. In contrast to the motor configuration (a) the currents in the drive windings do not only generate the motor torque but also cause shear forces which then need to be compensated by the bearing currents and thus lead to higher losses. This drawback can be overcome with an adequate control algorithm presented in [16]. With this, the currents applied to the drive windings only lead to the generation of the desired torque without influencing the stable levitation of the impeller. The subsequent calculation of the copper losses for this motor embodiment is carried out with regard to this optimized control scheme.

- Fig. 4(c) shows a three-phase BSM with nine separate windings and six claws building the drive and bearing systems and a two-pole ($p = 1$) PM rotor. In this configuration, the drive currents do not lead to a generation of bearing forces as it is the case for the previously described configuration.

- Fig. 4(d) shows a three-phase BSM with six coils in total that generate the torque and the axial forces in common.

The current rating is the same for all coils and again, a two-pole ($p = 1$) PM rotor is employed. The pitch winding configuration results in a more efficient torque generation and results in lower power losses compared to the motor embodiments (a) - (c) as will be shown later on.

- Fig. 4(e) shows a four-phase BSM with four concentrated coils and a four-pole ($p = 2$) PM rotor. In contrary to the previously described configurations, which are realised in temple motor design with each claw being placed in an orthogonal manner to the back iron, this motor is realised in disc shape with a homogenously orientated lamination of the iron sheets. The absence of intersections between the claws and the back iron in this configuration has a major impact on the total iron losses as well be shown later on. Although the motor has four phases, the torque generation is equivalent to a single phase motor. This design is the simplest solution that allows a bearingless operation [16]. Furthermore, it offers the benefit that only four power half-bridges are needed to generate the torque and levitation currents in the motor.

The advantage of the converter topologies (b), (c), and (d) is that two intelligent three-phase power modules can be applied. On the one hand this leads to a significantly higher compactness. On the other hand the manufacturing cost is lowered, since such power modules are used in a large variety of applications and thus produced in high numbers. This is as well the case for full-bridge modules. However, their manufacturing quantities are lower since it is not common to operate three phase ac-motors with full-bridges as it is the case for three phase power modules.

As an immanent property of the motor embodiments shown in Fig. 4(d) and (e) the rated current of all windings are showing the same value which leads to a good utilization of the power electronics. On the other hand, the concentrated coils contribute to a coupled and highly nonlinear force and torque generation and thus require a more complex control algorithm in order to safely operate the motor. In contrary, for the configurations depicted in Fig. 4 (a)-(c) separate winding systems for force and torque generation are used. This results in different current ratings for the drive and bearing windings and hence leads to an unbalanced utilisation of the semiconductor devices if identical three-phase power modules are utilised. In order to quantify these facts a detailed comparison will be carried out in the following.

IV. POWER LOSSES

The evaluation of the losses in the motor and the power electronics are indispensable for a thorough comparison of the motor concepts. The most important portion is the copper losses occurring in the windings of the motor. In order to receive a better comparability between motors (a)-(e) the same current densities in the drive and bearing windings, respectively, have been assumed as design criteria for the calculation of the windings. A further important portion is the speed dependant iron losses, which are evaluated for all the presented embodiments subsequently. In section C the equations for the calculation of the switching and forward losses occurring in the power semiconductors are presented. With this, the total losses will be comparatively evaluated for the five concepts.

A. Copper losses

The calculation of the copper losses in the motor phases is given by the following equation:

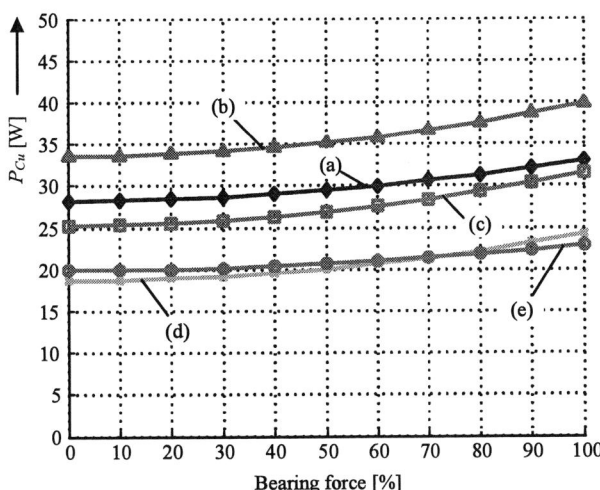

Fig. 5: Total copper losses subject to an increasing levitation force at the design point of 1200W drive power for the motor configurations (a)-(e) depicted in Fig. 4. 100% equals the maximum bearing force requested in a transient condition.

$$P_{Cu} = \sum_{i=1}^{m} R_i \cdot I_{i,rms}^2, \qquad (6)$$

with m as the number of winding phases of the subjected motor and R_i as the corresponding resistance value, which is calculated with

$$R_i = \frac{\rho_{Cu} \cdot l_w}{A_{Cu}}. \qquad (7)$$

Here, l_w stands for the average winding length of the drive or bearing winding, ρ_{Cu} the specific resistance of copper, and A_{Cu} the wire cross area.

Doing this for all of the motor embodiments presented in Fig. 4 for the load point specified in section III leads to a loss distribution as depicted in **Fig. 5**. Looking at the values at zero bearing force reveals that motor embodiment (d) only generates 56% of the losses occurring in configuration (b). While increasing the bearing force up to 100% (which equals 20N) this proportion stays almost the same. The largest increase in losses due to the generation of the levitation forces arises for motor (d). There, at maximum force the losses are increased by 30% compared to the load point with no currents in the bearing phases.

B. Iron losses

According to [17], the hysteresis losses of iron can be approximated under the assumption that the magnitude of the flux density $\hat{B}$ of an alternating field is in the range of 0.2-1.5T by the following equation:

$$P_{Hy} = c_{Hy} \cdot f_e \cdot \hat{B}^{1.6} \cdot m_{Fe}. \qquad (8)$$

The hysteresis losses are thus linearly dependant on a material constant c_{Hy}, the electrical frequency f_e of the motor, and the iron mass m_{Fe}, while the dependency on the flux density is of higher order. On the other hand, the eddy current losses [18] in the stator iron are given by

$$P_{Ed} = c_{Ed} \cdot f_e^2 \cdot \hat{B}^2 \cdot d_{Fe}^2 \cdot m_{Fe}, \qquad (9)$$

if the iron circuit is built up with isolated laminated sheets with a thickness of d_{Fe}.

However, for the evaluation of the hysteresis and eddy current losses the magnetic flux density in the iron path can not be assumed to have a homogeneous distribution, wherefore (8) and (9) can not be used directly. In order to calculate the iron losses accurately, the whole stator needs to be segmented into k parts with each having a constant flux density $\hat{B}_i$ and a mass $m_{Fe,i}$. The whole iron losses of each motor configuration can then be calculated according to

$$P_{Fe} = P_{Hy} + P_{Ed} =$$
$$= m_{Fe,i} \cdot \left(c_{Hy} \cdot f_e \cdot \sum_{i}^{k} \hat{B}_i^{1.6} + c_{Ed} \cdot f_e^2 \cdot \sum_{i}^{k} \hat{B}_i^2 \cdot d_{Fe}^2 \right). \qquad (10)$$

This correlation can also be written as

$$P_{Fe} = k_1 \cdot n_e + k_2 \cdot n_e^2 \qquad (11)$$

with k_1 being the linear and k_2 being the square loss factor, respectively and the motor speed n. These factors, which have been extracted from simulations and validated by measurements on experimental test setups, are compiled in Table I for all configurations.

TABLE I: IRON LOSS FACTORS

Configuration	(a)	(b),(c),(d)	(e)
k_1 [W/1000 rpm]	0.921	0.850	0.671
k_2 [W/(1000 rpm)^2]	0.645	0.602	0.153

It can be seen that especially the square loss factor (due to the eddy current losses) is significantly lower for configuration (e). The main reason is the fully radial construction of that setup with an equally orientated lamination of the iron sheets. In contrary, the concepts (a)-(d) are built in temple motor design, where the stator parts have to be linked in an orthogonal manner and a continuous lamination is not possible. Furthermore, setup (e) has a lower iron mass, which is, however, compensated by the higher electrical frequency (due to $p = 2$).

In total, as shown in Fig. 6, a clear advantage for the motor configuration (e) is given regarding the iron losses, especially for higher rotational speeds. However, due to dynamical limitations (as will be discussed later), this concept does not allow rotational speeds above 8000 rpm, which is a limiting factor for some applications.

C. Power Electronics losses

All of the presented motor configurations are operated with a

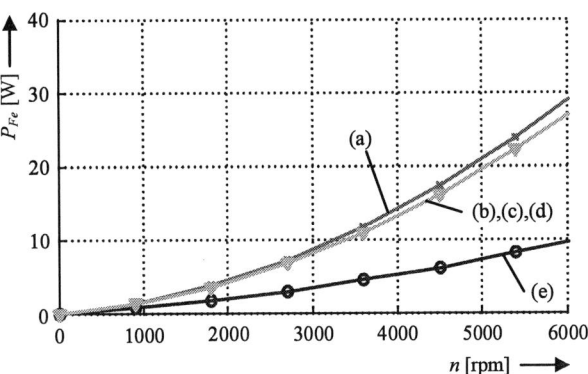

Fig. 6: Total iron losses P_{Fe} of the presented motor configurations (a)-(e) in dependence of the motor speed n.

symmetrical PWM switching pattern. Due to this and the utilization of similar components in the drive and bearing system of all configurations the switching losses can be evaluated based on the same switching loss energy data. The switching losses of a device are then given by integration of the loss energy over a $\pi/2$-wide interval.

$$P_{Sw} = f_s \cdot \frac{2}{\pi} \int_0^{\pi/2} w(i)\,d\varphi. \tag{12}$$

With the assumption of a linear dependency of the loss energy on the switched current

$$w(i) = k_i \cdot i \tag{13}$$

for a given voltage the calculation of the switching losses based on the component specific factors k_{on}, k_{off} and k_{rev} for the turn-on, the turn-off and the reverse recovery losses, respectively, can be utilised. These parameters, which have been evaluated by switching loss measurements on a three phase power module [19] account to k_{on}, = 22.54μJ/A, k_{off} = 11.43μJ/A and k_{rev} = 1.35μJ/A for a dc-link voltage of U_{dc} = 325V. For the sake of a fair comparison these parameters have also been used for the full-bridge topology (cf. Fig. 4(a), two-phase configuration) and the half-bridge topology (cf. Fig. 4(e), four-phase configuration). With this, the total switching losses per bridge leg are given by

$$P_{PE,Sw} = \frac{2 \cdot f_s}{\pi} \cdot (k_{on} + k_{off} + k_{rr}) \cdot \hat{I}. \tag{14}$$

The forward characteristics of the semiconductors can be approximated by a forward voltage drop and a forward resistance. The parameters $U_{CE,0}$ and r_{CE} for the IGBT, and $U_{F,0}$, r_F for the diode, respectively, have also been evaluated by measurements and account to $U_{CE,0}$ = 1.11V, r_{CE} = 77mΩ, $U_{F,0}$ = 1.05V and r_F = 83mΩ. With this, the conduction losses of any semiconductor can be derived by

$$P_{Fw,T} = U_{CE,0} \cdot I_{T,avg} + r_{CE,on} \cdot I_{T,rms}^2 \tag{15}$$

$$P_{Fw,D} = U_{F,0} \cdot I_{D,avg} + r_F \cdot I_{D,rms}^2 \tag{16}$$

The total power electronics losses in the half-bridges due to the bearing and drive currents can then be evaluated.

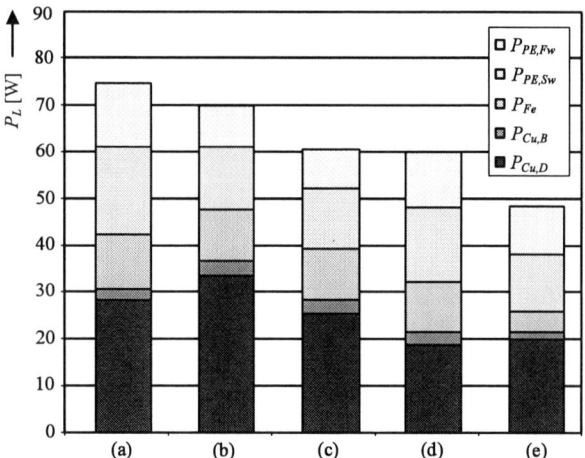

Fig. 7: Overall loss distribution for the motor and converter configurations depicted in Fig. 4(a)-(e). The shown losses are the copper losses in the drive system ($P_{Cu,D}$) and in the bearing system ($P_{Cu,B}$), the iron losses (P_{Fe}), the switching losses ($P_{PE,Sw}$) and the forward losses ($P_{PE,Fw}$) in the power electronics.

D. Total losses

With the help of the before-presented equations the overall losses P_L including the copper, the iron and the power electronics losses can be calculated. Their distribution is depicted in Fig. 7.

This comparison has been carried out for a rated drive power of P_{DR} = 1200W at 70% bearing force at a speed of n = 3600rpm, which is a typical operating point value. It can be seen that the total losses are lowest for configuration (e), followed by setup (d).

V. POWER ELECTRONICS REQUIREMENTS

Another important figure for the evaluation of the motor topologies is their Volt-Ampere (VA) power electronics requirement. In the past, a lot of research has been carried out on the VA requirements of different motors [20], [21]. Among different definitions the VA rating in terms of inverter peak voltage and rms current of the motor is the most suitable for the comparison of the evaluated BSM topologies.

The VA rating in a mathematical form is given by

$$P_{VA} = \sum_{i=1}^{m} \hat{U}_i \cdot I_{i,rms}, \tag{17}$$

with the rms current $I_{i,rms}$ and the peak phase voltage $\hat{U}_i$. When neglecting the resistive voltage drop across the drive or bearing winding the required value of the peak phase voltage $\hat{U}_i$ results in

$$\hat{U}_i = \sqrt{\left(\omega L_i \hat{I}_i\right)^2 + \hat{U}_{i,ind}^2}, \tag{18}$$

with the phase inductance L_i, the peak phase current $\hat{I}_i$ and the amplitude of the induced voltage $\hat{U}_{i,ind}$.

For the configurations (a)-(c), which feature separated drive and bearing systems, different peak phase voltage requirements can be calculated for both systems. However, due to the availability of only one common dc-link voltage of the converter, the higher voltage requirement (which occurs for the drive system) has been considered for both systems. Hence, the bearing system has a broad stability margin to compensate for external disturbances. For the configurations (d)-(e), which generate the levitation forces and the torque in the same windings, a certain stability margin for the bearing system has to be added explicitly in order to also guarantee a safe operation. Measurements on laboratory prototypes have shown that the voltage requirement has to be increased by a factor of 1.3 for the considered speed range, which is already taken into account for this comparative evaluation. For higher rotational speeds this margin would have to be increased even more and it was observed experimentally that for a dc-link voltage of 325V the configurations (d) and (e) cannot be operated above 8000rpm anymore. Additionally, for configuration (e) the maximum achievable speed is also limited by the digital control, which has to deal twice the electrical frequency due to $p = 2$ as compared to the other topologies.

The VA requirements scaled to the rated mechanical power are depicted in Fig. 8 for the different motor designs. One can see that the required bearing forces have significant impact on the VA requirement of the motor. This is especially the case for motor (d), where at 100% bearing force, which equals 20N, the VA requirement is nearly doubled as compared to zero force. If only the drive system is taken into

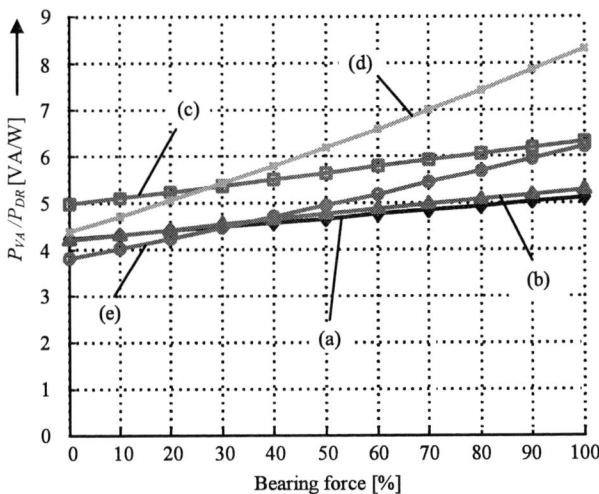

Fig. 8: Normalised VA requirement of the presented motor configurations (normalisation basis: rated drive power P_{DR} = 1200 W).

account (0% bearing force), configuration (e) demands the lowest VA requirement.

However, this condition does not arise in normal operation of a BSM since levitation forces are always needed in order to safely operate the system. For the considered operating point with 70% bearing forces, which is a typical value for pump applications, motor configurations (a) and (b) are the most efficient solutions regarding the necessary VA requirement of the converter.

VI. COST RELATED FACTORS

As mentioned in the beginning, BSM pump systems are getting more and more targeted for applications, where mass production becomes feasible. Therefore, additional cost-related factors must be taken into consideration for a complete comparison. In Table II a qualitative comparison of the concepts is given for these factors in addition to the previously discussed performance indices (power losses and VA requirement).

TABLE II
QUALITATIVE COMPARISON OF THE DIFFERENT BSM CONCEPTS

Motor	(a)	(b)	(c)	(d)	(e)
Power losses	−	−	✓	✓	+
VA requirement	+	+	✓	−	✓
Copper/Iron mass	−	−	−	−	+
Realisation effort	✓	✓	✓	✓	+
Control complexity	+	+	+	−	−
Scalability	+	+	+	−	−

As has been shown in section IV, configurations (d) and (e) are the favourable ones, if only the losses in the motor and the power electronics are considered. At the chosen operating point the overall losses occurring in configuration (e) are 65% of those resulting in configuration (a).

In terms of VA requirements, the motor setups (a) and (b) have been found to be preferable for applications, where significant bearing forces occur, e.g. for pumps. On the other hand, configuration (d) has the highest VA requirements and therefore leads to the largest power electronics volume.

The comparison of the copper and iron mass, which is re-

quired to realise the different motor embodiments, is lowest for configuration (e). The radial design of the iron circuit, which does not need additional vertical claws, as they are required in the temple motor configurations (a)-(d), results in an iron mass which is almost half as compared to the other concepts.

Looking at the embodiments from a manufacturing perspective and with this considering the necessary realisation effort reveals motor (e) as the most promising solution. This mainly emerges from the disc shape structure of this configuration, which offers certain production advantages. First, the four identical coils can directly be wound on the stator claws in one step, which simplifies the manufacturability. In addition, the iron circuit can be realised with a horizontally laminated iron stack. Due to the absence of additional vertical claws the manufacturing effort is clearly reduced compared to the temple motor configurations (a)-(d). Finally, the disc shape setup also offers the possibility of integrating the power electronics part in the motor while still keeping the thereby resulting total case volume in the range of the temple motor configurations without integrated power electronics. In addition, this greatly reduces the cabling effort. However, the sensor concept of motor (e) for the position detection of the impeller is the most difficult of all the presented configurations due to the shape of the iron circuit and the limited space that is available for the insertion of the sensors. Its design also influences the design of the pump impeller in terms of hydraulic efficiency. This fact also strongly influences the applicability of this concept in high pressure/flow applications. However, taking all the before-mentioned issues into account, the realization effort clearly is the lowest for configuration (e).

The control complexity of the presented motor embodiments highly depends on the chosen winding configuration. The fact that the currents, which are generating torque and levitation forces, are applied to the same coils in common for configurations (d) and (e), results in a more sophisticated control structure as it is the case for configurations (a)-(c). In the latter, the control of the drive and bearing system can be done independently. This results in a less complex control structure as for motors (d) and (e).

As explained in section V, the scalability of the motor configurations towards higher speeds and pressure is best for configurations (a)-(c). This results from the independent drive and bearing system, where due to the utilisation of the same dc-link voltage the bearing system usually features a large dynamical voltage margin to compensate for external disturbances. In contrast, for configurations (d) and (e) the stability margin for the bearing system has to be added explicitly in order to guarantee a safe operation. This voltage margin together with the maximum available dc-link voltage is the limiting factor for the maximum achievable drive speed for these concepts. In today's semiconductor applications a strong demand arises for pumps with high pressure ratings. For these applications motor configurations (d) and (e) cannot be considered as suitable solutions due to their limited speed capability.

Summing up, it can be stated that configuration (e) is highly interesting for future cost-sensitive bearingless motor applications such as pumps for plating industry, mixers for biotechnology processes, or heat and cooling pumps. However, the applicability of this configuration is limited to the low/medium speed range and is therefore not suitable for high-pressure applications. In this area, configuration (a) still seems to be the most preferable solution.

VII. Summary

In this paper two-, three- and four- phase bearingless slice motors (BSM) have been comparatively discussed based on performance indeces in order to find the most suitable motor embodiment for more cost sensitive application areas of next generation bearingless pump systems. The comparison has been carried out for a typical pump operating point (rated mechanical drive power 1200W, bearing forces in the range of 0-20N) and the performance indices have been defined as the occurring power losses in the converter and the motor, the power electronics VA requirements in order to achieve the operating point and cost related manufacturing issues (such as copper and iron masses, realisation effort in consideration of mass production, control complexity, and scalability towards higher speed and pressure ranges).

The comparison has not revealed a clear superior concept in all aspects, but has given a better insight to the specific attributes and possibilities of each concept. Generally, it can be stated that the three-phase motor configurations (b)-(d) do not show a clear advantage in none of the considered aspects, wherefore they will barely be selected as the next generation bearingless motor concept.

On the other hand, the four-phase motor configuration (e) seems to be a promising concept for future cost-sensitive applications in the low pressure and low/medium speed range, e.g. pumps for plating industry, mixers for biotechnology processes, or heat and cooling pumps. The advantages of this concept in terms of small iron and copper masses and easy manufacturability are mainly arising from its radial construction. In addition, the compact design allows the integration of the power electronics in the motor housing with a resulting volume comparable to that of the temple motor design without integrated electronics.

However, for high pressure applications in upcoming semiconductor applications, where speeds above 8000rpm are demanded, this concept cannot be considered anymore due to its inherent speed limitations. There, the standard two-phase topology (a) is still the most preferable solution due to the independent drive and bearing winding system.

VIII. References

[1] M. Neff, N. Barletta, R. Schöb, "Magnetically levitated centrifugal pump for highly pure and aggressive chemicals", *PCIM Conference* 2000, June 6-8, 2000, Nuremberg, Germany.

[2] R. Schöb, N. Barletta, J. Hahn, "The Bearingless centrifugal pump – A perfect example of a mechatronics system", *1st IFAC-Conference on Mechatronic Systems*, Darmstadt, Germany, 18-20 September 2000.

[3] M.T. Bartholet, T. Nussbaumer, P. Dirnberger, J.W. Kolar, "Novel converter concept for bearingless slice motor systems", IEEE Industry Applications Conference 2006, *Conference Record of the 41st IAS Annual Meeting*, October 2006, Tampa, USA, vol. 5, pp. 2496-2502.

[4] A. Chiba, R. Furuichi, Y. Aikawa, K. Shimada, Y. Takamoto and T. Fukao, "Stable operation of induction-type bearingless motors under loaded conditions", *IEEE Trans. Ind. Appl.*, vol. 33, no. 4, pp. 919-924, Jul./Aug. 1997.

[5] M. Ohsawa S. Mori and T. Satoh "Study of the induction type bearingless motor", *Proc. 7th Int. Symp. Magnetic Bearings*, Zurich, Switzerland, Aug. 2000, pp. 389-394.

[6] A. Chiba, T. Deido, T. Fukao and M.A.Rahman, "An analysis of bearingless AC motors", *IEEE Trans. Energy Convers.*, vol. 9, no. 1, pp.61-67, Mar. 1 94.

[7] W. K. S. Khoo, R. L. Fittro and S. D. Garvey, "AC polyphase self-bearing motors with a bridge configured winding", *Proc. 8th Int. Symp. Magnetic Bearings*, Mito, Japan, Aug. 2002, pp. 47-52.

[8] Y. Okada, K. Dejima and T. Ohishi, "Analysis and comparison of PM synchronous motor and induction motor type magnetic bearings", *IEEE Trans. Ind. Appl.*, vol. 31, no. 5, pp. 1047-1053, Sep./Oct. 1995.

[9] Y. Okada, S. Myamoto and T. Ohishi, "Levitation and torque control of internal permanent magnet type bearingless motor", *IEEE Trans. Contr. Syst. Technol.*, vol. 4, no. 5, pp. 565-571, Sep. 1996.

[10] S. Silber, W. Amrhein, P. Bosch, R. Schob, N. Barletta, „Design aspects of bearingless slice motors", *IEEE Trans on Mechatronics*, vol 10, pp. 611-617, Dec. 2005.

[11] Maxwell® 3D distributed by Ansoft Corporation, http://www.ansoft.com.

[12] K. Simonyi, Theoretische Elektrotechnik, Leipzig, Germany: Barth Verlagsgesellschaft GmbH, 1993.

[13] D. Schröder, Elektrische Antriebe – Grundlagen, 2. Auflage, Springer, 2000.

[14] Levitronix Pumps: http://www.levitronix.com.

[15] M.T. Bartholet, T. Nussbaumer, D. Krähenbühl, F. Zürcher, J.W. Kolar, „Modulation concepts for the control of a two-phase bearingless slice motor utilizing three-phase power modules", *IEEE Power conversion conference 2007*, Nagoya, Japan, 2007.

[16] S. Silber, "Beiträge zum lagerlosen Einphasenmotor, Dissertation, Johannes Kepler Universität, Linz, 2000.

[17] C. P. Steinmetz, "On the law of hysteresis", *Proc. IEEE*, vol. 72, pp. 196-221, Feb. 1984.

[18] C. Heck, "Magnetische Werkstoffe und ihre technische Anwendung, Dr. Alfred Hüthig, Heidelberg, 2. Auflage, 1975.

[19] International Rectifier: Integrated Power Hybrid IC for Appliance Motor Drive Applications, IRAM136-3063B, Preliminary Datasheet DR-2 (2007).

[20] M. Barnes, Ch. Pollock, "Power electronic converters for switched reluctance drives", *IEEE Trans. on Power Electron.*, vol. 13, pp. 1100-1111, Nov. 1998.

[21] T.J.E. Miller, "Switched Reluctance Motors and Their Control", Oxford, U.K: Magna Physics/Clarendon, 1993.

Compensation of Pole Position Estimation Error for Sensor-less IPMSM Drives with DC Link Current Detection

Hisao Kubota, Yusuke Shibano, Takayuki Kobayashi

School of Science and Technology, Meiji University, Japan

Abstract– The authors have proposed a method to estimate the pole position for sensor-less IPMSM drives with only DC link current measurement and use a triangular comparison PWM technique. Using the proposed method, the current control has not to be interrupted and the band-pass filter is not necessary. The proposed method needs the current detection at apexes of triangular carrier waves, precisely. So the effect of the detection timing delay and the effect of the dead time are investigated. A method to compensate the pole position estimation error is also proposed.

Index Terms—DC current detection, IPMSM, Sensorless drive, Shunt resistor

I. INTRODUCTION

In interior permanent magnet synchronous machine (IPMSM) drive systems, mechanical sensors such as rotary encoders are used to get the magnetic pole position information, because the stator currents of IPMSMs have to be controlled synchronized with the pole position. However, there are problems such that a mechanical sensor is expensive, is lacking in reliability, and increases placing space. Therefore various sensor-less control methods to get the magnetic pole position information with only current sensors have been proposed [1]-[8].

Some of them utilize the electro motive force which is basically proportional to the motor speed [1]-[4]. So, they cannot be applied to drives at low speeds or standstill. The others utilize the spatial saliency, which is the difference between *d* and *q* axis inductances. These methods are able to estimate the pole position at low speed region and standstill by superimposing high frequency components into the armature voltage commands [5]-[8]. Most of them need expensive Hall Effect current sensors to detect the motor phase currents, and need band-pass filters for extracting the high frequency components from the measured currents. The INFORM method proposed in [8] is applicable to the system with only an inexpensive DC link current sensor, but special PWM patterns are necessary and the current control has to be interrupted during supplying the pilot voltages.

The authors have proposed a method that estimates the pole position with only DC link current measurement and uses a triangular comparison PWM technique. The current control has not to be interrupted and band-pass filters are not necessary for the proposed method [9], [10].

The proposed method needs the current detection at apexes of triangular PWM carrier waves, precisely. So the effect of the detection timing delay and the effect of the dead time are investigated. A method to compensate the pole position estimation error is also proposed.

II. POLE POSITION ESTIMATION WITH ONLY DC LINK CURRENT MEASUREMENT AND THREE PHASE TRIANGULAR PWM CARRIER WAVE

Fig. 1 shows the frequency spectra of the output line to line voltages of a PWM inverter. Figs. 1(a) and 1(b) are the ones for the single phase triangular PWM carrier wave and for the three phase triangular PWM carrier waves, respectively. When the single phase carrier wave is used, the PWM carrier frequency components in the phase voltages are cancelled each other. While, the three phase carrier waves are used, the PWM carrier frequency components are dominant at the low modulation index region. So, these PWM carrier frequency components enable to detect the difference between d and q axis inductances.

An IPMSM is described by the following equation in the α–β stationary reference frame.

$$
\begin{bmatrix} v_\alpha \\ v_\beta \end{bmatrix} = R \begin{bmatrix} i_\alpha \\ i_\beta \end{bmatrix} + \begin{bmatrix} L_0 + L_1 \cos 2\theta & L_1 \sin 2\theta \\ L_1 \sin 2\theta & L_0 - L_1 \cos 2\theta \end{bmatrix} \frac{d}{dt} \begin{bmatrix} i_\alpha \\ i_\beta \end{bmatrix}
$$
$$
+ \frac{d}{dt} \left\{ \begin{bmatrix} L_0 + L_1 \cos 2\theta & L_1 \sin 2\theta \\ L_1 \sin 2\theta & L_0 - L_1 \cos 2\theta \end{bmatrix} \begin{bmatrix} i_\alpha \\ i_\beta \end{bmatrix} \right\}
$$
$$
+ \frac{d}{dt} \left\{ \psi \begin{bmatrix} -\sin \theta \\ \cos \theta \end{bmatrix} \right\}
$$

(1)

where $L_0 = (L_d + L_q)/2$, $L_1 = (L_d - L_q)/2$, θ: Pole position.
Considering the PWM carrier frequency component, the pole position is almost constant, so 1st, 3rd, and 4th terms are negligible. Then, the voltage equation for the PWM

978-1-4244-0644-9/07/$25.00 ©2007 IEEE

carrier frequency component is expressed by the following equation.

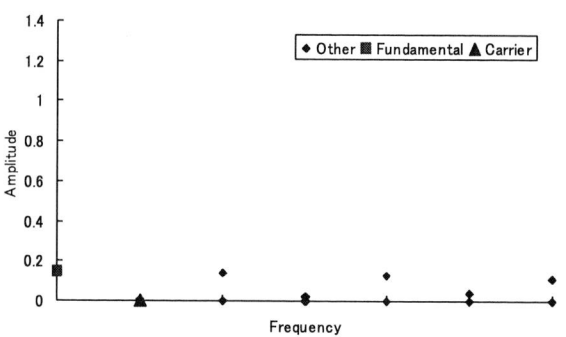

(a) Single Phase Triangular PWM Carrier Wave

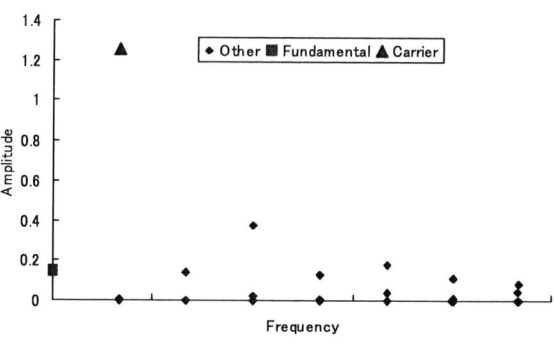

(b) Three Phase Triangular PWM Carrier Waves

Fig. 1. Frequency Spectra of PWM Inverter Output Voltage

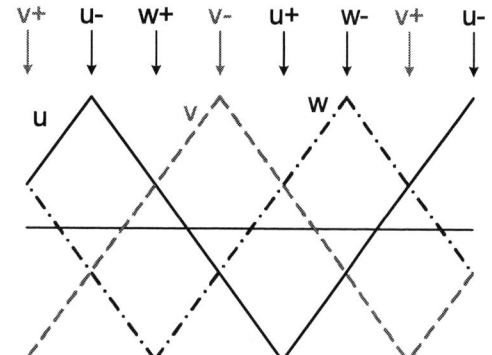

Fig. 2. Three Phase Triangular PWM Carrier Waves and Current Measurement Points

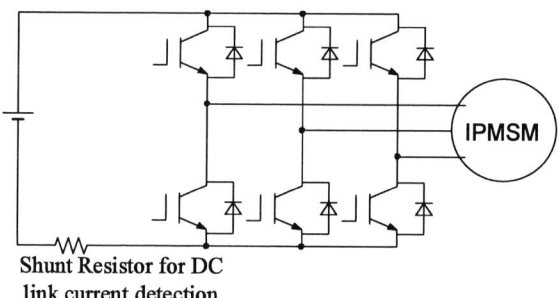

Shunt Resistor for DC link current detection

Fig. 3. IPMSM Drive System with Shunt Resistor in DC Link

$$\begin{bmatrix} v_{\alpha h} \\ v_{\beta h} \end{bmatrix} = \begin{bmatrix} L_0 + L_1 \cos 2\theta & L_1 \sin 2\theta \\ L_1 \sin 2\theta & L_0 - L_1 \cos 2\theta \end{bmatrix} \frac{d}{dt} \begin{bmatrix} i_{\alpha h} \\ i_{\beta h} \end{bmatrix} \quad (2)$$

where sub-script h means the PWM carrier frequency component.

When the three phase triangular PWM carrier waves shown in Fig.2 are used, the PWM carrier frequency components of the armature voltage are expressed by the following equation.

$$\begin{bmatrix} v_{\alpha h} \\ v_{\beta h} \end{bmatrix} = V_h \begin{bmatrix} \cos \omega_h t \\ \sin \omega_h t \end{bmatrix} \quad (3)$$

From equations (2) and (3), the PWM carrier frequency component of the armature current, $i_{\alpha h}$, is expressed by the following equation under steady state conditions.

$$i_{\alpha h} = \frac{V_h \left\{ (L_0 - L_1 \cos 2\theta) \sin \omega_h t + L_1 \sin 2\theta \cdot \cos \omega_h t \right\}}{\omega_h \left(L_0^2 - L_1^2 \right)} \quad (4)$$

To remove expensive Hall Effect current sensors, only the DC link current of the inverter is measured by the shunt resistor as shown in Fig.3. In order to estimate the pole position, the DC link current is measured at the points shown by arrows in Fig.2. The u-phase current is detected twice in a cycle of the PWM carrier waves at $\omega_h t = 0$ and $\omega_h t = \pi$. Therefore, the u-phase current at these two points are expressed by the following equation.

$$i_u = i_{uf} \pm \frac{V_{h1}}{\omega_h (L_0^2 - L_1^2)} L_1 \sin 2\theta \quad (5)$$

where i_{uf} is the fundamental component of the u-phase current, V_{h1} is the transferred value of V_h into the u-v-w three phase axis.

The difference between the u-phase currents at $\omega_h t = 0$ and $\omega_h t = \pi$ becomes

$$I_{uh} = \frac{2V_{h1}}{\omega_h (L_0^2 - L_1^2)} L_1 \sin 2\theta \quad (6)$$

The following equations are obtained about the phase v and w with the same procedures as the phase u.

$$I_{vh} = \frac{2V_{h1}}{\omega_h (L_0^2 - L_1^2)} L_1 \sin 2\left(\theta - \frac{2\pi}{3} \right) \quad (7)$$

$$I_{wh} = \frac{2V_{h1}}{\omega_h(L_0^2 - L_1^2)} L_1 \sin 2\left(\theta + \frac{2\pi}{3}\right) \tag{8}$$

From I_{uh}, I_{vh}, and I_{wh} , the following equation is obtained.

$$I_{uh} + aI_{vh} + a^2 I_{wh} = \frac{3}{2} \frac{V_{h1}}{\omega_h\left(L_0^2 - L_1^2\right)} \left(\sin 2\theta + j\cos 2\theta\right) \tag{9}$$

where $a = -\dfrac{1}{2} + j\dfrac{\sqrt{3}}{2}$.

Therefore, the estimated pole position, $\hat{\theta}$, can be calculated through the ratio of the real part and the imaginary part of Eq.(9) and the arctan function as follows.

$$\hat{\theta} = \frac{1}{2}\tan^{-1}\left\{\frac{\mathrm{Re}(I_{uh} + aI_{vh} + a^2 I_{wh})}{\mathrm{Im}(I_{uh} + aI_{vh} + a^2 I_{wh})}\right\} \tag{10}$$

III. THE EFFECT OF THE CURRENT DETECTION TIMING DELAY AND ITS COMPENSATION

The proposed method needs the current detection at tops and bottoms of the PWM carrier waves, precisely. So, the method is investigated about the effect of the current detection timing delay.

If the current detection timing is delayed, the u-phase current cannot be detected at $\omega_h t = 0$ and $\omega_h t = \pi$. So, Eq. (6) becomes the following equation.

$$
\begin{aligned}
I_{uh} &= i_{uh}(\omega_h t_1) - i_{uh}(\omega_h t_2) \\
&= \frac{V_{h1}}{\omega_h\left(L_0^2 - L_1^2\right)}\{(L_0 - L_1\cos 2\theta)(\sin\omega_h t_1 - \sin\omega_h t_2) \\
&\quad + L_1\sin 2\theta\cdot(\cos\omega_h t_1 - \cos\omega_h t_2)\}
\end{aligned}
\tag{11}
$$

where $\omega_h t_1$, $\omega_h t_2$ are points shifted a little from $\omega_h t = 0$ and $\omega_h t = \pi$, respectively. Fig. 4 shows the relationship of $\omega_h t_1$ and $\omega_h t_2$.

Eq. (11) is simplified to the following equation with the relationship $\sin\omega_h t_2 = \sin(\omega_h t_1 + \pi) = -\sin\delta$ and $\cos\omega_h t_2 = \cos(\omega_h t_1 + \pi) = -\cos\delta$.

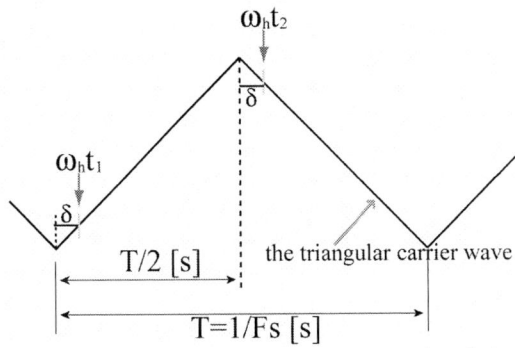

Fig. 4. Triangular PWM Carrier Wave and Delayed Detection Timing

TABLE I
TABLE I. RATINGS AND PARAMETERS OF TESTED IPMSM

Rated Power [kW]	1.5
Rated Voltage [V]	180
Rated Current [A]	6.1
Rated Frequency [Hz]	90
Rated Speed [min^{-1}]	1800
Pole Pairs	3
R [Ω]	1.57
L$_d$ [mH]	6.0
L$_q$ [mH]	18.0

$$I_{uh} = \frac{2V_{h1}\{(L_0 - L_1\cos 2\theta)\cdot\sin\delta + L_1\sin 2\theta\cdot\cos\delta\}}{\omega_h(L_0^2 - L_1^2)} \tag{12}$$

The equations about the phase v and w are obtained by the same procedures as the phase u.

$$I_{vh} = \frac{2V_{h1}[\{L_0 - L_1\cos 2(\theta - \frac{2}{3}\pi)\}\sin\delta + L_1\sin 2(\theta - \frac{2}{3}\pi)\cos\delta]}{\omega_h(L_0^2 - L_1^2)} \tag{13}$$

$$I_{wh} = \frac{2V_{h1}[\{L_0 - L_1\cos 2(\theta + \frac{2}{3}\pi)\}\sin\delta + L_1\sin 2(\theta + \frac{2}{3}\pi)\cos\delta]}{\omega_h(L_0^2 - L_1^2)} \tag{14}$$

Substituting Eqs.(12)-(14) into Eq.(10), following equation is obtained.

$$\hat{\theta} = \theta - \frac{1}{2}\delta \tag{15}$$

The estimated pole position includes some errors. The estimation error can be compensated by using the sum of I_{uh}, I_{vh}, and I_{wh}.

$$I_{DC} = \frac{1}{3}(I_{uh} + I_{vh} + I_{wh}) = \frac{2V_{h1}}{\omega_h(L_0^2 - L_1^2)}L_0\sin\delta \tag{16}$$

The compensated pole position, $\hat{\theta}'$, becomes as follows with consideration of the current detection timing delay.

$$
\begin{aligned}
\hat{\theta}' &= \frac{1}{2}\tan^{-1}\left\{\frac{\mathrm{Re}(I_{uh} + aI_{vh} + a^2 I_{wh})}{\mathrm{Im}(I_{uh} + aI_{vh} + a^2 I_{wh})}\right\} \\
&\quad + \frac{1}{2}\sin^{-1}\left\{\frac{I_{DC}}{2V_{h1}L_0}\cdot\omega_h(L_0^2 - L_1^2)\right\}
\end{aligned}
\tag{17}
$$

IV. EXPERIMENTAL RESULTS

The validity of the proposed method has been verified experimentally. The ratings and parameters of the tested IPMSM are shown in TABLE I. The period of the triangular carrier wave is set at 300 μs and the DC link current is measured at every 50 μs. A dc motor with a current controller is coupled with the IPMSM as a load.

A. Position Estimation and Estimation Error Compensation

The IPMSM is driven at constant speed, 18.8 rad/s (3 Hz) by a vector controller with a rotary encoder. Fig. 5(a) shows real and estimated pole positions when rated load torque is loaded. Differences between phase currents measured at tops and bottoms of PWM carrier wave contain some DC offset as shown in Fig. 5(b). The estimated pole position with the compensation is very close to the actual one. Table II shows the average position error at 18.8 rad/s.

B. Sensor-less Speed Control

In order to make a sensor-less speed control system, the motor speed is calculated using the PLL technique as shown in Fig. 6. Fig. 7 shows results for a speed step response. The speed reference is changed from 18.8 rad/s (3 Hz) to 31.4 rad/s (5 Hz). Fig. 8 shows results for a load step response. The load torque is changed from 0 N·m to 8 N·m, which is rated torque. The speed control system with the proposed position estimation works successfully.

V. CONCLUSIONS

A pole position estimation method for IPMSM sensor-less drives is proposed. It measures only DC link current, and enables to use an inexpensive shunt resistor instead of expensive Hall Effect current sensors. This method needs to detect currents at apexes of triangular PWM carrier waves. So, the effect of the detection timing delay is investigated. Furthermore, a method to compensate the effect is proposed. The validity of the proposed method is verified experimentally.

TABLE II Average Position Estimation Error
(Speed Reference: 18.8 rad/s)

	Average Position Estimation Error (rad)	
	Without Compensation	With Compensation
No Load	0.0404	-0.0626
Rated Torque (Motoring)	0.1553	-0.0485
Rated Torque (Regenerating)	0.0217	-0.09

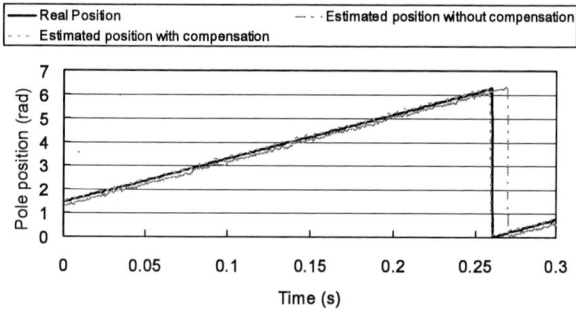

(a) Pole Position

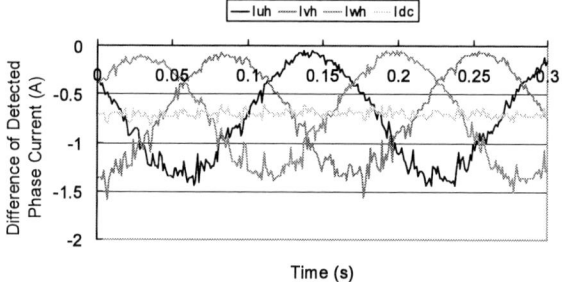

(b) Difference of Detected Phase Current and DC Offset

Fig. 5. Estimated Pole Position and Detected Currents
(18.8 rad/s (3 Hz), Rated Torque)

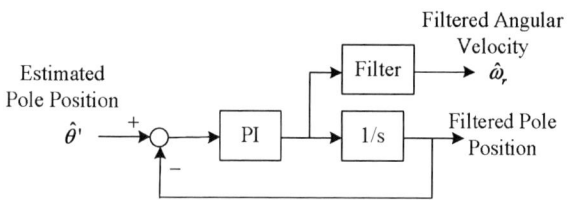

Fig. 6. Speed Estimator with PLL Technique

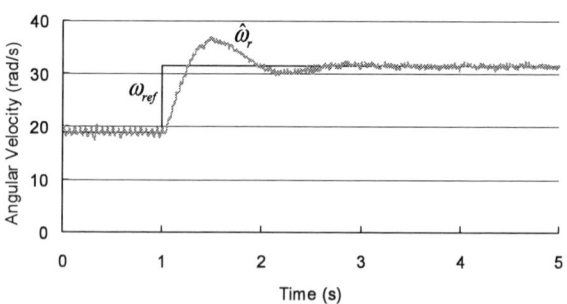

Fig. 7. Speed Step Response

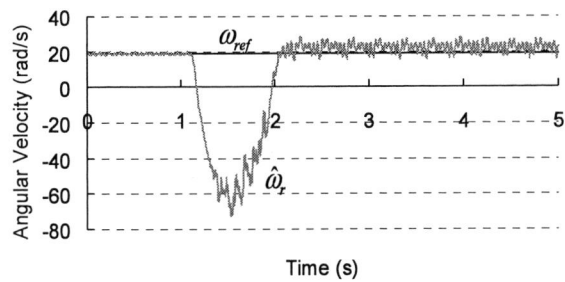

Fig. 8. Load Step Response

REFERENCES

[1] N. Matsui, and M. Shigyo, "Brushless DC Motor Control without Position and Speed Sensors," *IEEE Trans. on Ind. Appl.*, vol. 28, no.1, pp. 120-127, Jan./Feb. 1992.

[2] Z. Chen, M. Tomita, S. Doki, and S. Okuma, "An Extended Electromotive Force Model for Sensorless Control of Interior Permanent-magnet Synchronous Motors," *IEEE Trans. on Ind. Electron.*, vol. 50, no.2, pp. 288-295, Apr. 2003.

[3] G. Yang, R. Tomioka, M. Nakano, and T. Chin, "Position and Speed Sensorless Control of Brush-Less DC Motor Based on an adaptive Observer," *Trans. IEE Japan D*, vol. 113, no. 6, May. 1993, pp. 579 – 586. (in Japanese)

[4] S. Ichikawa, Z. Chen and M. Tomita, "Sensorless Controls of Salient-Pole Permanent Magnet Synchronous Motors Using Extended Electromotive Force Models," *Trans. IEE Japan D,* vol. 122, no.12, Feb. 2002, pp. 1088 – 1096. (in Japanese)

[5] M.J. Corley, and R.D. Lorenz, "Rotor Position and Velocity Estimation for a Salient-pole Permanent Magnet Synchronous Motor at Standstill and High Speeds," *IEEE Trans. on Ind. Appl.*, vol. 34, no. 4, pp. 784-789, Jul./Aug. 1998.

[6] J.-I. Ha, K. Ide, T. Sawa, and S.-K. Sul, "Sensorless Rotor Position Estimation of an Interior Permanent-magnet Motor from Initial States," *IEEE Trans. on Ind. Appl.*, vol. 39, no. 3, pp. 761-767, May/Jun. 2003.

[7] J. Oyama, T. Higuchi, T. Abe, D. Itoyama, K. Ogawa and M. Mamo "Improvement of Estimate Precision of Position Sensorless Control of IPM Motor using PWM Inverter Carrier Frequency Component," in *Proceedings of the 2002 Japan Industry Applications Society Conference*, 149, Vol. 1, pp. 583 – 588, 2002. (in Japanese)

[8] M. Schroedl, "Sensorless Control of AC Machines at Low Speed and Standstill based on the 'INFORM' Method," *1996 IEEE IAS Annual Meeting*, pp. 270-277, 1996.

[9] H. Kubota and S. Nakagawa, "Position Estimation of IPMSM with DC Link Current Measurement and Three Phase Triangular Carrier Wave," in *Proceedings of the 2005 International Power Electronics Conference (IPEC-Niigata2005)*, S72-4, pp. 2217-2220, 2005.

[10] T. Kobayashi, H. Kubota, and S. Nakagawa, "Investigation of IPMSM's Position Estimation Method in Very Low Speed Region with DC Link Current Detection," in *Proceedings of the 32nd Annual Conference of the IEEE Industrial Electronics Society (IEEE-IECON'06)*, TPC2, PF-006807, 2006.

A Novel Dual-Stator Hybrid Excited Synchronous Wind Generator

Liu Xiping[1,2], Lin Heyun[1], Yang Chengfeng[1], Fang Shuhua[1] and Guo Jian[1]

[1] School of Electrical Engineering, Southeast University, Nanjing 210018, P. R. China

[2] School of Mechanical and Electrical Engineering, Jiangxi University of Science and Technology, Ganzhou 341000, P. R. China

Abstract —This paper presents a novel dual-stator hybrid excited synchronous wind generator (DSHESG). The structure characteristics are presented and no-load magnetic field is computed by 3-D finite element method. Static characteristics including magnetic flux density, EMF, inductances are analyzed. A closed-loop control system for field windings using PWM method is built on the basis of mathematical model of DSHESG. Thanks to the field windings the air-gap magnetic flux can be easily controlled, and the output voltage of DSHESG can be kept at a constant value with the speed or load varying. The dual-stator machine can increase output voltage effectively as well. Tests are performed on the prototype machines to validate analysis results, and an excellent agreement is obtained.

Index Terms—Hybrid excited, Static characteristic, Air-gap flux, finite element method, Dual-stator

I. INTRODUCTION

Hybrid excited generator combines advantages of permanent magnet (PM) machines with the possibility of controlling air-gap magnetic flux easily by auxiliary windings, so it has good research prospects and value of application [1][2]. In some cases with low speed, such as wind power system, the diameter of electrical machine is larger and the power density is lower. In order to overcome these shortcomings, a novel dual-stator hybrid excited synchronous generator (DSHESG) is presented. Finite element analysis (FEA) and experimental results are compared for the prototype of machine and excellent agreement has been obtained. The output voltage can be effectively increased due to an inner stator. The transition processes between no-load and rated load are achieved by simulation. Based on the closed-loop control system for field windings, the satisfying simulation results are obtained when the speed or load of generator varies randomly.

II. STRUCTURE

Fig.1 shows the structure of DSHESG, which mainly includes permanent magnets (PM), claw-poles, field windings, out stator, inner stator etc. The PMs and claw-poles share with the outer stator, and the inner stator is fixed on shell of machine. There are two magnetic circuits in the DSHESG, one is called PM magnetic circuit mainly consisting of PMs, air-gap, cup rotor, laminated stator core and so on, the other is called DC field magnetic circuit mainly including claw poles, air-gap, laminated core of the outer stator and bracket of field windings, and they are in parallel independently. The PMs are fixed on the surface of rotor, and the outer PMs are in series with the inner PMs in magnetic circuit.

In the DSHESG, the magnetic motive force generated from PMs is constant, while that created from field windings is variable with different field current.

Fig.1. Structure of DSHESG

Table I presents the specification and design parameters of a DSHESG prototype. The slot number of outer stator and inner stator are 27 and 9 respectively.

TABLE I
SPECIFICATION AND DESIGN PARAMETERS OF A DSHESG

Item	Value	Item	Value
Rated speed (rpm)	400	Rated voltage (V)	120
Rated Phase current(A)	5	Pole number	8
PM flux density(T)	1.15	Core length of outer stator (mm)	164
PM Coercivity (kA/m)	835	Core length of inner stator (mm)	70

Fig.2 shows the distribution of armature windings, the dual-layer and short-pitch distributed windings are adopted for the outer stator. Additionally, the concentrated windings are adopted for the inner stator.

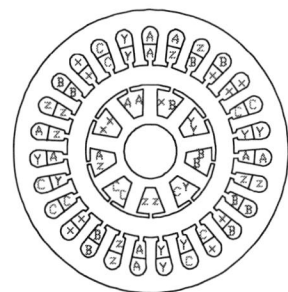

Fig.2. Distribution of armature windings

III. MAGNETIC FIELD ANALYSIS AND STATIC CHARACTERISTICS

A. Magnetic Field Analysis

Fig.3 shows 3-D FEM meshes of DSHESG, mainly including meshes of PM magnetic circuit (Fig.3.a) and DC field magnetic circuit (Fig.3.b).

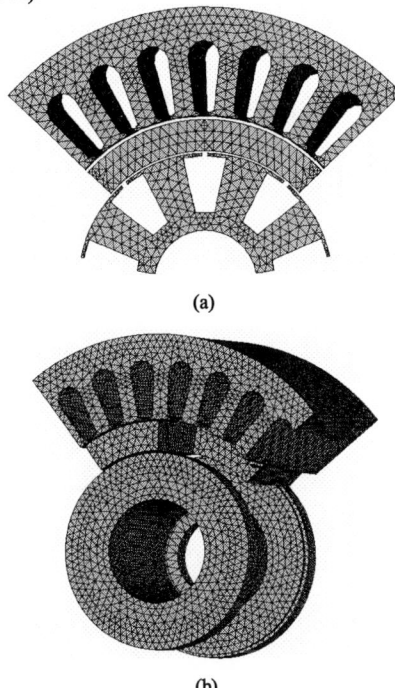

(a)

(b)

Fig.3. 3-D FEM meshes (a) PM circuit (b) DC field winding circuit

The air-gap magnetic flux distributions on the surface of claw poles and PMs during one pole distance based on the FEM model are shown in Fig.4. Comparing Fig.4.a with Fig.4.c, the air-gap magnetic flux distributions on the surface of claw poles are different obviously with different field current, while that on the surface of PMs has nearly no change. Additionally, the flux density on the surface of claw poles is nearly close to zero when the field current is 0 (Fig.4.b).

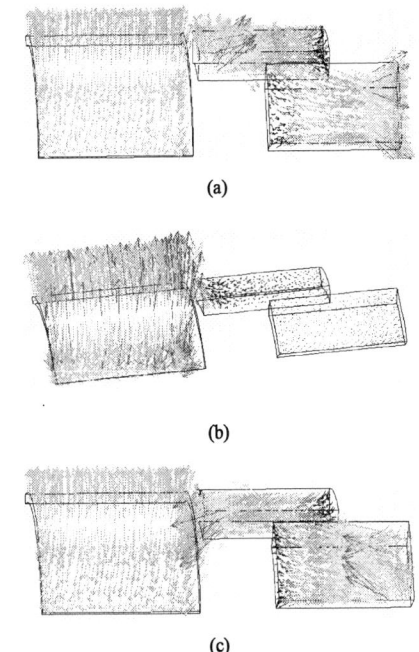

(a)

(b)

(c)

Fig.4.The distribution of magnetic flux distribution in air-gap (a) I=-0.5A (b) I=0 (c) I=0.5A

B. Static Characteristics

Calculations are carried out on the whole FEM structure model (see Fig.3). Fig.5 shows the flux linkage of dual-stator for one phase during one electrical period when the field current is 0, it shows the flux linkage can be increased about 23%.

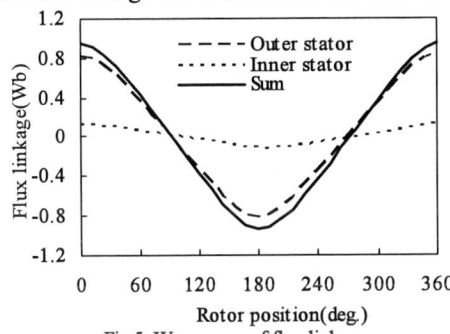

Fig.5. Wave curve of flux linkage

Based on 3-D FEM, waveforms of phase EMF and line-line EMF against rotor position during one electrical period are calculated at a speed of 400rpm, as shown in Fig.6 (a). Apparently, The FEA results agree well with that of experiments (see Fig.6 .b).

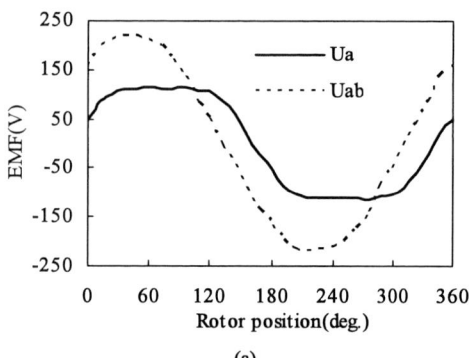

(a)

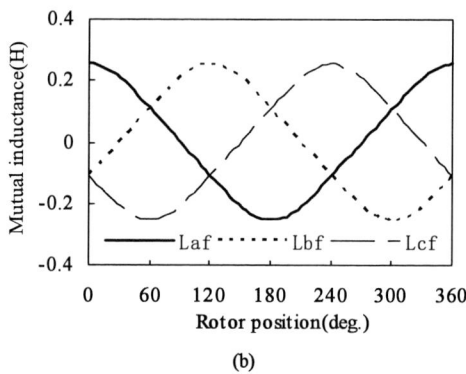

(b)

Fig.7. Wave curves of inductances (a) Self inductance of armature windings (b) Mutual inductance between armature windings and field windings

IV. MATHEMATICAL MODEL OF DSHESG

With the computed inductances, the flux linkages equations of armature windings and field windings can be expressed as

$$
\begin{bmatrix} \Psi_a \\ \Psi_b \\ \Psi_c \\ \Psi_f \end{bmatrix} = \begin{bmatrix} L_{aa} & M_{ab} & M_{ac} & M_{af} \\ M_{ba} & L_{bb} & M_{bc} & M_{bf} \\ M_{ca} & M_{cb} & L_{cc} & M_{cf} \\ M_{fa} & M_{fb} & M_{fc} & L_f \end{bmatrix} \begin{bmatrix} i_a \\ i_b \\ i_c \\ i_f \end{bmatrix} + \begin{bmatrix} \Psi_{pma} \\ \Psi_{pmb} \\ \Psi_{pmc} \\ 0 \end{bmatrix} \quad (1)
$$

where Ψ_{pm} is the flux linkages generated from PMs magnetic force in armature windings, i_a, i_b, i_c are three-phase currents, and i_f is field current, L_{aa}, L_{bb}, L_{cc} are three-phase self inductances, M_{ab} is mutual inductance between armature windings A and armature windings B, M_{fa} is mutual inductance between armature windings and field windings.

Based on equation (1), the EMF induced in the armature windings and field windings can be expressed as

$$
U = RI + \frac{d\Psi}{dt} \quad (2)
$$

where

$$
U = \begin{bmatrix} u_a & u_b & u_c & u_f \end{bmatrix}^T \qquad R = \begin{bmatrix} -R_a & & & \\ & -R_b & & \\ & & -R_c & \\ & & & R_f \end{bmatrix}
$$

$$
I = \begin{bmatrix} i_a & i_b & i_c & i_f \end{bmatrix}^T \qquad \Psi = \begin{bmatrix} \Psi_a & \Psi_b & \Psi_c & \Psi_f \end{bmatrix}^T
$$

V. SIMULATION AND EXPERIMENT

A. Prototype

Fig. 9 shows the prototype of the DSHESG based on the given parameters in Table I.

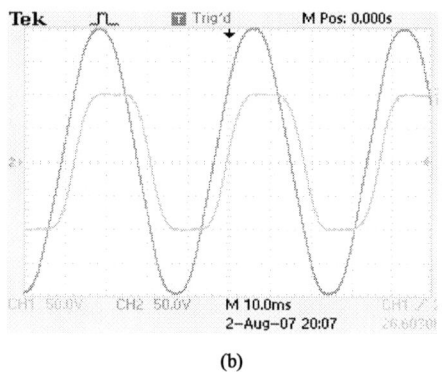

(b)

Fig.6. Wave curves of EMF (a) FEA (b) Experiment

According to flux linkage method, the inductances of armature windings and field windings are obtained. The FEA and experimental results of self inductance of armature windings are compared in Fig.7 (a), and they have good agreement each other. Wave curves of mutual inductance between armature windings and field windings against rotor position are shown in Fig.7 (b).

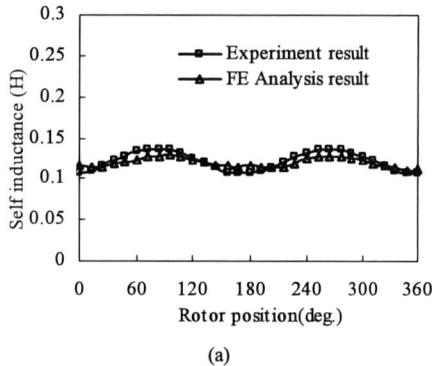

(a)

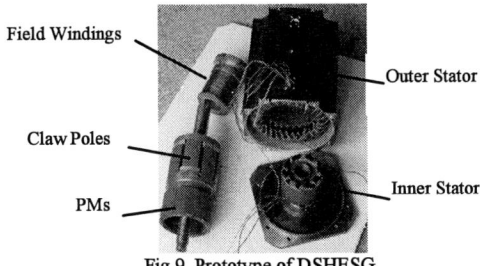

Fig.9. Prototype of DSHESG

The tests of DSHESG are carried out, and some experimental results are obtained.

B. Comparisons of Simulation and Experiment of Phase Voltage and Current

The simulation results are obtained based on the mathematical model. Fig.10 shows the comparisons results of phase voltage and current between the simulation and the experiment at a speed of 400rpm. They are agreed with each other very well.

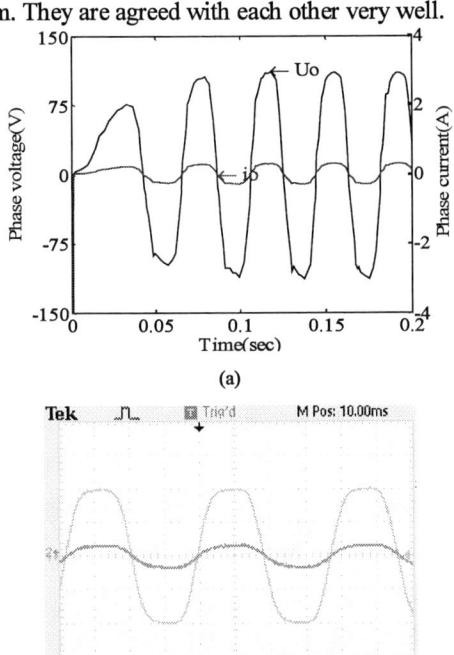

Fig.10. Wave curves of phase voltage and current (a) Simulation (b) Experiment

C. Transition Process between No-load and Load

Fig. 11 shows the transition processes between no-load and load. The time of simulation is 4s, and a load current is applied to armature windings at 2s suddenly. Fig. 11(a) is the transition process from no-load to load, it shows that the output voltage of generator is maintained at 80V after a short time. The output voltage and current are obtained by a full-bridge rectifier and

filter. The transition process from load to no-load is shown in Fig.11 (b), the output voltage reduces rapidly from the time of 2s, and then decreased to zero after a short time.

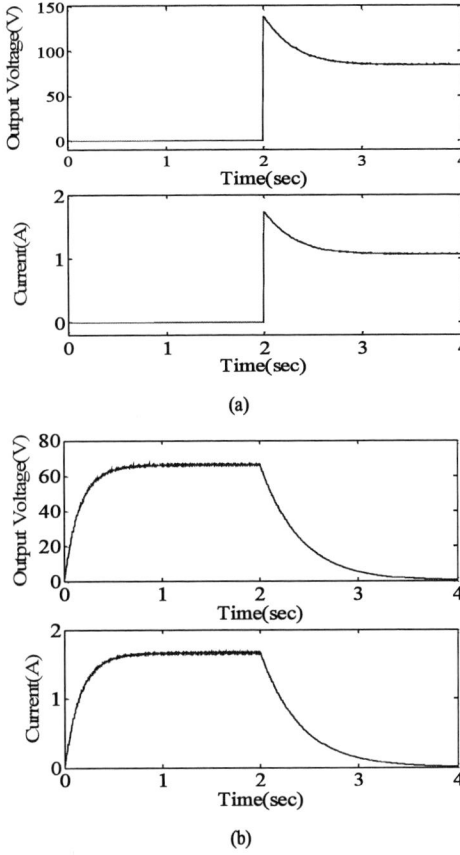

Fig.11. Wave curves of phase voltage and current (a) From no-load to load (b) From load to no-load

D. Closed-loop Control System

In order to obtain a constant output voltage when the speed or load of DSHSEG varies randomly, a closed-loop control system based on PWM method and PI regulator is built, as shown in Fig. 12. When the speed increases or the load current become lower, the output voltage will become higher than the given voltage. In order to make the output voltage close to the given voltage, the field current must be reduced. Contrarily, the field current must be increased to maintain the output voltage when the speed decreases or load current increases.

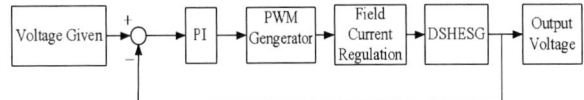

Fig.12.The closed-loop control system

The wave curves of output voltage and field current when the load current of generator varies at the rated speed of 400rpm are shown in Fig. 13(a). When the load current is increased suddenly at time of 5s, and the output voltage is increased to the given voltage again after short sag by regulating the field current to about 0.6A quickly.

Fig. 13(b) shows the simulation results when the speed of generator varies from 400rpm to 300rpm at 5s and the speed varies from 300rpm to 500rpm at 10s. Due to the closed loop control system, the output voltage can be also maintained at given voltage after a short time by regulating field current. The DSHSEG has a good dynamic performance and steady performance. The PWM method and PI regulator applied in the closed-loop control system are effective.

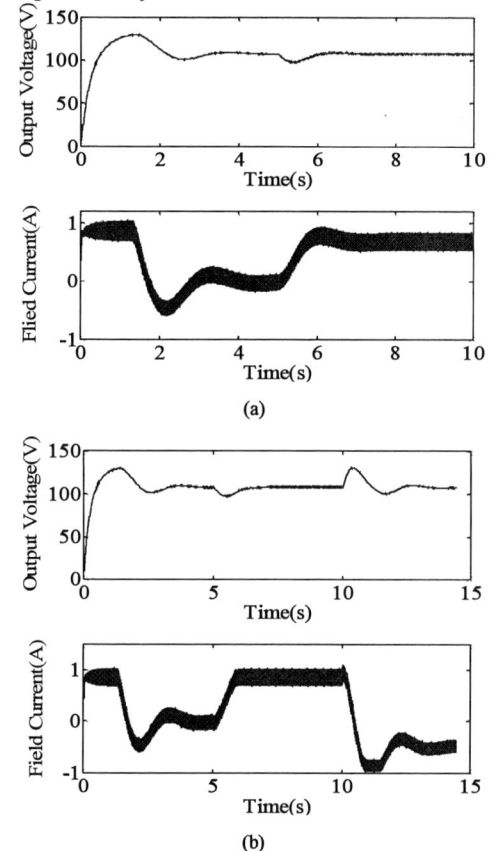

Fig.13 Simulation results of closed-loop control system (a) Load current varied (b) Rotor speed varied

VI. CONCLUSIONS

The paper describes the structure and operation principle of a novel dual-stator hybrid excited synchronous wind generator. A

prototype of generator was built, and its basic characteristics are computed by 3-D FEM. Based on mathematical equations of generator, the simulation model of closed-loop control system is built, and some useful results are obtained. The test results were obtained from the prototype of generator. From the FEM and experimental results, it is clear that the developed novel dual-stator hybrid excited synchronous wind generator has a good capacity of field-control, and its output voltage can be maintained at rated value by regulating field current. The structure of dual-stator can increase output voltage effectively as well.

ACKNOWLEDGMENT

Thanks the support of National Natural Science Foundation of P.R. China (NO. 50337030)

REFERENCES

[1] L.Video, M.Gabsi, "Homopolar and bipolar hybrid excitation synchronous machines" *Proc. of Int. Conf. on Electric Machines and Drives (IEMDC2005),* San Antonio, Texas(USA),May, 2005, pp. 1212-1218.

[2] Nobuyula Naoe, "Trial production of a hybrid excitation type synchronous machine," *Proc. of Int. Conf. on Electric Machines and Drives (IEMDC2001),* (France), 2001, pp. 545-547.

[3] Q.J.Wang and J.Chen, "The modeling and calculating of a new type hybrid claw pole alternator," *Proceedings of the CSEE,* vol.23, no.2, pp. 67-70, Feb.2003.

[4] Q.J.Wang and G.L.Li, "Investigation and calculations on 2 dimensional field and inductance of a hybrid claw pole alternator with PM excitation," *Transactions of China Electrotechnical Society,* vol.17, no.5, pp.1-5, May.2002.

[5] Amara Y.A new topology of hybrid synchronous machine. IEEE *Transactions on Industry Applications,* vol.37, no.5, pp.1273-1281, Sept./Oct.2001.

[6] J.H.Hu and J.B.Zou, "Finite element calculations of the saturation DQ-axes inductance for a direct drive PM synchronous motor considering cross-magnetization," *Proc. of 5th Int. Conf. on Power Electronics and Drive System (PEDS 2003),* Nov.2003,vol.1, pp. 677-681.

[7] Y. S.Chen, Z. Q. Zhu, "Calculation of d- and q-axis inductances of PM brushless ac machines accounting for skew," *IEEE Transactions on Magnetics,* vol.41, no.10, pp.3940-3942, Oct.2005.

[8] Q.J.Wang and Y.Y.Ni, "Computation of magnetic fields and inductances of a claw-pole alternator under load conditions," *Proceedings of the CSEE,* vol.24, no.3, pp. 91-95, 2004.

Application of Multi-level Multi-domain Modeling in the Design and Analysis of a PM Transverse Flux Motor with SMC Core

Youguang Guo*, Jianguo Zhu*, Dikai Liu*, Haiyan Lu**, and Shuhong Wang***

* Faculty of Engineering, University of Technology, Sydney, NSW 2007, Australia
** Faculty of Information Technology, University of Technology, Sydney, NSW 2007, Australia
*** State Key Laboratory of Electrical Insulation and Power Equipment, Faculty of Electrical Engineering,
Xi'an Jiaotong University, Xi'an, 710049, China

Abstract–This paper presents the design and analysis of a permanent magnet (PM) transverse flux motor with soft magnetic composite (SMC) core by applying multi-level multi-domain modeling. The design is conducted in two levels. The upper level is composed of a group of equations which describe the electrical and mechanical characteristics of the motor. The lower level consists of two domains: electromagnetic analysis and thermal calculation. The initial design, including structure, materials and major dimensions, is determined according to existing experience and empirical formulae. Then, optimization is carried out at the system level (the upper level) for the best motor performance by optimizing the structural dimensions. To successfully deal with such a multi-level multi–domain optimization problem, an effective modeling with both high computational accuracy and speed is required. For accurately computing the key motor parameters, such as back electromotive force, winding inductance and core loss, magnetic field finite element analysis is performed. The core loss in each element is stored for effective thermal calculation, and the winding inductance and back EMF are stored as a look-up table for effective analysis of the motor's dynamic performance. The presented approach is effective with good accuracy and reasonable computational speed.

Index Terms—Motor design, multi-level multi-domain modeling, permanent magnet (PM) transverse flux motor, soft magnetic composite (SMC) material.

I. INTRODUCTION

The design of electrical motors is a complex task involving many aspects such as performance prediction, parameter calculation, dimension and structure optimization, mechanical and thermal calculation, etc. A large amount of models and approaches have been developed by various researchers for these analyses and computations. For example, parameters can be computed with empirical formulae or magnetic field finite element analysis (FEA). The empirical formulae, based on many simplifications and assumptions, often produce large computational error and cause incorrect prediction of the motor performance, although the simulation speed is impressive. Any incorrect simulation may cause design failure and expensive design changes [1]. On the other hand, the approach based on numerical magnetic field

analysis can provide high computational accuracy, but the CPU time can be very long. The long computational time is a big problem for the design and analysis of a complex device such as an electrical machine, which is composed of multi-levels and multi-domains. Therefore, an effective modeling with both high accuracy and high speed is always desired by motor designers.

With the help of effective multi-level multi-domain modeling, this paper presents the design and analysis of a transverse flux motor (TFM) with soft magnetic composite (SMC) core and permanent magnet (PM) flux concentrating rotor. The core material, SOMALOY™ 500, is a relatively new SMC material developed by Höganäs AB, Sweden [2]. SMC materials possess many unique properties like magnetic isotropy and very low eddy current loss and hence are very suitable for applications of electrical machines with complex structures and three-dimensional (3-D) fluxes, such as claw pole motors and TFMs [3-4]. In a TFM, the magnetic field at the armature has significant component along any direction. Consequently, 3-D numerical analysis should be conducted to accurately determine the field distribution and key motor parameters.

A hierarchical two-level modeling approach is applied for the design of this TFM, as illustrated in Fig. 1. The upper level consists of a group of electrical and mechanical equations, which describe the motor characteristics. The lower level includes two domains: electromagnetic design and thermal calculation. Electromagnetic design is to mainly determine key motor parameters such as PM flux (defined as the flux of one phase winding produced by rotor PMs), back electromotive force (EMF), winding inductance and core loss, based on magnetic field FEAs. Thermal calculation is to predict the temperature rises of key locations such as winding and PMs, which must be within the limits for safe operation of the motor.

The initial design, including the structure, materials and dimensions, is determined based on existing experience and empirical formulae. Then, optimization is conducted at the system level for the best motor performances, such as the highest torque to cost ratio within a certain volume. To effectively couple the designs at different levels and different domains, the core loss at each element from electromagnetic design is

---

This work was partly supported by the Australian Research Council Discovery Project Grants (DP0773858).

stored as the input of thermal calculation, and the patterns of back EMF and winding inductance from electromagnetic analysis are stored for the motor performance prediction at the upper level. By this way, a good compromise between accuracy and speed can be achieved.

Another important domain, mechanical design, can be included in the lower level. In this paper, the mechanical strength is guaranteed by our experience, e.g. the minimal core dimension is controlled to be no less than 6 mm during the optimization process. It should be noted that more domains and levels, such as the mechanical design, could be easily added, leading to a more complex design architecture.

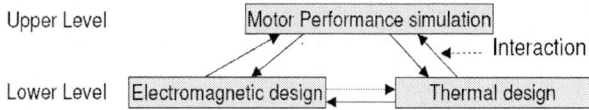

Fig. 1. Two-level hierarchical motor design architecture

II. PM TRANSVERSE FLUX MOTOR WITH SMC CORE

PM machines with transverse flux structure have attracted strong interest of research since Weh *et al.* proposed the first versions of TFMs in the 1980s [5]. Capable of producing very high specific torque provided that the number of poles is large, TFMs are naturally suitable for direct drive applications which request large torque at low rotational speed.

Generally, TFMs have complex structure with a large number of components, especially for the double-sided stator type, so it is very difficult to manufacture the cores by using the conventional laminated steels. TFMs have large leakage flux, which is 3-D in nature, in addition to the 3-D main flux. The flux component perpendicular to the lamination plane may cause excessive eddy currents. These problems have greatly limited the applications of TFMs, but they may be overcome by the development of new SMC materials [3-4]. The applications of SMCs in TFMs have been investigated by several researchers, and the results are quite promising [6-8].

Based on our previous experience with SMC motors [4, 8-9], this paper presents the design and analysis of a three-phase three-stack TFM with modified double-sided stator and PM flux concentrating rotor. SMC is used as the cores of both stator and rotor. Fig. 2 illustrates the magnetically relevant parts of one pole-pair of one stack (the single concentrated coil is not shown for clarity). Each stack forms a phase and three phases are axially stacked with the stator cores shifted by 120° electrical and the rotor PMs aligned. The rotor is supported by cantilevers on the shaft.

The main dimensions and parameters of the TFM include: 100 mm for the stator outer diameter, 56 mm for the stator and rotor inner diameter, 33 mm for the effective axial length of each stack, 9.5 mm for the PM radial length, and 20 poles. The motor is designed to operate with a brushless DC (BLDC) control scheme, delivering a power of 280 W at 3000 rev/min for compressor driving.

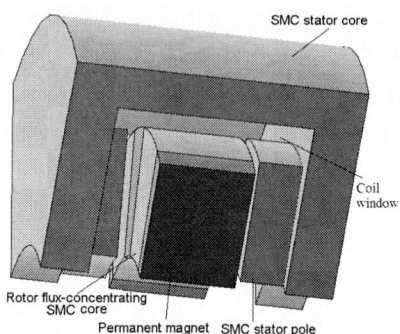

Fig. 2. Structure and FEA solution region of one pole-pair of one stack of a TFM

III. ELECTROMAGNETIC DESIGN

Electromagnetic analysis is conducted to determine the motor's key parameters. Due to the complex structure and 3-D field distribution, analytical and empirical formulae cannot provide accurate computation. Therefore, numerical field analysis like FEA is required, allowing the nonlinear properties of the materials and structural details to be considered. Because of the magnetic independence between stacks and the symmetry of the motor structure, it is only required to analyze the magnetic field in one pole-pair region of one phase, as shown in Fig. 2. At the two radial boundary planes, the magnetic scalar potential used to solve the magnetic field distribution obeys the so-called periodical boundary conditions:

$$\varphi_m\left(r, \Delta\theta, z\right) = \varphi_m\left(r, -\Delta\theta, -z\right) \tag{1}$$

where $\Delta\theta$=30° is the angle of one pole pitch. The original point of the cylindrical coordinate is located at the center of the motor.

From the numerical field distribution, many key motor parameters can be accurately obtained. For example, from the no-load field solutions, the curve of PM flux against rotor angle is found to be an almost perfect sinusoid with magnitude of ϕ_1=0.358 mWb, so the back EMF constant is calculated as 0.232 Vs/rad by

$$K_E = \frac{p}{2} N_s \frac{\phi_1}{\sqrt{2}} \tag{2}$$

where p=12 is the number of poles and N_s=153 is the number of turns of a phase winding.

With the numerical magnetic field solutions at no-load, the cogging torque can be calculated by using the virtual work method or Maxwell stress tensor method. It is found that this TFM has a negligible cogging torque with a peak value of 0.02 Nm, comparing to the rated output torque of 0.89 Nm. The major cogging torque component of one stack is the fundamental, which is cancelled by those of the other two stacks.

Winding inductance is another key parameter determining the motor performance. When the magnetic circuit is saturated, the motor performance depends on the incremental (differential) inductance rather than the secant (apparent) inductance [10]. In this paper, the winding inductance of the TFM is computed by a modified incremental energy method [11], which consists of the following steps: (1) Conduct a nonlinear analysis with the excitations of both rotor PMs and stator currents; (2) Calculate and save the incremental permeability in each element; (3) Conduct a linear analysis with the saved permeability and the excitation of a perturbed stator current Δi (from zero current) only; (4) Calculate the magnetic co-energy ΔW_c. Then, the winding incremental inductance can be obtained by

$$L_{inc} \approx \frac{2\Delta W_c}{(\Delta i)^2} \qquad (3)$$

Because each phase has an almost independent magnetic circuit, the mutual inductances between phase windings can be considered as zero. The average self incremental inductance of one phase winding is computed as 13.0 mH.

Unlike conventional electrical machines in which the copper loss is the dominant component of total power loss, an SMC machine has comparable core loss and copper loss, so accurate prediction of core loss is important in the motor design and analysis. It has been long known that core losses are caused by not only alternating but also rotating magnetic fields, and core losses caused by different patterns are very different [12]. In the TFM, when the rotor rotates, the flux density loci at certain location can be alternating (one-dimensional) with or without harmonics, circularly or elliptically rotating within a two-dimensional plane which may not be parallel to any axis, or even an irregular loop in a real 3-D space.

An improved method is applied for predicting the core losses of this 3-D flux SMC motor [13]. Different formulations are used for core loss prediction with alternating, circularly rotating, and elliptically rotating flux density vectors, respectively. The core loss of the TFM is computed as 16.3 W at 3000 rev/min. Due to the very small eddy current loss, the total core loss can be considered proportional to the rotor speed, or operational frequency.

Other parameters determined include: phase winding resistance of 2.304 Ω, copper loss of 15.6 W, and mechanical loss of 8.8 W.

IV. THERMAL DESIGN

Thermal design is crucial for the economical utilization of materials and safe operation of a motor. Commonly used methods for thermal analysis of electrical machines include lumped parameter thermal model [14], thermal network and thermal FEA [15]. The thermal network combines the accuracy of FEA and the speed of lumped parameter thermal model. As the core loss distribution has been obtained in each element, which can be directly input as the heat sources of thermal analysis, thermal FEA and a hybrid thermal model [16] are employed to analyze the temperature rise of the TFM. During the optimization process, the maximum temperatures of winding and PM are controlled as 139.5 °C and 120 °C, respectively.

V. MOTOR PERFORMANCE SIMULATION

A. Steady-State Characteristics

The motor can be operated with a BLDC drive scheme. At synchronous motor mode, the steady-state characteristic of the TFM can be predicted by an equivalent electrical circuit, as shown in Fig. 3, where E_1 is the root-mean-squared (RMS) value of induced back EMF, R_1 the phase winding resistance, ω_1 the operational angular frequency, and L_1 the synchronous inductance, which equals the self inductance of a phase winding plus half the mutual inductance between two phase windings. All these parameters have been obtained in the Electromagnetic Design.

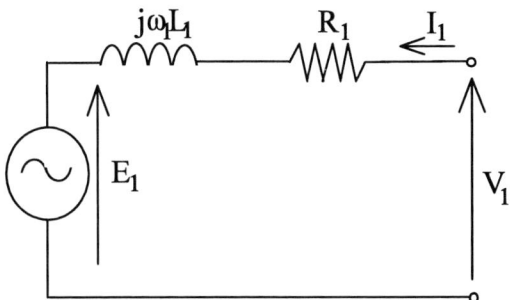

Fig. 3. Per-phase equivalent electrical circuit of PM synchronous motor

At the optimum BLDC control condition, the stator current I_1 is in phase with E_1, so that the electromagnetic power and torque can be calculated by

$$P_{em} = mE_1 I_1 \qquad (4)$$

$$T_{em} = \frac{P_{em}}{\omega_r} = K_T I_1 \qquad (5)$$

where ω_r is the rotor speed in mechanical rad/s, $K_T = mK_E$ is the torque constant, and $m=3$ is the number of phases. The RMS value of the back EMF is obtained by $E_1 = K_E \omega_r$.

For a given terminal voltage, V_1, the characteristic of rotor speed against electromagnetic torque can be derived as

$$\omega_r = \frac{\sqrt{\left(\frac{R_1 T_{em}}{m}\right)^2 + \left[\left(\frac{p}{2}L_1\frac{T_{em}}{K_T}\right)^2 + K_E^2\right]\left[V_1^2 - \left(\frac{R_1 T_{em}}{K_T}\right)^2\right]} - \frac{R_1 T_{em}}{m}}{\left(\frac{p}{2}L_1\frac{T_{em}}{K_T}\right)^2 + K_E^2} \qquad (6)$$

The output power P_{out}, output torque T_{out}, input power P_{in} and efficiency η of the total drive system can be calculated by

$$P_{out} = P_{em} - P_{Fe} - P_{mec} \qquad (7)$$

$$T_{out} = P_{out} / \omega_r \qquad (8)$$

$$P_{in} = P_{em} + P_{inv} + P_{cu} \qquad (9)$$

$$P_{cu} = 3I_1^2 R_1 \qquad (10)$$

$$\eta = P_{out} / P_{in} \qquad (11)$$

where P_{Fe} is the core loss, P_{mec} the mechanical loss, P_{inv} the inverter conduction loss, P_{cu} the copper loss, and ω_r the rotor angular speed in mechanical rad/s.

B. Dynamic Performance

The dynamic characteristics of the motor are determined by a group of equations as

$$v_j = r_j i_j + d\lambda_j / dt + e_j, \quad j = a, b, c \qquad (12)$$

$$\lambda_j = \sum_{k=a}^{c} L_{jk} i_k \qquad (13)$$

$$T_{em} = (\sum_{j=a}^{c} e_j i_j) / \omega_r + T_{cog} \qquad (14)$$

$$J\frac{d\omega_r}{dt} = T_{em} - T_L - \delta\omega_r \quad \text{and} \quad \frac{d\theta_r}{dt} = \omega_r \qquad (15)$$

where all the variables are of their conventional meanings.

For the symmetrical Y-connected 3 phase windings without central line, we have

$$i_a + i_b + i_c = 0 \qquad (16)$$

$$r_a = r_b = r_c \qquad (17)$$

$$L_{ab} = L_{ba}, \quad L_{bc} = L_{cb}, \quad L_{ca} = L_{ac} \qquad (18)$$

Based on (12)-(18), a complete simulation model can be built in Simulink environment [17]. This model can be applied to analyze the steady and dynamic performances of BLDC motors to predict if the design requirements can be met, e.g. the dynamic performance of start-up, i.e. whether or not the motor can reach the required steady speed under the full load when the rated DC voltage of the inverter is applied.

An optimization routine for the design of the PM transverse flux motor with SMC core has been set up, which considers multi-levels and multi-domains. The magnetic field FEAs, thermal analysis, performance calculation, and optimization searches are all implemented in a commercial comprehensive software package, ANSYS. It is found that the presented method is practical with both good accuracy and speed.

VI. Conclusions and Discussions

In this paper, a permanent magnet transverse flux motor with soft magnetic composite core is designed and analyzed by using a two-level modeling. The lower level consists of electromagnetic design domain and thermal analysis domain. The upper level consists of the performance simulation domain only and at this level optimization is carried out in close interaction with the lower level. Appropriate modeling is applied to achieve effective interaction between various levels and domains,

with both good computational accuracy and speed.

The design and analysis of a motor drive is a complex task. For a practical drive system, many aspects such as motor type and topology design, power electronic circuit design, controller system design, and dimension and loading design and material selection, should also be considered [18]. It should be noted that these aspects could be easily added to the hierarchical motor design architecture of Fig. 1, leading to a more complex multi-level multi-domain design problem.

References

[1] B. Knorr, D. Devarajan, D. Lin, P. Zhou, and S. Stanton, "Application of multi-level multi-domain modeling to a claw pole alternator," in *Proc. Annual Conf. of Society of Automotive Engineers*, USA, 2004, Paper 2004-01-0758.

[2] "Soft magnetic composites from Höganäs Metal Powders - SOMALOY™ 500," *Höganäs Product Guide*, 1997.

[3] "The latest development in soft magnetic composite technology," *SMC Update, Reports of Höganäs AB, Sweden,* 1997-2007. At http://www.hoganas.com/, see News then SMC Update.

[4] Y. G. Guo, J. G. Zhu, P. A. Watterson, and W. Wu, "Comparative study of 3-D flux electrical machines with soft magnetic composite core," *IEEE Trans. on Industry Applications*, vol. 39, no. 6, pp. 1696-1703, Nov. 2003.

[5] H. Weh and H. May, "Achievable force densities for permanent magnet excited machines in new configuration," in *Proc. Int. Conf. on Electrical Machines*, Munich, Germany, 1986, pp. 1107-1111.

[6] B. C. Mecrow, A. G. Jack, and C. P. Maddison, "Permanent magnet machines for high torque, low speed applications," in *Proc. Int. Conf. on Electrical Machines*, Vigo, Spain, 1996, pp. 461-466.

[7] G. Henneberger and M. Bork, "Development of a transverse flux traction motor in a direct drive system," in *Proc. Int. Conf. on Electrical Machines*, Helsinki, Finland, 2000, pp. 1457-1460.

[8] Y. G. Guo, J. G. Zhu, P. A. Watterson, and W. Wu, "Development of a permanent magnet transverse flux motor with soft magnetic composite core," *IEEE Trans. on Energy Conversion*, vol. 21, no. 2, pp. 426-434, June 2006.

[9] Y. G. Guo, J. G. Zhu, P. A. Watterson, and W. Wu, "Development of a claw pole permanent magnet motor with soft magnetic composite stator," *Australian J. Electrical & Electronic Eng.*, vol. 2, no. 1, pp. 21-30, 2005.

[10] M. Gyimesi and D. Ostergaard, "Inductance computation by incremental finite element analysis," *IEEE Trans. Magn.*, vol. 35, no. 3, pp. 1119-1122, May 1999.

[11] Y. G. Guo, J. G. Zhu, and H. Y. Lu, "Accurate determination of parameters of a claw pole motor with SMC stator core by finite element magnetic field analysis," *IEE Proceedings – Electric Power Application,* vol. 153, no. 4, pp. 568-574, July 2006.

[12] Y. G. Guo, J. G. Zhu, J. J. Zhong, and W. Wu, "Core losses in claw pole permanent magnet machines with soft magnetic composite core," *IEEE Trans. Magn.*, vol. 39, no. 5, pp. 3199-3201, Sep. 2003.

[13] Y. G. Guo, J. G. Zhu, Z. W. Lin, and J. J. Zhong, "Measurement and modeling of core losses of soft magnetic composites under 3D magnetic excitations in rotating motors," *IEEE Trans. Magn.*, vol. 41, no. 10, pp. 3925-3927, Oct. 2005.

[14] P. H. Mellor, D. Roberts, and D. R. Turner, "Lumped parameter thermal model for electrical machines of TFFC

design," *Proc. Inst. Elect. Eng. Pt. B*, vol. 138, no. 5, pp. 205-218, Sep. 1991.

[15] G. Cannistrà, G. Cannistrà, and M. S. Labini, "Thermal analysis in an induction machine using thermal network and finite element methods," in *Proc. IEE EMD Conf.*, 1991, pp. 300-304.

[16] Y. G. Guo, J. G. Zhu, and W. Wu, "Thermal analysis of SMC motors using a hybrid model with distributed heat sources," *IEEE Trans. Magn.,* vol. 41, no. 6, pp. 2124-2128, June 2005.

[17] Y. G. Guo, J. G. Zhu, J. X. Chen, and J. X. Jin, "Performance analysis of a permanent magnet claw pole SMC motor with brushless DC control scheme," in *Proc. Int. Power Electronics and Motion Control Conf.*, Shanghai, China, 13-16 Aug. 2006, Vol. 2. pp. 1-5.

[18] J. G. Zhu, D. K. Liu, and Y. G. Guo, "Application oriented system level optimum design method for advanced electrical drive systems," Australian Research Council Discovery Project, DP0773858, 2007-2009.

New Approximate 2DOF Digital Controller for DC-DC Converter with Second-Order Differential Characteristics

Eiji TAKEGAMI[1], Kohji HIGUCHI[2], Kazushi NAKANO[3]
Satoshi TOMIOKA[4], Kazushi WATANABE[5]
[1,4,5]DENSEI-LAMBDA K.K., 2701 Togawa, Settaya, Nagaoka 940-1195, Japan
email: e.takegami@densei-lambda.com, s.tomioka@densei-lambda.com
k18.watanabe@densei-lambda.com
[2,3]Dept. of Electronic Engineering, The University of Electro-Communications,
1-5-1 Chofu-ga-oka, Chofu, Tokyo 182-8585, Japan
email: higuchi@ee.uec.ac.jp, nakano@ee.uec.ac.jp

Abstract— Robust DC-DC converter which can cover extensive load changes and also input voltage changes with one controller is needed. Then the demand to suppressing output voltage change becomes still severer. We proposed an aproximate 2DOF digital controller which which realizes the start-up response and dynamic load response independently. The controller makes the control bandwidth wider, and at the same time makes a variation of the output voltage small at sudden changes of loads and the input voltages. In this paper, a new approximate 2DOF digital controller using additinal zeros is proposed. The new controller makes a variation of the output voltage more small at sudden changes of loads and the input voltages. This controller is actually implemented on a DSP and is connected to a DC-DC converter. Experimental studies demonstrate that this type of digital controller can satisfy given severe specifications.

I. INTRODUCTION

In many applications of DC-DC converters, loads cannot be specified in advance, i.e., their amplitudes are suddenly changed from the zero to the maximum rating. Generally, design conditions are changed for each load and then each controller is re-designed. Then, a so-called robust DC-DC converter which can cover such extensive load changes and also input voltage changes with one controller is needed.Then the demand to suppressing output voltage change becomes still severer. Analog control IC is used usually for the controller of DC-DC converter. Simple integral control etc. are performed with the analog control IC. Moreover, the application of the digital controller to DC-DC converter designed by the PID or root locus method etc. has been recently considered[1], [2]. However it is difficult to retain sufficient robustness of DC-DC converters by these techniques.

The authors proposed the method of designing a approximate 2-degree-of-freedom (2DOF) controller of DC-DC converter[3]. This controller realized a first-order differential transfer characteristics between equivalent disterbances and an output voltage. In this paper, we propose a new approximate 2DOF digital controller which realizes a second-order differential trasfer characteritics. The new controller makes a variation of the output voltage more

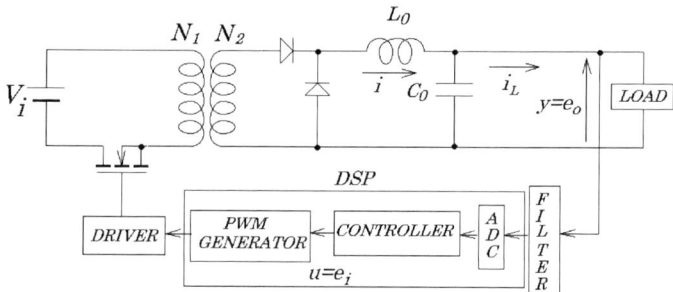

Fig. 1. DC-DC converter

small at sudden changes of resistive load and the input voltage. This characteristic is realized by introducing an additional zeros into a transfer function between equivalent disterbances and an output. A new DC-DC converter equipped the proposed controller in DSP is actually manufactured. Some simulations and experiments show that this new DC-DC converter can satisfy given severe specifications.

II. DC-DC CONVERTER

The DC-DC converter as shown in **Fig.1** has been manufactured. In order to realize the approximate 2DOF digital controller which satisfies given specifications, we use the DSP(TI TMS320LF2801). This DSP has a built-in AD converter and a PWM switching signal generating part. The triangular wave carrier is adopted for the PWM switching signal. The switching frequency is set at 400[KHz] and the peak-to-peak amplitude C_m is 125[V]. The LC circuit is a filter for removing carrier and switching noises. C_0 is 300[μF] and L_0 is 0.46[μH]. If the frequency of control signal u is smaller enough than that of the carrier, the state equation of the DC-DC converter at a resistive load in Fig.1 except for the controller in DSP can be expressed from the state equalizing method[4] as follows :

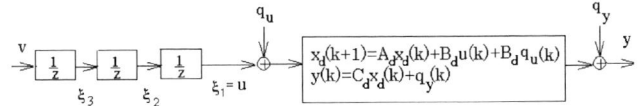

Fig. 2. Controlled object with input dead time $L_d(\leq T)$

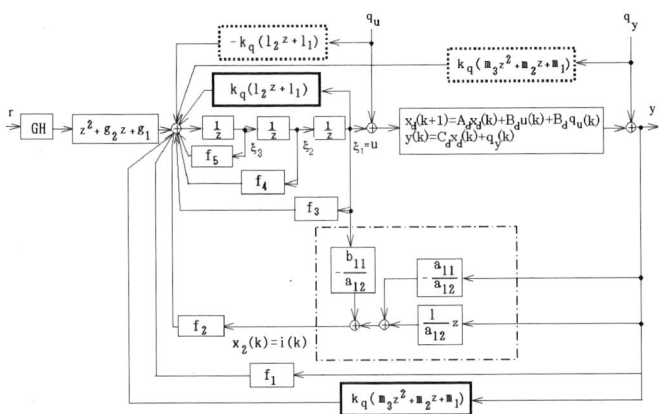

Fig. 3. Equivalent disturbances due to load variations (parameter variations) and model matching system with state feedback

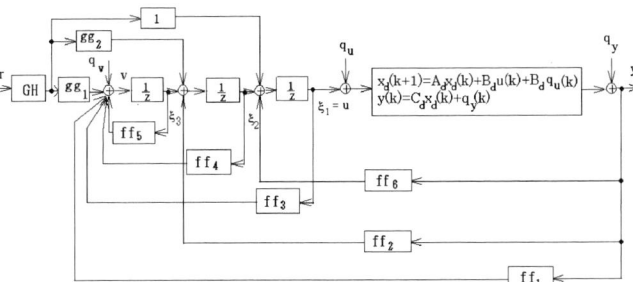

Fig. 4. Model matching system using only voltage (output) feedback

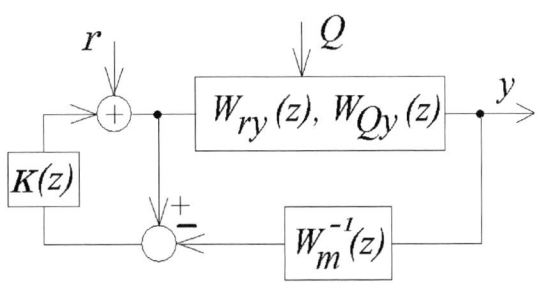

Fig. 5. System reconstituted with inverse system and filter

$$\begin{cases} \dot{x} = A_c x + B_c u + B_c q_u \\ y = Cx + q_y \end{cases} \quad (1)$$

where

$$x = \begin{bmatrix} e_o \\ i \end{bmatrix} \quad A_c = \begin{bmatrix} -\frac{1}{C_0 R_L} & \frac{1}{C_0} \\ -\frac{1}{L_0} & -\frac{R_0}{L_0} \end{bmatrix} \quad B_c = \begin{bmatrix} 0 \\ \frac{K_p}{L_0} \end{bmatrix}$$

$$C = \begin{bmatrix} 1 & 0 \end{bmatrix} \quad u = e_i \quad y = e_0 \quad K_p = -\frac{V_i N_2}{C_m N_1}$$

and R_0 is the total resistance of coil and ON resistance of FET, etc., and the value is $0.015[\Omega]$. Then the discrete-time state equation of the system (1) with a zero order hold is expressed as follows :

$$\begin{cases} x_d(k+1) = A_d x_d(k) + B_d u(k) + B_d q_u(k) \\ y(k) = Cx_d(k) + q_y(k) \end{cases} \quad (2)$$

where

$$A_d = \begin{bmatrix} a_{11} & a_{12} \\ a_{21} & a_{22} \end{bmatrix} = e^{A_c T} \quad B_d = \begin{bmatrix} b_{11} \\ b_{21} \end{bmatrix} = \int_0^T e^{A_c \tau} B_c d\tau$$

The transfer function of the system (2) is as follows:

$$G_p(z) = \frac{N_p(z)}{D_p(z)} \quad (3)$$

where

$$N_p(z) = b_{11}z + b_{21}a_{12} - a_{22}b_{11}$$
$$D_p(z) = z^2 - (a_{11} + a_{22})z + a_{11}a_{22} - a_{21}a_{12}$$

In practical use of a DC-DC converter, the characteristics of a startup transient response, a dynamic load response and an output response at input voltage sudden change are important. The DC-DC converter with the following

specifications 1-7 is designed and manufactured. Specs 1, 5 and 7 are new ones. These specs are more severe than $200[\mu F]$ and $50[mV]$ in references[3].

1. Input voltage V_i is 48[V] and output voltage e_o is 3.3[V].
2. Startup transient reponses are almost the same at resistive load and parallel load of resistance and capacity, where $0.165 \leq R_L < \infty[\Omega]$ and $0 \leq C_L \leq 300[\mu F]$.
3. The rising time of the startup transient reponse is smaller than $100[\mu s]$.
4. Against all the loads of spec.1, an over-shoot is not allowable in the startup transient response.
5. The dynamic load response is smaller than 25[mV] against 10[A] change of load current.
6. The specs. 2, 3, 4 and 5 are satisfied also to change of input voltage of ±20%.
7. Regulation of output voltage is amaller than 25[mV] at the sudden change of input voltage within ±20%.

The load changes for the controlled object and the input voltage change are considered as parameter changes in eq.(1). Such parameter changes can be transformed to equivalent disturbances q_u and q_y as shown in **Fig.3** even in discrete-time systems. Therefore, what is necessary is just to constitute the control systems whose pulse transfer functions from equivalent disturbances q_u and q_y to the output y become as small as possible in their amplitudes, in order to robustize or suppress the influence of these parameter changes and input voltage change.

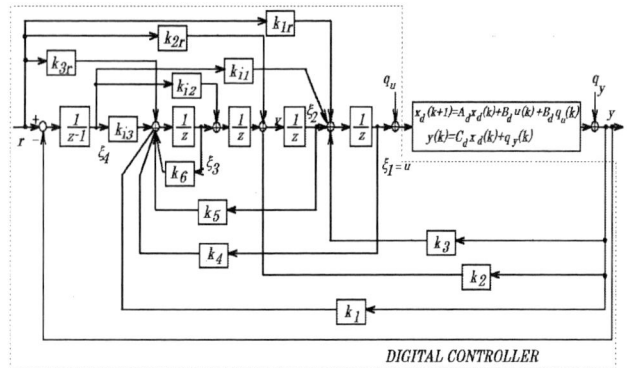

Fig. 6. Approximate 2DOF digital integral type control system

III. DESIGN METHOD OF APPROXIMATE 2DOF DIGITAL CONTROLLER WITH ADITIONAL ZEROS

A. Addional method of zeros

The following equation is obtained by repeating the difference of the output of eq.(2).

$$Y = O^* x_d(k) + U\bar{u}(k) + U\bar{q}_u + \bar{q}_y \qquad (4)$$

where

$$Y = \begin{bmatrix} y(k) \\ y(k+1) \\ y(k+2) \end{bmatrix} \quad O^* = \begin{bmatrix} C \\ CA_d \\ CA_d^2 \end{bmatrix} \quad U = \begin{bmatrix} 0 & 0 \\ CB_d & 0 \\ CA_dB_d & CB_d \end{bmatrix}$$

$$\bar{u} = \begin{bmatrix} u(k) \\ u(k+1) \end{bmatrix} \quad \bar{q}_u = \begin{bmatrix} q_u(k) \\ q_u(k+1) \end{bmatrix} \quad \bar{q}_y = \begin{bmatrix} q_y(k) \\ q_y(k+1) \\ q_y(k+1) \end{bmatrix}$$

If both sides of eq.(3) are mulutiplied by $\bar{I}_2$ from the left and solved about x_d, the following equation will be obtained.

$$\begin{aligned} x_d(k) &= (\bar{I}_2O^*)^{-1}\bar{I}_2Y - (\bar{I}_2O^*)^{-1}\bar{I}_2U\bar{u}(k) \\ &\quad - (\bar{I}_2O^*)^{-1}\bar{I}_2U\bar{q}_u - (\bar{I}_2O^*)^{-1}\bar{I}_2\bar{q}_y \end{aligned} \qquad (5)$$

where

$$\bar{I}_2 = \begin{bmatrix} 1 & 0 & 0 \\ 0 & 1 & 0 \end{bmatrix}$$

This equation is substituted into eq.(3), the following equation will be obtained.

$$\begin{aligned} & (I_3 - O^*(\bar{I}_2O^*)^{-1}\bar{I}_2)U\bar{q}_u + (I_3 - O^*(\bar{I}_2O^*)^{-1}\bar{I}_2)\bar{q}_y \\ &= (I_3 - O^*(\bar{I}_2O^*)^{-1}\bar{I}_2)Y - (I_3 - O^*(\bar{I}_2O^*)^{-1}\bar{I}_2)U\bar{u} \end{aligned} \qquad (6)$$

where I_3 is a 3×3 unit matrix. That is, $\bar{q}_u$ and $\bar{q}_y$ can be replaced to Y and $\bar{u}(k)$. Eq.(5) is transformed as follws:

$$\begin{aligned} & -(l_2z + l_1)q_u(k) + (z^2 + m_2z + m_1)q_y(k) \\ &= (l_2z + l_1)u(k) + (z^2 + m_2z + m_1)y(k) \end{aligned} \qquad (7)$$

where

$$\begin{bmatrix} l_1 & l_2 \end{bmatrix} = -(I_3 - O^*(\bar{I}_2O^*)^{-1}\bar{I}_2)U$$
$$\begin{bmatrix} m_1 & m_2 & m_3 \end{bmatrix} = (I_3 - O^*(\bar{I}_2O^*)^{-1}\bar{I}_2) \qquad (8)$$

The system of Fig.2 is constituted in consideration of a delay time for calculation etc., a estimation of current and a zeros addition. The state equation of Fig.2 can be expressed as follows:

$$\begin{cases} x_{dw}(k+1) = A_{dw}x_{dw}(k) + B_{dw}v(k) \\ y(k) = C_{dw}x_{dw}(k) \end{cases} \qquad (9)$$

where

$$x_{dw}(k) = \begin{bmatrix} x_d(k) \\ \xi_1(k) \\ \xi_2(k) \\ \xi_3(k) \end{bmatrix} \quad A_{dw} = \begin{bmatrix} A_d & B_d & 0 & 0 \\ 0 & 0 & 1 & 0 \\ 0 & 0 & 0 & 1 \\ 0 & 0 & 0 & 0 \end{bmatrix}$$

$$B_{dw} = \begin{bmatrix} 0 \\ 0 \\ 0 \\ 1 \end{bmatrix} \quad C_{dw} = \begin{bmatrix} C & 0 & 0 & 0 \end{bmatrix} \quad \xi_1(k) = u(k)$$

The following feedforwards from q_u, q_y and r, and state feedback are applied to the system of eq.(9) shown in Fig.(3).

$$\begin{aligned} v(k) &= -k_q(l_2z + l_1)q_u(k) + k_q(z^2 + m_2z + m_1)q_y(k) \\ &\quad + (z^2 + g_2z + g_1)r(k) + [f_1\ f_2\ f_3\ f_4\ f_5\ f_6]x_{dw}(k) \end{aligned} \qquad (10)$$

In Fig.3, the parts surrounded by dotted lines are the feedforward coefficients from q_u and q_y and the part surrounded by a chain line is the estimated part of current. From eq.(5), the feedforward of eq.(7) are changed as follws:

$$\begin{aligned} v(k) &= k_q(l_2z + l_1)u(k) + k_q(z^2 + m_2z + m_1)y(k) \\ &\quad + (z^2 + g_2z + g_1)r(k) + [f_1\ f_2\ f_3\ f_4\ f_5\ f_6]x_{dw}(k) \end{aligned} \qquad (11)$$

That is, the parts surrounded by the dotted lines are replaced by the parts surrounded by solid lines from u and y. The system except for the parts surrounded by the dotted lines in Fig.3 can be transformed equivalently as shown in Fig.4. In Fig.4,

$$\begin{aligned} ff_1 &= f_1 - f_2(a_{11}/a12) + k_qm_1 + f_5(k_qm_2 + f_2/a_{12}) \\ &\quad + f_5^2k_qm_3 + (f_4 + k_ql_2)k_qm_3 \\ ff_2 &= k_qm_2 + f_2/a_{12} + f_5k_qm_3 \\ ff_3 &= f_3 - f_2(b_{11}/a_{12}) + k_ql_1 \\ ff_4 &= f_4 + k_ql_2 \\ ff_5 &= f_5 \\ ff_6 &= k_qm_3 \end{aligned} \qquad (12)$$

The transfer functions between r and y, q_u and y, and q_y and y in Fig.4 are as follows:

$$W_{ry}(z) = N_{ry}(z)/D(z) \qquad (13)$$
$$W_{q_uy}(z) = N_{q_uy}(z)/D(z) \qquad (14)$$
$$WW_{q_yy}(z) = N_{q_yy}(z)/D(z) \qquad (15)$$

where

$$N_{ry}(z) = GH(z^2 + g_1z + g_0)(b_{11}z + b_{21}a_{12} - a_{22}b_{11})$$

282

$$N_{q_u y}(z) = N_{qz} N_p$$
$$N_{q_y y}(z) = N_{qz} D_p$$
$$\begin{aligned} N_{qz} &= (a_{12}z^3 - a_{12}f_5 z^2 + (a_{12}b_{11}k_q - a_{12}f_4)z \\ &- f_3 a_{12} + f_2 b_{11} - a_{12}a_{22}b_{11}k_q + a_{12}^2 b_{21}k_q) \end{aligned}$$
$$\begin{aligned} D(z) &= z^5 + (-f_5 - a_{22} - a_{11})z^4 + (a_{11}f_5 \\ &+ a_{11}a_{22} - a_{21}a_{12} + a_{22}f_5 - f_4)z^3 \\ &+ (a_{21}a_{12}f_5 - f_3 + a_{11}f_4 - a_{11}a_{22}f_5 \\ &+ a_{22}2f_4)z^2 + (a_{22}f_3 + a_{21}a_{12}f_4 + a_{11}f_3 \\ &- b_{11}f_1 - f_2 b_{21} - a_{11}a_{22}f_4)z + f_2 a_{11}b_{21} \\ &- a_{21}f_2 b_{11} + a_{21}a_{12}f_3 + f_1 a_{22}b_{11} \\ &- f_1 a_{12}b_{21}21 - a_{11}a_{22}f_3 \end{aligned}$$

From $D(z)$, the pole of the overall system can arrange arbitrarily by f_1, f_2, f_3, f_4 and f_5. From $N_{ry}(z)$, two zeros of r-y can arrange arbitrarily by g_0 and g_1. Moreover, from common N_{qz} in $N_{q_x}(s)$ and $N_{q_y y}(s)$, one zeros of $q_u - y$ and $q_y - y$ can be arbitrarily arranged at the same place by k_q. That is, one zeros can be arbitrarily added to W_{Qy}.

B. Design method

First, the transfer function between the reference input r and the output y is specified as follows:

$$W_{ry}(z) = N_H(z)/D_H(z) \tag{16}$$

where

$$\begin{aligned} N_H(z) &= (1 + H_1)(1 + H_2)(1 + H_3)(z - n_1)(z - n_2) \\ &\times (z + H_4)(z + H_5) \\ D_H(z) &= (1 - n_1)(1 - n_2)(z + H_1)(z + H_2)(z + H_3) \\ &\times (z + H_4)(z + H_5) \end{aligned}$$

where, n_1 and n_2 are the zeros for the discrete-time control object (2). $[f_1\ f_2\ f_3\ f_4\ f_5]$, $[g1\ g2]$ and GH are determined so that $W_{ry}(z)$ becomes eq.(16). And the transfer function among the disturbance inputs q_u, q_y and the output y is specified as follows:

$$W_{q_u y}(z)) = (z - 1)\bar{N}_{qz}N_p/D_H(z) \tag{17}$$
$$W_{q_y y}(z)) = (z - 1)\bar{N}_{qz}D_p/D_H(z) \tag{18}$$

Here the zeros are placed at 1 by setting k_q as the solution of $N_{qz}(1) = 0$. The k_q becomes as follows:

$$k_q = \frac{-a_{12} + a_{12}f_5 + a_{12}f_4 + f_3 a_{12} - f_2 b_{11}}{a_{12}(b_{11} - a_{22}b_{11} + b_{21}a_{12})} \tag{19}$$

$\bar{N}_{qz}$ is the remaining zeros of N_{qz} which cannot be placed arbitrarily.

It shall be specified that the relation of H_1 and H_2, H_3 becomes $|H_1| \gg |Re(H_2)|$, $|H_1| \gg |Re(H_3)|$. Then $W_{ry}(z)$ can be approximated to the following first-order model:

$$W_{ry}(z) \approx W_m(z) = \frac{1 + H_1}{z + H_1} \tag{20}$$

This target characteristics $W_{ry}(z) \approx W_m(z)$ is specified so that it satisfies the specs.3 and 4.

The system added the inverse system and the filter to the system in Fig.4 is constituted as shown in **Fig.5**. In Fig.5, the transfer function $K(z)$ becomes

$$K(z) = \frac{k_z}{z - 1 + k_z} \tag{21}$$

The transfer functions between $r - y$, $q_u - y$ and $q_y - y$ of the system in Fig.5 are given by

$$y = \frac{1 + H_1}{z + H_1} \frac{z - 1 + k_z}{z - 1 + k_z W_s(z)} W_s(z)r \tag{22}$$
$$y = \frac{(z - 1)^2}{z - 1 + k_z} \frac{z - 1 + k_z}{z - 1 + k_z W_s(z)} \frac{\bar{N}_{qz}N_p}{D_H(z)} q_u y \tag{23}$$
$$y = \frac{(z - 1)^2}{z - 1 + k_z} \frac{z - 1 + k_z}{z - 1 + k_z W_s(z)} \frac{\bar{N}_{qz}D_p}{D_H(z)} q_y y \tag{24}$$

where

$$W_s(z) = \frac{(1 + H_2)(1 + H_3)(z - n_1)(z - n_2)}{(z + H_2)(z + H_3)(1 - n_1)(1 - n_2)} \tag{25}$$

Here, if $W_s(z) \approx 1$, then eqs.(22), (23) and (24) become, respectively,

$$y \approx \frac{1 + H_1}{z + H_1}r \tag{26}$$
$$y \approx \frac{(z - 1)^2}{z - 1 + k_z} \frac{\bar{N}_{qz}N_p}{D_H(z)} q_u y \tag{27}$$
$$y \approx \frac{(z - 1)^2}{z - 1 + k_z} \frac{\bar{N}_{qz}D_p}{D_H(z)} q_y y \tag{28}$$

From eqs.(26), (27) and (28), it turns out that the characteristics from r to y can be specified with H_1, and the characteristics from q_u and $q_y y$ to y can be independently specified with k_z. That is, the system in Fig.5 is an approximate 2DOF, and its sensitivity against disturbance becomes lower with the increase of k_z.

If an equivalent conversion of the controller in Fig.5 is carried out, the approximate 2DOF digital integral-type control systems will be obtained as shown in **Fig.6**. In Fig.6, the parameters of the controller are as follows:

$$\begin{aligned} k_1 &= (f_1 - f_2 a_{11}/a_{12} + k_q(-a_{11}^2 - a_{21}a_{12} + (a_{12}a_{11} \\ &+ a_{12}a_{22})a_{11}/a_{12} + f_5(-k_q(a_{12}a_{11} \\ &+ a_{12}a_{22})/a_{12} + f_2/a_{12}) + f_5^2 k_q + (f_4 - k_q b_{11})k) \\ &- GHk_z(g0 + (g1 + f_5)f_5 + f_4 - k_q b_{11})/(1 + H2) \\ k_2 &= k_q - GHk_z/(1 + H2) \\ k_3 &= -k_q(a_{11}a_{12} + a_{12}a_{22})/a_{12} + f_2/a_{12} + f_5 k_q \\ &- (g1 + f_5)GHk_z/(1 + H2) \\ k_4 &= f_3 - f_2 b_{11}/a_{12} + k_q((a_{12}a_{11} \\ &+ a_{12}a_{22})b_{11}/a_{12} - a_{11}b_{11} - b_{21}a_{12}) \\ k_5 &= f_4 - k_q b_{11} \\ k_6 &= f_5 \\ k_i 1 &= (g0 + (g1 + f_5)f_5 + f_4 - k_q b_{11})GHk_z \\ k_i 2 &= (g1 + f_5)GHk_z \end{aligned}$$

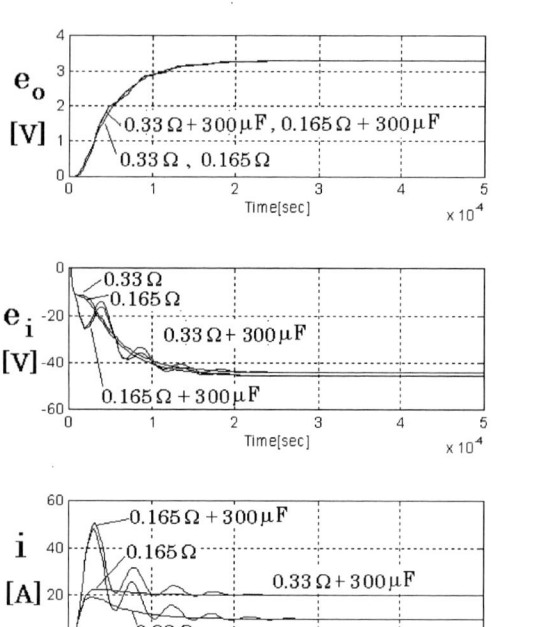

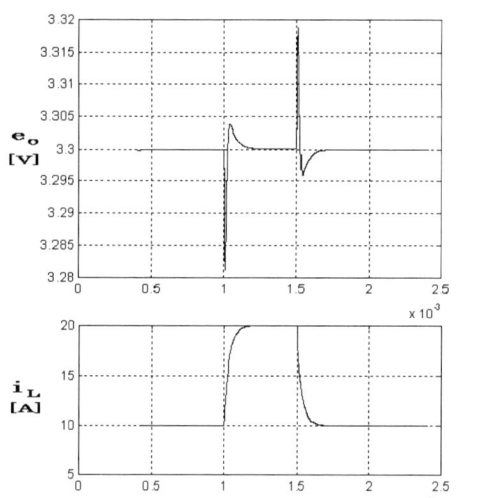

Fig. 7. Simulation results of startup responses at various loads

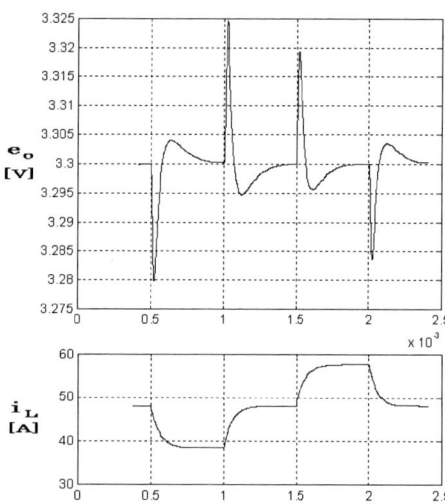

Fig. 9. Simulation result of dynamic load response at resistive load

Fig. 10. The manufactured new quarter brick DC-DC converter

ponse, H_1 and H_4 are specified as

$$
\begin{aligned}
H_1 &= -0.94 \quad H_2 = -0.18 + 0.2i \quad H_3 = -0.18 - 0.2i \\
H_4 &= -0.2 + 0.3i \quad H_5 = -0.2 - 0.3i \\
k_z &= 0.17 \quad k_q = 548.31
\end{aligned}
\tag{30}
$$

Then the parameters of controller become as follows:

$$
\begin{aligned}
k_1 &= 861.85 \quad k_2 = 584.67 \quad k_3 = -1061.2 \\
k_4 &= 0.72688 \quad k_5 = 0.45193 \quad k_6 = -0.178811 \\
k_{i1} &= -1.4955 \quad k_{i2} = 1.2629 \quad k_{i3} = -2.1818
\end{aligned}
\tag{31}
$$

It must be better that k_{r1} and k_{r2} are set to 0, since the characteristics of the control system hardly changes in this case.

The simulation results of the startup responses are shown in **Fig.7**. From the output voltage $y = e_o$ in this figure, it turns out that the specifications are satisfied. It is checked that almost the same simulation results as Fig.7 are obtained when the input voltage V_i is changed by ±20%. The simulation result of the dynamic load responses is shown in **Fig.8**. Fig.8 is the result at resistive load and the value is changed as $R_L = 0.33 \leftrightarrow 0.165[\Omega]$. It is checked that almost the same simulation result as Fig.8 is obtained at parallel load of resistance ($R_L = 0.33 \leftrightarrow 0.165[\Omega]$) and capacity ($C_L = 200[\mu F]$). **Fig.9** shows the output response at resistive load $R_L = 0.33[\Omega]$ when input voltage changed

Fig. 8. Simulation result of dynamic load response at resistive load

$$
\begin{aligned}
k_i 3 &= GHk_z \\
k_1 r &= GH \\
k_2 r &= (g1 + f_5)GH \\
k_3 r &= (g0 + (g1 + f_5)f_5 + f_4 - k_{b11})GH
\end{aligned}
\tag{29}
$$

IV. EXPERIMENTAL STUDIES

The sampling period T are set at $2.5[\mu s]$ and the input dead time L_d is about $0.999T[\mu s]$. The nominal value of R_L is $0.33[\Omega]$. We design a control system so that all the specifications are satisfied. First of all, in order to satisfy the specification on the rising time of startup transient re-

284

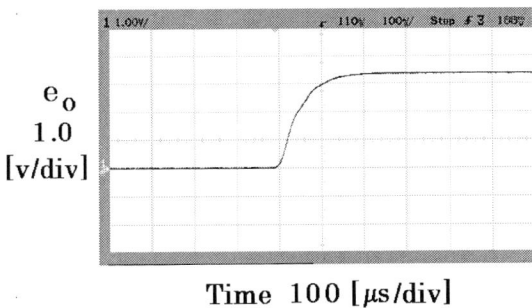

Time 100 [μs/div]

Fig. 11.   Experimental result of startup response at resisitive load
($R_L = 0.33[\Omega]$)

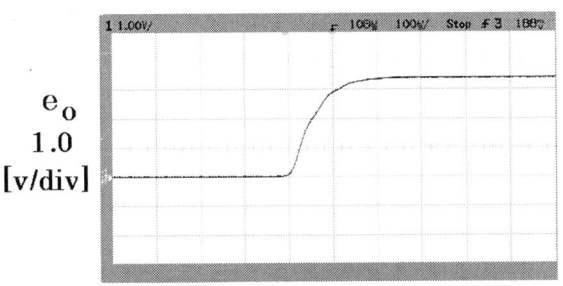

Time 100 [μs/div]

Fig. 12.   Experimental result of startup response at resistive load
($R_L = 0.165[\Omega]$)

suddenly $48 \to 38 \to 48 \to 58 \to 48[V]$. It turns out that
all the specifications are satisfied.

The manufactured new DC-DC converter built-in DSP
is shown in **Fig.10**. This fits in a quarter brick size (37mm
x 58mm x 8mm) one. Experimental results when the digital
controller with the parameters of eq.(31) is equipping
in the DSP shown in **Figs.11-14**. Fig.11 shows a startup
response at the resisitive load $R_L = 0.33[\Omega]$. Fig.12 shows
a startup response atresistive load $R_L = 0.33[\Omega]$. From
$y = e_o$ in these figure, it turns out that almost the same
exprimental results as the simulation ones in Fig.9 are obtained
and the specifications are satisfied. It is checked that
the specifications are satisfied when the input voltage V_i is
changed by ±20%. Fig.13 shows the dynamic load response
at the parallel load of resistance ($R_L = 0.33 \leftrightarrow 0.165[\Omega]$
)and capacity ($C_L = 300[\mu F]$). It turns out that almost
the same exprimental results as the simulation results in
Fig.10 are obtained. It is checked that almost the same

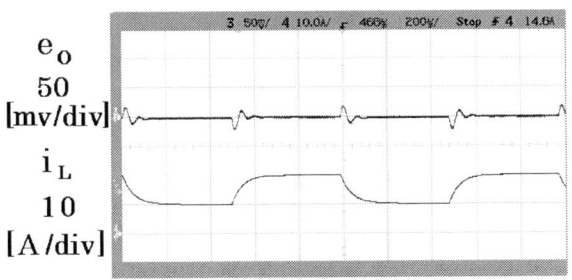

Time 200 [μs/div]

Fig. 13.   Experimental result of dynamic load response at resistive
load ($R_L = 0.33 \leftrightarrow 0.165[\Omega]$, $C_L = 300[\mu F]$)

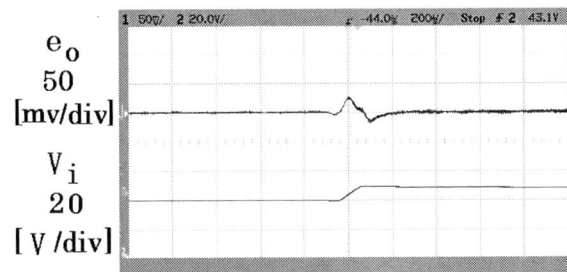

Time 200 [μs/div]

Fig. 14.   Experimental result of output reponse at parallel load of
resistance ($R_L = 0.33[\Omega]$) and capacity ($C_L = 300[\mu F]$)
when input voltage changing suddenly from 38[V] to 48[V]

exprimental results as Fig.13 are obtained at the resistive
load ($R_L = 0.33 \leftrightarrow 0.165[\Omega]$). Although the load current
changed suddenly from 20 [A] to 10 [A] or reverse,
the output voltage change is very small and is suppressed
within about 25[mV]. Fig.14 shows the output response
at parallel load of resistance ($R_L = 0.33[\Omega]$) and capacity
($C_L = 300[\mu F]$) when input voltage changed suddenly from
38[V] to 48[V]. It turns out that almost the same exprimental
results as the simulation results in Fig.11 are obtained.
It is checked that almost the same exprimental results as
Fig.9 are obtained when input voltage changed suddenly
from 48[V] to 58[V]. It turns out that the specifications are
satisfied. We checked by expriments that all other specifications
are satisfied.

V. CONCLUSION

In this paper, the concept of controller of a DC-DC converter
to attain good robustness against extensive load
changes and input voltage change was given. The proposed
digital controller was implemented on the DSP(TI
TMS320LF2801). The new DC-DC converter built-in
this DSP was manufactured. It was shown from experiments
that a sufficiently robust digital controller is realizable.
The characteristics of the dynamic load response and
the output response against sudden input voltage change
were improved by using the proposed method for approximate
2DOF digital controller with additional zeros. This
fact demonstrates the usefulness and practicality of our
method. The future work is to design a digital controller
robust enough, when (LC+LC) circuits etc. are used as
filters for removal of switching and carrier noises.

REFERENCES

[1]   L. Guo, J. Y. Hung, and R. M. Nelms, "Digital controller Design
for Buck and Boost Converters Using Root Locus", *IEEE
IECON'2003*, 1864/1869 (2003).

[2]   H. Guo, Y. Shiroishi and O. Ichinokura, "Digital PI Controller
for High Frequency Switching DC/DC Converters Based on
FPGA", *IEEE INTELEC'03*, 536/541 (2003).

[3]   K. Higuchi, K. Nakano, T.Kajikawa, E.Takegami, S.Tomioka,
K.Watanabe, "Robust Control of DC-DC Converter by High-Order
Approximate 2-Degree-of-Freedom Digital Controller",
IEEE IECON'2004, (CD-ROM), 2004.

[4]   H. Fukuda and M. Nakaoka, "State-Vector Feedback Controlled-based
100kHz Carrier PWM Power Conditioning Amplifier and
Its High-Precision Current-Tracking Scheme", *IEEE IECON'93*,
pp. 1105/1110(1993).

High Accuracy CMOS Current Sensing Circuit for Current Mode Control Buck Converter

Yuang-Shung Lee* and Chih-Jen Hsu**

* Graduate Institute of Applied Science and Engineering, Fu Jen Catholic University, Taiwan, R.O.C.
** Department of Electronic Engineering, Fu Jen Catholic University, Taiwan, R.O.C.

Abstract– **A current mode control （CMC） integrated circuit （IC） with accuracy current sensing circuit （CSC） for buck converter is presented in this paper. With the proposed accurate integrated current sensor, the sensed inductor current, combined with the internal ramp signal, can be used for current mode DC-DC converter feedback control. The proposed CSC does not need an operational amplifier to implement, and has been fabricated with a standard 0.35μ m CMOS process. Simulation results show that the switching converter can be operated up to 1MHz. Which is suitable for signal cell lithium-ion battery supply applications, the power efficiency is over 85% for supply voltage from 2.5V to 5V and the output current is 200 mA. The performance of the proposed circuit is the good compared to the other circuits.**

Index Terms—**Current mode control, Current sensing circuit, DC/DC Converter**

I. INTRODUCTION

In today's consumer electronics market, battery powered portable electronic devices are in great demand. Portable high frequency and low voltage DC/DC converters to efficiently generate low-voltage supplies from a single cell battery source to maximize the system run time. To decrease the size and weight of these devices, minimization of the power module is essential. As a result, the trend is to focus on the CMOS converter implementations with low power consumption.

It is well-known that the current mode control （CMC） DC-DC buck converter has the advantages of automatic over-current protection, better stability, better line regulation and faster dynamic responses compared with the voltage-mode control [1]. It is used for overcurrent protection and current-mode feedback control. Many different current sensing schemes have been developed and implemented to sense the inductor current [2-5]. Other current sensing approaches include using a series resistor, power MOSFET on-resistance and even an integrator [4]. However, these schemes have their limitations such as high power dissipation, process dependence, control difficulty and high implementation complexity. Therefore, an accurate current sensing circuit （CSC） is necessary for all CMC switching mode power supplies (SMPS).

Fig. 1 illustrates a simplified buck converter structure with CMC integrated circuit (IC). This converter is composed of a power stage and a feedback network. The current sensing circuit function block provides an accurate integrated current sensor. This function block can be used for current mode DC-DC converter feedback control. The power stage is the synchronous rectifier buck converter contains a switching element, consisting of a power PMOSFET, NMOSFET transistors and an output filter, that is constructed using an inductor L1 and a filtering capacitor C with equivalent series resistor (ESR). The output voltage is scaled down to βV_o to compare with the reference voltage V_{ref} before feeding into the compensator. The compensator, compensation ramp and sensed inductor current signal output will pass through the modulator and digital control block to define the duty ratio d(t). The duty ratio d(t) controls the on-time and off-time duration of the power MOSFET switches. The negative feedback is achieved to control the duty ratio perturbation to regulate the buck converter output voltage V_o. A soft start circuit is designed to control the transient inrush current of the inductor in CMC buck converter.

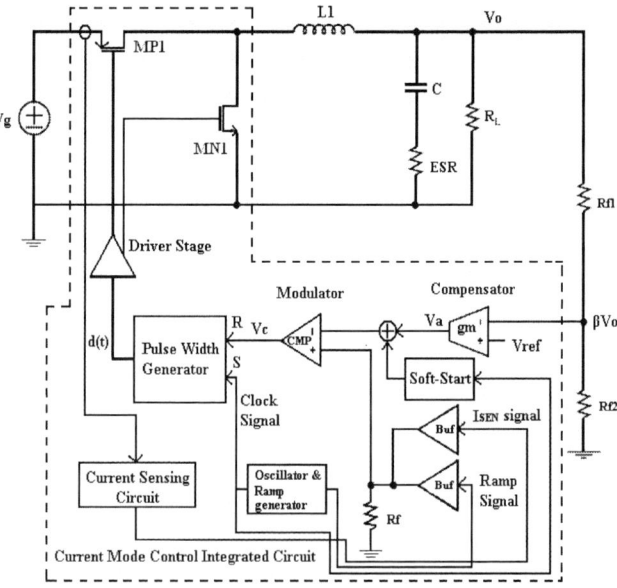

Fig. 1 Structure of buck converter with CMC IC

Therefore, a high accuracy CSC for CMC DC-DC buck converter with a Taiwan Semiconductor

978-1-4244-0644-9/07/$25.00 ©2007 IEEE   286

Manufacturing Company（TSMC）0.35 μ m polycide 2P4M CMOS process is proposed. The proposed circuit presents the low average power consumption, the high current-sensing accuracy and the low number of MOSFETs, since it does not need an operational amplifier or BICMOS process treated as a voltage mirror for the current-sensing scheme. The existing CSC is addressed in section II. The operating principle of the proposed on-chip CSC is described in section III. The simulation result is included in section IV. Section V states our conclusions.

II. EXISTING CSC SCHEMES

A high side current sensing amplifier is referenced in the U. S. as Patent 5,627,494 [2]. The high side current sensing amplifier is comprised of a feedback resistor, a current sensing resistor, an amplifier and a Darlington transistor pair. The feedback resistor has first and second inputs coupled to a non-inverting amplifier input. The Darlington transistor pair has a collector coupled to the non-inverting amplifier input. The transistor base is coupled to the amplifier output. The current sensing resistor is coupled between the emitter of the Darlington transistor pair and ground. A differential voltage is applied across the first feedback resistor input and the amplifier inverting input. The Darlington transistor pair converts the amplifier output voltage into a feedback current for generating a voltage across the feedback resistor. Under stable conditions the voltage across the feedback resistor is equal to the differential voltage. However, the amplifier must used diodes to provide a bias voltage and Darlington pair to generate the current sensing signal. They cannot be used in the low voltage IC design condition.

There are other simple sensing circuits implemented using the standard CMOS process [3]. A circuit and a related method to sense the DC/DC converter inductor current have been achieved. The inductor current is sensed by generating a voltage drop across a fully integrated sensing resistor. The voltage drop is proportional to the inductor current in the pass device by supplying a fraction of the inductor current out of a source-follower, which matches the power MOSFET stage. The source-follower source is connected to a sense-resistor that also connected to the same supply as the power MOSFET stage. The passed current of the power MOSFET generates the voltage drop, and feedback into the inductor current for minimizing the power loss. Mirroring amplification and voltage drop offset correction across the sense-resistor are performed using a single matching pair of source-followers. However, the CMOS current mirror cannot provide a good current mirror and thus the accuracy of the current sensing circuit is seriously degraded.

There is another simple sensing circuit implemented using the standard CMOS process. The CMOS high

accuracy current sensing circuit depends mainly on the voltage at VA and VB, as shown in Fig. 2 [4,5]. To achieve high accuracy, an operational amplifier is used to force the voltage at nodes A and B into the same voltage level. The two current sources I1 and I2 have small and equal magnitude pulling the current from nodes VA and VB. An output current, I_o, that flows through the power transistor MP1 is mirrored to the sensing transistor MP2. Any change in VA will force a similar change in VB due to the virtual short-circuit provided by the operational amplifier. Thus, the drain-to-source voltage V_{DS} of transistors MP1 and MP2 are nearly the same, as well as their drain current density. Nevertheless, the transistor MP1 and MP2 are scaled so that power transistor MP1 on the output side of the circuit has an aspect ratio that is much greater than that of transistor MP2 on the sensing side. As a result, the current I_{sense} on the sensing side is much smaller than, and proportional to, the current I_o on the output side. The output sensing current I_{sense}, which passes through the internal sensing resistor R, is the difference between the sensing current I_{sense} flowing through the small biasing current source. Consequently, the current I_{sense} flowing through the internal sensing resistor R is proportional to and substantially much smaller than current I_o flowing through the load. However, this scheme needs an operating amplifier to implement the voltage mirror, which will lead to higher power consumption and a number of MOSFET devices.

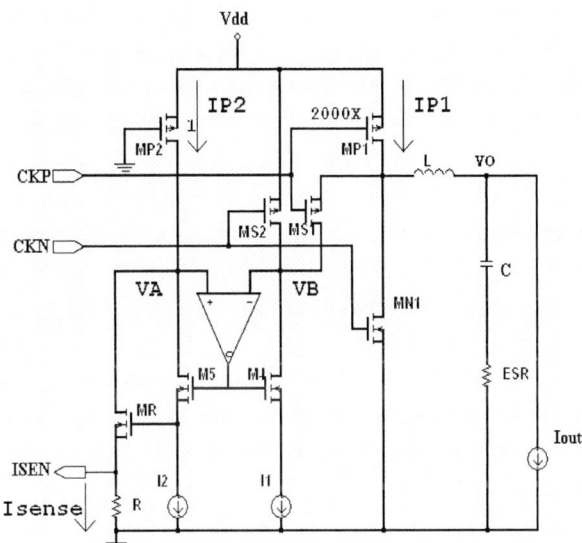

Fig. 2 The CMOS CSC for buck converter

III. PROPOSED CMC INTEGRATED CIRCUIT

Figure 3a shows the proposed current-sensing circuit. This scheme is modified from the high side current sense amplifier in [2] and the current sensing circuit for buck converter in [3], which can control the output voltage and manage the load current of the buck converter. The CSC of the buck converter is performed by adding a PMOSFET source follower M3 to the PMOSFET power device MP1. A sense-

287

resistor R_{SEN} is connected to the source follower M3 and to the supply voltage VDD. The source of the power device is also connected to VDD. The drain of the power device MP1 and the drain of the source follower M3 are connected to the drain of the synchronous device MN1, which is connected to the buck converter external inductor L1. In a preferred embodiment, $1MH_Z$ switching frequency is used.

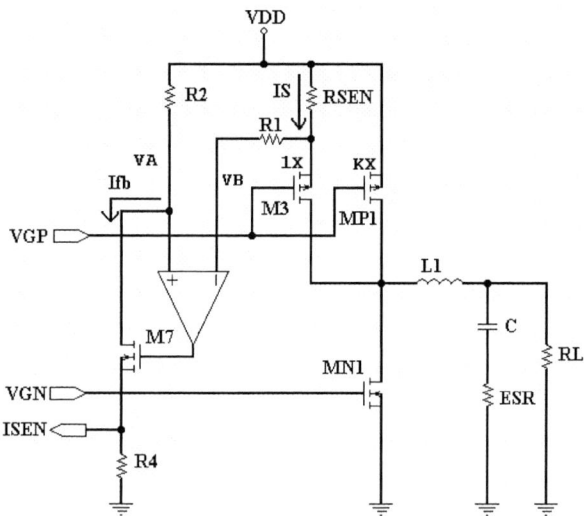

Fig. 3a The proposed integrated current sensing circuit

The complete schematic is shows in Fig. 3b, the voltage drop across the sense-resistor R_{SEN} is mirrored with a second pair of PMOSFET source followers M4 and M5, from one side of M4 and M5 pair to the other side. The common gate stage is comprised of PMOS transistors M4 and M5 and resistors R1 and R2. Transistors M4 and M5 have equal conductive width/length so that they conduct equal currents with equal voltage bias. The current mirror stage is comprised of NMOS transistors M8, M9, M10 and M11. The current mirror stage mirrors the current of NMOS transistors M11 to NMOS M10. The NMOS transistors M8 and M9 are cascade devices that provide high output impedance to the drain of PMOS transistor M8. However, cascade device will limit the V_{out} maximum value and waste one PMOS threshold voltage in the swing, so that the M8 and M9 are biased at the edge of the triode region [6].

In the current mode control buck converter application, the current sensing voltage ISEN is useful in the control feedback loop only during the on-state. Thus, only the signal from the power transistor MP1 and mirror transistor M3 during turn-on is useful, as a result, there will be a small current flowing through the sensing resistor R_{SEN}. The voltage to current output stage provides current feedback to maintain a condition where nodes VA and VB are at an equal voltage. When a current I_S flows in R_{SEN}, in the direction of the arrow, it unbalances the current mirror produced by the two MOSFETs M4 and M5, this unbalance causes the third

MOSFET M7 to be forward biased that a current I_{fb} flows in this transistor. The current I_{fb} increases the voltage drop across the resistor R4 until the sources of the two transistors M4 and M5 are returned to the same voltage. As VA equals to VB, it follows:

$$I_{fb} \times R_4 = I_S \times R_{SEN} = V_{SEN} \qquad (1)$$

This corresponds to the voltage across resistor R1 being equal to the combined voltage of the differential voltage V_{SEN} and the voltage across resistor R2. The current of PMOS transistors M4 and M5 is approximately I_1 in this stable condition. A feedback current I_{fb} and the I_1 current of the PMOS transistor M5 combine to generate a voltage drop across resistor R1 to bring the current sensing circuit into balance. Hence, the current IM7 flowing through the internal resistor R4 is proportional to and substantially smaller than the current Io flowing through the load.

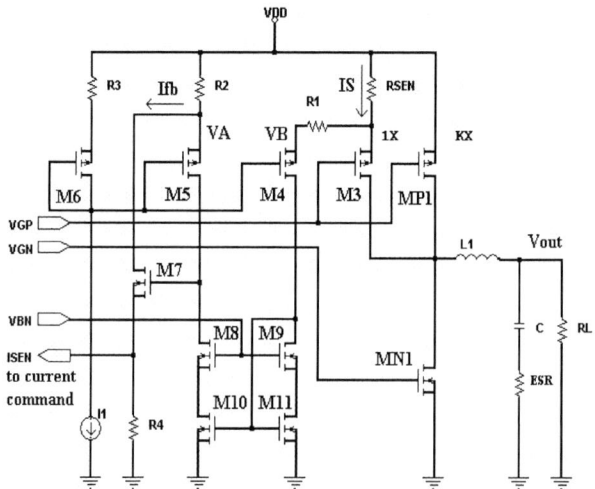

Fig. 3b The complete schematic integrated current sensing circuit

To provide a low-cost, low-power and fully integrated CMOS power module for portable applications, a current-mode DC/DC buck converter with a novel current sensor for feedback control was fabricated using the Taiwan Semiconductor Manufacturing Company （TSMC） 0.35 μ m polycide CMOS process. This converter was implemented using a synchronous rectifier that the forward-bias diode voltage drop is eliminated and the overall efficiency is enhanced. All power MOSFETs, feedback control circuits, and current-sensing circuits are fabricated on-chip. Only one off-chip inductor and one off-chip capacitor are needed at the power stage, and no off-chip inductor current-sensing is needed. This reduces the number of I/O pins and the number of reactive components needed for the converter. Moreover, it can be designed for an input voltage from 2.5V to 5V with the load current below 200 mA, which is suitable for portable electronic devices powered using a single-cell lithium-ion battery [7, 8].

IV. SIMULATION RESULTS

The current sensing circuit simulations are based on the circuits shown in Fig. 3b. Ideal gate signals are used to drive the power transistors and the two switches, MP1 and MN1 for the DC/DC converter PWM control. The simulation results for the sensing signal and the scale inductor current are shown in Fig 4, Fig 5, Fig 6 and Fig 7. The value of inductor L1 is $6.8\,\mu$ H. The capacitance of the output capacitor C is $47\,\mu$ F with ESR=5mΩ and R_L =5 Ω, respectively. The ideal gate driver's switching frequency is 1MH$_Z$ and the duty cycle is 50%. When the supply voltage V_{DD} is 3.6V, Fig. 4 shows that the inductor peak current is 436 mA at 247.5 μ sec. Fig. 5 shows that the current sense circuit signal is 440 mV at 247.5 μ sec. The accuracy of the current-sensing is 99.0%. When the supply voltage V_{DD} is 2.5V, the inductor peak current is 316mA at 246.7 μ sec, as shown in Fig. 6. In Fig. 7, the current sense circuit signal is 320mV at 246.7 μ sec. The current-sensing accuracy is up to 98%.

The system design simulations are based on the circuits shown in Fig. 1. The converter is supplied with the input voltage of 3.6V and the switching frequency of 1MH$_Z$. The simulation results for the current-sensing and error-amplifier output signal are shown in Fig. 8.

The efficiency with respect to the input voltage is shown in Fig. 9 under the output voltage of 1.8V and the output current of 200mA. The maximum efficiency is 93.4% at 4.4V input voltage. There are two major power dissipations, conduction loss and switching loss, in the switching mode power supplies and they depend on the size of the power transistor. For PWM and synchronous rectification control, switching loss is dominant at light load and conduction loss is dominant at heavy load conditions. Figure 9 shows that conduction loss is dominant when the input voltage is larger than 4.4V and the efficiency is decreased.

The performance simulation results of the current sensing circuits are summarized in Table 1. The proposed circuit has the lowest average power consumption, the highest current-sensing accuracy and the lowest number of MOSFETs since it does not need an operational amplifier as a current mirror. The proposed circuit performance is better than the other CMOS current sensing circuits.

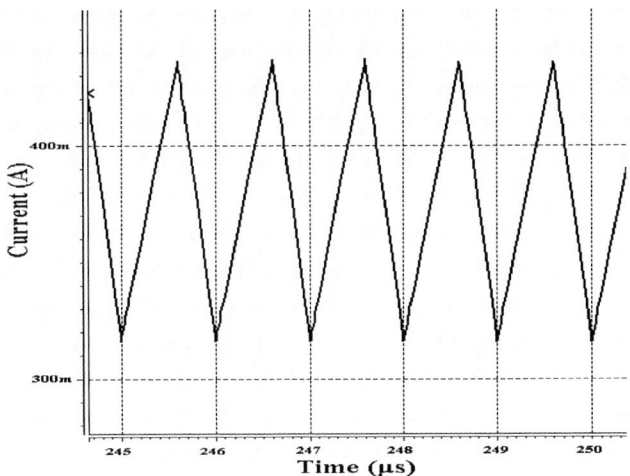

Fig. 4 The waveform of the inductor current on supply voltage is 3.6V

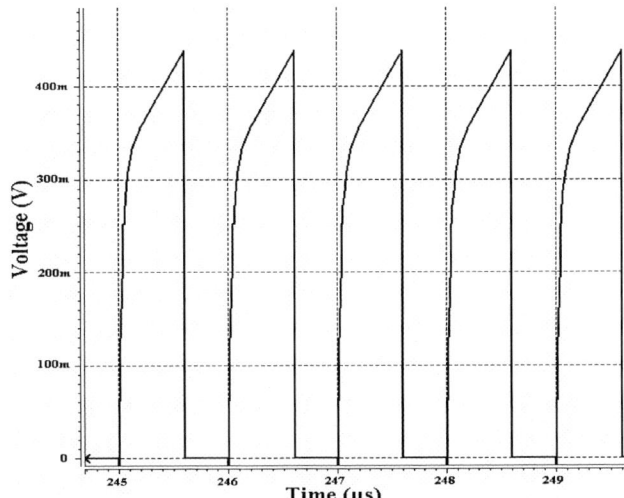

Fig. 5 The waveform of the current sensing signal on supply voltage is 3.6V

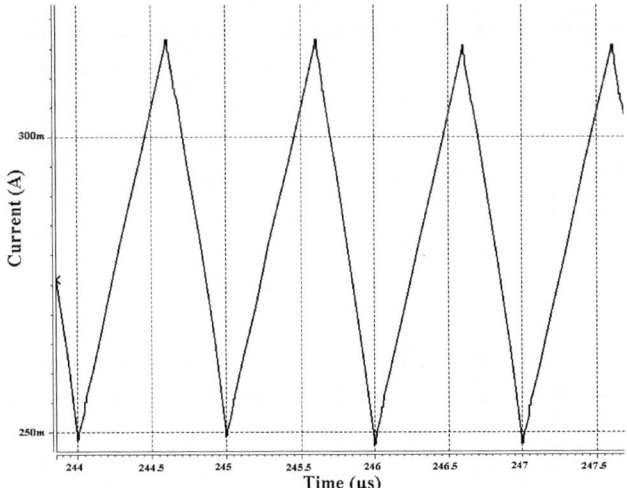

Fig. 6 The waveform of the inductor current on supply voltage is 2.5V

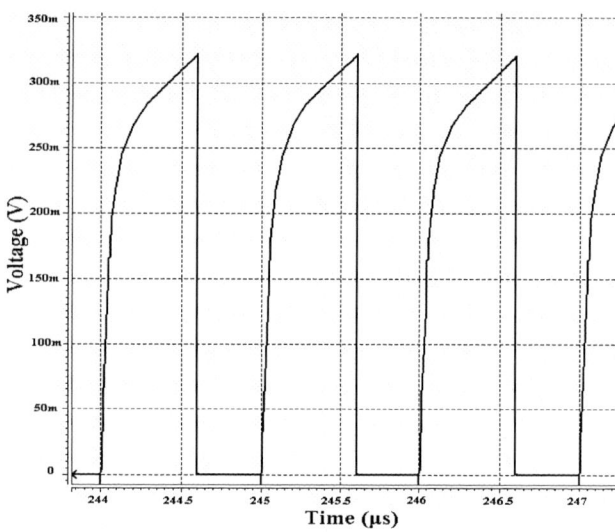

Fig. 7 The waveform of the current sensing signal on supply voltage is 2.5V

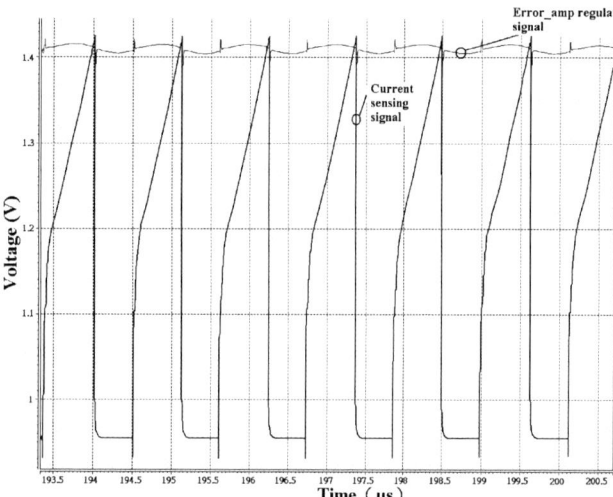

Fig. 8 The waveform of the regulated signal on output voltage is 1.8V

V. CONCLUSIONS

The proposed current mode control IC with the current sensing scheme was addressed including the operation, design issues, circuit implementation and simulation results. The results show that the performance of the proposed current sensing technique is more accurate and simpler than that of conversional methods. In addition, this technique is not strongly dependent on other parameter variations, such as frequency, temperature and processes. This CMC IC with the CSC can be simply fabricated using any standard CMOS process. From the proposed CMC integrated circuit simulation results, the proposed current sensing scheme can successfully operate up to 1MHz. The accuracy of proposed current-sensing circuit is 98%. This work will be supported by certification conducted at the National Chip Implementation Center, National Science Council of Taiwan.

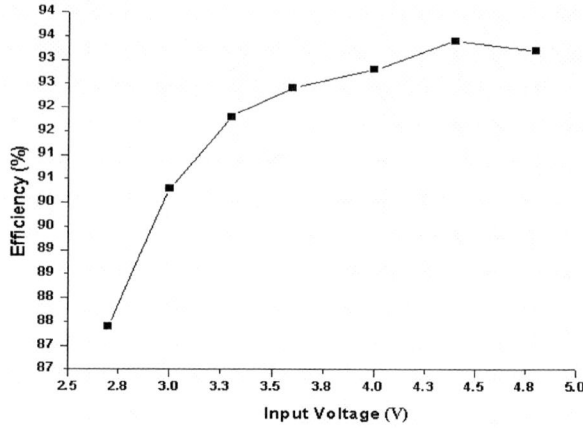

Fig. 9 Efficiency with respect to input voltage at V_o=1.8v and I_o=200mA

TABLE 1
FEATURE OF THE CSC FOR CMC DC-DC BUCK CONVERTER

	Proposed Circuit
Technology	Taiwan Semiconductor Manufacturing Company 0.35 Polycide 2P4M CMOS Process
Numbers of Core Circuit's MOSFET	9
Operating Voltage Range	2.5V-5V
Accuracy of Current - Sensing	98%
Need the OP Amp to as a Voltage Mirror	No
Maximum Operating Frequency	1MHz
Maximum Efficiency	93.4%

ACKNOWLEDGMENTS

The design tools in this work were supported by the National Chip Implementation Center, National Science Council of Taiwan.

REFERENCES

[1] R. W. Erickson and D. Maksimovic, *Fundamentals of Power Electronics*.

[2] T. A. Somerville, "High Side Current Sense Amplifier," US Patent 5,627,494, May 6, 1997.

[3] E. Marschalkowski and J. Malcolm, "Current Sensing Circuit for DC/DC Buck Converters," US Patent 6,992,473, Jan. 31, 2006.

[4] C. F. Lee and P. K. T. Mok, "On-Chip Current Sensing Technique for CMOS Monolithic Switch-Mode Power Converter," *IEEE Int. Symp. Circuits and Systems*, vol. 5, Scottsdale, AZ, May 2002, pp. 265-268.

[5] C. F. Lee and P. K. T. Mok, "A Monolithic Current-Mode CMOS DC-DC Converter with On-Chip Current-Sensing Technique," *IEEE J. Solid-State Circuits*, vol. 39, Jan. 2004, pp. 3-14.

[6] B. Razavi, Design of Analog CMOS Integrated Circuits, Boston, MA: McGraw-Hill, 2001.

[7] Yuang-Shung Lee and Guo-Tian Cheng, "Quasi-Resonant Zero Current Switching Bi-directional Converter for Battery Equalization Applications," *IEEE Transaction on Power Electronics*, vol. 21, no. 5, Sept. 2006, pp. 1213-1224.

[8] Yuang-Shung Lee and Ming-Wang Cheng, "Intelligent Control Battery Equalization for Series Connected Lithium-Ion Battery Strings," *IEEE Transaction on Industrial Electronics*, vol.52, no. 5, Oct. 2005, pp. 1297-1307.

Small Signal Analysis of a dual-switch forward Converter with non-ideal transformer in Current-Programmed Control

Weiping Zhang , Yuzhou Lei, Xiaoqiang Zhang, and Yuanchao Liu

College of Information Engineering, North China Univ. of Tech., Beijing, P.R. China

Tel (Fax): 86-10-88802880 E-mail:zwp@ncut.edu.cn

Abstract—A small signal analysis approach of a dual forward converter with non-ideal transformer in current-programmed control (CPC) has been investigated in this paper. A small-signal model has been derived to determine the small-signal characteristics of the converter. By employing this model, the following significant features have been revealed or explained: (1). If the artificial ramp slop is properly selected, the ideal line-to-output transfer function of the converter becomes zero. That is to say, the output voltage fluctuation of causing from a perturbation of the DC bus voltage can be substantially restrained; (2). The CPC can greatly reduce the Q-factor and make the converter to become a low Q-factor system; (3). The current-programmed control can extend the bandwidth; (4).The magnetizing inductor plays an important role in the converter, so we must consider it when we design the control circuit.

Index Terms--Current-Programmed Control , Dual-switch Forward Converter，Small-Signal Model

I. INTRODUCTION

Forward converter is one of the most widely used topology in high power converters. Dual-switch forward converters overcome the following disadvantages that a Forward converter suffers: (1) reduce switching high voltage stress, and each switch only withstands DC input voltage; (2) magnetic reset circuit is simple and can ensure a reliable magnetic transformer reset; (3) each bridge consists of one switch with a diode in series, and there is no danger that both bridges get through at same time [2]. Therefore, the dual-switch forward converter has advantages that other converters can not reach.

Under normal circumstances, the transformer in this kind of circuit is regarded as an ideal component, but the reality is not the case. Fig.1 gives out a dual-switch forward converter topology that considers magnetizing inductance. The other circuit elements are ideal. The circuit works at CCM mode and the duty cycle D<0.5.

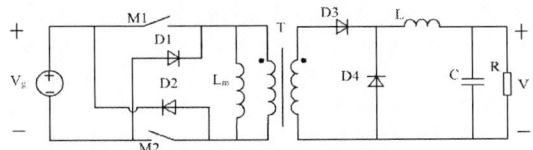

Fig. 1. Dual-switch forward converter

Project supported by Natural Science Foundation of China (No 50477054);Project supported by Beijing Natural Science Foundation of China (No. 4052011)

II. MODELING OF THE CPC DUAL-SWITCH FORWARD CONVERTER

Fig.2 illustrates the relationship between input current equivalent value and the control current [1][2]. m_1, m_{21}, m_{22}, m_a respectively are the slope of input current equivalent value when the switches are on, the slope of input current equivalent value, the slope of input current equivalent value while the magnetizing inductance current is not zero and the switches are off, and the slope of the artificial slope compensation. <i(t)>dTs is the average value of the input current equivalent value while the switches are on; <i'(t)>dTs is the average value of the input current equivalent value while the magnetizing inductance current is not zero and the switches are off; <i''(t)>dTs is input current equivalent value in the rest time of a cycle.

The average value of the input current equivalent value can be expressed as,

$$\left\langle i(t) \right\rangle_{T_s} = \left\langle i_c(t) \right\rangle_{T_s} - m_a dT_s - 0.5 dm_1 dT_s \tag{1}$$
$$-0.5 dm_{21} dT_s - 0.5(1-2d)m_{22}(1-2d)T_s$$

Where,
$$m_1 = v_g / L_m + (v_g / n - v)/(nL),$$
$$m_{21} = v_g / L_m + v / nL, m_{22} = v /(nL)$$

To find out a small-signal current programmed controller, there are the followings assumptions:

(1) The small-signal perturbations are introduced:

$$\left\langle i(t) \right\rangle_{T_s} = I + \hat{i}(t), \left\langle i_c(t) \right\rangle_{T_s} = I_C + \hat{i}_c(t), d = D + \hat{d}(t),$$

$$m_1 = M_1 + \hat{m}_1(t), m_{21} = M_{21} + \hat{m}_{21}(t), m_{22} = M_{22} + \hat{m}_{22}(t)$$

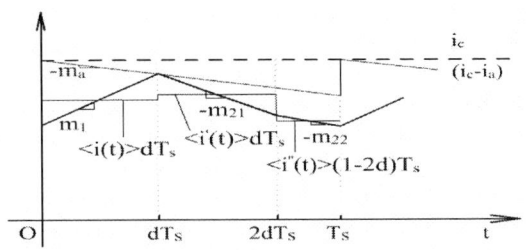

Fig.2.The relationship between input current equivalent value and control current

$$\hat{i}\left(t\right)=\hat{i}_m\left(t\right)+\hat{i}_L\left(t\right)/n,\hat{m}_1$$

where, $=\hat{v}_g/L_m+(\hat{v}_g/n-\hat{v})/(nL)$,

$$\hat{m}_{21}=\hat{v}_g/L_m+\hat{v}/(nL),\hat{m}_{22}=\hat{v}/(nL),$$

(2)The m_a does not have any perturbation, then $\hat{m}_a=M_a$.

(3) Use of the equilibrium relationship

$$M_1D=DM_{21}+2M_{22}(1-2D)$$

and neglect of the higher-order ac terms.

Based on the above assumptions, the small-signal model for CPC can be found as

$$\hat{d}(t)=F_m\begin{bmatrix}\hat{i}_C(t)-\hat{i}_m(t)-\hat{i}_L(t)/n\\-F_g\hat{v}_{g2}-F_v\hat{v}(t)\end{bmatrix}\qquad(2)$$

where, $F_m=\dfrac{1}{M_aT_s}$, $F_g=\dfrac{D^2T_s}{2}(\dfrac{2}{L_m}+\dfrac{1}{n^2L})$,

$$F_v=(1-2D)^2T_s/(2nL)$$

$\hat{i}(t)$ is a perturbation of a summation of the average inductor current and the magnetizing current. We can apply the average small-signal ac model of forward converter to study the dynamic properties. The small-signal model for CPC- dual-forward converter has been derived in this paper, shown in Fig.3.

III. SMALL-SIGNAL CHARACTERISTICS

By using the small-signal model derived in previous section, the transformer functions can be solved to express the characteristics of CPC dual-switch forward converter. The output voltage $\hat{v}(s)$, inductor current $\hat{i}_L(t)$ and the magnetizing current $\hat{i}_m(t)$ can be expressed as a function of the duty-cycle $\hat{d}$ and input voltage $\hat{v}_g$, using the transformer function G_{vd},G_{vg}, G_{id},G_{ig} and G_{md},G_{mg},

$$\hat{v}(s)=G_{vd}\hat{d}(s)+G_{vg}\hat{v}_g(s)\qquad(3)$$

$$\hat{i}_L(s)=G_{id}\hat{d}(s)+G_{ig}\hat{v}_g(s)\qquad(4)$$

$$\hat{i}_m(s)=G_{md}\hat{d}(s)+G_{mg}\hat{v}_g(s)\qquad(5)$$

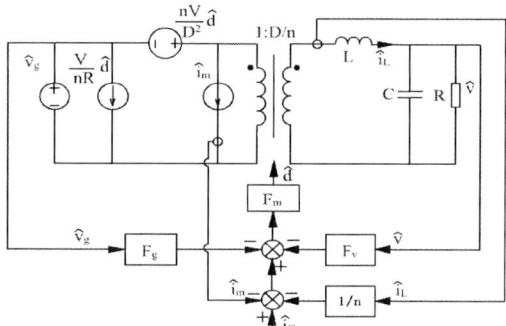

Fig. 3. Small-signal model for CPC- dual-forward converter

$$G_{vd}(s)=\dfrac{\hat{v}(s)}{\hat{d}(s)}\Big|_{\hat{v}_g(s)=0}=\dfrac{V}{D}H(s),$$

$$G_{vg}(s)=\dfrac{\hat{v}(s)}{\hat{v}_g(s)}\Big|_{\hat{d}(s)=0}=\dfrac{D}{n}H(s),$$

Where, $G_{id}(s)=\dfrac{\hat{i}_L(s)}{\hat{d}(s)}\Big|_{\hat{v}_g(s)=0}=\dfrac{V}{DR}(1+sRC)H(s),$

$$G_{ig}(s)=\dfrac{\hat{i}_L(s)}{\hat{v}_g(s)}\Big|_{\hat{d}(s)=0}=\dfrac{D}{R}(1+sRC)H(s),$$

$$G_{md}(s)=\dfrac{\hat{i}_m(s)}{\hat{d}(s)}\Big|_{\hat{v}_g(s)=0}=\dfrac{V_gT_s}{L_m},$$

$$G_{mg}(s)=\dfrac{\hat{i}_m(s)}{\hat{v}_g(s)}\Big|_{\hat{d}(s)=0}=\dfrac{DT_s}{L_m},$$

$$H(s)=\dfrac{1}{1+sL/R+s^2LC}.$$

The solution of (2) ~ (5) leads to the following desired result:

The CPC-dual-forward converter control to output transfer function is as the following

$$A_P(s)=\dfrac{\hat{v}(s)}{\hat{i}_c(s)}\Big|_{\hat{v}_g(s)=0}$$

$$=\dfrac{F_mG_{vd}}{1+F_m(G_{md}+G_{id}/n+F_vG_{vd})}\qquad(6)$$

$$=\dfrac{G_{co}}{1+\dfrac{s}{\omega_cQ_c}+\dfrac{s^2}{\omega_c^2}}$$

The line-to-output transfer function is

$$A(s)=\dfrac{\hat{v}_2(s)}{\hat{v}_g(s)}\Big|_{\hat{i}_c(s)=0}$$

$$=\dfrac{G_{vg}-F_mF_gG_{vd}}{1+F_m(G_{md}+G_{id}/n+F_vG_{vd})}\qquad(7)$$

$$=\dfrac{G_{go}}{1+\dfrac{s}{\omega_cQ_c}+\dfrac{s^2}{\omega_c^2}}$$

$$G_{go}=\dfrac{D}{n}\dfrac{1-\dfrac{M_2'}{2M_a}}{A},\omega_c=\omega_0\sqrt{\dfrac{A}{(1+\dfrac{nF_mT_sV}{DL_m})}},$$

Where, $Q_c=Q_0\dfrac{\sqrt{A}\sqrt{1+\dfrac{nF_mT_sV}{DL_m}}}{1+\dfrac{F_mVRC}{nDL}+\dfrac{nF_mVLT_s}{DL_m}},Q_0=R\sqrt{C/L},$

$$G_{co}=\dfrac{VF_m}{DA},A=1+\dfrac{F_mV}{nDR}+\dfrac{F_mF_vV}{D}+\dfrac{nF_mVLT_s}{DRL_m},$$

$$\omega_o=\dfrac{1}{\sqrt{LC}},M_2'=\dfrac{2nV}{L_m}+\dfrac{V}{nL}.$$

We make the following discussion to describe the important behaviors of CPC dual-switch forward converter:

(1). If selection of artificial ramp slop is that $M_a = 0.5M_2'$, the ideal line-to-output transformer function $A(s)$ becomes zero. Therefore, the CPC has ability to eliminate low frequency output voltage ripple causing from the input voltage fluctuation. In the engineering design, the reasonable selection is that $M_a = 0.75m_2'$. In this case, the current-programmed control has a good ability to restrain low frequency output ripple; (2). It can be seen from the expression of Q_c that $Q_c < Q_o$. The current-programmed control can greatly reduce the Q-factor and make it to become a low Q-factor system; (3).Because $\omega_c > \omega_o$, the current-programmed control can extend the bandwidth[1][2].

Because CPC can greatly reduce the Q-factor in generally and make it to become a low Q-factor system, consequently, for larger F_m and smaller F_v, the poles become real and well separated in magnitude, ω_{p1} and ω_{p2}, $\omega_{p1} < \omega_{p2}$

$$\omega_{p1} = \frac{\omega_c}{Q_c}\frac{1-\sqrt{1-4Q_c^2}}{2}$$

$$\omega_{p2} = \frac{\omega_c}{Q_c}\frac{1+\sqrt{1-4Q_c^2}}{2}$$

Based on above discussion, the approximating formula of control-to-output transformer function is

$$A_P(s) \approx \frac{G_{co}}{(1+j\dfrac{\omega}{\omega_{p1}})(1+j\dfrac{\omega}{\omega_{p2}})} \qquad (8)$$

The line-to-output transfer function is

$$A(s) \approx \frac{G_{go}}{(1+j\dfrac{\omega}{\omega_{p1}})(1+j\dfrac{\omega}{\omega_{p2}})} \qquad (9)$$

IV. THE SIMULATION RESULTS

The frequency responses of control-to-output transformer functions, operating with CPC and duty-cycle control are shown in Fig.4. It can be seen that, for duty cycle control, the transformer function G_{vd} exhibits resonant dual-pole response. The substantial damping introduced by CPC leads to essentially a dual-pole response in current-programmed control-to-output transformer function $A_p(S)$ and $A_p'(S)$. $A_p(S)$ considers the role of magnetizing inductance but $A_p'(S)$ does not.

The frequency responses of line-to-output transformer function are simulated and illustrated in Fig.5. G_{vg} for duty-cycle control, $A(s)$ and $A'(s)$ with CPC. The magnitude of the line-to-output transformer function $A(s)$ and $A'(s)$ with CPC is significantly reduced and

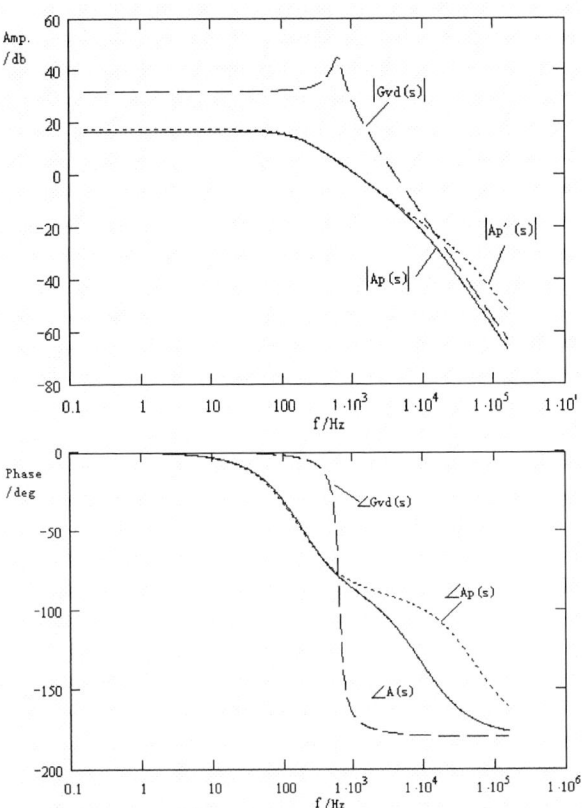

Fig. 4 Comparison of CPC with duty-cycle control for control-to-output frequency response

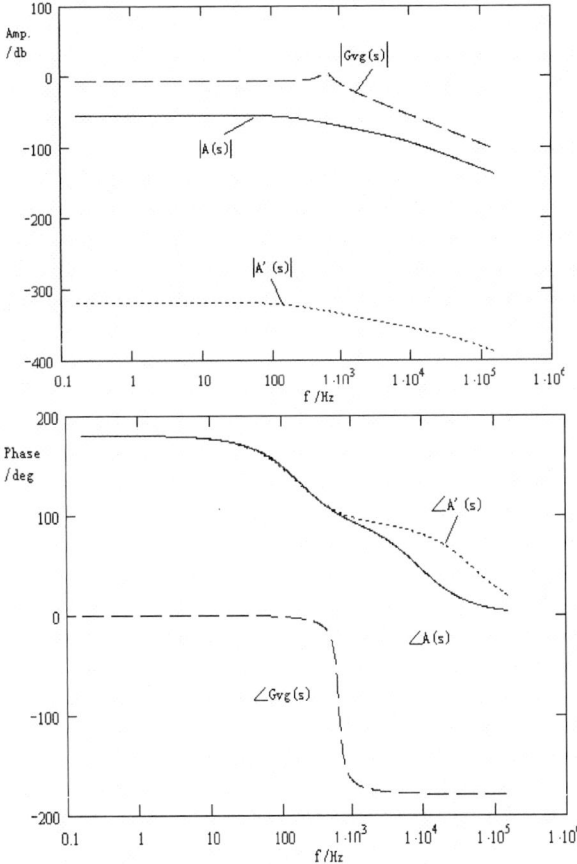

Fig.5 Comparison of CPC with duty-cycle control for line-to-output frequency response

TABLE I
COMPARISON OF PARAMETERS TRANSFER FUNCTIONS WITH AND WITHOUT L_M

L_m	G_{co}	G_{go}	Q_c	ω_c
5.6mH	6.875	$-1.796*10^{-3}$	0.132	$8.002*10^3$
$+\infty$	7.393	$-1.081*10^{-12}$	0.054	$1.811*10^4$

simulation result exhibits that more than 30dB additional attenuation has been obtained in low frequency area. That is to say that the current-programmed control has a great ability to reduce the low frequency ripple at the output voltage.

But there are obvious differences between the $A_p(S)$ - $A_p'(S)$ and $A(s) - A'(s)$, which we can see from Table I, Fig.4 and Fig.5. From the above analysis, we can get that this is caused by magnetizing inductance. So, we can not get rid of the role of magnetizing inductance when we design the control loop of the converter.

V. CONCLUSIONS

By employing small-signal analysis of CPC dual-switch forward converter, the following important conclusions have made out in this paper: (1). If the artificial ramp slop is properly selected, the ideal line-to-output transfer function of the converter becomes zero. That is to say, the output voltage fluctuation of causing from a perturbation of the DC bus voltage can be substantially restrained; (2). The CPC can greatly reduce the Q-factor and make the converter to become a low Q-factor system; (3). The current-programmed control can extend the bandwidth; (4).The magnetizing inductor plays an important role in the converter, so we must consider it when we design the control circuit.

REFERENCES

[1] Weiping Zhang, *Modeling and Design of the Control Loops of Switching Regulators*, 2006.1, CEPP.

[2] Zhang Weiping, Cheng Yaai, Guan Xiaohan, Zhang Dongyan, *Small-Signal Analysis of CPC-Single Stage AC/DC Converter with PFC*, TENCON 2006. 2006 IEEE Region 10 Conference, 14-17 Nov. 2006, pp1-4.

[3] Weiping Zhang, *A single stage AC/DC converter with resonant model PFC*, IEEE PESC, 2004, pp.1799-1802.

[4] A. Fontan, S. Ollero, E. de la Cruz, J. Sebastian, *Peak Current Mode Control Applied to the Forward Converter with Active clamp*, Volume 1, May. 1998, pp.45 -51.

[5] Byung-Il Kwon, *Forward Converter Analysis by the Method of Coupling Electromagnetic Field with Hysteresis and Circuit Equations*, IEEE TRANSACTIONS ON MAGNETICS, VOL. 36, NO. 4, JULY 2000, pp1426-1430.

[6] Hua Yang, *Modeling of Peak Current Control Mode for Forward Converter*, Research and Design of Power Supply Technique,2005.8.

High Frequency Transformer Designs for Improving Cross Regulation in Multiple-Output Flyback Converters

Kusumal Chalermyanont, Pairote Sangampai, Anuwat Prasertsit, Surapon Theinmontri
Faculty of Engineering, Prince of Songkla University, Hatyai, Songkla, Thailand
Email: kusumal.c@psu.ac.th, Eng_road@hotmail.com, anuwat.p@psu.ac.th, surapon.t@psu.ac.th

Abstract - **The cross regulation in multiple-output flyback converters is affected by leakage inductances of the transformers. The suitable winding arrangements of the transformers will lead to the improvement of cross regulation. This paper presents the comparative study of various winding arrangements in the three winding flyback transformers. The cross regulation model is considered in form of the resistance matrix related to the leakage inductance parameters. The design guidelines for the transformer for improving cross regulation are given based on the experimental results.**

Keywords: **Cross regulation, multiple-output flyback converters, leakage inductance**

I. Introduction

The multiple-output switching converter is generally used in low power applications for electronic equipments that need different levels of output voltages such as computers. One of the most favorite used configurations is a multiple-output flyback converter due to its low component counted and cost-effective structure.

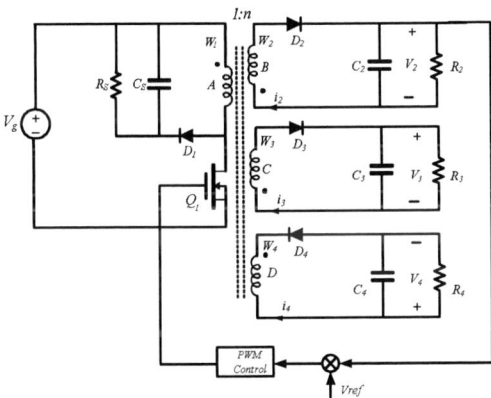

Fig. 1 The three-output flyback converter where W_1 is a primary winding, W_2 is a main secondary winding and W_3 and W_4 are auxiliary windings.

Typically, as shown in Fig. 1, the main output voltage of the converter is directly sensed and regulated to the desired value by the feedback network. The auxiliary outputs are not directly regulated, but they are fixed with respect to the principle output voltage by coupling between the transformer windings which is called cross regulation. In the past, the cross regulation analytical models and the methods for improving the cross regulation in multiple-output converters are proposed in many literatures [2, 3, 4 and 5]. However, it still lacks general sense because they are only considered for

one simple magnetic structure or simplifying assumption [3, 4 and 5]. The effects of various magnetic structures of high frequency transformers on cross regulation and the comparison of transformer winding arrangements that effect the cross regulation in a multiple-output flyback converter are presented in [6] and [7]. However, they are shown only one method of winding arrangement.

In this paper, three methods of transformer winding arrangements are presented in comparison. Each winding arrangement is reasonably performed in such a way to control leakage inductance in order to improve cross regulation in a multiple output flyback converter. The experimental results are illustrated and the transformer design guidelines are concluded as simple rules to improve cross regulation in multiple-output flyback converter

II. Analytical model of cross regulation in a multiple flyback converter

a. The Extended Cantilever Model and the N-port Model

In this paper, the extended cantilever magnetic model as shown in Fig. 2 is used to analyze the behaviors and cross regulation mechanisms in the multiple-output flyback converter. The extended cantilever model has $n(n+1)/2$ independent parameters including leakage inductances. L_{11} is a self-inductance and l_{jk} is a leakage inductance between winding j^{th} and k^{th}. All parameters of the model can be directly measured by using a two port technique. Additionally, it is also closely related to the N-port description of transformer which is useful in computer simulation and simple analysis of the multiple-output flyback converter.

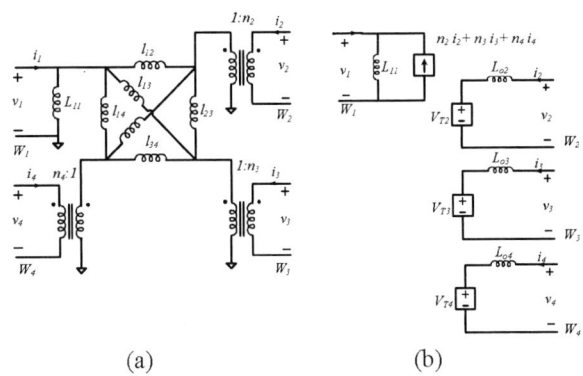

(a)   (b)

Fig. 2 (a) The extended cantilever model and (b) the N-port model.

According to the analysis of the multiple-output flyback converter using the N-port model, the change rate of the

978-1-4244-0644-9/07/$25.00 ©2007 IEEE   295

secondary output current i_k of the multiple-output flyback converter is given by

$$\frac{di_k}{dt} = \frac{1}{n_k l_{1k}} v_1 + \frac{1}{n_k n_{k+1} l_{k(k+1)}} v_{k+1} + \ldots + \frac{1}{n_k n_N l_{kN}} v_N - \frac{1}{L_{ok}} v_k \quad (5)$$

and the rate of change of the output current i_k can also be determined in the term of the average DC output current I_k [5] as

$$\frac{di_k}{dt} = \frac{2f}{(1-D)^2}(I_k - (1-D)i_k(t_c)) \quad (6)$$

By combining (5) and (6), the DC output current I_k is written as a linear combination of the input and output voltages as

$$\frac{2f}{(1-D)^2}(I_k - (1-D)i_k(t_c)) = \frac{1}{n_k l_{1k}} v_1 + \frac{1}{n_k n_{k+1} l_{k(k+1)}} v_{k+1} + \ldots + \frac{1}{n_k n_N l_{kN}} v_N - \frac{1}{L_{ok}} v_k \quad (7)$$

It can be noticed from (7) that the coefficients of any output voltage V_N beside the considered output voltage V_k are functions of the leakage inductances between the secindary winding N and the secondary winding k (l_{kN}). Therefore, if those leakage inductances are relatively large in values, the change of the output current I_k does not significantly affect the change in the output voltages. It is in turn to improve cross regulation in the multiple-output flyback converter.

III. The winding arrangements of high frequency transformers

a. Winding arrangement designs to improve cross regulation in a multiple-output converter

Based on the results of [7] and the conclusion in a previous section, cross regulation in a multiple-output converter can be improved by setting the values of specific leakage inductance parameters. Leakage inductance values between a primary and secondary windings (l_{1k}) must be minimized to magnetic flux and leakage inductance values among secondaries (l_{jk}) must be maximized to reduce the effect of magnetic flux between windings when the load change.

As we know, the value of leakage inductance depends on the winding geometry in a transformer core. Changing position of the winding will change the value of the effective leakage inductance [1]. For example, the smallest value of leakage inductance is given when two windings are tightly coupled while the largest value of the leakage inductance is happened when two windings are put far way to each other. Therefore, three different winding methods (Stack, Sandwich, and Interleave) are chosen to construct with the ETD49/25/16 ferrite core and foil copper winding with 1.5-mm thickness. The winding arrangement in each method is assemble based on the required values of the leakage inductance parameters to improve cross regulation.

The prototype three-output flyback converter is constructed with the specification of input voltage of 40V main output voltage of 5V, and auxiliary output voltages of 12V and -12V. The converter operates at switching frequency of 100 kHz and duty ratio of 0.5. Fig. 3 shows experimental winding geometries of three different winding arrangement where A presents a primary winding of 40V (W_1), B presents a main secondary of 5V (W_2), C and D present auxiliary secondaries of 12V (W_3) and -12V (W_4).

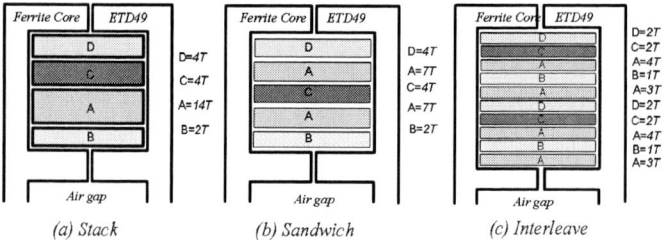

(a) Stack   (b) Sandwich   (c) Interleave

Fig. 3 Three different winding geometries in a flyback transformers.

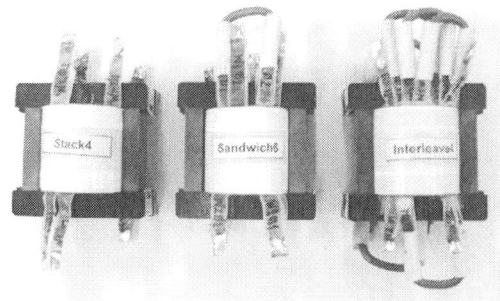

Fig.4 Three different experimental high frequency transformers.

b. Measurement of leakage inductance values

Measured values of leakage inductances of experimental Stack, Sandwich and Interleave winding arrangements are shown in Table 1 using the method mentioned previously.

Table1. Summary of flyback transformer parameter measurements

Inductance parameter	Measure values		
	Stack	Sandwich	Interleave
n_2	0.15	0.14	0.14
n_3	0.3	0.29	0.29
n_4	0.3	0.29	0.29
L_{11}	294 μH	304μH	328 μH
l_{12}	6 μH	5.3 μH	11.2 μH
l_{13}	3 μH	3.1 μH	2.9 μH
l_{14}	9.6 μH	3.2 μH	3.3 μH
l_{23}	59 μH	187 μH	129 μH
l_{24}	82 μH	196 μH	136 μH
l_{34}	3 μH	38 μH	16 μH

By considering the measured values of leakage inductances in each winding arrangement, Sandwich method gives the largest values of secondary leakage inductances (l_{23}, l_{24}, and l_{34}) while the Stack method gives the smallest values of the secondary leakage inductances. According to the required values to improve cross regulation in the converter, one can predict that the converter with the Sandwich arrangement transformer will give the better cross regulation than that of the Stack transformer. Secondary output current waveforms of the experimental transformers with the main output current I_2 of 3.5A and the auxiliary output currents I_3 and I_4 of 1A are shown in Fig.5-7 Slopes of the main output current of the experimental transformers are all positive while Slopes of the auxiliary output currents are all negative.

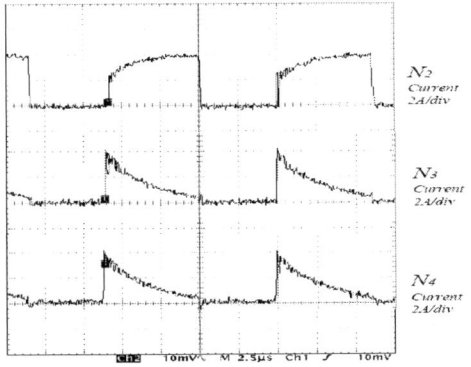

Fig.5 Experimental waveforms in the Stack arrangement transformer

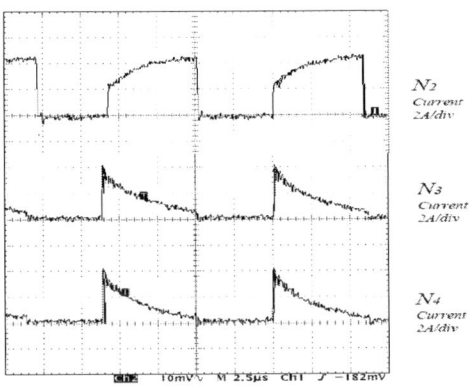

Fig.6 Experimental waveforms in the Sandwich arrangement transformer

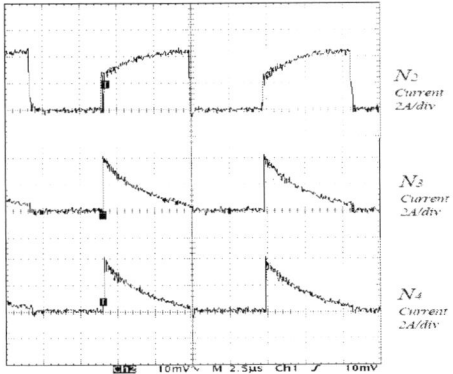

Fig.7 Experimental waveforms in the Interleave arrangement transformer

The N-port model which is related to measured values of leakage inductance parameters in the extended cantilever model and simulation waveforms of the experimental transformers (Stack model) are illustrated in Fig.8-9 as an example.

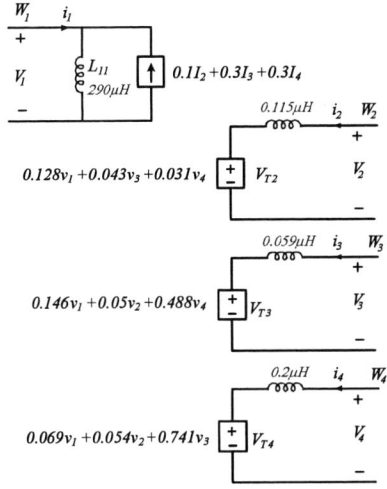

Fig.8 The N-port model of the Stack arrangement transformer

The simulated output current waveforms perfectly agree with the experimental results.

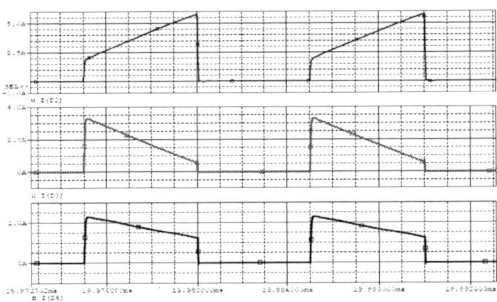

Fig. 9 Simulated waveforms of the N-port model for flyback converter.

IV. The cross-regulation results of the experimental transformers in a multiple output flyback converter

Each experimental transformer shown in Fig. 4 is placed to the prototype three-output flyback converter. Cross regulation of the converter is determined by the change of the output voltages while varying the main output current form 0.5A to 3.5A and maintaining other output currents constant at 1A. The change of the output voltage (ΔV) can be expressed as a function of the change of output current (ΔI) in a matrix form as $\Delta \mathbf{V} = -\mathbf{R}\Delta\mathbf{I}$ where $\mathbf{R}$ is the effective resistance matrix. For the converter operates in continuous conduction mode (CCM) at all operating points, the changes of output voltages when varying the main output current of each designed transformer are comparatively illustrated in Fig. 10.

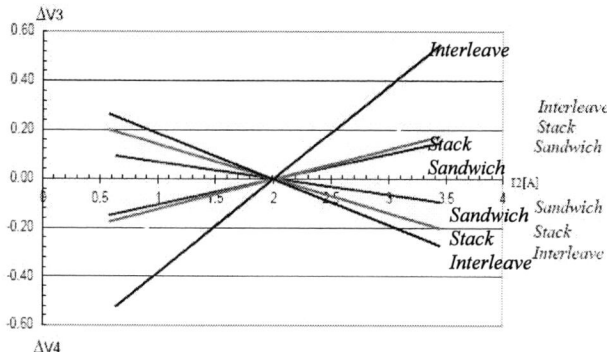

Fig. 10 The comparison of cross regulation in multiple output flyback converter using different winding arrangements

Elements in the effective resistance matrix **R** can be obtained as slopes of the changes of the output voltages over the output currents. The effective resistance matrix of each designed transformers can be determined as in (8), (9) and (10). The cross regulation in the auxiliary output voltages is considered from non-diagonal elements of the matrix. The small values are required for the improved cross regulation.

$$Stack = \mathbf{R}_{expriment} = \begin{bmatrix} 0.35 & 0.15 & 0.14 \\ 0.12 & 0.66 & 0.41 \\ -0.13 & -0.21 & -0.79 \end{bmatrix} \Omega \quad (8)$$

$$Sandwich = \mathbf{R}_{expriment} = \begin{bmatrix} 0.36 & 0.13 & 0.12 \\ 0.11 & 1.20 & 0.66 \\ -0.09 & -0.21 & -0.59 \end{bmatrix} \Omega \quad (9)$$

$$Interleave = \mathbf{R}_{expriment} = \begin{bmatrix} 0.47 & 0.21 & 0.25 \\ 0.36 & 1.2 & 0.81 \\ -0.17 & -0.32 & -0.89 \end{bmatrix} \Omega \quad (10)$$

From experimental results, the transformer with Sandwich winding arrangement is the best among other designs. Since the cross regulation at auxiliary output voltages V_3 and V_4 has been affected by the leakage inductance parameters between windings W_3 and W_4 and the main output winding W_2 (l_{23} and l_{24}). Therefore, controlling the leakage inductance values among those windings will be able to control the cross regulation. In the Sandwich winding arrangement, the auxiliary windings are placed away from the main winding. The corresponding leakage inductance values of the Sandwich winding arrangement in Table 1 are large that results in the small values in non-diagonal elements in the matrix. Hence, based on the experimental results, the transformer design guidelines for improving cross regulation in a multiple-output flyback converter can be given as follows:

- The leakage inductances between primary and secondary windings (l_{12}, l_{13} and l_{14}) should be small to minimize magnetic flux.

- The leakage inductances among secondary windings (l_{23}, l_{24}, and l_{34}) should be maximized to reduce the effect of magnetic flux between windings when the load changes.
- The main output winding should be positioned near transformers core to reduce the leakage inductance.
- The primary and secondary windings should be in proximity while each secondary windings should be placed away as much as possible.

V.Conclusion

The cross regulation in multiple-output flyback converters is affected by leakage inductances of the transformer. The suitable winding arrangements of the transformer will lead to the improvement in cross regulation. This paper presents the comparative study of three different winding arrangements in three winding flyback transformer. The cross regulation model is considered in the form of resistance matrix that is related to leakage inductance parameters. The design guidelines of transformers for improving cross regulation are given based on the experimental results.

References

[1] Dauhajre, "Modelling and Estimation of Leakage Phenomena in Magnetic Circuits," Ph.D.Thesis, California Institute of Technology, Pasadena, California, April 1986
[2] Erickson and D. Maksimovic, "A Multiple-Winding Magnetics Model Having Directly Measurable Parameters," *IEIE Power Electronics Specialists Conference, May 199*8
[3] Chuanwen Ji,K.Mark Smith,Jr " Cross Regulation in Flyback Converters: Analytic Model ," *IEIE Transactions Power Electronics Conference, 1999*
[4] D. Maksimovic and R. Erickson, "Modeling of Cross Regulation in Multiple-Output Flyback Converters," *IEIE Applied Power Electronics Conference, 1999*
[5] Kusumal Changtong " Magnetic Modeling for improving cross-regulation in multiple output flyback converters" Ph.D. Thesis, University of Colorado at Boulder,1999.
[6] Pairote Sangampai,Kusumal chalermyanont, Anuwat prasertsit and Surapon theinmontri " The Study the Effects of High Frequency Transformer Core Types for Improve Cross-Regulation in Multiple-Output Flyback Converters " PSU-Engineering Conference ,Thailand (*PEC 5*), May2007
[7] Pairote Sangampai,Kusumal chalermyanont, Anuwat prasertsit and Surapon theinmontri "The Comparative Study and Design Guidelines of a High Frequency Transformer for Improving Cross Regulation in Multiple-Output Flyback Converters"ECTI-CON 2007 Mea Fah Luang University Chaiang Rai ,Thailand

Operation of a wye Connected Three-Level Active Power Filter under Non-ideal Conditions

H.B. Zhang   A.M.Massoud   S.J.Finney   B.W.Williams   T.C. Lim   H. Hotait

Abstract___ **A three-level wye shunt active power filter based on a predictive current controller and line voltage space vector modulation is presented and analyzed. This control scheme treats the non-ideal load as a pure resistive load and compensates the reactive power and harmonics under both ideal and non-ideal conditions, thus the power quality at point of common couplings is improved. Experimental results are presented to demonstrate the effectiveness of this approach, and also reveal the limitation of wye structure APF.**

Keywords___ **Active Power Filters, predictive current controller, line voltage space vector.**

I. INTRODUCTION

For many years, shunt active power filters (Shunt APF) have been researched with the aim to compensate reactive power and eliminate harmonics, thus to improve the power quality at the point of common coupling (PCC). Since the 1990s, attention has been paid to the shunt APF operating under non-ideal conditions [1]-[4].

The non-ideal power system conditions may have distorted or unbalanced mains voltages, or both. Fig. 1 defines the terms for a three-phase system model.

Huibin zhang is with Department of Electronic and Electrical Engineering, University of Strathclyde, UK (email: hzhang@eee.strath.ac.uk)

A.M.Massoud is with Department of Electronic and Electrical Engineering, University of Strathclyde, UK. (email: ahmed.massoud@eee.strath.ac.uk)

S.J.Finney is with Department of Electronic and Electrical Engineering, University of Strathclyde, UK (email: s.finney@eee.strath.ac.uk)

B.W.Williams is with Department of Electronic and Electrical Engineering, University of Strathclyde, UK. (email: B.W.Williams@eee.strath.ac.uk)

T.C. Lim is with Department of Electronic and Electrical Engineering, University of Strathclyde, UK. UK.(email: ceetcl@eee.strath.ac.uk)

H. Hotait is with Department of Electronic and Electrical Engineering, University of Strathclyde, UK, UK. (email: hadi.hotait@eee.strath.ac.uk)

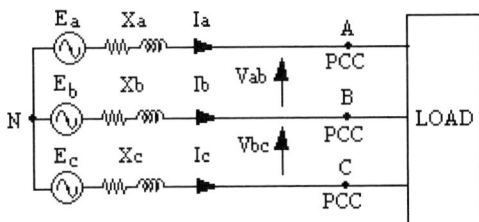

Figure 1: A three-phase three-wire system

In Fig. 1, if the load is non-linear, distorted currents $i_x (x = a, b, c)$ are drawn from the source; with the impedance $X_i (i = a, b, c)$ along the transmission lines, the distorted currents cause distorted voltages before the PCC. Furthermore, when a non-linear load is unbalanced, there will be unbalanced and distorted voltages at the PCC, which further affect other loads parallel connected at the PCC.

In this paper, a three-level shunt active power filter operating with distorted line voltage quantities is assessed. The shunt APF synthesizes the output demand voltages and through interface inductors, the output voltages force the source side current to be sinusoidal and synchronized with 1st positive component of the phase voltages at the PCC. Thus the harmonics and reactive power drawn by the load are compensated. The reference line voltage vector of the shunt APF is obtained from a predictive current controller, and SVM transformed from line voltage space vectors to obtain the timing parameters. This control scheme employs the virtues of both predictive control and SVM. Predictive control produces a constant switching frequency, lower current ripple [5]; while SVM, is readily implemented in a digital processor. Three-level APF reduces dv/dt and THD, allows the increase of connection voltage, thus is adaptable in medium power applications. The validity of the proposed design is confirmed by experimentation.

II. APF CONTROL

Regardless of the conditions of load or source, one operational option for the shunt APF is to make the source currents sinusoidal and synchronized to the fundamental component of the phase voltages at the PCC. Thus the reactive power and harmonics of each phase can be compensated. Furthermore, the resultant sinusoidal source currents cause sinusoidal voltage drop along line impedances thus reducing the adverse effects on other loads connected to the PCC.

Fig. 2 shows the cascaded three-level APF used for compensation, while the nonlinear load consists of a three-phase diode bridge rectifier plus a single-phase bridge to unbalance the load. Without the APF, the currents drawn from source are distorted and unbalanced, thus distorting and unbalancing voltage drops across the line impedances.

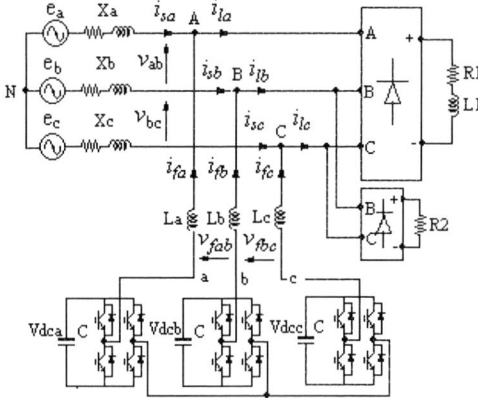

Figure 2: A wye three-level APF with nonideal three-phase system

A. Capacitor Balance Control

The APF compensates non linear currents by forcing the input currents i_s (a, b, c) to track a set of sinusoidal reference currents, in phase with fundamental component of the supply phase voltage.

The magnitude of the reference current is controlled such that the real power from the supply matches that of the load and the APF losses. This is achieved by controlling the magnitude of the input current according to the error between the APF capacitor voltage and a fixed reference, as shown in Fig. 3.

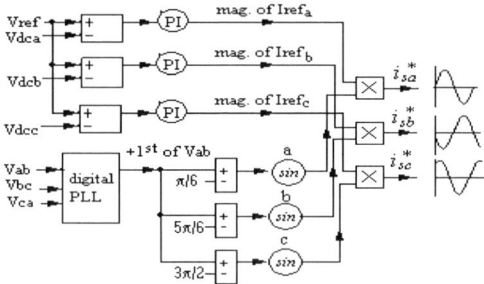

Figure 3: Control diagram for obtaining the reference currents

The first positive sequence component of the phase voltage lags the line voltage by $\pi/6$. After subtraction of $\pi/6$, the output of a digital PLL locks to the first positive component of the corresponding line voltage which is fed into a sinusoidal generator to give the phase for reference current i_{sa}^*. By $2\pi/3$ and $4\pi/3$ displacements, the phases for reference currents i_{sb}^* and i_{sc}^* are obtained.

Three PI controllers are used to obtain the corresponding current magnitudes, one for each phase, and the outputs of the PI controllers and sinusoidal generators are multiplied to yield the instantaneous source current references i_{sa}^*, i_{sb}^* and i_{sc}^*.

B. Predictive Current Controller

Using Kirchhoff's Voltage Law, and with the presence of the reference source currents, the predictive current controllers are obtained.

In loop b-a-A-B-b of Fig. 2,

$$v_{fab} = v_{ab} + L_a \times \frac{di_{fa}}{dt} - L_b \times \frac{di_{fb}}{dt} \qquad (5)$$

where v_{fab} is the output voltage between leg a and leg b of the APF, v_{ab} is the line voltage between phase A and B at the PCC, and L_a, L_b are the inductances of the corresponding interface inductors.

Writing (5) in a discrete form

$$V_{fab(n)} = V_{ab(n)} + L_a \times \left(\frac{i_{fa(n+1)} - i_{fa(n)}}{T_s}\right) - L_b \times \left(\frac{i_{fa(n+1)} - i_{fa(n)}}{T_s}\right) \qquad (6)$$

where T_s is the switching period, $i_{fx}(n+1)$ and $i_{fx}(n)$, $(x=a, b)$ are two samples from the time interval T_s.

To compensate the instantaneous source current $i_{sa}(n)$ giving $i_{sa}(n)^*$, the compensating current through interface inductor for phase A should be $i_{sa}(n)- i_{sa}(n)^*$, then the voltage drop across the inductor is $L_a \times (i_{sa(n)} - i_{sa(n)}^*)/T_s$. Similarly for phase B these quantities are $i_{sb}(n)-i_{sb}(n)^*$, and $L_b \times (i_{sb(n)} - i_{sb(n)}^*)/T_s$ respectively. Substituting the two voltage terms into (6) gives:

$$V_{fab(n)} = V_{ab(n)} + L_a \times \left(\frac{i_{a(n)} - i_{sa(n)}^*}{T_s}\right) - L_b \times \left(\frac{i_{b(n)} - i_{sb(n)}^*}{T_s}\right) \qquad (7)$$

Equation (7) shows the demand output line voltage between legs a and b of the APF, its relation to the corresponding line voltage V_{ab}, and the voltages across the applicable interface inductors.

Similarly:

$$V_{fbc(n)} = V_{bc(n)} + L_b \times \left(\frac{i_{b(n)} - i_{sb(n)}^*}{T_s}\right) - L_c \times \left(\frac{i_{c(n)} - i_{sc(n)}^*}{T_s}\right) \qquad (8)$$

$$V_{fca(n)} = V_{ca(n)} + L_c \times \left(\frac{i_{c(n)} - i_{sc(n)}^*}{T_s}\right) - L_a \times \left(\frac{i_{a(n)} - i_{sa(n)}^*}{T_s}\right) \qquad (9)$$

300

Equations (7), (8), and (9) show the demand output line voltages of the APF in order to force the source currents to be the predicted references, sinusoidal and synchronized with the first positive sequence component of the phase voltages.

III. LINE VOLTAGE SPACE VECTOR MODULATION

In this case, the reference space vectors are line voltage vectors, thus the space vector modulation corresponding to line voltages results.

For the three-level APF, in the $\alpha - \beta$ plane there are 27 phase voltage space vectors (V_a V_b V_c) as shown in Fig. 4, and 19 line voltage space vectors (V_{ab} V_{bc} V_{ca}) obtained from V_a -V_b, V_b -V_c, and V_c -V_a, as shown in Fig.5.

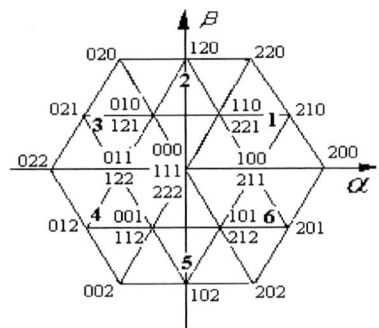

Figure 4: $\alpha - \beta$ plane of phase voltage vectors

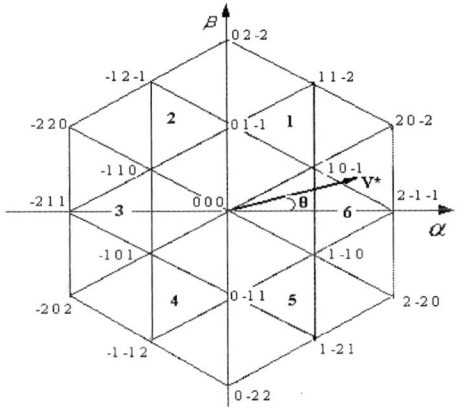

Figure 5: $\alpha - \beta$ plane of line voltage vectors

After V_{ab} V_{bc} V_{ca} are obtained from (7) (8) and (9), the V_α, V_β components of reference voltage vector are obtained:

$$\begin{bmatrix} V_\alpha \\ V_\beta \end{bmatrix} = \frac{2}{3} \begin{bmatrix} 1 & -\frac{1}{2} & -\frac{1}{2} \\ 0 & \frac{\sqrt{3}}{2} & -\frac{\sqrt{3}}{2} \end{bmatrix} \begin{bmatrix} V_{ab} \\ V_{bc} \\ V_{ca} \end{bmatrix} \quad (12)$$

The magnitude V* and phase angle θ of the reference vector are:

$$V^* = \sqrt{V_\alpha^2 + V_\beta^2} \quad (13)$$

$$\tan \theta = \frac{V_\beta}{V_\alpha} \quad (14)$$

Compared with the phase vectors as shown in figure 4, the magnitude of line voltage vectors are $\sqrt{3}$ times that of phase vectors and lead them by $\pi/6$. Due to this feature, timing parameters for line voltage vectors can be readily obtained by adopting the usual algorithm for phase vectors after the following transformations:

$$V_p^* = V^* / \sqrt{3} \quad (15)$$

$$\theta_p^* = \begin{cases} \theta + \frac{11\pi}{6} & (2k\pi \le \theta \le 2k\pi + \frac{\pi}{6}) \\ \theta - \frac{\pi}{6} & (2k+\frac{1}{6})\pi \le \theta \le 2(k+1)\pi \end{cases} \quad (16)$$

where V_p^* and θ_p^* are the magnitude and phase angle of the target vector in the hexagon transformed from phase voltage vectors.

Note that at any time the sum of line voltages is zero, so there are no zero components associated with line voltage vectors, that is, line voltage SVM is always two-dimensional.

IV. POWER FLOW IN THE SYSTEM

The proposed control scheme aims at compensating the reactive power and harmonics drawn by the non-linear load. Fig. 6 shows the power flow after the APF is activated.

The source would only supply real power, p_a, p_b and p_c, demanded by the load. When the load is balanced, then the real power supplied from the balanced source is balanced. With the wye connection of APF, to maintain a constant capacitor voltage, no real power flows between each leg of the APF. Thus when the load is unbalanced, the real power supplied by the source corresponds to the real power drawn by each phase of the load. The reactive powers, q_a, q_b and q_c, drawn by each phase of the load,

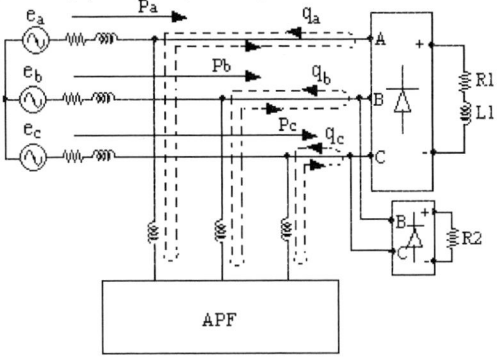

Figure 6: Power flow in the system after the activation of APF

301

circulate between the corresponding leg of the APF and the load.

V. EXPERIMENTAL RESULTS

The experimental platform is as in Fig. 2. An Infineon TriCore ™ TC1796 implements the control scheme. Three experimental scenarios are considered, first an ideal source without impedance X before the PCC and the load is balanced (without the single phase bridge load); then with balanced impedance and finally with unbalanced source impedance while the load is unbalanced in both cases.

The fixed system parameters are listed in Table I:

Table I: Parameters for Experiment

E_a, E_b, E_c	110 Vac (rms)
R_1	110 Ω
R_2	250 Ω
L_1	8 mH
L_a, L_b, L_c	20 mH
C	470 μF
V_{dc}	130 Vdc
Switching frequency	12 kHz

A. Ideal Mains and Balanced Load

In this case the load is the three-phase diode bridge. There are notches in the line voltages due to the commutation of diodes, as shown in figure 7. Source currents drawn by the balanced nonlinear load are distorted and balanced. After APF activation, the source currents become sinusoidal and balanced, as shown in Fig. 8.

Balanced source currents result because the load draws balanced real power from each phase. Outputs of PI controllers monitoring the voltage of capacitor provide equal quantities for the reference currents of each phase.

Fig. 9 compares the FFT analysis of the source current of phase A, before and after the APF operates. The characteristic harmonics of six pulse inverter, for example, the 5th, 7th, 11th, 13th, are significantly decreased.

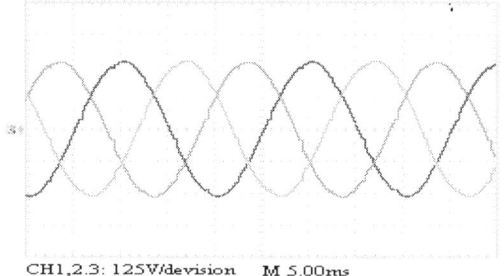

CH1,2.3: 125V/devision   M 5.00ms

Figure 7: line voltages without source impedance.

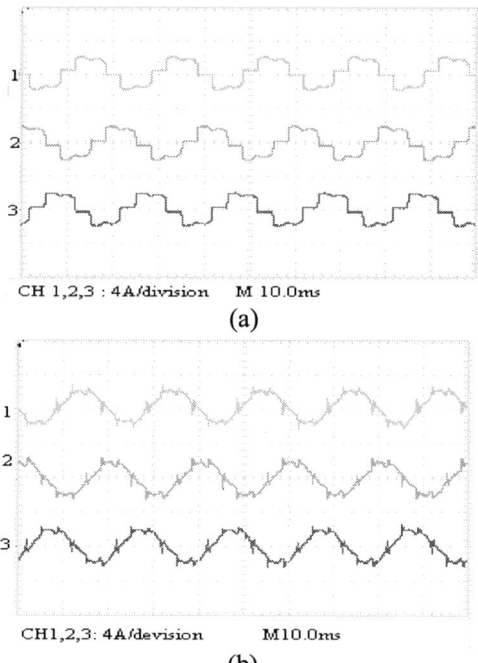

CH 1,2,3 : 4A/division   M 10.0ms

(a)

CH1,2,3: 4A/devision   M10.0ms

(b)

Figure 8: Source currents with ideal ac mains
(a) Before APF activation.
(b) After APF activation.

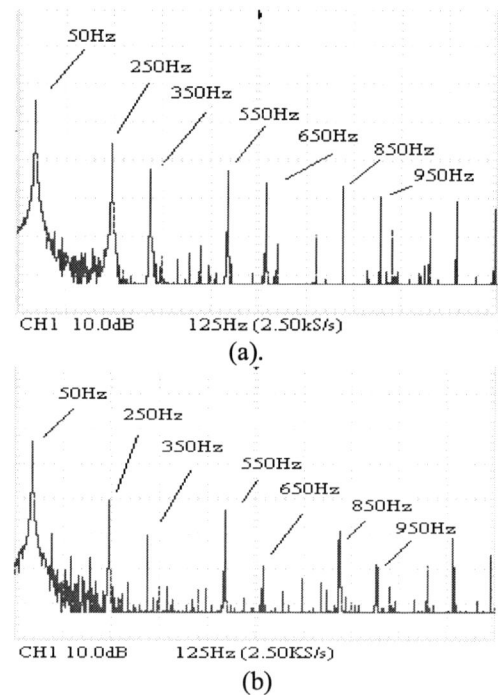

CH1 10.0dB   125Hz (2.50kS/s)

(a).

CH1 10.0dB   125Hz (2.50KS/s)

(b)

Figure 9: FFT analysis of source current
(a) Before the APF activation;
(b) After APF activation.

B. Balanced Source Impedance and unbalanced load.

Balanced impedances X ($R = 5\Omega, L = 5mH$) are added to the source side, and the loads are unbalanced due to the connection of a single phase diode bridge rectifier load between phases b and c. Before APF activation, the unbalanced and distorted load current cause distorted and unbalanced voltage drops along the line impedance before the PCC. The unbalanced and distorted voltage drops affect the load. The source currents before APF activation are shown in Fig.10(a). The currents in Fig. 10(a) also show that the connection of the single phase load between phase b and c draws distorted and unbalanced currents.

After APF activation, the source currents are forced to be sinusoidal, as shown in Fig. 10(b), but unbalanced. The phase b and c carry the same load, although the corresponding waveforms in Fig 10(a) are different, the compensated currents are the same magnitude which reveals that two phases draw the same real power.

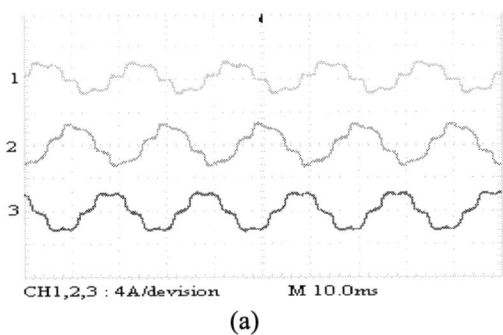

CH1,2,3 : 4A/devision   M 10.0ms

(a)

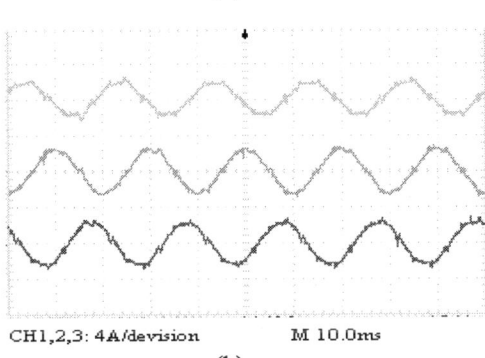

CH1,2,3: 4A/devision   M 10.0ms

(b)

Figure 10: Source currents in the case of balanced source impedances
(a) Before APF Activation.
(b) After APF activation.

C. Unbalanced Source Impedance and unbalanced load

The source impedances are unbalanced, as shown in Table II:

Table II: Unbalanced Source Impedance

R_a	$3.5\,\Omega$	L_a	5mH
R_b	$5.5\,\Omega$	L_b	5mH
R_c	$7\,\Omega$	L_c	5mH

The line voltages and source currents before APF activation are shown in Fig. 11.

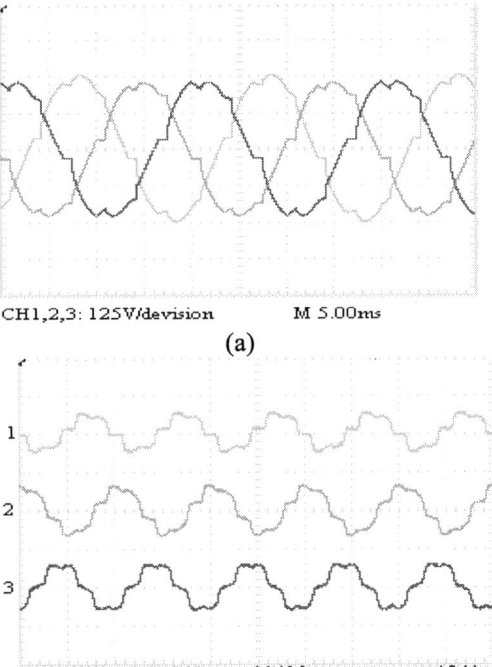

CH1,2,3: 125V/division   M 5.00ms

(a)

CH1,2,3: 4A/devision   M 10.0 ms

(b)

Figure 11: Unbalanced source impedances and loads conditions
(a) Line voltages before APF activation.
(b) Source currents before APF activation.

Under such conditions the line voltages become further distorted and unbalanced. Compared with ideal balanced conditions, the imbalanced extent has increased to 9.78%. After APF activation, source currents are driven to be sinusoidal and related to the real power of each phase. Furthermore, due to the sinusoidal voltage drop across the source impedance, although now the voltage at the PCC incorporates a high frequency voltage developed across the source capacitance, the line voltages distortion is decreased to 6.12%, as shown in Fig. 12(a). This benefits other loads connected in parallel to the nonlinear load at the PCC.

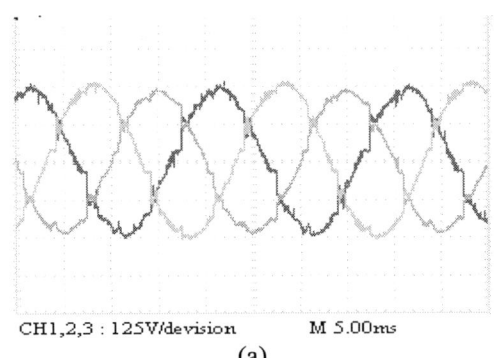

CH1,2,3 : 125V/devision   M 5.00ms

(a)

303

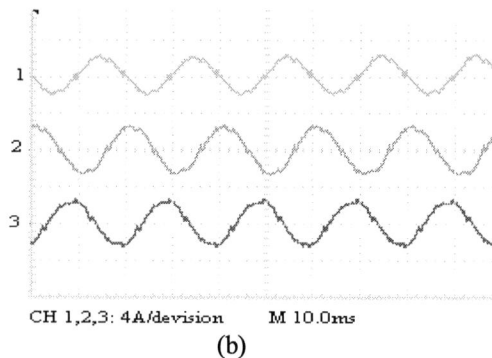

CH 1,2,3: 4A/devision   M 10.0ms

(b)

Figure 12: Unbalanced source impedances and
loads conditions
(a) Line voltages after the APF activation.
(b) Source currents after APF activation.

VI. CONCLUSION

A three-level shunt APF control scheme is presented which operates under non-ideal conditions. Based on a predictive current controller and LVSVM, the controller performs under both ideal and non-ideal conditions. Experimental results demonstrate that the control scheme works in the presence of distorted source voltages, compensates reactive power and harmonics, and thus improves the power quality at the PCC. With unbalanced loads, the wye structure APF drives the source currents to be sinusoidal but without the ability to balance the real power drawn from sources. This suggests that both the compensating algorithm and hardware structure should be considered when designing an effective APF.

REFERENCES

[1] M.Machmoum, N.Bruyant, "Control methods for three-phase active power filters under non-ideal mains voltages," Power System Technology, 2000. Proceedings , Volume 3, pp.1613 – 1618,.

[2] V Soares, P Verdelho and G Marques, "A control method for active power filters under unbalanced non-ideal conditions", 'Power Electronics and Variable Speed Drives', Conference Publication No. 429, Sep. 1996

[3] Engin Özdemir, Murat Kale, Sule Özdemir, "A novel control method for active power filter under non-ideal mains voltage," Proceedings of IEEE CCA, 2003. Vol. 2, pp.931 – 936, June 2003.

[4] G.D. Marques, "A comparison of active power filter control methods in unbalanced and non-sinusoidal conditions," IECON '98, Vol. 1, pp. 444-449.

[5] A.M.Massoud, S.J.Finney, B.W.Williams, "Predictive Current Control of a Shunt Active Power Filter," PESC'04, Vol.5, pp 3567-3572.

Application of GPRS Techniques for Wide-Area Power Quality Monitoring

Shun-Yu Chan*   Jen-Hao Teng
Member IEEE
Department of Electrical Engineering,
Cheng-Shiu University,
Kaohsiung, Taiwan

David Chang   Li-Yuan Chin*

Department of Electrical Engineering,
I-Shou University,
Kaohsiung, Taiwan

Abstract: This paper tries to use USB and Personal Digital Assistant to develop a novel Power Quality (PQ) monitoring platform and then integrates GPRS technique into the proposed PQ platform to realize wide-area PQ monitoring. The works of this paper can be divided into three parts. First, a small-scale PQ monitoring platform with appropriately designed I/O interfaces and peripherals is designed and implemented. Next, a GPRS module which can be integrated into the designed PQ monitoring platform is developed. Finally, a web server with well-designed database used to record abnormal PQ data is designed. The proposed GPRS based wide-area PQ monitoring system can be used for PQ monitoring with minimum cost and maximum efficiency. All the functions implemented in this paper can realize the novel, real-time and wide-area PQ monitoring. Experimental results demonstrate the validity of the proposed system.

Keywords: Power Quality Monitoring Platform, Total Harmonic Distortion, GPRS, Personal Digital Assistant

I. INTRODUCTION

A good understanding of Power Quality (PQ) can help power companies choose the correct instruments to improve PQ and obtain the benefits from upgrading the supply circuits. Therefore, many companies perform PQ survey at specific sites to gather the real-time PQ information. PQ monitoring and recording becomes one of the major services of the power companies for their customers. Besides, the era of deregulation has brought more attentions for PQ. It is possible to provide additional services to some customers on an optional basis and to charge for those services. For example, several competing distribution companies might base their prices on the level of PQ provided [1-5].

Wide-area power quality monitoring is one of the most important and difficult issues for utilities; therefore, how to monitor, display and analyze the PQ data as effective and efficient as possible is still an on-going work. The key in realizing the wide-area PQ monitoring is the communication. Wire communication requires all meters being connected to the remote terminal units. It has the advantage of high reliability; however, the construction cost is huge. Wireless communication is suitable for wide-area power quality monitoring. But the traditional wireless communication techniques, such as radio station, have the limitation of low data rate, uncertain time-delay and low reliability. The new wireless communication techniques such as Bluetooth, IEEE 802.11 and GSM/GPRS (Group Special Mobile, GSM/General Packet Radio Service, GPRS) provide more reliable and efficient data exchange techniques; therefore, they will be more suitable for constructing a wide-area PQ monitoring system. Due to the communication distance and wide spread of communication cell, GSM/GPRS is a better choice for wide-area PQ monitoring.

In order to realize the wide-area monitoring, some new hardware interfaces and mobile devices such as USB and PDA (Personal Digital Assistant, PDA) should be utilized for PQ monitoring and recording. Most of instruments still use RS-232 as major data transmission interface. However, due to its lower transmission rate, RS232 is difficult to be used for equipments with high-speed control and monitoring requirement. Recently, personal computers already have built-in USB interface. Comparing to RS-232 interface, USB has the characteristics of faster transmission rate and plug and play [6]. Therefore, USB-based interface can be used for PQ monitoring that has high-speed data transmission requirement. Recently, since the memory and processing speed of those PDA devices become more powerful, they now can run complex programs and can be used for graphical and multimedia applications. Those newly designed PDAs can perform more complex functions such as education, data collection, intelligent appliances and industrial automation etc., which are previously available only via desktop computers [7, 8]. With PDA, a tiny and mobile PQ analyzer can be developed.

This paper tries to use USB and PDA to develop a novel PQ monitoring platform and then integrates GPRS technique into the proposed platform to realize wide-area PQ monitoring. The works of this paper can be divided into three parts. First, a small-scale PQ monitoring platform with appropriately designed I/O interfaces and peripherals is designed and implemented. The PQ analyzing functions implemented in the proposed system include 1) system voltage and current calculation, 2) system frequency and power factor calculation, 3) harmonic spectrum and Total Harmonic Distortion (THD) calculation and 4) abnormal event recorder etc. Next, a GPRS module which can be integrated into the designed PQ monitoring platform is developed. Finally, a web server with well-designed database used to record abnormal PQ data is designed. The proposed GPRS based wide-area PQ monitoring system can be used for remote PQ monitoring with minimum cost and maximum efficiency. All the functions implemented in this paper can realize the real-time and wide-area PQ monitoring.

II. HARDWARE AND FIRMWARE ARCHITECTURE

2.1 Hardware Architecture

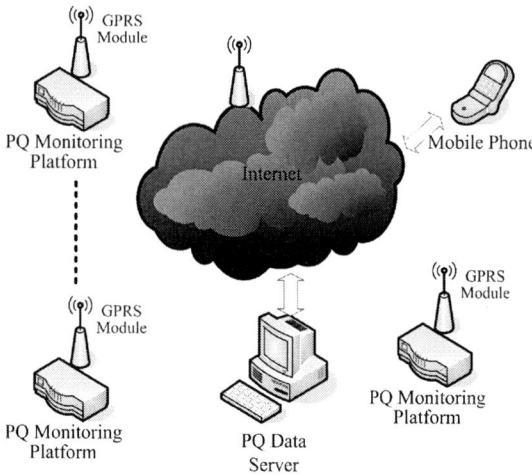

Fig. 1: Configuration of the proposed GPRS-based wide-area PQ monitoring system

Fig. 1 is the hardware configuration of the proposed GPRS-based PQ monitoring system. From Fig. 1, the whole system consists of PQ monitoring platforms embedded with the GPRS communication function and a PQ data server used to save the abnormal PQ events. Fig. 2 shows the hardware architecture of the proposed PQ monitoring platform. The periodic PQ monitoring and recording is accomplished in the local PQ monitoring platform. Only the abnormal PQ event and its measured and calculated data are transmitted to the PQ data server. The communication between local PQ monitoring platforms and PQ data server can be achieved by GPRS. The hardware of the proposed PQ monitoring platform can be divided into two parts–the C8051F320 MCU [9] and the circuit for signal level-scaling and shifting. The C8051F320 manufactured by Silicon Laboratories Co. Ltd. features a configurable unified 2304 bytes internal RAM and 16k bytes flash memory, an I^2C serial interface, one UART, four timers, 25 programmable I/O ports, 10-bit 200 ksps 17 channels single-ended/ differential Analog to Digital Conversion (ADC) and a USB controller.

The circuit for signal level scaling and shifting part is used to scale and shift the signal level to the permitted level of ADC. The MCU is used to obtain the voltage and current signals, calculate the PQ indices and then transmit the waveform data and PQ indices to PC or PDA by USB and RS232, respectively. The USB and PC based PQ analyzer can provide a long-term PQ monitoring and recording. The RS232 and PDA based PQ analyzer can be used for PQ monitoring with tiny volume and mobility. When an abnormal PQ event occurred, for example, the PQ indices exceed the predetermined limit, a message will be sent to operator's mobile phone by the short message service of GSM and the PQ data will also be transmitted to the PQ data server by GPRS. The user can access the PQ data server via the Internet and obtain the abnormal PQ data without extra infrastructure.

2.2 Firmware Designed in MCU

In order to achieve the goals set in this paper, appropriate firmware should be designed for MCU. Fig. 3 shows the flowchart of MCU firmware. The "MCU setting and initialization" shown in Fig. 3 is used to set the parameters of MCU, enable the I/O pins for data acquisition, select the communication interface and choose the sampling rate for ADC etc. The "Acquire voltage/current waveform" is used to acquire voltage and current signals at the specific PQ monitoring site. After those ADC data were acquired, "Power quality analyzing" is used to calculate the PQ indices. The flowchart of "Power quality analyzing" is shown in Fig. 4. The sampling rate used in this paper is 7680 sampled data per second. The PQ analyzing functions implemented in the proposed system include Root Means Square (RMS) of acquired voltage and current, system frequency, power factor, harmonic spectrum and THD etc. The harmonic spectrum can be obtained by Discrete Fourier Transform (DFT). The Fast Fourier Transform (FFT) is an efficient method for DFT calculation. The result of FFT is identical to the result of the DFT, but a number of redundant calculations can be eliminated to allow faster computation; therefore, FFT is implemented in this paper. The FFT implemented in this paper is Radix-2 Decimation-in-Time algorithm [10].

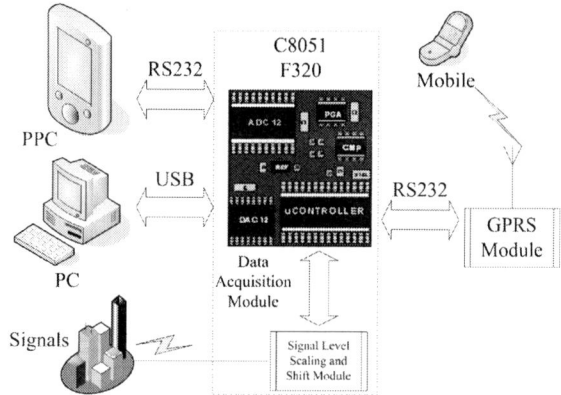

Fig. 2: Configuration of the PQ monitoring platform

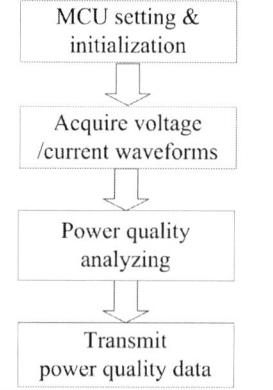

Fig. 3: The flowchart of MCU firmware

"Transmit power quality data" is used to transmit the acquired voltage and current data and the calculated PQ indices to PC, PDA and mobile phone by USB, RS232 and GSM/GPRS module, respectively. Fig. 5 shows the flowchart of data transmission between the

306

C8051F320 and GSM/GPRS module. From Fig. 5, it can be seen that the MCU will send some AT commands [11, 12] to GSM/GPRS module if an abnormal event occurred. A message including the abnormal event location, voltage frequency, voltage RMS and voltage THD will then be sent to operator's mobile phone by the short message service of GSM. The interface shown in the mobile phone is designed as Fig. 6. The detailed PQ data will also be transmitted to the PQ data server by GPRS. If the PQ data are received by PC and PDA, then user-friendly HMIs (Human Machine Interface, HMI) are designed to display and record those data.

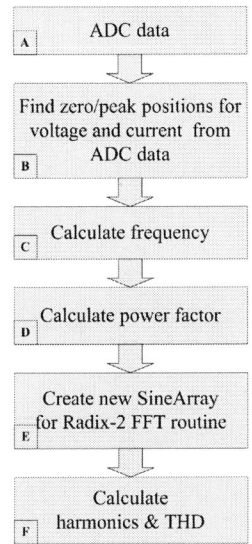

Fig. 4: The flowchart for power quality analyzing

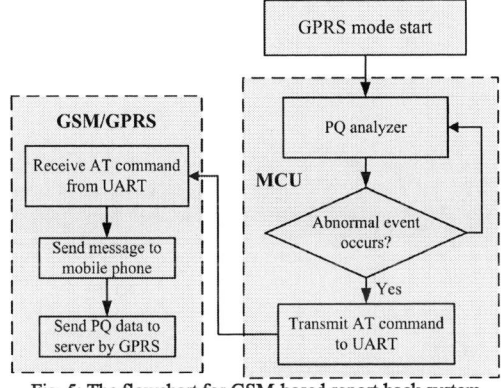

Fig. 5: The flowchart for GSM-based report back system

III HMIs FOR THE PROPOSED PQ MONITORING PLATFORM

Two HMIs designed for PC and PDA are developed. The PC-based HMI can fully make use of the high speed transmission rate of USB, and therefore, it can be used for real-time PQ monitoring and abnormal event recording. The PDA-based HMI can achieve the PQ monitoring with tiny volume and mobility requirements. Microsoft eMbedded VB (eVB) and eMbedded VC++ (eVC) are the popular development tools for Pocket PCs (PPCs), the PDAs run on Windows CE OS. eVB enables programmers to develop Windows CE-based applications using an integrated development environment similar to that used in developing desktop VB applications. Therefore, it is used by most

programmers who are familiar with VB programming. The main difference between eVB and VB is that the objects supported in the eVB are less than VB's due to the memory and performance limitation of PPCs. In this paper, PPC and eVB are used to develop the proposed HMI due to the simplicity and wide usage of eVB [13, 14].

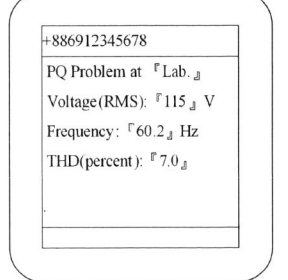

Fig. 6: The information shown in mobile phone

Figs. 7 are the PPC-based HMI for the proposed PQ monitoring platform. The HMI is composed of three sub-interfaces: Main, Monitor and RS232 setting interfaces. Main interface shows the PQ data including frequency, RMS and THD (%) etc. Monitor interface displays the voltage waveform and harmonic spectrum. The RS232 setting interface is used to set the parameters such as baud rate, data bits, stop bits and parity etc, and then connect to the MCU via RS232.

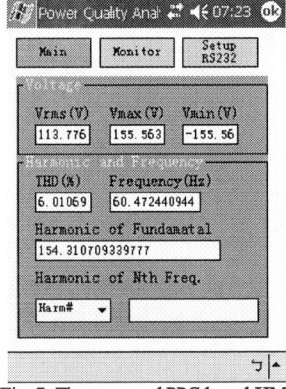

Fig. 7: The proposed PPC based HMI

IV. WIDE-AREA PQ MONITORING SYSTEM

Fig. 8 shows the configuration of the proposed wide-area PQ monitoring system. When an abnormal PQ event occurred, for example, the PQ indices exceed the predetermined limit, a message will be sent to operator's mobile phone by the short message service of GSM and the PQ data will also be transmitted to the PQ data server by GPRS. The user can access the PQ data server via the Internet and obtain the PQ data. Fig. 9 shows the architecture of the PQ data server. From Fig. 9, it can be seen that the PQ data server is composed of the TCP/IP programs used to receive the PQ data transmitted from GPRS, the interactive web pages used to display the abnormal PQ events to operators, and the database used to store the abnormal PQ data. Fig. 10 shows the flowchart of TCP/IP programs designed by Microsoft Visual Basic. From Fig. 10, it can be seen that when an abnormal PQ event occurred, the PQ data acquired and calculated will be prepared and then transmitted to GPRS module by MCU. The TCP/IP programs will initialize TCP/IP socket, setup TCP/IP configuration, establish socket binding and listening and then wait for the client connection. After the client connection, the PQ data will be received from Internet and be saved in the database. Fig. 11 shows the data table of abnormal PQ database. It can be seen that the IP address of PQ monitoring platform, data/time of abnormal event and the PQ data including maximum voltage, minimum voltage, RMS, THD and harmonic spectrum etc are all saved in the database.

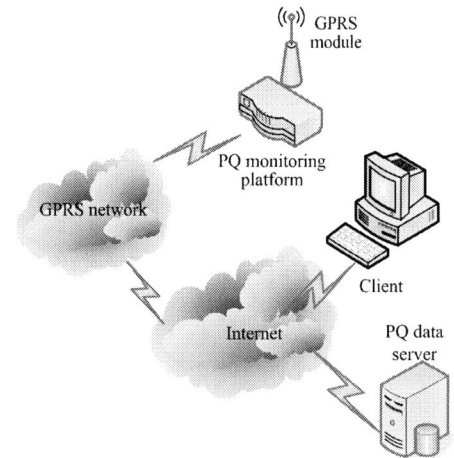

Fig. 8: Wide-area PQ monitoring system

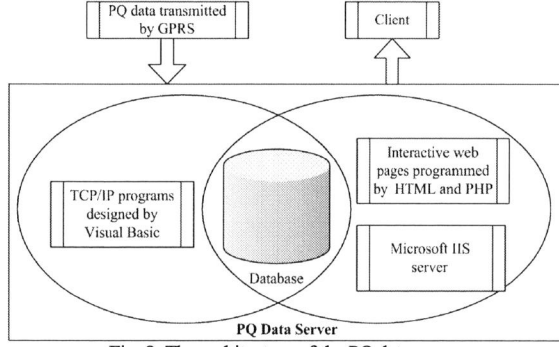

Fig. 9: The architecture of the PQ data server

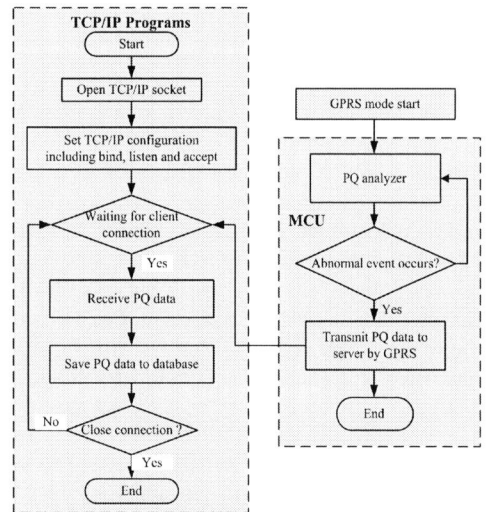

Fig. 10: The Flowchart of TCP/IP Programs

The interactive web pages are designed by PHP and HTML on the Microsoft IIS server. Fig. 12 shows the flowchart of the web pages. Fig. 13 shows the home page of the abnormal PQ information center. From Fig. 13, it can be seen that the "top.php" will display the total abnormal events and the "main.php" will display the five newest abnormal events. Users can select the PQ event which they want to observe, the detailed PQ information for this event will then be shown.

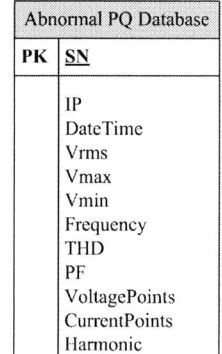

Fig. 11: Data table of abnormal PQ database

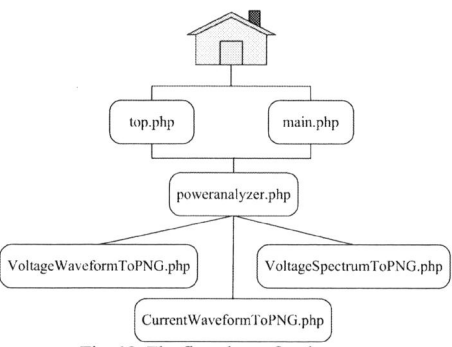

Fig. 12: The flowchart of web pages

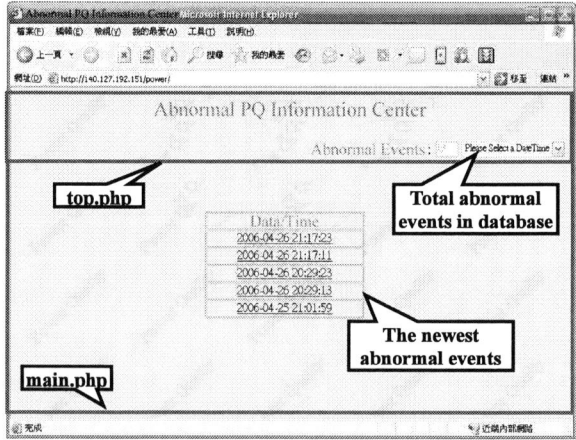

Fig. 13: The home page of abnormal PQ information center

V. EXPERIMENTAL RESULTS

Experiments have been conducted to make sure that the proposed system can be applied to wide-area PQ monitoring. However, only the execution procedure and some experimental results are shown here due to limited space. For PDA mode, the communication parameters between PPC and C8051F320 are baud rate 19200 bps, data bits 8, stop bits 1 and no parity. Figs. 7 and 14 show the measured PQ data, waveform and spectrum at the specific PQ monitoring site. If the proposed system is set in GSM/GPRS mode and the THD alarm value is set as 6%. When an abnormal event occurred, a message as shown in Fig. 15 will be received in the operator's mobile phone. Operators can then connect to the PQ data server and the detailed PQ information as shown in Fig. 16 can be displayed. From the experimental results, it can be seen that the proposed system has successfully integrated the widely used techniques including USB, PDA and GSM/GPRS into developing a wide-area PQ monitoring system.

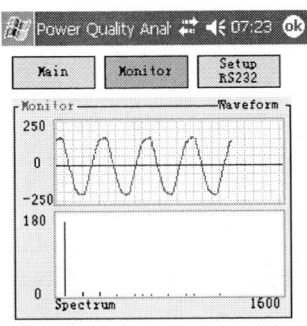

Fig. 14: The interface for waveform and spectrum

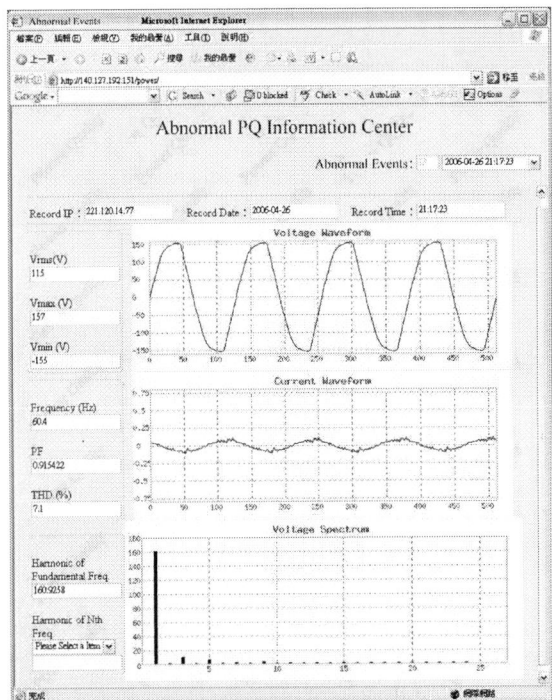

Fig. 16: The abnormal PQ information

Fig. 15: The message received by the system operator

VI. DISCUSSIONS AND CONCLUSIONS

This paper used USB and PDA to develop a PQ monitoring platform and then integrated GPRS technique into the proposed platform to realize wide-area PQ monitoring. A web server with well-designed database used to record the abnormal PQ data was designed. The proposed GPRS based wide-area PQ monitoring system can be used for remote PQ monitoring and recording with minimum cost and maximum efficiency. All the functions implemented in this paper can realize the real-time and wide-area PQ monitoring. Experimental results demonstrated the validity of the proposed system.

VII. ACKNOWLEDGEMENTS

This work was sponsored by National Science Council, Taiwan, under research grant NSC 95-2221-E-230-034.

VIII. REFERENCES

[1] Heydt, G.T.; "Electric power quality: a tutorial introduction," IEEE Computer Applications in Power, Vol. 11, Issue 1, Jan. 1998, pp.15 – 19

[2] Baker, P.P., Short, T.A., and Burns, C.M. "Power quality monitoring of a distribution system," IEEE Trans. on Power Delivery, Vol. 9, No. 2, April 1994, pp. 1136-1142.

[3] Brent M. Hughes, John S. Chan, and Don O. Kvoal, "Distribution customer power quality experience," IEEE Trans. on Industry Application, Vol. 29, No. 6, Nov. 1993, pp. 1204-1211.

[4] Erich W. Gunther, and Harshad Mehta, "A survey of distribution system power quality – preliminary results," IEEE Trans. on power Delivery, Vol. 10, No. 1, Jan. 1995, pp.322-329.

[5] Chen, S.; "Open design of networked power quality monitoring systems," IEEE Trans. on Instrumentation and Measurement, Vol. 53, Issue 2, April 2004, pp. 597 – 601

[6] Universal Serial Bus Specification Revision 2.0, http://www.usb.org/

[7] Myers, B.A.; Beigl, M.; "Handheld computing," Computer, Vol. 36, Issue: 9, Sep. 2003, pp. 27-29

[8] Kiely, D.; "Wanted: programmers for handheld devices," IEEE Computer, Vol. 34, Issue 5, May 2001, pp. 12-14

[9] C8051F32x Data Sheet, http://www.silabs.com/, Silicon Laboratories, 2003

[10] Oppenheim, A.V. and Schafer, R.W., "Discrete-time signal processing," 2nd Ed. Prentice Hall, Inc., Upper Saddle River, New Jersey, 1999

[11] User Manual for GSM/GPRS Module (EGD-01), http://www.eup.com.tw

[12] AT command set, http://www.control.com.sg/docs/ SIM100_AT_Command_Set.pdf

[13] David Chang, Ming-Jieh Cheng, Shun-Yu Chan and Jen-Hao Teng, "Development of a Novel Power Quality Monitoring and Report-Back System," IEEE TENCON Hong Kong 2006, PE 2.2

[14] Douglas Boling, Programming Microsoft Windows CE .Net, 3rd Edition

Design and Development of Autotransformer Based 24-Pulse AC-DC Converter fed Induction Motor Drive

Bhim Singh, *Senior Member, IEEE,* Vipin Garg, *Member, IEEE* and, G.Bhuvaneswari, *Senior Member, IEEE*
Department of Electrical Engineering, Indian Institute of Technology, Delhi, New Delhi, India -110016

Abstract-- **This paper deals with a reduced rating delta-polygon autotransformer based twenty-four-pulse ac-dc converter feeding variable frequency induction motor drives (VFIMD's) for improving power quality at the point of common coupling (PCC). The proposed twenty four-pulse ac-dc converter is realized on the principle of dc ripple re-injection technique for harmonic mitigation. The design of the proposed autotransformer is developed alongwith the necessary modifications required for making it suitable for retrofit applications, where presently a 6-pulse diode bridge rectifier is used. The effect of load variation on VFIMD is also studied to demonstrate the effectiveness of the proposed ac-dc converter. The extensive tests are conducted on the developed prototype autotransformer based ac-dc converters. Different power quality indices of 6-pulse, 12-pulse and proposed 24-pulse ac-dc converters are obtained from simulation and verified from experimental results.**

Index Terms—Autotransformer, multipulse AC-DC converter, pulse multiplication, power quality improvement, VFIMD.

I. INTRODUCTION

The revolution in solid state power conversion has opened an era for enhanced use of variable frequency induction motor drives (VFIMD's) in various applications such as air conditioning, blowers, fans, pumps for waste water treatment plants, cement industry etc. These VFIMD's are now a days used generally in vector control mode [1] due to their inherent advantages. These voltage source inverter (VSI) based induction motor drives are generally fed from the 6-pulse diode bridge rectifier, which results in injection of current harmonics in to the ac mains. These current harmonics while propogating through the source impedance result in voltage distortion at the point of common coupling (PCC), thereby affecting the nearby consumers.

There has been an impetus on research in power quality interface due to stringent power quality regulation and strict limits on current and voltage distortion imposed by various standards such as IEC 61000-3-2 [2] and IEEE 519-1992 [3]. This has led to the consistent research in innovating various configurations of ac-dc converters for mitigating these harmonics [4]. Different techniques for power quality improvement such as use of multipulse converters [5-14], use of multiphase converters [15] and use of dc ripple re-injection technique [16-17] for pulse multiplication have been reported in the literature. These techniques for harmonic mitigation are found to be rugged, reliable and energy efficient over their counterparts. These methods use two or more converters, where the harmonics generated by one converter are cancelled by other converter, by proper phase shift. The autotransformer based converters result in reduction in rating of the magnetics,

as the transformer transfers only a small portion of the total kVA of the VFIMD. Different topologies of 12-pulse ac-dc converters have been reported in the literature for harmonic mitigation [5-11]. But these configurations fail to comply with above mentioned standards. Increasing the number of pulses further results in improvement in various power quality indices, but alongwith the additional cost of different converters and increased system complexity. Autotransformer based configurations of 18-pulse ac-dc converters have also been reported in the literature [12-15]. The Total Harmonic Distortion (THD) of ac input current in an 18-pulse ac-dc converter is reported to be 8.6% at full load, which may further deteriorate at light load [14]. Moreover, the rating of magnetics required in an 18-pulse converter is of the order of 55% [15], which is quite high. To achieve an improved performance in terms of various power quality indices at a reduced cost, a 24-pulse converter employing dc ripple re-injection technique for pulse doubling is designed and developed in this work.

This paper presents a novel reduced rating autotransformer based twenty-four-pulse ac-dc converter suitable for retrofit applications (where presently a six-pulse diode bridge rectifier is used, as shown in Fig.1, referred as Topology 'A'). The design procedure of the autotransformer is given for making it suitable for retrofit applications [18]. The dc ripple re-injection technique used for harmonic mitigation needs two additional diodes alongwith a suitable tapped interphase reactor for increasing the number of pulses. The proposed ac-dc converter is able to provide almost unity power factor in wide operating range of the VFIMD. A set of tabulated results giving the comparison of different power quality indices such as total harmonic distortion (THD) and crest factor of ac mains current (CF), power factor (PF), displacement factor (DPF) and distortion factor (DF), THD of supply voltage at PCC is presented for a VFIMD fed from an existing 6-pulse ac-dc converter, 12-pulse ac-dc converter and proposed 24-pulse ac-dc converter. Various tests are conducted on the prototype autotransformers developed in the laboratory alongwith the interphase reactors and the Zero Sequence Blocking Transformer (ZSBT). The experimental results validate the simulation results for both 12-pulse and proposed 24-pulse ac-dc converters.

II. ANALYSIS AND DESIGN OF THE PROPOSED AUTOTRANSFORMER

The minimum phase shift required for proper harmonic elimination is given by [5]:

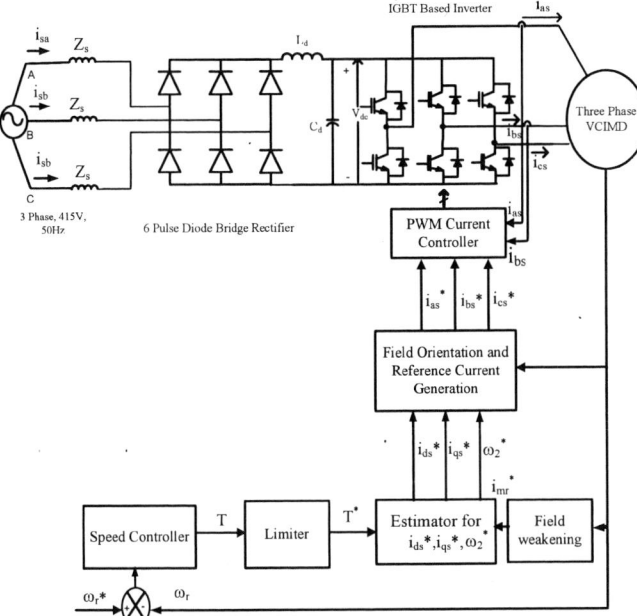

Fig.1 Six-pulse diode bridge rectifier fed vector controlled induction motor drive. (Topology A).

Phase shift = 60^v/ Number of six-pulse converters

For achieving 12-pulse ac-dc conversion, the phase shift between the two sets of voltages may be either of 0^0 and 30^0 or $\pm 15^0$ with respect to the supply voltages. In this work the autotransformer based on $\pm 15^0$ has been considered to reduce the size of the magnetics.

From the supply voltages, two sets of 3-phase voltages (phase shifted through $+15^0$ and -15^0) are produced. The number of turns required for $+15^0$ and -15^0 phase shift are calculated as follows. Fig. 2 shows the winding connection diagram of the proposed autotransformer. Fig. 3 shows the corresponding phasor diagram of different phase voltages.

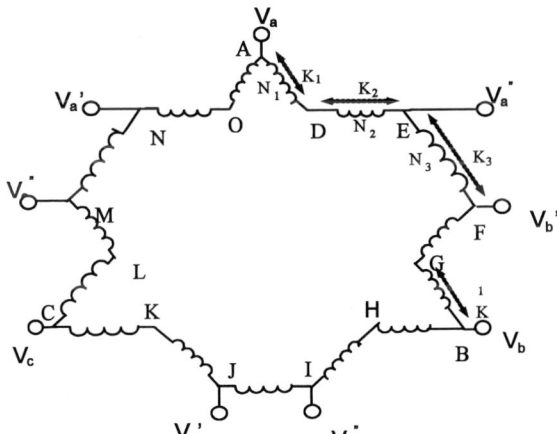

Fig. 2 Proposed autotransformer winding connection diagram.

Consider phase 'a' voltages in Fig.2 as:

$$V_a' = V_a + K_1 V_{ca} + K_2 V_{bc} \tag{1}$$
$$V_a'' = V_a + K_1 V_{ab} + K_2 V_{bc} \tag{2}$$

Considering the following set of voltages:

$$V_a = V\angle 0^0, \ V_b = V\angle -120^0, \ V_c = V\angle 120^0, \ V_{ab} = 1.732V\angle 30^0,$$
$$V_{bc} = 1.732V\angle 90^0, \ V_{ca} = 1.732V\angle -30^0 \tag{3}$$
$$V_a' = V\angle 15^0, \ V_b' = V\angle -105^0, \ V_c' = V\angle 135^0 \tag{4}$$
$$V_a'' = V\angle -15^0, \ V_b'' = V\angle -135^0, \ V_c'' = V\angle 105^0 \tag{5}$$

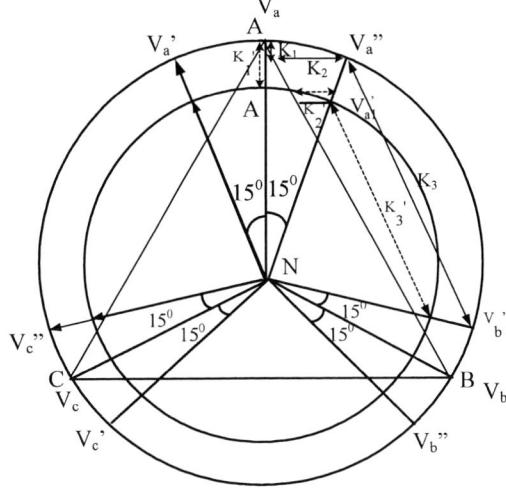

Fig. 3 Phasor diagram of voltages in the proposed autotransformer connection alongwith modifications for retrofit arrangement.

where, v is the rms value of phase voltage.

Using above equations K_1 and K_2 can be calculated. These equations result in $K_1 = 0.0227$ and $K_2 = 0.138$ for the desired phase shift in autotransformer.

The angle between DB and DE in Fig.2 is 60^0, similarly angle between FG and GD is 60^0. Thus, from Fig.2, it can be written as :

$$K_1 + K_2\cos 60^0 + K_3 + K_2\cos 60^0 + K_1 = 1 \tag{6}$$
$$2K_1 + K_2 + K_3 = 1 \tag{7}$$

Thus $K_3 = 0.862$

A phase shifted voltage (e.g. V_a') is obtained by tapping a portion (0.0227) of line voltage V_{ca} and connecting one end of an approximately (0.138) of line voltage (e.g. V_{bc}) to this tap. Thus the autotransformer can be designed with these known values of winding constants i.e. K_1, K_2 and K_3.

With this transformer arrangement the dc link voltage obtained is slightly higher than that of a 6-pulse diode bridge rectifier output voltage. To make the proposed ac-dc converter suitable for retrofit applications, the transformer design has been modified to make the dc link voltage same as that of 6-pulse diode bridge rectifier. Fig.3 shows the generalized phasor diagram for achieving different voltage ratios from the autotransformer by simply varying the tapping positions on the windings. The different tap positions for retrofit arrangement have been shown on inner circle. This ensures that both the output voltages are still having the required phase shift of $\pm 15^0$ (for achieving the twelve-pulse operation). By following the above procedure, for same dc link voltage as that of 6-pulse diode bridge rectifier, the values of K_1', K_2' and K_3' are as $K_1' = 0.042066$, $K_2' = 0.12838$ and $K_3' = 0.78746$ where K_1', K_2' and K_3' are the new constants. Thus, by simply changing the transformer winding tapping, as shown in Fig.3, the same dc link voltage as that of 6-pulse diode bridge rectifier is obtained. Fig.4 shows the 12-pulse ac-dc converter fed VCIMD with the proposed autotransformer, referred as Topology 'B'.

The kVA rating of the transformer is calculated as [5]:

$$kVA = 0.5\sum V_{winding} \ I_{winding} \tag{8}$$

The kVA rating of the interphase transformer is also calculated using the above relationship.

311

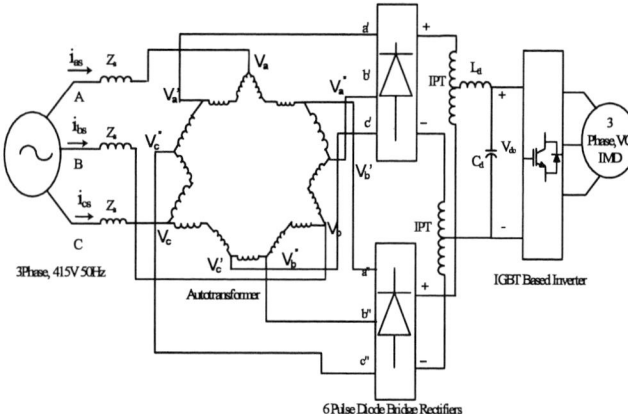

Fig.4 Proposed autotransformer based 12-pulse converter (with phase shift of $+15^0$ and -15^0) fed VCIMD. (Topology B).

III. DC RIPPLE RE-INJECTION FOR TWELVE-PULSE AC-DC CONVERTERS

This technique is used for increasing the number of rectification pulses (without much additions in the system layout) resulting in harmonic reduction on both ac mains as well as dc bus. Fig.5 shows the proposed autotransformer based 24-pulse ac-dc converter fed VFIMD, referred as Topology 'C' and for retrofit applications, it is referred as Topology 'D'.. This configuration needs one zero sequence blocking transformer (ZSBT) to ensure independent operation of the two diode rectifier bridges. It exhibits high impedance to zero sequence currents, resulting in symmetrical conduction for each diode of the bridges and also results in equal current sharing in the output. An interphase reactor tapped suitably to achieve pulse doubling is connected at the output of the ZSBT.

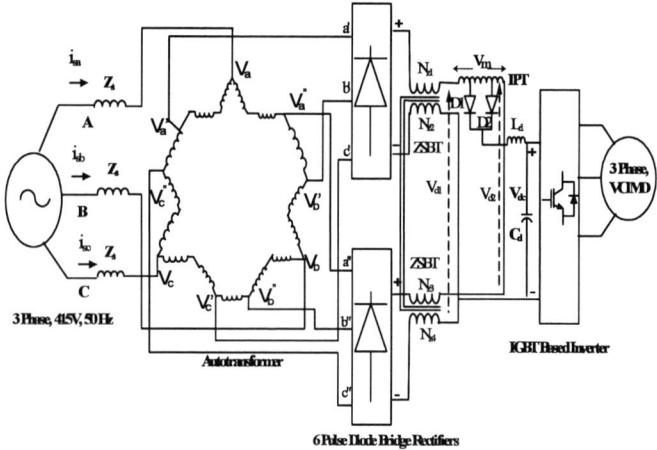

Fig.5 Proposed 24-pulse ac-dc converter with dc ripple re-injection scheme fed VCIMD (Topology C and D).

A. Design of Interphase Transformer

The schematic diagram of pulse doubling scheme for diode bridge rectifiers as shown in Fig.5 is realized based on dc ripple re-injection. The operation of interphase transformer is shown in Fig.6. The voltage appearing across the interphase transformer v_m (shown in Fig.7) is an ac voltage ripple of six times the source frequency, resulting in smaller size, weight

and volume of the transformer. Depending upon the polarity of the impressed voltage across the interphase transformer, diodes D_1 or D_2 conduct, result in pulse multiplication. When the voltage v_m is positive, diode D_1 becomes forward biased (Fig.6a) and the full load current flows through this diode. From the MMF relationship of the interphase transformer, it can be written as:

$$i_{o1} = (0.5 + K) i_o \qquad (9)$$
$$i_{o2} = (0.5 - K) i_o \qquad (10)$$

Similarly, when the voltage v_m is negative, diode D2 becomes forward biased (Fig.6b) and the full load current flows through this diode. From the MMF relationship of the interphase transformer, it can be written as:

$$i_{o1} = (0.5 - K) i_o \qquad (11)$$
$$i_{o2} = (0.5 + K) i_o \qquad (12)$$

It is observed that the polarity of the voltage v_m modulates the output currents i_{o1} and i_{o2}. The value of K has been chosen to be 0.26 to eliminate the harmonics upto 23rd order.

The turn's ratio of the interphase reactor is given by

$$N_{i1}/N_o = 0.26, \ N_{i1} + N_{i2} + N_{i3} = N_o \text{ and } N_{i1} = N_{i3} \qquad (13)$$

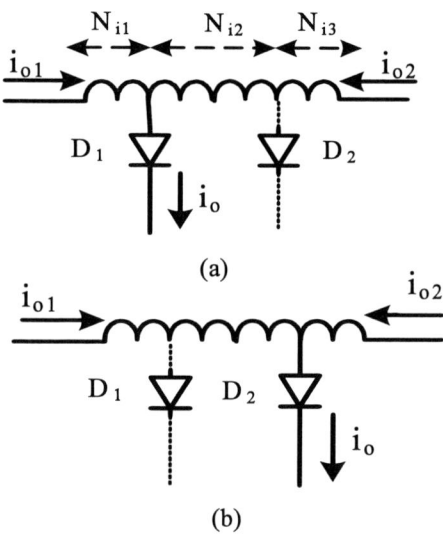

Fig.6 Operation of interphase transformer.

B. Zero Sequence Blocking Transformer

The zero sequence blocking transformer (ZSBT) results in independent operation of the two rectifier bridges, thus eliminating the unwanted conducting sequence of the rectifier diodes. ZSBT offers very high impedance for zero sequence current components. The voltage waveform across ZSBT (V_{ZSBT}), shown in Fig.7 contains only triplen frequency components resulting in smaller size, weight and volume of the transformer.

IV. MODELLING OF VECTOR CONTROLLED INDUCTION MOTOR DRIVE

Fig.1 shows the schematic diagram of a 6-pulse ac-dc converter fed indirect vector controlled induction motor drive, referred as Topology 'A'. The motor is controlled in vector

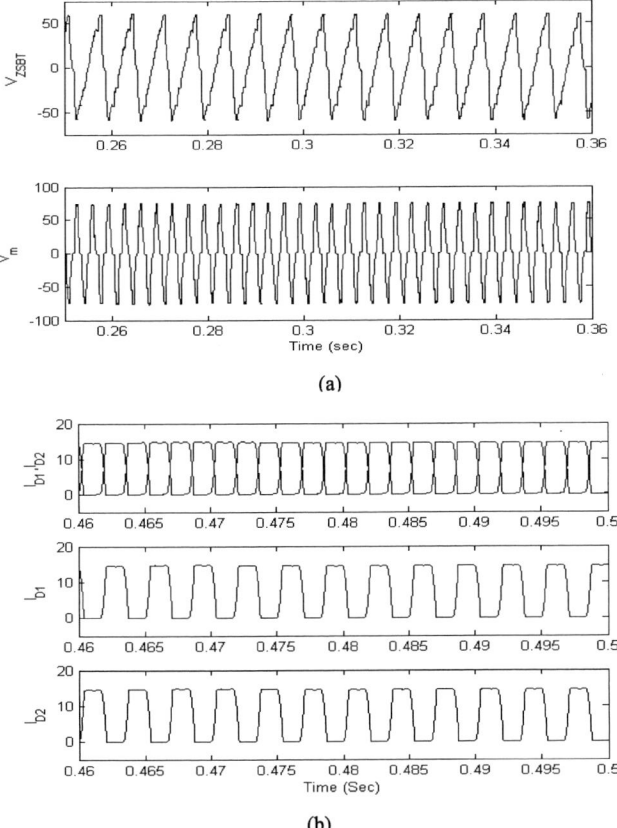

Fig.7 Different waveforms (a) V_{ZSBT} across ZSBT and V_m across interphase reactor, (b) Diodes D_1 and D_2 current waveforms.

control mode using indirect vector control technique and connected as load on the proposed converter. To realize the vector control of an induction motor, two currents of motor phases namely i_{as} and i_{bs} and the motor speed signal (ω_r) are sensed. The closed loop PI speed controller compares the reference speed (ω_r^*) with motor speed (ω_r) and generates reference torque $T_{(n)}^*$ (after limiting it to a suitable value).

$$T_{(n)}^* = T_{(n-1)}^* + K_p \{\omega_{e(n)} - \omega_{e(n-1)}\} + K_I \omega_{e(n)} \qquad (14)$$
$$\omega_{e(n)} = \omega_{r(n)}^* - \omega_{r(n)} \qquad (15)$$

where, $T_{(n)}^*$ and $T_{(n-1)}^*$ are the output of the PI controller (after limiting it to a suitable value) and $\omega_{e(n)}$ and $\omega_{e(n-1)}$ refer to speed error at the nth and $(n-1)^{th}$ instants. K_p and K_I are the proportional and integral gain constants.

The exciting current is governed by the rotor speed of the induction motor and is expressed as:

$$i_{mr} = I_{mr} \qquad \text{if } \omega_r < \omega_s \qquad (16)$$
$$i_{mr} = I_{mr} (\omega_r / \omega_s) \qquad \text{if } \omega_r > \omega_s \qquad (17)$$

where I_{mr} is the rated exciting current and ω_s is the base synchronous speed of the induction motor.

The flux control signal (i_{mr}) alongwith $T_{(n)}^*$ are fed to the vector controller, which computes the flux producing component of current (i_{ds}^*), torque component of current (i_{qs}^*), slip speed (ω_2^*) and the flux angle (ψ) as given below:

$$i_{ds}^* = i_{mr} + \tau_r (\Delta i_{mr}/\Delta t) \qquad (18)$$
$$i_{qs}^* = T^* /(K_m i_{mr}) \qquad (19)$$
$$\omega_2^* = i_{qs}^* /(\tau_r i_{mr}) \qquad (20)$$
$$\Psi_{(n)} = \Psi_{(n-1)} + (\omega_2^* + \omega_r) \Delta t \qquad (21)$$

where K_m is a constant and it depends on motor parameters,

$\Psi_{(n)}$ and $\Psi_{(n-1)}$ are the value of rotor flux angles at n^{th} and $(n-1)^{th}$ instants respectively and Δt is the sampling time.

These currents (i_{ds}^*, i_{qs}^*) in synchronously rotating frame are converted to stationary frame three phase currents (i_{as}^*, i_{bs}^*, i_{cs}^*) as given below:

$$i_{as}^* = -i_{qs}^* \sin\Psi + i_{ds}^* \cos\Psi \qquad (22)$$
$$i_{bs}^* = \{-\cos\Psi + \sqrt{3}\sin\Psi\}i_{ds}^*(1/2) + \{Sin\Psi + \sqrt{3}\cos\Psi\}i_{qs}^*(1/2) \qquad (23)$$
$$i_{cs}^* = -(i_{as}^* + i_{bs}^*) \qquad (24)$$

These reference currents (i_{as}^*, i_{bs}^* and i_{cs}^*) generated by the vector controller are compared with the sensed motor currents (i_{as}, i_{bs} and i_{cs}). The calculated current errors are:

$$i_{ke} = i_{ks}^* - i_{ks}, \text{ where k = a,b,c} \qquad (25)$$

These current errors are amplified and fed to the PWM current controller, which controls the duty ratio of different switches in VSI. The VSI generates the PWM voltages being fed to the motor to develop the torque required to maintain the rotor speed equal to the reference speed.

V. MATLAB BASED SIMULATION

The proposed ac-dc converter feeding VFIMD is simulated in MATLAB environment alongwith Simulink and Power System Blockset (PSB) toolboxes. Fig.8 shows the MATLAB model of a vector controlled induction motor drive. The VFIMD consists of a 10 hp, 415V induction motor drive controlled using indirect vector control technique (the detailed parameters are given in Appendix). Fig.9 shows the MATLAB model of the proposed ac-dc converter to improve various power quality indices. The source impedance is kept at a practical value of 3% in all the simulations.

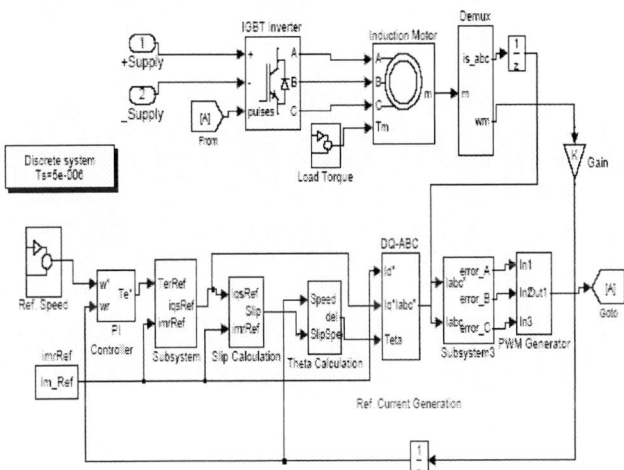

Fig.8 MATLAB block diagram of VCIMD.

VI. EXPERIMENTATION

The simulated results are verified on a test bench consisting of the newly designed and developed autotransformers alongwith small rating interphase reactor and a ZSBT as shown in Fig.5. Three single phase autotransformers are designed and wound in the laboratory as per the design details given below:

Flux density = 0.8 Tesla, Current density = 2.3A/mm^2, Core Size = 8 No., Area of cross section of core = 3225mm^2

313

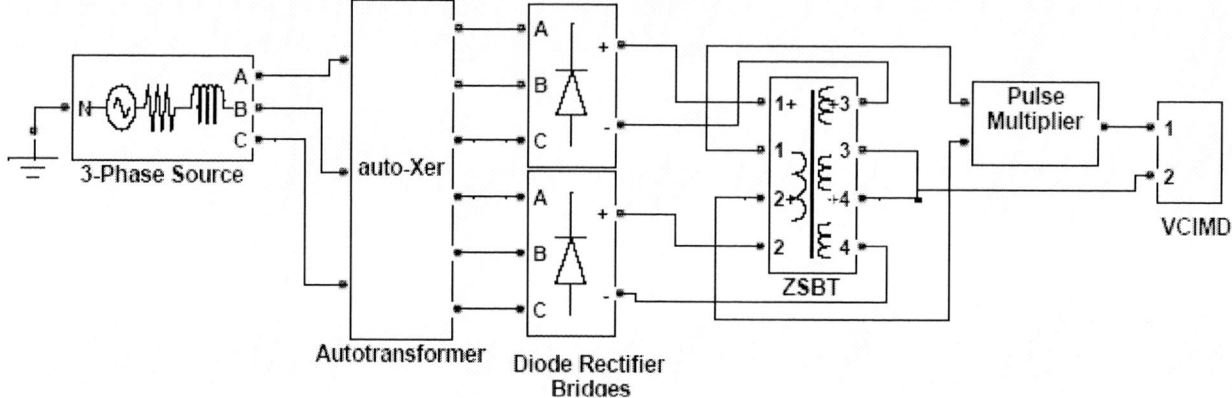

Fig.9 MATLAB block diagram of proposed 24-pulse ac-dc converter with dc ripple re-injection scheme fed VCIMD (Topology 'D').

(50.8mm x 63.5mm). E-Laminations: Length =184.1mm, width = 171.4mm, I-lamination: Length = 171.4mm, width = 50.8mm. Number of turns of different windings (shown in Fig.2) and the conductor cross section for both 12-pulse and 24-pulse autotransformers are given below:
Autotransformer for AC-DC converter: N_1 = 24 (SWG = 13), N_2 = 79 (SWG = 13), N_3 = 548 (SWG = 21).

Similarly, the interphase transformers of small ratings are designed and fabricated. The flux density is taken as 0.8Wb/m^2 and the current density is considered as 2.3A/mm^2. The interphase transformer is wound using core of size 3 No. with E-I laminations of size (76mm X127mm) and (127mm x 19mm) respectively. Based on the voltage across different windings, the number of turns are calculated and based on the current flowing through different windings, the gauge of wire is calculated and these are given in Table I.

TABLE I
Interphase transformer winding details

Winding	Number of turns	Gauge of wire
N_{i1}	35	11
N_{i2}	67	18
N_{i3}	35	11

The ZSBT is designed and wound on core size 8 No. The dimensions of E-laminations used are (123mm X 185mm) and that of I-laminations are (185mm X 25mm). The number of turns (shown in Fig.5) are calculated and are given as:
$N_{z1} = N_{z2} = N_{z3} = N_{z4} = 105$.
Based on the current flowing through different windings, the gauge of wire used in all the windings is taken as 11.
Various tests are carried out at three-phase line voltage of 230V AC input and with an equivalent resistive load. The test results are recorded using Fluke make power analyzer model 43B on the developed prototype of the converter.

VII. RESULTS AND DISCUSSION

The proposed twenty-four-pulse ac-dc converter has been designed and modeled alongwith the VFIMD for retrofit applications where presently a 6-pulse diode bridge rectifier is being used. Fig.10 shows the supply current waveform alongwith its harmonic spectrum at full load, showing THD of ac mains current as 31.3%, which deteriorates to 62.2%

at light load (20%) as shown in Table-II. Moreover, the power factor at full load is 0.935, which deteriorates to 0.807, (as shown in Table-II) as the load is reduced to 20%. Thus there is sufficient scope as well the need for improving the power quality at ac mains using some ac-dc converter which can easily replace the existing 6-pulse converter.

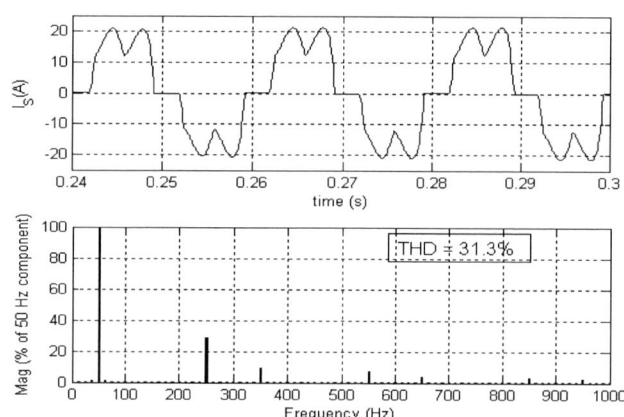

Fig. 10 AC mains current waveform of VCIMD fed by 6-pulse diode rectifier along with its harmonic spectrum at full load (Topology 'A').

A. Performance of Twelve-Pulse AC-DC Converter Fed VFIMD

To improve the power quality indices, a novel autotransformer for 12-pulse ac-dc converter fed VFIMD has been designed, modeled and simulated, referred as Topology 'B'. Fig. 11 shows the waveform of the supply current alongwith its harmonic spectrum at full load and the THD of supply current at full load is 9.23% and that at light load is 16.3%, whereas the power factor under these conditions is 0.981 and 0.977 respectively, thus not complying with IEEE Standard 519 [3]. Moreover, the rating of magnetics is 28.04%, as shown in Table III and which is on a higher side.

B. Performance of Proposed Twenty-Four-Pulse AC-DC Converter Fed VFIMD

The dc ripple re-injection technique has been incorporated in the above configuration to achieve the twenty-four-pulse rectification (Topology 'C'). In this Topology 'C', the dc link voltage is higher than that of a 6-pulse diode bridge rectifier, as given in Table-II. The autotransformer has been redesigned for making it suitable for retrofit applications,

TABLE II
Comparison of power quality parameters of a VCIMD fed from different converters

Sr. No.	Topology	THD V_s (%) Full Load	I_s (A)		Total Harmonic Distortion Factor of I_s (%)		Distortion Factor (DF)		Displacement Factor (DPF)		Power Factor (PF)		DC Link Voltage(V) Average	
			Full Load	Light Load (20%)	Full Load	Light Load (20%)	Full Load	Light Load (20%)	Full Load	Light Load (20%)	Full Load	Light Load (20%)	Full Load	Light Load (20%)
1.	A	6.76	14.35	4.35	31.3	62.2	.954	.849	.979	.950	.935	.807	547	555
3.	B	3.78	11.4	2.37	9.23	16.3	.995	.987	.985	.989	.981	.977	566	575
4.	C	3.76	11.36	2.34	4.51	5.97	.999	.998	.997	.995	.996	.995	563	566
5.	D	3.57	11.49	2.35	4.19	5.68	.999	.998	.997	.995	.995	.993	546	550

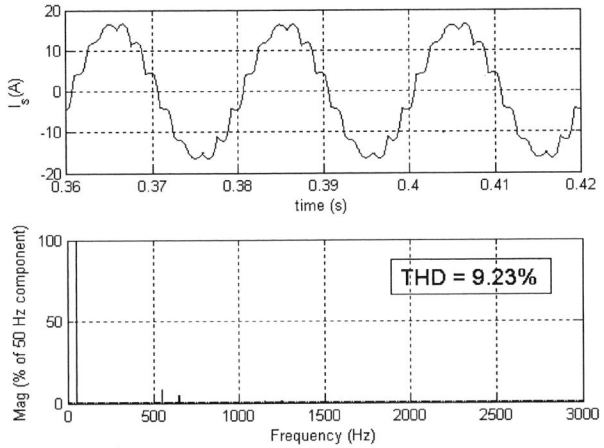

Fig. 11 AC mains current waveform alongwith its harmonic spectrum at full load for Topology 'B'.

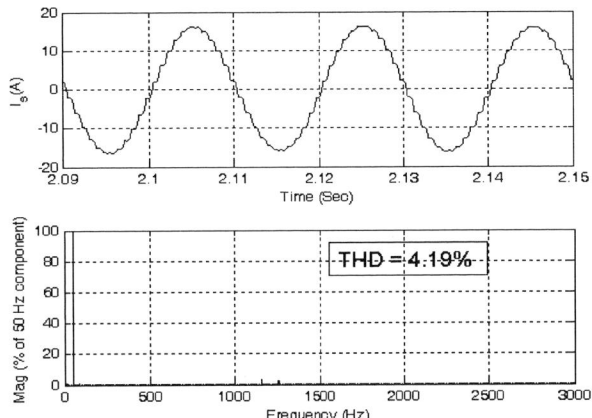

Fig. 13 AC mains current waveform alongwith its harmonic spectrum at full load for Topology 'D'.

resulting in Topology 'D', which is exactly similar to Topology 'C', except the minor difference in the number of turns in autotransformer windings. It can be observed that the presented design technique is capable of giving same dc link voltage as that of a six-pulse ac-dc converter.

Fig.12 shows the performance of the proposed 24-pulse ac-dc converter fed VFIMD under steady state and load perturbation. The set of curves consists of supply voltage v_s, supply current i_s, motor speed w_r(R/s), motor winding currents i_{abc}(A), developed torque T_e(N-m) and dc link voltage v_{dc} (V).

The supply current waveform at full load alongwith its harmonic spectrum is shown in Fig.13 for Topology 'D', which shows that the THD of ac mains current is 4.19% and the power factor obtained is 0.995. Fig.14 shows the supply current waveform alongwith its harmonic spectrum under light load condition (20%). At light load condition, the THD of ac mains current is 5.68% and the power factor is 0.993, showing the improvement in these indices.

To demonstrate the effect of load variation on the VFIMD, the load has been varied and the variation of different power quality indices has been shown in Table IV. It is seen that the proposed ac-dc converter is able to perform satisfactorily

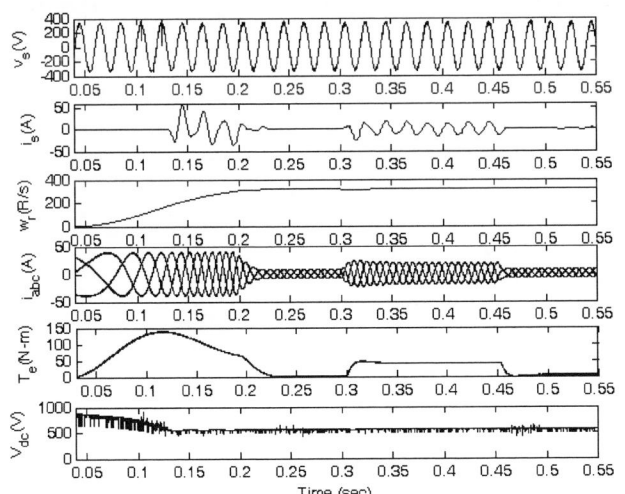

Fig. 12 Dynamic response of proposed ac-dc converter (Topology 'D') fed VCIMD with load perturbation.

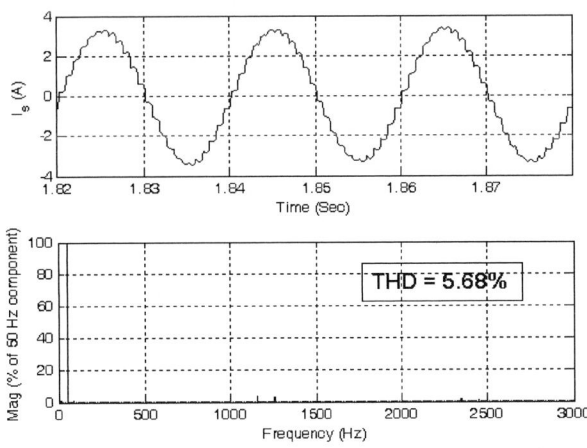

Fig. 14 AC mains current waveform alongwith its harmonic spectrum at light load (20%) for Topology 'D'.

TABLE III
Comparison of rating of magnetics in different converter fed VFIMD

Sr. No	Topol ogy	Transformer Rating (kVA)	Interphase Transformer rating (kVA)	ZSBT Rating (kVA)	Rating of magnetics % of drive rating
1	A	0	0.0	0.0	0.0
2.	B	1.807	1.12	0.0	28.04
3.	C,D	1.33	0.131	0.588	19.63

TABLE IV
Comparison of power quality indices of proposed 24-pulse harmonic mitigator fed VFIMD under varying loads.

Load (%)	THD (%)		CF of I_s	DF	DPF	PF	RF	V_{dc} (V)
	I_s	V_t						
20	5.68	1.61	1.40	.998	.995	.993	3.72	550
40	5.28	2.09	1.40	.998	.996	.995	3.67	550
60	4.86	2.64	1.40	.998	.996	.995	3.25	548
80	4.51	3.17	1.40	.999	.996	.995	2.16	547
100	4.19	3.57	1.40	.999	.996	.995	1.72	546

under load variation on VFIMD with almost unity power factor in the wide operating range of the drive and THD of supply current is always less than 8%. This is within the IEEE Standard 519 [3] limits for SCR >20. A comparison of different power quality indices of a VFIMD fed from different ac-dc converters is shown in Table I, showing the improvement in these indices with the proposed ac-dc converter. On the magnetics front also, there is reduction in rating, as it needs an autotransformer of 1.33kVA, interphase transformer of 0.131kVA and ZSBT of 0.588kVA, totaling the magnetics rating to 19.63% of the drive rating, as given in Table –III.

C. Experimental Performance of Proposed 12-Pulse and 24--Pulse AC-DC Converters Fed VFIMD (Topologies 'B' and 'D')

Various tests have been carried out on the developed prototypes for Topologies 'B' and 'D', shown in Figs.15-19. The waveform of supply voltage and current alongwith the harmonic spectrum of ac mains current at full load in Topology 'B' is shown in Fig. 15 and at light load (20%) it is shown in Fig.16. Similarly, the waveform of supply voltage and current alongwith the harmonic spectrum of ac mains current at full load in Topology 'D' is shown in Fig.17 and at light load (20%) it is shown in Fig.18.The effect of load variation on different power quality indices in Topology 'D' is shown in Table-V.

TABLE V
Experimental comparison of power quality indices in proposed 24- pulse harmonic mitigator under varying loads

Supply Current I_s (A)	THD (%)		CF of I_s	DF	DPF	PF	V_{dc} (V)
	I_s	V_s					
2.50	7.7	1.9	1.4	.997	0.99	0.987	302
3.66	7.5	2.0	1.4	.998	1.00	0.988	298
4.46	7.0	2.1	1.4	.999	1.00	0.999	297
6.33	6.1	2.1	1.4	.999	1.00	0.999	295
8.16	5.2	2.1	1.4	.999	1.00	0.999	293
10.67	4.6	2.0	1.4	.999	1.00	0.999	290

It can be observed that the test results show similar trend as simulated results, thus validating the developed design procedure and simulation model of proposed 12-pulse and 24-pulse ac-dc converters.

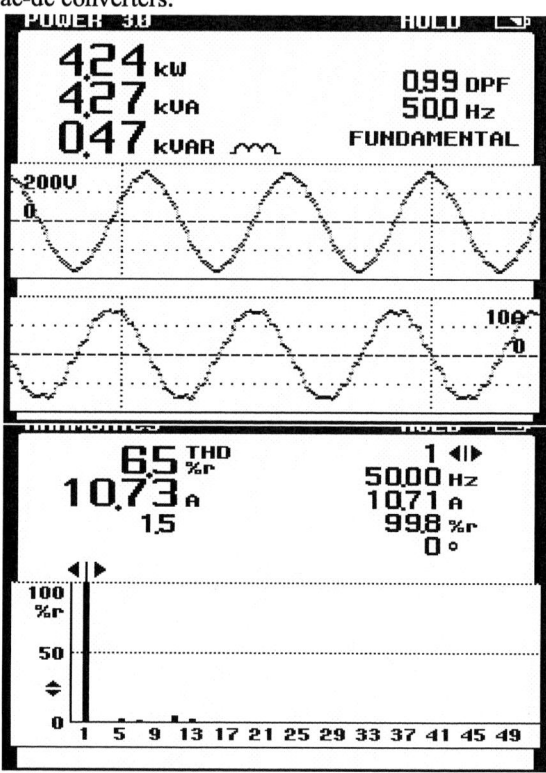

Fig. 15 Experimental results of AC mains voltage and current waveforms alongwith harmonic spectrum of ac mains current for Topology 'B' at full load.

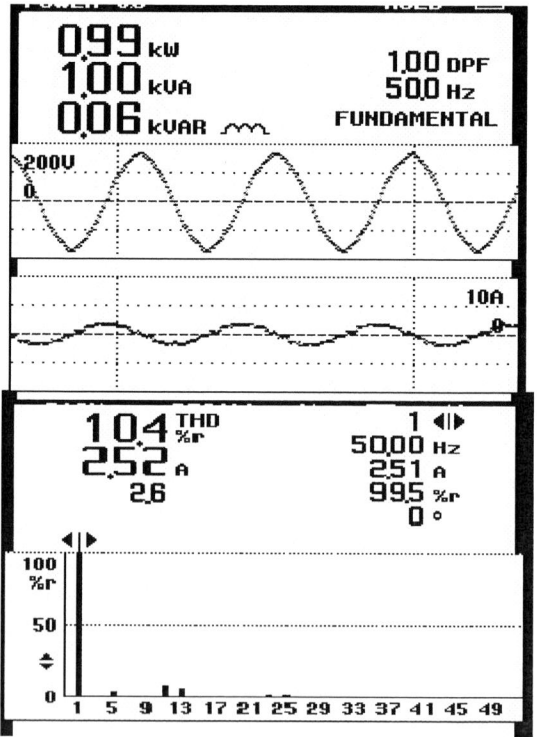

Fig. 16 Experimental results of AC mains voltage and current waveforms alongwith harmonic spectrum of ac mains current for Topology 'B' at light load (20% of full load).

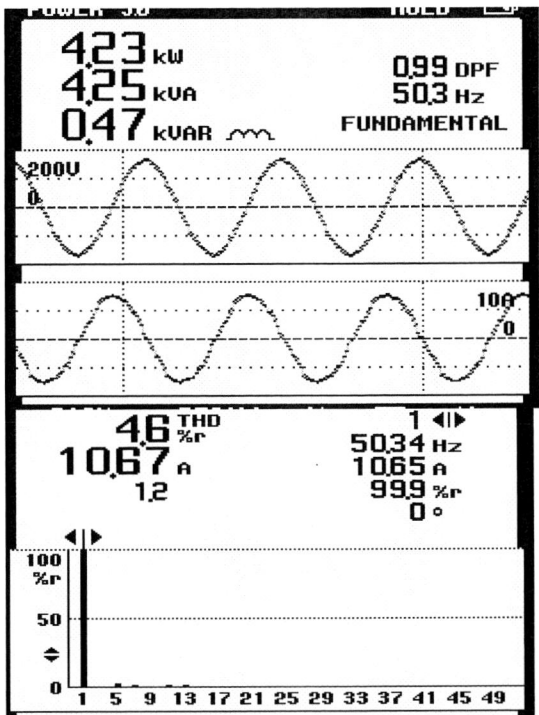

Fig. 17   Experimental results of AC mains voltage and current waveforms alongwith   harmonic spectrum of ac mains current for Topology 'D' at full load.

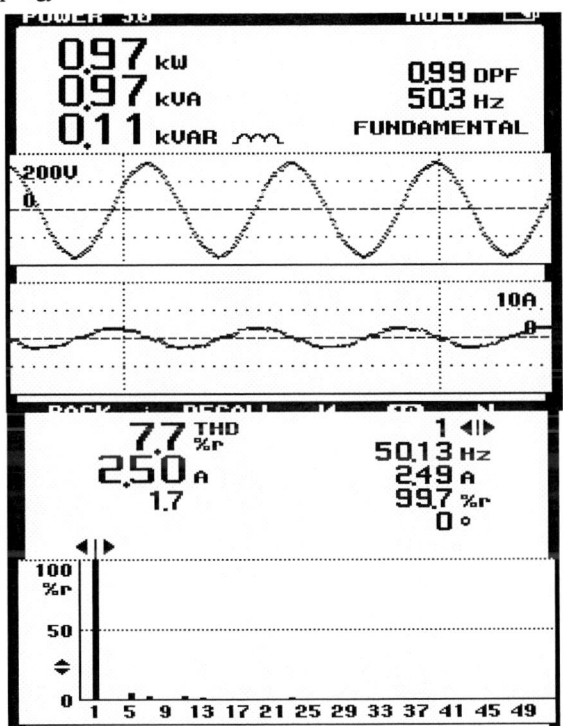

Fig. 18   Experimental results of AC mains voltage and current waveforms alongwith   harmonic spectrum of ac mains current for Topology 'D' at light load (20% of full load).

VIII. CONCLUSIONS

The design and modeling of a novel autotransformer based twenty four-pulse ac-dc converter with a VFIMD load has been presented for retrofit applications. Pulse multiplication has been achieved using dc ripple re-injection technique in the twelve-pulse ac-dc converter. The proposed twenty-four-pulse ac-dc converter has resulted in reduction in rating of the magnetics, leading to the saving in weight, size, volume and finally the overall cost of the drive. There has been remarkable improvement in the THD and rms values of ac mains current as well the power factor with almost close to unity power factor in the wide operating range of the drive.

IX. APPENDIX

Motor and Controller Specifications
    Three-Phase Squirrel Cage Induction Motor –10 hp (7.5kW), 3-Phase, 4 Pole, Y- connected, 415 V, 50 Hz, rated current = 14.5A, R_s = 1.0 ohms, R_r = 0.76 ohms, X_{ls} = 0.77 ohms, X_{lr} = 0.77 ohms, X_m = 18.84 ohms, J= 0.1 kg-m^2
PI Speed Controller: K_p = 7.0, K_i = 0.1
DC link parameters: L_d = 0.002H, C_d = 2200µF.
Magnetics ratings: Autotransformer Rating 1.33kVA, Interphase Transformer 0.13kVA, ZSBT Rating 0.588kVA..

X. REFERENCES

[1] P.Vas, *Sensorless vector and direct torque control*, Oxford University Press, 1998.

[2] *Limits for harmonic current emissions*, International Electrotechnical Commission Standard 61000-3-2, 2004.

[3] *IEEE Guide for harmonic control and reactive compensation of Static Power Converters*, IEEE Standard 519-1992.

[4] M.H. Shwehdi, A.H. Mantawy and H.H Al-Bekhit, "Solving the harmonic problems produced from the use of adjustable speed drives in industrial oil pumping field," *Proc. Int. Conf. Powercon*, Oct.2002, vol.1,pp. 86-92.

[5] D. A. Paice, *Power Electronic Converter Harmonics: Multipulse Methods for Clean Power*, New York, IEEE Press, 1996.

[6] D.A. Paice, "Multipulse converter system", U.S. Patent No. 4876634, Oct. 24,1989.

[7] P.W. Hammond, "Autotransformer", U.S. Patent No. 5619407, 8 April, 1997.

[8] D.A.Paice, "Transformers for multipulse AC/DC converters", U.S. Patent No. 6101113, 8 August 2000.

[9] S.Martinius, B.Halimi and P.A.Cahono, "A transformer connection for multipulse rectifier applications," *Proc. Int. Conf. Powercon*, Oct.2002, vol.2, pp. 1021-1024.

[10] G. R. Kamath, B. Runyan and Richard Wood, "A compact autotransformer based 12-pulse rectifier circuit," Proc. *IEEE IECON, Conf.* 2001, pp. 1344-1349.

[11] Steffan Hansen, Uffe Borup and Frede Blaabjerg, "Quasi 12-pulse rectifier for adjustable speed drives," *Proc. IEEE APEC Conf.* 2001, pp. 806-812.

[12]   G. R. Kamath, D. Benson and R. Wood, "A novel autotransformer based 18-pulse rectifier circuit," in *Proc., IEEE IECON'03, 2003*, pp. 1122-1127.

[13]   K.Oguchi and T.Yamada, "Novel 18-pulse diode rectifier circuit with non-isolated phase shifting transformers," IEE Proc. Electric Power Applications Vol.14, No.1, Jan.1997. pp.1-5.

[14] F.J.M.de Seixas and I.Barbi, "A 12 kW three-phase low THD rectifier with high frequency isolation and regulated dc output," *IEEE Trans. on Power Electronics*, Vol.19, No.2, March 2004, pp.371-377.

[15] D.A. Paice, "Simplified wye connected 3-phase to 9-phase autotransformer," U.S. Patent No. 6,525,951 B1, Feb. 25, 2003.

[16] Shota Miyairi, Shoji Iida, Kiyoshi Nakata and Shideo Masukawa, "New method for reducing harmonics involved in input and output of rectifier with interhase transformer," *IEEE Trans. on Industry Applications*, Vol.22, No.5, Oct. /Nov.1986, pp. 790-797.

[[17]   M.E.Villablanca and J.A. Arrilaga, "Pulse multiplication in parallel converters by multi tap control of interphase reactor," *IEE Proceeding-B*, Vol.139, No.1, Jan.1992, pp 13-20.

[[18]   V. Kumar, "Power quality improvements at ac mains in variable frequency induction motor drives," Ph.D. Dissertation, Indian Institute of Technology, Delhi, New Delhi, India, 2006.

Power Quality Monitoring System Using Real-Time Operating System

Krisda Yingkayun*, Suttichai Premrudeepreechacharn**, Kosol Oranpiroj*
*Department of Electrical Engineering, Rajamangala University of Technology Lanna,
128 Huaykeaw Road. T.Changpuek A.Muang Chiang Mai, Thailand 50300
**Department of Electrical Engineering, Chiang Mai University,
239 Huaykeaw Road T.Sutep A.Muang Chiang Mai, Thailand 50200

Abstract--This paper represents the methodology for monitoring continuously of power quality (PQ) by using real-time operating system (RTOS). The monitoring system acquired analog signals and send data via Ethernet network based on RTOS programs. After obtained signals, the monitoring system then send PQ information to the personal computer. The applied protocol is the user datagram protocol (UDP) to send PQ information via Ethernet. The architecture and software implementation are mentioned in the paper. An application test was created to test the throughput of the communication between the measurement point and the target computer.

Keyword-- power quality (PQ), real-time operating system (RTOS), user datagram protocol (UDP).

I. INTRODUCTION

During the recent year a concern with power quality is increasing. The problems of power quality such as voltage, current or frequency deviation cause a failure or malfunction of end-user equipment [1]. In order to monitor the power quality, it's essential to measure voltage, current, frequency, harmonic distortion and waveform. But power quality problems are not restricted to harmonic distortion. The standard IEEE 1159 classified various electromagnetic phenomena in power system voltage (which are related to power quality problems), namely: impulses, oscillations, sags, swell, interruptions, undervoltages, overvoltages, DC offset, harmonics, interharmonics, notches, noise, flicker and frequency variation [2]. Various researchers have implemented power quality measurements[3]-[9]. Some researcher applied data acquisition board to monitor power quality based on PC [4] or use DSP processor to monitoring power quality in real time [5].

Nowadays, Ethernet is widely use in embedded system, which applied to the microcontroller. Ethernet chip is low-cost and easy to apply. To implement the power monitoring system, which takes raw measurement data and additional information for power quality diagnosis, a power quality monitoring using Ethernet communication is proposed. The proposed main-architecture consists of AT89C51ED2 microcontroller from Atmel Corporation, ENC28J60 Ethernet chip, MCP3304 analog to digital converter (ADC) from Microchip Technology Inc. and DS1302 trickle-charge

timekeeping chip from Maxim Integrated Products. The program monitor can execute the tasks in real time based on real-time operation system (RTOS) programming. All measurement data transmitted to remote computer via Ethernet networks. Power quality information processed and stored in remote computer. The power monitoring hardware supported single-phase, three-phase voltage and current measurements. This paper suggest a power monitoring system that monitor the power quality continuously, which is measured at many places, easy to separate or combine power quality monitoring system depend on the number of remote computer. The computer received packets, which Ethernet chip send from measurement points and process power quality factors. However, this paper focused in the first part of the complete system. Fig. 1 shows the block diagram of PQ monitoring system.

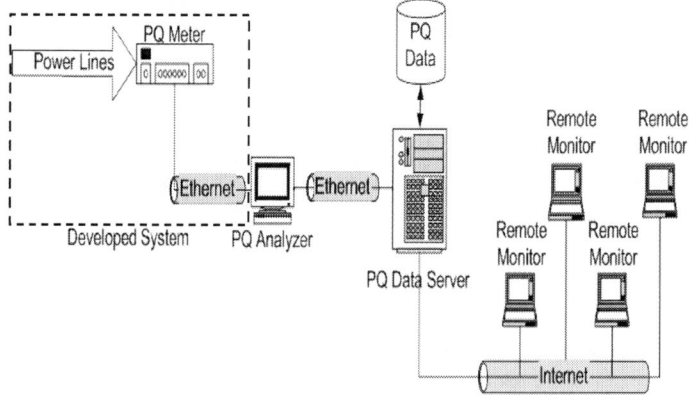

Figure 1. Block diagram of the PQ monitoring system

II. SYSTEM ARCHITECTURE

The power quality monitoring system is designed by considering the high-speed microcontroller, a data acquisition device using an ADC, signal-conditioning module, real time clock component and Ethernet chip. The monitoring system has several advantages such as high-accuracy measurements, low-cost and low power consumption. Fig. 2 shows the architecture of the PQ meter in terms of a block diagram. The prototype uses signal conditioning module to isolate AC signals from the monitoring system.

978-1-4244-0644-9/07/$25.00 ©2007 IEEE

The Microchip MCP3304 13-bit A/D converters feature full differential inputs and low power consumption in a small package that is ideal for battery powered systems and remote data acquisition applications. It's sampling rate up to 100 ksps with 5V supply voltage. The Atmel AT89C51ED2 is an 80C52 Compatible high-speed microcontroller up to 60 MHz. The Microchip ENC28J60 is a stand-alone Ethernet controller with SPI™ (Serial Peripheral Interface) interface, speeds up to 10 Mb/s and compatible with standard IEEE 802.3. The DS1302 Trickle Charge Timekeeping Chip contains a real time clock/calendar. It communicates with a microprocessor via a simple serial interface. The real time clock/calendar provides seconds, minutes, hours, day, date, month, and year information, which represent timebase.

After sampling A/D data, read time and date into memory. The microcontroller packed all measured data included time and date from memory into UDP packet, then send to remote computer. The computer received the UDP packets and the prototype hardware controlled by computer. Fig. 3 shows the block diagram of the PQ meter. All sampling data transmitted to the remote computer. The remote computer stored the incoming data to harddisk or external storage.

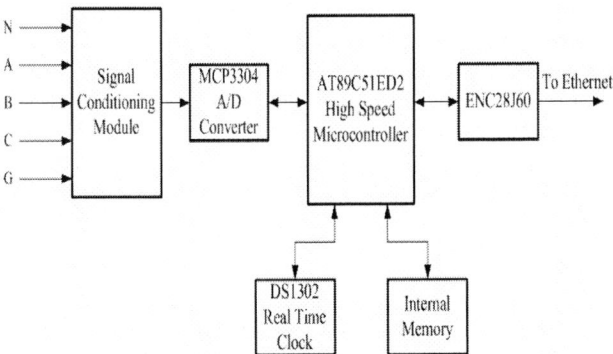

Figure 2. Block diagram of the PQ meter

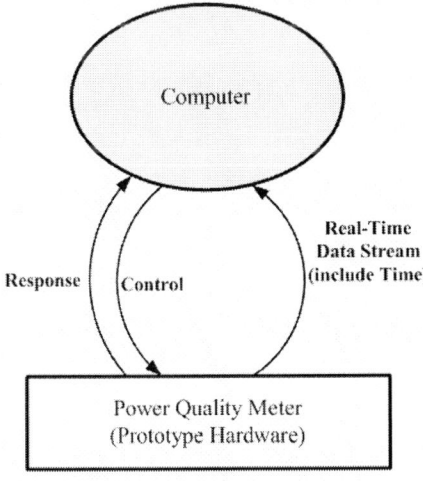

Figure 3. Intercommunication principal

III. SOFTWARE DESIGN

A. Program Types

Various programmers using multitask programming to control the tasks. However, the heart of programming based on RTX-51 real time kernel from keil software. RTX51 is a real-time kernel for the 8051 family of microcontrollers that is designed to solve two problems common to embedded programs as follows:

1) Multitasking: several operations must execute simultaneously.
2) Real-time control: operations must execute within a defined period of time.

Fig. 4 illustrated the example flowchart of multitask programs. The functions (or tasks) are executed in order, one after another. For example, if the task3 executes for a long time, the loop may take to long to get back around to the task2. If the task2 receives data and data may be lost. Of course, the task2 may be called more often in the loop to correct this issue, but eventually this technique will not work.

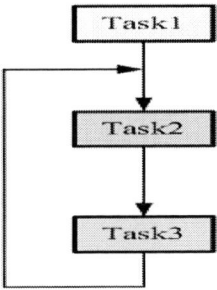

Figure 4. Example block diagram of multitask programs

The multitask programs written in keil C51, which compared with RTOS programs in term of speed and performance. For example:

```
void main (void)
{
   serial_init( );
   do{
       get_adc( );
       get_time( );
       pack_data( );
       ethernet_send( );
       check_serialport( );
       process_command( );
   }while(1);
}
```

Fig. 5 illustrated the system flowchart of RTOS programs, which applied in project. The tasks (or programs) managed by the task switcher. The task switcher is the RTOS program which executed the created tasks.

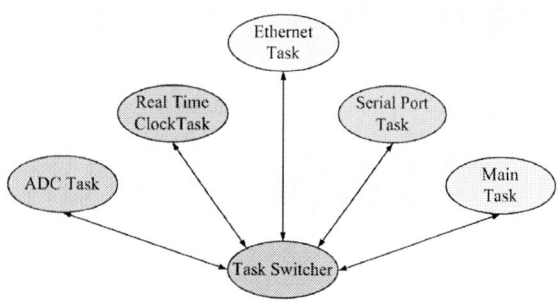

Figure 5. System flowchart of RTOS programs

Each task executed simultaneously and handled by RTOS programs. RTOS programs that written in keil C51.
For example:

```
void ethernet_task (void) _task_ 1
{
/* This task processes Ethernet communication */
}
void serial_task (void) _task_ 2
{
/* This task processes serial port communication */
}
void time_task (void) _task_ 3
{
/* This task processes date and time */
}
void atod_task (void) _task_ 4
{
/* This task processes analog to digital signals */
}
void startup_task (void) _task_ 0
{
os_create_task (1);   /* Create ethernet Task */
os_create_task (2);   /* Create serial port Task */
os_create_task (3);   /* Create date and time Task */
os_create_task (4); /*Create analog to digital signal Task */
os_delete_task (0);   /* Delete the Startup Task */
}
```

B. Package structures

To defines the stream of data that send over the Ethernet networks. The stream consists of sampled A/D data and timebase, which included to UDP data and then the Ethernet task packed into packets and send to the remote computer. The structure of the stream as the following shows.

```
struct datetime  /* Date and Time Structure */
{
  char Hundred;
  char Second;
```

```
  char Minute;
  char Hour;
  char Date;
  char Month;
  char Year;
};
struct adc_channel  /* Signals structure */
{
    int ch0;
    int ch1;
    int ch2;
    int ch3;
    int ch4;
    int ch5;
    int ch6;
    int ch7;
};
struct ethernet_stream /* Ethernet Data Structure */
{
    struct adc_channel ad;
    struct datetime dt;
};
```

C. Protocol packet design

Power quality meter is transmitted in UDP protocol packet [10], which the measurements data packed and carried on Ethernet networks. UDP is a protocol within the TCP/IP (Transmission Control Protocol/Internet Protocol) [11] protocol suite that is used in place of TCP when a reliable delivery is not required. There is less processing of UDP packets than there is for TCP. UDP is widely used for streaming audio and video, voice over IP (VoIP) and videoconferencing, because there is no time to retransmit erroneous or dropped packets. Fig. 6 shown the UDP packet within Ethernet frame.

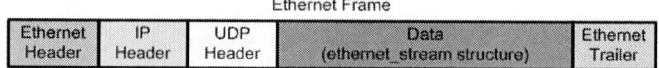

Figure 6. UDP within Ethernet frame

All measurement data packed within Ethernet frame and transmitted trough the Ethernet networks.

IV. EXPERIMENTAL RESULT

First of all, the monitoring hardware must be setting up the time to the real-time clock chip and assign the network configuration to the Ethernet chip via serial port. After setting up, the monitoring hardware is ready to measurement. The monitoring system sends the Ethernet packet out to the remote computer. The first experiment used the multitask programs to controlled the operation of the monitoring hardware and another experiment used the RTOS programs. The processes were executed in following steps.

1. Sampling 8-channel analog to digital signals.
2. Read date and time data.
3. Send out the Ethernet packets to the remote computer.

Fig. 7 shows program the incoming packets that sent from the monitoring hardware.

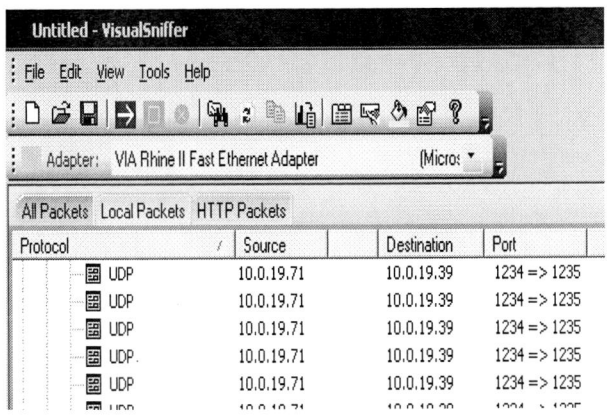

Figure 7. Example of incoming packets to PC

Fig. 8 shows the packet contents of UDP protocol, which applied for data transmission.

```
0000 : 00 01 29 4b 65 3b 00 50 22 00 4a ec 08 00 45 00
0001 : 00 34 6f a4 00 00 80 11 90 a7 0a 00 13 47 0a 00
0002 : 13 27 04 d2 04 d3 00 20 dc 13 e8 03 f7 03 d0 07
0003 : ef 09 54 07 34 ff da 06 49 fe 50 47 28 17 15 04
0004 : 07 00
```

Figure 8. UDP packet contents

After testing the function by counting the packets. Table 1 presents the experimental results of multitask program and RTOS programs.

TABLE 1
THE EXAMPLE OF EXPERIMENTAL RESULT

Experiment Number	Number of Packets (Multitask)	Number of Packets RTOS
1	67	113
2	66	110
3	64	108

As shown in Table I. It is confirmed that data is transmitted, which using RTOS programs can transmits data faster than using multitasking program. All the experiments are based on identical hardware.

V. CONCLUSION

In this paper, power quality monitoring system using real-time operating is a method which separate functions (or tasks) of the monitoring system independently. The benefit of RTOS programs, which executes the multiple-tasks simultaneously. The experimental results clearly show this. The development hardware can operate functionally. The next phase of the monitoring system based on PC, which manipulate all packets converted to power quality analysis.

REFERENCES

[1] Dugan R.C., *Electric Power Quality,* 2nd ed. Scottsdale AZ: Stars in a Circle Publication, 1996.
[2] R. C. Dugan. M. F. Macgranaghan, H. W. Beaty, *Electrical Power System Quality*, MacGraw-Hill, 1996.
[3] L.F. Auler and R. d'Amore, "Power Quality Monitoring and Control Using Ethernet Networks," *IEEE International Conference on Harmonics and Quality of Power, 10th*, vol. 1, pp. 208-213, 2002.
[4] J. Batista, J. L. Alfonso, and J. S. Martins, "Low-Cost Power Quality Monitor Based on a PC," *IEEE International Symposium on Industrial Electronics.* vol. 1, pp 323-328, 2003.
[5] A. Rahim bin Abdullah and A. Zuri bin Sha'ameri, "Real-Time Power Quality Monitoring System Based On TMS320CV5416 DSP Processor," *IEEE International Conference on Power Electronics and Drive Systems*, vol. 2, pp. 1668-1672, November 2005.
[6] D. Hong, J. Lee, and J. Choi, "Power Quality Monitoring System Using Power Line Communication," *IEEE International Conference on Information, Communications and Signal Processing*, 5th, pp. 931-935, December 2005.
[7] A. So, N. Tse, W. L. Chan, and L. L. Lai, "A Low-Cost Power Quality Meter for Utility and Consumer Assessments," *IEEE International Conference on Electric Utility Deregulation and Restructuring and Power Technologies*, pp. 96-100, April 2000.
[8] G. H. Yang, and B. Y. Wen, "A Device for Power Quality Monitoring Based on ARM and DSP," *IEEE Conference on Industrial Electronics and Applications*,1st, pp. 1-5, May 2006.
[9] M. E. Salem, A. Mohamed, S.Abd. Samad, and R. Mohamed, "Development of a DSP-Based Power Quality Monitoring Instrument for Real-Time Detection of Power Disturbances," *IEEE International Conference on Power Electronics and Drives Systems*, vol. 1, pp. 304-307, January 2006.
[10] User Datagram Protocol, RFC 768.
[11] Transmission Control Protocol, RFC 793.

Technology Performance Comparison of Triacs Subjected to Fast Transient Voltages

L. Gonthier, A. Passal

STMicroelectronics, IMS / ASD & IPAD, Rue Pierre et Marie Curie – BP 7155, France

Abstract – **This paper presents an experimental comparison of several Triac devices under immunity tests, as described in the IEC 61000-4-4 standard. After a short reminder of the different Triac technologies available today (TOP, MESA and PLANAR technologies), the IEC 61000-4-4 test procedure to compare the devices is explained. The immunity results are discussed according to the devices' technology and the gate current sensibilities. A discussion about relevance of dV/dt parameter and die area is carried on to differentiate the devices in term of immunity capability.**

I. INTRODUCTION

Home appliances such as washing machines, refrigerators and dishwashers integrate a lot of low power loads such as valves, door lock systems, dispensers and drain pumps. These loads are usually powered by the mains in ON / OFF mode, and are mostly controlled by Triacs.

The direct connection of the silicon switch to the mains, through the load, requires that these devices must withstand the line transients to make the system compliant with the international Electromagnetic Compatibility (EMC) standards. The silicon devices are then subjected to fast transient voltages, as described in the IEC/EN 61000-4-4 standard.

The immunity of several Triac technologies is evaluated here experimentally. Several guide-lines can then be pointed out to design high immunity appliances.

II. TRIAC TECHNOLOGIES AND IMMUNITY

A. Triac silicon structures

Triac devices are by far the most used silicon devices to drive AC loads directly connected on the mains voltage. Triacs are widely used because they present three major advantages:
1. Low forward voltage drop
2. Turn-on by gate current pulse
3. High voltage immunity

Advantages 1 and 2 come from the active silicon structure of the Triac which is based on four alternatively doped silicon layers (N-P-N-P). These four layers implement two bipolar transistors which are coupled to maintain ON state even if the gate current is removed (cf. Fig. 1).

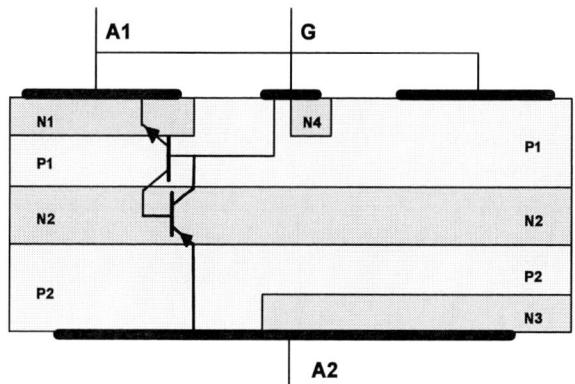

Fig. 1. Active simplified structure of a Triac

Advantage 3 comes from the fact that the voltage is withstood by the silicon thickness. It's then quite easy to reach breakdown levels above 1 kV. The main issue is then the junction termination. Several low cost technologies are used since the 70's which used a glass passivation. For example the "MESA" technology (cf. Fig. 2) uses glass on both sides of the die [1][2][3] to passivate both junctions, which hold both forward and reverse high voltages (referred to "Cathode" or "A1" or "COM" polarity). The "TOP" glass technology [1][2][3] uses only one glass layer (cf. Fig. 3). This avoids having some glass at the bottom of the die, and so it is easier to put such dies in every kind of package.

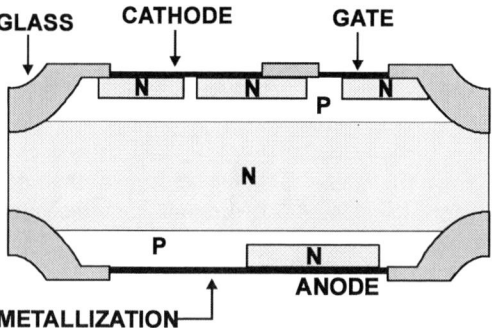

Fig. 2. "MESA" glass technology

978-1-4244-0644-9/07/$25.00 ©2007 IEEE

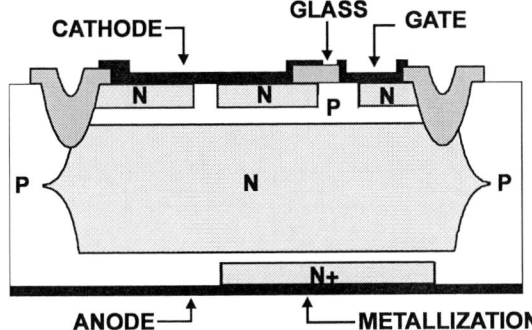

Fig. 3. "TOP" glass technology

Both TOP and MESA glass technologies are very cost effective. Indeed, the voltage capability of glass is very high; this helps to achieve high voltage devices with a limited periphery area. For sure, MESA glass technology is the cheaper one as this technology uses less silicon area to withstand reverse voltage. For TOP technology, the reverse PN junction is terminated on the upper side on the die thanks to the deep P well. Such dies are bigger for the same active area than MESA dies. This is the reason why TOP technology is mainly used for low current Triacs.

Unfortunately it is not possible to ensure a good operation of die with glass passivation when the voltage exceeds its maximum allowed value (VDRM or VRRM parameters). Indeed, if the voltage reaches the breakdown value, a current will flow through the die periphery, causing heat dissipation between the glass-silicon interface. This heat could cause mechanical stress and damage this interface. The device could then be damaged.

To develop switches able to work up to their breakdown voltage, a planar technology has to be implemented. Such a technology uses photolithography to terminate the PN junction at the top of the die, and oxide passivation instead of glass (cf. Fig. 4). There is no more glass-silicon interface issue. ACST devices use this kind of technology [4].

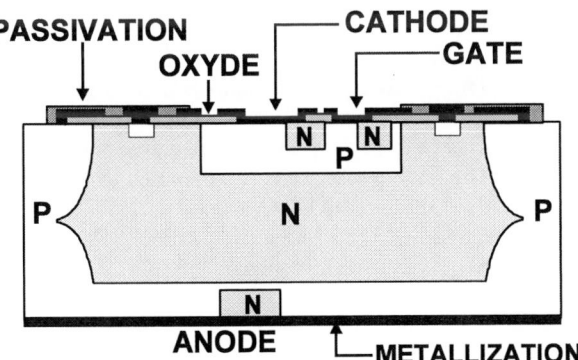

Fig. 4. PLANAR technology

B. Sensibility and dV/dt immunity

Our experimental comparison on immunity level has been performed on different devices which target the same application. This application is the control of valves or pumps in white goods. The load current is usually less than 1 A. Anyway, appliances designers use, to control these loads, 1 A, 2 A and even 4 A Triacs. The use of a high current rating device is usually motivated by desire for highly immune and robust systems.

To rationally compare the different devices, we have selected devices with the same voltage and sensibility ratings. The voltage rating is given by the VDRM and VRRM parameters (800 V in this paper). The sensibility is given for Triacs by the Igt parameter which is the minimum gate current that should be applied in order to turn-on the device. The devices we have chosen all have a 10 mA maximum Igt.

The Triac sensibility can not be reduced too much. Indeed, an appliance designer has to work with the sensitivity (supply consumption) and immunity compromise. If the Igt is too low, the device will not be able to withstand too high dV/dt rates across its power terminals (A1 and A2 for Triacs, or OUT and COM for ACST). Triac manufacturers then give a dV/dt parameter which is measured at the maximum junction temperature. Results of Table 1 are given for dV/dt tests performed at a 125° C junction temperature and a 400 V applied voltage. It should be noted that some values are lower than values given in constructors datasheet for the devices which are specified at 110° C instead of 125° C. This is a normal result as the dV/dt immunity decreases with the temperature.

Table 1 gives the Igt and dV/dt parameters that we measured for the different samples we used during IEC 61000-4-4 tests, and for the different bias polarities.

The different selected P/N are:
- the Z0409N: a TOP glass 4 A Triac
- the T410-800: a MESA glass 4 A Triac
- the ACST2: a PLANAR 2 A Triac
- the COMP.1A: a PLANAR 1 A Triac
- the COMP.2A: a PLANAR 2 A Triac

P/N	Techno.	Sample	Igt (mA)			dV/dt (V/µs) @ 125° C	
			Q1	Q2	Q3	Positive	Negative
Z0409N	TOP	1	1.02	3.37	4.69	118	94
		2	0.90	2.90	3.49	140	135
		3	1.02	2.99	4.08	120	95
		4	0.98	2.72	3.69	160	138
T410-800	MESA	1	1.57	5.82	3.43	200	450
		2	1.67	7.14	3.69	180	710
		3	1.52	5.43	3.31	180	525
		4	1.65	6.99	3.64	170	630
ACST2	PLANAR	1	3.54	6.20	6.83	>5000	>5000
		2	3.48	6.20	6.82	>5000	>5000
		3	3.45	6.20	6.83	>5000	>5000
		4	3.47	6.20	6.73	>5000	>5000
COMP.1A	PLANAR	1	4.89	4.05	5.10	500	470
		2	4.85	4.28	5.22	580	500
		3	4.86	4.05	5.05	530	430
		4	4.85	4.16	5.34	540	560
COMP.2A	PLANAR	1	3.54	6.20	6.83	>5000	>5000
		2	3.48	6.20	6.82	>5000	>5000
		3	3.45	6.20	6.83	>5000	>5000
		4	3.47	6.20	6.73	>5000	>5000

TABLE 1:   Measured Igt and dV/dt parameters of tested samples
(Igt is measured with standard 30 Ω/ 12 V circuit and dV/dt with standard test circuit: open gate and 400 V applied peak voltage)

III. IEC 61000-4-4 TESTS

A. Test procedure

To compare the immunity capabilities of the different technologies, we have performed tests as described in the IEC 61000-4-4 standard [5]. European appliance manufacturers usually require that their equipment will not present any malfunction for burst levels up to 2 kV.

Here, we increase the burst level by 0.1kV steps to reach at least one device spurious triggering within one minute. To easily compare only the Triacs performances, we used a dedicated board featuring all the AC switches, and their gate connected to their A1 terminals through gate resistors. This allows us to be sure that the spurious triggering comes from the switch itself and not from any control circuit. The spurious triggering is detected when the light bulb, connected in series with one of the switch, is switched on. Indeed, as a Triac will latch, usually the on time will last several ms. This is well long enough for a human eye to detect it.

The tests are carried out in the following conditions:
- Printed circuit board 10 cm above reference plane.
- A mains S14K300 varistor (refer to "V" on Fig. 5) is connected to the mains input on the PCB
- The board embeds four Triacs (or ACSTs).

- Each Triac-A2 (or ACST OUT) terminal is linked to a 25 W light bulb (resistive loads are chosen to get easier device turn-on as dI/dt is higher than with inductive load).
- Each gate is connected to A1 or COM terminals respectively, for Triac and ACST, through a 220 Ohm resistor (refer to "Rg" on Fig. 5) to be free of spurious firings coming from any control circuit.
- No snubber circuits are added across the Triacs or ACSTs.
- Ambient temperature: 25° C
- The board is plugged into a L-N plug which is disturbed by a burst generator (bursts are coupled to N or to L, for positive or negative bias)
- The burst generator is programmed as required in the IEC 61000-4-4 standard (15 ms burst duration, 3 Hz burst frequency, 5 kHz spikes frequency, one minute test duration).

N.B: only results with coupling modes to L and N are presented here as the other coupling modes are usually less stressful. For example, a simultaneous coupling of the burst to L, N and the PE (Protective Erath, which is not connected to our board) will lead to the less stressful test.

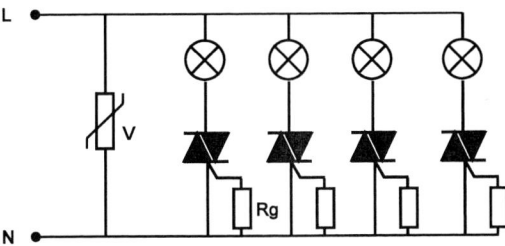

Fig. 5.   IEC 61000-4-4 test schematic

B. Impact of a varistor on mains input

The IEC 61000-4-4 standard as been developed mainly to protect equipment from fast voltage transients coming from bad mains connection, electrostatic discharge or inductive load disconnection. Designers usually think that the impact of the burst is mainly due to the high dV/dt, and so due to the capacitive currents circulating through all parasitic capacitors. For sure, this effect is one of the major ones, by applying bursts can also cause some spurious device turn-on due to too high voltage across the switches. For example, Fig. 6 shows the voltage and current waveforms of an ACST2 used in the previously described board but without any varistor, for a 1 kV burst. We see that when the voltage spikes are applied at peak mains voltage, the device turns on, and turns off for the next ZCS (Zero Current Switching) point. If we zoom in this oscillogram with a 0.4 µs/div time scale, we clearly see that the triggering comes from an overvoltage. Indeed, Fig. 7 shows that the voltage across the ACST2 reaches more than 800 V, and then the device turns on in breakover mode.

Such a turn-on can be allowed with ACST2 devices, contrary to other technologies and even contrary to other planar devices from other companies. For ACST2 there will not be any risk of device failure, as long as the applied current remains below the guaranteed limit. But, applying such bursts without any voltage protection on mains input could lead to devices failure with TOP or MESA glass technologies. This is the reason why we put a varistor on the board we used for IEC 61000-4-4 tests.

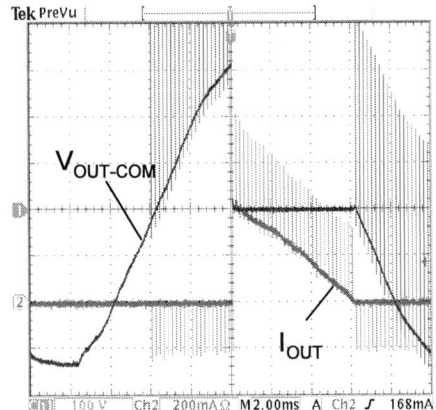

Fig. 6.   ACST2 behaviour during 1 kV burst
(without input varistor)

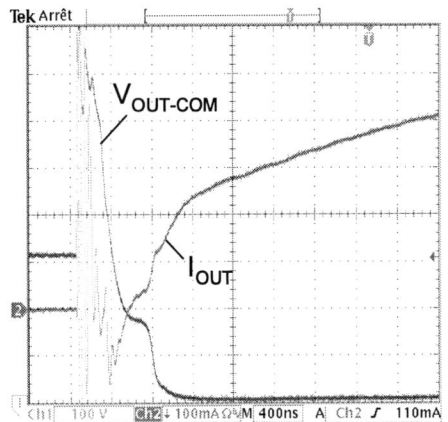

Fig. 7.   ACST2 turn-on zoom
(without input varistor)

C. technology impact on immunity levels

Immunity tests results according to P/N are given in Fig. 8. This figure gives the maximum burst levels that the different devices were able to withstand before turn-on. We see that most of the devices present highly dispersed results. Even if these tests have been performed with devices with a low dispersion on Igt levels (for a same P/N, refer to Table 1), their immunity levels can vary with a 1 to almost 3 scale. The coupling mode and burst polarity also have an impact on the immunity capability of a sample. This impact can vary between 20 to 60%. To conclude, only the minimum value of the "maximum burst level" range, given in Fig. 8, has to be taken into-account. Indeed, this minimum level corresponds to the worst-case operation among all the different coupling cases covered by the IEC 61000-4-4 standard.

It should be also noted that the maximum capability of our burst generator is 4.5 kV. That means that we didn't succeed in reaching the maximum immunity level of most ACST2 devices.

These results give several interesting information:

- The PLANAR technology allows the highest level to be reached, but the different companies don't reach the same level.
- ACST2 supports levels two times higher that its closest competitor (COMP.2A).
- The 2 A device (COMP.2A) in planar technology is approximately as immune as the 4A Triac in MESA glass technology.
- A device which presents a good dV/dt immunity (around 500 V/us for COMP.1A) can only be as immune as a 4A TOP device with a dV/dt capability close to 100 V/us

About the last point, it should be noted that the COMP.1A device presents a die area which is approximately 20% lower than the T410-800. And generally speaking it can be said that the bigger the die area is, the higher the immunity level is. Only the ACST2, which presents the highest immunity level, presents also the highest "immunity-level-to-die-area" ratio.

325

Indeed, its die area is almost two times smaller than the T410-800 for a two times higher immunity.

Compared to COMP.2A, this "immunity-area" ratio is also 40% higher. To reach approximately the same immunity level (2.4 KV in average, compare to 3 kV with ACST2), a 47 nF - 100 Ohm snubber circuit has to be added across the COMP.2A device.

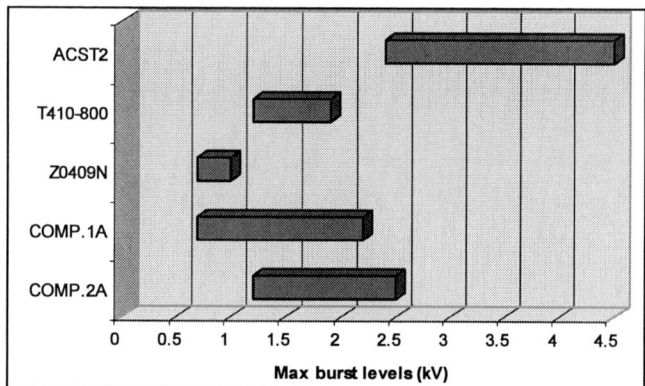

Fig. 8. P/N rankings towards IEC 61000-4-4 tests

D. Sensibility impact

According to previous results, it is difficult to conclude anything about impact of device sensibility on immunity. We have even seen that some devices with good dV/dt capability can present poor immunity level. But, for sure, such conclusion is difficult because we have compared here components from different processes.

To better analyze the sensibility impact, we have also performed some tests of other Triacs coming from the same process. We took some T405-800 devices which are manufactured with the same process as the T410-800. The T405-800 device is guaranteed to present Igt currents less than 5 mA. The samples we've tested present measured Igt levels approximately two times lower than the T410 devices we've tested, and this for the three quadrants. It means that we measured an Igt level of approximately 0.75 mA, 2.30 mA and 1.75 mA respectively for quadrants 1, 2 and 3. The maximum burst level that these devices were able to withstand was around 1 kV, i.e. 30% lower than the T410-800. This result highlights that sensibility has also a strong impact on immunity, as is well known. But this statement is only valid if one compares two devices from the same process.

V. CONCLUSION

This paper has given experimental benchmarking analysis among different alternating current switch offerings. The impacts of silicon process technologies and sensibility levels have been discussed. It has thus been shown that sensibility and dV/dt parameters are not relevant enough to evaluate the immunity rating of different devices. A lower Igt will lead,

for sure, to a less immune device but only if we compare devices from a same silicon process. It has been furthermore shown that the die area has also a big impact on immunity. Indeed, a bigger die area will lead to a higher device capacitance. This will help to absorb energy coming from the bursts, and thus reducing the dV/dt across the device. On the other hand, when dV/dt immunity of Triacs is measured, the voltage rate is forced across the device. So a bigger die area will lead to a higher capacitive current. So a bigger device, with same sensitivity level, will withstand a lower dV/dt level.

PLANAR technology seems to offer devices with the highest immunity. However, Z04 and T410 devices, respectively produced with a TOP and MESA technologies, offer the same minimum level of immunity as the tested PLANAR devices. ACST2 presents the highest immunity level compared to other technologies or compared to other planar devices from other companies. ACST2 can then be used without any snubber circuit with a high immunity to bursts and is the solution for applications requiring a high immunity level.

VI. REFERENCES

[1]. J. Arnould, P. Merle, "Dispositifs de l'Electronique de Puissance", Vol. 2, HERMES edition, 1992, ISBN 2-86601-308-5.

[2]. STMicroelectronics, "SCRs, Triacs and AC Switches Quality and Reliability", SCRs, Triacs and AC Switches databook, 4th Edition, 2001.

[3]. B. Morillon, "Study of Aluminum Thermomigration in Silicon for the Industrial Implementation of Isolation Walls in Bidirectioanl Power Devices", PhD Dissertation, Institut National des Sciences Appliquées de Toulouse, Toulouse (France), July 2002.

[4]. STMicroelectronics, "ACST2 series – AC Switch family", datasheet, Rev.1, March 2007.

[5]. L. Gonthier, "ACS and Triacs Comparison towards Fast Voltage Transients", conf. Power Conversion Intelligent Motion, PCIM 2000, Nuremberg, Germany, 2000, Proc. CD-ROM.

On-line Junction Temperature Measurement of CoolMOS Devices

Andreas Koenig, Thomas Plum, Peter Fidler and Rik W. De Doncker
Institute for Power Electronics and Electrical Drives
RWTH Aachen University
Jaegerstrasse 17-19, 52066 Aachen, Germany
Phone: +49 241 8096959
Email: ak@isea.rwth-aachen.de

Abstract— To operate power electronic devices at high ambient temperatures it has to be ensured that the maximum specified junction temperature is not exceeded at any time during operation. This paper presents a method to calculate the actual junction temperature by measuring voltage and current at a power MOSFET during operation of a converter. Based on this temperature, the actual transferred power of a converter can be controlled to ensure a safe operation within the specified temperature limits.

I. INTRODUCTION

Nowadays, power electronic devices are operated at high temperatures - either to operate electronic systems at high ambient temperatures (e.g. geothermal exploration, power systems, combustion engines, etc.) or to minimize the volume of the heat sink. In both cases it has to be ensured that the junction temperature remains below the maximum specified junction temperature. Operating the devices at a higher junction temperature could result in a reduced reliability or even in a thermal runaway of the semiconductor which finally leads to the destruction of the device.

To prevent these failure modes, there are several methods to ensure a safe operation. One possibility is to measure the heatsink temperature and based on these data the junction temperature is calculated using a thermal model of the system. The disadvantage of this method is that the result depends on the accuracy of the system model. Furthermore, the influence of a fluid based direct cooling of the semiconductors (e.g. by operating the devices in an insulating liquid) is very difficult to consider in the model. Another possibility is to oversize the heatsink to ensure a safety margin leading to an increased volume of the system.

To avoid these problems, a direct measurement of the junction temperature during operation would be advantageous. If the actual temperature is available, the control unit of a power converter could regulate the actual transmitted power in dependence of the junction temperature. In general, by reducing the transmitted power the power loss in the semiconductor devices is also reduced which leads to lower thermal losses and consequently a lower junction temperature. This control method enables the converter to operate at a lower power level when the maximum junction temperature is reached instead of switching off the whole system. A known approach to measure the temperature of the semiconductor devices is the inspection of an opened power module with an infrared camera [1]. Since infrared cameras are very expensive and bulky and operating a converter with open modules is problematic, this method can only be performed in a laboratory environment. Furthermore, this method is not applicable for encapsulated semiconductor devices (such as TO247, etc.).

In this paper, an innovative method is proposed which enables temperature sensing of a CoolMOS chip by only measuring electrical parameters of the device.

II. FUNDAMENTALS

In several applications MOSFETs are employed. In 1999, CoolMOS devices appeared in the market. These devices belong to the group of superjunction MOSFETs and can be characterized by a fast switching behavior und very low on-state losses [2], [3]. To evaluate the junction temperature of a device, a temperature dependent characteristic can be taken into account [4]. The idea of the proposed junction temperature measurement method is to determine the junction temperature by measuring the on-state resistance during operation. As depicted in Fig. 1, the on-state resistance strongly depends on the junction temperature and changes its value between $20°C$ and $120°C$ by a factor of about two. If this value is available for the control circuitry in a converter, the transferred power could be controlled depending on the junction temperature.

The on-state resistance of a MOSFET device comprises of the sum of several resistances (e.g. resistance of the package, source layer, channel, accumulation layer, n^--Layer and substrate). The composition of these resistances on the total on-state resistance for a power MOSFET is depicted in Table I. The composition of the resistances of a CoolMOS device is similarly.

It can easily be recognized, that the on-state resistance is mainly determined by the resistance R_{epi} of the n^--

978-1-4244-0644-9/07/$25.00 ©2007 IEEE

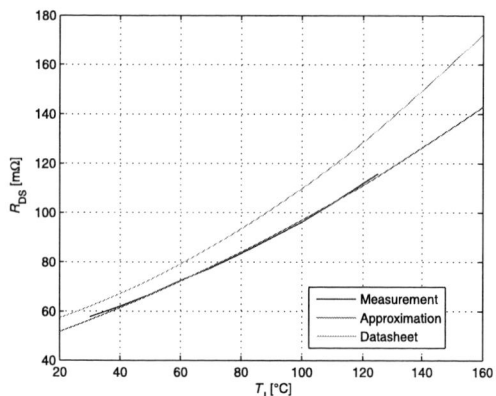

Fig. 1. On-state resistance of a SPW47N60C3 CoolMOS manufactured by Infineon

Variable	Description	Value
R_S	package	0,5 %
R_{n+}	source layer	0,5 %
R_{CH}	channel	1,5 %
R_a	accumulation layer	0,5 %
R_{epi}	n^--layer	96,5 %
R_{sub}	substrate	0,5 %

TABLE I

COMPOSITION OF THE R_{DS} OF A 600V POWER MOSFET [3]

layer. The resistance is antiproportional to the mobility of electrons μ_n:

$$R_{epi} \propto \rho \qquad (1)$$

$$\rho = \frac{1}{q\mu_n(T)N_D} \qquad (2)$$

Thereby determines ρ the specific resistance, q the elementary charge, T the temperature and N_D the donor concentration. The only temperature dependent variable in this equation is the mobility μ_n. Taking the following temperature dependent mobility model into account, the mobility can be calculated to [5]:

$$\mu_n(T) = \mu_n(300K) \left(\frac{T}{300K} \right)^{-x} \qquad (3)$$

Recapitulating, the resistance R_{epi} is proportional to:

$$R_{epi} \propto \left(\frac{T}{300K} \right)^x \qquad (4)$$

If the value x is set to 2.6 by a curve fitting procedure the resistance R_{epi} equals the measured value of R_{DS} with a very good accuracy. In Fig. 1 the measured on-state resistance R_{DS} and the approximated curve are depicted. Furthermore, the specified on-state resistance specified in the datasheet is given.

In this frist approach the influence of the channel resistance is neglected. In order to take the influence of this resistance into account the dependency of the current on the on-state resistance has to be included.

III. MEASUREMENT PRINCIPLE

The goal of the measurement principle is to determine the on-state resistance during the normal operation of the converter. This can be realized by simultaneously measuring U_{DS} and I_D while the MOSFET is switched on. As shown in Fig. 1, the resistance value corresponds to only one junction temperature. The calculated resistance must be compared with a look-up table in order to evaluate the corresponding junction temperature.

With this measurement method only an averaged temperature of the complete semiconductor can be determined. Hot spot temperatures can not be detected.

A. Hardware

Since the drain-source voltage U_{DS} of a MOSFET normally amounts several hundred volts during blocking and only several hundred millivolts during the conduction period, measuring this voltage is quite a challenge and requires special measurement circuits. The goal of this circuit is to cut off all voltages over a certain limit of several volts (e.g. 5 V).

In this approach the topology depcited in Fig. 2 was chosen: This topology is basically an emitter follower [6], [7]. Its main components are a detector MOSFET (T) and a resistor R_1. This circuit cuts off all voltages above a certain value of only a few volts. The gate connector of the detector MOSFET (T) is connected to a constant voltage of 9 V DC. During blocking of the power MOSFET (marked in Fig. 2 as DUT), the detector MOSFET (T) is almost switched off. During on-state of the power MOSFET (DUT), the detector MOSFET (T) is turned on and the voltage at the resistor R_1 equals the drain-source voltage of the power MOSFET (DUT).

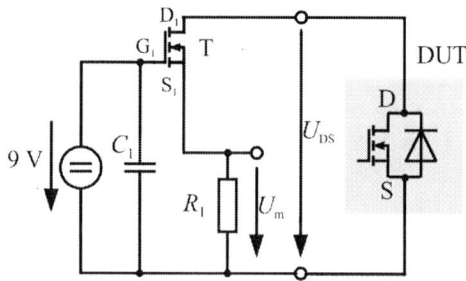

Fig. 2. Emitter-follower to measure U_{DS} of a MOSFET

Using this circuit, it can be ensured that the voltage U_m never exceeds an upper limit of several volts (which equals approximately the gate voltage of the detector MOSFET). The voltage U_m can now be connected to a DSP or microcontroller and can be measured very accurately without overdriving any operational amplifiers. Without this auxiliary circuit, an accurate measurement of the voltage would not be possible.

To design this measurement circuit it has to be ensured that the basic behaviour and the electric characteristics of the power MOSFET (DUT) are not influenced. Therefore a fast detector MOSFET device (T) with low parasitic capacitances (e.g. drain-source capacitance) has to be chosen. Certainly it must also be specified for the desired operating voltage. To ensure a safe operation of this device it should be characterized by the same blocking voltage as the power MOSFET (DUT).

The resistance R_1 should be large enough to limit the power dissipation within the detector MOSFET (T). During blocking of the power MOSFET (DUT) the detector MOSFET (T) is almost completely turned off. That means, that the voltage U_m equals the gate voltage U_{G1} applied to the detector MOSFET (T) minus the threshold voltage U_{th} of the MOSFET. The current which flows in this status can be calculated to:

$$I_{R1} = \frac{U_{G1} - U_{th}}{R_1} \qquad (5)$$

The current I_{R1} also flows through the detector MOSFET (T). Since the voltage drop at the resistor R_1 can be neglected compared to the blocking voltage during blocking of the power MOSFET (DUT), the power loss during blocking can be calculated to:

$$P_{loss} = \frac{U_{G1}}{R_1} \cdot U_{DSmax} \qquad (6)$$

During on-state of the power MOSFET (DUT) the drain-source voltage U_{DS} is lower than the gate voltage U_{G1}. The detector MOSFET (T) is turned on. The voltage U_{R1} equals U_{DS} in this case.

The drain current I_D can be measured very easily with a shunt resistor or with a closed loop current sensor (e.g. LEM). The measured value has to be preprocessed and can than be easily connected to a DSP or microcontroller for further handling.

B. Measuring Method

By dividing U_{DS} by I_D the actual R_{DS} can be calculated. The measured resistance has to be compared with the look-up table in which the dependency between resistance and corresponding junction temperature is stored. There are two methods on which the data in the look-up table are based:

1) Since the data available in the datasheets of the power devices are only worst case estimations and not typical values, the correlation between the R_{DS} and the junction temperature can be calibrated by an experiment and stored within a look-up table.
2) The minimal values for the on-state resistance specified in the datasheet are stored in the look-up table. Assuming that the resistance R_{DS} of a real device

is higher or at least equal in comparison to the specified resistance, the actual junction temperature is always lower in the device than determined by the measurement with this method.

In order to determine the actual junction temperature possibility 1 should be addressed. To calibrate the measurement system the on-state resistance R_{DS} of the power MOSFET (DUT) hs to be determined in dependency of the temperature. Therefore, the junction of the MOSFET device has to be heated up to a defined temperature (e.g. 50°C). With a short current pulse the resistance $R_{DS}(50°C)$ can be determined. At least three measurements are required to determine the characteristic $R_{DS}(T)$ over the total temperature range. The current pulse must be kept short in order to prevent a self heating of the junction during calibration since this affects the accuracy of the measurement.

Furthermore, it has to be ensured that the junction is heated to a defined temperature. This can be realized by placing the whole heat sink into a fluid to distribute the heat equally.

IV. REALIZATION

A prototype has been developed to evaluate the measurement method during operation. A DC/DC-converter module (single active bridge) has been chosen. This converter topology consists of an input capacitor, an inverter bridge, a transformer, an output rectifier and an output capacitor. As an example, the load is represented by a resistor. The circuit diagram is depicted in Fig. 4. A picture of the realized prototype is shown in Fig. 5.

The input inverter is operated with a phase shift of 180°and a duty cycle of approximately 50%. The resulting current in the transformer is triangular at nominal load conditions. The transformer's input voltage and current are depicted in Fig. 3. In the middle of the conduction period of a switch the current and the voltage are measured. Based on this value the on-state resistance R_{DS} is calculated and afterwards comapred to a look-up table to determine the junction temperature.

In this prototype, a microcontroller from Texas Instruments was chosen which generates the switching signals, evaluates the voltage and current signals and calculates the resistance R_{DS}. The calculated resistances are transmitted via a RS232 interface to a PC where the measured data is visualized.

V. MEASUREMENT RESULTS

To present this measurent principle the DC/DC-converter was operated transmitting a power of 3 kW. The CoolMOS devices were mounted on a standard heatsink. The heatsink was cooled by convection cooling. The test period was 120 min. Besides the junction temperature measured with the presented measurement principle the heatsink temperature was logged. The heat sink temperature was measured with a standard temperature sensor (thermocouple).

The characteristics are depicted in Fig. 6. The figure shows the temperature in dependency of the time. It can be clearly recognized that the difference between juntion temperature and heatsink increases at higher temperatures. The reason for this phenomena are higher losses in the semiconductor devices caused by higher on-state resistances. Since the thermal resistance between junction and heatsink stays constant, higher power losses result in a higher temperature difference.

VI. Accuracy Estimation

The accuracy of this measurement method depends on several factors. First, the accuracy of the sensors to acquire voltage and current is important. Second, because only the dominant influence of the n^--layer is regarded in the measurement system and not a possible current dependent influence of the channel resistance the accuracy is restricted.

To check the accuracy of this measurement method, the following experiment has been performed: a power MOSFET device was immersed in a thermal fluid at a temperature of 100°C to ensure a constant temperature within the device. Every two seconds a short current pulse was generated (10 to 20 μs) and the drain-source voltage was measured. It is important to keep the current pulse in this experiment short to prevent a self heating of the junction which would lead to erratic results. Since the junction temperature is always constant the on-state resistance R_{DS} should also be constant for all measurements. The experiment took several hours and about 6000 measurements were performed. To estimate the deviation from the expectation the results were statistically analyzed. Figure 7 shows the Gaussian distribution. The expectation is $R_{DS}(100°C) = 102.2\ \Omega$. The standard deviation is 0.7 Ω which means that 68% of the measured values devitate from the expectation within 0.7 Ω. As a comparison, a standard Gaussian distribution with $\mu = 102.2\ \Omega$ and $\sigma = 0.7\ \Omega$ is depicted. At this temperature the deviation of 0.7 Ω corresponds to a temperature of 1K. In order to optimize the accuracy, a sequence of several measured values can be averaged which results in a reduction of σ. Since the thermal time constants are mostly large in

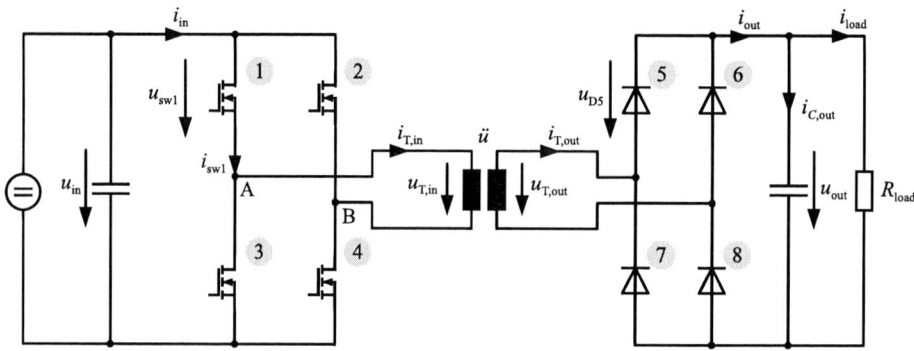

Fig. 4. DC/DC-Converter module (Single Phase Single Active Bridge, SAB1)

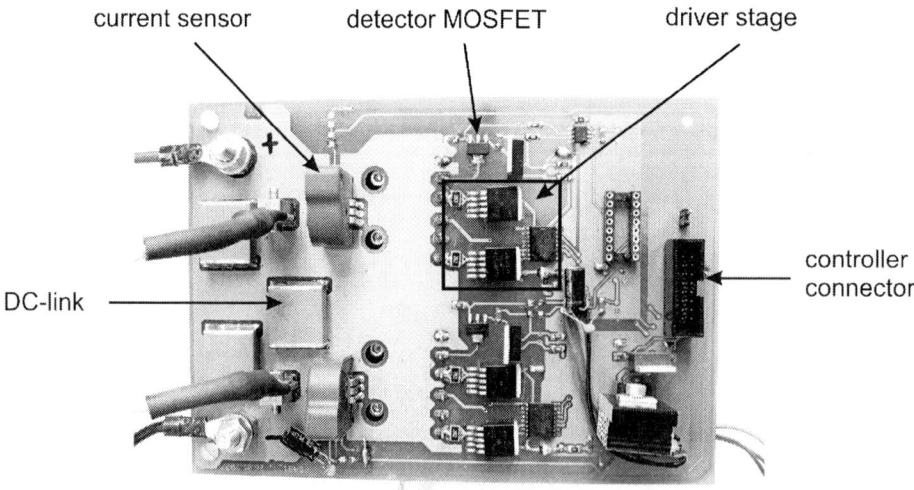

Fig. 5. Inverter section with sensing electronics

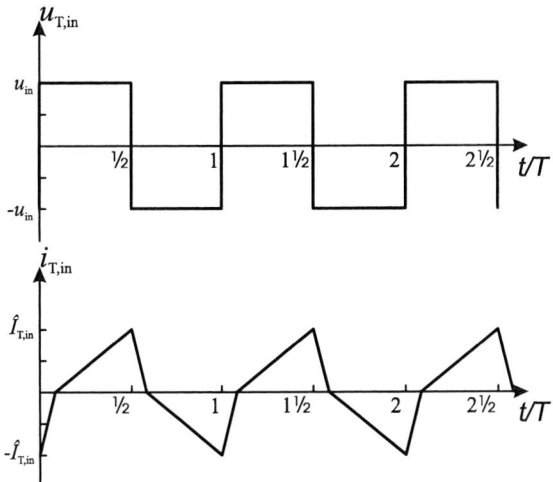

Fig. 3. DC/DC-converter (SAB1) waveforms

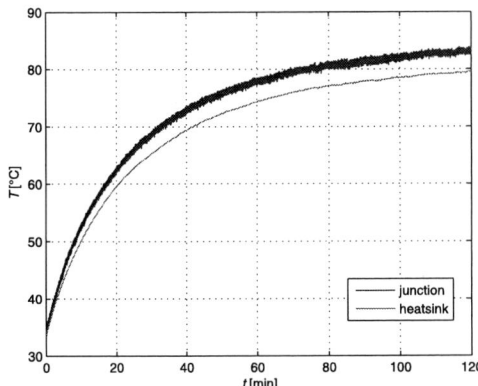

Fig. 6. Junction temperature during operation of DC/DC-converter

comparison to the switching frequency, an averaging of ten values for instance is tolerable.

VII. CONCLUSION

With this method, an accurate measurement of the junction temperature can be established, where no thermal model and other thermal resistances have to be regarded. This method can be employed either in applications where the devices are operated at their thermal boundaries to ensure a safe operation at any time. Furthermore, this method could be used in order to compare different topologies with each other in terms of semiconductor losses (e.g. resonant and non-resonant converters).

VIII. FUTURE WORK

As mentioned before the dependency of the on-state resistance from the drain current is not regarded since the influence of the channel resistance is neglected. Figure 8 shows the dependency of the on-state resistance on the drain current. In this figure the drain current I_D is plotted on the x-axis and the corresponding resistance R_{DS} on

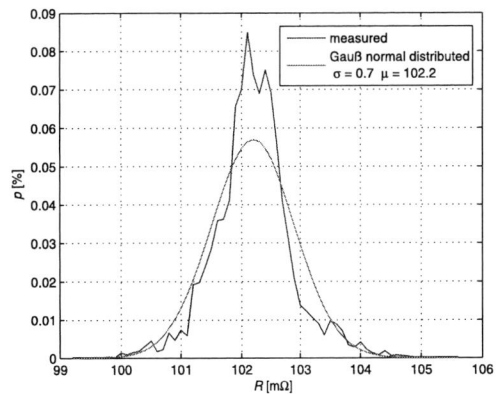

Fig. 7. Gaussian distribution

the y-axis for two temperatures (25°C and 150°C). It can be clearly seen that the gradient of the curve depends on the temperature. For lower temperatures the gradient is almost negligible. In contrast to this, the gradient for higher temperatures is significant.

In the approach presented in this paper the effect can be neglected because the system is calibrated at the same current level (e.g. 5 A) which occurs at the sampling point during operation of the converter. In this case the dependency of the on-state resistance on the current does not effect the measurement.

If a temperature measurement is desired which can measure the temperature over the entire current range of a device, the dependency of the on-state resistance depending on the drain current also has to be entered in the look-up table.

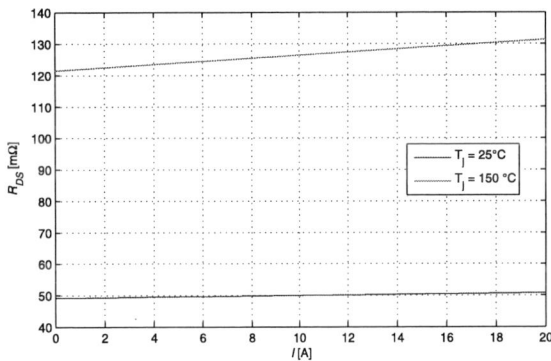

Fig. 8. Influence of temperature on channel resistance

REFERENCES

[1] T. Bruckner and S. Bernet, "Estimation and measurement of junction temperatures in a three-level voltage source converter," *2005. Fourtieth IAS Annual Meeting. Conference Record of the 2005 Industry Applications Conference*, vol. 1, pp. 106–114, Oct. 2005.

[2] L. Lorenz and M. Maerz, "CoolMOSŹ - A new approach towards high efficient power supplies," in *Proceedings of the 39th PCIM*, 1999, pp. 25–33.

[3] J. Lutz, *Halbleiter-Leistungsbauelemente, Physik, Eigenschaften, Zuverlaessigkeit.* Springer Verlag, 2006.

[4] F. R. Lappe, *Leistungselektronik-MeSStechnik.* Verlag Technik GmbH, Berlin - München, 1993.

[5] C. Lombardi, S. Manzini, A. Saporito, and M. Vanzi, "A physically based mobility model for numerical simulation of nonplanar devices," *IEEE Transactions on Computer-Aided Design of Integrated Circuits and Systems*, vol. 7, no. 11, pp. 1164–1171, Nov. 1988.

[6] R. W. D. Doncker, "Zero-voltage crossing detector for soft-switching devices," U.S. Patent 5,166,549, 1992.

[7] P. Koellensperger, J. von Bloh, S. Schroder, and R. W. De Doncker, "A GCT-driver optimized for soft-switching high-power inverters with short circuit protection," *2004. PESC 04. 2004 IEEE 35th Annual Power Electronics Specialists Conference*, vol. 1, pp. 105–111, June 2004.

[8] B. Cyril, B. Dominique, M. Herv, A. Bruno, E. Rene, and B. Pascal, "Towards a sensorless current and temperature monitoring in MOSFET-based h-bridge," *2003. PESC '03. 2003 IEEE 34th Annual Power Electronics Specialist Conference*, vol. 2, pp. 901–906, June 2003.

Analytical Design of High-Power MTO Thyristors

Thomas Plum and Rik W. De Doncker
Institute for Power Electronics and Electrical Drives
RWTH Aachen University
Jaegerstrasse 17-19, 52066 Aachen, Germany
Phone: +49 241 8096942
Email: pl@isea.rwth-aachen.de

Abstract— The design of semiconductor devices is usually performed with finite element methods. In this paper an analytical approach for the design of a high-power MOS Turn-Off Thyristor (MTO) is presented. The model enables the calculation of on-state voltage drop and turn-off losses analytically. The results are compared to a finite-element (FE) model of the MTO. The analytical model offers a high degree of accuracy together with fast calculation times and can therefore be used to find an optimized device design for a given application.

I. INTRODUCTION

The typical design process of semiconductor devices consists of two major steps. During the first phase, basic device properties as doping and thickness of the voltage sustaining base layer are designed analytically. During the second phase the device is simulated with finite element software.

The simulative approach has several major advantages: A basic simulation model can be implemented rather quickly. In addition, more effects can be included in the simulation and the device can be modeled with a high degree of accuracy. On the other hand, such a design process is time consuming, because the finite element calculation of a turn-off process of a device can last several hours even on a modern computer system. To find an optimized solution a large number of simulations have to be performed so that the total simulation time can be unacceptably. In addition, the influences of all the modeled effects on the results are hard to seperate from each other.

In this paper, a different design process based on analytical calculations is performed. Therefore, a rather complex analytical model is implemented, with which all basic characteristics of the device like on-state voltage drop and switching losses can be calculated. Once the model has been implemented, the characteristics of the device can be derived quickly and thereby a time efficient optimization process can be realized.

II. MOS-TURN-OFF THYRISTOR

The MOS-Turn-Off Thyristor (MTO) was first proposed by D. E. Piccone et al. [1]. Similar to a classical thyristor, the device is turned on by injecting a current pulse in the on-gate whereas the turn-off process is realized by voltage control. Fig. 1 depicts the basic device structure

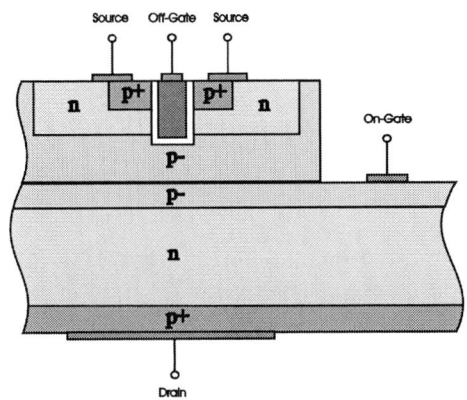

Fig. 1. Device Structure of a MOS Turn-Off Thyristor Cell

of a MTO-thyristor cell. In parallel to the gate-cathode junction, a low resistive switch is connected which is realized by a MOSFET cell. To turn off the device, the MOSFET structure is turned on. Ideally, the cathode current instantaneously commutates to the MOSFET thus bypassing the cathode side n-emitter. Consequently, the device turns off with unity gain similar to the open base pnp-transistor of an IGBT. Hence, similar to the IGCT, the turn-off process can be characterized as switching the device from the thyristor in the transistor mode. More details concerning the device operation can be found in [2].

MTO thyristors are used for high-power applications. Therefore, it is advantageously to build the MTO as a disc-type device. A method for integrating the MOS structures in disc-type devices is the silicon-silicon bonding technique. Details concerning the manufacturing of such devices can be found in [3] and [4].

III. ANALYTICAL DESIGN PROCESS

The analytical design process consists of four different phases. In the first phase the width and doping of the voltage blocking base layer is designed. Afterwards the carrier distribution in the low doped gate and base regions is analytically derived. Based on these results, the on-state voltage drop and the switching losses can be calculated. This model can be used to find an optimized device design for a specific application within a short calculation time.

978-1-4244-0644-9/07/$25.00 ©2007 IEEE

A. Blocking State

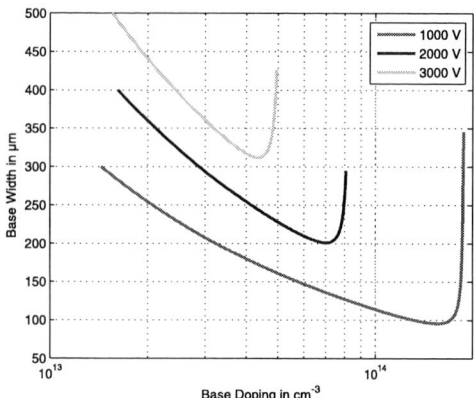

Fig. 2. Base Width as a function of Base Doping for different Blocking Voltages

The blocking capability of the MTO is limited by avalanche generation. During the off-state of the device, the MTO can be represented by an open base pnp-transistor. Breakdown occurs when the following well-known equation is fulfilled:

$$M_p \cdot \alpha_{pnp} = 1 \qquad (1)$$

Where M_p is the multiplication factor of the holes and α_{pnp} is the common base current gain of the pnp-transistor. In contrast to the traditional transistor theory α_{pnp} is dominated by the emitter effiency γ of the anode side emitter rather than by the transport factor.

$$\alpha_{pnp} = \gamma = \frac{J_{pB}}{J_{pB} + J_{nA}} = \frac{1}{1 + \frac{J_{nA}}{J_{pB}}} \qquad (2)$$

with J_{pB}: hole current density in the undepleted base region; J_{nA}: electron current density in the anode region. The emitter efficiency can be calculated by assuming low injection condition in the base and in the anode region of the MTO during blocking. By further assuming that the minority carrier diffusion lengths in the anode and in the base region are much larger than the undepleted widths of the corresponding regions, a constant carrier gradient exists. Thereby, a simple expression for the emitter efficiency can be derived:

$$\gamma = \frac{1}{1 + \frac{D_n}{D_p} \cdot \frac{L_B \cdot N_D}{q_A}} \qquad (3)$$

With: D_n, D_p: Diffusion coefficients of electrons and holes; L_B: undepleted width of the base region; N_D: donor concentration in the base region; $q_A = N_A \cdot L_A$: majority carrier charge of the anode region.

M_p can be calculated via the ionization rates of electron and holes according to [5]. Combining equations 1 and 3 can then be used to compute the base thickness as a function of the doping concentration in the base region.

The results are depicted in Fig. 2 for three different blocking voltages. It can be seen that the base width is initally reduced with increasing base doping while the maximum field strength is well below the limit where impact ionisation occurs. Once impact ionisation occurs, M_p increases and the base width has to be enlarged to keep the emitter effiency low.

For each blocking voltage an optimum base doping and base thickness can be determined. For a non-punch through (NPT) design of a 3.3 kV MTO the base thickness can be calculated to be at least $380 \mu m$ with a base doping of $3.8 \cdot 10^{13} cm^{-3}$. In the following, a total width of the base and gate region is assumed to be $400 \mu m$.

B. Carrier Distribution in Gate and Base Layer

The on-state and switching characteristics are strongly influenced by the shape of the carrier distribution in the gate and base layer of the MTO. The carrier distribution is analytically derived in this section.

The lifetimes of electrons and holes are assumed to be high as no lifetime killing is applied to the device. Therefore, the recombination current can be neglected leading to constant electron and hole currents in the low doped regions for steady state. The ratio between the electron current density J_n and the total current density J_{tot} is one important design parameter and is abbreviated by α in the following:

$$\alpha = \frac{J_n}{J_{tot}} \qquad (4)$$

Assuming high injection condition, the carrier distribution of electrons and holes in the gate and the base layer can be derived. A mobility degradation model due to carrier-carrier scattering is included in the calculations: [6], [7]

$$D_n = \frac{a_n}{n + b_n} \qquad (5)$$

a_n and b_n are constants and are determined by curve fitting to existing mobility data. An equivalent equation is used to describe the carrier-carrier scattering of holes with the constants a_p and b_p.

Starting with the definitions from electron and hole currents and eleminating the electrical field a differential equation for the electron concentration results which can be solved by integration. Finally, the following equation for the electron concentration results:

$$n(x) = -k + (n(0) + k) \cdot e^{\frac{x}{l}} \qquad (6)$$

where k and l are constants which only depend on α, J_{tot} and the fitting parameters of the carrier-carrier scattering model. As can be seen from equation 6, the shape of the carrier concentration only depends on two parameters: one is the concentration of the electrons in the base layer on the anode side $n(0)$ and the other parameter is α the ratio between electron current and total current.

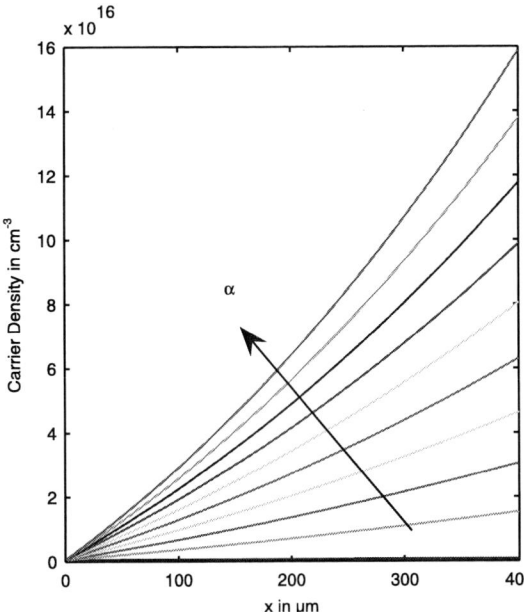

Fig. 3. Carrier Density Distribution for increasing α

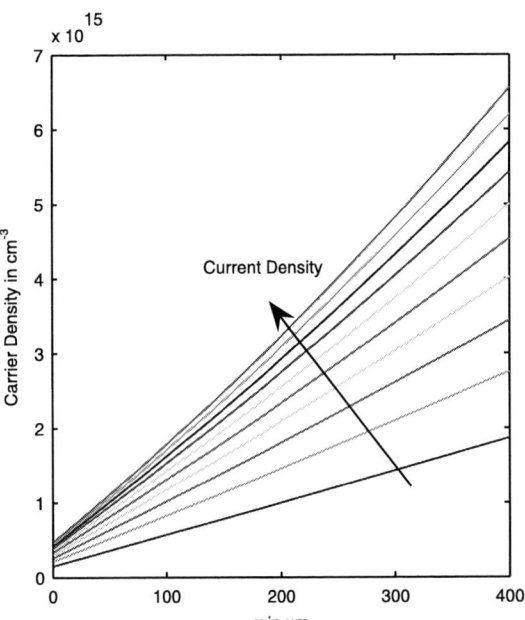

Fig. 4. Carrier Concentration in the Base and Gate Region for increasing Current Density

Fig. 3 depicts the carrier density distribution in the base and gate layer of the MTO for a fixed $n(0)$ and increasing α. It can be seen that the curve becomes steeper for increasing α. In addition, the minimal carrier concentration is located at the anode side of the device rather than somewhere in the middle resulting in minimized switching losses. The carrier distribution in the base and gate layer of the device can be adjusted by appropriate emitters. The emitter design is performed according to [8]. Assuming that the emitter regions are in low injection condition during device operation the following equation can be derived for the electron current density $J_n(n(0))$ on the anode side of the base region:

$$J_n(n(0)) = q \cdot \frac{D_n}{q_A} \cdot n(0)^2 \qquad (7)$$

A similar equation can also be set up on the cathode side of the MTO. With the help of this equations the carrier distribution derived in the previous paragraph can be realized. The carrier concentration in base und gate layer with increasing current density is exemplarily shown in Fig. 4. Here $x = 0$ corresponds to the anode-base junction of the MTO. At $x = 400\mu m$ the junction between gate and cathode is located.

C. On-State Voltage Drop

Once the carrier concentration within the device is known, the on-state voltage drop can be calculated. In a first step, the electrical field is determined and the voltage drop which occurs across the gate and base layer can be derived by integrating the electrical field across these regions. The junction voltages are calculated by the

well known Boltzmann equations. To verify the results obtained by this model, a comparison between a finite element simulation realized with the software package ISE TCAD and the analytical model is depicted in Fig. 5. It can be seen that the analytical model describes the on-state characteristics of the MTO accurately.

D. Switching Losses

The switching losses of the MTO are derived assuming ideally hard-switching conditions. In the first phase the voltage across the device is increasing while the full load current is flowing through the device. During the second phase, the current is going to zero while the full blocking voltage is applied. The swichting losses E_{off} are derived by inspecting the follwing integral during device turn-off:

$$E_{off} = \int_{t=0}^{t=t_{off}} u(t) \cdot i(t) dt = \int_{t=0}^{t=t_{off}} u \cdot \mathrm{dq} \qquad (8)$$

with: $t = 0$: beginning of turn-off process; $u(t), i(t)$: voltage over and current through the device; t_{off}: time instant when the current has decayed to the steady state.

Equation 8 will be analyzed separatley in the two phases voltage rise time and current fall time.

1) Voltage Rise Time: The turn-off process is initiated by switching on the MOSFET-cell. Thereby the emitter on the cathode side is immediately bypassed. The whole turn-off process is only characterized by the characteristics of the open-base transistor formed by the gate, base and anode region.

At the beginning, the gate region is discharged from holes while the on-state voltage across the device remains

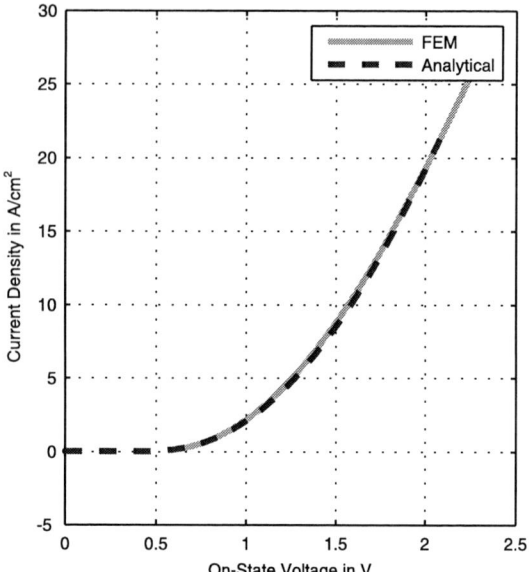

Fig. 5. Comparison of the On-State Characteristics derived by FE-Simulations and the Analytical Model

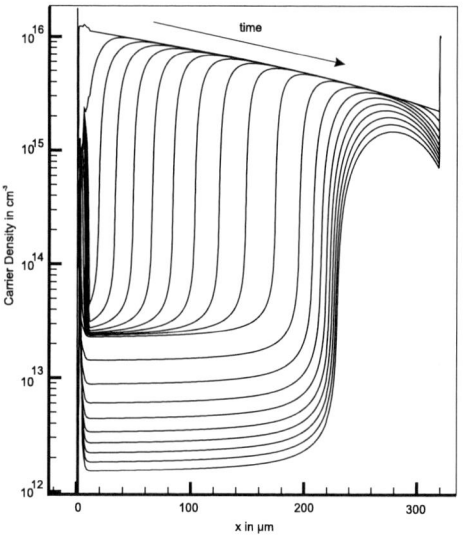

Fig. 6. Hole Distribution in a MTO during Turn-Off (FE-Simluation)

low so that the losses generated during this period can be neclegted. Once the gate-base junction is free of mobile carriers the device starts to block voltage. This voltage can be expressed in terms of the width of the depletion layer in the base region x as follows:

$$ u(x) = 0.5 \cdot \frac{q \cdot N_D + \frac{J_{tot}}{v_{sat}}}{\epsilon} \cdot x^2 \qquad (9) $$

with: $u(x)$: blocking voltage for a given depletion layer width x; q: elementary charge; v_{sat}: saturation velocity of holes; $\epsilon = \epsilon_r \cdot \epsilon_0$: dielectric constant in silicon.

The difference between the hole current injected on the

anode side $(1 - \alpha) \cdot J_{tot}$ and the hole current extracted on the gate side of the device J_{tot} is removing the holes stored in the base region. As can be seen from Fig. 6 the hole gradient near the depletion layer is steep, so that the charge extracted can be assumed to have a rectangular shape: $dq = q \cdot p(x)dx$.

Combining all these equations an equation to determine the switching losses during the voltage rise time E_{rt} can be derived.

$$ E_{rt} = \frac{q \cdot (N_D + \frac{J_{tot}}{v_{sat}})}{2 \cdot \epsilon \cdot \alpha} \cdot \int_{x=0}^{x=x(U_{max}, J_{tot})} x^2 \cdot p(x)dx \quad (10) $$

2) Current Fall Time: As the voltage remains constant in that phase, eq. 8 can be simplified. Additionally, it can be seen from FE simulations that the hole current injected by the anode side emitter falls rapidly to zero during the current fall time. Therefore, the following simple equation can be used to calculate the losses occuring during the current fall time E_{ft} quite accuratley:

$$ E_{ft} = U_{max} \cdot \int j_{tot}(t)dt = q \cdot U_{max} \int_{x(U_{max}, J_{tot})}^{w_{base}} p(x)dx \qquad (11) $$

where $x(U_{max}, J_{tot})$ is the width of the depletion region at the time instance when both the maximal voltage and the full current is applied to the device. w_{base} is the total base width.

E. Technology Curve

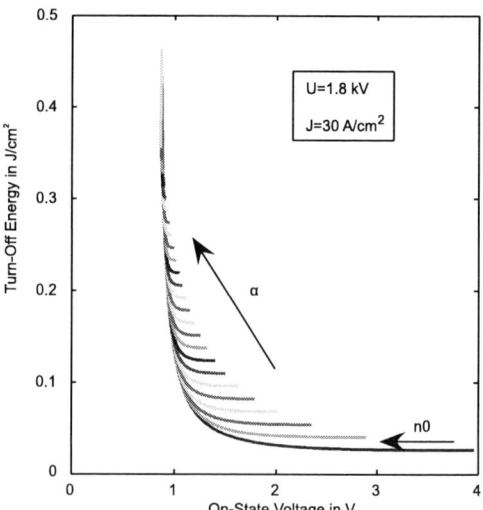

Fig. 7. Technology Curve of a 3.3 kV MTO

With the help of the analytical model described in the previous paragraphs the technology curve for an arbitrary MTO can be calculated quickly. The design only depends on three design parameters: the blocking voltage, the concentration of electron and holes on the anode side of the base region $n(0)$ and the design parameter α describing the fraction of electron current to the whole current flow in the device. A technology curve for the 3.3 kV MTO is shown in Fig. 7.

IV. Validation

FE Simulation	Analytical Modell	Error in %
$U_{on} = 1.12V$	$U_{on} = 1.14V$	1.8%
$E_{off} = 82.3mJ$	$E_{off} = 80.9mJ$	1.7%
$U_{on} = 1.51V$	$U_{on} = 1.54V$	2.0%
$E_{off} = 38.4mJ$	$E_{off} = 37.8mJ$	1.6%
$U_{on} = 2.06V$	$U_{on} = 1.99V$	3.4%
$E_{off} = 23.9mJ$	$E_{off} = 25.4mJ$	6.3%

TABLE I

Camparison between Analytical Model and FE Simulation

The on-state voltage and the swichting losses obtained with the analytical model were compared with a FE simulation of a MTO cell in ISE TCAD. The FE simulation includes the effects of high-field drift-velocity saturation for electrons and holes and the mobility degradation due to carrier-carrier scattering. The results for three different points in the technology curve are depicted in Tab. I.

The maximum error in the calculation of the on-state voltage drop U_{on} is 3.4% and for the swichting losses the maximum error is 6.3%. The maximum difference occurs for the device with the highest on-state voltage drop. This can be explained by the effect that the high-injection conditions in the gate layer are not fullfilled anymore. The model can be easily extended to also cover this effect and thereby eliminating this error.

V. Conclusion

In this paper, an analytical model of MTO thyristors is developed which enables an accurate determination of the on-state and switching losses in hard switching operation. The maximum error compared to a FE simulation is well below 10%. The simulation time for a complete technology curve is less than one minute on a modern PC. The model can be used during the design process for an application specific thyristor device. In the future, the model can be extended to also cover effects of recombination and lifetime killing.

References

[1] D. E. Piccone, R. W. De Doncker, J. A. Barrow, and W. H. Tobin, "The MTO thyristor-a new high power bipolar MOS thyristor," in *Industry Applications Conference, 1996. Thirty-First IAS Annual Meeting, IAS '96., Conference Record of the 1996 IEEE*, vol. 3, San Diego, CA, Oct. 1996, pp. 1472–1473.

[2] R. W. De Doncker, D. Detjen, and T. Plum, "New concepts for high-power BIMOS-devices," in *11th International Conference EPE-PEMC*, 2004.

[3] D. Detjen, S. Schroder, T. Plum, and R. W. De Doncker, "Novel MTO-design based on silicon-silicon bonding," in *Power Electronics Congress, 2002. Technical Proceedings. CIEP 2002. VIII IEEE International*, Oct. 2002, pp. 27–32.

[4] D. Detjen, T. Plum, S. Schroder, and R. W. De Doncker, "Characterization and modeling of bonded hydrophobic interfaces for high-power BIMOS-devices," in *Industry Applications Conference, 2003. 38th IAS Annual Meeting. Conference Record of the*, vol. 2, Oct. 2003, pp. 1236–1243.

[5] S. Sze, *Physics of Semiconductor Devices*. John Wiley & Sons, 1969.

[6] M. Naito, H. Matsuzaki, and T. Ogawa, "High current characteristics of asymmetrical p-i-n diodes having low forward voltage drops," *IEEE Transactions on Electron Devices*, vol. 23, no. 8, pp. 945–949, Aug. 1976.

[7] A. Nakagawa, "Theoretical investigation of silicon limit characteristics of IGBT," in *Power Semiconductor Devices and IC's, 2006 IEEE International Symposium on*, June 2006, pp. 1–4.

[8] J. Lutz, *Halbleiter-Leistungsbauelemente, Physik, Eigenschaften, Zuverlaessigkeit*. Springer Verlag, 2006.

A Novel Gate Driver with Output Voltage Having Double Source Voltage

K. I. Hwu[1], Y. T. Yau[2]

[1]Center for Power Electronics Technology, National Taipei University of Technology, Taiwan

[2]Industrial Technology Research Institute, Taiwan

eaglehwu@ntut.edu.tw

Abstract- **This paper presents a gate driver, which has output voltage with double source voltage. Such a gate driver can reduce the number of voltage sources required to power a complicated system, such as a servo system. And hence, the space for the overall system can be decreased. The detailed operating principles are illustrated and some experimental results are provided to verify the effectiveness of the proposed topology.**

I. INTRODUCTION

For the power electronics circuits to be considered, the analog control chip plays an important role, is composed of peripheral protection circuits and feedback compensation circuits containing OPs, comparators, etc, and generally includes switch drivers. Since the analog integrated circuit (IC) is manufactured mainly by high-voltage process so that it can operate under high-voltage conditions from 10V to 20V. However, since the power supply system is getting more and more complicated today, including monitor, communications, power management, the analog control does not meet such requirements. Consequently, the digital control for the power supply based on FPGA, DSP, uP or specially-defined ASIC will become a trend in the next years.

But there are some problems existing in the components of the digital IC. In order to reduce the product cost, the circuits for the digital control are getting more and more miniature and hence the required working voltage is getting lower and lower, limited to 3.3V or less. Although reducing the working voltage of the digital IC facilitates reducing the dissipated power and accelerating the operating velocity, the gate driving voltage for the MOSFET outside the digital IC can not be reduced. For example, the voltage of 3.3V can not drive the 5V gate of MOSFET. Besides, the more the gate driving voltage is reduced, the more the gate capacitance is required, and hence it is more difficult in driving the MOSFET to be fully turned on. Furthermore, if the logic circuit and the gate driving circuit are separated, then two different working sources are indispensable, implying that more discrete components are used and hence the high-density integration of the digital circuit is not easy to realize. Besides, Fig. 1 shows the conventional gate driver. In past years, there are many researches on the efficiency [1-4] and level shift [5, 6] of the gate driver but few researches on how to boost voltage at a given voltage source. Since the industrial electrical product is getting smaller and smaller, its response is getting faster and faster, and hence the required working voltage is getting lower and lower and the gate driver having the voltage boosting capability is indispensable in the future.

Based on the mentioned above, a new gate driver is presented with the output voltage having double source voltage. In this paper, the input voltage is set to 5V and hence the output voltage is 10V. Such a drive possesses the capability of the fast driving speed and is easy to realize. That is to say, only single voltage source is utilized not only to provide energy for the digital IC but also to offer higher voltage to drive the MOSFET. By doing so, the overall system can be simplified and compact.

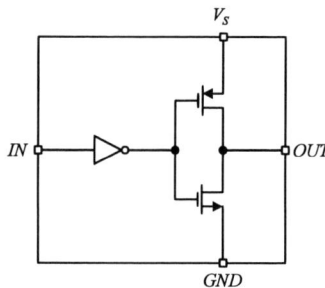

Fig. 1. Conventional gate driver.

II. CIRCUIT TOPOLOGY FOR THE PROPOSED GATE DRIVER

The circuit for the proposed gate driver is presented in Fig. 2. Such a circuit with only one positive-voltage source given, used to source or sink the gate of the MOSFET, is mainly established by one p-channel MOSFET Q_1 and one n-channel MOSFET Q_2. Besides, one capacitor C_1 and one diode D_1 are utilized to form the voltage-clamping circuit, which controls the level of the output voltage, double source voltage or ground. The selection of the current direction, responsible for the path of charging or discharging of the gate of the MOSFET switch, is controlled by one p-channel MOSFETG Q_3 and one n-channel MOSFET Q_4. In this case, the gate of the n-channel MOSFET to be driven is modeled by one capacitor C_{gs}.

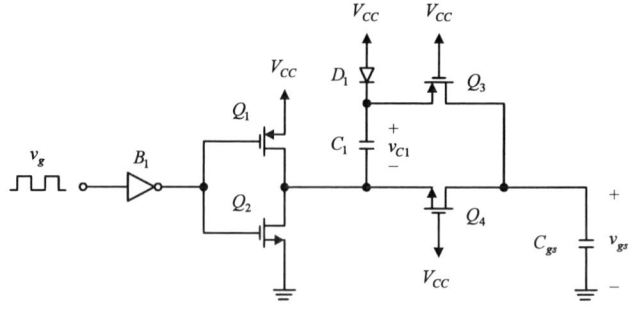

Fig. 2. Proposed gate driver.

III. Operating Principle of the Proposed Gate Driver

Before entering into this section, it is assumed that the voltage drop across switches and diodes are negligible. There are two operating modes for the proposed gate driver to drive the gate of the n-channel MOSFET, modeled by one capacitor C_{gs}.

Mode 1:

In Fig. 2, as the level of the PWM signal v_g is low, the n-channel MOSFET Q_2 is turned on and hence the voltage across the capacitor C_1 is charged to V_{CC}. Besides, due to forward bias of the diode D_1, the p-channel MOSFET Q_3 is turned off because the voltage across the terminals G and S of this MOSFET is zero. At the same time, the n-channel MOSFET Q_4 is turned on because the voltage across the terminals G and S of this MOSFET is V_{CC}. Therefore, the output voltage of this driver is grounded, thereby rendering the power flowing via Q_4 through Q_2 to discharge C_{gs}.

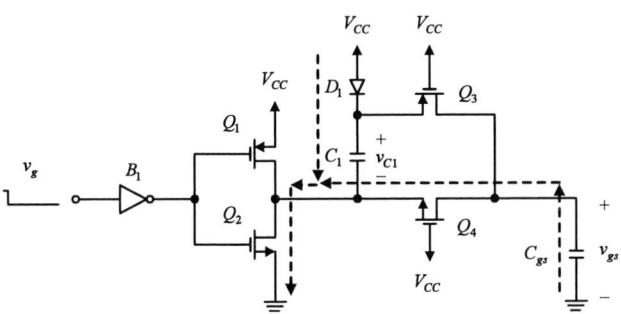

Fig. 3. Power flow of the proposed gate driver with C_{gs} discharged.

Mode 2:

In Fig. 3, as the level of the PWM signal v_g is high, the p-channel MOSFET Q_1 is turned on and hence the voltage across the terminals D and S of this MOSFET is zero. Besides, the voltage at the negative terminal of the capacitor C_1 is V_{CC}, and hence the voltage at the positive terminal of the capacitor C_1 is $2V_{CC}$, thereby making the diode D_1 reverse biased. At the same time, the n-channel MOSFET Q_3 is turned on because the voltage across the terminals G and S of this MOSFET is $-V_{CC}$. Therefore, the output voltage of this driver is $2V_{CC}$, thus rendering the power flowing via Q_1 through Q_3 to charge C_{gs}.

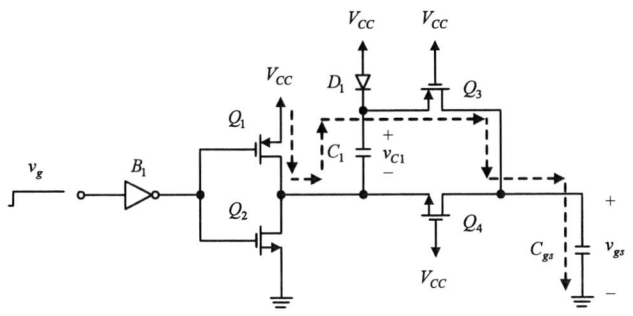

Fig. 4. Power flow of the proposed gate driver with C_{gs} charged.

IV. Simulation and Experimental Results

The specifications of the components in the proposed gate driver are as follows: (i) V_{CC} is set to 5V; (ii) v_g is set to a square wave; (iii) the value of C_1 is set to 4.7μF ; (iv) the value of C_{gs} is chosen to be 22000pF ; (v) the output buffer switches of MIC4420 are used as Q_1 and Q_2; (vi) the product names of Q_3 and Q_4 are IRLML5103 and NDS335N respectively; and (vii) the product name of D_1 is 1N5819.

The following simulation results shown in Figs. 5 to 8 are based on Fig. 2 at the switching frequencies of 1kHz, 10kHz 100kHz and 500kHz, respectively. And after this, under the same conditions, the experimental results are shown in Figs. 9 to 12. Besides, Figs. 13 and 14 show the measured waveforms relevant to the rising period and the falling period respectively. It is obvious that there are output voltages having double source voltage to drive the n-channel MOSFET under the proposed gate driver with the positive-voltage source given.

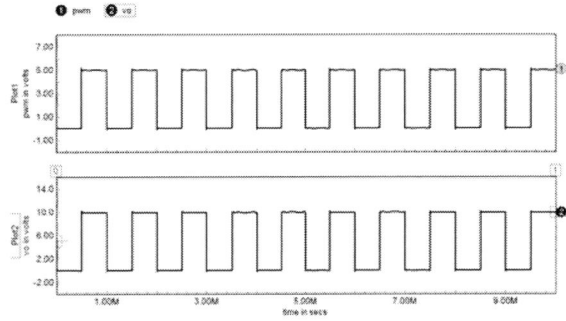

Fig. 5. Simulation results at the switching frequency of 1kHz: (a) the input PWM signal v_g at the top; and (b) the voltage v_{gs} across C_{gs} at the bottom.

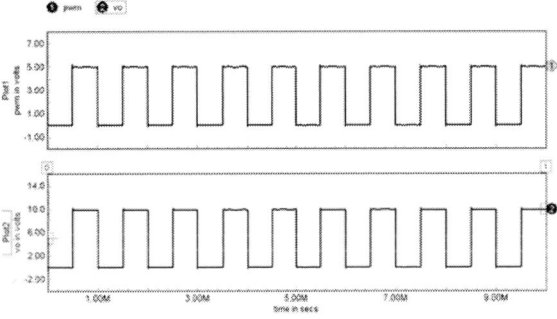

Fig. 6. Simulation results at the switching frequency of 10kHz: (a) the input PWM signal v_g at the top; and (b) the voltage v_{gs} across C_{gs} at the bottom.

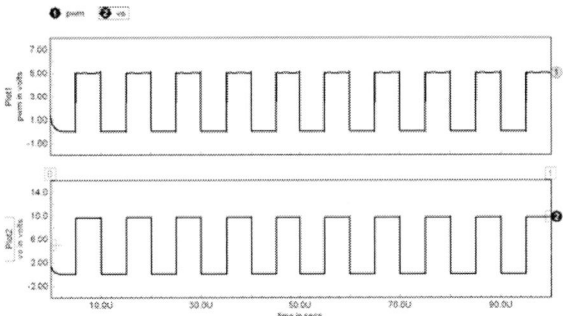

Fig. 7. Simulation results at the switching frequency of 100kHz: (a) the input PWM signal v_g at the top; and (b) the voltage v_{gs} across C_{gs} at the bottom.

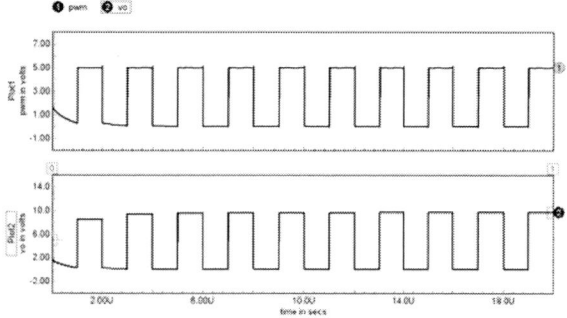

Fig. 8. Simulation results at the switching frequency of 500kHz: (a) the input PWM signal v_g at the top; and (b) the voltage v_{gs} across C_{gs} at the bottom.

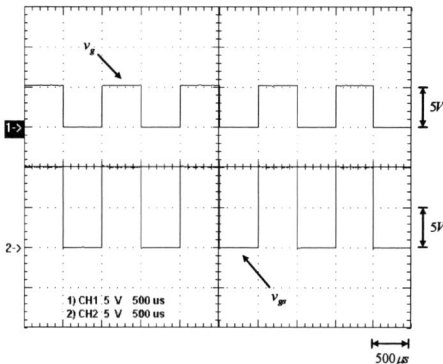

Fig. 9. Measured results at the switching frequency of 1kHz: (a) the input PWM signal v_g at the top; and (b) the voltage v_{gs} across C_{gs} at the bottom.

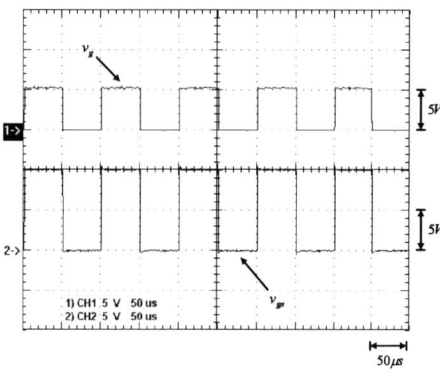

Fig. 10. Measured results at the switching frequency of 10kHz: (a) the input PWM signal v_g at the top; and (b) the voltage v_{gs} across C_{gs} at the bottom.

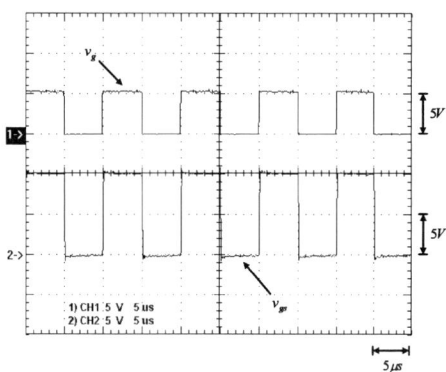

Fig. 11. Measured results at the switching frequency of 100kHz: (a) the input PWM signal v_g at the top; and (b) the voltage v_{gs} across C_{gs} at the bottom.

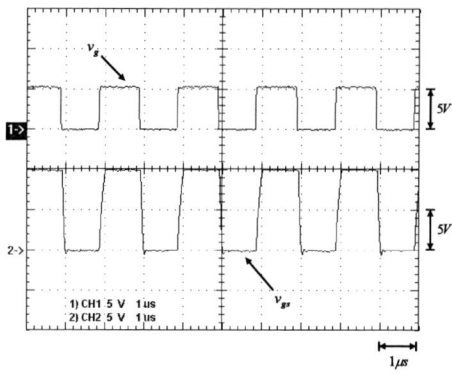

Fig. 12. Measured results at the switching frequency of 500kHz: (a) the input PWM signal v_g at the top; and (b) the voltage v_{gs} across C_{gs} at the bottom.

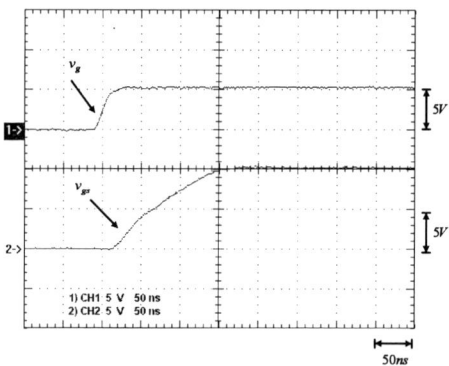

Fig. 13. During the rising period under the value of $C_{gs} = 22000pF$: (a) the input PWM signal v_g at the top; and (b) the voltage v_{gs} across C_{gs} at the bottom.

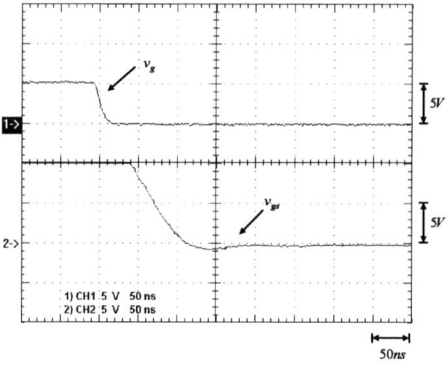

Fig. 14. During the falling period under the value of $C_{gs} = 22000pF$: (a) the input PWM signal v_g at the top; and (b) the voltage v_{gs} across C_{gs} at the bottom.

CONCLUSION

In this paper, a novel gate driver with a single positive-voltage source offers a positive output voltage with double source voltage. By doing so, the number of voltage sources is reduced in powering the digital control IC and the gate driver. By the way, the proposed gate driver can be also applied to the IGBT.

REFERENCES

[1] Kaiwei Yao and F. C. Lee, "A novel resonant gate driver for high frequency synchronous buck converter," *IEEE Trans. Power Electron.*, vol. 17, no. 2, pp. 180-186, 2002.

[2] Yuhui Chen, F. C. Lee, L. Amoroso and Ho-Pu Wu, "A resonant MOSFET gate driver with complete energy recovery," *IEEE IPEMC'00*, vol. 1, pp. 402-406, 2000.

[3] Yuhui Chen, F. C. Lee, L. Amoroso and Ho-Pu Wu, "A resonant MOSFET gate driver with efficient energy recovery," *IEEE Trans. Power Electron.*, vol. 19, no. 2, pp. 470-477, 2004.

[4] I. D. de Vries, "A resonant power MOSFET/IGBT gate driver," *IEEE APEC'02*, vol. 1, pp. 179-185, 2002.

[5] Sihong Park and T. M. Jahns, "A self-boost charge pump topology for a gate drive high-side power supply," *IEEE Trans. Power Electron.*, vol. 20, no. 2, pp. 300-307, 2005.

[6] R. L. Nerone, "A novel MOSFET gate driver for the complementary Class D converter," *IEEE APEC'99*, vol. 2, pp. 760-763, 1999.

Effects of Internal Feedback and Gate-Drive Signal on the Turn-off Loss of MOSFET ZVS

Youthana Kulvitit*, Puckapon Opanuruk*, and Tanvaa Tansatit**

* Dept. of Electrical Eng., Chulalongkorn University. 254 Phyathai Rd., Bangkok 10330 Thailand.
e-mail: youthana.k@chula.ac.th
** Dept. of Medicine., Chulalongkorn University 1873 Rama IV Rd., Bangkok 10330 Thailand.
e-mail: tansatit@yahoo.com

Abstract--This paper studies the effects of internal feedback through device's parasitic capacitances and inductances associated with gate-drive signal on the turn-off loss of MOSFET operates as zero-voltage switch in Class-D resonant inverter. Expressions relating parameters of Miller feedback and gate-drive circuit to Miller voltage were derived. Impeding turn-off and re-turn-on of conduction channel are responsible for higher turn-off loss. Switching trajectory, energy transferred into the MOSFET in each switching cycle, as well as turn-off loss for different gate-drive signal were calculated using data obtained from computer simulations. The results are compared with those calculated from experimental data. Voltage dependent parasitic capacitances of the MOSFET were estimated from experimental data and used for the calculation channel current.

Index Terms-- Channel current, gate-drive signal, impeding turn-off, Miller feedback, re-turn-on, turn-off loss.

I. INTRODUCTION

The main losses and loss mechanism in MOSFET zero-voltage switch (ZVS) were examined in [1]. High speed MOSFET gate-drive circuits were presented in [2]. The paper [1] derived design equations to predict the proper magnitude of dead time and load current to allow zero-voltage switching of the devices in Class-D RF power amplifier. Both square wave gate drive and sinusoidal resonant gate drive are examined with more preference and emphasis on sinusoidal resonant gate drive. Proper design of dead time eliminates loss caused by capacitor discharge. Loss due to voltage and current crossover was mentioned without detailed investigation. Evaluation of interconnection parasitic inductances on MOSFET switching characteristics was present in [3]. In this paper, effects of Miller feedback and source-current feedback were examined. Miller feedback is well-known for its enhancement of gate-drive charge and loss. The finding of this paper discloses the linkage between the internal feedback, gate-drive signal (v_{SIG}) and turn-off loss. Turn-off loss is minimal if v_{SIG} is so designed that, the gate-to-source voltage (v_{GS}) decreases monotonically and cross gate threshold voltage (V_{Th}) only once in each

turn-off operation. Apart from Miller feedback, source-current feedback can exacerbate turn-off loss by reducing the negative gate current (i_G) even further that may cause v_{GS} as well as channel current (i_{Ch}) to rise or ring. The rising or ringing of i_{Ch} while drain-to-source voltage (v_{DS}) is high substantially increases turn-off loss. This paper shows that, in order to reduce turn-off loss, not only dead time but also amplitude and shape of the negative gate voltage and current must be properly controlled.

II. CIRCUIT OPERATIONS

To study the effects of internal feedback, amplitude and shape of the negative gate voltage and current on the turn-off loss of MOSFETs operate as zero-voltage switchs in a Class-D resonant inverter, the switching transition between the two devices should be examined carefully. For Class-D resonant inverter as in Fig. 1, there are two switching operations in each switching cycle. As the two switching operations are equivalent, only one switching operation will be presented. There are 5 different switching intervals in each switching operation. In the derivation of circuit equations, drain-lead and source-lead resistances (R_D and R_S) are neglected. The gate-lead resistance (R_{Gi}) and output resistance of gate drive circuit (R_{Gate}) are lumped together as gate circuit resistance (R_G). The effects of gate-lead inductances (L_G) are negligible in

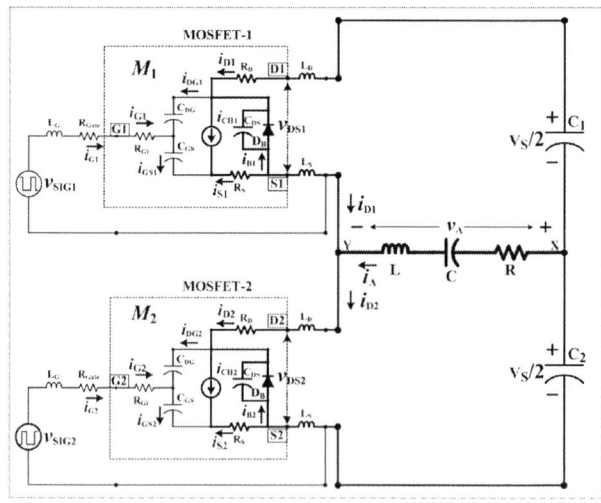

Fig. 1. Detailed circuit of a class-D resonant inverter.

The authors wish to thank National Electronics and Computer Technology Center for their financial support.

978-1-4244-0644-9/07/$25.00 ©2007 IEEE

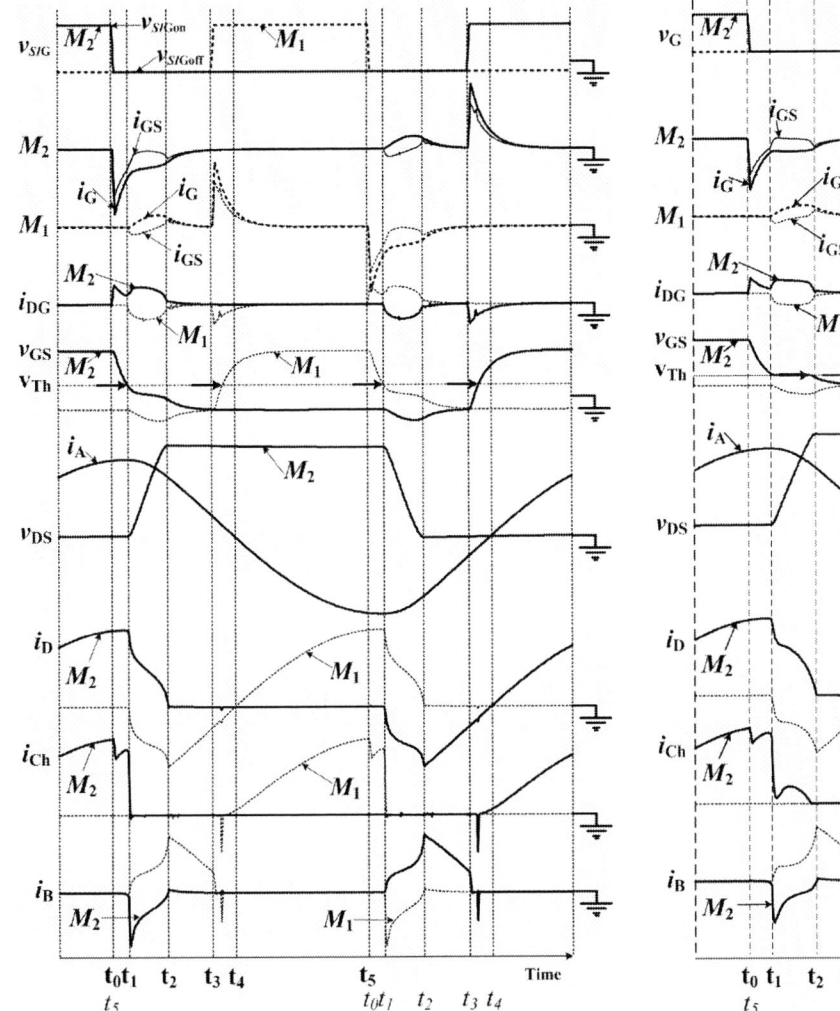

Fig. 2. Simulated waveforms when V_{SIGoff} = -2.5 V and all the parasitic inductances are excluded.

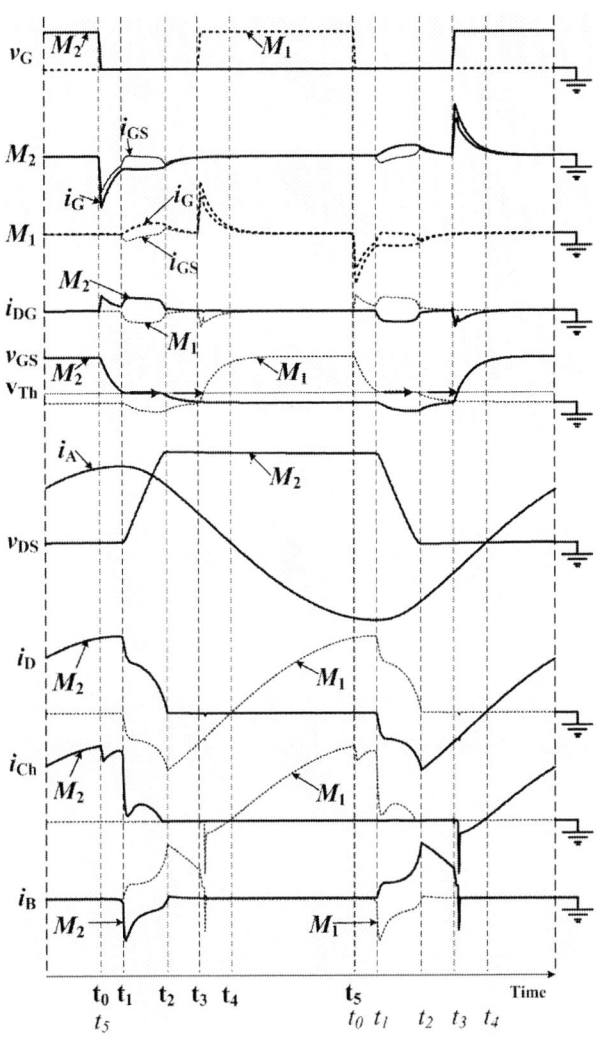

Fig. 3. Simulated waveforms when V_{SIGoff} = 0 V and all the parasitic inductances are excluded.

almost all intervals except in the transition period of v_{GS}. Due to the high rate of change of gate current (i_G) in the transition period, the induced voltage across L_G could be comparable to other voltage components in the circuit.

In order to alleviate the complexity of Miller voltage (V_{Mi}) estimation, L_G was neglected in the theoretical calculations but included in the circuit for computer simulations. To reduce discrepancies between theoretical calculations and experimental results, measures were taken to minimize L_G so that its effects are minimal. All the corresponding parasitic inductances of the upper and lower switches are assumed to be equal. Devices' parasitic capacitances are voltage dependent. Fig. 2 shows simulated waveforms of circuit's voltages and currents when proper gate-drive signal for minimum turn-off loss is used. Fig. 3 shows similar waveforms when less appropriate gate-drive signal is used. For the analysis, internal drain-lead and source-lead parasitic inductances were lumped together with circuit-lead inductances and were push out of the MOSFET model. As parasitic inductances generally cause voltage and current ringing that may obscure the switching characteristics of the

MOSFETs, in the explanation of MOSFETs' operations, waveforms were obtained from computer simulations of an inverter circuit without parasitic inductances.

Simulation results of an inverter circuit with drain-lead and source-lead inductances (L_G , L_D and L_S) are presented in Fig. 6 and Fig. 7. An estimated output resistance of gate-drive IC (Ro) was also included in the Model. The operations of an inverter circuit in each interval are as follows:

Interval 1 $[t_0 < t < t_1]$: Prior to the beginning of this interval: v_{GS1} is lower than threshold voltage (V_{Th}), M_1 is off, drain current $i_{D1} \cong 0$; v_{GS2} is equal to positive gate-drive voltage (V_{SIGon}) which is higher than V_{Th}, M_2 is on, all positive load current i_A passes through M_2, $i_{D2} = i_A$, $v_{DS1} = V_{DC}$, $v_{DS2} \cong 0$. At t = t_0 : M_2's gate-drive signal v_{SIG2} changes from positive gate-drive voltage (V_{SIGon}) to negative gate-drive voltage (V_{SIGoff}); gate current i_{G2} as well as gate-to-source current (i_{GS2}) goes negative while drain-to-gate current (i_{DG2}) goes positive. The negative current -i_{GS2} discharges C_{GS2}, while positive current i_{DG2} charges C_{DG2}. Gate-source voltage v_{GS2} is monotonically

343

decreasing but its value is still higher than both Miller and threshold voltage. MOSFET M_2 is operating in the ohmic region and conducting most of the load current. A large part of the drain current i_{D2} flows through its conduction channel. About 15%–20% of the drain current passes through and charges C_{DG}. During this interval, inverter voltages and currents can be expressed by the following equations:

$$v_{DS1} \approx V_{DC}, \quad v_{DS2} \approx 0, \quad i_{D1} \approx 0, \quad i_{D2} \approx i_A \quad (1)$$

$$V_{SIGoff} - L_G \cdot \frac{di_{G2}}{dt} - R_G \cdot i_{G2} - v_{GS2} - L_S \cdot \frac{d(i_{D2}+i_{G2})}{dt} = 0 \quad (2)$$

$$as \; v_{DS} \approx constant, \; i_{G2} \approx C_{ISS2}\frac{dv_{GS2}}{dt}, \; v_{GS2}(t_0) = V_{SIGon} \quad (3)$$

In this interval, as the rate of change of drain current is low and can be neglected, induced voltage across L_S and L_G can be grouped together. Substituting (3) into (2)

$$\left(L_G + L_S\right)\cdot C_{ISS2}\frac{d^2 v_{GS2}}{dt^2} + R_G \cdot C_{ISS2}\frac{dv_{GS2}}{dt} + v_{GS2} = V_{SIGoff} \quad (4)$$

Second order differential equation (4) can be solved using standard technique if all the circuit components are linear. The solution of (4) depends on damping effect of R_G. Nevertheless, as v_{DS2} changes from small negative value to small positive value, more than tenfold variation of C_{DG} is common [4]. Solving equation (4) with nonlinear variation of C_{ISS2} is not an easy task. If gate circuit is well damped, (4) could be approximated by first order differential equation as in equation (5)

$$R_G \cdot C_{ISS2}\frac{dv_{GS2}}{dt} + v_{GS2} = V_{SIGoff} \quad (5)$$

Solution of the first order differential equation (5) with average value of C_{ISS2} is

$$v_{GS2} = V_{SIGoff} + \left(V_{SIGon} - V_{SIGoff}\right)\cdot exp\left(\frac{-(t-t_0)}{R_G \cdot C_{ISS2}}\right) \quad (6)$$

$$i_{G2} = -\left[\frac{V_{SIGon} - V_{SIGoff}}{R_G}\right]\cdot exp\left[\frac{-(t-t_0)}{R_G \cdot C_{ISS2}}\right] \quad (7)$$

The simulated waveforms of v_{GS2} and i_{G2} in the first interval of Fig. 2 and Fig. 3 correspond to equations (6) and (7). The agreement is natural, as parasitic inductances were not included in the circuit model according to the condition for the derivation of equations (6) and (7).

This interval ends at $t = t_1$ when drain-to-gate feedback takes effect. Charging current of C_{DG2} (i_{DG2}), due to drain-to-source voltage variation, reduces discharging current of C_{GS2} ($-i_{GS2}$) and hence the discharging rate of C_{GS2}. As discharging current of C_{GS2} is not completely reduced to zero for $V_{SIGoff} = $ -2.5 V, existence of Miller voltage V_{Mi} and Miller plateau in Fig. 2 is not obvious as that for $V_{SIGoff} = 0$ in Fig. 3. Discharging time (t_1-t_0) of v_{GS2} from V_{SIGon} to V_{Mi} for $V_{SIGoff} = $ -2.5 V in Fig. 2 is shorter than that for $V_{SIGoff} = 0$ in Fig. 3.

Interval 2 $\left[t_1 < t < t_2\right]$: At $t = t_1$: v_{GS2} is discharged to Miller voltage V_{Mi}. The current carrying capability of M_2's conduction channel decreases. The channel current is totally reduced to zero for $V_{SIGoff} = $ -2.5 V in Fig. 2 or somewhere between i_A and zero for $V_{SIGoff} = 0$ V in Fig. 3. Part or all of the load current diverted from M_2's conduction channel is shared among the charging current of C_{OSS2} (i_{D2}) and discharging current of C_{OSS1} ($-i_{D1}$). The discharging current of C_{OSS1} is shared among discharging current of both C_{DS1} (i_{B1}) and C_{DG1} ($-i_{DG1}$). The negative current $-i_{DG1}$ is shared among positive gate current (i_{G1}) and discharging current of C_{GS1} ($-i_{GS1}$). Positive gate current (i_{G1}) does not cause any harmful effect to the operation of Q_1 as it passes through C_{DG1} and does not cause v_{GS1} to increase. On the contrary, the discharging current of C_{GS1} ($-i_{GS1}$) renders v_{GS1} to an even higher negative value as can be seen in Fig. 2 and Fig. 3. Charging current of C_{OSS2} transfers electrical energy to store in C_{OSS2} and renders v_{DS2} to increase while discharging current of C_{OSS1} restores electrical energy from C_{OSS1} and renders v_{DS1} to decrease. The power circulating between C_{OSS1} and C_{OSS2} does not contribute to switching loss and can be considered as reactive power. M_2's output voltage and current variations are fed back to its gate circuit through drain-to-gate capacitor C_{DG} and source-lead parasitic inductance L_S. For practical value of parasitic components, Miller feedback through C_{DG} is usually dominant. It can impede M_2's conduction channel from being completely turned off, as in Fig.3 until the feedback ceases. The impeding turn-off of the conduction channel generally increases turn-off loss. When feedback through L_S takes effect, channel current ringing is common. The re-turn-on of the conduction channel caused by ringing also contributes to the higher turn-off loss. During interval 2, voltages and currents of the inverter can be calculated and approximated by the following equations:

Gate circuit:

$$V_{SIGoff} - R_G \cdot \left[C_{GS2}\frac{dv_{GS2}}{dt} - C_{DG2}\frac{dv_{DG2}}{dt}\right] - v_{GS2} - L_s \cdot \frac{d(i_{D2}+i_{G2})}{dt} = 0 \quad (8)$$

$$\frac{dv_{GS2}}{dt} = \frac{V_{SIGoff} - v_{GS2}}{R_G \cdot C_{ISS2}} + \frac{C_{DG2}}{C_{ISS2}}\cdot\frac{dv_{DS2}}{dt} - \frac{L_s}{R_G \cdot C_{ISS2}}\cdot\frac{d(i_{D2}+i_{G2})}{dt} \quad (9)$$

The derivative of gate-source voltage in (9) shows that, v_{GS2} is controlled by gating signal V_{SIGoff}. The variations of output voltage dv_{DS2}/dt (Miller feedback) and drain current di_{D2}/dt are fed back through C_{DG2} and L_S respectively. The feedback of v_{DS} through C_{DG} is a shunt-shunt feedback while the feedback of i_D through L_S is a series-series feedback [5]. The derivative of v_{GS2} is zero when v_{GS2} cross Miller voltage. Miller voltage is obtained from (9) by setting $dv_{GS2}/dt = 0$ as in (10)

$$v_{GS2} = V_{Mi} = V_{SIGoff} + R_G \cdot C_{GD2}\frac{dv_{DS2}}{dt} - L_s \cdot \frac{d(i_{D2}+i_{G2})}{dt} \quad (10)$$

$$v_{DS2} = v_{GS2} + v_{DG2} = V_{Mi} + v_{DG2} \quad (11)$$

$$i_{G2} = -C_{DG2}\cdot\frac{dv_{DG2}}{dt} = -C_{DG2}\cdot\frac{dv_{DS2}}{dt} \quad (12)$$

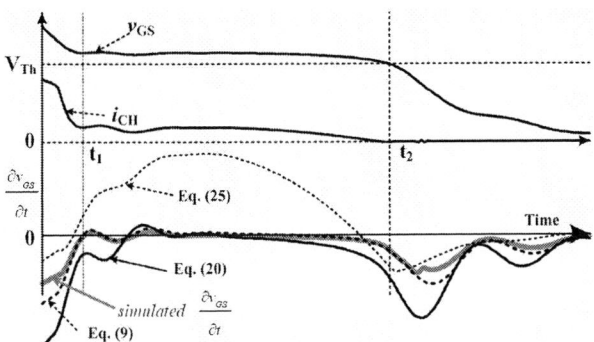

Fig. 4. v_{GS2} and dv_{GS2}/dt during turn-off period.

Equation (10) shows the parameters affecting Miller voltage $\mathbf{V}_{Mi}$. Fig. 4 Shows v_{GS2}, i_{CH} and dv_{GS2}/dt calculated from equations (9), (20) and (25).

Output circuit:

$$V_{DC} = v_{DS1} + (L_D + L_S) \cdot \frac{di_{D1}}{dt} + v_{DS2} + (L_D + L_S) \cdot \frac{di_{D2}}{dt} + L_S \cdot \frac{d(i_{G1} + i_{G2})}{dt} \quad (13)$$

$$i_{D1} = C_{OSS1} \frac{dv_{DS1}}{dt}, \quad i_{D2} \cong i_{CH2} + C_{OSS2} \frac{dv_{DS2}}{dt} \quad (14)$$

$$i_A = i_{D2} - i_{D1} = i_{CH2} + C_{OSS2} \frac{dv_{DS2}}{dt} - C_{OSS1} \frac{dv_{DS1}}{dt} \quad (15)$$

Despite the origin of Miller feedback is due to drain-to-source voltage variation, the voltage variation itself is dependent on load current i_A and channel current i_{CH}. So, Miller voltage will be calculated as a function of load current i_A and channel current i_{CH} at the turn-off instant. Load current can be calculated from load circuit, while channel current of a MOSFET operating in constant current region is a parabolic function of gate-source voltage (v_{GS}) as in (16)

$$i_{CH2} \cong K(V_{GS2} - V_{Th})^2 \cong K(V_{Mi} - V_{Th})^2 \quad (16)$$

Closed-form solutions of (10) to (16) including load-circuit equations are arduous or impossible. A less accurate but feasible approximate solution will be derived and can be used as rough estimation of the influence of affecting parameters. The Miller voltage ($\mathbf{V}_{Mi}$) may be estimated by the following approximations:

A. First Approximation

If induced voltages of parasitic inductances are small compare to V_{DC} and drain to source voltages, we obtain

$$V_{DC} \cong v_{DS1} + v_{DS2} \quad (17)$$

$$\frac{dv_{DS1}}{dt} = -\frac{dv_{DS2}}{dt} \quad (18)$$

Equation (17) is generally true except for the very beginning and ending parts of turn-off process. Despite the validity of (17), (18) may be erroneous if high frequency ringing is presented as can be seen in Fig. 6

and Fig. 7. Substitute (18) into (15) and rearrange the equation:

$$\frac{dv_{DS2}}{dt} = \frac{(i_A - i_{CH2})}{(C_{OSS2} + C_{OSS1})} \quad (19)$$

By neglecting feedback through L_S and substituting (16) and (19) into (9) we obtain:

$$\frac{dv_{GS2}}{dt} = \frac{V_{SIGoff} - v_{GS2}}{R_G \cdot C_{ISS2}} + \frac{C_{DG2}}{C_{ISS2}} \cdot \frac{\left[i_A - K(V_{GS} - V_{Th})^2 \right]}{(C_{OSS2} + C_{OSS1})} \quad (20)$$

Miller voltage is obtained from (20) by setting $dv_{GS2}/dt = 0$ and $v_{GS2} = V_{Mi}$

$$V_{Mi} = V_{SIGoff} + R_G \cdot C_{DG2} \cdot \frac{\left[i_A - K(V_{Mi} - V_{Th})^2 \right]}{(C_{OSS2} + C_{OSS1})} \quad (21)$$

Rearranging (21)

$$a \cdot V_{Mi}^2 + (1 - 2aV_{Th})V_{Mi} + \left[aV_{Th}^2 - \left(V_{SIGoff} + D \cdot i_A \right) \right] = 0 \quad (22)$$

$$D = \frac{R_G \cdot C_{DG2}}{(C_{OSS2} + C_{OSS1})}, \quad a = D \cdot K, \quad (23)$$

Miller voltage can be obtained by solving quadratic equation (22). If the amplitude of feedback current through C_{DG} is equal or higher than that of negative gate current at the turn-off instant, Miller voltage exists, one of the positive real roots of equation (22) is higher than V_{Th}. Complex roots of equation (22) imply the absence of Miller voltage.

B. Second Approximation:

For minimum turn-off loss, $i_{CH2} = 0$, then

$$\frac{dv_{DS2}}{dt} = \frac{i_A}{(C_{OSS2} + C_{OSS1})} \quad (24)$$

Substitute (24) into (9) and neglect drain current feedback through L_S we obtain

$$\frac{dv_{GS2}}{dt} = \frac{V_{SIGoff} - v_{GS2}}{R_G \cdot C_{ISS2}} + \frac{C_{DG2}}{C_{ISS2}} \cdot \frac{i_A}{(C_{OSS2} + C_{OSS1})} \quad (25)$$

Miller voltage V_{Mi} is obtained from (25) by setting $dv_{GS2}/dt = 0$

$$V_{Mi} = V_{SIGoff} + R_G \cdot \frac{C_{DG2}}{(C_{OSS2} + C_{OSS1})} i_A \quad (26)$$

It can be seen that zero crossing of all three curves of dv_{GS2}/dt in Fig. 4 occur around the point where slopes of v_{GS2} are zero.

The MOSFET will not be re-turned on if $\mathbf{V}_{Mi}$ in (26) is lower than threshold voltage (V_{Th}) as in equation (27)

$$V_{Mi} = V_{SIGoff} + R_G \cdot \frac{C_{GD2}}{(C_{OSS2} + C_{OSS1})} i_A < V_{Th} \quad (27)$$

345

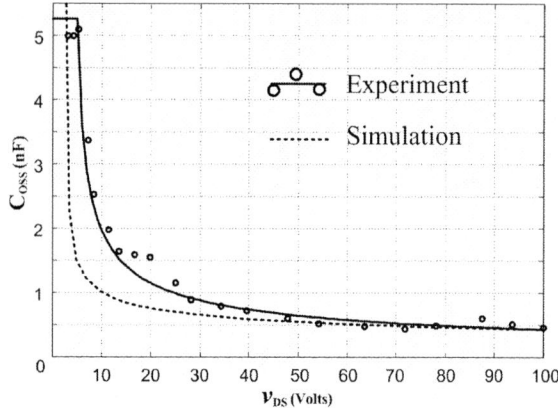

Fig. 5. Voltage dependent characteristics of C_{OSS} calculated from simulation and experimental data.

Due to voltage dependent of C_{OSS} as shown in Fig. 5, absent of re-turn-on is guaranteed if (27) is valid at the turn-off instant for the entire range of load current i_A and v_{DS}. To avoid conduction channel re-turn-on and hence higher turn-off loss, gate drive designer should maximize negative V_{SIGoff} and minimize parasitic-capacitor ratio ($C_{DG}/(C_{OSS2}+C_{OSS1})$) as well as gate-circuit resistance R_G. The parasitic-capacitance ratio is normally optimized by devices manufacturer. The capacitance ratio may be reduced, if appropriate, by adding external capacitors across the drain-to-source terminals. Fig. 3 reveals the occurrence of conduction channel re-turn-on which is the consequence of the violation of equation (27). During the transition period, When Miller feedback takes effect, charging current of C_{DG} passing through the gate circuit resistance also contributes to the turn-off loss. Miller feedback increases losses in gate-drive circuit by increasing effective gate charge and hence average discharging current. Energy loss in the gate circuit during turn-off process increases as the voltage ratio V_{Mi}/V_{DC} increases.

For an appropriate dead time of gate-drive signal and lagging power factor load, a large part of positive load current will completely discharge C_{OSS1} and transfer all the stored energy to other parts of the circuit before M_1 is turned on. Other part of the load current continues to charge C_{OSS2} up to V_{DC} or higher. This interval end when the voltage across C_{OSS1} is reduced to zero at $t = t_2$.

Interval 3 $[t_2 < t < t_3]$: At $t = t_2$: C_{OSS1} is discharged to a small negative voltage, anti-parallel diode D_1 is turned on clamping v_{DS1} at approximately zero. Part of the positive load current (i_{B1}) passing through clamping diode D_1 feeds the energy stored in load's reactive components back to the dc source. The other part of the load current passing through parasitic inductances connected in series with M_2 subsides as the stored energy is transferred to other parts of the circuit including dc source and load. Due to voltage dependent characteristic of C_{OSS} as in Fig. 5, The C_{OSS1} will carry more current than C_{OSS2} because voltage across C_{OSS1} is approximately zero while voltage across C_{OSS2} is approximately equal V_{DC}. The subsidence of the current i_{D2} generally causes voltage and current

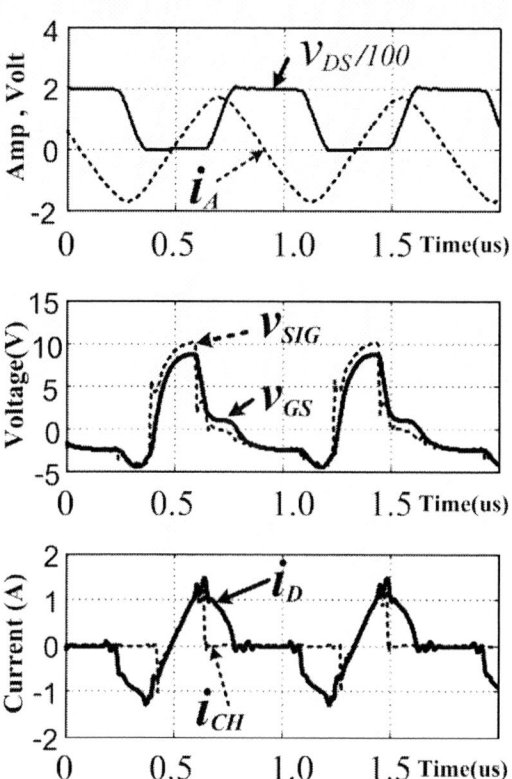

Fig. 6. Simulated waveforms for V_{SIGoff} = -2.5 V. Estimated gate-driver IC output resistance Ro and parasitic inductances of the output circuit included.

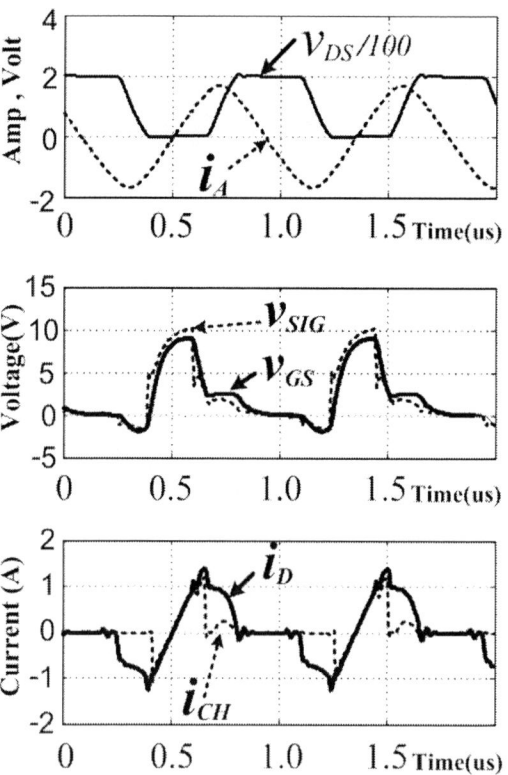

Fig. 7. Simulated waveforms for V_{SIGoff} = 0 V Estimated gate-driver IC output resistance Ro and parasitic inductances of the output circuit included.

346

ringing as can be seen in Fig. 6 and Fig. 7. The voltage dependent characteristic of C_{OSS} is very desirable for the reduction of voltage and current ringing, because it largely reduces the amplitude of drain current of the MOSFET just turned off as can be seen in Fig. 2 and Fig 3. Circuit equation during the ringing period (28) is obtained from (13) by neglecting v_{DS1} and di_{G2}/dt

$$V_{DC} = (L_D + L_S) \cdot \frac{di_{D1}}{dt} + v_{DS2} + (L_D + L_S) \cdot \frac{di_{D2}}{dt} \quad (28)$$

After t_2, v_{GS2} is generally discharged to lower than V_{Th} and i_{CH2} is reduced to zero, then

$$i_{D2} \cong C_{OSS2} \frac{dv_{DS2}}{dt}, \quad \frac{di_{D2}}{dt} \cong C_{OSS2} \frac{d^2 v_{DS2}}{dt^2} \quad (29)$$

$$i_A = i_{D2} - i_{D1} \quad , \quad \frac{di_{D1}}{dt} = \frac{di_{D2}}{dt} - \frac{di_A}{dt} \quad (30)$$

$$V_{DC} = 2(L_D + L_S) \cdot C_{OSS2} \frac{d^2 v_{DS2}}{dt^2} + v_{DS2} - (L_D + L_S) \cdot \frac{di_A}{dt} \quad (31)$$

$$2(L_D + L_S) \cdot C_{OSS2} \frac{d^2 v_{DS2}}{dt^2} + v_{DS2} - \left[(L_D + L_S) \frac{di_A}{dt} + V_{DC} \right] = 0 \quad (32)$$

Equation (32) describes the ringing characteristic of v_{DS2}. Absence of damping in the equation is normal because all the resistive components are neglected.

The ringing voltage continues to feedback to the gate circuit of M_2. Nevertheless, the rate of change of feedback voltage and current are lower than those occurred during the transition period. During this period, voltage and current feedback to the gate circuit has negligible effect on the turn-off loss. Gate circuit equation (9) can be approximated by equation (5). The initial voltage at t_2 $v_{GS2}(t_2)$ is approximately equal to threshold voltage V_{Th}. The negative gate current continues to discharged v_{GS2} to its final voltage V_{SIGoff}, v_{GS2} and i_{G2} decrease exponentially according to (33) and (34) respectively.

$$v_{GS2} = V_{SIGoff} + \left(V_{Th} - V_{SIGoff} \right) \exp\left(\frac{-(t-t_2)}{R_G \cdot C_{ISS2}} \right) \quad (33)$$

$$i_{G2} = -\left[\frac{\left(V_{Th} - V_{SIGoff} \right)}{R_G} \right] \left[\exp\left(\frac{-(t-t_2)}{R_G \cdot C_{ISS2}} \right) \right] \quad (34)$$

This interval ends at t_3 when positive gate voltage v_{SIG1} is applied to the gate circuit of M_1. The interval between t_0 to t_3 is defined as dead time of gate-drive signal.

Interval 4 $[t_3 < t < t_4]$: At $t = t_3$, positive gate voltage v_{SIG1} is applied to the gate circuit of M_1. C_{GS1} is charge up by positive gate current i_{G1}. The positive gate current diverts part or all of body diode current (i_{B1}). When v_{GS1} crosses V_{Th}, M_1 is turned on. The small negative drain-to-source voltage across C_{DS1}—the forward bias voltage of D1—is discharged. The discharging surge current can be observed in both channel current ($-i_{CH}$) and body current ($-i_{B1}$). The positive load current i_A passing through the anti-parallel diode D_1 is mostly diverted to the conduction channel as well as gate circuit and turning off the anti-parallel diode D_1. As there is no Miller feedback when

M_1 is turned on, Miller plateau does not exist during the turn-on period. Input circuit capacitance C_{ISS1} is charged up exponentially to its final value V_{SIGon} while i_A is decreasing. This interval end at $t = t_4$ when i_A decreases to zero.

Interval 5 $[t_4 < t < t_5]$: At $t = t_4$: Load current i_A crosses zero and changes direction, $v_{GS2} < V_{Th}$, M_2 is off, drain current $i_{D2} \cong 0$; $v_{GS1} = V_{SIGon} > V_{Th}$, M_1 is on. Load current i_A flows through conduction channel of M_1 from drain to source. The anti-parallel diode D_1 is reverse biased and turned off. This interval end at $t = t_5$ when gate drive signal v_{SIG1} changes from high to low and marks the beginning of first interval for turn-off process of M_1. The turn-off process of M_1 is similar to that of M_2 except the roles of M_1 and M2 are interchanged.

III. SIMULATION AND ANALYSIS

Due to voltage dependent characteristic of the output capacitance C_{OSS}, the system is nonlinear. Both input and output circuits are described by second or higher order differential equations with multiple feedback paths. Closed-form solutions are arduous. Computer simulation and numerical analysis can render more information to examine the behavior the switching devices and circuit. Circuit equations were used as guides to identify the parameters affecting circuit operation. Approximate solution of V_{Mi} gives a rough estimation of the influence of the affecting parameters. As the switching loss is partly caused by device's lead resistances, all the lead resistances are included in the simulation model. The value of circuit components and devices' parameters are given in Table I.

In order to examine the effect of amplitude and shape of negative gate current on the switching loss, switching trajectories of both i_D vs. v_{DS} and i_{CH} vs. v_{DS} for two values of V_{SIGoff} were shown in Fig. 8 and Fig. 9. As V_{SIGoff} has influence on both the amplitude and fall-time of gate current, V_{SIGoff} will be used as a control parameter. The trajectory of i_D vs. v_{DS} in the first quadrant indicates that, energy is transferred into the MOSFET during the turn-off process. Part of the energy is stored in C_{OSS} while the other part is transferred to the conduction channel. When C_{OSS} is discharged, switching trajectory of i_D vs. v_{DS} in the fourth quadrant indicates that, the stored energy is retrieved from the device and transferred to other parts of the circuit. The trajectory of i_{CH} vs. v_D

TABLE I. VALUES OF CIRCUIT COMPONENTS AND DEVICES' PARAMETERS

LOAD	V_{DC} V	L uH	C nF	R Ω	L_D nH	L_S nH
Circuit	200	18	2.4	740	10	50
GATE	V_{SIGon}	V_{SIGoff}	R_O Ω	R_{Gate} Ω	L_G nH	f MHz
Circuit	10 V	varied	7	3	20	1.18
MOS	Level	L μm	W μm	K_P	V_{TO}	I_S pA
FET	3	2	1.9	1.1E-5	2.3V	96.3
Model	C_{GS0} nF		C_{GD0} pF		C_{BD} nF	
	1.95		85		3.0	
Resis-	R_{Gi} Ω		R_D Ω		R_S Ω	
tances	1.556		2		1	

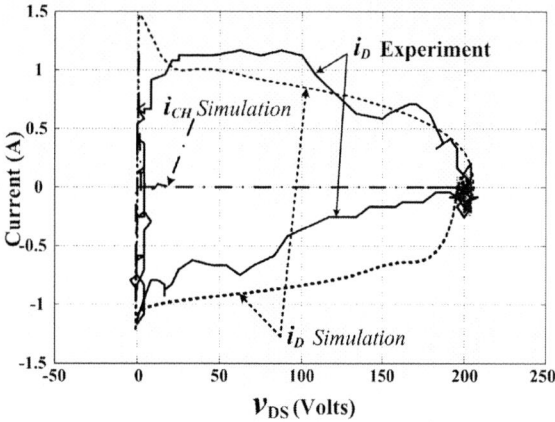

Fig. 8. Switching trajectories of i_D vs. v_{DS} and i_{CH} vs. v_{DS} for V_{SIGoff} = -2.5V

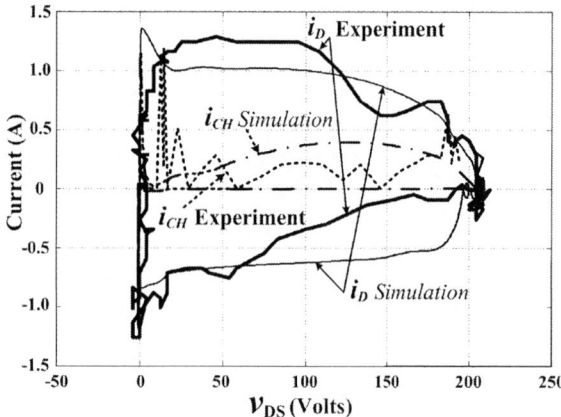

Fig. 9. Switching trajectories of i_D vs. v_{DS} and i_{CH} vs. v_{DS} for V_{SIGoff} = 0.

shows that, there is energy transferred into the conduction channel of the MOSFET during the turn-off process. Absence of i_{CH} vs. v_{DS} trajectory in the fourth quadrant indicates that, the energy transferred into the conduction channel is dissipated as heat in the conduction channel. As the two dimensional switching trajectories do not contain time information, energy transferred into and out of the MOSFET can not be retrieved. During the energy transfer process, certain amount of energy is loss in the resistances of the connection leads. Not all the stored energy can be retrieved. Nonzero channel current during the transition period contributes to higher turn-off loss. It can be seen from Fig. 8 and Fig. 9 that, as the V_{SIGoff} becomes more negative, the energy loss in the conduction channel decreases. Energy loss in the conduction channel is largely reduced for V_{SIGoff}= - 2.5 V.

Fig. 12 shows the energy transferred into the MOSFET as a function of time for different gate drive signals (dotted line). The transferred energy is calculated from data of circuit simulation. The difference of transferred energy between the beginning and the end of turn-off process is the turn-off loss per switching operation. When V_{SIGoff} is reduced from 0 V to -2.5 V energy loss per switching operation is reduced from 4.9 μJ to 2.0 μJ.

IV. EXPERIMENTAL RESULTS

A class-D resonant inverter with 1.17 MHz operating frequency had been built. Load circuit and dead time of gate-drive signal were properly designed so that the MOSFET is turned on at zero voltage. Values of circuit components and circuit parameters are similar to those used for computer simulations. Experimental waveforms for two values of V_{SIGoff} are present in Fig. 10 and Fig.11. General appearances of the experimental waveforms in Fig. 10 and Fig. 11 are similar to those of simulated waveforms. Discrepancies are noticeable, particularly the decaying rate of drain current at the turn-off instant. The difference of the decaying rate is believed to be a consequence of the different decaying rate of C_{OSS} with the increasing drain-to-source voltage. Fig. 12 shows the energy transferred into the MOSFET as a function of time for different gate-drive signals (solid line). The transferred energy is calculated from experimental data. When V_{SIGoff} is reduced from 0 V to -2.5 V energy loss per switching cycle is reduced from 8.1 μJ to 4.5 μJ.

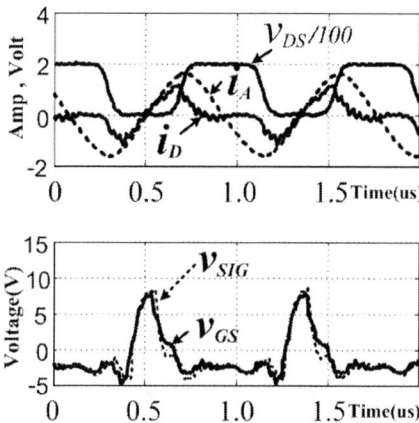

Fig. 10. Experimental waveforms for V_{SIGoff} = -2.5 V. Estimated gate-driver IC output resistance Ro and parasitic inductances of the output circuit included.

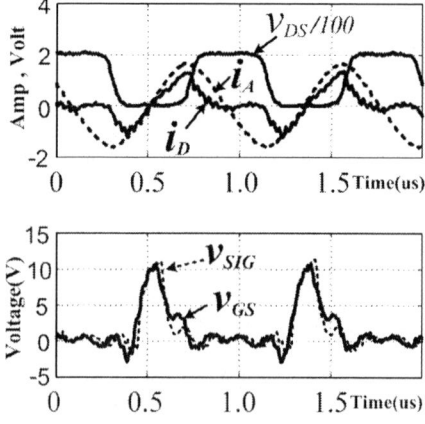

Fig. 11. Experimental waveforms for V_{SIGoff} = 0 V Estimated gate-driver IC output resistance Ro and parasitic inductances of the output circuit included.

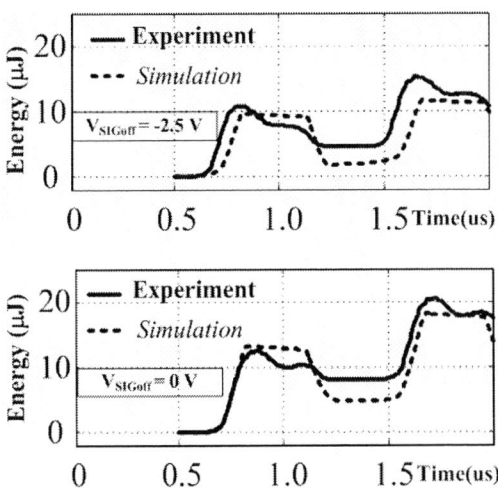

Fig. 12. shows simulation and experimental waveforms of voltage, current and energy transferred into the MOSFET.

V. CONCLUSION

Switching characteristics of MOSFET ZVS in a Class-D resonant inverter were investigated in great details especially during the turn-off transition period. Due to the complexity of the analysis, only simplified circuit equations were derived. Feedback of output voltage and current to the gate circuit were investigated and found to be the origin of impeding turn-off and re-turn-on of conduction channel. A simplified expression of Miller voltage as a function of circuit components and variables were derived. The simplified expression identifies the parameters affecting Miller voltage, while approximate solution may be used as a rough estimation of the influence of the affecting parameters. Turn-off loss for two gate-drive signals calculated from both simulation and experimental data are given. Voltage-dependent capacitances were estimated and included in the numerical calculations. Effects of voltage dependent capacitances on the switching characteristics are indicated.

REFERENCES

[1] S.A. El-Hamamsy, "Design of High-Efficiency RF Class-D Power Amplifier," IEEE Trans. Power Electron., vol. 9, no. 3, pp. 297-308, May 1994.

[2] Laszlo Balogh, "Design and Application Guide for High Speed MOSFET Gate drive Circuits," Power Supply Design Seminar SEM-1400, Topic 2, Texas Instruments Literature No. SLUP169.

[3] Y.Xiao,H. Shah, T.P. Chow, and R.J. Gutmann, "Analysis and Modeling and Experimental Evaluation of Interconnect Parasitic Inductance on MOSFET Switching Characteristics," .

[4] Motorola Power MOSFET Transistor Device Data. Motorola, Inc 1994, pp 2-6-2.

[5] A.S.Sedra, K.C.Smith, "Microelectronics Circuits" Fourth Edition.

A Novel Bridge Type FCL Based on Single Controllable Switch

Wanmin Fei, Yanli Zhang and Qi Wang

School of Electrical and Automation Engineering, Nanjing Normal University, 78 Bancang Street, Nanjing, P.R.China

Abstract—A novel bridge type FCL (Fault Current Limiter) with single controllable switch is proposed in this paper. The topologies and control strategies of the new FCL for single-phase and three-phase power systems are investigated in detail. The advantage of the new FCL includes simplicity in circuit and control method, low cost and high reliability. Simulations based on Psim 6 are carried out. A three-phase experimental model is constructed. Simulation and experiment results proved the validity of the new FCL.

Index Terms--Power systems faults, Fault current limiters, Power semiconductor devices, Power electronics.

I. INTRODUCTION

For highly reliable power supply, Fault Current Limiter (FCL) is becoming an essential part in the modern power system. The conventional technology used today to clear the fault is based on circuit breaker (CB) with over current relay. The typical operational time delay of practical circuit breaker ranges from few cycles to several seconds. During this time, only the system impedance can limit the fault current and the fault current may stress the system equipment to the mechanical, and in certain cases, also to the thermal limits. Current limiting device is required to be introduced into the power system for limiting the fault current before it rising to its full prospective value. Advantages of using a fault current limiter include reduced fault level of the supply and smaller voltage sag during a short-circuit fault. These will avoid upgrading switchgears during system expansion and improve the power quality delivered to customers.

Several kinds of FCL have been proposed in recent years. Superconductor-based FCL has advantages such as compact design, high efficiency even including refrigeration penalty, and is a most promising Limiter, but its cost is too high. Reactor parallel connected with GTO scheme proposed by EPRI may arouse oscillation and operating over-voltage, and requires very high operating speed, its reliability is low. The solid state bridge type FCL based on power electronics has advantages over others in multi-operation capability, lower voltage drop, lower cost and higher reliability[4]~[7].

II. HALF-CONTROLLABLE-BRIDGE TYPE FCL AND ITS DROWBACKS

Reference [8] proposes a self-turn-off device based bridge type FCL, its topology is shown by Fig 1 and 2. The single-phase structure is shown by Fig 1, which is composed of a half-controlled single-phase rectifier, a

DC reactor L1, a limiting reactor L2 and a ZnO arrestor. In normal situation, T1 and T2 are in full conduction status, because the voltage across L1 can only be one direction, the current in L1 keeps increasing until reaching the peak value of load current after power is on. In steady state, ignore the loss of L1 and the on-state voltage drop of D1, D2 and T1, T2, the current in L1 keeps constant and the voltage of L1 is zero. The existence of the FCL has almost no harmful effect on the power systems. When short circuit fault happens, the rectifier, DC reactor L1 and limiting reactor L2 are inserted automatically to limit the fault current. The source voltage applied through the rectifier on L1 and results in a rapid increasing of current in L1 because the inductance of L1 is small. When current in L1 exceed the preset value, turn off self-turn-off devices T1 and T2, and L1 current freewheels through D1 and D2 afterwards. Because the inductance of L1 is reduced greatly and little magnetic energy is stored in it, the current of L1 fall to zero in about one cycle. Then fault current will be limited by the limiting reactor L2 and is finally cut off by the circuit breaker S1. A ZnO arrestor is employed to eliminate the operating over-voltage.

The three-phase structure for three-phase three-wire power systems is shown in Fig 2, which is more compact and economical than the scheme of using three single-phase FCLs to constitute a three-phase FCL. For three-phase three-wire systems shown in Fig 2, there are two short circuit fault modes: two phase and three phase short circuit. Because the system can not work after a two-phase or three-phase short circuit fault happens, so the control method is very simple: remove the gating signals of T1-T4 when a short circuit fault happens and is

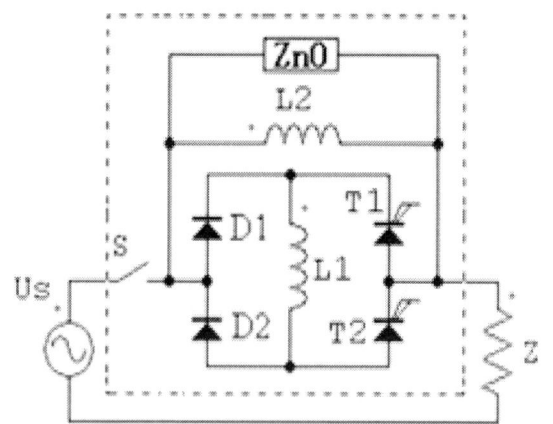

Fig. 1. Single phase half-controllable-bridge type FCL

detected regardless of the fault mode. The current of L1

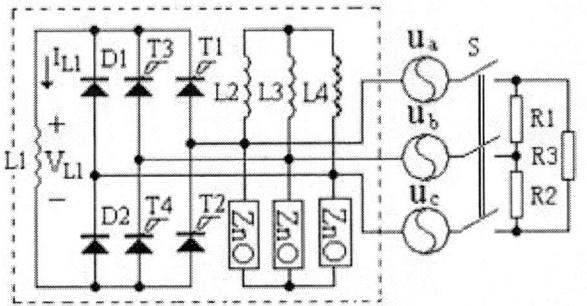

Fig. 2. Half-controllable-bridge type FCL for three-phase three-wire power systems.

freewheels through D1 and D2 afterwards. The current of L1 fall to zero in about one cycle. Then fault current will be limited by the bypass reactors. Turn off the circuit breaker when the fault needs to be cut off finally.

The control method and control circuit of half-controlled bridge type FCL shown by Fig 1 and 2 is much simpler than the full controlled bridge type FCL, but it still has more than one self-turn-off device and all the self-turn-off devices are turned off simultaneity after the short circuit fault being detected. All the self-turn-off devices work as one switch in the common path of the L1 current indeed. So the topology, control method and control circuit can be farther simplified.

In this paper, a single-controllable switch based bridge type FCL has been proposed, its main circuit is simplified, the control method and control circuit are simpler, the cost is reduced and the reliability can be increased greatly. Simulation and experiment are carried out which proved the practicability and validity of the new FCL.

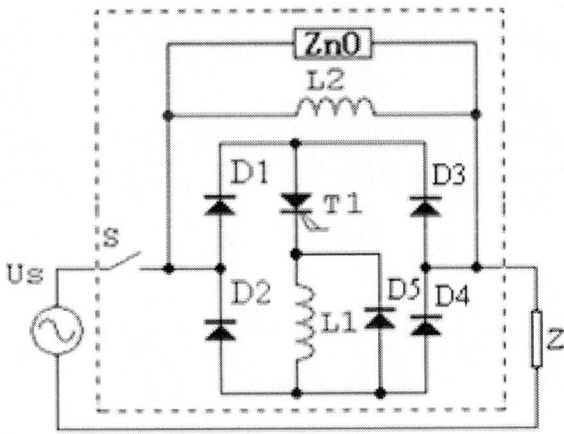

Fig. 3. Topology of the proposed single-phase FCL.

III. TOPOLOGY AND CONTROL METHOD OF THE NEW FCL

Topology of the new FCL is shown by Fig 3 and 4. Fig 3 shows the single-phase structure. Fig 4 shows the three-phase structure for three-phase three-wire power systems.

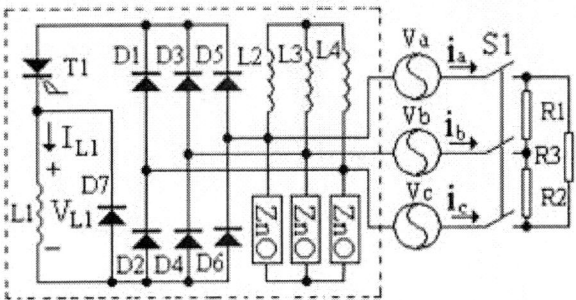

Fig. 4. FCL based on single controllable switch for three-phase three-wire power systems

The differences between FCL shown by Fig 3 and 4 from that shown by Fig 1 and 2 include two parts: a) the rectifier bridge in the FCL shown by Fig 3 and 4 is composed of diodes which are much cheaper and simpler than controllable devices; b) a controllable switch is used in the dc side of the rectifier and a diode is employed for current continuation. There is only one controllable switch in the new FCL, so the cost of the main circuit can be decreased. Only one group of gate-driving and protection circuit is needed, so the control method and control circuit are simplified greatly. The reliability can be enhanced. Both reverse conducting device and reverse blocking device can be selected for the controllable switch T1 in the new FCL because of the location of the controllable switch in the main circuit.

Take three-phase structure for three-phase three-wire power system to describe the principle of the new FCL. In normal status, the controllable switch T1 keeps in full conduction. After a short magnetizing process, current in L1 reaches the peak value of load current and then keeps constant if the loss of L1 and on-state voltage drops of diodes and T1 are ignored. There is no voltage and power consumption on the FCL.

When a two-phase short circuit fault happens, take phase a and b for instance, the fault point serves as a new neutral point of power supply, the reverse of fault phase line voltage will applied through the rectifier on L1 and result in a rapid increasing of L1 current from the peak value of load current. When a three-phase short circuit fault happens, the fault point serves as a new neutral point, a reversed three-phase voltage of power supply apply on L1 and result in a rapid increasing of L1 current. When L1 current exceed the preset value, turn off T1 no matter two-phase or three-phase short circuit fault, then L1 current freewheels through D7 and falls to zero in about one cycle. From the moment of fault, currents are produced in the limiting reactors L2, L3 and L4 according to the fault mode. When two-phase fault occurs, take phase a and b for instance, currents are produced in L2 and L3. When three-phase fault occurs, currents are produced in L2, L3 and L4. After T1 is turned off, limiting reactor L2, L3 and L4 take the role of current limiting completely. The fault currents are finally cut off by circuit breaker S1, the over-voltage excited by the operation will be absorbed by the ZnO arrestors.

351

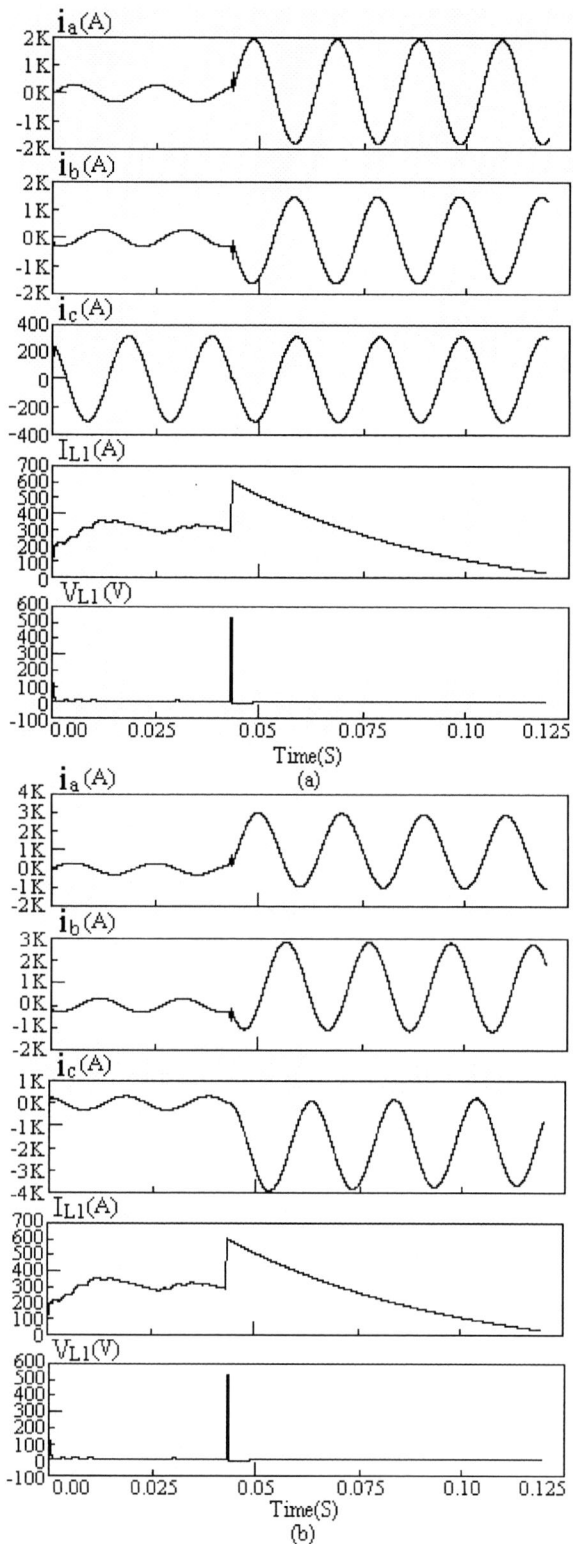

Fig. 5. Simulation waveforms of the proposed three-phase FCL

IV. SIMULATION

In order to verify the validity of the proposed FCL, Simulations based on PSIM 6.0 for the circuit shown by Fig. 4 are accomplished. The simulation waveforms are shown in Fig.5, where ia, ib, and ic is three phase source current respectively. I_{L1} and V_{L1} is current and voltage of

DC reactor L1 respectively. Fig 5(a) shows the simulation waveforms of the new FCL at two phase fault mode and Fig 5(b) shows that at three phase fault mode. Parameters the simulations based on are listed: phase voltage 220V/50Hz; rated load current 220A; load resistance 1Ω; inductance of DC reactor L1=0.6mH, the internal resistor of L1 is 0.05ohm; inductances of bypass reactor L2=L3=L4=0.5mH; preset short circuit fault current is 7 times of the rated load current. Power is turned on at t=0, short circuit fault occur at t=0.0428s. The interval from the fault moment to the time T1 is turned off is 200 microsecond. Voltage drop of diodes D1-D7 and T1 is 1.5V and 2.0V respectively.

V. EXPERIMENT

To further verify the validity of the new FCL, an experimental model of the proposal FCL shown by Fig 6 is constituted. Fig 6(a) shows the photo of the experimental model. Fig 6(b) shows the main circuit

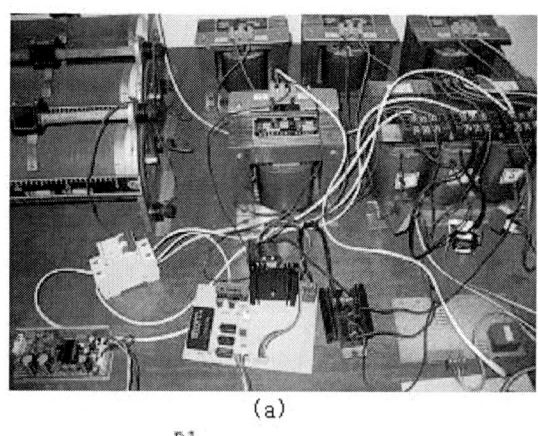

(a)

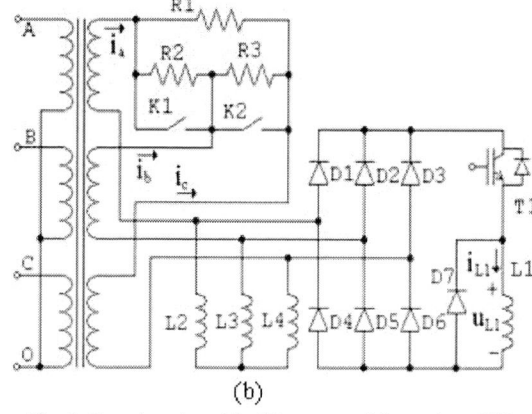

(b)

Fig. 6. Experiment model of the proposed three-phase FCL

principle. In order to establish a three-phase three-wire power system model, the primary windings of the transformer TR1 are Y-type linked and the secondary windings are linked as shown by Fig6(b). Fault current is preset to 7A. Self-turn-off device IGBT with a reverse diode is adopted. L1=15mH, L2=L3=L4=17mH, R1=R2=R3=50ohm, K1 and K2 are used for realization of short circuit fault.

Experiment waveforms are shown in Fig 7. Fig 7(a) shows two fault-phase currents under two-phase short circuit fault mode. Fig 7(b) shows one fault-phase current

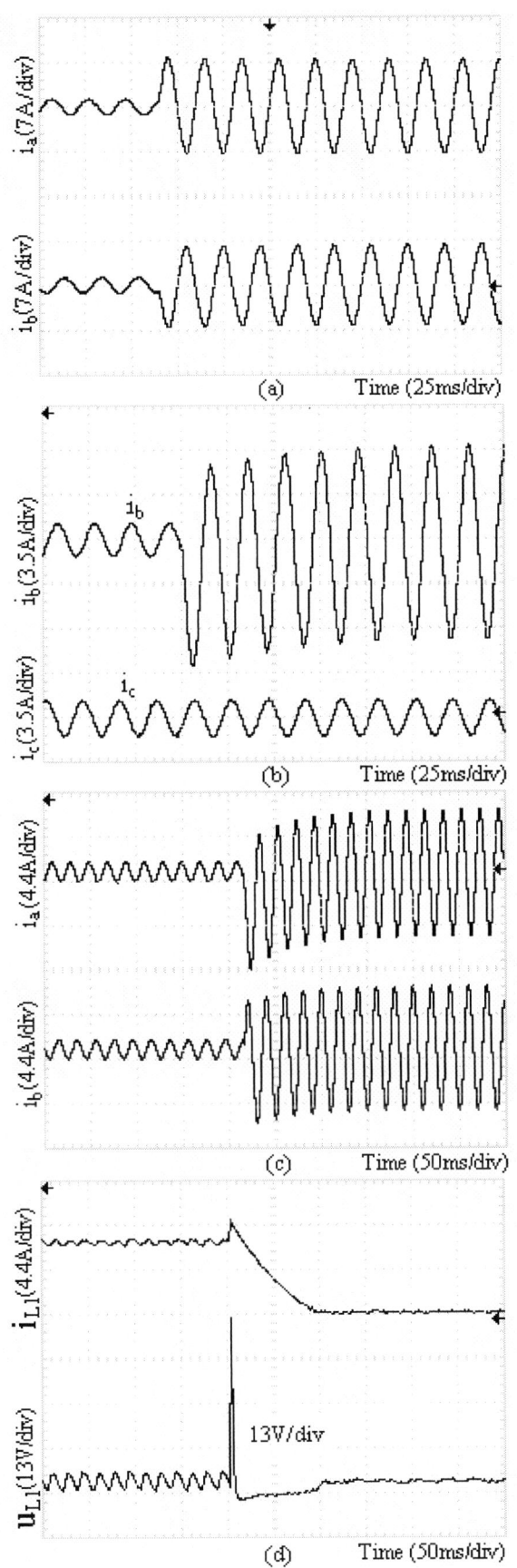

Fig. 7. Experiment results of the proposed three-phase FCL.

and one normal phase current under two-phase fault

mode. Waveforms of Fig 7(a) and (b) are from two tests and have no time relation. Fig 7(c) is two phase currents under three-phase fault mode. Fig 7(d) is the waveforms of voltage and current of L1 under three-phase fault mode. It can be seen from the waveforms that the current can be limited to a preset value after short circuit fault happen. Current of L1 falls to zero in about one cycle after T1 is turned off. The fault current limiting effect is excellent and the proposed FCL is practicable.

VI. CONCLUSION

A single-controllable-switch-based bridge type FCL has been proposed. Topology and control method are described in detail. Compared with other FCL, the new one has some advantages such as low cost, simple control circuit and control method, high reliability and so on. Simulations with Psim 6 are carried out. An experimental model of the proposal three-phase FCL is constructed and investigated in detail. The experimental results agree strictly to the simulation waveform. The fault current limiting effect is excellent.

REFERENCES

[1] Kenji Yasuda, Ataru Ichinose, Akio Kimura et al. Research & Development of Superconducting Fault Current Limiter in Japan, IEEE Transactions on Applied Superconductivity, Vol. 15, No. 2, June 2005, pp:1978-1981.

[2] H. Shimizu, Y. Yokomizu, and T. Matsumura. Comparison of Fundamental Performance of Different Type of Fault Current Limiters With Two Air Core Coils, IEEE Transactions on Applied Superconductivity, Vol. 14, No. 2, June 2004, pp:807-810.

[3] M.M.R. Ahmed, G.A. Putrus, Ran Li, Xiao Lejun. Harmonic Analysis and Improvement of a New Solid-state Fault Current Limiter. IEEE TRANSACTIONS ON INDUSTRY APPLICATIONS, VOL. 40, NO. 4, JULY/AUGUST 2004, Pages: 1012~1019

[4] Wanmin Fei, Zhenyu Lu, Lingyan Tan et al. Research of fault-current limiter for three-phase four-wire power system, IEEE Power Electronics Specialists Conference, 2002, Vol. 2, pp:709-712

[5] Wu Zhao-lin. A Type of Short Circuit Protection Circuit. Chinese Patent, ZL 96123001.0

[6] Wu Zhao-lin. Three Phase Short Circuit Current Limiting Transformer, Chinese Patent, ZL 00206596.7

[7] Zhengyu Lu, Daozhuo Jiang and Zhaolin Wu. A New Topology of Fault-Current Limiter and Its Parameters Optimization. IEEE Power Electronics Specialists Conference, Mexico:2003, pp:462-465

[8] Wanmin Fei, Yanli Zhang and Zhaojuan Meng. A Novel Solid-State Bridge Type FCL for Three-Phase Three-Wire Power Systems, IEEE Applied Electronics Conference and Exposition, Anaheim, California, U.S.A., 2007, Vol.2, pp:1084-1088

A Novel Isolation Power Supply for Gating Multiple Devices in FACTS Equipment

Yanli Zhang*, Wanmin Fei *, and Zhengyu Lu**

* School of Electrical and Automation Engineering, Nanjing Normal University, 78 Bancang Street, Nanjing, P.R. China
** College of Electrical Engineering, Zhejiang University, 38 Zheda Road, Hangzhou, P.R. China

Abstract--A new kind of power supply with multiple isolated outputs for gating semiconductor devices in FACTS equipments is developed. The main parts of the power supply include a DC current source, a bridge-type inverter, a series of output transformers and shunt voltage regulators. A single turn primary winding is used in output transformers. The isolation voltage among all outputs and input can be considerable high if the primary winding is made by high-voltage cable. The number and locations of the transformers can be arranged quite easily to meet the requirements of the main system. Compared with other power supply used in same purpose, the proposal one has the advantage in size, weight, cost, convenience, efficiency and reliability. With the increase of voltage level and number of devices in the FACTS equipment, the proposal power supply will be more advantageous.

Index Terms--Isolation, Power Semiconductor Device, Multiple outputs, Power supply, FACTS equipments.

I. INTRODUCTION

As increasing development of high-voltage and high-power power electronics equipments, such as FACTS equipments, multiple devices usually are connected in series to sustain a high operation voltage. Recently, the multi-level technology has been used more and more in FACTS equipments. It makes the device driving circuits become more complex, especially for the problem of high-voltage isolation. The main circuit of short-circuit fault-current limiter used in power systems, a typical example of high-voltage equipment, is shown in Fig.1. It mainly consists of a three-phase thyristor rectifier, a dc limiting reactor L1 and three bypass reactors L2, L3 and L4[1]. The limiter is most likely to be used in 15kV power systems [2]. Each thyristor arm in the limiter must consist of eight 6kV thyristors connected in series. The total number of thyristors used in the limiter will be 64. Namely, at least 60 isolated power supplies are needed for gating the thyristors. Hence, development of a new power supply for the limiter and other power electronic equipments is essential.

Dusan M. Raonic proposed a SCR self-supplied gate driver by means of the snubber capacitor as an energy storage element to eliminate the need of one isolated power supply for almost each power switch [3], but can not be used here because there is no voltage across the rectifier bridge in normal situation. Liuchen Chang developed a driver board power supply for high-power IGBTs used in three-phase inverter, which has four isolated outputs with one multi-winding transformer [4]. However, the increasing of winding number and isolation voltage would result in the transformer high complexity and bulkiness. Lothar Heinemann et al proposed a power supply with very high insulation requirements for IGBT gate-drivers by the use of a transformer with special structure, which has only one output [5]. Hence, it is inevitably too bulky and too inconvenient if the number of isolated auxiliary power source exceeds ten or more. Dejan VASIC et al introduced a MOSFET & IGBT gate drive insulated by a piezoelectric transformer. However, there are some drawbacks with the piezoelectric transformer: the insulation voltage cannot be high enough for ultra-high voltage application; the energy transmission is limited (usually less than 10W); and the cost is too expensive [6].

Based on a Chinese Patent [7], a novel high frequency switch power supply with many isolated outputs used for gating devices in FACTS equipment is developed. The main idea and topology of the main circuit are described. The characteristics and structure of the transformer are discussed. Experiments to verify the validity of the power supply are accomplished.

II. MAIN IDEA OF PROPOSED CONVERTER

The primary requirements for the power supply include multiple outputs and high isolation voltage among the outputs and input. Transformer is almost the best device for electric energy transmitting and isolation. But a high isolation voltage between the primary winding and secondary winding of an ordinary transformer would result in a great increase in its volume and weight. So a new kind of transformer is adopted, which has a single-turn primary winding. The isolation voltage between primary winding and secondary winding can be considerable high if the primary winding is made by a high-voltage cable. Because the voltage drop of the output transformer is small, N output transformers can be connected in series to obtain N isolated outputs. An ac current source is advantageous over ac voltage source for driving of series connected output transformers. The

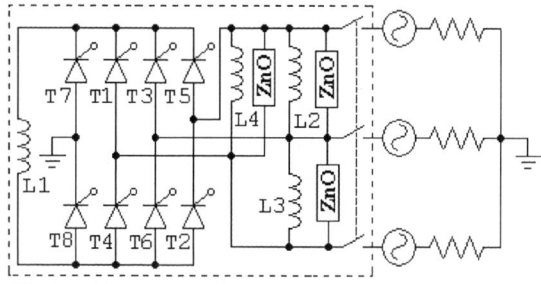

Fig. 1. Bridge Type FCL for three-phase four-wire power systems

978-1-4244-0644-9/07/$25.00 ©2007 IEEE

number of output transformers can be changed easily to

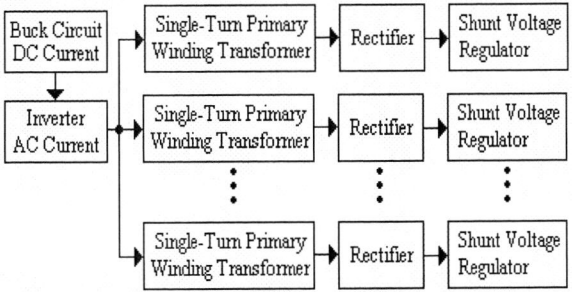

Fig. 2. Diagram of the proposed power supply

meet the requirements of the main system. Increase in the frequency of the ac current source can decrease the volume and magnetizing current of the transformer. So the main idea is producing a high-frequency ac current and using this current to drive a series of output transformers with single-turn primary winding. Fig.2 shows the diagram of the proposed power supply. In order to produce a high frequency ac current, a buck circuit and a single-phase inverter are used. The buck circuit produces a constant dc current. The inverter converts the constant dc current to a high-frequency ac current. By the output transformer, a high-frequency ac current source is isolated. Then, the ac current is rectified to be a dc current. Because a current source is employed to provide the energy, shunt voltage regulator is adopted to convert the dc current into a stable dc voltage.

III. DESIGN CONSIDERATIONS

A. Topology of the main circuit

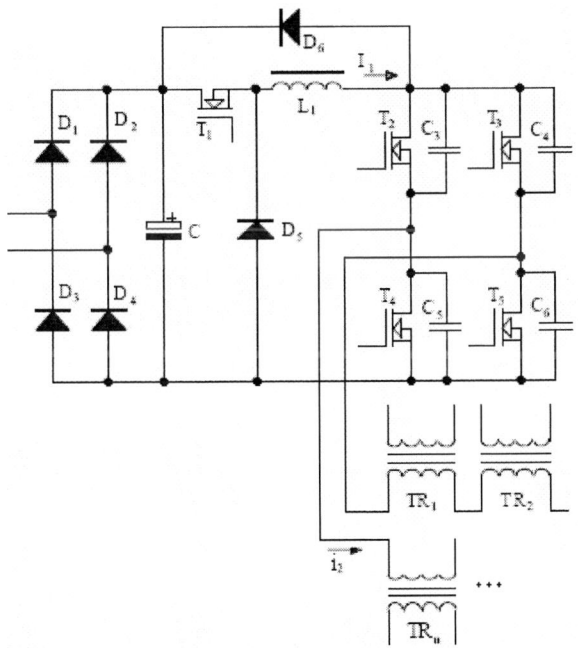

Fig. 3. Main circuit of the proposed power supply

As shown by fig.3, the main circuit of the proposed power includes a buck circuit and a single-phase bridge type inverter. The buck circuit is composed of T1, D5 and L1. Controlled by a current- loop, the buck circuit

generates a constant dc current I1. The single-phase inverter converts the constant DC current I1 to a high frequency alternative square-wave current i2. T2 and T5 (or T3 and T4) are controlled by one signal and turned on and off simultaneously, by which the square-wave current is produced. TR1, TR2, ~TRn are output transformers with special structure to be explained in following content. In order to minimize the size of the proposed power supply, the switching frequency should be high enough.

B. Output transformers

Each output transformer uses a ferrite toroid and a single turn in the primary winding. Each secondary winding offers an ac current i2. The number of outputs of the proposed power supply can easily be increased or decreased. One power supply can afford great number of isolated outputs if the current I1 and the output voltage of the rectifier are great enough. The location of every output of the proposed power supply can be arranged conveniently to meet the requirements of the devices in the main systems. Because the devices are used in high voltage FACTS equipment, isolation voltage between any two gating circuits must be high enough. The isolation voltage among input and all outputs can be considerable high if the primary winding is made by high-voltage cable. Because the number of the primary winding is 1, the magnetic conductivity of the magnetizer must be excellent, the length of the magnetic-core should be as short as possible and the section of the magnetic-core should be great enough so as to obtain a good electro-magnetic coupling result and to reduce the magnetizing current.

C. Shunt voltage regulators

Shunt voltage regulators are output circuits of isolation power source. Fig.4 shows the principle of the circuit. The rectifier composed of D8~D11 convert alternative

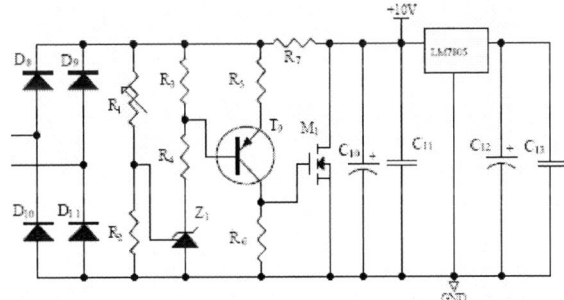

Fig. 4. Secondary circuit of the proposed power supply

current i2 to DC current. Circuit composed of resistor R1~R7, shunt regulator Z1, transistor T7 and MOSFET M1 convert DC current to a stable voltage.

IV. EXPERIMENT RESULTS

A power supply based on the principle described above is accomplished, which has 27 isolated outputs. Capacity of each output is 10W and isolation voltage between outputs is 40kV. The dc current I1 is 4A. Current in the secondary winding of output transformer is

Fig. 5. Main circuit and controller of the proposed power supply

Fig. 6. Secondary circuit and SCR triger

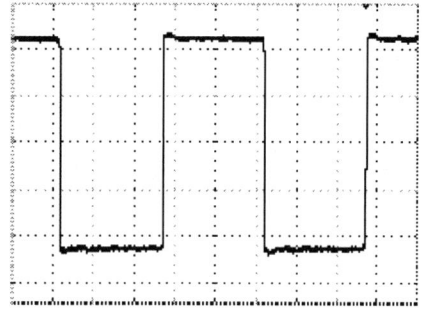

Horizontal:5V/div; Vertical:2uS/div

Fig. 7. Voltage waveform of isolation transformer

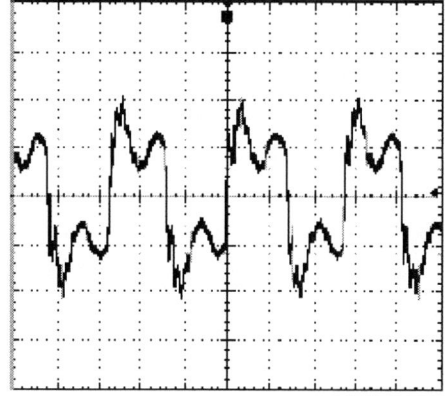

Horizontal: 2A/div; Vertical: 4uS/div

Fig. 8. Current waveform of isolation transformer

1A. Voltage across capacitor C10 and C11 in Fig 4 is

10V. The waveform of output voltage and current in the secondary side of output transformer is shown by Fig. 7 and Fig. 8 respectively. The power supply has already been used in the demo system of a SCR-based bridge type fault-current limiter shown by Fig. 1.

V. CONCLUSION

A novel Power supply with multiple isolated outputs for driving of thyristors used in FACTS equipments is developed. The number of outputs of the proposed power supply is great enough for driving all the thyristors used in the limiter. Using ultra-high voltage cable to make the common one turn primary winding of all the output transformers, the isolation voltage between its outputs can be made very high. The number and locations of its outputs can be changed very easily to meet the requirements of power electronics devices in the FACTS equipment. Compared with a power supply made in other ways with the same capacity, same number of outputs and same isolation voltage, the proposal power supply has the advantage of smaller in size, lighter in weight, lower cost, higher efficiency and reliability. The proposed power supply is well suited for high power application such as large motor drive systems where a great number of power electronic devices are involved.

REFERENCES

[1] Zhengyu Lu, Daozhuo Jiang and Zhaolin Wu. A New Topology of Fault-Current Limiter and Its Parameters Optimization. IEEE Power Electronics Specialists Conference, PESC03. Mexico:2003, pp.462-465

[2] P. K. Smith, P. G. Slade, M. sarkozi et al. Solid State Distribution Current Limiter and Circuit Breaker: Application Requirements and control Strategies, IEEE Transactions on Power Delivery, Vol.8, No.3, July 1993:1155~1164

[3] Dusan M. Raonic. SCR Self-Supplied Gate r for Medium-Voltage Application with Capacitor as Storage Element. IEEE Transactions on Industry Applications, Volume 36, No.1, January/February 2000: Page(s): 212-216

[4] LiuChen Chang. Development of a Driver Board Power Supply for High Power IGBTs Used in Three Phase Inverter. IEEE AES Systems Magazine, August 1996 Pages: 24-28

[5] Lothar Heinemann, Jochen Mast, Guntram Scheible et al. Power Supply for Very High Insulation Requirements in IGBT Gate-Drivers. The 1998 IEEE Industry Applications Conference. Thirty-Third IAS Annual Meeting, Volume: 2 , 1998,Page(s): 1562 –1566

[6] Dejan VASIC, François COSTA, Emmanuel SARRAUTE. A New MOSFET & IGBT Gate Drive Insulated By A Piezoelectric Transformer. Power Electronics Specialists Conference, 2001. PESC. 2001 IEEE 32nd Annual , Volume: 3 , Page(s): 1479 -1484

[7] Lu ZhengYu, Jiang DaoZhuo, Wu ZhaoLin, Wu GuoYan. Power supply with multiple isolated outputs applicable to power electronic device. Chinese Patent: ZL 01 1 45559.4

Voltage and Frequency Controller for Parallel Operated Isolated Asynchronous Generators

Bhim Singh, *Senior Member IEEE and* Gaurav Kumar Kasal

Dept. of Electrical Engineering, Indian Institute of Technology Delhi, New Delhi 110016 India

(e-mail: bhimsinghr@gmail.com , gauravkasal@gmail.com).

Abstract-- **This paper deals with a voltage and frequency controller for parallel operated isolated asynchronous generators used in constant power applications such as driven by uncontrolled pico hydro turbines. The proposed controller is having capability of controlling the voltage and frequency in decoupled manner. For controlling the voltage, a static compensator (STATCOM) is used as a reactive power compensator along with harmonic eliminator and a load balancer while for controlling the frequency; an electronic load controller (ELC) is used to regulate the total active power at the generators terminals. The STATCOM is realized using IGBTs (Insulated gate bipolar junction transistors) based voltage source converter (VSC), and a capacitor as an energy storage element at its DC link, while an ELC consists of a diode bridge rectifier, a chopper switch and an auxiliary load. The proposed generating system along with its controller is modeled in MATLAB using SIMULINK and PSB (Power System Block Sets) toolboxes. The simulated results are presented to demonstrate the capability of an isolated generating system with the proposed controller.**

Index Terms-- Isolated Asynchronous Generator, Parallel Operation, Voltage and Frequency Controller.

I. INTRODUCTION

Asynchronous generators (AGs), with their low maintenance and simplified control, appear to be an effective solution for small hydro and wind power plants. For their simplicity, robustness, and small size per generated kW, the AGs are more favored for such applications [1-4]. Recently, there is a global growing use of isolated asynchronous generators (IAGs) in pico hydro generation, especially for supplying electric power in remote areas where utility lines are uneconomical to install due to terrain, the right-of-way difficulties or the environment concerns. But poor voltage and frequency regulation with speed and load variations are the major drawbacks for their commercialization, therefore different voltage and frequency control schemes are proposed for IAGs in constant power applications driven by uncontrolled pico hydro turbines and wind power applications at different sites [5-12].

In this paper a new voltage and frequency control scheme is proposed for parallel operated isolated asynchronous generators in constant power applications driven by uncontrolled pico hydro turbines. Analysis of such a system is considered relevant because with increased load demand and full utilization of the generated power, parallel operation of generators is required and in comparison to other electric generators, asynchronous generators is not having problems of synchronization as no need of synchronizing torque and no problem due to hunting. Although number of attempts have been made for modeling and steady state analysis of parallel operated isolated asynchronous generators [13-16] and their voltage control [17].

The proposed controller is a combination of a STATCOM for controlling the voltage and an ELC for controlling the active power which in turn maintains the system frequency constant. Basic principle of the controller operation is that for regulating the voltage during load variation, a continuous and varying reactive power demand of the generators and loads is fulfilled by the STATCOM, while an ELC absorbs the additional generated power which is not consumed by the consumer loads so that total generated powers at the generators terminals remain constant which in turn regulates the system frequency.

II. SYSTEM CONFIGURATION AND CONTROL STRATEGY

Fig.1 shows the system configuration of parallel operated IAGs, with individual excitation capacitors, proposed voltage and frequency controller and consumer loads. The value of excitation capacitors for both generators is selected for generating the rated voltage at no-load, the additional reactive power requirement of the load and generators is met by the STATCOM which acts as a source of lagging or leading current to maintain the constant terminal voltage and it also functions as a harmonic eliminator and a load balancer. The STATCOM consists of a voltage source converter (VSC), DC bus capacitor and AC inductors. The output of the VSC is connected through the AC filtering inductors to the point of common coupling (PCC). The DC bus capacitor is used as an energy storage device and provides self-supporting DC bus of STATCOM. For regulating the frequency, an ELC is used which consists of a diode bridge rectifier and a chopper at DC link with auxiliary load which absorbs the additional generated power not consumed by the consumer loads by varying the duty cycle of the chopper, so that in this way the controller effectively regulates the generated voltage and frequency of the proposed electrical system.

Fig.2 shows the control strategy of the proposed controller for parallel operated isolated asynchronous generators. The control strategy for the STATCOM is based on the generating

978-1-4244-0644-9/07/$25.00 ©2007 IEEE

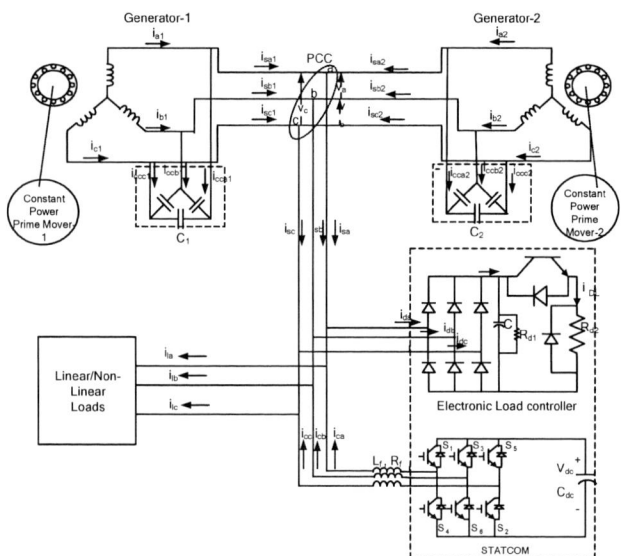

Fig.1 Schematic diagram of VF controller for an isolated asynchronous generators.

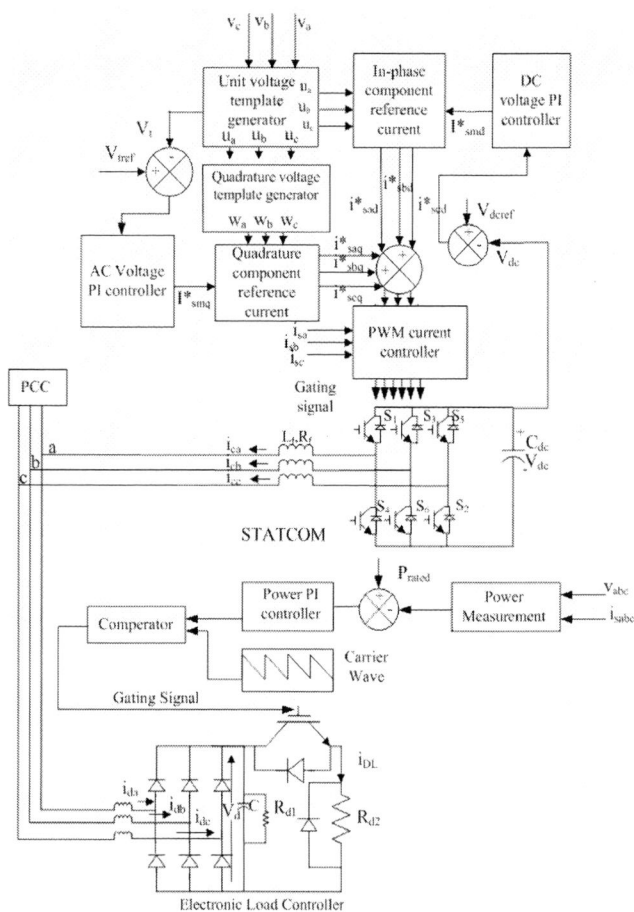

Fig. 2 Control scheme for the proposed voltage and frequency controller.

reference source current (i^*_{sa}, i^*_{sb} and i^*_{sc}) for voltage regulation as well harmonic current compensation and load balancing. These reference source currents are having two components one is the quadrature component (i^*_{saq}, i^*_{sbq} and i^*_{scq}) for reactive power compensation while other one is in phase component (i^*_{sad}, i^*_{sbd} and i^*_{scd}) for maintaining the DC link voltage constant. For generating the in-phase components of the currents, sinusoidal in-phase unit vectors (u_a, u_b, u_c) are estimated by dividing the voltages v_a, v_b and v_c by their amplitude V_t. To provide self supporting DC bus of STATCOM, its DC voltage is sensed and compared with DC reference voltage. The voltage error is processed in a PI controller. The output of the PI controller (I_{smd}^*) decides the amplitude of active current. Multiplication of in-phase unit vectors (u_a, u_b and u_c) with output of the PI controller (I_{smd}^*) yields the in-phase component of the reference source currents (i_{sad}^*, i_{sbd}^* and i_{scd}^*). For generating the quadrature component of the reference source currents another set of quadrature unit vectors (w_a, w_b, w_c) are estimated from in-phase unit vectors. To regulate the AC terminal voltage (V_t), it is sensed and compared with the reference voltage. The voltage error is processed in the PI controller. The output of the PI controller (I_{smq}^*) for an AC voltage control loop decides the amplitude of reactive current to be generated by the STATCOM. Multiplication of quadrature unit templates (w_a, w_b and w_c) with the output of the PI based AC voltage controller (I_{smq}^*) yields the quadrature component of the reference source currents (i_{saq}^*, i_{sbq}^* and i_{scq}^*). The instantaneous sum of quadrature and in-phase components provides the reference source currents (i_{sa}^*, i_{sb}^* and i_{sc}^*), which are compared with the sensed line currents (i_{sa}, i_{sb} and i_{sc}). These current error signals are amplified and used in the hystresis controller to generate the gating signals for IGBTs of the VSC.

For controlling the chopper of an electronic load controller, estimated power (P_{gen}) is compared with reference power (total rated power of the generators P_{rated}) and then output of the power PI controller is compared with saw-tooth carrier wave resulting in the PWM output of varying duty cycle for the IGBT of the ELC.

III. CONTROL ALGORITHM

Basic equations of control scheme of the proposed controller for IAGs are developed into two sections. Section 'A' which describes the equations of controlling the "STATCOM" while section 'B' deals with the chopper control of electronic load controller (ELC).

A. Control Algorithm for STATCOM

Different components of IAGs-STATCOM system shown in Figs. 1-2 are modeled as follows.

Three-phase voltages at the IAGs terminals (v_a, v_b and v_c) are considered sinusoidal and hence their amplitude is computed as:

$$V_t = \sqrt{(2/3)\,(v_a^2 + v_b^2 + v_c^2)} \qquad (1)$$

The unit template in phase with v_a, v_b and v_c are derived as:

$$u_a = v_a/V_t;\ u_b = v_b/V_t;\ u_c = v_c/V_t \qquad (2)$$

The unit template in quadrature with v_a, v_b and v_c may be derived using a quadrature transformation of the in-phase unit templates u_a, u_b and u_c as:

$$w_a = -u_b / \sqrt{3} + u_c / \sqrt{3} \qquad (3)$$

$$w_b = \sqrt{3}\, u_a / 2 + (u_b - u_c) / 2\sqrt{3} \qquad (4)$$

$$w_c = -\sqrt{3}\, u_a / 2 + (u_b - u_c) / 2\sqrt{3} \qquad (5)$$

1) Quadrature Component of Reference Source Currents

The AC voltage error $V_{er(n)}$ at the nth sampling instant is:

358

$$V_{er(n)} = V_{tref(n)} - V_{t(n)} \qquad (6)$$

where $V_{tref(n)}$ is the amplitude of reference AC terminal voltage and $V_{t(n)}$ is the amplitude of the sensed three-phase AC voltage at the IAGs terminals at n^{th} instant.

The output of the PI controller ($I^*_{smq(n)}$) for maintaining AC terminal voltage constant at the nth sampling instant is expressed as:

$$I^*_{smq(n)} = I^*_{smq(n-1)} + K_{pa} \{ V_{er(n)} - V_{er(n-1)}\} + K_{ia} V_{er(n)} \qquad (7)$$

where K_{pa} and K_{ia} are the proportional and integral gain constants of the proportional integral (PI) controller. $V_{er(n)}$ and $V_{er(n-1)}$ are the voltage errors in n^{th} and $(n-1)^{th}$ instant and $I^*_{smq(n-1)}$ is the amplitude of quadrature component of the reference source current at $(n-1)^{th}$ instant.

The quadrature components of reference source currents are computed as:

$$i^*_{saq} = I^*_{smq} w_a; \; i^*_{sbq} = I^*_{smq} w_b; \; i^*_{scq} = I^*_{smq} w_c \qquad (8)$$

2) In-Phase Component of Reference Source Currents

The error in DC bus voltage of STATCOM ($V_{dcer(n)}$) at n^{th} sampling instant is:

$$V_{dcer(n)} = V_{dcref(n)} - V_{dc(n)} \qquad (9)$$

where $V_{dcref(n)}$ is the reference DC voltage and $V_{dc(n)}$ is the sensed DC link voltage of the STATCOM. The output of the PI controller for maintaining DC bus voltage of the STATCOM at the nth sampling instant is expressed as:

$$I^*_{smd(n)} = I^*_{smd(n-1)} + K_{pd} \{ V_{dcer(n)} - V_{dcer(n-1)}\} + K_{id} V_{dcer(n)} \qquad (10)$$

where $I^*_{smd(n)}$ is considered as the amplitude of active source current. K_{pd} and K_{id} are the proportional and integral gain constants of the DC bus PI voltage controller.

In-phase components of reference source currents are computed as:

$$i^*_{sad} = I^*_{smd} u_a; \; i^*_{sbd} = I^*_{smd} u_b; \; i^*_{scd} = I^*_{smd} u_c \qquad (11)$$

3) Reference Source Currents

These are sum of in-phase and quadrature components of the reference source currents as:

$$i^*_{sa} = i^*_{saq} + i^*_{sad} \qquad (12)$$

$$i^*_{sb} = i^*_{sbq} + i^*_{sbd} \qquad (13)$$

$$i^*_{sc} = i^*_{scq} + i^*_{scd} \qquad (14)$$

4) PWM Current Controller

The reference currents (i^*_{sa}, i^*_{sb} and i^*_{sc}) are compared with the sensed source currents (i_{sa}, i_{sb} and i_{sc}). The ON/OFF switching patterns of the gate drive signals to the IGBTs are generated from the PWM current controller. The current errors are computed as:

$$i_{saerr} = i^*_{sa} - i_{sa} \qquad (15)$$

$$i_{sberr} = i^*_{sb} - i_{sb} \qquad (16)$$

$$i_{scerr} = i^*_{sc} - i_{sc} \qquad (17)$$

These current error signals are amplified and then fed to the hystresis controller for generating the switching signals of IGBT of VSC.

B. Control Algorithm for ELC

To maintain the generated power constant at generators terminal, the measured power (P_{gen}) is compared with generators rated power (P_r) considered as reference power and power error $P_{er(n)}$ at n^{th} sampling instant is calculated as:

$$P_{er(n)} = P_{r(n)} - P_{gen(n)} \qquad (18)$$

where $P_{r(n)}$ is the reference power and is taken equal to rated power. $P_{g(n)}$ is the computed power at the nth sampling instant and it is computed as:

$$P_{gen} = (1/\sqrt{3}) (v_a i_a + v_b i_b + v_c i_c) \qquad (19)$$

where v_a, v_b and v_c are line voltages.

The output of the power PI controller at the nth sampling instant is expressed as:

$$P^*_{con(n)} = P^*_{con(n-1)} + K_{pp} \{ P_{er(n)} - P_{er(n-1)}\} + K_{pi} P_{er(n)} \qquad (20)$$

where K_{pp} and K_{pi} are the proportional and integral gain constants of the power controller. The PI controller output ($P^*_{con(n)}$) is compared with the triangular carrier (P_{tri}) waveform and output is fed to the gate of the chopper switch (IGBT) in ELC of the controller.

If $P^*_{con(n)} > P_{tri}$, then SD = 1 and if $P^*_{con(n)} < P_{tri}$, then SD = 0 The SD is the switching function used for generating the gating pulse of IGBT of the chopper of ELC.

IV. MATLAB BASED MODELING

The proposed controller for parallel operated isolated asynchronous generators are modeled and simulated for supplying balanced/unbalanced, linear/ non-linear loads on MATLAB along with Simulink and PSB toolboxes. Fig. 3 shows the complete model of the proposed electrical generating system while Figs. 4(a)-4 (b) demonstrate the corresponding subsystems for VSC control of STATCOM and chopper control on the DC side of ELC. The proposed system is modeled using 15kW, 415V, 50Hz, Y-connected and 7.5kW, 415 V, 50Hz, Y-connected asynchronous machines. Two delta connected capacitor banks of rating of 9kVAR and 5kVAR are used for generating the rated voltage at no load. The STATCOM is realized with IGBT based VSC using universal bridge with capacitor at DC link, while an ELC is realized using diode bridge rectifier and an IGBT based chopper switch with an auxiliary load at DC bus. Values of different parameters are given in Appendix. Both linear and non-linear loads are considered here to demonstrate the capability of the proposed controller for parallel operated isolated asynchronous generators. Simulation is carried out in discrete mode at 5e-6 step size with ode23tb (stiff/TR-BDF-2) solver on MATLAB version of 7.1 considering magnetizing saturation characteristics of the IAGs.

V. RESULTS AND DISCUSSION

The performance of the controller for parallel operated isolated asynchronous generators system feeding linear/ non-linear, balanced/unbalanced loads is simulated and transient waveforms of the generators voltage (v_{abc}) and generators currents (i_{abc1}, i_{abc2}), individual excitation capacitors currents (i_{cca1}, i_{cca2}), load currents (i_{labc}), STATCOM currents (i_{cabc}), ELC current (i_{da}), amplitude of terminal voltage (v_t), the DC link voltage (V_{dc}), generators speed (ω_1, ω_2) frequency and (f) and variation of generators power (P_{gen1}, P_{gen2}) consumer and auxilliary load powers (P_{load} and $P_{auxiliary}$) etc are shown in Figs. 5-6. Detailed parameters of asynchronous machines are given in Appendix.

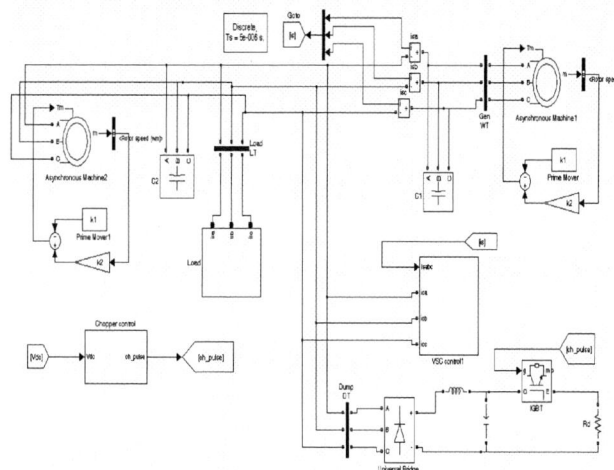

Fig. 3 MATLAB based simulation model of the electrical system with controller.

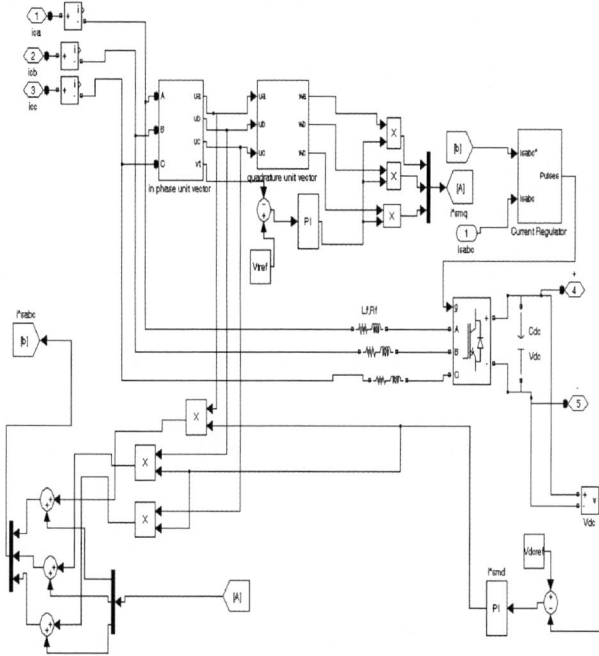

Fig. 4 (a) Subsystem of VSC control of STATCOM.

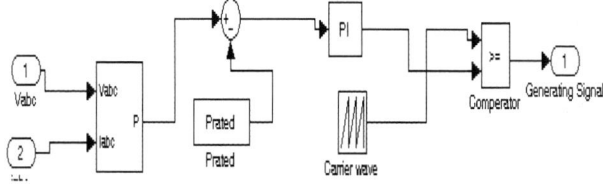

Fig. 4 (b) Subsystem of chopper control of ELC.

A. Performance of the System Feeding Linear Loads

Fig. 5 shows the performance of the proposed system with balanced/unbalanced resistive loads. At 2.6 s a balanced delta connected full load of 22.5kW is applied then the powers is drawn by a auxiliary load (P_{dump}) reduces to almost zero value. With opening of one phase at 2.75 s, and another phase at 2.85 s, the load becomes unbalanced and the load balancing aspect of the controller is demonstrated along with voltage and frequency regulation. At 2.95s, the consumer load is fully removed; the auxiliary load absorbs the full active power

generated by the generators, which shows that the controller maintains the constant value of the voltage and frequency of the generators.

B. Performance of the System Feeding Non-Linear Loads

Similarly Fig. 6 demonstrates the capability of the proposed controller for parallel operated IAGs with balanced-unbalanced nonlinear loads. At 2.6 s around 80% consumer load is applied and it is observed that the ELC current (i_{da}) is reduced for maintaining constant power at the generator terminal which in turn regulates the system frequency. At 2.8 s one phase of the load is opened and the load becomes unbalanced but the controller balances the load on the generators by maintaining balanced voltage and current at the generators terminal which shows its load balancing aspects. At 2.9 s 'b' phase of the load is again closed and it observed that the ELC current again reduces for maintaining constant power, while at 3.0 s the load is removed and the ELC absorbs the full power to regulate the frequency while regulation of voltage is achieved by the STATCOM.

C. Power Quality Issues

Table-I demonstrates the performance of the controller as a harmonic eliminator. Here it is observed that due to non-linear behavior of the ELC, it injects the harmonics currents in the system which is compensated by the STATCOM along with voltage control. At 80% of non-linear consumer loads, the ELC draws very less current and the load current which is having total harmonic distortion (THD) of around 22.29%, is compensated by the STATCOM and THD of terminal voltage, generator-1 current and generator-2 current are observed order of 0.38%, 1.23% and 0.64% respectively. At single phase non-linear load, the load current THD of 36.71% is observed. Under this condition, the THD of terminal voltage, generator-1 current and generator-2 current are observed order of 0.83%, 2.69% and 1.65% respectively which is much less than 5%, the limit imposed by IEEE – 519 Standard.

TABLE I
TOTAL HARMONIC DISTORTION UNDER DIFFERENT NON-LINEAR LOAD

S.N	Condition of Load	% Total Harmonioc Distortion (THD)			
		v_a	i_{a1}	i_{a2}	i_{lc}
1	Balanced Non-linear Load	0.38	1.23	0.64	22.29
2	Un-Balanced Non-linear Load	0.83	2.69	1.65	36.71

VI. CONCLUSION

The performance of the proposed controller is demonstrated for parallel operated isolated asynchronous generators in constant power applications driven by uncontrolled pico hydro turbines. It is observed that the controller is having capability of voltage and frequency regulation along with harmonic compensation and load balancing. In comparison to other voltage and frequency controllers the proposed controller controls the active and reactive currents in decoupled manner, so that the cost and rating of the VSC of STATCOM is comparatively reduced.

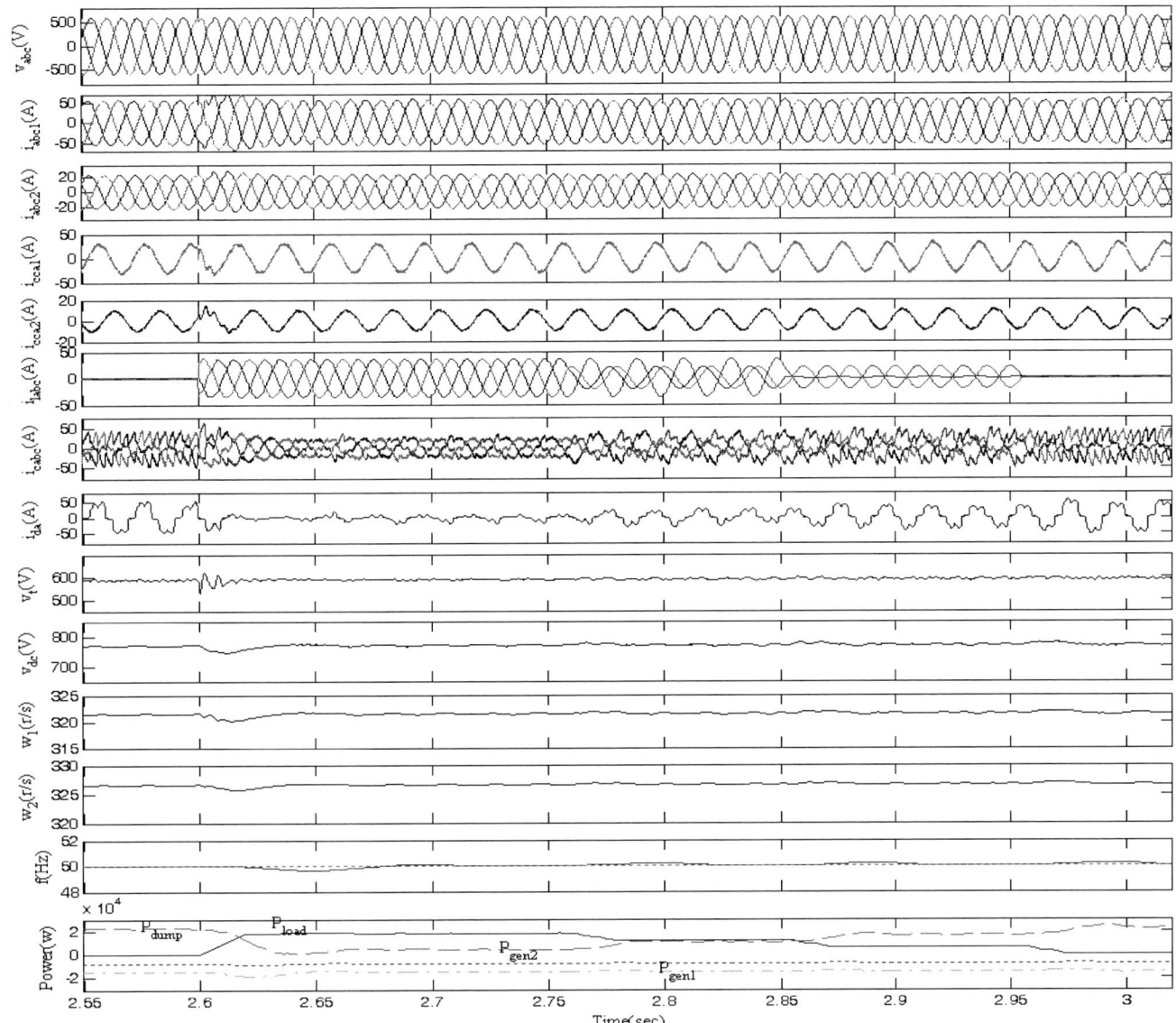

Fig. 5 Transient waveforms of parallel operated IAGs with controller feeding linear load.

VII. APPENDIX

A. The parameters of 15kW, 415V, 50Hz, Y-connected, 4-pole asynchronous machine are given below.

$R_s = 0.58\Omega$, $R_r = 0.78\Omega$, $X_{lr} = X_{ls} = 2.52\Omega$, $J = 0.23$kg-m^2,

$L_m = 0.22$ $I_m < 2.8$

$L_m = 0.0001I_m^2 - 0.0138I_m + 0.27$ $2.8 < I_m < 14.2$

$L_m = 0.1$ $I_m > 14.2$

B. The parameters of 7.5kW, 415V, 50Hz, Y-connected, 4-pole induction machine are given below.

$R_s = 1\Omega$, $R_r = 0.77\Omega$, $X_{lr} = X_{ls} = 1.5\Omega$, $J = 0.1384$ kg-m^2

$L_m = 0.134$ $(I_m < 3.16)$

$L_m = 9e-5I_m^2 - 0.0087I_m + 0.1643$ $(3.16 < I_m < 12.72)$

$L_m = 0.068$ $(I_m > 12.72)$

C. Controller Parameters

STATCOM Parameters:

$L_f = 5$mH, $R_f = 0.1\Omega$, and $C_{dc} = 6000\mu$F

$K_{pa} = 0.13$, $K_{ia} = 0.032$ $K_{pd} = 0.05$, $Ki_d = 0.01$

ELC Parameters:

$R_{d1} = 5$kΩ, $R_{d2} = 5\Omega$, $C = 1000$ μF

D. Consumer Loads

Resistive load 7.5kW single phase loads

Non-linear load   18kW with 1000μF capacitor and 1mH inductor at DC end of diode bridge rectifier

E. Prime Mover Characteristics for 15kW Machine

$T_{sh} = K_1 - K_2 \omega_r$, $K_1 = 6500$, $K_2 = 40$

F. Prime Mover Characteristics for 7.5 kW Machine

$T_{sh} = K_1 - K_2 \omega_r$, $K_1 = 3300$, $K_2 = 10$

REFERENCES

[1] R.C. Bansal, "Three phase self excited induction generators: an overview," *IEEE Trans. on Energy Conversion*, vol 20, no. 2, pp. 292-299, June 2005.

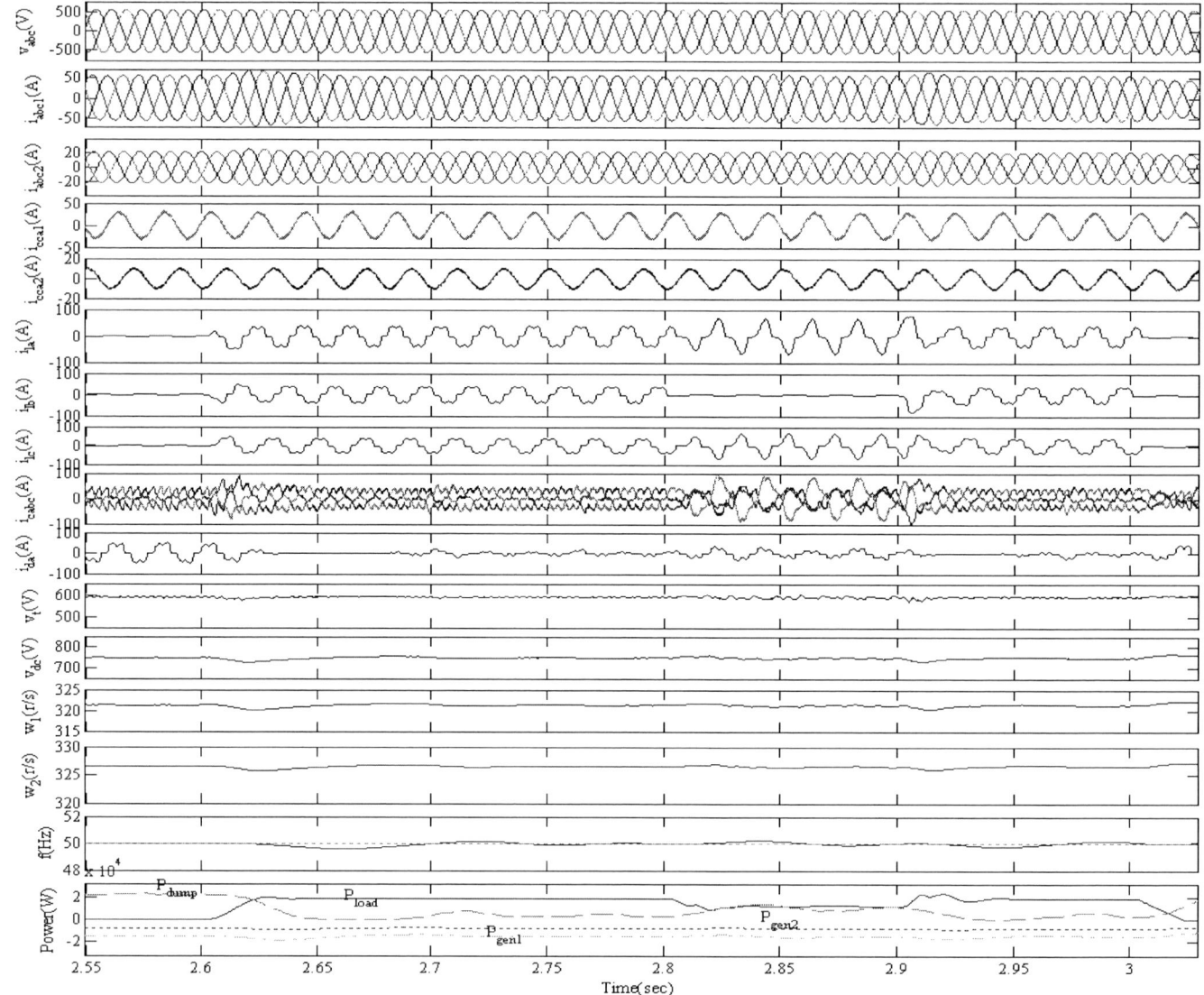

Fig. 6 Transient waveforms of parallel operated IAGs with controller feeding non-linear load.

[2] S.N. Bhadra, D. Kastha, S.Banerjee, "Wind Electrical Systems" 1st Ed. Oxford University Press, New Delhi 2004.

[3] G.K.Singh, "Self-excited induction generator research- a survey" *Electrical Power System Research,* vol 69, no. 2-3, pp 107-114, May 2004.

[4] M. Godoy Simoes, Felix A. Farret, "Renewable Energy Systems," 1st Ed. CRC Press Florida, 2004.

[5] D. Joshi, K.S Sandhu and M.K Soni. "Constant Voltage Constant Frequency Operation for a Self-Excited Induction Generator," *IEEE Transactions on Energy Conversion,* vol 21, no.1, pp.228 – 234, March 2006.

[6] E. Suarez and G. Bortolotto, "Voltage – frequency control of a self excited induction generator," *IEEE Trans Energy Conversion,* vol. 14, no. 3 pp.394-401, Sep 1999.

[7] Woei-Luen Chen and Yuan-Yih Hsu, "Controller design for an induction generator driven by variable-speed wind turbine," *IEEE Transactions on Energy Conversion,* vol 21, no.3, pp.625 – 635, Sept 2006.

[8] T.F. Chan, K.A. Nigim and L.L.Lai, "Voltage and frequency control of self excited slip ring induction generators" *IEEE Trans. on Energy Conversion,* Vol. 19, No. 1, pp. 81-87, March 2004.

[9] R. Bonert and S. Rajakaruna, "Self-excited induction generator with excellent voltage and frequency control," *IEE Proc.-Gener. Transm. Distrib.,* vol. 145, no. 1, pp. 33-39, January 1998.

[10] E. G. Marra and J. A. Pomilio, "Self excited induction generator controlled by a VS-PWM bi-directional converter for rural application,"

IEEE Trans. on Industry Applications, vol. 35, no. 4, pp. 877-883, July/August 1999.

[11] B.Singh, S.S. Murthy and Sushma Gupta, "A voltage and frequency controller for self-excited induction generators" *Electric Power Components and Systems,* vol., 34, pp 141-157, 2006.

[12] Luiz A.C. Lopes and Rogerio G. Almeida, "Wind-driven induction generator with voltage and frequency regulated by a reduced rating voltage source inverter" *IEEE Trans. on Energy Conversion,* vol. 21, no. 2, pp. 297-304, June 2006.

[13] D. B. Watson and I.P. Milner, "Autonomous and parallel operation of self excited induction generator," *International Journal of Electrical Engineering Education,* vol. 22, pp. 365-374, 1985.

[14] F.A. Farret, B. Palle and M.G. Simoes, "Full expandable model of parallel self excited induction generator" *IEE Proc- Electr. Power Appl.,* vol. 152, no.1, pp. 96-102, 2005.

[15] Chandan Chakraborty, S.N. Bhadra, Muneaki Ishida and A.K. Chattopadhyay, "Performance of parallel operated self excited induction generators with the variation of machine parameters," in *Proc. of IEEE Conference on Power Electronics and Drive Systems,* July. 1999, pp. 86-91.

[16] L. Wang and C.H. Lee, "Dynamic analysis of parallel operated self excited induction generator feeding an induction motor load," *IEEE Trans. Energy Conversion,* vol. 14, vol.3, pp 479-485, Sept 1999.

[17] A.H. Al-Bahrani and N.H. Malik, "Voltage control of parallel operated self excited induction generators," *IEEE Trans. Energy Conversion,* vol.8, no.2, pp. 236-242, June 1993

DSP controlled Semiconductor based High-Voltage Source

F. Martin*, T. Leibfried*, O. Kerz* and K. Mössner*

*University of Karlsruhe, Institute of Electric Energy Systems and High-Voltage Technology (IEH)
Kaiserstrasse 12, 76128 Karlsruhe, Germany

Abstract-- **For testing electrical components the test voltage has to achieve special requirements. Tests showed that it's not possible to fulfill these requirements using a semiconductor based frequency converter with a conventional static pulse pattern. This paper deals with the development of an appropriate closed loop control unit and its implementation on a Digital Signal Processor (DSP). The focus is on the synthesis of the control, and the transfer to a digital and time-discrete system for the DSP as well as the obtained results after the application in a frequency converter with semiconductor valves.**

Index Terms-- **Closed Loop Control, DSP, Semiconductor based Power Supply, Frequency Converter**

I. INTRODUCTION

The requirements on HV test voltages are defined in IEC 60060-1 [11]. In addition there is the demand for HV tests with high sinusoidal AC voltages of higher or lower frequency than line frequency. So far amplitude and frequency variable sinusoidal output voltages have been generated with classic motor-generator (m.-g.) sets. In accordance with the proceeding technical development in the area of electronic power semiconductors the maximum off-state voltages and on-state currents have rapidly increased in the recent years. As a consequence semiconductor elements are not only used for the known purposes. A new challenge is to replace the standard m.-g. sets with static frequency converters based on power semiconductor elements.

Even non-sinusoidal alternating voltage shapes of a wide variety can be generated by static frequency converters, if the converters are controlled in an appropriate way. The non-sinusoidal voltages may be important for insulation research for future power systems.

Testing electrical equipment for power generation, transmission and distribution, high voltages have to be generated by a step-up transformer between the converter and the test object. The sinusoidal voltage at the output of this transformer is distorted with a multitude of harmonics caused by the pulse pattern of the inverter. Due to this, the voltage has to be smoothed with a reactance or a multistage sinus filter. However, this would result in a remarkable power loss for higher power ratings of the static frequency converter. On the other hand the static pulse pattern does not adapt to different loads, and thus the total harmonic distortion (THD) for a sinusoidal voltage is higher than 5 %, the value demanded by IEC 60060-1 [11].

To prevent these issues a novel control strategy for static frequency converters was developed to meet the requirements. This paper deals with the automatic control, the realization on a DSP development system and the im-

plementation in a semiconductor based converter. Additionally the paper presents results from a real system with different loads and voltage forms providing a total harmonic distortion factor smaller than 5 %.

II. CONTROL SYNTHESIS

The basic arrangement of the inverter with a passive LC-filter is shown in Fig. 1 [2].

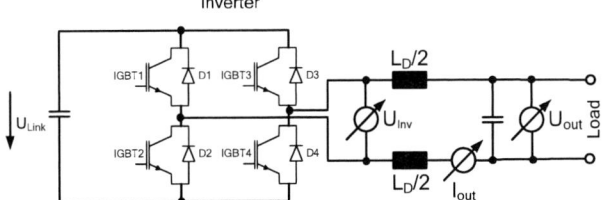

Fig. 1: Frequency inverter

There are many ways to approach the design of a control of the output voltage. A first solution results in controlling the known pulse-width modulation by a variation of the control voltage. However simulations show that this control is too slow, and due to this there is no significant improvement of the output voltage. Another attempt is the immediate control providing the switching signals for the valves of the converter directly without pulse width modulation. This principle offers a multitude of possible realizations, which have to be examined carefully. The intention of the synthesis is to develop an appropriate closed-loop control structure which is stable and stationary correct, sufficiently damped, but also quick enough.

The fundamental idea for the control is based on a simple comparison between a setpoint and the actual value applied to the output voltage and the output current. However, if this approach was realized the converter output would permanently seesaw between $+U_{link}$ for $U_{ref} < U_{actual}$ and $-U_{link}$ for $U_{ref} > U_{actual}$. This easy control would not be stationary correct as well as tend to oscillations and instabilities. Due to this a third possible state has to be implemented: the free-wheel-state, at which the output of the converter is 0 V. Therefore a tolerance zone around the reference voltage has to be defined.

Different control structures were examined to find the most suitable scheme for a closed-loop control. During the simulation the cascade control, shown in Fig. 2, proved to be the best conception. Here the control of the output current is superimposed by the control of the output voltage.

978-1-4244-0644-9/07/$25.00 ©2007 IEEE

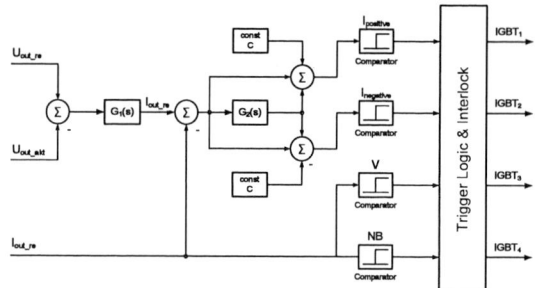

Fig. 2: Cascade control for one phase of the inverter

III. DSP IMPLEMENTATION

The development of the control was followed by a simulation of the whole system including the converter link and the converter itself controlled by the new closed loop control. The simulation results with different loads were very good, and thus an already existing single phase pulse-width modulated converter was modified and fit out with the new closed loop control.

There are two possibilities for the hardware realization: An analog circuit design on one hand or an implementation with a DSP on the other hand. The advantages of the digital solution are:

- Possible automatic adaptation of the control parameters during service
- Data transfer via CAN bus system
- Use of DSP internal libraries for ideal sinusoidal nominal value generation
- Easy realization of modifications with program code in comparison to a new circuit design necessary to change the analog control

An appropriate DSP has to fulfill the high demands set by the control design and the used IGBTs: The allowed switching frequency for the IGBTs is 10 kHz, but the minimum on- and off-state time for an IGBT valve is only 5 µs. Hence for the best output voltage shape the control, respectively the DSP, has to deliver new results at least every 5 µs. This requires a fast processor, which can execute an A/D conversion of the actual value of the output voltage and the output current and the subsequent processing of the closed loop control in 5 µs. Furthermore the DSP should be equipped with auxiliary features e.g. analog-digital-converter, CAN bus interface, general-purpose I/O ports and on-chip flash memory.

The Texas Instruments TMS320F2812 proved to be qualified for the application as a digital control with the properties mentioned above. For developing purposes it was used on a Spectrum Digital Inc. eZdsp™ F2812 evaluation board. The 150 MHz clock of the CPU is sufficient to control the four IGBT valves of the H-Bridge inverter, and it includes multiple additional tools and libraries needed for control applications in general.

A. Implementation

In a first step the elements of the time-continuous control had to be transformed to a time-discrete system. Table 1 shows the allocation of analog and digital control elements; T is the cycle time of the algorithm, Yi respec-

tively is the output at discrete equidistant points of time $t_i = T \cdot i$, $i = 1,2,3,\dots$ [1].

TABLE 1: ANALOG AND DIGITAL CONTROL ELEMENTS

	Time Continuous	Time Discrete
P-element	$Y(t) = K \cdot X(t)$	$Y_i = K \cdot X_i$
Σ-element	$Y(t) = X_1(t) + X_2(t)$	$Y_i = X_{1i} + X_{2i}$
I-element	$Y = K \int_0^t X(\tau)d\tau + Y(0)$	$Y_i = Y_{i-1} + K \cdot T \cdot X_i$
PT1-element	$T_1 \dfrac{dY}{dt} + Y = X$	$Y_i = Y_{i-1} + \dfrac{T}{T_1+T} \cdot (X_i - Y_{i-1})$

Apart from calculating the control stages the software must execute the initialization of the DSP and its peripherals, must trigger the A/D conversion, arrange the communication between the control unit and the user interface SIEMENS OP17 via eCAN and Profibus DP as well as provide the nominal sinusoidal value for the setpoint of the output voltage. Hence the program code was developed corresponding to the structure shown in Fig. 3.

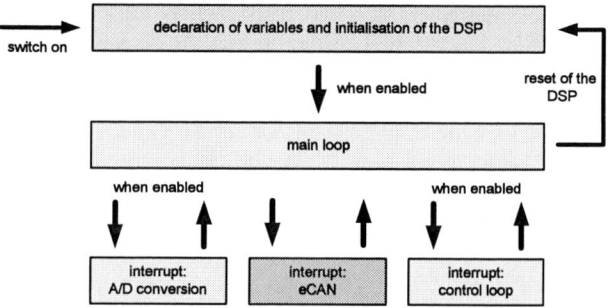

Fig. 3: Defined program structure for the DSP

During the initialization of the DSP

- the boot-loading software is executed.
- the system clock and the peripheral clock is set.
- the required interrupts are enabled and the interrupt service routines (ISR) are declared.
- the A/D converter is powered up.
- the values of the control constants are assigned.

The interrupt "A/D conversion" is called periodically every 5 µs and hence it is the start of one program loop. During the interrupt the actual value of the output current and the output voltage is A/D-converted and the digital setpoint of the output voltage is calculated with the DSPs own high precision sinus generator library.

Triggered by the end of the A/D conversion the interrupt "control loop" is started. In the corresponding service routine the control scheme shown in Fig. 2 is applied to the output of the A/D conversion and the generated setpoint value. Depending on the result the four logical output signals are set high or low and transmitted to the trigger logic and interlock unit via four assigned output pins.

A further interrupt is needed to arrange communication via CAN: This interrupt is triggered by a new eCAN message but then is suppressed for 50 ms. This time limit is necessary to reduce the length of one program cycle but it's little enough to capture the new desired values set by the operator.

B. Communication

As the DSP does not provide a Profibus-DP-interface, an intermediate step is necessary for the communication between the users interface and the DSP. This task is done by a microcontroller allocated to the central control and measurement unit. The bidirectional communication concept is given in Fig. 4 [2], [4], [5], [6].

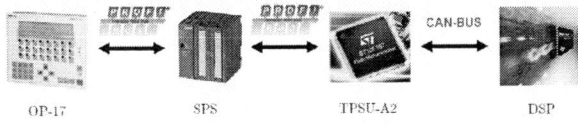

Fig. 4: Communication concept between the user interface and the DSP

The proportional and integral values for the voltage and current control as well as the free-wheel constant for the current control can be set with the user interface. Additionally the desired amplitude and frequency value of the output voltage is transmitted to the DSP.

C. Digital Nominal Value Generation

The TMS320F2812 can process the various signal generator libraries from Texas Instruments which include not only sinusoidal shapes but as well offer other signal forms. Due to the desired frequency band between 1 Hz and 250 Hz a high precision sinus generator should be used. 8 bits, i.e. 256 values are stored in a databank and 15 bits are used for linear interpolation. The frequency is generated with a 32 bit modulo counter. The minimum frequency step is calculated by

$$\Delta f = \frac{f_{max}}{step_{max}} \qquad (1)$$

with

$$step_{max} = 2^{32} \cdot f_{max} \cdot T_{ISR} \overset{!}{<} 2^{32}. \qquad (2)$$

fmax = 250 Hz is the maximum desired output frequency and TISR = 5 µs is the cycle time for the interrupt service routine. The TI library demands stepmax to be less than 232. The resultant minimum frequency step Δf = 10-27 Hz.

D. Adaptation Circuit Board

After the F2812 was programmed in C with the aid of TIs Code Composer Studio (CCS), the code was optimized concerning the cycle time, and the control was tested in a laboratory test setup. The input signals were simulated by external signal generators, and the output signals were displayed on an oscilloscope screen. For the actual operation connected to the measurement unit and the trigger logic of the converter an analog adaptation circuit board had to be designed to accomplish the following tasks:

- adaptation of the measured voltages to the front-end voltage level of the DSPs A/D converter (0 V to +3 V)
- overvoltage protection for the A/D converter
- adaptation of the output signals to the 5 V-TTL-level for the trigger logic and interlock
- CAN bus module for the communication

E. Complete Control System

The prototype of the static frequency converter used to generate testing voltages for high voltage applications still is operated with the DSP running on the evaluation board, as development and further investigation is much easier to realize compared to stand alone solutions with an own redesigned circuit board for the DSP. Fig. 5 shows the DSP based inverter control system, which can be plugged into the central control unit shown in Fig. 6.

Fig. 5: Control system with evaluation board and analog adaptation circuit

Fig. 6: Central control, feedback-control and measurement unit for the converter

IV. RESULTS

Tests were carried out with every possible combinations of resistive, inductive and capacitive loads. Feeding a capacitive load the output voltage approaches an ideal sinusoidal shape as the additional capacitance in parallel increases the smoothing capacity at the output of the converter. Table 2 shows the summery of the oscilloscope screenshots. The maximum tested output power was 45 kVA.

TABLE 2: OVERVIEW FOR THE OSCILLOSCOPE SCREENSHOTS

Load	1,1 Ω	2 mH	1,1 Ω + 2 mH
Voltage	212 V	128 V	174 V
Current	196 A	202 A	137 A
THD	0,6 %	1,3 %	1,2 %
Fig. No.	7	9	11

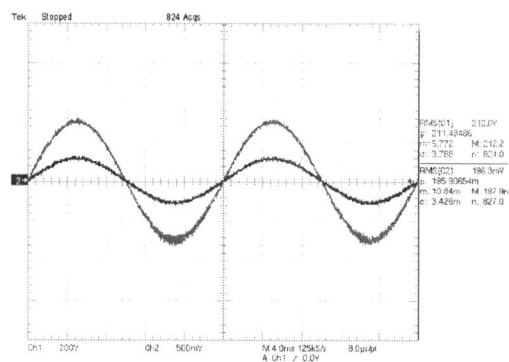

Fig. 7: Oscilloscope screenshot of a resistive load (voltage = brown, current = blue)

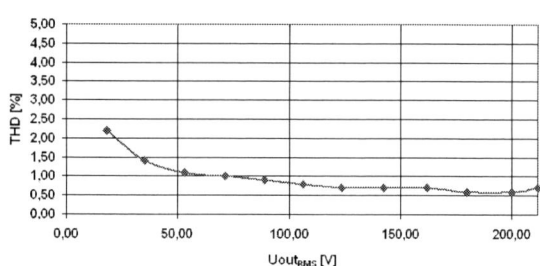

Fig. 8: Measuring results for the THD, resistive load

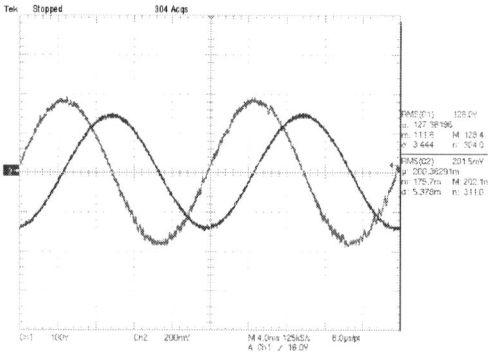

Fig. 9: Oscilloscope screenshot of an inductive load (voltage = brown, current = blue)

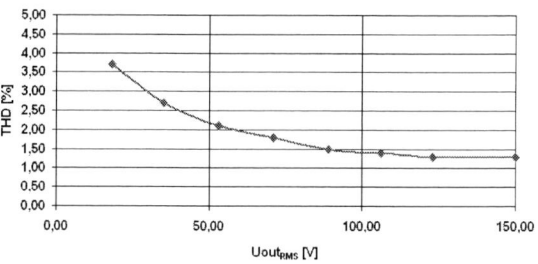

Fig. 10: Measuring results for the THD, inductive load

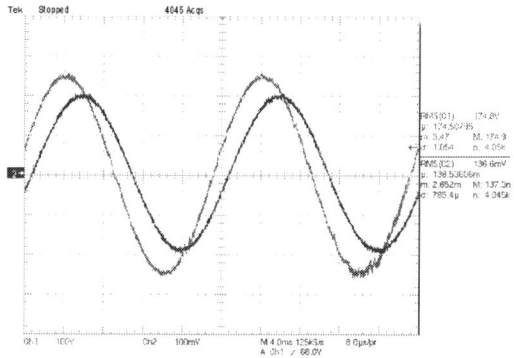

Fig. 11: Oscilloscope screenshot of a resistive/inductive load (voltage = brown, current = blue)

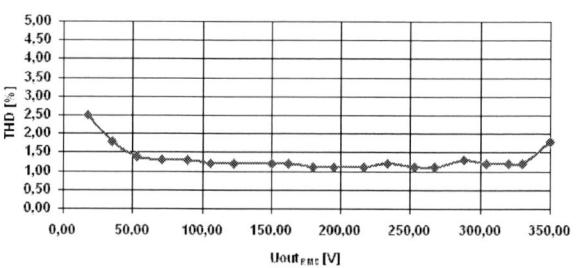

Fig. 12: Measuring results for the THD, resistive/inductive load

Connecting a non linear load at the output of the converter the control can still form a sinusoidal output voltage. Fig. 13 shows an example with a saturated transformer during an induced voltage test with line frequency.

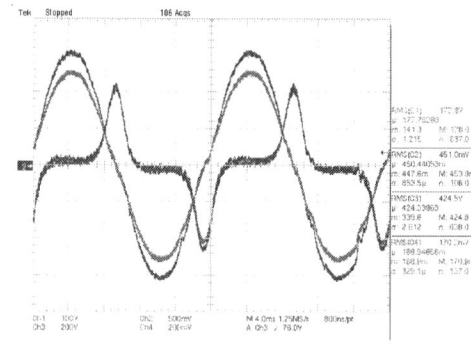

Fig. 13: Oscilloscope screenshot for non linear loads

The current crest factor of Channel 3 at the output of the converter is 110 A / 45 A = 2.44. The according THD of the voltage in CH 1 at the output of the converter is with 4,5 % still below the limit of 5 %. CH 2 and CH 4 are the voltage and the current at the secondary winding of the step-up transformer.

Fig. 14 shows the variety of the closed-loop control: It is possible to generate every voltage shape e.g. rectangular or triangular voltage shapes, as the control forces the output voltage to follow its reference value.

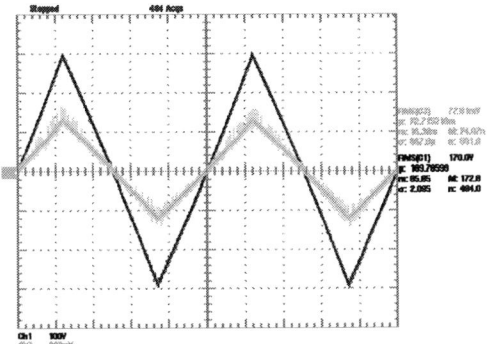

Fig. 14: Triangular waveform, resistive load (voltage = blue, current = green)

V. APPLICATIONS

The different sinusoidal and non-sinusoidal voltage forms open up a wide field of testing applications. The maximum RMS-value of the voltage and current at the output of the single phase static frequency converter presented in this paper is 380 V respectively 250 A. Thus for high voltage testing applications a step-up transformer has to be installed between the static frequency converter and the device under test to obtain any required voltage level.

To ensure testing conditions according to IEC 60060-3 [12] respectively IEC 60076-3 [13] the experience gained with the single phase system influenced the design of the first mobile test system for on-site testing of power transformers in the world. The static frequency converter, which was assembled for this task, provides a three phase output with a line to line voltage of 650 V. The maximum output voltage of the entire system is 350 kV for applied voltage tests; due to the transformation ratio of the step-up transformer the testing level for no load tests and induced voltage tests is limited to 75 kV. With this configuration this system has been successfully used for various on-site tests in the meantime [7], [8], [9] and [10].

VI. CONCLUSION

After the design of the closed-loop control the performance of the entire system, including the power line, rectifier, link, inverter and different loads, was successfully simulated. Subsequently the DSP was programmed and the hardware was developed to implement the digital control in a real system. In the course of this development an existing pulse-width modulated single phase inverter was modified to a static frequency converter with a controlled output voltage.

Measurements during the operation of the modified converter show, that an output voltage with a total harmonic distortion factor less than 5 % can always be obtained (to a certain level even with non linear loads). Thus the direct control of the IGBTs, which was developed for a frequency and amplitude variable generation of test voltage, guarantees testing with output voltages conform to international standards for all the examined loads. An adaptation of the control parameters to different loads is hardly necessary, as universal settings could be found providing satisfying results for every type of load.

Meanwhile the first industrial application a mobile on-site test system for power transformers has been successfully established at the market.

REFERENCES

[1] Föllinger, O. Regelungstechnik: Einführung in die Methoden und ihre Anwendungen. *Hüthig*, 1994

[2] Meyer, M. Leistungselektronik Einführung Grundlagen Überblick, *Springer Verlag*, 1990

[3] Giessler, W. Simatic S7, SPS-Einsatzprojektierung und –Programmierung, *VDE Verlag Berlin*

[4] Dobling, G. Signalprozessoren, Schlembach Fachverlag, Wilburgstetten

[5] Lawrenz, W. CAN, *Hütig Verlag*, Heidelberg

[6] Weigmann, J. Dezentralisierung mit Profibus-DP/DPV1. Publicis Corporate Publishing, Erlangen

[7] Werle, P. "On-Site Tests of Power Transformers" *Nord-IS*", 2007

[8] ABB Mobiles Testsystem für Transformatoren ABB Kundenmagazin "Connect", 2006, 04, 12

[9] Winter, A.; Thiede, A.; Stephan, U.; Werle, P.; Scheil, K.; Steiger, M. & Vogel, M. „Mobile on-site test system for off-line tests and diagnostics at power transformers" *ETG Fachtagung Diagnostik elektrischer Betriebsmittel*, VDE, 2006

[10] Winter, A.; Coors, P. & Stephan, U. A Mobile Transformer Test System Based on a Static Frequency Converter *HIGHVOLT Kolloquium*, 2007, 137-142

Standards:

[11] IEC 60060-1 "High-voltage test techniques. Part 1: General definitions and test requirements" 1989

[12] IEC 60060-3 "High-voltage test techniques. Part 3: Definitions and requirements for on-site testing" 1989

[13] IEC 60076-3: Power transformers – Part 3: Insulationlevels, dielectric tests and external clearances in air. (second edition 2000-03)

Open Switch Fault Diagnosis for a Doubly-Fed Induction Generator

W. Sae-Kok and D M Grant

Department of Electronic and Electrical Engineering, University of Strathclyde,
204 George Street, Glasgow, United Kingdom, G1 1XW

Abstract-- This paper addresses the analysis and detection of open switch faults in back-to-back PWM converters used in doubly-fed induction generators (DFIG). Several methods have previously been proposed to detect open switch faults in either the machine-side or line-side converter. The operating conditions that can cause possible false alarms with these methods are investigated. The proposed method detects open switch faults more reliably than any of the existing methods and hence improves overall system reliability. The performance of the existing methods and proposed methods has been verified by both simulation and experiment.

Index Terms-- DFIG, Doubly-Fed Induction Generator, Open Switch Fault Diagnosis.

I. INTRODUCTION

Wind turbine generators have become extensively used in many countries to generate power to the grid. Many types of generators may be used with the turbine to supply the energy. The doubly-fed induction generator (DFIG) is widely used with off-shore wind turbines, and because of the difficulty in accessing such turbines, health monitoring and fault diagnosis become important to help schedule maintenance and minimize unforeseen failure.

A DFIG consists of a wound-rotor induction machine, back-to-back PWM converters, three-phase filter and a three-phase transformer. As shown in Fig.1, the stator is connected directly to the grid whereas the rotor is connected to the grid via the back-to-back PWM converters, filter and transformer. Compared with other parts, the back-to-back converters are the least reliable parts of this generator. The faults that can occur in a conventional induction motor drive, as listed in [2], can also occur in this generator.

This paper focuses on open switch faults occurring in either the machine-side or line-side converter and presents the effects of open switch faults on doubly-fed induction generator variables. The existing methods which have been proposed to detect open switch faults for electrical drives are then briefly discussed as well as the advantages and drawbacks of each method when applied to the DFIG. This paper presents a new method to detect open-switch faults on the DFIG which overcomes the problems of the existing methods. The experimental setup and simulation method are described in detail.

II. EFFECT OF OPEN SWITCH FAULTS ON DOUBLY-FED INDUCTION GENERATOR VARIABLES

The effects of open switch faults can be broadly classified into local effects and global effects. A local effect is the effect on the current flowing through the faulty converter whereas a global effect is an effect on other system variables. In this paper, in the case of global effects, only the significant variables such as stator current, electromagnetic torque, active power and reactive power will be discussed. Considering local effects, once one switch of a phase leg of a three phase voltage source converter (VSC) is faulty, the current of the faulty phase loses one half cycle every period, either positive or negative depending on whether the top or bottom switch is faulty. The current waveform of the faulty phase is similar to the current waveform of a half-wave rectifier. According to the assumption that the summation of all three-phase currents is zero $(i_a + i_b + i_e = 0)$, the other two healthy phase currents have DC offset. This local effect is presented in Fig 2.

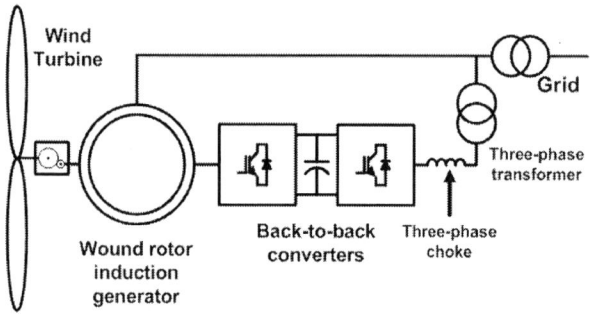

Fig.1. The topology of a doubly-fed induction generator used with a wind turbine

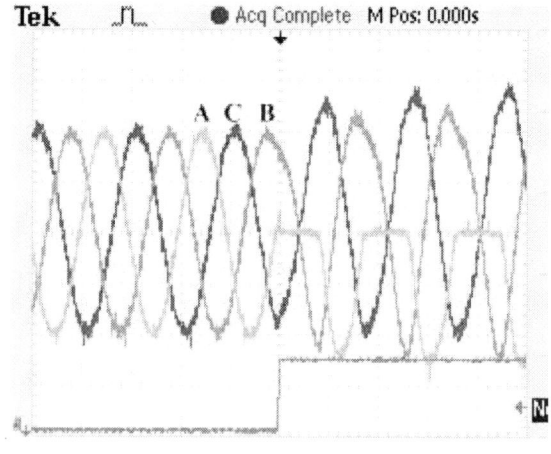

(a)

978-1-4244-0644-9/07/$25.00 ©2007 IEEE

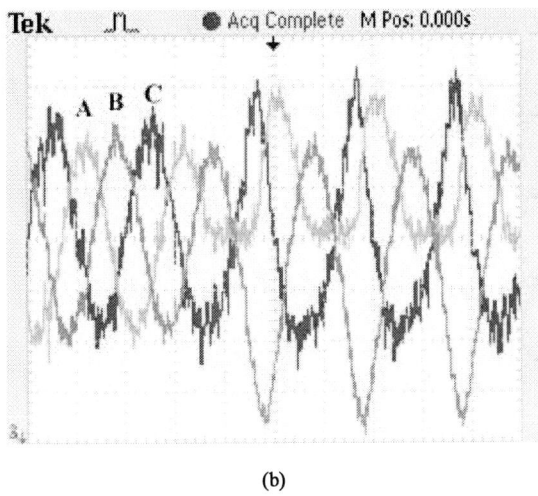

(b)

Fig. 2. Waveforms of (a) the rotor currents with one open switch fault of the machine-side converter open and (b) the line-side converter currents with one open switch fault of the line-side converter open.

In the case of global effects, although an open switch fault does not immediately damage the converter or cause any protection system to trip, it can cause severe damage to the mechanical parts of the generator and the turbine. Generally, the operating speed of the DFIG is in the range of 33% above and below synchronous speed [1]. The frequency of the rotor current referred to a stationary reference frame will then vary from 0 to 33% of line frequency. Fig. 3 shows the effect of an open switch fault at phase A of the machine-side converter under super synchronous generation.

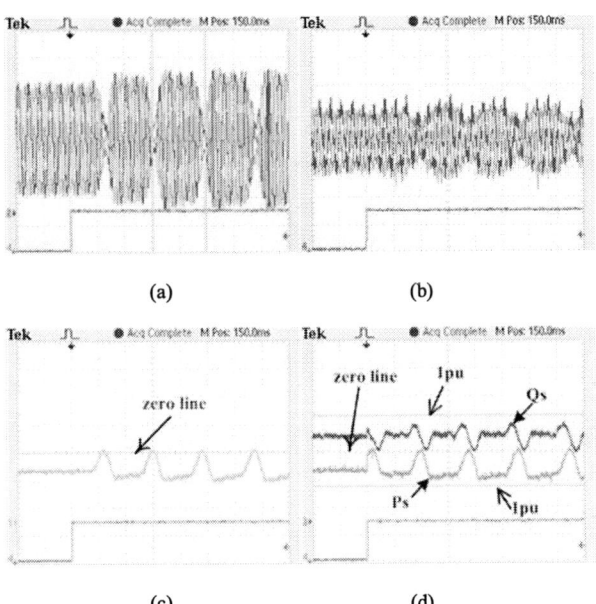

(a)   (b)

(c)   (d)

Fig.3. The effect of an open switch fault at the machine-side converter on (a) stator currents, (b) line-side converter currents (c) electromagnetic torque, (d) stator active and reactive power,

Considering the machine-side converter, according to Fig.3 a fault in this converter causes torque oscillations which may cause low frequency speed oscillations, depending on their amplitude and the inertia of the turbine. Any high frequency oscillation, having a period

much shorter than the time constant of the turbine, is largely suppressed. The oscillation frequency caused by rotor current may be quite low and can cause low frequency vibration of the generator shaft, the gear box and the turbine. As is shown in Fig. 4b the magnitude of the rotor frequency component (10 Hz.) is quite high under open switch fault condition in the machine-side converter running at 20 % above synchronous speed, whereas under normal operating condition the magnitude of rotor frequency component is almost zero. Moreover, as shown in Fig. 3d, an open switch fault in the machine-side converter can also cause a low frequency oscillation of the stator active and reactive power of the generator.

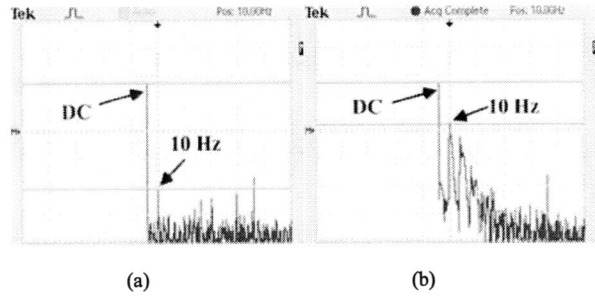

(a)   (b)

Fig.4. The spectrum of torque (a) under healthy condition and (b) under one open switch fault condition at the machine-side converter

Considering the line-side converter, an open switch fault has less effect than a fault in the machine-side converter. As shown in Fig. 5, the effect on generator variables is negligible. There is no significant oscillation appearing on the stator active power and reactive power signal as well as stator current. However, as presented in Fig. 2b, once the open switch fault appears in the line-side converter, there is a DC component appearing on all three phase currents of the line-side converter. This can cause saturation in the inductor filter and the transformer.

Generally, the cost of the converter is low compared with the generator, turbine and other power equipment, but once a fault appears in the converter, damage may occur to the more expensive parts of the turbine. For this reason, fault diagnosis for the power electronic converters becomes essential.

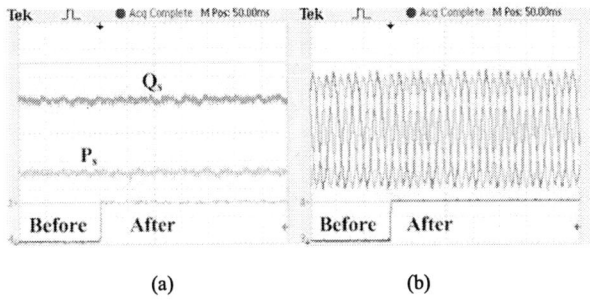

(a)   (b)

Fig.5 The effect of an open switch fault at the line-side converter on (a) stator active and reactive power, (b) stator currents

III. SOME OPEN SWITCH FAULT DIAGNOSIS METHODS

There have been several studies on open switch fault diagnosis. The diagnosis methods considered in this paper are signal processing based methods, and no model-based method is considered. This is because model-based methods require adequate knowledge of machine parameters [10]. The methods considered are as follows:

- Park's Vector Method [3]
- Slope Method [4, 5]
- Control Deviation Method [6]
- Normalized DC Current Method [7]
- Modified Normalized DC Current Method [8, 9]
- Simple DC Current Method [8, 9].

The advantages and drawbacks of the methods above are discussed in [8, 9]. The most significant problem identified was false alarms, and this was partially resolved by introducing a fixed dead time for each method [8]. However, this was not able to prevent false alarms under every operating condition. The fixed dead time used in the line-side converter cannot be used in the machine-side converter which has variable fundamental frequency. This is one of the weaknesses of the existing fault detection methods.

With the exception of the Slope Method, each method is based on calculation of the DC component and/or fundamental component of either the space vector of the three-phase currents or of all three phase currents, and this is used as a diagnostic index. A recursive moving average technique is employed to achieve the calculation of these indices. However, in case of the machine-side converter, the frequency of the current varies through zero frequency. Under this condition, when the speed of the generator passes synchronous speed, only the Slope Method and Control Deviation Method do not show false alarms. The Park's Vector Method, Normalized DC Current Method, Modified Normalized DC Current Method and Simple DC Current Method show false alarms under this condition, because of the nature of their moving average calculation. This requires the previous instantaneous data for the present moving average value. In case of N samples per cycle, the present average value needs the present value and N-1 previous values for calculation. When the machine passes synchronous speed the current becomes DC. Depending on the rate of change of speed the fundamental component of the current is zero or almost zero for a certain time. This causes the average and normalized average values to become very high after the speed passes synchronous speed.

The Control Deviation Method does not suffer from false alarms under these conditions because there is no fundamental component in the command current, but it has false alarms under transient conditions and under low current conditions. Moreover, the fault detection algorithm must be integrated into the control algorithm.

The Slope Method did not perform well for fault detection as described in [8, 9]. It has a longer detection time compared with other methods because, under healthy conditions, the Slope Method has some values in the tolerance range of faulty states [8].

As illustrated in [8, 9], the Modified Normalized DC Current Method is the most effective of the existing methods. It was first proposed by Abramik et al. [7] and was later modified by Rothenhagen et al. [8, 9]. This method uses the moving average value of the line current as a diagnostic variable. To make this variable independent of load, the moving average value is normalized by the fundamental component of the line current by means of the Discrete Fourier Transform (DFT), as shown in the equations below.

$$\mu_v = \frac{1}{N} \sum_{k=1}^{N} i_v(k\tau) \tag{1}$$

$$i_{1,v} = a_{1,v} \cos\left(\frac{2\pi}{T} k\tau\right) + b_{1,v} \sin\left(\frac{2\pi}{T} k\tau\right) \tag{2}$$

$$\gamma_v = \frac{\mu_v}{i_{1,v}} \tag{3}$$

$$a_{1,v} = \frac{2}{N} \sum_{k=1}^{N} i_v(k\tau) \cos\left(\frac{2\pi k}{N}\right) \tag{4}$$

$$b_{1,v} = \frac{2}{N} \sum_{k=1}^{N} i_v(k\tau) \sin\left(\frac{2\pi k}{N}\right) \tag{5}$$

$$v \in [a, b, c] \tag{6}$$

$$\frac{1}{f} = N\tau \tag{7}$$

where k is 1, 2, 3...64, N is 64, and γ_v is a diagnostic variable for the Modified Normalized DC Current Method of each phase.

Under healthy conditions, the value of γ_v is always within the threshold of 0.45[8, 9], but under faulty conditions, the value of γ_v of the faulty phase exceeds the threshold. In the case where the value of γ_v exceeds the threshold in more than one phase, the phase with the highest absolute of γ_v is the faulty phase.

This method seems to work well for any frequency except when the frequency of the current signal is close to zero or passes zero. When applied to the detection of an open switch fault in the machine side-converter of the DFIG, if the rotor speed passes synchronous speed the normalized DC current value exceeds the threshold of 0.45 as shown in Fig. 6. Fig.6a and 6c show the simulation and experimental rotor current waveform respectively passing synchronous speed from sub-synchronous speed region to super-synchronous speed region. In Fig. 6d, the Modified Normalized DC Current Method had a delay of one half cycle introduced, but this method still shows a false alarm. A longer delay time might overcome this problem but the wind speed can cause the DFIG run at speeds around synchronous speed and hence cause unavoidable false alarms. Moreover, a long delay time will increase the time to detect an open switch fault at other frequencies. This method also has a relatively long computational time because of the sinusoidal calculations and multiplications of the DFT. Nevertheless, this is a minor problem if the DSP technology provides high computation speed.

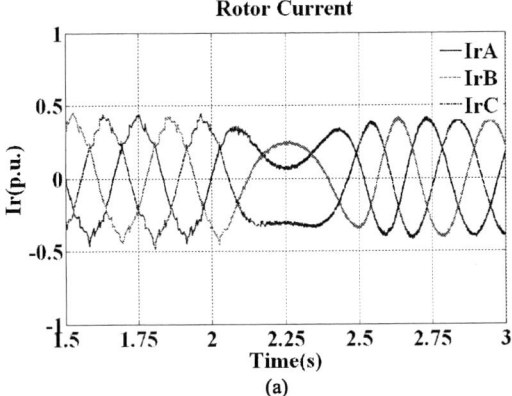

Rotor Current

(a)

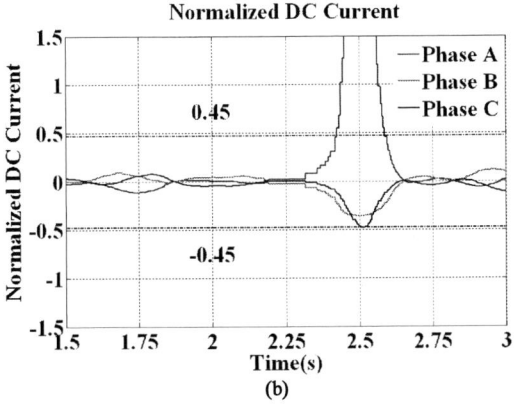

Normalized DC Current

(b)

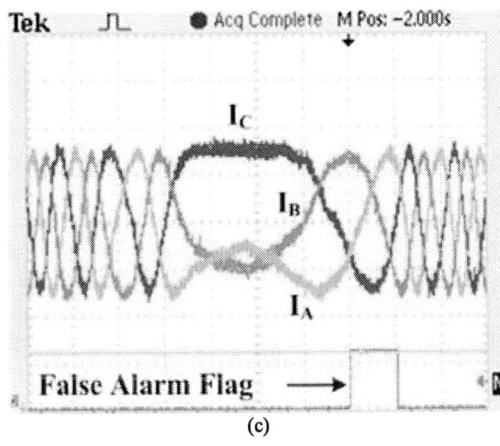

(c)

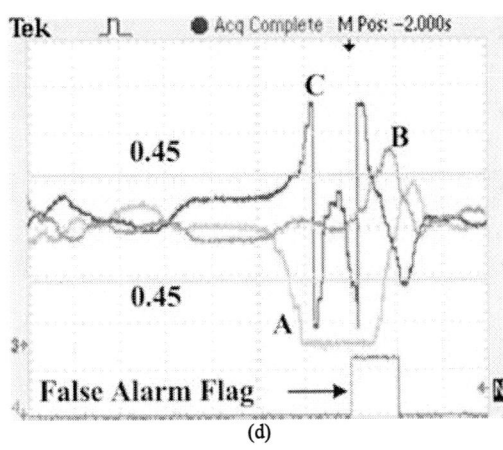

(d)

Fig. 6. Simulation results (a and b) and experimental results (c and d) showing a condition that causes a false alarm (a),(c) rotor current, (b),(d) normalized DC current

IV. PROPOSED METHODS

Since the objective of introducing fault detection is to improve reliability, any false alarm must be avoided. Therefore, in this section, an alternative method to detect and locate the fault, and overcome the false alarm problem, is presented.

Considering the Normalized DC Current Method [7] and the Modified Normalized DC Current Method [8, 9], as explained earlier, these methods are time consuming and need DFT calculations. To reduce the burden on the DSP, an alternative diagnostic variable is proposed.

This new method, named the Absolute Normalized DC Current Method, uses the average of the absolute value of each phase current, instead of the fundamental component, as a normalizing variable. Under faulty conditions, the faulty phase current has only one half cycle, so the normalized average value is equal to 1. Using this method only the calculation of the absolute value is required. This method requires less computational time and has smaller code size for fault detection. A threshold of 0.7 was found to be suitable from both simulation and experiment. The equations used to calculate the diagnostic variable for this method are (1) and the following equations.

$$\lambda_v = \frac{1}{N} \sum_{k=1}^{N} |i_v|(k\tau) \qquad (8)$$

$$\xi_v = \frac{\mu_v}{\lambda_v} \qquad (9)$$

where ξ_v is a diagnostic variable for the Absolute Normalized DC Current Method of each phase.

The values of ξ_v after an open switch fault in the machine-side converter are shown in Fig.7. Here, phase A top switch is open and therefore it loses positive half cycles. The value of ξ_v of phase A exceeds the threshold and saturates at 1 whereas the value of ξ_v of phase B and C still lie within the threshold. This picture proves that ξ_v can be used as a diagnostic variable in the same way as γ_v.

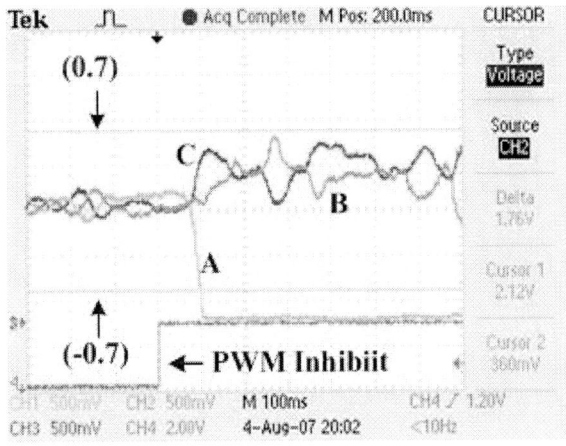

Fig.7. The value of the diagnostic variable of the Absolute Normalized DC Current method under an open switch fault at the machine-side converter at 10 Hz

Due to the periodic nature of ξ_v, if the DFIG operates at or around synchronous speed, the value of ξ_v may exceed the threshold for at least 1/3 cycle. Therefore, to avoid false alarms, a delay of one half cycle ($N/2$ point) is introduced. During this delay, if the ξ_v of one of the other phases also exceeds the threshold, the phase fault flag is set to zero. Therefore, no false alarms appear although the speed of the generator swings slightly around synchronous speed. With this algorithm, the problem of operating the DFIG around synchronous speed or passing through synchronous speed is completely solved.

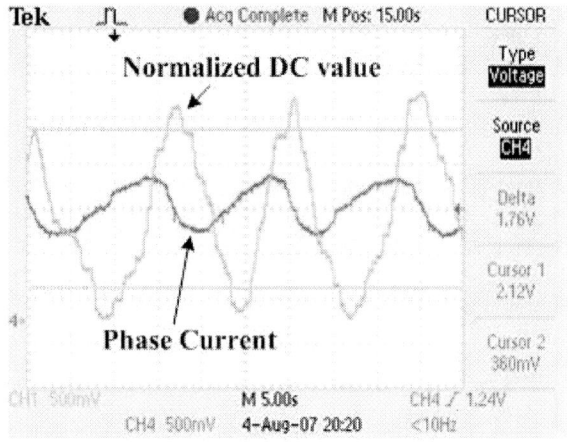

Fig.8. The picture show the periodic nature of the ξ_v

This paper also proposes an alternative open switch fault detection method named the Sampling Point Comparison Method. According to the sampling technique proposed by Abramik *et al.* [7], a signal is sampled N times in one period with a fixed angular step between samples. Dividing a unit circle into 4 sectors, and considering only the first sector as shown in Fig.9, each sector contains 16 samples with the value of $\sin(2m\pi/N)$, where m is 1,2,3,...,16 and N is 1,2,3,...,64.

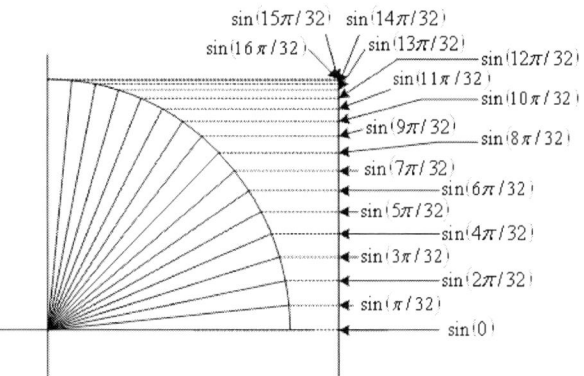

Fig. 9. The first sector of a unit circle used for calculation of the faulty band for current during a faulty half cycle

The idea of this method is to check the number of samples lying in a "faulty band" and also in the positive

and negative half cycles of the signal during faulty conditions. The faulty band ranges between the positive and negative product of the moving average value of the peak current vector and the sine value of the unit circle. The peak current can be obtained from the amplitude of the $\alpha - \beta$ components of the three-phase current vector. The value of $\sin(2(2\pi/N))$ was found from both simulation and experiment to be a suitable multiplier to set the faulty band range.

In this paper, the number of sampling points in one cycle is 64. The number of samples lying in the faulty band is included in the number of samples in both positive and negative half cycles. The number of samples is stored in a fault accumulator, a positive accumulator and a negative accumulator respectively. Once a fault appears in the converter, if the fault accumulator and either the positive or negative accumulator exceed the fault threshold, the faulty phase is detected. Then the sign of the moving average value of the faulty phase is checked to locate the faulty switch. This method can be graphically explained in Fig.10.

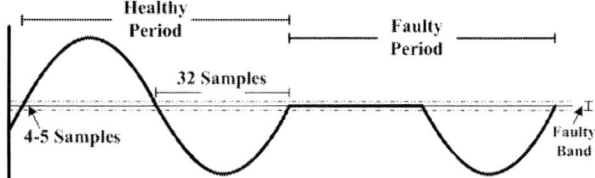

Fig. 10. The graphical explanation of the algorithm to detect an open switch fault by the Sampling Point Comparison Method

The advantage of this method is fast fault detection time because, under faulty conditions, the faulty current signal loses one half cycle. During this condition, the number of samples within the faulty band will exceed the limit (4 or 5 samples). If the threshold for the fault accumulator is set to a value close to the limit e.g. 8 counts, and the threshold for both positive and negative accumulators is set to N/2 + 8 counts, the minimum detection time is approximately 1/8 cycle which is much faster than using the Modified Normalized DC Current Method which relies on the time that the value γ_v exceeds the threshold. Another advantage is that it can be used to detect two open switch faults in the same pole. Under two open switch fault condition, the faulty phase current become zero which means that the faulty current always stays within the faulty band. Therefore, the fault can be detected within 1 cycle.

From experiments, if the number of samples in the fault accumulator exceeds 24 and the number of samples in the positive or negative accumulator exceeds 56, then a fault is recognized. These threshold values were determined by extensive tests at different speeds and torques.

If both the Sampling Point Comparison Method and the Absolute Normalized DC Current Method are combined together, the combined method can detect both one open switch and two open switch faults and has false alarm suppression capability. Furthermore, this method can also be applied to the four quadrant electrical motor drive when the motor changes the direction.

V. EXPERIMENTAL SETUP

The schematic of the experimental system and a picture of the experimental test rig are shown in Fig. 11 and 12 respectively.

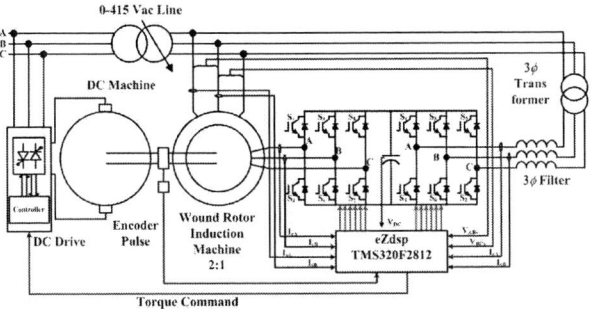

Fig.11. Schematic of experimental system

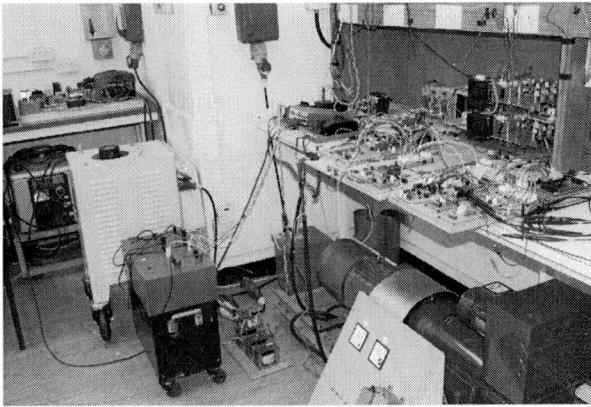

Fig.12. Experimental test rig

The converter switching frequency of the system is 5.1 kHz. A single DSP, the TMS320F2812 on an eZdspF2812 board, is used for both control and fault diagnosis. The program to detect and locate the open switch fault is performed every switching cycle.

In this paper, the rotor flux producing current is controlled to be zero to avoid the consumption of reactive power and the line-side converter is controlled to have unity power factor [11]. Any gate firing signal may be disabled internally by software to create an open-switch fault. The fault flag is made available via a digital-to-analog converter channel. The system is tested under three different conditions - super-synchronous speed generation, sub-synchronous speed generation and synchronous speed generation. The torque applied to the generator follows a wind turbine characteristic. Relevant signals are supplied by the DSP via D/A converter channels for oscilloscope display.

VI. SIMULATION AND EXPERIMENTAL RESULTS

This section starts with the comparison of simulation results from the Sampling Point Comparison Method and the Modified Normalized DC Current Method with half

cycle delay. In the case of the Modified Normalized DC Current Method, the delay time is set to one half cycle to try to avoid false alarms. As shown in Fig.13, the detection time of the open switch fault by the Sampling Point Comparison Method is shorter than the time to detect the open switch fault by the Modified Normalized DC Current Method.

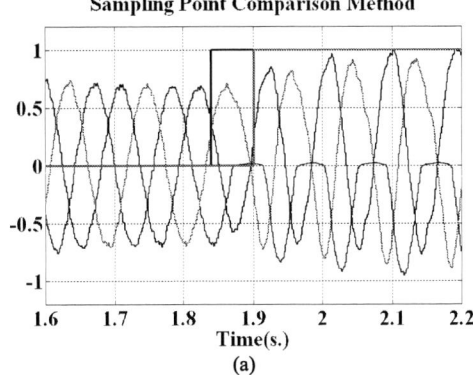

(a)

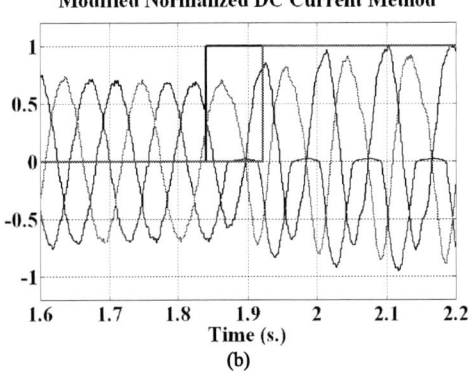

(b)

Fig.13. Simulation results of open switch fault detection for the machine-side converter by (a) the Sampling Point Comparison Method, (b) the Modified Normalized DC Current Method with half cycle delay

Fig.14 and 15 show experimental results from the Absolute Normalized DC Current Method with false alarm suppression algorithm and the Modified Normalized DC Current Method with delay. The open switch fault detection times are essentially the same. This comparison proves that the Absolute Normalized Current Method can be used instead of the Modified Normalized Current Method.

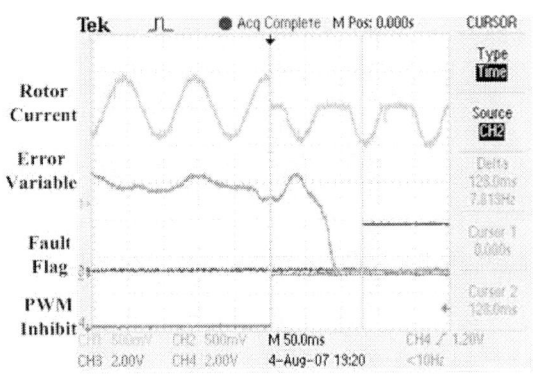

(a)

373

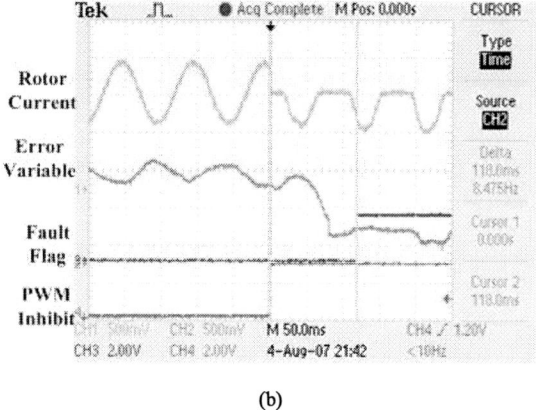

(b)

Fig.14 Experimental results of open switch fault detection for the machine-side converter by (a) the Absolute Normalized DC Current Method, (b) the Modified Normalized DC Current Method with half cycle delay

Since each converter has its own fault detection element, the delay time for the Modified Normalized DC Current Method for each converter is set to match its operating condition. For the line-side converter, the delay of 1/4 cycle is set, to follow the published reference [8].

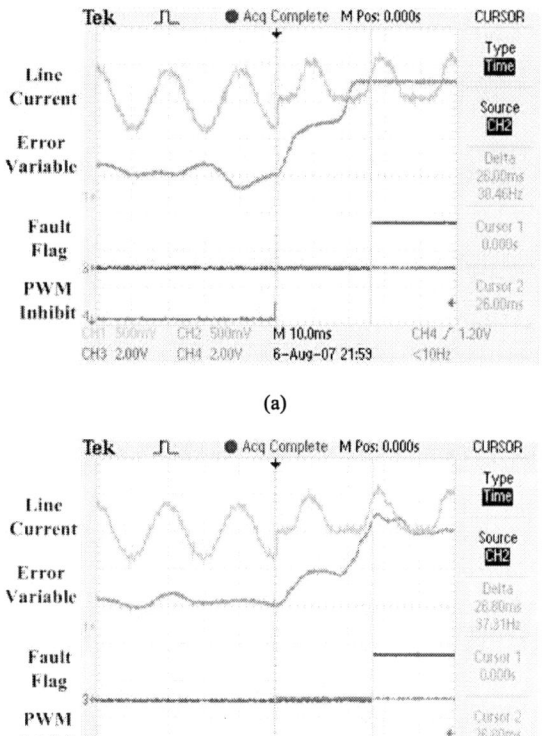

(a)

(b)

Fig.15 Experimental results of open switch fault detection for the line-side converter by (a) the Absolute Normalized DC Current Method, (b) the Modified Normalized DC Current Method with 1/4 cycle delay

Fig.16 shows the time to detect an open switch fault by the Sampling Point Comparison Method for the machine-side converter. The fault is made at the same point as in Fig.14. Here, the time to detect the open switch fault is

faster than using the Modified Normalize DC Current Method. The detection time depends on the point on the wave at which the fault occurs.

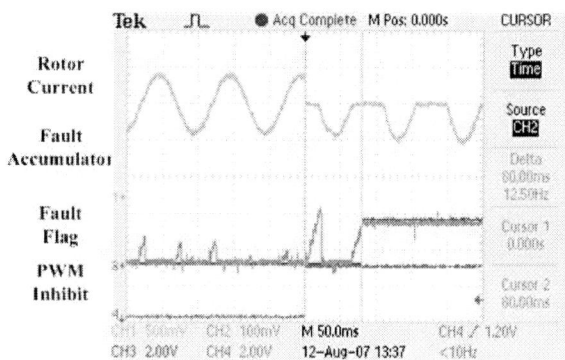

Fig.16.   Experimental results of open switch fault detection for the machine-side converter by the Sampling Point Comparison Method

Fig.17 shows the rotor current waveform, ξ_v, the sample count when ξ_v exceeds the threshold, and the false alarm flag under operation around synchronous speed.

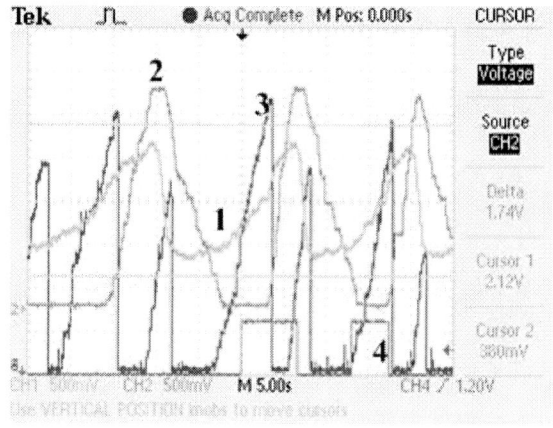

(a)

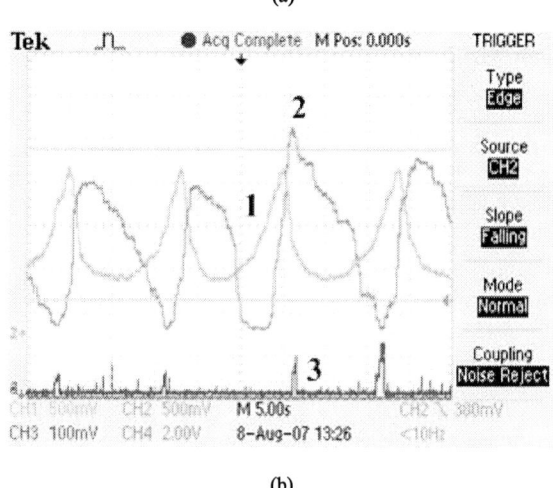

(b)

Fig.17 Experimental results when the DFIG operates around synchronous speed (a) without false alarm suppression and (b) with false alarm suppression (1) rotor current, (2) the value of ξ_v, (3) analog display of sample count when ξ_v exceeds the threshold and (4) false alarm flag.

In Fig.17a, the system is operating without the false alarm suppression algorithm. The value of the sample count when ξ_v exceeds the threshold is high and it exceeds the limit set in the software (32 counts) causing false alarms to appear. Conversely, in Fig. 17b, the false alarm suppression algorithm is added into the software. Although ξ_v exceeds the limit, the sample count is still low because of the algorithm explained in section IV and no false alarm flag appears.

VII. CONCLUSIONS

The existing fault diagnosis methods presented in previous research were tested with a stator-fed electrical machine and were not studied when the signal frequency passes through zero. Therefore, when applied to the DFIG, the rotor frequency of which varies from 0 to a limited frequency (approximately 10-20 Hz) such methods reduce the reliability of the system by showing false alarms. However, as discussed before, false alarms must be avoided.

This paper proposes two alternative fault detection methods, namely the Absolute Normalized DC Current Method and the Sampling Point Comparison Method. The former method provides similar fault detection capability while requiring less computational time than the Modified Normalized DC Current Method. The latter method provides fast fault detection time having a simple algorithm and it can also detect two open switch faults in the same leg.

As described earlier, for the machine-side converter, existing methods suffer from false alarms especially when the signal frequency passes or swings around 0. Therefore this paper proposes an algorithm to suppress false alarms to apply with the Absolute Normalized DC Current Method.

The combined method, including the false alarm suppression algorithm, increases the performance of open switch fault detection by its ability to detect two open switch faults and makes the method immune to any condition that causes false alarms. The combined method can also be applied to four quadrant electrical drives when the motor changes direction.

REFERENCES

[1] S. Muller, M. Deicke, Rik W, "Doubly Fed Induction Generator Systems for Wind Turbines A viable Alternative to Adjust Speed over a Wide Range at Minimal Cost," *IEEE Industry Application Magazine*, May/June 2002.

[2] D. Katsha, B.K. Bose, "Investigation of Fault Modes of Voltage-Fed Inverter System for Induction Motor Drive," *IEEE Trans. Industry Applications*, vol. 30, No.4, July/August 1994

[3] A. M. S. Mendes, A. J. Marques Cardoso, "Fault Diagnosis in a Rectifier-Inverter System Used in Variable Speed AC Drive, by Park's Vector Approach," *Proceedings of the 1999 EPE 8th European Conference on Power Electronics and Applications*, pp. P.1-P.9.

[4] R. Peuget, S. Courtine, J. P. Rognon, "Two Knowledge-Based Approaches to Fault Detection and Isolation on a PWM Inverter," *Proceedings of the 1997 IEEE*

International Conference on Control Applications, pp. 565-570.

[5] R. Peuget, S. Courtine, J. P. Rognon, "Fault Detection and Isolation on a PWM Inverter by Knowledge-Based Model," *IEEE Trans. Industry Applications*, vol. 34, no. 6, Nov./Dec. 1998, pp. 1318-1326.

[6] C. Kral, K. Kafka, "Power Electronics Monitoring for a Controlled Voltage Source Inverter," *Proceedings IEEE Power Electronics Specialists Conference*, Vol.1, pp.213-217

[7] S. Abramik, W. Sleszynski, J. Nieznanski, H. Piquet, "A Diagnostic Method for On-line Fault Detection and Localization in VSI-Fed AC Drive," *Proceedings of the 2003 EPE 10th European Conference on Power Electronics and Applications*, pp. P.1-P.8.

[8] K. Rothenhagen, F.W. Fuchs, "Performance of Diagnosis Methods for IGBT Open Circuit Faults in Voltage Source Active Rectifiers," *Proceedings of the 2004 IEEE 35th Power Electronics Specialists Conference*, pp. 4348-4354.

[9] K. Rothenhagen, F.W. Fuchs, "Performance of Diagnosis Methods for IGBT Open Circuit Faults in Three Phase Voltage Source Inverters for AC Variable Speed Drives," *Proceedings of the 2005 EPE 12th European Conference on Power Electronics and Applications*, pp. P.1-P.7.

[10] Wojciech Sleszynski, Janusz Nieznanski, Artur Cichoqski, "Real-Time Fault Detection and Localization in Vector-Controlled Induction Motor Drives," *Proceedings of the 2005 EPE 12th European Conference on Power Electronics and Applications*, pp. P.1-P.7.

[11] R. Pena, J. C. Clare, and G. M. Asher, "Doubly fed induction generator using back-to-back PWM converters and its application to variable-speed wind-energy generation," *Proc. Inst. Elect. Eng.*, pt. B, vol.143, pp. 231–241, May 1996.

Rapid Analysis & Design Methodologies of High-Frequency LCLC Resonant Inverter as Electrodeless Fluorescent Lamp Ballast

Yong-Ann Ang, David Stone, Chris Bingham, Martin Foster

Dept of Electronic Engineering, University of Sheffield, Mappin Street, Sheffield, S1 3JD. UK. (0)114 2225046.
d.a.stone@sheffield.ac.uk

Abstract - The papers presents methodologies for the analysis of 4th-order LCLC resonant power converters operating at 2.63MHz as fluorescent lamp ballasts, where high frequency operation facilitates capacitive discharge into the tube, with near resonance operation at high load quality factor enabling high efficiency. State-variable dynamic descriptions of the converter are employed to rapidly determine the steady-state cyclic behaviour of the ballast during nominal operation. Simulation and experimental measurements from a prototype ballast circuit driving a 60cm, 8W T5 fluorescent lamp are also included.

I. INTRODUCTION

Fluorescent lighting takes a major role in today's lighting requirements (with around 1.2 billion units being produced per year) due to benefits afforded by crisp white light output compared to traditional incandescent and high intensity discharge lamps. Fluorescent lamps also provide a higher Lumens/Watt output, and higher efficiency, particularly when excited at high frequencies, typically 30-50kHz, by virtue of there being insufficient time between each half cycle of the supply for a significant number of mercury ions in the discharge to re-combine (and thereby necessitating a re-strike), as occurs with standard mains frequency excitation, for instance.

The affect of this non-linear frequency dependence of lamp voltage and current can be clearly seen from a comparison of Figs. 1(a) & 1(b), which show the voltage vs. current relationship for an 8W, 60cm, T5 fluorescent tube excited by 50Hz and 50kHz input voltage, respectively.

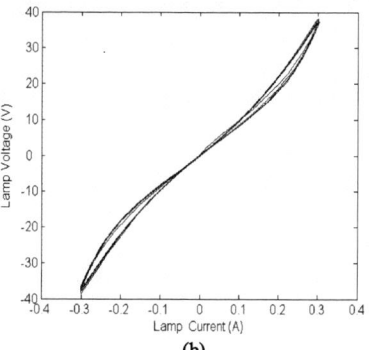

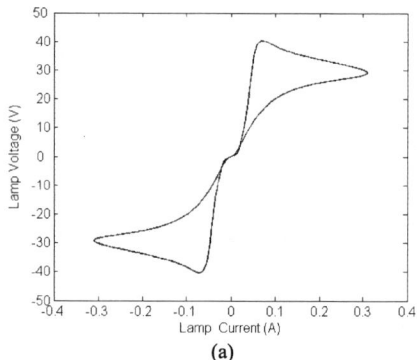

(a)

Figure 1 Lamp voltage vs. current (a) 50Hz excitation (b) 50kHz excitation

The relative 'loop area' shows that less re-combination occurs within the lamp between each half cycle of the input voltage when high frequency excitation is employed. Typically, fluorescent lamps are constructed with an oxide coated, tungsten filament electrode. Passing current through the electrode prior to striking to heat it (termed pre-heating) lowers the electrode work function, thereby allowing electrons to be emitted more readily. This consequently allows the lamp to strike at lower voltage than would normally be required, which in-turn reduces damage to the electrode from ion bombardment during the ignition event. Despite electrode pre-heating, however, the most common lamp failure mechanism is due to breakdown of the coating on the electrodes, giving rise to a blackening at one end of the tube and the lamp subsequently acting with similar characteristics to that of a gas diode. To circumvent this problem, and ultimately increase lamp lifetime, electrode-less lamps have been developed with various techniques being employed to sustain the arc viz. RF induction and capacitive discharge [1], and usually require excitation frequencies in the MHz range. Along with increasing lifetime, electrode-less excitation also removes the loss associated with the electrode heating, and therefore encourages higher operating efficiency. Such lamps (for example the GE Genura 23W commercially available induction coupled lamp) are increasingly becoming a preferred candidate for inaccessible environments viz: high ceiling sports halls requiring low maintenance etc. A demonstrator capacitively coupled lamp, 8W, 60cm, T5 fluorescent tube coupled to a ballast via copper tape applied to the outside of the tube, is shown in Fig. 2(a). To provide

978-1-4244-0644-9/07/$25.00 ©2007 IEEE

ignition and sustained light output from the lamp, an electronic ballast must develop sufficiently high voltage, typically between 400V and 1000V for striking, and subsequently provide current limiting to promote stable operation. The latter issue is a key motivator for adopting resonant converters for fluorescent lamp ballasts, since, after striking, the lamp exhibits a negative incremental impedance characteristic, as shown in Fig. 2(b) & (c) for an 8W lamp with an ignition voltage of 270V. The series impedance of resonant converters naturally acts to counter this destabilizing characteristic and encourage steady continuous operation. When operating without electrodes, the requirement for high frequency excitation also has the additional advantage of reducing the volume envelope requirements of the reactive components of the ballast, although this is at the expense of significantly complicating the design of the ballast since circuit behaviour can become dominated by parasitic elements.

Electronic ballasts with half-bridge series resonant inverters are relatively straightforward to design, and have been widely reported e.g [2, 3, 4], along with the more complicated 3rd-order LCC inverter variants [5]. More generically, however, for applications that are battery powered or require battery backup facilities, for instance, a low DC input voltage must be 'boosted' for lamp ignition through the incorporation of a step-up transformer [6].

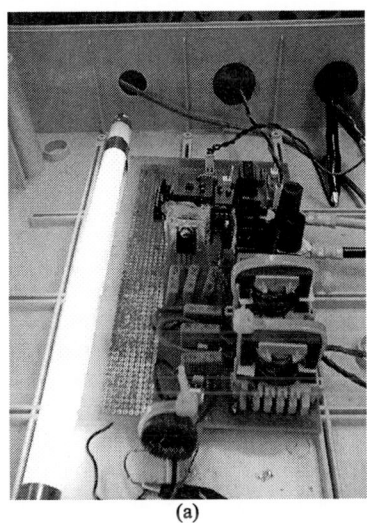
(a)

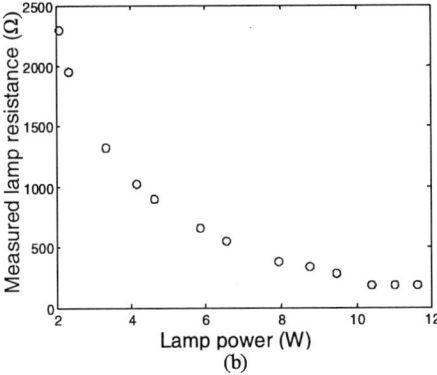
(b)

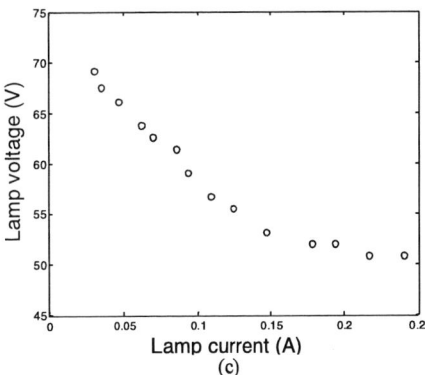
(c)

Figure 2 Capacitively coupled fluorescent lamp (a) Lamp and inverter (b) Measured lamp resistance against power (c) Measured voltage and current characteristics

The subsequent effects of the high-frequency transformer's magnetising inductance and inter-winding capacitance, are then best represented by a model of a 4th-order LCLC resonant inverter. The widespread adoption of such high-order resonant inverters, however, has been impeded by the higher peak electrical stresses to which individual electronic components are exposed (compared to hard-switched inverter counterparts), and the lack of suitable design methodologies that can provide an accurate and rapid analysis of the circuit at the design stage; particularly those that consider the significant effects that parasitic resistances, capacitances and inductances have on the resonant tank behaviour.

These difficulties have prompted researchers to investigate techniques for the reliable analysis and design of resonant inverters. Time domain mathematical models used to describe compact fluorescent ballasts have previously been reported in [7, 8], where theoretical results of series-resonant and series-parallel LCC electronic ballasts are shown to provide good agreement with measured data, at the expense of requiring significant computation overhead. Moreoften therefore, designers return to Fundamental Mode Analysis (FMA) [9, 10] for simplifying and speeding-up the design and analysis process, at the expense of neglecting the important harmonic and sub-harmonic content of the circuit voltages and currents.

Here then, the paper considers *cyclic analysis* [11, 12, 13, 14] as a candidate technique for the rapid analysis and design of 4th-order LCLC resonant inverters, suitable for use as electronic ballasts for example, including a treatment of parasitic effects. The main aim of the full paper is therefore to address the lack of suitable design methodologies for high order, high frequency resonant converters, the lamp providing a suitable candidate application for exploitation and verification of the work. By suitable manipulation, the resulting models are then employed to analytically predict the voltage and current stresses on the resonant components, with measurements from a prototype 8W, 4th-order, capacitively coupled ballast used to demonstrate the accuracy of the model predictions.

II. ANALYSIS OF ELECTRODE-LESS FLOURESCENT-LAMP BALLAST

As previously discussed, suitable 4th-order electronic ballasts must provide sufficient voltage to promote ignition, and a current limiting capability thereafter. Before the lamp is ignited, it can be assumed that the resistance presented to the output of the ballast is infinite i.e. the output stage is open circuit. Consequently, the resonant inverter behaves as a tank circuit with a high effective Q to facilitate ionization of the gas within the tube. Of note, is that the minimum capacitive discharge voltage decreases with increasing frequency, when the capacitive coupling reactance becomes small. Once gaseous breakdown has occurred, the resistance decreases as the lamp conducts current. From Fig.2(b), the nominal resistance of the lamp during normal (8W) operation can, for this case, be estimated to be $\cong 470\Omega$. An equivalent circuit of the system therefore consists of a 4th-order LCLC resonant inverter loaded by the lamp resistance, R_{lamp}, see Fig.3.

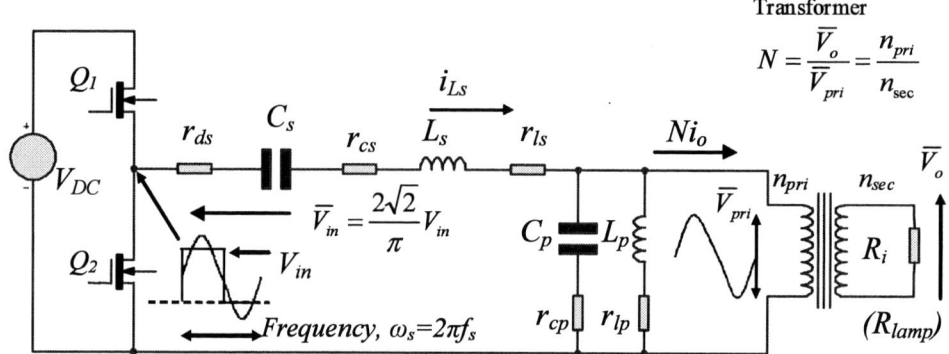

Figure 3 LCLC resonant inverter for fluorescent lamp ballast (R_{lamp})

State Variable Modelling

Figure 3 depicts the structure of a 4th-order LCLC resonant power inverter. In high frequency inverters, such as those for fluorescent lamp ballast applications, the parallel resonant components, L_p and C_p, are designed to be the magnetising inductance and parasitic capacitance of a step-up transformer, whilst the series resonant component, L_s, takes advantage of the transformer leakage inductance. In this way, the high order circuit is achieved with few additional passive components. Parasitic circuit elements have also been included in Fig. 3 for completeness. A state-variable dynamic model of the circuit can be derived by considering the resonant tank components and power switches:

$$\frac{dV_{Cp}}{dt} = \frac{i_{Ls} - i_{Lp} - N \cdot i_o}{C_p}$$

$$\frac{dV_{Cs}}{dt} = \frac{i_{Ls}}{C_s}$$

$$\frac{di_{Lp}}{dt} = \frac{V_{Cp} - i_{Lp} \cdot (r_{cp} + \eta_p) + r_{cp} \cdot i_{Ls} - N \cdot r_{cp} \cdot i_o}{L_p}$$

$$\frac{di_{Ls}}{dt} = \frac{V_{in} - V_{Cs} - V_{Cp} - i_{Ls} \cdot (r_{ds} + r_{cp} + r_{cs} + r_{ls}) + r_{cp} \cdot i_{Lp} + N \cdot r_{cp} \cdot i_o}{L_s}$$

$$(1)$$

with the output voltage V_o and load current i_o given by:

$$V_o = N \cdot \left[V_{Cp} + r_{cp} \cdot (i_{Ls} - i_{Lp} - N \cdot i_o)\right], \quad i_o = N \cdot \left[\frac{V_{Cp}}{(R_i + N \cdot r_{cp})} + \frac{r_{cp} \cdot (i_{Ls} - i_{Lp})}{(R_i + N \cdot r_{cp})}\right] \quad (2)$$

The proposed state-variable model is therefore given by:

$$
\begin{bmatrix} \dot{V}_{Cp} \\ \dot{V}_{Cs} \\ \dot{i}_{Lp} \\ \dot{i}_{Ls} \end{bmatrix} =
\begin{bmatrix}
0 & 0 & -\frac{1}{C_p} & \frac{1}{C_p} \\
0 & 0 & 0 & \frac{1}{C_s} \\
\frac{1}{L_p} & 0 & -\left(\frac{r_{cp} + r_{lp}}{L_p}\right) & \frac{r_{cp}}{L_p} \\
-\frac{1}{L_s} & -\frac{1}{L_s} & \frac{r_{cp}}{L_s} & -\left(\frac{r_{ds} + r_{cs} + r_{cp} + r_{ls}}{L_s}\right)
\end{bmatrix} \cdot
\begin{bmatrix} V_{Cp} \\ V_{Cs} \\ i_{Lp} \\ i_{Ls} \end{bmatrix} +
\begin{bmatrix} -N\left(\frac{i_o}{C_p}\right) \\ 0 \\ -N\left(\frac{r_{cp} \cdot i_o}{L_p}\right) \\ \frac{V_{in}}{L_s} + N\left(\frac{r_{cp} \cdot i_o}{L_s}\right) \end{bmatrix}
$$

$$(3)$$

To demonstrate the accuracy of the model, an experimental 4th-order LCLC inverter (V_{DC}=12V) is considered, as required for example for battery backup lighting applications. A step-up toroidal transformer with primary to secondary turn ratio n=0.5 is incorporated to demonstrated the ability of the higher order resonant circuit to take advantage of the transformer magnetising inductance. The resonant frequency of the tank circuit is highly sensitive to variations in component values, and whilst C_s and C_p can normally be assumed to be within standard component tolerances, the inter-turn capacitance (C_T) of L_s and L_p must be accommodated. For example, the inter-turn capacitance of L_p is measured as $C_T \approx 30pF$ at self-resonance. The resonant inductor and the inter-turn capacitance constitute a parallel resonant circuit, which resembles an equivalent inductance when excited below the combined resonant frequency. A method of accommodating these effects is to define L_s and L_p as frequency-dependent equivalent inductances, see Fig. 4.

The frequency-dependent equivalent inductance, L_{peq}, is then obtained from the parallel combination of L_p & C_T:

$$Z_{eq} = \frac{j\omega_s L_p}{1 - \omega_s^2 L_p C_T} = j\omega_s L_{peq}, \quad \text{where } L_{peq} = \frac{L_p}{1 - \omega_s^2 L_p C_T} \quad (4)$$

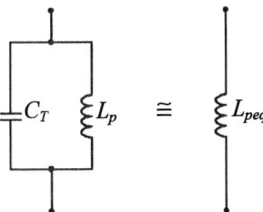

Figure 4 Equivalent circuit of resonant inductor and stray capacitance

After inclusion of the equivalent inductance (4) in the dynamic model, the measured output voltage of the inverter is compared with that predicted from the proposed state-variable model, and Spice simulations, over a range of operating frequencies and output loads, including the open-circuit condition are shown in Fig. 5. It can be seen that the proposed model provides commensurate accuracy with results obtained from Spice.

III. PRINCIPLES OF CYCLIC ANALYSIS

Although the presented state-variable description has been shown to accurately model the behaviour of the 4th order inverter, the computation time remains prohibitive due to the requirement for integration, and impedes the use of such models as an interactive design tool. This drawback can be abated to some degree however by only considering the steady-state behaviour of the circuit, thereby allowing analytical solutions from the state equations to be obtained for investigative and design validation purposes.

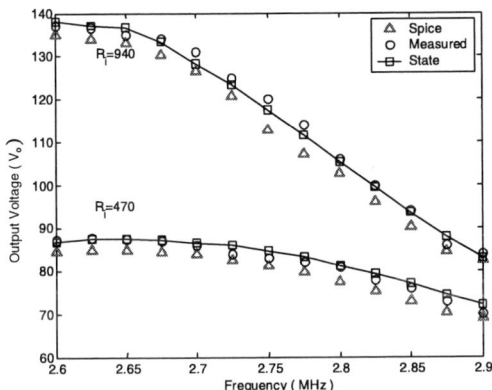

Figure 5 Output voltage of the 4th-order resonant inverter at R_i= 470Ω and, R_i= 940Ω

In particular, it is now shown that cyclic-modelling [11, 13, 14] provides a convenient methodology to facilitate the rapid solution of the steady state voltages and currents for the 4th-order ballast.

Cyclic Modes
An electronic circuit is said to operate in a cyclic-mode when the state vector $x(t)$ at any time t is equal to $x(t+nT)$, where T is the switching period of the converter and n is an integer, i.e. $x(t+nT)=x(t)$. For a resonant inverter, each cycle is comprised of multiple operating modes, m_i, each dependent on the state of the input voltage. When

considering operation in a cyclic-mode, a system of piecewise linear (state-space) equations that describe the inverter in each operational mode during a cycle can be derived, viz.:

$$\dot{x}_i = \mathbf{A}_i x_i + \mathbf{B}_i \qquad (5)$$

where x_i is the state vector, A_i represents the dynamics and B_i is the excitation matrix during the ith-operating mode. For the ith mode, (5) can be solved analytically to give:

$$x_i(t) = e^{\mathbf{A}_i t} x_i(t_0) + \int_0^t e^{\mathbf{A}_i (t-\tau)} \mathbf{B}_i \, d\tau = \Phi_i x_i(t_0) + \Gamma_i \qquad (6)$$

where $\Phi_i = \Phi(t,t_0) = e^{\mathbf{A}_i t}$, $\Gamma_i = \int_0^t e^{\mathbf{A}_i (t-\tau)} \mathbf{B}_i \, d\tau$, and $x_i(t_0)$ are the initial conditions for the ith mode. By noting that the time during which the circuit operates in the ith mode is $d_i T$, where d_i is the duty, the complete solution for the system can be obtained by employing the state vector at time $d_i T$ as the initial condition for the subsequent dynamics of the $(i+1)^{th}$ mode. However, the need to evaluate the integral in (6) is a key cause of the computational overhead when analysing the system in this manner. By combining A_i and B_i to form an augmented dynamics matrix, (7), the integration overhead can be eliminated at the expense of obtaining only the 'cyclic' steady-state description:

$$\frac{d}{dt}\left(\frac{x_i(t)}{1}\right) = \left(\begin{array}{c|c} \mathbf{A}_i & \mathbf{B}_i \\ \hline 0 & 0 \end{array}\right)\left(\frac{x_i(t)}{1}\right), \quad \text{or} \quad \frac{d}{dt}\hat{x}_i(t) = \hat{\mathbf{A}}_i \hat{x}_i(t) \qquad (7)$$

Now, if mode 1 corresponds to the time period between t_0 and t_1, and mode 2 corresponds to the time period between t_1 and t_2 the solution for the state vector at the transition time between modes 1 & 2, t_1, is given by $\hat{x}_1(t_1) = e^{\hat{\mathbf{A}}_1 d_1 T} \hat{x}_1(t_0) = \hat{\Phi}_1 \hat{x}_1(t_0)$. Similarly, the state vector at the transition time between modes 2 and 3, t_2, is $\hat{x}_2(t_2) = e^{\hat{\mathbf{A}}_2 d_2 T} \hat{x}_2(t_0) = \hat{\Phi}_2 \hat{\Phi}_1 \hat{x}_1(t_0)$. In general, for the mth mode:

$$\hat{x}(t_m) = \hat{\Phi}_m \hat{\Phi}_{m-1} \cdots \hat{\Phi}_1 \hat{x}(t_0) = \hat{\Phi}_{tot} \hat{x}(t_0) \qquad (8)$$

where $\hat{\Phi}_i = \left(\begin{array}{c|c} \Phi_i & \Gamma_i \\ \hline 0 & 1 \end{array}\right)$, and $\hat{x}(t_m)$ is the state-vector at time t_m for an initial condition $\hat{x}(t_0)$, and, therefore, by definition of the cyclic mode, is equivalent to the initial condition for the cyclic solution. Since behaviour in the cyclic mode necessarily presumes periodic steady-state operation, the initial condition for operation in a cyclic mode is given by,

$$x_{per}(t_0) = \left(I^n - \Phi_{tot}\right)^{-1}\Gamma_{tot} = \hat{x}(t_0) \qquad (9)$$

from which the initial condition for the voltage and current of the inverter and, together with (8) can be used to determine the voltage and current state at subsequent times.

IV. OPERATION OF LCLC INVERTER IN A CYCLIC MODE

For operation above resonance, which is the norm, analysis of the behaviour of the 4th-order LCLC inverter identifies 6 modes of operation within each periodic cycle.

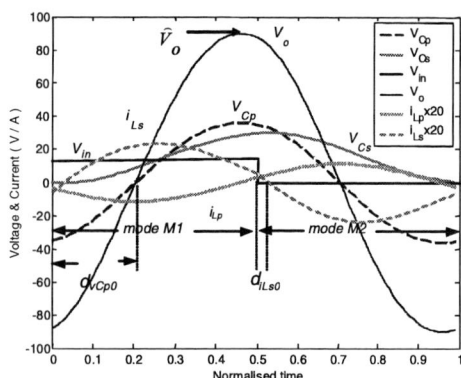

Figure 6 Voltage and current waveforms and dominant operating modes of the 4th-order resonant inverter

These are defined with respect to the polarity of the input voltage, V_{in}, and the state of the series resonant inductor current i_{Ls} and the parallel resonant capacitor voltage V_{Cp}. However, for analysis purposes two dominant modes, $M1$ and $M2$ can be identified in each cycle with respect to the polarity of V_{in}, as shown in Fig. 6. A state-variable description of the circuit behaviour can represented by the dynamics matrices $\mathbf{A}_i$ and the input excitation matrices, $\mathbf{B}_i$, for each mode. In mode $M1$, $V_{in}>0$ and the piecewise linear state-equation is defined from the matrices:

$$\mathbf{A}_1 = \begin{bmatrix} -\dfrac{N^2}{C_p(R_i+N\cdot r_{cp})} & 0 & -\dfrac{1}{C_p}+\dfrac{N^2\cdot r_{cp}}{C_p(R_i+N\cdot r_{cp})} & \dfrac{1}{C_p}-\dfrac{N^2\cdot r_{cp}}{C_p(R_i+N\cdot r_{cp})} \\ 0 & 0 & 0 & \dfrac{1}{C_s} \\ \dfrac{1}{L_p}-\dfrac{N^2 r_{cp}}{L_p(R_i+N\cdot r_{cp})} & 0 & -\left(\dfrac{r_{cp}+r_{ip}}{L_p}\right) & \dfrac{r_{cp}}{L_p} \\ -\dfrac{1}{L_s}+\dfrac{N^2\cdot r_{cp}}{L_s(R_i+N\cdot r_{cp})} & -\dfrac{1}{L_s} & \dfrac{r_{cp}}{L_s}-\dfrac{N^2\cdot r_{cp}^2}{L_s(R_i+N\cdot r_{cp})} & -\left(\dfrac{r_{ds}+r_{cs}+r_{cp}+r_{ls}}{L_s}\right)+\dfrac{N^2\cdot r_{cp}^2}{L_s(R_i+N\cdot r_{cp})} \end{bmatrix}$$

$$\text{and, } \mathbf{B}_1 = \begin{bmatrix} 0 \\ 0 \\ 0 \\ \dfrac{V_{in}}{L_s} \end{bmatrix}$$

(10)

Due to symmetry, the dynamics matrix $\mathbf{A}_2$ and the input matrix, $\mathbf{B}_2$, for inverter operation in mode $M2$ ($V_{in}=0$) is given by:

$$\mathbf{A}_2=\mathbf{A}_1 \text{ and, } \mathbf{B}_2=\mathbf{0}_{4x1} \qquad (11)$$

Since only two dominant modes are considered, depending on the polarity of the input voltage, symmetry dictates that the duty of each mode is $0.5T_s$, where T_s is the period for one cycle i.e. $d_1 = 0.5$, $d_2 = 0.5$. Substituting (10, 11) together with the duties, into (8, 9) therefore provides the initial operating condition of the circuit in steady state and an analytical solution for the circuit behaviour. An example phase portrait of the resonant circuit voltages and currents in steady state prior to ignition and during normal operation is shown in Fig. 7. It is notable that the computation time to obtain the analytical steady-state cyclic solution of the state vector is ~1/10000× that required from Spice and other integration based simulation packages.

VI. CONCLUSIONS

State-variable dynamic descriptions of a 4th-order resonant power inverter, used as fluorescent lamp ballast, have been presented.

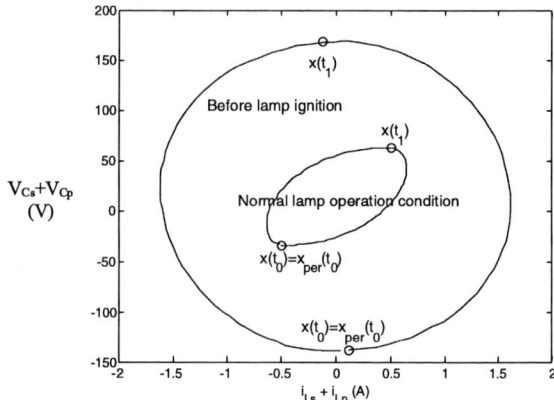

Figure 7 Steady state cyclic trajectory of the 4th-order resonant inverter during normal lamp operation, and prior to ignition of lamp

The model is subsequently used to obtain a steady-state cyclic description of the circuit, and the derivation of analytical formulae to calculate the electrical stresses on the resonant tank components. The accuracy of the proposed techniques has been shown by comparisons with both Spice simulations, and from measurements of an experimental 8W, capacitively coupled, fluorescent lamp, with good agreement being demonstrated. It is notable that, whilst the state-variable dynamic model requires a commensurate computational overhead to that of Spice simulations, the presented cyclic analysis method is typically $10^4\times$ faster.

VII REFERENCES

[1] D. O. Wharmby, 'Electrodeless lamps for lighting: a review,' IEE Proceedings-A, vol. 140, pp. 465-473 Nov. 1993

[2] M. K. Kazimierczuk and W. Szaraniec, 'Electronics Ballast for Fluorescent Lamps,' IEEE Trans. on Power Electronics, vol. 8, pp. 384-395, Oct. 1993

[3] C. Chang, J. Chang, G. W. Bruning, 'Analysis of the Self-Oscillating Series Resonant Inverters for Electronics Ballasts,' IEEE Trans. on Power Electronics, vol. 14, pp. 533-540, May 1999

[4] S. Y. R. Hui, L. M. Lee, H. Chung, Y. K. Ho, 'An Electronics ballast with Wide Dimming High PF, and Low EMI,' IEEE Trans. on Power Electronics, vol. 16, pp. 465-471, July 2001

[5] M. C. Cosby and R. M. Nelms, 'A Resonant Inverter for Electronics Ballast Applications,' IEEE Trans. on Industrial Electronics, vol. 41, pp. 418-425, Aug. 1994

[6] C. S. Moo, W. M. Chen, and H. K. Hsieh, 'Electronic Ballast with Piezoelectric Transformer for Cold Cathode Fluorescent Lamps,' IEE Procs. Electric Power Applications, vol. 150, pp. 278-282, March 2003

[7] L. R. Nerone, 'A Mathematical Model of the Class D Converter for Compact Fluorescent Ballast,' IEEE Trans. on Power Electronics, vol. 10, pp. 708-715, Nov. 1995

[8] S. Yaakov, M. Shvartsas, J. Lester, 'A Behavioural SPICE Compatible Model of an Electrodeless Fluorescent Lamp,' APEC, 2002, pp. 948-954

[9] J. Alonso, C. Blanco, E. Lopez, A. J. Calleja, and M. Rico, 'Analysis, Design, and Optimization of the LCC Resonant Inverter as a High-Intensity Discharge lamp Ballast,' IEEE Trans. on Power Electronics, vol. 13, pp. 573-585, May 1998

[10] M. K. Kazimierczuk, Resonant Power Converters. New York : John Wiley and Son, 1995

[11]   H. R. Visser and P. P. J. Borch, 'Modelling of Periodically Switching Networks,' PESC 91 Records 22nd IEEE Power Electronics Specialists Conference, 1991, pp. 67-73

[12]   Y. A. Ang, D. A. Stone, C. M. Bingham and M. P. Foster, 'Analysis and Design of High-frequency LCLC Converters for Electrodeless Fluorescent Lamp Ballast,' in press for PEMD 2004

[13]   M. P. Foster, H. I. Sewell, C. M. Bingham, D. A. Stone, D. Hente & D. Howe: 'Cyclic-averaging for high-speed analysis of resonant converters.' IEEE Transactions on Power Electronics, vol. 18, pp. 985-993, July 2003.

[14]   Y. A. Ang, M. P. Foster, C. M. Bingham, D. A. Stone, H. I. Sewell, D. Howe: 'Analysis of 4th-order LCLC Resonant Power Converters.' IEE Electric Power Applications, vol. 151, pp. 169-181, March 2004

Analysis and Control of Dual-Output LCLC Resonant Converters, and the Impact of Leakage Inductance

Y. Ang, C. M. Bingham, M. P. Foster, D. A. Stone
Department of EEE, The University of Sheffield, Sheffield, UK.
E-mail d.a.stone@sheffield.ac.uk

Abstract-The analysis, design and control of 4th-order LCLC voltage-output series-parallel resonant converters (SPRCs) for the provision of multiple regulated outputs, is described. Specifically, state-variable concepts are developed to establish operating mode boundaries with which to describe the internal behaviour of dual-output resonant converters, and the impact of output leakage inductance. The resulting models are compared with those obtained from SPICE simulations and measurements from a prototype power supply under closed loop control to verify the analysis, modeling and control predictions.

I. INTRODUCTION

To-date, several approaches have been explored to address cross-regulation, complexity and overall circuit performance issues of multi-output converters, the solutions being divided into three distinct categories. The first regulates a single primary output using closed-loop feedback, with the auxiliary outputs being semi-regulated and, therefore, subject to cross-regulation error. The second category achieves precise post-regulation of each output by using either linear regulators or hard-switched dc-dc converters. However, although relatively straightforward to design, such circuits are rarely used in practice due to cost constraints. The third category is specific to applications which require only two regulated outputs, as is commonly found in signal processing and microprocessor based systems. They avoid the need for post-regulation by utilising two closed-loop feedback configurations. A 3rd-order LLC converter with two independently controlled outputs was reported in [1]. However, optimum performance characteristics have yet to be forthcoming, primarily due to the significant complexity associated with the highly non-linear behaviour between the various outputs as a function of load. Nevertheless, it is a solution that broadly falls within this third category that is the subject of this paper. Specifically, dual-output resonant LCLC converters, are considered, with control of each output being achieved by switching the power devices asymmetrically over each half switching cycle using a combination of PWM and frequency control.

II. DUAL O/P LCLC-SPRC MODEL

A half-bridge LCLC-SPRC with two outputs is shown in Fig. 1(a). To achieve zero-voltage switching, the converter is assumed to operate on the negative gradient of the input-output frequency characteristic, above the primary resonant peak. When operating in this region, the resulting waveforms can be sub-divided into two distinct time intervals, viz. intervals 1 and 2, as depicted in Fig. 1(b):

Interval 1: Clamping of the parallel capacitor voltage. Here, the combined series inductor L_s (Fig.1(c)) and capacitor C_s provide resonant behaviour whilst the voltage across L_p and C_p is clamped by the output voltage. As the current through the series inductor, L_s, decays to zero, C_p begins to contribute to resonant behaviour, and operation enters the second interval.

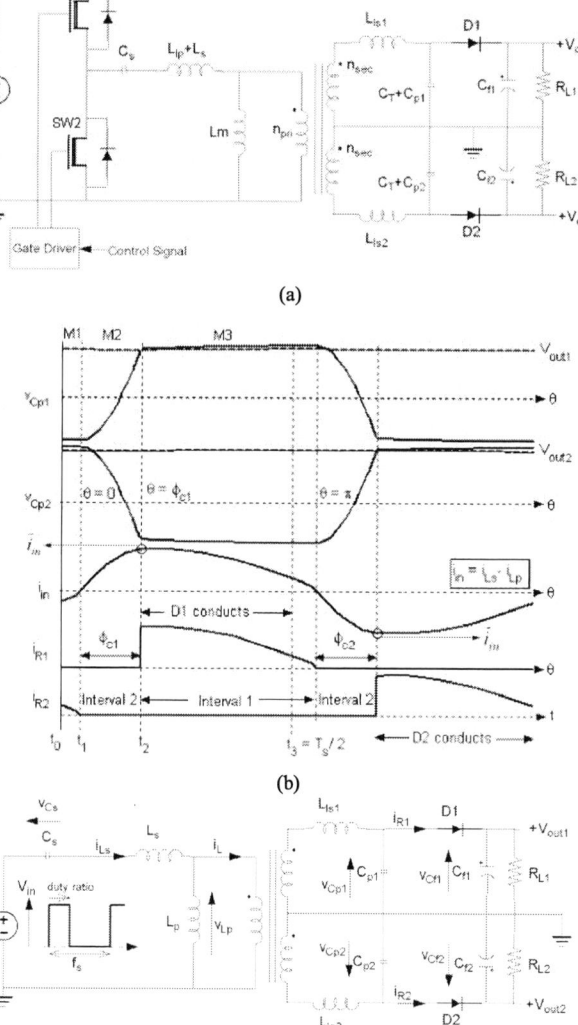

Fig.1 Dual-load 4th-order resonant converter (a) schematic (b) typical operating waveforms (c) simplified circuit

Interval 2: Decoupling of the rectifier and output filter. Here, all the tank components contribute to resonant behaviour, with the rectifier effectively becoming reverse biased. Current into both the high- and low-side diodes remains zero, and the parallel capacitors are charged until their voltage is clamped at either $+V_{out1}$ or $-V_{out2}$, thereby providing the boundary at the end of this time interval. During each half-cycle of operation, three Modes, M1, M2, and M3 can be identified, as shown in Fig. 1(b).

Circuit Mode M1 ($t_0 \le t < t_1$). At the start of M1, SW2 is turned off at t_0 and SW1 turned on. The series inductor current, i_{Ls}, is negative and flows through the internal diode of SW1, thereby facilitating ZVS of SW1. Also during this period, i_{Ls} allows D2 to conduct and transfer energy to support V_{out2}, whilst the voltage on C_{p2} is clamped to V_{out2}. All the rectifier current, therefore, flows to the load. At the end of M1, the rectifier current i_{R2} has decayed to zero, and both the high side and low side diodes, and the output filter, are effectively decoupled from the resonant tank.

Circuit Mode M2 ($t_1 \le t < t_2$). Here, the series resonant inductor current i_{Ls} becomes positive. Since SW1 is turned on during M1, current flow is now through SW1. Initial conditions for this mode are that i_{Ls}=0 and $v_{cp2}= V_{out2}$. The inductor current i_{Ls} and parallel resonant capacitor voltages take on a sinusoidal characteristic. Since the outputs are effectively disconnected from the tank, both C_{p1} and C_{p2} contribute to resonant behaviour. Both rectifier currents are zero, and the converter outputs are in an 'idle' state, with energy being supplied solely by the charge on the filter capacitors. By initially neglecting the rectifier on-state voltage, and noting that the effective parallel resonant capacitance C_p is the sum of the shunt network capacitances C_{p1} and C_{p2}, v_{Cp1} during the capacitor charging period is described by:

$$v_{Cp1}(t) = v_{Cp1}(t_1) + \frac{1}{C_p}\int_{t_1}^{t_2} \hat{i}_{in}\sin(2\pi f_s t)dt \tag{1}$$

where $\breve{i}_{in} = -(i_{Ls} - i_{Lp})$. Evaluating (1) with initial conditions $v_R(t_1) = v_{Cp1}(t_1) = -V_{out2}$ yields:

$$v_{Cp1}(t_2) = -V_{out2} + \hat{i}_{in} \times \frac{1 - \cos(2\pi f_s(t_2 - t_1))}{2\pi f_s C_p} \tag{2}$$

The boundary for the end of the capacitor charging period is $v_{Cp1}(t_2) = +V_{out1}$, which yields the rectifier non-conduction angle, ϕ_{c1}, associated with a positive polarity of current, i_R, through the high side rectifier:

$$t_2 - t_1 = \frac{1}{2\pi f_s} \times \cos^{-1}(\phi_{c1}) \quad \phi_{c1} = \cos^{-1}\left(1 - \frac{2\pi f_s C_p v_{tot}}{\hat{i}_{in}}\right) \tag{3}$$

where $v_{tot} = V_{out1} + V_{out2}$.

Circuit Mode M3 ($t_2 \le t < T_s/2$). At $t = t_2$, D1 becomes forward biased whilst D2 reverse biased. The rectifier diode current i_{R2} remains zero throughout the duration of M3, and D1 clamps the capacitor voltage v_{cp1} to $+V_{out1}$ until i_{Ls} decays to zero, at which time the second half cycle of operation commences.

For 50% duty-cycle excitation, the 2nd half-cycle of operation is the mirror image of the first. However, for asymmetrical excitation, the output rectifier diode (D2) non-conduction angle, associated with the series resonant inductor current being of negative polarity, is given by:

$$\phi_{c2} = \cos^{-1}\left(1 - \frac{2\pi f_s C_p v_{tot}}{\breve{i}_{in}}\right) \tag{4}$$

where $\hat{i}_{in} = i_{Ls} - i_{Lp}$. The voltage, v_{cp1}, across the parallel resonant capacitor can, therefore, be expressed as a function of the angle θ —see Fig. 1(b):

$$v_{Cp1}(\theta) = \begin{cases} -V_{out2} & \text{for } \theta = 0 \ldots \phi_{c1} \\ -V_{out2} + \frac{\hat{i}_{in}}{2\pi f_s C_p}\times(1-\cos(\theta)) \\ \quad + V_{out1} & \text{for } \theta = \phi_{c1}\cdots\pi \\ V_{out1} - \frac{\hat{i}_{in}}{2\pi f_s C_p}\times(1-\cos(\theta)) \\ \quad -V_{out2} & \text{for } \theta = \pi \ldots \pi + \phi_{c2} \\ \quad & \text{for } \theta = \pi + \phi_{c2} \ldots 2\pi \end{cases} \tag{5}$$

Under steady-state conditions, the mean output current i_{out1}, flowing through D1 towards the output filter and load, can be determined from the mean current flowing through the rectifier when it is of positive polarity. Since this occurs during the interval $\phi_{c1} \le \theta < \pi$, i_{out1} is given by:

$$i_{out1} = \frac{1}{2\pi} \times \int_{\phi_{c1}}^{\pi} \hat{i}_{in}\sin(\theta)d\theta \tag{6}$$

Substituting (3) into (6) and evaluating the integral provides the solution for i_{out1}:

$$i_{out1} = \frac{\hat{i}_{in}}{2\pi}\times(1 + \cos(\phi_{c1})) = \frac{\hat{i}_{in} - \pi f_s C_p v_{tot}}{\pi} \tag{7}$$

Simple mathematical manipulation of (3) and (7) then gives the corresponding rectifier non-conduction angle ϕ_{c1}:

$$\phi_{c1} = \cos^{-1}\left(\frac{\pi i_{out1} - \pi f_s C_p v_{tot}}{\pi i_{out1} + \pi f_s C_p v_{tot}}\right) \tag{8}$$

V_{out1} is determined by assuming the output filter capacitance C_f is sufficiently large to impart negligible output voltage ripple. In this case:

$$V_{out1} = i_{out1}R_{L1} = \frac{\hat{i}_{in}R_{L1}}{2\pi}\times(1 + \cos(\phi_{c1})) = \frac{R_{L1}(\hat{i}_{in} - \pi f_s C_p v_{tot})}{\pi} = \frac{R_{L1}}{\pi}\times\frac{\hat{i}_{in} - \pi f_s C_p V_{out2}}{1 + R_L f_s C_p} \tag{9}$$

Equations (6) to (9) can be further manipulated to provide the complementary D2 non-conduction angle, ϕ_{c2}, and the output current, i_{out2}, and output voltage V_{out2}, as follows:

383

$$i_{out2} = \frac{\tilde{i}_{in}}{2\pi} \times (1 + \cos(\phi_{c2})), \quad V_{out2} = \frac{R_{L2}}{\pi} \times \frac{\tilde{i}_{in} - \pi f_s C_p V_{out1}}{1 + R_L f_s C_p} \qquad (10)$$

III. STATE-VARIABLE ANALYSIS

A state-variable model describing the behaviour of the dual-output converter can be obtained by considering the electrical network shown in Fig. 1(c) and separating the dynamics into 'fast' and 'slow' sub-systems, with their interaction related by a set of coupling equations. The fast sub-system is considered to describe the dynamics of the resonant tank and power switches:

$$\frac{dv_{Cs}}{dt} = \frac{i_{Ls}}{C_s}, \quad \frac{di_{Ls}}{dt} = \frac{V_{in} - v_{Cs} - v_{Lp}}{L_s}, \quad \frac{di_{Lp}}{dt} = \frac{v_{Lp}}{L_p}$$

$$\frac{dv_{Cp1}}{dt} = \frac{i_{Ls} - i_{Lp} - i_{R1} - i_{Cp2} - i_{R2}}{C_{p1}}, \quad \frac{dv_{Cp2}}{dt} = \frac{i_{Ls} - i_{Lp} - i_{R2} - i_{Cp1} - i_{R1}}{C_{p2}}$$

$$(11)$$

The output filter dynamics are described by:

$$\frac{dv_{cf1}}{dt} = \frac{i_{R1}}{C_{f1}} - \frac{v_{cf1}}{C_{f1}R_{L1}}, \quad \frac{dv_{cf2}}{dt} = \frac{i_{R2}}{C_{f2}} - \frac{v_{cf2}}{C_{f2}R_{L2}} \qquad (12)$$

As discussed, during interval $t_1 \to t_2$ (see Fig. 1(b)) v_{cp1} is clamped to v_{cf1} during the positive half-cycle, and conversely, to $-v_{cf2}$ during the negative half-cycle, due to the action of the diodes. By noting that there will be negligible current flowing through C_p during these periods, the rectifier input voltage is dependent on the direction of the current leaving the resonant tank inductances, i.e. $i_L = i_{Ls} - i_{Lp}$. The relevant coupling terms are, therefore, obtained by equating voltages at either side of the rectifier for each respective half-cycle:

$$v_{Cp1} = \text{sgn}(i_L)(V_{out1} + v_{diode}) = \text{sgn}(i_L)(v_{cf1} + v_{diode}) \qquad (13)$$

$$v_{Cp2} = \text{sgn}(i_L)(V_{out2} + v_{diode}) = \text{sgn}(i_L)(v_{cf2} + v_{diode})$$

Assuming a constant rectifier voltage, (12) can be manipulated to give:

$$\frac{dv_{Cp1}}{dt} = \text{sgn}(i_L)\frac{dv_{cf1}}{dt}, \quad \frac{dv_{Cp2}}{dt} = \text{sgn}(i_L)\frac{dv_{cf2}}{dt} \qquad (14)$$

Considering the rectifier current, i_{R2}, to be zero during the positive half-cycle of the parallel capacitor voltage, the rectifier current i_{R1}, is given from:

$$\frac{i_L - i_{R1} - i_{Cp2} - i_{R2}}{C_{p1}} = \text{sgn}(i_L)\left(\frac{i_{R1}}{C_{f1}} - \frac{v_{cf1}}{C_{f1}R_{L1}}\right) \qquad (15)$$

$$\therefore i_{R1} = \frac{C_{p1}C_{f1}}{\text{sgn}(i_L)C_{p1} + C_{f1}}\left(\frac{i_L - i_{Cp2} - i_{R2}}{C_{p1}} + \frac{\text{sgn}(i_L)v_{cf1}}{C_{f1}R_{L1}}\right)$$

This leads to the following coupling equations which describe the rectifier currents within each half of a switching cycle:

$$i_{R1} = \begin{cases} \dfrac{C_{p1}C_{f1}}{\text{sgn}(i_L)C_{p1} + C_{f1}}\left(\dfrac{i_L - i_{Cp2} - i_{R2}}{C_{p1}} + \dfrac{\text{sgn}(i_L)v_{cf1}}{C_{f1}R_{L1}}\right) & \text{for } v_{Cp1} = V_{out1} + v_{diode} \\ 0 & \text{for } v_{Cp1} < V_{out1} + v_{diode} \end{cases}$$

$$i_{R2} = \begin{cases} \dfrac{C_{p2}C_{f2}}{\text{sgn}(i_L)C_{p2} + C_{f2}}\left(\dfrac{i_L - i_{Cp1} - i_{R1}}{C_{p2}} + \dfrac{\text{sgn}(i_L)v_{cf2}}{C_{f2}R_{L2}}\right) & \text{for } v_{Cp2} = V_{out2} + v_{diode} \\ 0 & \text{for } v_{Cp2} < V_{out2} + v_{diode} \end{cases}$$

$$(16)$$

Notably, the voltage across L_p can be considered a reflection of the voltages across C_{p1} and C_{p2}, and the state vector for the parallel inductor current in the fast sub-system (see (11)) simplifies to $v_{Lp} = v_{Cp}$. The state-variable equations for the parallel resonant capacitor voltage (11) can be simplified to:

$$\frac{dv_{Cp1}}{dt} = \frac{i_{Ls} - i_{Lp} - i_R}{2C_{p1}}, \quad \frac{dv_{Cp2}}{dt} = \frac{i_{Ls} - i_{Lp} - i_R}{2C_{p2}} \qquad (17)$$

The complete state-variable model of the dual load converter (excluding the effects of output leakage inductances) is, therefore, given by:

$$\dot{\mathbf{x}} = \begin{bmatrix} 0^{3\times3} & \mathbf{A}_1 & 0^{2\times3} \\ \mathbf{A}_2 & 0^{2\times2} & 0^{2\times2} \\ 0^{2\times3} & 0^{2\times2} & \mathbf{A}_3 \end{bmatrix}\mathbf{x} + \mathbf{B} \qquad (18)$$

where

$$\mathbf{x} = \begin{bmatrix} v_{Cp1} & v_{Cp2} & v_{Cs} & i_{Lp} & i_{Ls} & v_{Cf1} & v_{Cf2} \end{bmatrix}^T$$

$$\mathbf{A}_1 = \begin{bmatrix} -\dfrac{1}{2C_{p1}} & \dfrac{1}{2C_{p1}} \\ -\dfrac{1}{2C_{p2}} & \dfrac{1}{2C_{p2}} \\ 0 & \dfrac{1}{C_s} \end{bmatrix}, \quad \mathbf{A}_2 = \begin{bmatrix} \dfrac{1}{L_p} & 0 & 0 \\ -\dfrac{1}{L_s} & 0 & -\dfrac{1}{L_s} \end{bmatrix}, \quad \mathbf{A}_3 = \begin{bmatrix} -\dfrac{1}{C_f R_{L1}} & 0 \\ 0 & -\dfrac{1}{C_{f2} R_{L2}} \end{bmatrix}$$

$$\mathbf{B} = \begin{bmatrix} -\dfrac{i_R}{2C_{p1}} & -\dfrac{i_R}{2C_{p2}} & 0^{1\times2} & \dfrac{V_{in}}{L_s} & \dfrac{i_{R1}}{C_{f1}} & \dfrac{i_{R2}}{C_{f2}} \end{bmatrix}^T$$

$$(19)$$

The model can be used to investigate the behaviour of dual load converters when subject to asymmetrical input excitation. By way of example, the parameters of a candidate converter are given in Table I when supplied from a 30V dc link. A plot of the resulting steady-state output voltage characteristics of the converter, V_{out1} and V_{out2}, as a function of switching frequency and duty-cycle ratio is given in Fig. 2. It is evident that for operation above resonance, the sum of the output voltages applied to the loads increases as the operating frequency tends to the effective resonant frequency, for fixed values of duty-cycle ratio. Furthermore, for a 50% duty-cycle, giving symmetric square-wave excitation of the tank, the converter delivers identical voltages to both the high side and low side outputs, for a fixed operating frequency, as expected. For a given operating frequency, a decrease in the duty-cycle ratio, from 50%, is seen to deliver more energy from the resonant tank to energize output V_{out1}, thereby yielding a correspondingly higher output voltage and power, and vice-versa. It is, therefore, clear that for balanced loads, the voltage and power distribution to each output can be independently influenced by a suitable choice of duty ratio and switching frequency.

TABLE I
CONVERTER MODEL PARAMETERS

Parameters	Value
Characteristic impedance (Ω)	2.5
Resonant inductance ratio, L_n	0.01
Resonant frequency, f_o (kHz)	130
Resonant capacitance ratio, C_n	0.03
Series load quality factor, Q_{op1}	6

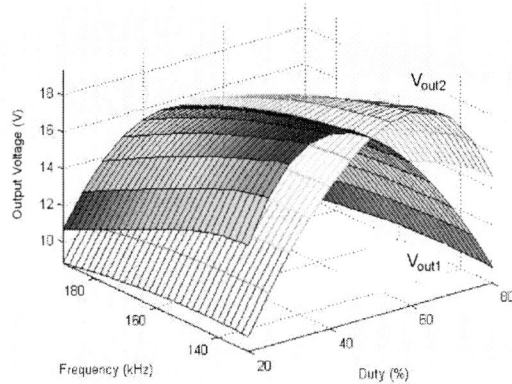

Fig.2 Variation of output voltage distribution with switching frequency and duty-cycle ratio

IV. IMPACT OF OUTPUT LEAKAGE INDUCTANCE

Although resonant converters are often designed to operate at relatively high frequencies, typically in the 200-500 kHz range, designers must still consider the impact of low levels of leakage inductance on converter performance. This is particularly evident for converters with multiple outputs, where the presence of such effects significantly complicates the analysis, particularly when determining the state of the parallel resonant inductor current, i_{Lp}, since the voltage seen across L_p cannot be assumed to be directly related to v_{Cp}, as a result of the voltages across the leakage inductances, L_{ls1} and L_{ls2}. The model must, therefore, be augmented with v_{Lp} to allow a solution for i_{Lp}. The converter can be separated into fast and slow sub-systems, Fig. 3 showing a model of the resonant converter 'fast' sub-system. Whilst the rectifier is omitted in the state-variable representation, its influence on the fast sub-system is accommodated through the addition of current sources, as shown in Fig. 3. This step is justified by noting that the interaction between the fast and slow sub-systems is solely based on coupling equations consisting of the characteristics of the rectifier output currents i_{R1} and i_{R2}. The slow sub-system describes the behaviour of the high side and low side rectifier outputs and the capacitive output filters and loads, v_{Cp1} and v_{Cp2} being considered to be the inputs to the high side and low side output sub-systems, respectively.

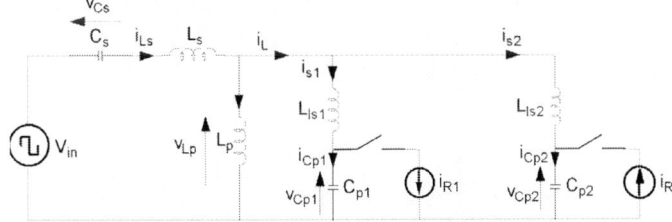

Fig. 3 State-variable representation of the fast sub-system.

The dynamics of the fast sub-system, therefore, consist of a set of state-variables whose value at time $t = t_0$, together with the input for all $t > t_0$, completely determines the behaviour of the system for any time $t > t_0$. An observable canonical state-space realization is, therefore, considered, as it allows the impact of parasitic elements to be readily included in the formulation, albeit at the expense of losing some of the physical

significance of the state-variables. The equivalent circuit in Fig. 3 is analyzed by considering each voltage and current source independently and using the principal of superposition to obtain the effective dynamic description for i_{Lp}:

$$\frac{di_{Lp}}{dt} = \frac{v_{Lp}}{L_p} = \frac{v_{Lp\_vi} + v_{Lp\_iR1} + v_{Lp\_iR2}}{L_p} = \frac{y_1 + y_2 - y_3}{L_p}$$

$$\dot{x}_1 = A_o x_1 + B_{o1} V_{in} \quad y_1 = v_{Lp\_vi} = C_o x_1 + D_{o1} V_{in}$$

$$\dot{x}_2 = A_o x_2 + B_{o2} i_{R2} \quad y_2 = v_{Lp\_iR1} = C_o x_2 + D_{o2} i_{R2}$$

$$\dot{x}_3 = A_o x_3 + B_{o2} i_{R2} \quad y_3 = v_{Lp\_iR2} = C_o x_3 + D_{o3} i_{R2}$$

where

$$A_o = \begin{bmatrix} 0 & 0 & 0 & 0 & 0 & -\dfrac{a_{13}}{a_{10}} \\ 1 & 0 & 0 & 0 & 0 & 0 \\ 0 & 1 & 0 & 0 & 0 & -\dfrac{a_{12}}{a_{10}} \\ 0 & 0 & 1 & 0 & 0 & 0 \\ 0 & 0 & 0 & 1 & 0 & -\dfrac{a_{11}}{a_{10}} \\ 0 & 0 & 0 & 0 & 1 & 0 \end{bmatrix} ; \quad B_{c1} = \begin{bmatrix} -\dfrac{b_{10}}{a_{10}} \times \dfrac{a_{13}}{a_{10}} \\ 0 \\ -\dfrac{b_{10}}{a_{10}} \times \dfrac{a_{12}}{a_{10}} + \dfrac{b_{11}}{a_{10}} \\ 0 \\ -\dfrac{b_{10}}{a_{10}} \times \dfrac{a_{11}}{a_{10}} + \dfrac{b_{12}}{a_{10}} \\ 0 \end{bmatrix} ; \quad B_{c2} = \begin{bmatrix} \dfrac{b_{22}}{a_{10}} \\ 0 \\ \dfrac{b_{21}}{a_{10}} \\ 0 \\ \dfrac{b_{20}}{a_{10}} \\ 0 \end{bmatrix} ;$$

$$B_{c3} = -B_{c2} \quad C_o = \begin{bmatrix} 0^{5 \times 1} & 1 \end{bmatrix} ; \quad D_{c1} = \begin{bmatrix} 1 \end{bmatrix} ; \quad D_{c2} = D_{c3} = \begin{bmatrix} 0 \end{bmatrix}$$

$$b_{10} = L_p L_{ls1} L_{ls2} \quad b_{11} = \frac{L_p L_{ls1}}{C_{p2}} + \frac{L_p L_{ls2}}{C_{p1}} \quad b_{12} = \frac{L_p}{C_{p1} C_{p2}}$$

$$a_{10} = L_p L_{ls1} L_{ls2} + L_s L_p L_{ls2} + L_s L_p L_{ls1} + L_s L_p L_{ls2}$$

$$a_{11} = \frac{L_s L_{ls1}}{C_{p2}} + \frac{L_s L_{ls1}}{C_{p1}} + \frac{L_s L_p}{C_{p2}} + \frac{L_s L_p}{C_{p1}} + \frac{L_{ls1} L_{ls2}}{C_s} + \frac{L_p L_{ls2}}{C_s} + \frac{L_p L_{ls1}}{C_s} + \frac{L_p L_{ls1}}{C_{p2}} + \frac{L_p L_{ls2}}{C_{p1}}$$

$$a_{12} = \frac{L_s}{C_{p1} C_{p2}} + \frac{L_{ls1}}{C_s C_{p2}} + \frac{L_{ls1}}{C_s C_{p1}} + \frac{L_p}{C_s C_{p2}} + \frac{L_p}{C_s C_{p1}} + \frac{L_p}{C_{p1} C_{p2}} \quad a_{13} = \frac{1}{C_s C_{p1} C_{p2}}$$

$$b_{20} = L_s L_p L_{ls2} \quad b_{21} = \frac{L_p L_{ls2}}{C_s} + \frac{L_s L_p}{C_{p2}} \quad b_{22} = \frac{L_s}{C_s C_{p2}}$$

Adding the state vectors yields,

$$\dot{x}_1 + \dot{x}_2 + \dot{x}_3 = A_o(x_1 + x_2 + x_3) + \begin{bmatrix} B_{o1} & B_{o2} & B_{o3} \end{bmatrix} \begin{bmatrix} V_{in} \\ i_{R1} \\ i_{R2} \end{bmatrix}$$

$$y_1 + y_3 + y_3 = v_{Lp\_vi} + v_{Lp\_iR1} + v_{Lp\_iR2} = C_o(x_1 + x_2 + x_3) + \begin{bmatrix} D_{o1} & 0 & 0 \end{bmatrix} \begin{bmatrix} V_{in} \\ i_{R1} \\ i_{R2} \end{bmatrix}$$

Now, by defining, $\begin{matrix} x = x_1 + x_2 + x_3 \\ y = y_1 + y_3 + y_3 = v_{Lp} \end{matrix}$ the observable canonical state-space equation simplifies to,

$$\dot{x} = A_o(x_1 + x_2 + x_3) + \begin{bmatrix} B_{o1} & B_{o2} & B_{o3} \end{bmatrix} \begin{bmatrix} V_{in} \\ i_{R1} \\ i_{R2} \end{bmatrix}$$

$$y = v_{Lp} = C_o(x_1 + x_2 + x_3) + \begin{bmatrix} D_{o1} & 0 & 0 \end{bmatrix} \begin{bmatrix} V_{in} \\ i_{R1} \\ i_{R2} \end{bmatrix}$$

Substituting the output state, together with the coupling equations, leads to a steady-state solution of the model. The fast and slow sub-system models can also be combined and used for implementation in simulation environments, eg. MATLAB®/SIMULINK.

V. EXPERIMENTAL RESULTS

Measured results have been obtained on an experimental converter with a step-down capability, the measured component values being given in Table II. A ferrite core, 3F3, suitable for high frequency applications, was used for both the transformer core and the resonant inductor. Since the transformer leakage inductances are dependent on the winding arrangement, the secondaries were bifilar wound adjacent to the core, beneath the primary winding, so as to reduce secondary leakage flux.

TABLE II
PROTOTYPE DUAL OUTPUT CONVERTER COMPONENT VALUES

Parameter	Value
DC link input voltage, v_{DC} (V)	15
Series resonant inductances, L_s (μH)	0.85
Series resonant capacitances, C_s (μF)	1.5
High side parallel resonant capacitances, C_{p1} (μF)	0.116
Low side parallel resonant capacitances, C_{p2} (μF)	0.116
Load resistance, R_L (Ω)	4
Filter capacitance, C_f (μF)	100
Magnetising inductance, L_m (μH)	109
Transformer turns ratio	1
Transformer output leakage inductance, L_{ls} (μH)	0.1
Transformer primary leakage inductance, L_{lp} (μH)	0.7

The effective parallel resonant inductor is designed to be on the transformer primary side, such that L_p utilises the magnetising inductance, L_m, of the transformer. The effective series inductance comprises of the series inductor, L_s, and the primary leakage inductance, L_{lp}, and was measured as 1.55μH. A comparison of the output voltage obtained from the state variable model, simulated to steady-state, with SPICE simulation results and measurements, is given in Fig. 4. As is clearly evident, the correlation between the theoretical predictions and the experimental data is extremely good. Moreover, a comparison of simulated and measured characteristics, when duty-cycle ratio control is employed, has also been obtained at an operating frequency of f_s=150 kHz. The results are shown in Fig. 5 from which it can be seen that the proposed state-variable model provides sufficient accuracy for design and analysis purposes, the maximum error being <5%.

VI. CLOSED-LOOP CONTROL

The structure of the control methodology, which employs two decoupled feedback loops for independent control of frequency and duty-cycle ratio, is shown in Fig. 6. Parameters for the decoupled SISO controllers are chosen for good transient response and disturbance rejection. The digital compensator is tuned to respond quickly to variations of V_{out1}, whilst the controller reacting to variations of V_{out2}, acts relatively slowly, thereby allowing an effective decoupling of the control loops, for simplicity. The prototype converter controller is shown in Fig. 7. The converter (see Table II for parameters) is required to provide regulated +5V and +3.3V outputs from a DC link input voltage in the range 15V to 20V. Figure 8 shows the steady-state error between the reference voltages, V_{ref1} and V_{ref2}, and the resulting measured output voltages of the converter, over the specified range of DC link input voltages (15V to 20V) with a 5Ω load.

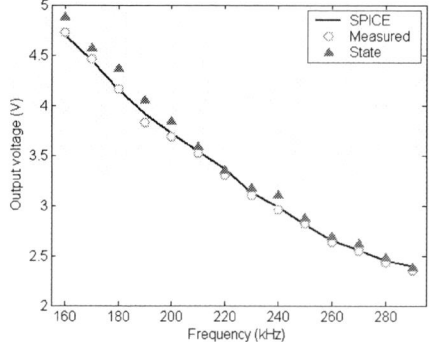

Fig. 4 Output voltage vs switching frequency

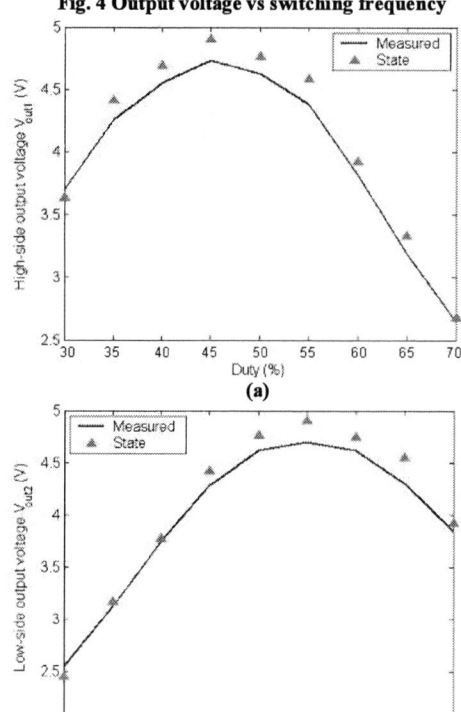

Fig. 5 Control characteristic curves for f_s=150 kHz (a) high side output, and (b) low side output.

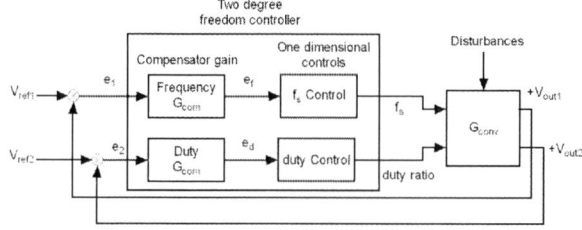

Fig. 6 Closed-loop control of the dual-load converter.

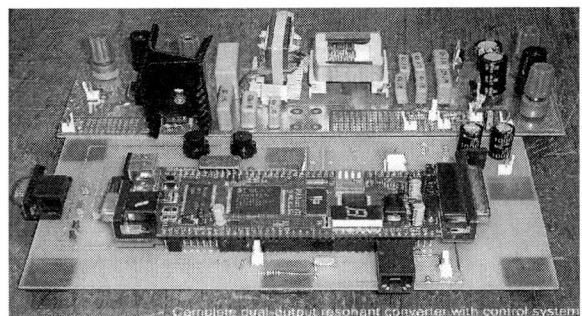

Fig. 7 Digital control of dual-output converter.

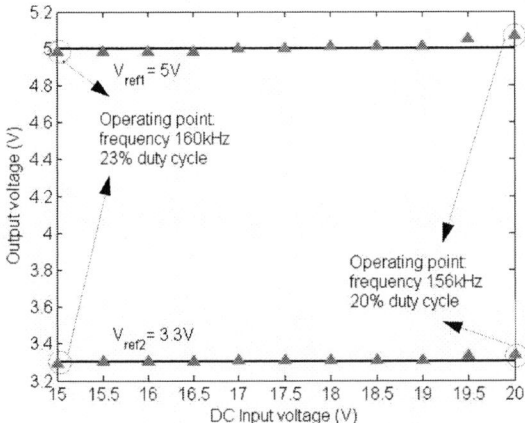

Fig. 8 Output voltage regulation vs. input voltage

It can be seen that the maximum regulation error for both outputs is <5%. Finally, Fig. 9 shows the response of the converter resulting from transient start-up conditions, for a range of applied input voltages and dual output voltage ratios. It can be seen that the converter voltages converge rapidly to the reference values, with an initial overshoot of ~10%, corresponding to an equivalent 2nd-order damping ratio of $\zeta \approx 0.6$.

VII. CONCLUSIONS

The characteristics of dual-load, 4th-order LCLC voltage-output resonant converters, have been explored. From the output voltage distribution derived from an example converter, the impact of the converter design parameters, and output leakage inductance, on the attainable output voltage ratio, has been investigated. A comparison of measurements from an experimental converter, capable of delivering 5V and 3.3V, with predictions from the derived state-variable model and SPICE simulations, shows that the state-variable model provides accurate predictions of output voltage under steady-state conditions. Moreover, a basic control scheme to allow reliable regulation of both outputs, has been realised, with steady-state measurements showing independent regulation using a combination of duty-cycle and frequency control, and good start-up transient behaviour on both outputs under a range of operating conditions.

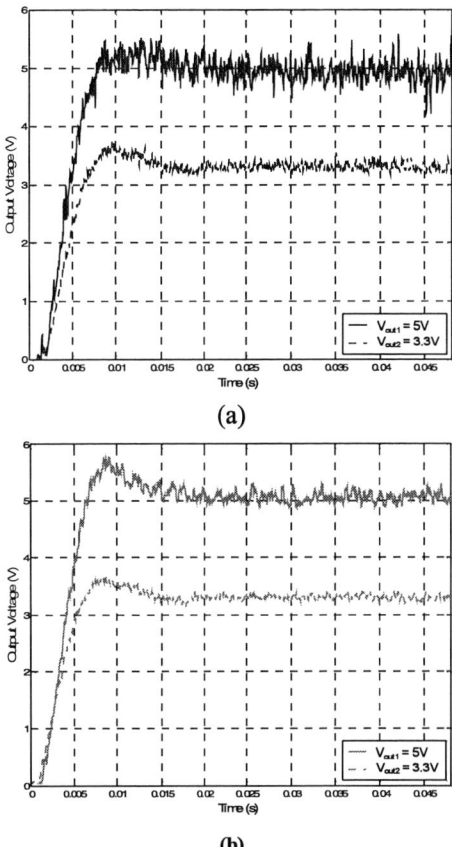

(a)

(b)

Fig. 9 Start-up transient response for various dual output voltage ratios and DC-link input voltages (a) v_{DC}=15V, V_{out1}=5V, V_{out2}=3.3V; (b) v_{DC}=20V, V_{out1}=5V, V_{out2}=3.3V

REFERENCES

[1] R. Elfrich and T. Duerbaum, "A new load resonant dual-output converter", *IEEE Power Electronics Specialists Conference Rec.*, 2002, pp. 1319-1324.

[2] S. Glozman and S. Ben-Yaakov, "Dynamic interaction of high frequency electronic ballasts and fluorescent lamps", *IEEE Power Electronics Specialists Conference Proc.*, 2000, pp. 1363-1368.

[3] J. A. Ferreira, "A series resonant converter for arc-striking applications", *IEEE Trans. on Industrial Electronics*, **45**, 1998, pp. 585-595.

[4] J. P. Agrawal, "Determination of cross regulation in multi output resonant converters", *IEEE Trans. on Aerospace and Electronic Systems*, 36, 2000, pp.760-772.

[5] J.P. Agrawal and I. Batarseh, "Improving the dynamic modeling and static cross regulation in multi-output resonant converters", *IEEE Applied Power Electronics Conference Proc.*, 1993, pp. 65-70.

[6] I. Batarseh and C. Q. Lee, "Multi-output LLCC-type parallel resonant converter", *IEEE Industrial Electronics Society Conference Proc.*, 1990, pp. 850-856.

[7] Y. Ang, C. M. Bingham, M. P. Foster, D. A. Stone and D. Howe, "Design orientated analysis of 4th-order LCLC converters with capacitive output filter", *IEE Proc. - Electric Power Applications*, **152**, 2005, pp. 310-322.

A Novel QR ZCS Switched-Capacitor Bidirectional Converter

Yuang-Shung Lee*, Yi-Pin Ko*, and Chien-An Chi**

* Graduate Institute of Applied Science and Engineering, Fu Jen Catholic University, Taipei, Taiwan
** Department of Electronic Engineering, Fu Jen Catholic University, Taipei, Taiwan

Abstract– The proposed quasi-resonant (QR) zero current switching (ZCS) switched-capacitor (SC) converter is a new type of bi-directional power flow control conversion scheme. They are able to provide the voltage conversion ratios from 2 versus 1/2 (double-mode / half-mode) to n versus 1/n (n-mode / 1/n-mode) by adding a different number of switched-capacitors and power MOSFET switches with a small series connected resonant inductor for forward and reverse schemes. The low current stress and balance resonance current are the advantage of the proposed quasi resonant switched-capacitor converter. The principle of operation, theoretical analysis of the proposed bi-directional power conversion scheme is described in detail with circuit model analysis. Simulation and experimental results are carried out to verify the performance of the new type ZCS SC bi-directional QR converters.

Index Terms– Quasi–resonant converter, Zero current switching, Switch-capacitor converter, Bi-directional converter.

I. INTRODUCTION

Traditional switching mode DC/DC converter use magnetic components and the size of the transformer are large in the converter. The switched-capacitor DC/DC converters do not requires any inductors and only need capacitors and MOSFET switches. However, the switching-capacitor DC/DC converters produce larger switching loss and high current stress at the high switching frequency [1]. Therefore, the quasi-resonant switched-capacitor DC/DC converter can operate at constant switching frequency with zero-current or zero-voltage switching (ZCS or ZVS) has been used to reduce the switching losses in the converters to overcome the above-mentioned disadvantages [2-7].

The bi-directional power control scheme have been used to the uninterruptible power supplies (UPS), battery charging and discharging system, and dual voltage vehicle power system [6-9]. A switched capacitor based bi-directional DC/DC converter providing the capability of step-down / step-up voltage is proposed in [8,9]. The countermeasures of input/output current pulsating peaks of the bi-direction SC converter are suggested but the

switching losses and the overall converter efficiency are not significantly improved.

This paper presents a new type of switched-capacitor step-up/step-down DC/DC quasi-resonant bi-directional converter that can be designed to operate at non-inverting mode and constant frequency for obtained low switching losses. The high switching current stresses can also be reduced under the bi-directional power flow control schemes [10]. The proposed converters have topologies for a number of voltage conversion ratios from $2 / \frac{1}{2}$ to $n / \frac{1}{n}$ under various control strategy of the switched-capacitor networks. The advantages of the proposed quasi-resonant ZCS switched-capacitor DC/DC converter are low weight, small volume, high efficiency, low current stress, balance resonance current, and different multiple output voltages. Analysis and experimental results are used to verify and validate the new concept of the proposed bi-directional converters. The operating principle of the proposed converter topology is described in section II. The converter circuit analysis under the forward and reverse power flow control are included in section III. Simulation and experimental results are discussed in section IV. Section V states the conclusions.

II. DESCRIPTION OF THE PROPOSED TOPOLOGY

Fig. 1 shows the circuit configuration of the proposed non-inverting type 4-mode / 1/4-mode (quadruple-mode/ quarter-mode) zero current switching switched-capacitor bi-directional converter that was developed based on the ZCS SC QR converter. Each of the circuits uses 2n MOSFET switches (n = 4) and only a very small inductor series connected with the switched-capacitors are needed to construct the resonant tanks in the converter. A resonant inductor L_r is connected in series with a set of the switching-capacitors comprising of C_{1b}, C_{2b} and C_{3b} to achieve a resonant characteristic when each of the switches Q_{NS} (contains Q_{N1}, Q_{N2}, Q_{N3} and Q_{N4}) or Q_{PS} (contains Q_{P1}, Q_{P2}, Q_{P3} and Q_{P4}) are switched on during the operating interval. The switches can be designed to switch on and off at the zero-current state while the L_r–C_r resonant current is rising and falling to zero to achieve zero current switching for reducing power losses of the MOSFET switches. Switches Q_{N1} or Q_{P1} can control the forward power flow from source V_1 to the output source V_2 as a quadruple-mode converter (i.e. $V_2 = 4V_1$) shown as Fig.2. Figs. 2(a)-2(d) show the equivalent circuit of the proposed ZCS SC QR bi-directional DC/DC converter during the various operation intervals under forward

---

This research work was supported by National Science Council of Taiwan under grant NSC95-2475-E-030-002-URD and Fu-Jen Catholic University 409531040426

978-1-4244-0644-9/07/$25.00 ©2007 IEEE

power flow control scheme. On the other hand, switches Q_{N2}, Q_{N3}, Q_{N4}, or Q_{P2}, Q_{P3}, Q_{P4} can control the reverse power flow from the source V_2 to the other source V_1 as a quarter-mode converter (i.e. $V_1 = 1/4\ V_2$) shown as Fig.3. Figs. 3(a)–3(d) show the equivalent circuit of the proposed bi-directional converter during the various operation intervals under the reverse power flow control scheme. When the resonant current increases to a peak value and decreases to zero current, it cannot reverse into negative current because there is a diode in the circuit loop of the converter, which ceases the current reversing. The detail circuit analysis and demonstration of the non-inverting type 4-mode / 1/4-mode ZCS SC bi-directional converter is illustrated in the following sessions.

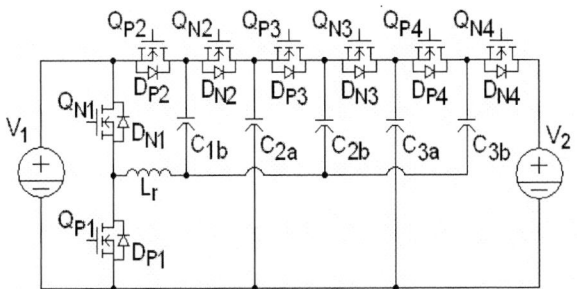

Fig. 1 Quadruple-mode / quarter-mode ZCS SC QR bi-directional converter

III. CONVERTER ANALYSIS

A. Forward Power Flow Scheme

Figs. 2(a)-(d) show the equivalent circuit of the proposed non-inverting type 4-mode/1/4-mode ZCS SC bi-directional converter with forward power flow control. The inductor L_r is small, C_{1b}, C_{2b} and C_{3b} are used to the resonant tank, the capacitors of C_{2a}, C_{3a}, and C_{4a} are larger enough to storage energy, Let $C_{1b} = C_{2b} = C_{3b} = C_r$ and all elements are ideal. The operation mode analysis of the converter is described as follow:

State I [t_0-t_1]

Switch Q_{N1} is turned on, and Q_{P1} is turned off during this interval shown as Fig. 2(a). The energy of the source V_1 that series connected with C_{1b}, C_{2b} and C_{3b} is charged to C_{2a}, C_{3a}, and C_{4a} respectively. The state equations in this stage are

$$V_1 = V_{C(k+1)a} - L_r \frac{di_{Lr}}{dt} - V_{Ckb} \qquad (1)$$

$$i_{Lr} = (n-1)C_r \frac{dV_{Ckb}}{dt} \qquad (2)$$

The solutions of (1) and (2) are:

$$V_{Ckb} = V_{c(k+1)a} - V_1 + \left(V_{ckb0} + V_1 - V_{c(k+1)a}\right)\cos\omega_0(t-t_0) \qquad (3.a)$$

$$\cong kV_1 + \frac{1}{2}nI_0T_sZ_0\omega_0\cos\omega_0(t-t_0) \qquad (3.b)$$

$$i_{Lr} = -(n-1)C_r\omega_0\left(V_{ckb0} + V_1 - V_{c(k+1)a}\right)\sin\omega_0(t-t_0) \qquad (4.a)$$

$$\cong -\frac{1}{2}nI_0T_s\omega_0\sin\omega_0(t-t_0) \qquad (4.b)$$

Using input and output energy conservation, equations (3.a) and (3.b) can be simplified as (4.a) and (4.b), respectively. Where k = 1, 2, 3, $V_{C4a} = V_o$, and

$$\omega_0 = \sqrt{\frac{1}{L_r(n-1)C_r}} = \frac{2\pi}{T_0} \quad , \quad Z_0 = \sqrt{\frac{L_r}{C_r(n-1)}}$$

State II [t_1-t_2]

All switches are turned off shown as Fig. 2(b), the stored energy in C_{4a} is discharged to the load, and the equations of the state are:

$$V_{Ckb} = kV_1 - \frac{1}{2}nI_0T_sZ_0\omega_0 \qquad (5)$$

$$i_{Lr} = 0 \qquad (6)$$

State III [t_2-t_3]

Switch Q_{P1} is turned on, and Q_{N1} is turned off shown as Fig. 2(c). The energy of the source V_1 is charged to C_{1b}, the energy of C_{2a} is charged to C_{2b}, and C_{3a} is charged to C_{3b}. In this stage, the state equations are:

$$V_{Cka} = L_r\frac{di_{Lr}}{dt} + V_{Ckb} \qquad (7)$$

$$i_{Lr} = (n-1)C_r\frac{dV_{Ckb}}{dt} \qquad (8)$$

The solutions of (7) and (8) are:

$$V_{Ckb} = V_{cka} + (V_{ckb2} - V_{cka})\cos\omega_0(t-t_2) \qquad (9.a)$$

$$\cong kV_1 - \frac{1}{2}nI_0T_sZ_0\omega_0\cos\omega_0(t-t_2) \qquad (9.b)$$

$$i_{Lr} = -(n-1)C_r\omega_0(V_{ckb2} - V_{cka})\sin\omega_0(t-t_2) \qquad (10.a)$$

$$\cong \frac{1}{2}nI_0T_s\omega_0\sin\omega_0(t-t_2) \qquad (10.b)$$

Using input and output energy conversation, equations (9.a) and (10.a) can be simplified as (9.b) and (10.b), respectively. Where k = 1, 2, 3, and $V_{C1a}=V_1$.

State IV [t_3-t_4]

Switches Q_{N1} and Q_{P1} are turned off. The stored energy in C_{4a} is discharges to the load shown as Fig. 2(d), the state equations are:

$$V_{Ckb} = kV_1 + \frac{1}{2}nI_0T_sZ_0\omega_0 \qquad (11)$$

$$i_{Lr} = 0 \qquad (12)$$

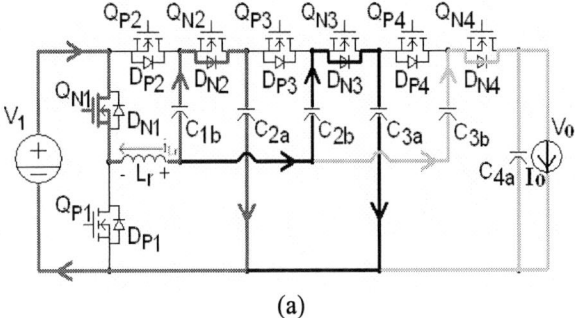

(a)

389

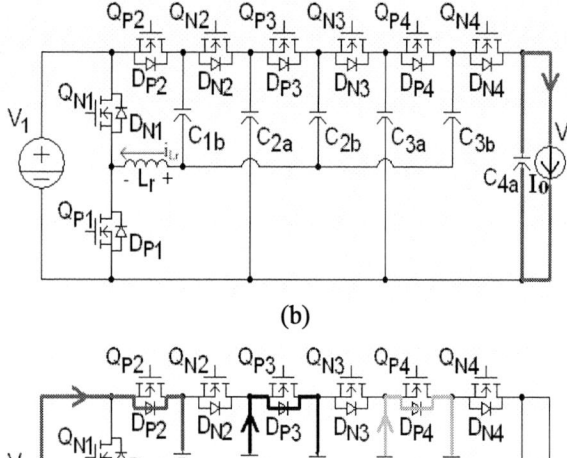

(b)

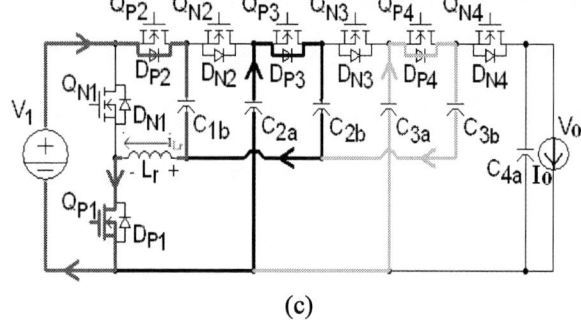

(c)

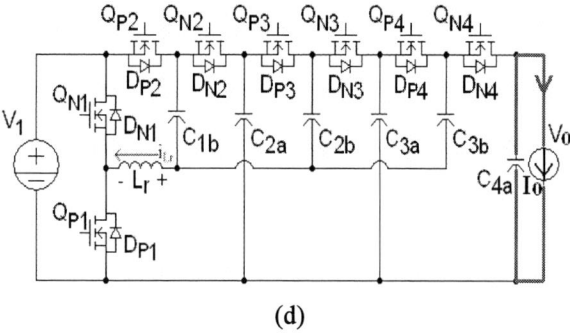

(d)

Fig. 2 Equivalent circuit of various operation stages for the proposed ZCS SC QR bi-directional DC/DC converter under forward power flow. (a) State I [t_0-t_1] (b) State II [t_1-t_2] (c) State III [t_2-t_3] (d) State IV [t_3-t_4]

B. Reverse Power Flow Scheme

Figs. 3(a)-(d) show the equivalent circuit of the proposed non-inverting type 4-mode / 1/4-mode ZCS SC bi-directional converter with reverse power flow control, and the analysis of the each equivalent circuit is the same as the forward power flow control.

State I [t_0-t_1]

During this interval, switches Q_{N2}, Q_{N3} and Q_{N4} are turned on shown as Fig. 3(a), the state equations are

$$V_O = V_{C(k+1)a} - L_r \frac{di_{Lr}}{dt} - V_{Ckb} \quad (13)$$

$$i_{Lr} = (n-1)C_r \frac{dV_{Ckb}}{dt} \quad (14)$$

The solutions of (13) and (14) are:

$$V_{Cb} = (V_o - V_{c(k+1)a}) + (V_{ckb} + V_{c(k+1)a} - V_o)\cos(t - t_0) \quad (15.a)$$

$$\cong kV_O - \frac{1}{2}nI_0T_SZ_0\omega_0\cos\omega_0(t-t_0) \quad (15.b)$$

$$i_{Lr} = -(n-1)C_r\omega_0(V_{ckb} + V_{c(k+1)a} - V_o)\sin\omega_0(t-t_0) \quad (16.a)$$

$$\cong \frac{1}{2}nI_0T_S\omega_0\sin\omega_0(t-t_0) \quad (16.b)$$

Using the method in part A, the equations (15.a) and (15.b) can be simplified as (16.a) and (16.b), respectively. Where k = 1, 2, 3, and V_{C4a}= V_2 .

State II [t_1-t_2]

All switched are turned off and the stored energy in C_{1a} is discharged to the load shown as Fig. 3(b), the equations of this state are:

$$V_{Ckb} = kV_O + \frac{1}{2}nI_0T_SZ_0\omega_0 \quad (17)$$

$$i_{Lr} = 0 \quad (18)$$

State III [t_2-t_3]

The switches Q_{P2}, Q_{P3} and Q_{P4} are turned on shown as Fig. 3(c), the state equations are

$$V_{Cka} = L_r\frac{di_{Lr}}{dt} + V_{Ckb} \quad (19)$$

$$i_{Lr} = (n-1)C_r\frac{dV_{Ckb}}{dt} \quad (20)$$

The solutions of equation (19) and (20) are:

$$V_{Ckb} = V_{cka} + (V_{ckb2} - V_{cka})\cos\omega_0(t-t_2) \quad (21.a)$$

$$\cong kV_O + \frac{1}{2}nI_0T_SZ_0\omega_0\cos\omega_0(t-t_2) \quad (21.b)$$

$$i_{Lr} = -(n-1)C_r\omega_0(V_{ckb2} - V_{cka})\sin\omega_0(t-t_2) \quad (22.a)$$

$$\cong -\frac{1}{2}nI_0T_S\omega_0\sin\omega_0(t-t_2) \quad (22.b)$$

With the same method as before, equation (21.a) and (22.a) can be simplified as (21.b) and (22.b), respectively. Where k = 1, 2, 3, and V_{C1a}= V_o .

State IV [t_3-t_4]

All switched are turned off and the stored energy in C_{1a} is discharged to the load shown as Fig. 3(d), the equations of this state are:

$$V_{Ckb} = kV_O - \frac{1}{2}nI_0T_SZ_0\omega_0 \quad (23)$$

$$i_{Lr} = 0 \quad (24)$$

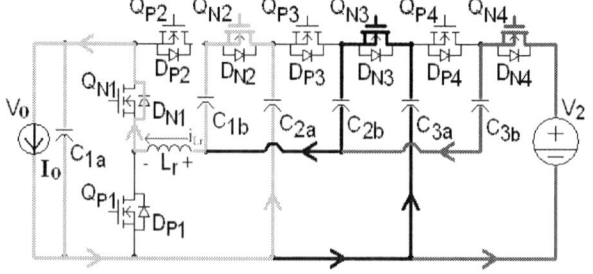

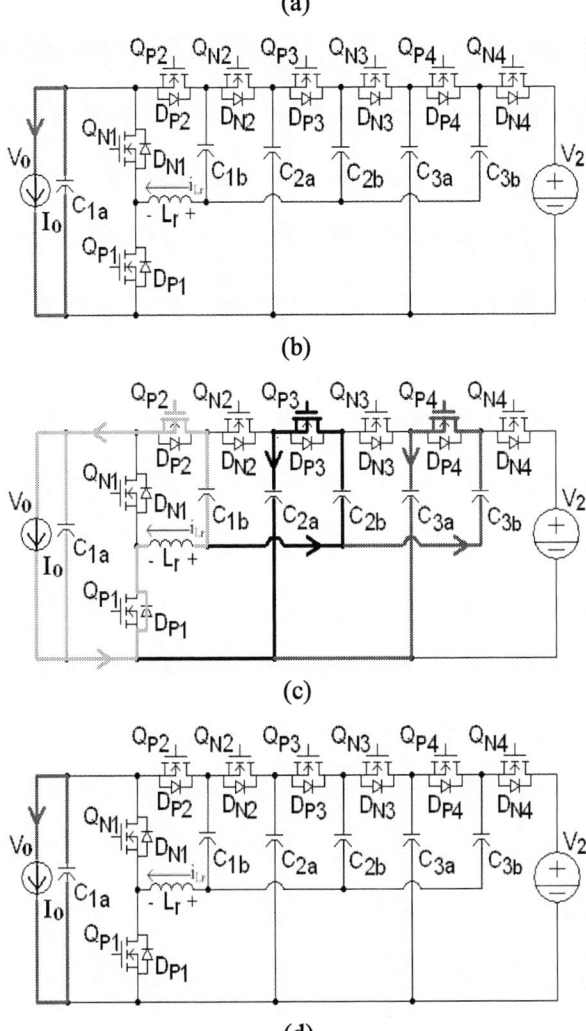

Fig. 3 Equivalent circuit of various operation stages for the proposed ZCS SC QR bi-directional DC/DC converter under reverse power flow. (a) State I [t_0-t_1] (b) State II [t_1-t_2] (c) State III [t_2-t_3] (d) State IV [t_3-t_4]

According to the forward and reverse power flow control shown as Figs. 2(a)-(d) and Figs. 3(a)-(d) in this paper, Table 1 shows the switching sequence of the MOSFET switches and diodes in the bi-directional converter for the various operation intervals under the forward and reverse power flow control schemes, respectively. The switching sequence of semiconductor switches in the proposed converter for the various operation intervals under forward power flow control is showed in the second column of Table 1. And the other sequence of the same converter under reverse power flow control is showed in the third column of Table1.

The proposed converter topology can be extended to design as an n-mode / $\frac{1}{n}$ -mode zero-current switching switched-capacitor DC/DC converter shown as Fig. 4.

IV. SIMULATION AND EXPERIMENTAL RESULTS

In order to validate the performance of the proposed ZCS SC bi-directional DC/DC converter, a P-Spice simulations and experiments are carried out for the proposed quadruple-mode / quarter-mode non-inverting ZCS SC DC/DC converter. The designed parameters are list as follows: MOSFET switches are CEP8060L, Schottky diodes are SBR2060, $L_r = 1\mu H$, $C_r = 0.22\mu F$, $f_s = 150$ kHz, the duty ratio and input/output voltages are, $D_1 = 0.45$ and $D_2 = 0.45$, $V_1 = 20$ (V) and $V_2 = 80$ (V) for the forward and reverse power flow control, respectively. Figs. 5(a) and 5(b) show the simulation waveforms of quadruple-mode / quarter-mode ZCS SC QR DC/DC converter for the forward and reverse power flow control, respectively. Figs. 6(a) and 6(b) show the corresponding experimental results of the same converters. Figs. 7(a) and 7(b) show the relation of the output voltage and the converter efficiency with respect to the output power. From the Figs. 5-7, several observation of the proposed ZCS SC bi-directional converter can be summarized as follows:

· The power MOSFETs of the proposed bi-directional converter are turned on and turned off in the zero current state. The total switching losses of the MOSFETs in the converter can be significantly reduced compared with the conventional bi-directional SC converter.

· The conductive EMI emission into the power source and the peak current stresses of the MOSFETs are significantly reduced compared with the conventional hard switching switched-capacitor bi-directional converter.

· The resonant inductor current is symmetry in the proposed bi-directional converters. Such that, there is no dc component current existed in the resonant inductor to avoid the saturation problem.

· The proposed bi-directional converter can be extended to design a non-inverting type n-mode / $\frac{1}{n}$ mode ZCS SC bi-directional converter for the high frequency and high voltage conversion ratio power supply applications.

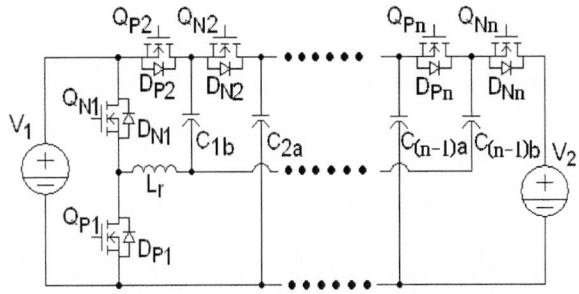

Fig. 4 n-mode / $\frac{1}{n}$ -mode ZCS SC QR bi-directional converter

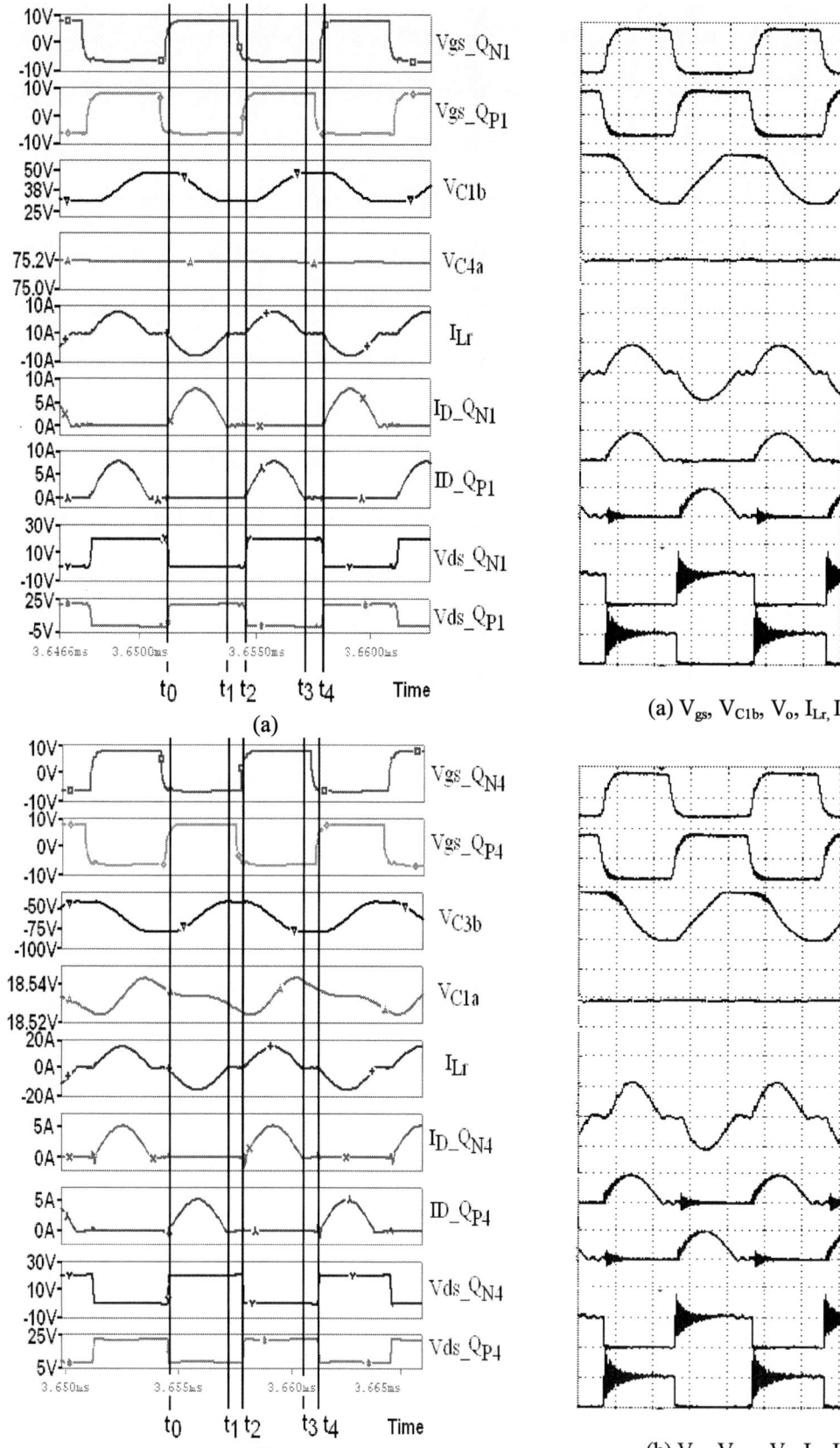

(a) V_{gs}, V_{C1b}, V_o, I_{Lr}, I_d, V_{ds} waveforms

(b) V_{gs}, V_{C3b}, V_o, I_{Lr}, I_d, V_{ds} waveforms

Fig. 5 Typical switching and simulation waveforms of quadruple-mode/quarter-mode ZCS SC QR DC/DC converter for (a) forward (b) reverse power flow control

Fig. 6 Experimental results of the proposed ZCS SC bi-directional converter for (a) forward (b) reverse power flow control

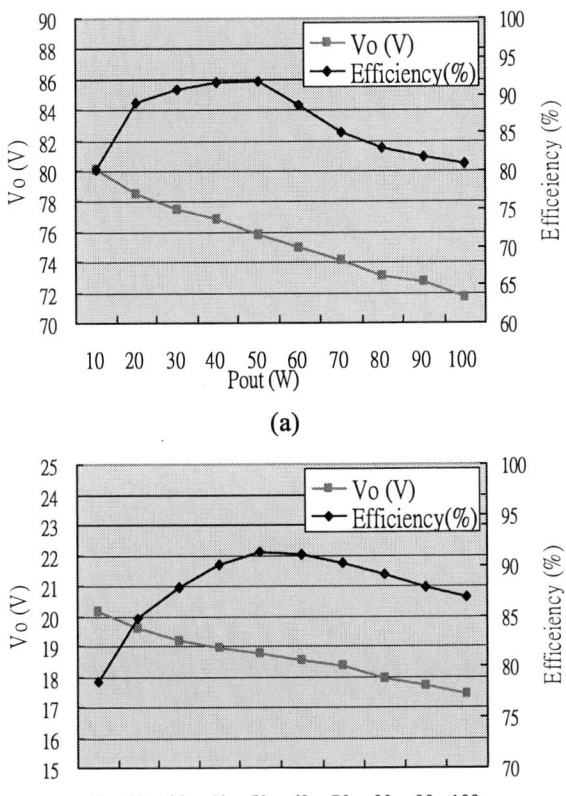

(a)

(b)

Fig. 7 Experiment results of the output voltage and efficiency versus output power for (a) forward (b) reverse power flow control

TABLE1
SWITCHING SEQUENCES OF SWITCHES AND DIODES OF PROPOSED CONVERTER UNDER VARIOUS OPERATING MODES

Intervals	Forward power flow control				Reverse power flow control			
	$[t_0\text{-}t_1]$	$[t_1\text{-}t_2]$	$[t_2\text{-}t_3]$	$[t_3\text{-}t_4]$	$[t_0\text{-}t_1]$	$[t_1\text{-}t_2]$	$[t_2\text{-}t_3]$	$[t_3\text{-}t_4]$
Q_{P1}			ON					
D_{P1}							ON	
Q_{N1}	ON							
D_{N1}					ON			
Q_{P2}							ON	
D_{P2}			ON					
Q_{N2}					ON			
D_{N2}	ON							
Q_{P3}							ON	
D_{P3}			ON					
Q_{N3}					ON			
D_{N3}	ON							
Q_{P4}							ON	
D_{P4}			ON					
Q_{N4}					ON			
D_{N4}	ON							

V. CONCLUSIONS

A novel QR ZCS switched-capacitor bi-directional DC/DC converter is proposed. According to the simulations and experimental results, a non-inverting type 4-mode/1/4-mode ZCS SC bi-directional converters have been developed. In the power converter, the switching losses will be decreased, the MOSFET current stress will be reduced, the total efficiency will be increased and easily extended to design as an n-mode/$\frac{1}{n}$-mode zero current switching SC QR DC/DC converter. The proposed converter is suitable for high frequency, high efficiency and high voltage conversion ratio power supply applications.

REFERENCES

[1] K. W. E. Cheng," Zero-Current-Switching Switched-Capacitor Converters," *IEE Proceedings Electric Power Applications*, Vol. 148, pp. 403-409, 2001.

[2] K. K. Law, K. W. E. Cheng, and Y. P. Benny Yeung," Design and Analysis of Switched-Capacitor-Based Step-Up Resonant Converters," *IEEE Transaction on Circuits and Systems*, Vol. 52, pp.943-948, 2005.

[3] Yuang-Shung Lee, Yin-Yuan Chiu, and Ming-Wang Cheng," ZCS Switched-Capacitor Bi-directional Quasi-Resonant Converters," *IEEE PEDS05*, pp. 867-871, 2005

[4] Y. P. B. Yeung and K. W. E. Cheng," Unified Analysis of Switched -Capacitor Resonant Converters," *IEEE Transaction on Industrial Electronics*, Vol. 51, No. 4, pp. 864-873, 2004.

[5] Y. P. B. Yeung, K. W. E. Cheng and D. Sutanto," Multiple and Fractional Voltage Conversion Rations for Switched-Capacitor Resonant Converters," *IEEE PESC01, 32nd Annual Meeting*, Vol. 3, pp. 1289-1294, 2001.

[6] H. Li, F. Z. Peng and J. Lawler," Modeling, Simulation, and Experimental Verification of Soft-Switched Bi-directional DC-DC Converters," *IEEE APEC01, Sixteenth Annual Meeting*, Vol. 2, pp. 736-742, 2001.

[7] Y. S. Lee and G. T. Cheng," ZCS Bi-directional DC-to-DC Converter Application in Battery Equalization for Electric Vehicle," *IEEE PESC04, 35th Annual Meeting, Aachen, Germany*, pp. 2766-2772, 2004.

[8] H. S. H. Chung and W. C. Chow," Development of A Switched-Capacitor DC/DC Converter with Bidirectional Power Flow," *IEEE Trans. Circuits and Systems-I: Fundamental Theory and Applications*, Vol. 47, No. 9, pp. 1383-1389, 2000.

[9] H. S. H. Chung and W. L. Cheung," Generalized Structure of Bi-directional Switched-Capacitor DC/DC Converter," *IEEE Trans. Circuits and Systems-I: Fundamental Theory and Applications*, Vol. 50, No. 6, pp. 743-753, 2003.

[10] Y. P. B. Yeung, K. W. E. Cheng, D. Sutanto and S. L. HO," Zero-Current Switching Switched-Capacitor Quasiresonant Step-Down Converter," *IEE Proc. Electr. Power Appl.*, Vol. 149, pp. 111-121, 2002.

Analysis of a Half – Bridge Inverter for a Small-Size Induction Cooker Using Positive-Negative Phase-Shift Control under ZVS and NON-ZVS Operation

P. Achara* , P. Viriya* and K. Matsuse**

* Dept. of Electrical Engineering, Faculty of Engineering, King Mongkut's Institute of Technology Ladkrabang,
Bangkok, 10520, Thailand, Tel. 662-7373000 EXT. 3515, 3516 Fax. 662-3264550, E-Mail : kpviriya@kmitl.ac.th
** School of Science and Technology, Meiji University, 1-1-1 Higashimita, Tama-ku, Kawasaki-shi 214, Japan,
Tel. +81-44-934-7293, Fax. 03(3296)4339, E-Mail : matsuse@ics.meiji.ac.jp

Abstract—This paper presents a detailed analysis of circuit operation under ZVS and NON-ZVS conditions in a high-frequency half-bridge inverter for a small-size and low-voltage induction cooker, using the principle of positive-negative phase-shift control over a wide control range both in positive and negative directions. A variety of modes of circuit operation with the voltage and current equations during phase-shift power control under the operating conditions of ZVS and NON-ZVS are analyzed as a first step and the output voltage and current waveforms are obtained by MATLAB program. These waveforms will be analyzed by Fourier analysis which can lead further to the calculation of ac output power P_o, dc input power P_d, and hence the conversion efficiency η of the half-bridge inverter. The analysis results shows that the control ranges of ac output power P_o and dc input power P_d are limited by the occurrence of NON-ZVS operating condition, which changes according to the switching frequency f_s.

Index Terms— Half-bridge, Induction cooker, Power control, Phase-shift, ZVS, NON-ZVS

I. Introduction

The concept of this research paper is achieved by interchanging between the following two different ideas. One is from the idea ① presented in the research paper [1-4] concerning a single-phase full-bridge phase-shift control under ZVS and NON-ZVS for induction heating which is suited for high-voltage and high-power application, while the other is from the idea ② presented in the research paper [5-7] concerning a single-phase half-bridge frequency control under ZVS for induction cooking which is suited for low-voltage and low-power application. So, by interchanging or mixing these two ideas or features into one, another new concept of the research topic for a half-bridge inverter to be presented in this paper can be achieved as shown by an idea in Fig. 1.

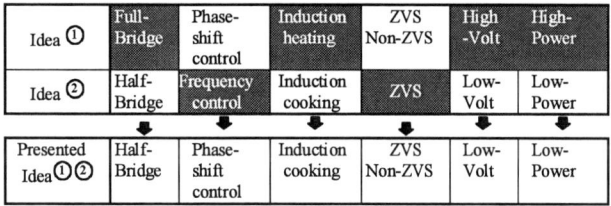

Fig. 1 The idea to achieve the concept of this paper

The half-bridge inverter which is obtained by this idea has a very close relation to the full-bridge inverter. This can be understood by considering first the circuit operation of the full-bridge phase-shift inverter with its operating output voltage and current waveforms illustrated in the upper part of Fig. 2. The circuit operation in this case consists of two repeated periods of modes of circuit operation over one cycle; that is, powering (P), free wheeling (F) and regenerating (R) for

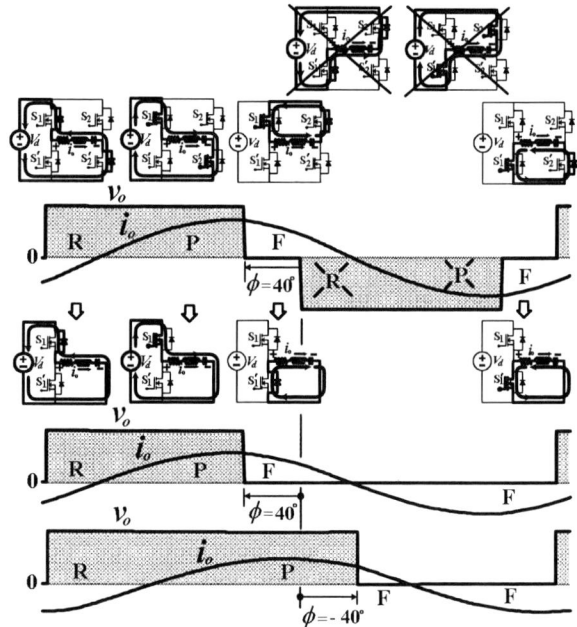

Fig. 2 Circuit operation of a half-bridge inverter related to that of a phase-shift full-bridge inverter

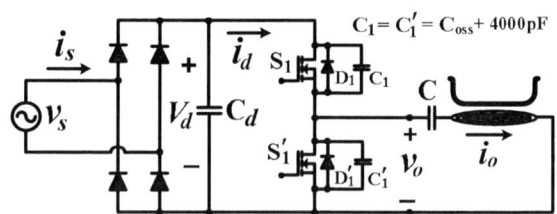

Fig. 3 A half-bridge inverter for a small-size induction cooker

two periods. When transforming into another case of a half-bridge inverter with its output voltage reduced to half of that of the case for full-bridge, two active modes

978-1-4244-0644-9/07/$25.00 ©2007 IEEE

of powering (P) and regenerating (R) from the total six modes are to be eliminated, and there are only four modes of circuit operation remaining for this case, as shown by an idea in Fig 2. The idea circuit operation in this case can be made possible, using the actual half-bridge inverter circuit of Fig.3, where the dc input-side is supplied with a dc smoothed rectified voltage. Such a half-bridge inverter with a wide control range of phase-shift ϕ both in positive and negative directions will be used in this paper. The inverter will be analyzed in details for various electrical quantities during phase-shift control both in positive direction from $0°$ to $180°$ and in negative direction from $0°$ to $-180°$, taking ZVS and NON-ZVS into consideration, since the phase-shift control range in both directions will be limited when ZVS circuit operation becomes NON-ZVS operation. These quantities are, for examples, ac output voltage v_o, ac output power P_o, dc input power P_d, and inverter efficiency η, etc.

II. ANALYSIS OF CIRCUIT OPERATION

First, we show a variety of modes of circuit operation for the main power circuit which are illustrated in Fig. 4 for the case of phase-shift control ($\phi = 0 \sim 180°$ and $\phi = 0 \sim -180°$). The circuit operation in one cycle of output voltage and current waveforms v_o, i_o is shown in 6 modes under the case of ZVS operation (Modes ①②③①'②③'). Fig. 5 shows the linear variation of the phase difference between the fundamental output voltage $v_{o,1}$ and the front edge of square wave output voltage v_o during phase-shift control both in positive and negative directions. It can be seen that with positive

phase-shift from $0°$ to $+180°$, the phase angle of the fundamental output voltage $v_{o,1}$ will move away from the font edge of output voltage v_o with an increasing leading angle. This makes the zero-crossing of output current i_o move toward the font edge of output voltage v_o, where NON-ZVS operation may occur, but in case of negative phase-shift control, the NON-ZVS operation may occur at the tailing edge of square wave voltage v_o, and in case of zero phase-shift control, NON-ZVS operation may occur both at the front and tailing edges, especially when the operating frequency is not high enough. In Fig. 6, we also show the circuit operation in one cycle of output voltage and current waveforms v_o, i_o in another 6 modes under the cases of the following NON-ZVS operation : (1) Modes ②③①②③❹ for phase-shift $\phi = 72°$ can cause a breakdown to the upper switch , by considering Mode ❹. (2) Modes ②③❹②③❹ for phase-shift $\phi = 0°$ can cause a breakdown to the lower switch, by considering Mode ❹ and to the upper switch, by considering Mode ❹. (3) Modes ①②③❹ ②③ for phase-shift $\phi = -72°$ can cause a breakdown to the lower switch, by considering Mode ❹. So, there are three possibilities for the half-bridge inverter switches to become breakdown (the upper switch, the lower switch, or both).

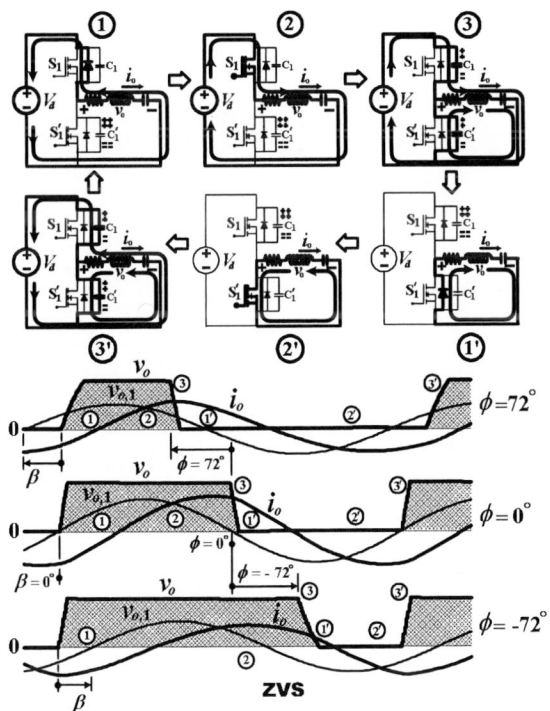

Fig. 4   Circuit operation for the case of ZVS at phase-shift

$\phi = 72°$, $0°$ and $-72°$

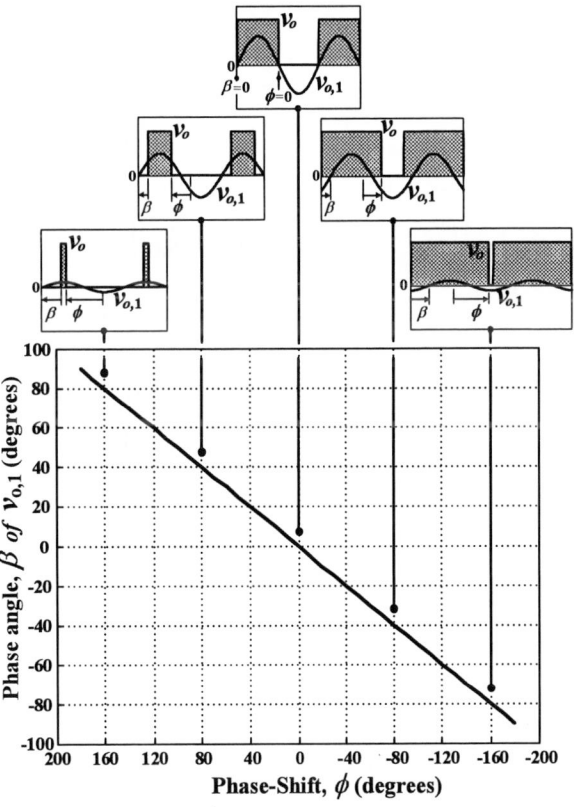

Fig. 5 Linear variation of phase angle β of $v_{o,1}$ vs. phase-

shift angle ϕ

395

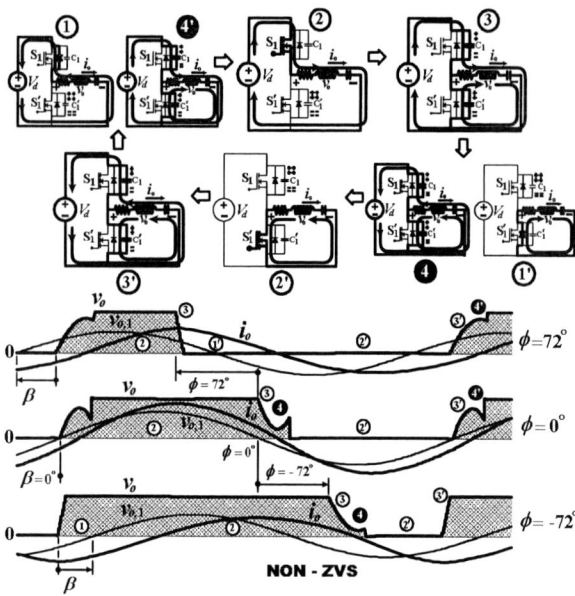

Fig. 6 Circuit operation for the case of NON-ZVS at phase-shift
$\phi = 72°$, $0°$ and $-72°$

III. VOLTAGE AND CURRENT EQUATIONS IN EACH MODE OF CIRCUIT OPERATION

From these modes of circuit operation, various equations of output voltage v_o and output current i_o can be also calculated and obtained in the following four cases of different equations :

Case (1) Equations ① ② ③ ①′ ②′ ③′ for ZVS operation

Case (2) Equations ② ③ ①′ ②′ ③′ ❹ for $\phi > 0°$

Case (3) Equations ② ③ ❹ ②′ ③′ ❹′ for $\phi = 0°$

Case (4) Equations ① ② ③ ❹′ ②′ ③′ for $\phi < 0°$

$$v_o = V_d$$
$$i_o = e^{-\alpha t}\left[\left(\frac{V_d - V - \alpha LI}{\omega_1 L}\right)\sin\omega_1 t + I\cos\omega_1 t\right] \qquad \text{① ②}$$

$$v_o = \frac{1}{C_1'(\alpha^2 + \omega_2^2)}\left[e^{-\alpha t}(A\sin\omega_2 t + B\cos\omega_2 t) + D\right] + V_1'$$
$$i_o = \frac{e^{-\alpha t}}{\omega_2}\left[\left(\frac{V_d - 2V - V_1 + V_1'}{2L} - \alpha I\right)\sin\omega_2 t + \omega_2 I\cos\omega_2 t\right] \qquad \text{③ ❹}$$

$$v_o = 0$$
$$i_o = e^{-\alpha t}\left[\left(\frac{-V - \alpha LI}{\omega_1 L}\right)\sin\omega_1 t + I\cos\omega_1 t\right] \qquad \text{①′ ②′}$$

$$v_o = \frac{1}{C_1'(\alpha^2 + \omega_2^2)}\left[e^{-\alpha t}(A\sin\omega_2 t + B\cos\omega_2 t) + D\right] + V_1'$$
$$i_o = \frac{e^{-\alpha t}}{\omega_2}\left[\left(\frac{V_d - 2V - V_1 + V_1'}{2L} - \alpha I\right)\sin\omega_2 t + \omega_2 I\cos\omega_2 t\right] \qquad \text{③′ ❹′}$$

Where ;

V_1 : the initial value of voltage V_{C_1}

V_1' : the initial value of voltage V_{C_i}

I : the initial value of load current i_o in each mode of circuit operation

V : the initial value of load capacitor voltage in each mode of circuit operation

$$\alpha = \frac{R}{2L}$$

$$\omega_1 = \sqrt{\frac{1}{LC} - \left(\frac{R}{2L}\right)^2}$$

$$\omega_2 = \sqrt{\left(\frac{1}{LC} + \frac{1}{2LC_{ds}}\right) - \left(\frac{R}{2L}\right)^2}$$

$$\begin{aligned}A = \{&-RC_{ds}(-V_d + V_1 + V_1')(\alpha^2 + \omega_2^2) - LI(\alpha^2 + \omega_2^2)\\ &+\alpha(C_{ds}/C)(-V_d + V_1 + V_1') + \alpha(V + V_1')\\ &+\alpha LC_{ds}(-V_d + V_1 + V_1')(\alpha^2 + \omega_2^2)\}(1/2L\omega_2)\end{aligned}$$

$$\begin{aligned}B = \{&(C_{ds}/C)(-V_d + V_1 + V_1') + (V + V_1')\\ &-LC_{ds}(-V_d + V_1 + V_1')(\alpha^2 + \omega_2^2)\}(1/2L)\end{aligned}$$

$$D = \{-(C_{ds}/C)(-V_d + V_1 + V_1') - (V + V_1')\}(1/2L)$$

where the above equations in cases (2), (3) and (4) are those of NON-ZVS. Fig. 7 shows the calculated and experimental results of output voltage and current waveforms under ZVS operation, using the equations in case (1) with the use of MATLAB program. The calculated waveforms are obtained with phase-shift $\phi = 72°, 0°, -72°$ at switching frequency of 29 kHz. It can be observed that the peak value of output current waveforms becomes the highest at phase-shift $\phi = 0°$ and then become decreasing with the increase or decrease of phase-shift from $0°$ to $72°$ or $0°$ to $-72°$, respectively. Moreover, at only the mid-point of phase-shift $\phi = 0°$, the output current i_o can be obtained with almost a sinusoidal waveform. These calculated waveforms with

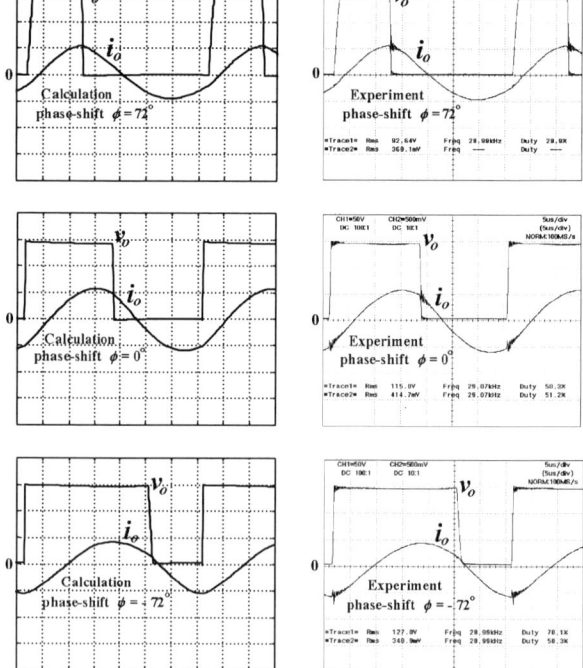

Fig. 7 Calculated and experimental output voltage and current waveforms under ZVS operating condition at 29 kHz : 50 V/div , 5 A/div , 5 µs/div

396

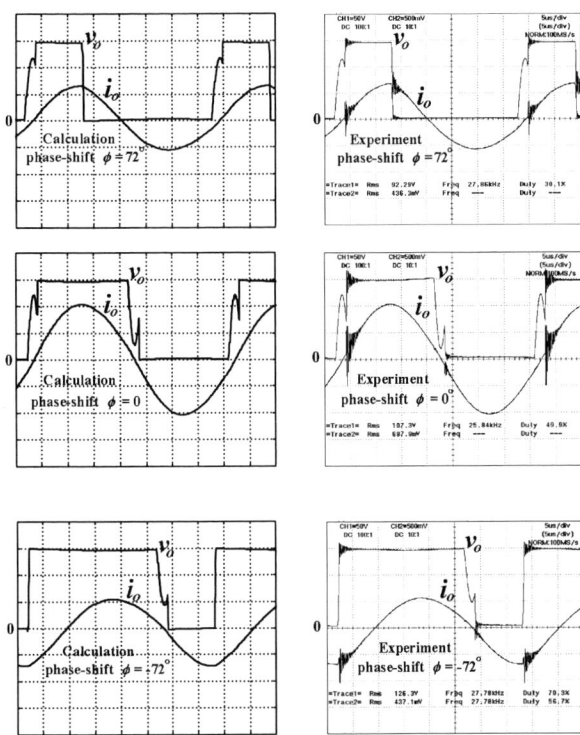

Fig. 8 Calculated and experimental output voltage and current waveforms under **NON-ZVS** operating condition : 50 V/div , 5 A/div , 5 µs/div

the principle of circuit operation are also verified by comparison with the experimental ones. Fig. 8 shows the calculated and experimental results of output voltage and current waveforms under NON-ZVS operation, using the equations in cases (2), (3) and (4) with the use of MATLAB program. The calculated waveforms are obtained with phase-shift $\phi = 72°, 0°, -72°$ at switching frequency around 28 kHz, 26 kHz , and 28 kHz, respectively. These calculated waveforms with the principle of circuit operation are also verified by comparison with the experimental ones. Also , it is observed that at different phase-shift control, NON-ZVS operation occurs at different position of the output voltage waveform (front edge at $\phi = 72°$: breakdown to the upper switch, tailing edge at $\phi = -72°$: breakdown to the lower switch,   and $\phi = 0°$: breakdown to both switches).

IV. Analysis of Output Power P_o , DC Input Power P_d and Efficiency

The calculated output voltage v_o of Fig. 7 can be analyzed into various component waveforms as shown in (5), by Fourier analysis. Then, applying these component waveforms as the input voltage to the RLC load equivalent circuit, the current equation i_o can be obtained as shown in (6). Again, the voltage v_o in (5) can be used to find out the rms values of output voltage in terms of ac-dc components $(V_{o,rms(ac,dc)})$, dc

component $(V_{o,rms(dc)})$, ac component $(V_{o,rms(ac)})$, and fundamental component $(V_{o,rms(1)})$ as shown in Fig. 9.

$$
\begin{aligned}
v_o = &\left(\frac{1}{2} - \frac{\phi}{2\pi}\right)V_d + \frac{V_d}{\pi}\left[\sin(\pi-\phi)\cos\omega_s t + (1-\cos(\pi-\phi))\sin\omega_s t\right] \\
&+ \frac{V_d}{2\pi}\left[\sin 2(\pi-\phi)\cos 2\omega_s t + (1-\cos 2(\pi-\phi))\sin 2\omega_s t\right] \\
&+ \frac{V_d}{3\pi}\left[\sin 3(\pi-\phi)\cos 3\omega_s t + (1-\cos 3(\pi-\phi))\sin 3\omega_s t\right] + \ldots
\end{aligned} \tag{5}
$$

$$
\begin{aligned}
i_o = &\frac{V_d}{\pi Z_1}\left[\sin(\pi-\phi)\cos(\omega_s t - \theta_1)\right. \\
&\left. + \{1-\cos(\pi-\phi)\}\sin(\omega_s t - \theta_1)\right] \\
&+ \frac{V_d}{2\pi Z_2}\left[\sin(2\pi-2\phi)\cos(2\omega_s t - \theta_2)\right. \\
&\left. + \{1-\cos(2\pi-2\phi)\}\sin(2\omega_s t - \theta_2)\right] \\
&+ \frac{V_d}{3\pi Z_3}\left[\sin(3\pi-3\phi)\cos(3\omega_s t - \theta_3)\right. \\
&\left. + \{1-\cos(3\pi-3\phi)\}\sin(3\omega_s t - \theta_3)\right] + \ldots
\end{aligned} \tag{6}
$$

$$
\cos\theta_n = \cos\left(\tan^{-1}\left(\frac{\omega_s L - (1/\omega_s C)}{R}\right)\right),\ \cos\left(\tan^{-1}\left(\frac{2\omega_s L - (1/2\omega_s C)}{R}\right)\right), \tag{7}
$$
$$
\cos\left(\tan^{-1}\left(\frac{3\omega_s L - (1/3\omega_s C)}{R}\right)\right),\ \ldots
$$

$$
P_o = V_{o,1}I_{o,1}\cos\theta_1 + V_{o,2}I_{o,2}\cos\theta_2 + V_{o,3}I_{o,3}\cos\theta_3 + \ldots \tag{8}
$$

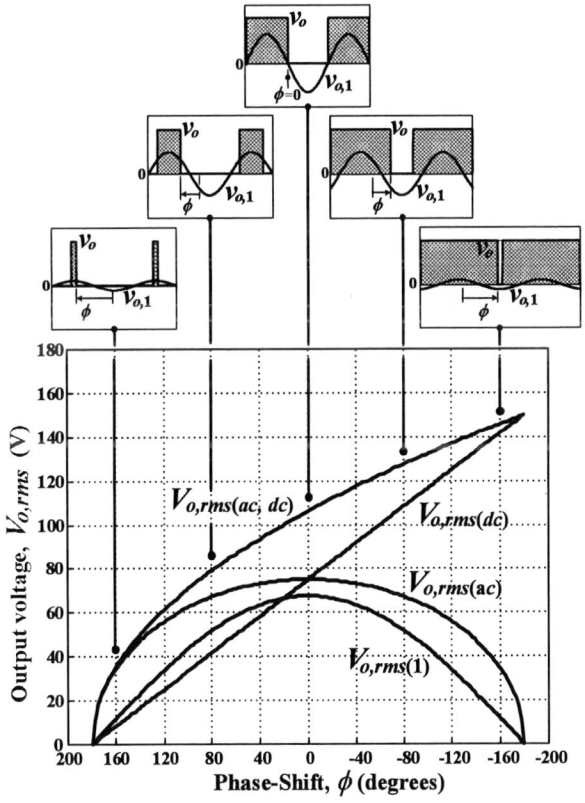

Fig. 9 Variation of various output voltages $V_{o,rms(ac,dc)}$, $V_{o,rms(dc)}$, $V_{o,rms(ac)}$ and $V_{o,rms(1)}$ vs phase-shift ϕ

These equations can also lead to the calculation of ac output power P_o, using the definition of P_o in (8). Fig. 10 shows the output voltage and current waveforms v_o, i_o with the phase difference angle of each harmonic. With phase-shift $\phi = +72°$ and $\phi = -72°$ there will be harmonic order of 1, 2, 3, 4, 5, . . . and with phase-shift $\phi = 0$ there will be harmonic order of 1, 3, 5, ... From this harmonic content, it can be seen that from the second harmonic order upward, each pair of output voltage and current waveforms will have the phase difference angle

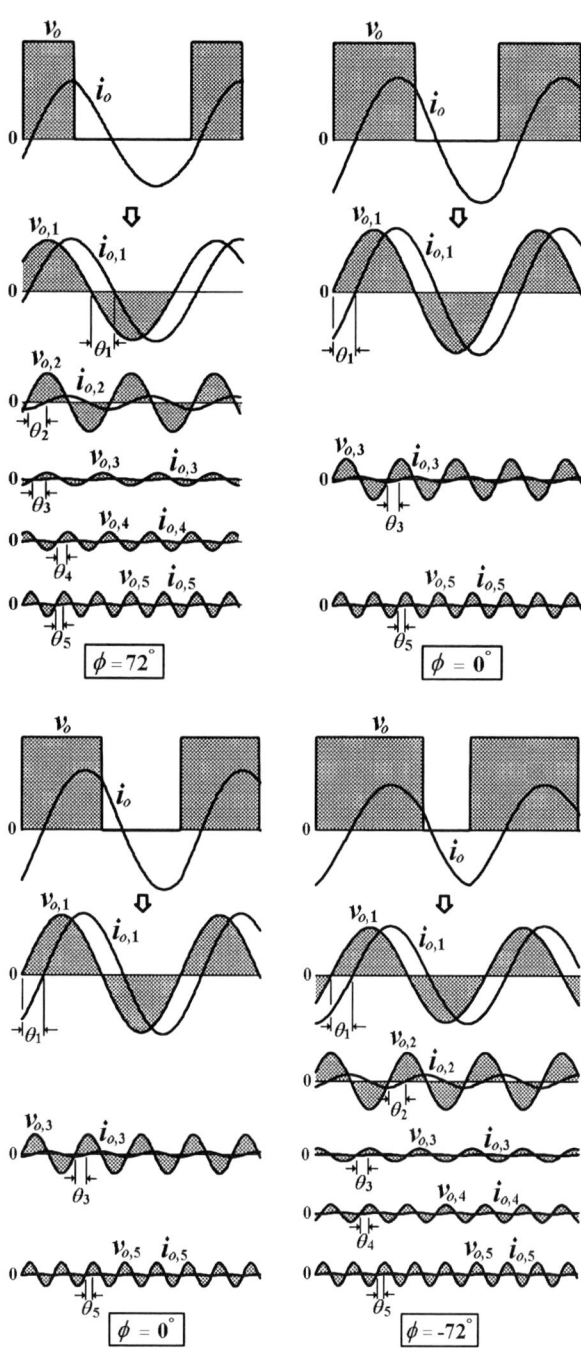

Fig. 10 Output voltage and current waveform v_o, i_o with the phase difference angle of each harmonic

almost equal to $90°$, since the RLC equivalent circuit of the load now becomes equivalent to almost a pure inductive reactance and consequently can not generate any output power . So, it is quite reasonable to calculate approximately the output power from the fundamental pair of output voltage and current waveforms without considering other harmonic orders. The calculated result and the experimental one are shown for comparison in the same graph of Fig. 11 with each switching frequency held constant at 27 kHz, 28 kHz, 29 kHz, 30 kHz and 31 kHz, during change of phase-shift ϕ from $0°$ to both the positive and negative directions. From the starting point of phase-shift ($\phi = 0°$) to both the positive and negative directions, it is observed that there is a certain limitation for the control range of phase-shift for each characteristic curve of output power P_o under a constant operating frequency. This can be understood by considering first the waveforms v_o, i_o at zero phase-shift ($\phi = 0°$) and switching frequency $f_s = 27\text{kHz}$ and also the waveforms v_o, i_o at the same zero phase-shift but at switching frequency $f_s = 31\text{ kHz}$. It can be seen that the phase difference between the output voltage and current waveforms v_o, i_o for these two cases are quite different. Higher switching frequency of 31 kHz can results in a larger phase difference due to lower level of output power. So, higher switching frequency can result in a wider control range of phase-shift under ZVS operation and when NON-ZVS operation is encountered the control range begin to be terminated.

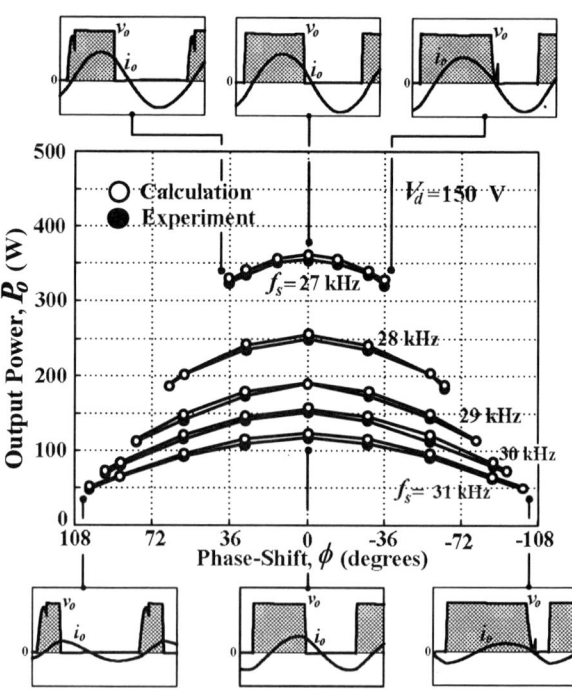

Fig. 11 AC output power P_o vs phase-shift ϕ at switching frequency $f_s = 27$ kHz, 28 kHz, 29 kHz, 30 kHz and 31 kHz

For the calculation of dc input current I_d, the output current i_o can be used again to calculate the dc input current I_d, since the current flow on the dc input side for a certain time duration is the same as that on the ac output side. The calculated result of this dc input current I_d is obtained as shown by an equation in (9) and is also obtained as shown in a graph of Fig. 12.

$$I_d = \frac{V_d}{2\pi^2 Z_1}\Big[\sin(\pi-\phi)\{\sin(\pi-\phi-\theta_1)-\sin(-\theta_1)\}$$
$$+\{1-\cos(\pi-\phi)\}\{-\cos(\pi-\phi-\theta_1)+\cos(-\theta_1)\}\Big]$$
$$+\frac{V_d}{2\times2^2\pi^2 Z_2}\Big[\sin(2\pi-2\phi)\{\sin(2\pi-2\phi-\theta_2)-\sin(-\theta_2)\}$$
$$+\{1-\cos(2\pi-2\phi)\}\{-\cos(2\pi-2\phi-\theta_2)+\cos(-\theta_2)\}\Big]$$
$$+\frac{V_d}{2\times3^2\pi^2 Z_3}\Big[\sin(3\pi-3\phi)\{\sin(3\pi-3\phi-\theta_3)-\sin(-\theta_3)\}$$
$$+\{1-\cos(3\pi-3\phi)\}\{-\cos(3\pi-3\phi-\theta_3)+\cos(-\theta_3)\}\Big]+\cdots$$
(9)

$$P_d = V_d I_d \qquad (10)$$

Where $Z_n = \sqrt{R^2+\left(n\omega_s L-\dfrac{1}{n\omega_s C}\right)^2}$

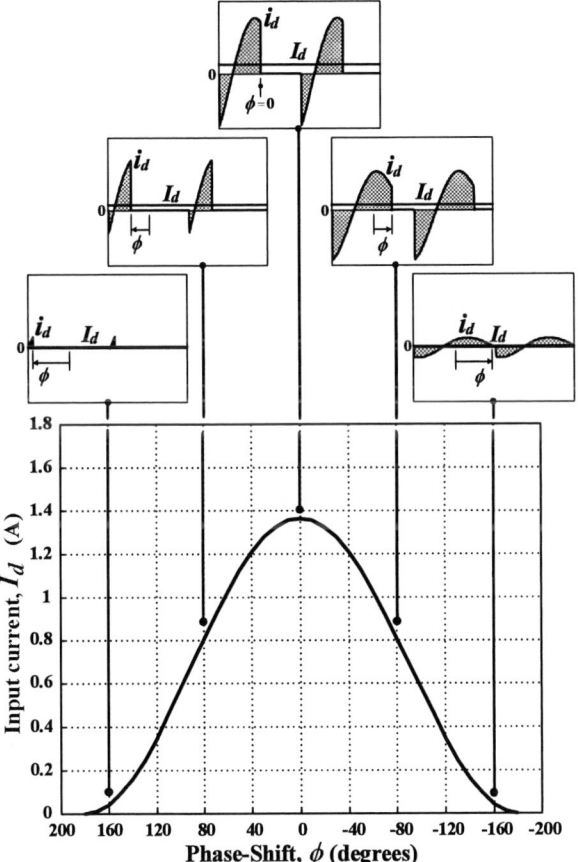

Fig. 12 Variation of dc input current I_d vs phase-shift ϕ at switching frequency f_s = 29 kHz

The dc input current I_d can be also used to calculate the dc input power P_d. The calculated result with the experimental one are also shown in the graph of Fig. 13 with each switching frequency held constant at 27 kHz, 28 kHz, 29 kHz, 30 kHz and 31 kHz, during change

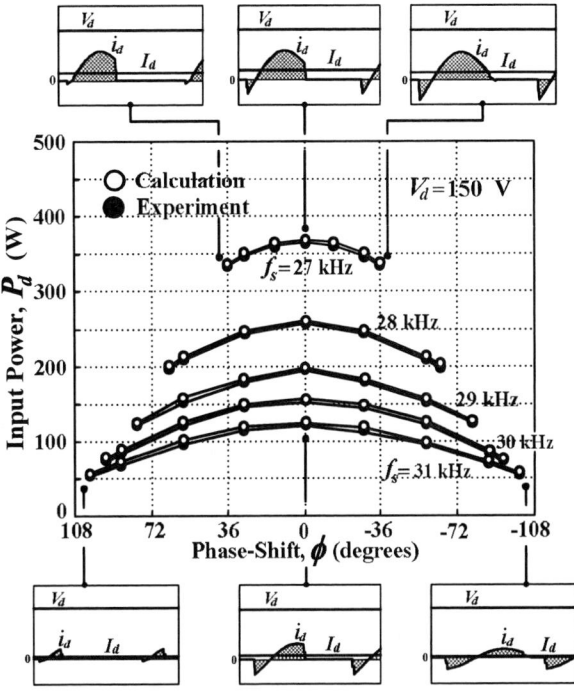

Fig. 13 DC input power P_d vs phase-shift ϕ at switching frequency f_s = 27 kHz, 28 kHz, 29 kHz, 30 kHz and 31 kHz

of phase-shift ϕ from 0° to both the positive and negative directions. From the starting point of phase-shift ($\phi = 0°$) to both the positive and negative directions, it is observed that there is a certain limitation for the control range of phase-shift for each characteristic curve of dc input power P_d under each constant operating frequency.

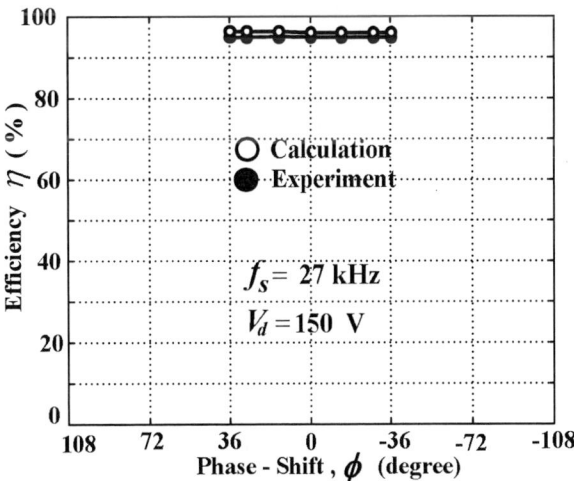

Fig. 14 Inverter efficiency vs phase-shift ϕ at switching frequency f_s = 27 kHz

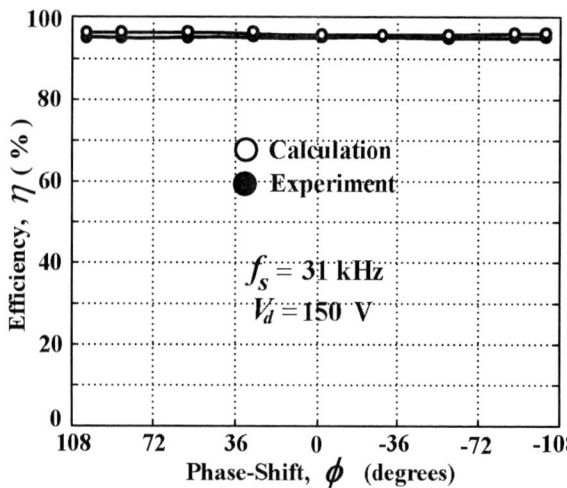

Fig. 15 Inverter efficiency vs phase-shift ϕ at switching frequency $f_S = 31$ kHz

Then, the ratio of output power P_o and dc input power P_d makes possible the calculation of the half-bridge inverter efficiency η vs. phase-shift ϕ which is plotted in various curves, each of which is held at constant frequency of 27 kHz, 28 kHz, 29 kHz, 30 kHz and 31 kHz. The calculated results are shown as some examples in Figs. 14 and 15 for 27 kHz and 31 kHz respectively and they are verified by comparing with the experimental ones in the same figure. The result shows that the efficiency is almost constant at 96 % over the whole control range of phase-shift ϕ which is not constant but the control range will change according to the switching frequency; that is, a wider control range of phase-shift ϕ will be obtained with a higher operating frequency.

V. CONCLUSION

The detailed analysis of circuit operation under ZVS and NON-ZVS switching conditions in a high-frequency half-bridge inverter for a small-size and low-voltage induction cooker, using the principle of positive-negative phase-shift control has been already presented both theoretically and experimentally. There are four main important points to be concluded here as follows :

1. In case of the positive phase-shift control of output voltage v_o, the phase angle of fundamental output voltage $v_{o,1}$ will lead the front edge of square wave voltage v_o. This makes the zero-crossing of output current i_o move toward the front edge, where NON-ZVS operation may occur. Oppositely, in case of the negative phase-shift control, the NON-ZVS operation may occur at the tailing edge of square wave voltage v_o, and in case of zero phase-shift control, NON-ZVS operation may occur both at the front and tailing edges, especially when the operating frequency is not high enough. All these three cases of the occurrences of NON-ZVS operation

can be avoided by increasing the operating frequency in order to move the zero-crossing of current i_o away from the front and tailing edges.

2. In phase-shift control, the maximum ac rms output voltage will be obtained at phase-shift $\phi = 0$ and when phase-shift angle ϕ is increased away from zero degree, the ac rms output voltage $V_{o,rms(ac)}$ and consequently the output current i_o and finally the output power P_o will become decreasing symmetrically both in positive and negative phase-shift control.

3. Similarly, when phase-shift angle ϕ is increased away from zero degree, the dc input current I_d and consequently the dc input power P_d will become decreasing symmetrically both in positive and negative phase-shift control.

4. During phase-shift control, there is a certain limitation for the control range of phase-shift for each characteristic curve of output power P_o under a constant operating frequency. Higher switching frequency can results in a larger phase difference due to lower level of output power. So, higher switching frequency can result in a wider control range of phase-shift under ZVS operation and when NON-ZVS operation is encountered the control range begins to be terminated. At this point, if further increase of phase-shift is required, this is also possible by increasing the switching frequency to a higher value and when NON-ZVS is encountered the same process can be repeated again.

REFERENCES

[1] P. Viriya , N. Yongyuth , I. Miki and K. Matsuse "Analysis of Circuit Operation under ZVS and NON-ZVS Conditions in Phase-Shift Inverter for Induction Heating," *IEEJ Trans. IA.*, vol. 126, no. 5, pp. 560-567, May 2006.

[2] P. Viriya , N. Yongyuth and K. Matsuse "Analysis of Transition Mode from Phase Shift to Zero-Phase Shift Under ZVS and NON-ZVS Operation for Induction Heating Inverter," *Proc. Power Conversion Conf. (PCC), Nagoya, Japan,* April 2007, pp. 1512-1519.

[3] L. Grajales, J. A. Sabate, K. R. Wang, W. A. Tabisz, and F. C. Lee, "Design of a 10 kW, 500 kHz Phase-Shift Controlled Series-Resonant Inverter for Induction Heating," *Proc. IEEE Industry Applications Soc. Annu. Meeting, Toronto, Canada,* 1993, pp. 843-849.

[4] J. M. Burdio, L. A. Barragan, F. Monterde, D. Navarro, and J. Acero, "Asymmetrical Voltage-Cancellation Control for Full-Bridge Series Resonant Inverter," *IEEE Trans. Power Electron.*, vol. 19, no. 2, pp. 461-469, Mar. 2004.

[5] P. Viriya , S. Sittichok and K. Matsuse , "Analysis of high-frequency induction cooker with variable frequency power control," *Proc. Power Conversion Conf. (PCC), Osaka, Japan,* 2002, pp. 423-428.

[6] J. A. Sabate, R. W. Farrington, M. M. Jovanovic, and F. C. Lee, "Effect of Switch Capacitance on Zero-Voltage Switching of Resonant Converters," *Proc. Applied Power Electron. Conf.,* 1992, pp. 213-220.

[7] H. Ogiwara, M. Itoi and M. Nakaoka, " PWM – controlled soft – switching SEPP high – frequency inverter for induction – heating application" , *IEE Proc. – Electr. Power Appl,* vol. 151, no. 4, pp. 404-413, July 2004.

Adaptive Phase Control Method for Load Variation of Resonant Converter with Piezoelectric Transformer

S. T. Yun*, J. M. Sim*, J. H. Park**, S. J. Choi** and B. H. Cho**

* System Engineering & Integration Team, Korea Aerospace Research Institute, 115 Gwahangno Yuseong, Daejeon 305-333, Korea

** School of Electrical Engineering and Computer Science, Seoul National Univ, #042 San 56-1 Shillim-dong Kwanak-ku Seoul 151-744, Korea

Abstract—this paper proposes a new method eliminating external effect from (load, temperature) variation on a Piezoelectric Transformer (PT) converter by utilizing the phase-locked loop (PLL) modulation technique. The PLL control scheme involves detecting phase difference between injected sinusoidal signal and demodulated terminal voltage of PT. Generally, frequency control method decreases the efficiency of PT in wide input range. And, duty cycle control method has a problem of insufficient operating range for the load variation. In other to overcome these problems, this paper introduces a two loop control (PWM and PLL) with PT which provides wide ZVS condition nearly independent of the load variation, and excellent line regulation for the universal input voltage and load range. The performance of the proposed control method is verified by experimental result with 40 [W] DC/DC converter hardware prototypes.

Index Terms—piezoelectric transformer, pulse with modulation, phase locked loop, DC/DC converter.

I. INTRODUCTION

Recently, demands for the development of compact, lightweight power supplies with higher power density and higher efficiency have been increased. Since Piezoelectric Transformer (PT) was emerged in device and material industry, it has been suggested as a viable alternative to the magnetic transformer in some applications. PT has some advantages such as low profile and mechanical energy transfer with little electromagnetic interface (EMI). Also, PT can provide high voltage stepping ratio with good isolation and requires no copper windings saving copper usage especially for large voltage conversion differences [1, 2]. Conventional control of PT converter has mainly two-way. One is the pulse frequency modulation (PFM) control method and the other is the pulse width modulation (PWM) control method [3, 4]. As shown in fig.1 (a), it is known that the maximum PT efficiency can be obtained when it operates near the resonant frequency of the PT. As the operating frequency moves further from the resonant frequency, the PT efficiency decreases dramatically due to the increase of the circulating current. Thus a DC/DC PT converter using PFM control method for a wide input voltage variation reveals a poor efficiency. As for another approach, PWM control, since PT's resonant frequency moves according to the load

conditions, fixed frequency PWM control is in effective for the load variation [5, 6]. This PT characteristic is shown in fig.1 (b). Therefore, this paper proposes a new control method tracking resonant point against load variation by phase locked loop modulating method.

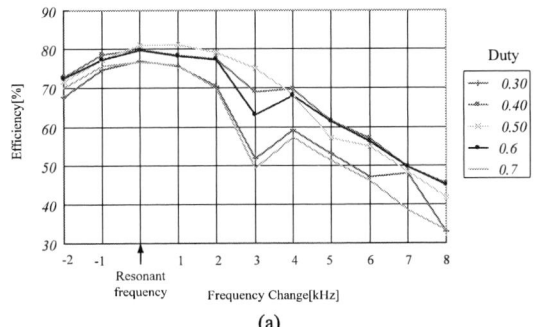

(a)

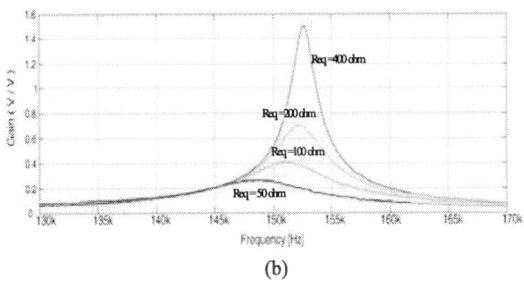

(b)

Fig. 1. (a) Efficiency curves of a PT converter according to duty cycle and operating frequency (b) Gain curve with load change in frequency sweep

II. THE PROPOSED APPROACH

A. PWM Piezoelectric converter

Figure.2 shows the circuit diagram of the PWM PT converter with an active-clamp circuit. Switches S1 and S2 are turned on and off alternately with a short dead time. During the dead time (see fig.3), the energy stored in L_p, charges and discharges the output capacitors of the switches and the input capacitor, C_{d1}, of the PT. The amplitude of voltage, V_1, depends on the duty ratio of the main switch. Thus, output voltage, V_O can be controlled by the duty ratio with a fixed frequency. But, as shown fig. 1(b) the efficiency of PT is dramatically change. Therefore, for load variation, frequency control of PWM PT converter is needed.

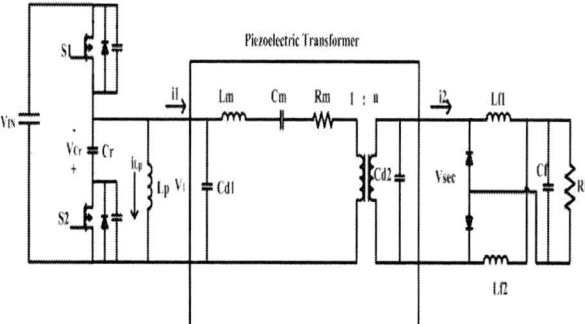

Fig. 2. PWM Resonant PT converter [5].

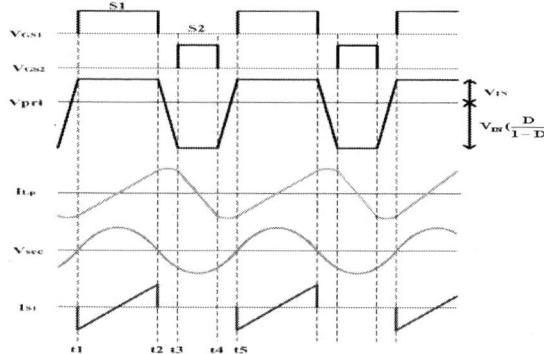

Fig. 3. Waveform of the PWM converter in Fig. 2.

B. Phase locked loop control with modulation

As mentioned above, the resonant peak gain frequency of PT is changed by load condition. For efficient power transfer through PT, the converter needs a peak gain tracking control. Thus, it is unsatisfactory to fixed-frequency PWM PT converter itself. As shown in Fig. 3, the primary waveform is asymmetry. Hence, primary voltage and secondary voltage direct comparing method makes huge error for using phase information. Since there are input and output capacitances in parallel, it is inaccurate to sense the resonant branch current, directly. Therefore, primary voltage and current sensing is not suitable for piezoelectric transformer application. Finally, it is difficult to sense phase difference information because the switching frequency of the typical PT resonant converter is relatively high. Hence, the conventional phase control approaches of the resonant converter are unsuitable for this application.

This paper proposes a new scheme for sensing phase delay by using a low frequency injection method. Figure. 4,5 describes the concept of modulating method. When a constant low frequency sinusoidal voltage is injected to saw tooth waveform, it is modulated with the switching frequency. Hence, fundamental switching frequency is divided by $f_{sw}+f_{sig}$ and $f_{sw}-f_{sig}$, where f_{sw} is fundamental switching frequency, f_{sig} is frequency of injected sinusoidal voltage. The modulated signal is shown in Eq (1).

$$V_{pri} \cong V_{in} \cdot \frac{2}{\pi(1-D)} \cdot \sin(\pi D) \cdot \sin(\omega_s \pm \omega_{sig}) \qquad (1)$$

where V_{pri} is PT primary voltage, V_{in} is DC input voltage, D is duty.

Once modulated signal is passed the PT, phase delay is occurred by its frequency component. Because PT act like band-pass filter, based on equivalent circuit model of fig.7, second voltage can approximate (2). And, demodulation of V_2' holds phase delay at frequency of $\omega_s \pm \omega_{sig}$.

$$V_2' = \frac{V_{pri}}{\|M(\varpi_{sw} \pm \varpi_{sig})\|} \cdot \cos(k(\varpi_{sw} \pm \varpi_{sig})t - \angle M(\varpi_{sw} \pm \varpi_{sig}))$$

$$(2)$$

where $M = \dfrac{sC_m R_{eq}}{N^2(1+sC_{d2}R_{eq})(s^2 L_m C_m + sC_m R_m + 1) + sC_m R_{eq}}$,

$$\omega_{rs} = \frac{1}{\sqrt{L_m \cdot C_m}} \qquad , \qquad R_{eq}' = \frac{R_{eq}/N^2}{1+(\omega C_{d2}R_{eq})^2} ,$$

$$C_{d2}' = N^2 \frac{1+(\omega C_{d2}R_{eq})^2}{\omega C_{d2}R_{eq}^2}, \quad Q_m = \frac{1}{\omega_{rs} \cdot C_m \cdot R_m}, \quad Q = \omega_{rs} \cdot C_{d2} \cdot R_{eq}$$

Consequently, comparison between the injected low frequency reference signal and demodulation signal of secondary voltage (the signal frequency is the same as the reference) can be a good information for phase control. Because the frequency of the injected signal is already known, only one point sensing (secondary voltage) is required. Also, due to low frequency of the injected signal, high bandwidth sensor is unnecessary for this method. Fig. 7 shows the gain and phase relationships of between V_{Sig} and V_{de} under different load condition from full load to 10% load condition, where V_{Sig} is injected constant amplitude sinusoidal signal and V_{de} is demodulation signal of V_2'. The frequency of injection signal in hardware experiment is 4 kHz. However, as shown in Fig. 7, when R_L is equal to 50Ω, the angle of V_{de}/V_{sig} at the resonant frequency is approximately -45°. Based on Eq(3), (4) and (5), as R_{eq}' increases, the phase angle of V_{de}/V_{sig} is decreased.

$$C_o = n^2 C_{d2} \cdot \frac{1+(\omega \cdot C_{d2} \cdot R_{eq})^2}{(\omega \cdot C_{d2} \cdot R_{eq})^2} \qquad (3)$$

$$C_{eq} = \frac{C_m \cdot C_o}{C_m + C_o} \qquad (4)$$

$$f_m = \frac{1}{2\pi\sqrt{L_m \cdot C_{eq}}} \qquad (5)$$

Therefore, from the fig.7, it is shown that the phase difference between V_{Sig} and V_{de} at the resonant frequency (f_m) has some variations. Hence, compensation of variation of phase difference is needed. Based on Eq (2),

(3),(4) and (5), Fig. 8 can be plotted to show the difference between the resonant frequency and the frequency at the fixed 45°phase delay from full load to 10% load condition. As shown in this curve, up to 40% of the load, the frequency difference is linearly increased. Hence, if the variations of load condition pass over the 40% load condition, the frequency difference is approximately saturated to 1 kHz. To compensate this difference, sensing load current and error term of phase difference is used to phase lock loop, Phase reference function is defined As follows:

$$\begin{cases} \text{phase reference voltage} + \alpha \cdot \varepsilon & \text{above 40\% load} \\ \text{phase reference voltage} + \beta & \text{below 40\% load} \end{cases}$$

where α is linear gain, ε is phase error, β is constant gain.

Due to phase reference voltage determines phase lock delay, changing this term makes tracking resonant frequency for load change. Applying this phase reference function; fig. 9 can be plotted, which shows compensated relationship under varying load condition. Accordingly, while the load (R_{eq}) changes from 50Ω to 500Ω, the phase V_{de}/V_{sig} at the resonant frequency is fixed at -45°. These results show that proposed PLL algorithm is adequate for tracking resonant frequency of PT.

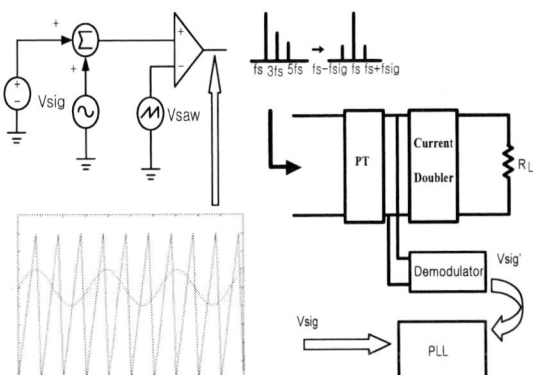

Fig. 4. Concept of phase-modulation control.

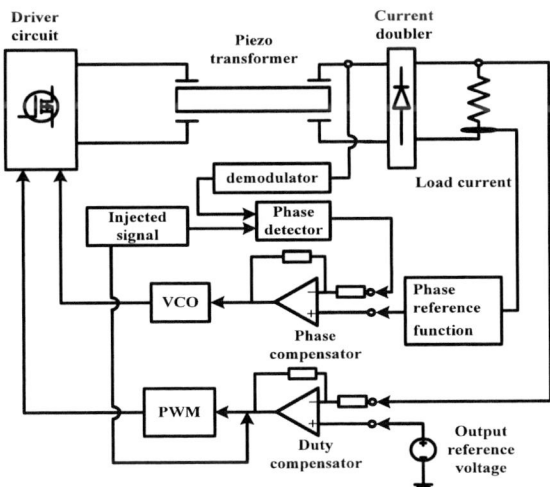

Fig. 5. Low frequency injection and Modulation control diagram.

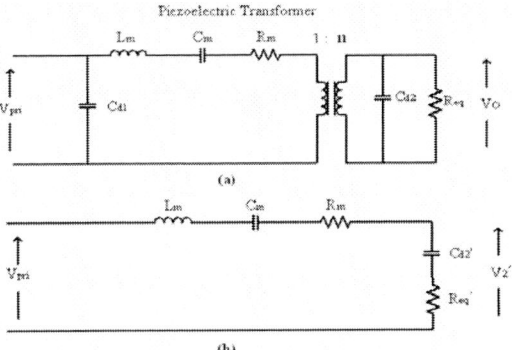

Fig. 6. (a) The equivalent circuit model of resonant converter with PT (b). The simplified circuit of (a).

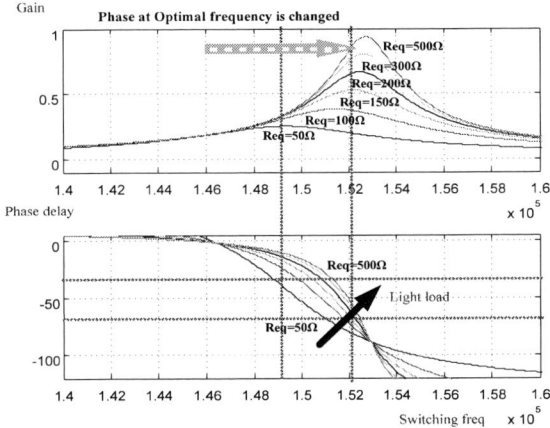

Fig. 7. Gain and phase of V_{de}/V_{sig} with difference load condition.

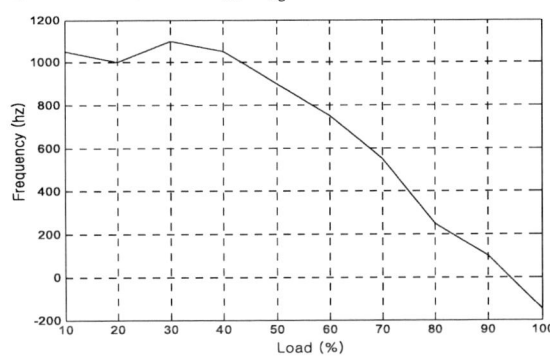

Fig. 8. Frequency difference between optimal frequency and fixed 45° phase frequency with difference load condition.

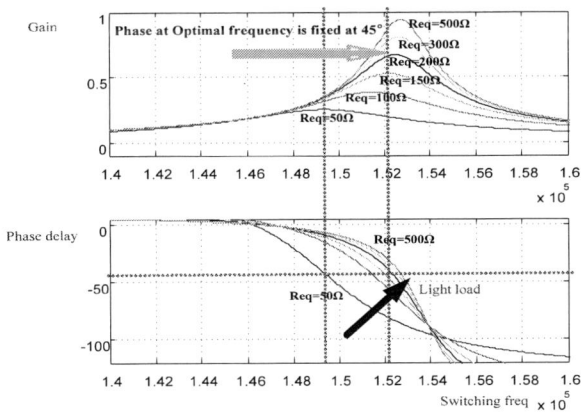

Fig. 9. Compensation of gain and phase of V_{de}/V_{sig} under different load condition.

C. PWM and PLL control modeling by using Extended Describing Function modeling and experimental result

In order to design control loop, small-signal modeling of PT converter is derived. Extended describing function (EDF) is used for the PWM control modeling of the PT resonant converter. Because PT performs band-pass filter for resonant frequency, fundamental approach of EDF is suitable for the analysis of PWM PT converter. Fig. 10 shows the modeling and experimental results of duty to output voltage bode plot. And, Fig. 11 shows the frequency to phase modeling. The measure experiment of model was made by 4194 network analyzer. The modeling result is very similar to the experiment one, and it is utilized for the compensator design.

To verify the proposed control method, load change experiment from 10% to 100% (0.2A~2A) was performed and the result is shown in fig 12. The experimental proto type converter is 40W, and regulation voltage is 20V. Because the load changes from light to heavy, resonant frequency is decreased from 154 kHz to 147 kHz and the controlled duty ratio is increased from 0.2 to 0.7. This experimental result represents that the injected modulation method can used for Phase control. Figure 12 A, B shows the phase difference between V_{Sig} and V_{de} for light and heavy load condition. As mentioned above, proposed method eliminate external effect from load variation on a PT converter

Figure.13 shows the efficiency of injected PLL with PWM control, PFM control and PWM control for wide load variation. This result shows that PWM only control method can't used for wide load range variation and PFM only control dramatically decreased efficiency for wide load range condition. Otherwise, proposed two loop control method presents good performance for wide load range.

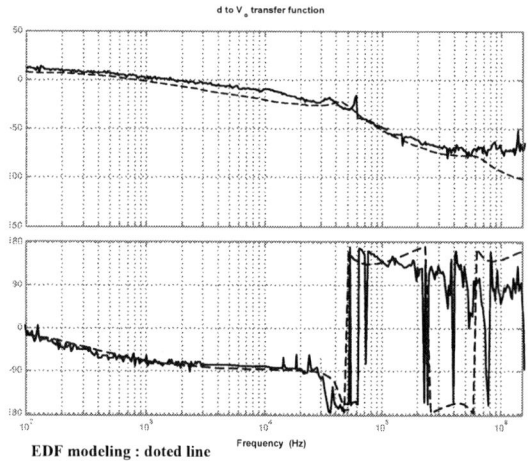

EDF modeling : doted line

Network analyzer experiment result : line

Fig. 10. Simulation and experiment small signal modeling result of duty to output voltage.

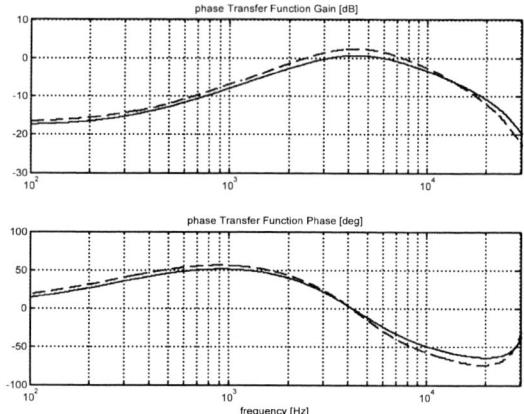

EDF modeling : doted line

Network analyzer experiment result : line

Fig. 11. Simulation and experiment small signal modeling result frequency to phase .

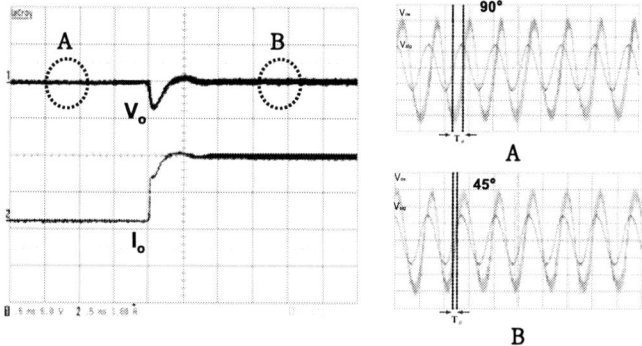

Fig. 12. Transient response during load variation (0.2-2A).
(A) Phase difference waveform between V_{Sig} and V_{de} during A period.
(B) Phase difference waveform between V_{Sig} and V_{de} during B period.

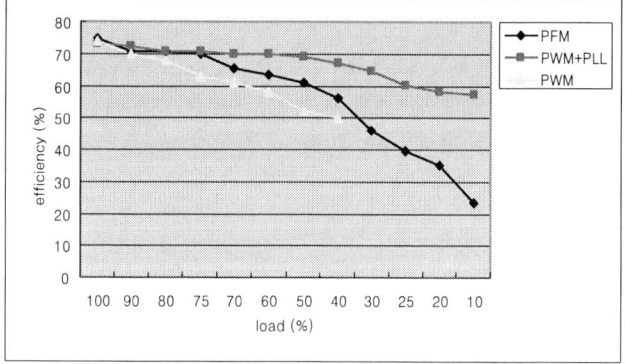

Fig. 13. The efficiency comparison of PT converter with different load condition.

III. CONCLUSIONS

This paper proposed a new resonant tracking algorithm using phase difference information between reference signal and PT secondary signal and PWM and PLL two loop control. The proposed control is verified by experimental result with 40 [W] DC/DC converter hardware prototypes. The experimental results show an efficiency of 59~75%, which is higher than PWM, PFM only control methods with wide load variation. Based on the analysis, it is confirmed that the proposed circuit can be a good solution for the universal off-line input voltage and load specifications

In conclusion, the proposed method has advantage in size, efficiency, and sensing implementation and wide input voltage and load variation.

REFERENCES

[1] G. Ivensky, I. Zafrany and S. Ben-Yaakov, *"Generic operational characteristics of piezoelectric transformers,"* PESC '00. June 2000, pp. 1657 – 1662.

[2] S. Ben-Yaakov and S. Lineykin, *"Frequency tracking to maximum power of piezoelectric transformer HV converters under load variations,"* PESC '02, June 2002, pp. 657 – 662.

[3] T Zaitsu, T Shigehisa, T Inoue, M Shoyama, T Ninomiya, *"Piezoelectric transformer converter with frequency control,"* INTELEC '95, Oct. 1995, pp. 175 – 180.

[4] T. Zaitsu, T. Shigehisa, M. Shoyama and T. Ninomiya, *"Piezoelectric Transformer Converter with PWM Control,"* APEC '96, March 1996, pp. 279-283.

[5] M.H. Ryu, S.J. Choi, B.H. Cho, *"A New Piezoelectric Transformer Driving Topology for Universal Input PWM Control AC/DC Adapter,"* APEC, 2006.

[6] S. Hamamura and D. Kurose, *"New control method of piezoelectric transformer converter by PWM and PFM for wide range of input voltage,"* CIEP '00, Oct. 2000, pp. 3 – 8.

Adaptation of Motor Parameters in Sensorless PMSM Drives

Antti Piippo, Marko Hinkkanen, and Jorma Luomi
Power Electronics Laboratory
Helsinki University of Technology
P.O. Box 3000, FI-02015 TKK, Finland

Abstract—The paper proposes an on-line method for the estimation of the stator resistance and the permanent magnet flux in sensorless permanent magnet synchronous motor drives. An adaptive observer augmented with a high-frequency signal injection technique is used for sensorless control. The observer contains excess information that is not used for the speed and position estimation. This information is used for the adaptation of the motor parameters. At low speeds, the stator resistance is estimated, whereas at medium and high speeds, the permanent magnet flux is estimated. Steady-state analysis and small-signal analysis are carried out to investigate the parameter estimation, and adaptation mechanisms are defined for the parameters. The convergence of the parameter estimates is shown by simulations and laboratory experiments. The stator resistance adaptation works down to zero speed in sensorless control.

I. INTRODUCTION

Permanent magnet synchronous machines (PMSMs) are used in many high-performance applications. For vector control of PMSMs, information on the rotor position is required. In sensorless control, the methods for estimating the rotor speed and position can be classified into two categories: fundamental-excitation methods [1], [2] and signal injection methods [3], [4]. The methods can also be combined by changing the estimation method as the rotor speed varies [5], [6].

The fundamental-excitation methods used for sensorless control are based on models of the PMSM. Hence, the electrical parameters are needed for the speed and position estimation [7]. The errors in the stator resistance estimate result in an incorrect back-emf estimate and, consequently, impaired position estimation accuracy. The operation can also become unstable at low speeds in a loaded condition. The detuned estimate of the permanent magnet (PM) flux results in incorrectly estimated electromagnetic torque [8], and also impairs the position estimation accuracy. Errors in the d- and q-axis inductances of a salient PMSM also affect the estimation and the torque production, and can degrade the current control performance.

The stator resistance and the PM flux depend on the motor temperature, and thus change rather slowly. On the other hand, magnetic saturation decreases the inductances, which thus depend on the load condition. The inductances can be modeled as functions of the stator flux or the stator current, but an estimation scheme is required for the stator resistance and the PM flux. The rotor back-emf is proportional to the PM flux and the resistive voltage drop to the stator resistance. At medium and high speeds, the effect of the PM flux estimation error is more significant than that of the stator resistance estimation

error. On the other hand, the back-emf is small at low speeds, and the stator resistance estimate plays an important role in the estimation.

Several methods have been proposed to improve the performance of a PMSM drive by estimating the electrical parameters. In [7], an MRAS scheme is used for the on-line estimation of the stator resistance and the PM flux with position measurement. Reactive power feedback is used for estimating the PM flux in [9]. The stator current estimation error and a neural network can be used for estimating both the PM flux and the stator resistance [10]. The stator inductances and the PM flux are estimated using the steady-state voltage equations and the flux harmonics, respectively in [11]. A DC-current signal is injected to detect the resistive voltage drop for the resistance estimation in [12], and the PM flux linkage is estimated by taking it as an additional state of an extended Kalman filter in [13].

Some parameter estimation schemes have also been developed for sensorless control methods. In [7], an MRAS scheme is applied for the stator resistance estimation. A parameter estimator is added to two position estimation methods for estimating the stator resistance and the PM flux in [14]. In [15], these parameters are estimated using both the steady-state motor equations and the response to an alternating current signal. In [14], [15], the convergence of the estimated parameters to their actual values is not shown. [16] proposes a method where the resistance and the inductances of a salient PMSM are extracted from an extended EMF model. Three electrical parameters are estimated simultaneously, but the behavior of the stator resistance estimate is not convincing.

This paper proposes a method for the on-line estimation of the stator resistance and the PM flux in sensorless control. The method is based on a speed-adaptive observer that is augmented with a high-frequency (HF) signal injection technique at low speeds [17]. The excess information available in the observer is used for the adaptation of the parameters. At medium and high speeds, the PM flux is estimated from the d-axis current estimation error. At low speeds, the stator resistance is estimated from a speed correction term produced by the signal injection method. The sensitivity of the d-axis current estimation error and the speed correction term to the parameter errors are investigated by means of steady-state and small-signal analyses, and adaptation laws are designed for the estimation of the parameters. The stability and the convergence of the parameter estimators are investigated by means of simulations and laboratory experiments. The resistance adaptation is shown to work down to zero speed in sensorless control.

978-1-4244-0644-9/07/$25.00 ©2007 IEEE

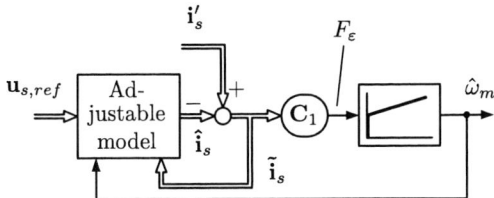

Fig. 1. Block diagram of the adaptive observer.

II. PMSM MODEL

The PMSM is modeled in the d-q reference frame fixed to the rotor. The d axis is oriented along the PM flux, whose angle in the stator reference frame is θ_m in electrical radians. The stator voltage equation is

$$\mathbf{u}_s = R_s \mathbf{i}_s + \dot{\boldsymbol{\psi}}_s + \omega_m \mathbf{J} \boldsymbol{\psi}_s \tag{1}$$

where $\mathbf{u}_s = [\, u_d \ u_q \,]^T$ is the stator voltage, $\mathbf{i}_s = [\, i_d \ i_q \,]^T$ the stator current, $\boldsymbol{\psi}_s = [\, \psi_d \ \psi_q \,]^T$ the stator flux, R_s the stator resistance, $\omega_m = \dot{\theta}_m$ the electrical angular speed of the rotor, and

$$\mathbf{J} = \begin{bmatrix} 0 & -1 \\ 1 & 0 \end{bmatrix}$$

The stator flux is

$$\boldsymbol{\psi}_s = \mathbf{L} \mathbf{i}_s + \boldsymbol{\psi}_{pm} \tag{2}$$

where $\boldsymbol{\psi}_{pm} = [\, \psi_{pm} \ 0 \,]^T$ is the PM flux and

$$\mathbf{L} = \begin{bmatrix} L_d & 0 \\ 0 & L_q \end{bmatrix}$$

is the inductance matrix, L_d and L_q being the direct- and quadrature-axis inductances, respectively. The electromagnetic torque is given by

$$T_e = \frac{3p}{2} \boldsymbol{\psi}_s^T \mathbf{J}^T \mathbf{i}_s \tag{3}$$

where p is the number of pole pairs.

III. SPEED AND POSITION ESTIMATION

A. Adaptive Observer

An adaptive observer [17] is used for the estimation of the stator current, rotor speed, and rotor position. The speed and position estimation is based on the estimation error between two different models; the actual motor can be considered as a reference model and the observer—including the rotor speed estimate $\hat{\omega}_m$—as an adjustable model. The error term used in a speed adaptation mechanism is based on the estimation error of the stator current. The estimated rotor speed, obtained using the adaptation mechanism, is fed back to the adjustable model.

The adaptive observer is formulated in the estimated rotor reference frame. The block diagram of the adaptive observer is shown in Fig. 1. The adjustable model is based on (1) and (2), and defined by

$$\dot{\hat{\boldsymbol{\psi}}}_s = \mathbf{u}_{s,ref} - \hat{R}_s \hat{\mathbf{i}}_s - \hat{\omega}_m \mathbf{J} \hat{\boldsymbol{\psi}}_s + \boldsymbol{\lambda} \tilde{\mathbf{i}}_s \tag{4}$$

where estimated quantities are marked by $\hat{}$ and $\mathbf{u}_{s,ref}$ is the stator voltage reference. The estimate of the stator current and the estimation error of the stator current are

$$\hat{\mathbf{i}}_s = \hat{\mathbf{L}}^{-1}(\hat{\boldsymbol{\psi}}_s - \hat{\boldsymbol{\psi}}_{pm}) \tag{5}$$

$$\tilde{\mathbf{i}}_s = \mathbf{i}'_s - \hat{\mathbf{i}}_s \tag{6}$$

respectively, where $\mathbf{i}'_s$ is the measured stator current expressed in the estimated rotor reference frame. The feedback gain matrix $\boldsymbol{\lambda}$ is proportional to the rotor speed up to the nominal speed [17].

The speed adaptation is based on an error term

$$F_\varepsilon = \mathbf{C}_1 \tilde{\mathbf{i}}_s \tag{7}$$

where $\mathbf{C}_1 = [\, 0 \ \hat{L}_q \,]$, i.e., the current error in the estimated q direction is used for adaptation. The estimate of the electrical angular speed of the rotor is obtained using a PI speed adaptation mechanism

$$\hat{\omega}_m = -k_p F_\varepsilon - k_i \int F_\varepsilon dt \tag{8}$$

where k_p and k_i are nonnegative gains. The gain selection is described in [17]. The estimate $\hat{\theta}_m$ for the rotor position is evaluated by integrating $\hat{\omega}_m$.

B. High-Frequency Signal Injection

The adaptive observer described above is augmented with a HF signal injection method to stabilize the speed and position estimation at low speeds [17]. By using the HF signal injection method with an alternating voltage u_c as a carrier excitation signal [18], an error signal $\varepsilon \approx 2K_\varepsilon \tilde{\theta}_m$ proportional to the position estimation error $\tilde{\theta}_m = \theta_m - \hat{\theta}_m$ is obtained, K_ε being the signal injection gain. The error signal is used for correcting the estimated position by influencing the direction of the stator flux estimate of the adjustable model. For the combined observer, the adjustable model (4) is modified to

$$\dot{\hat{\boldsymbol{\psi}}}_s = \mathbf{u}_{s,ref} - \hat{R}_s \hat{\mathbf{i}}_s - (\hat{\omega}_m - \omega_\varepsilon) \mathbf{J} \hat{\boldsymbol{\psi}}_s + \boldsymbol{\lambda} \tilde{\mathbf{i}}_s \tag{9}$$

where

$$\omega_\varepsilon = \gamma_p \varepsilon + \gamma_i \int \varepsilon dt \tag{10}$$

is the speed correction term, γ_p and γ_i being the gains of the PI mechanism driving the error signal ε to zero. In accordance with [6], these gains are selected as

$$\gamma_p = \frac{\alpha_i}{2K_\varepsilon}, \quad \gamma_i = \frac{\alpha_i^2}{6K_\varepsilon} \tag{11}$$

where α_i is the approximate bandwidth of the PI mechanism.

At low speeds, the combined observer relies both on the signal injection method and on the adaptive observer. The influence of the HF signal injection is decreased linearly as the speed increases by decreasing both the HF excitation voltage and the bandwidth α_i. At speeds higher than a threshold speed ω_Δ, the estimation is based only on the adaptive observer.

IV. PARAMETER ADAPTATION

The current estimation error $\tilde{\mathbf{i}}_s$ of the adaptive observer contains information that can be used for the parameter adaptation. In [7], [10], the components of $\tilde{\mathbf{i}}_s$ are used for the adaptation of two parameters in a PMSM drive equipped with a motion sensor. Solving the parameters from the steady-state voltage equations as in [11] would require filtering to prevent incorrect operation in transients. It is to be noted that the PMSM dynamics offer only two degrees of freedom. Since the component $\tilde{i}_q$ is used for the speed estimation in this paper, one parameter can thus be adjusted using $\tilde{i}_d$.

In low-speed operation, the HF signal injection method provides additional information through the speed correction term ω_ε. Instead of solving the parameters directly from the response to the injected signal [12], ω_ε is used here for the parameter adaptation. Hence, if ω_ε differs from zero in steady state or if the current estimation error $\tilde{i}_d$ is nonzero, the motor parameter estimates are inaccurate and the variables ω_ε and $\tilde{i}_d$ can be driven to zero by adjusting the parameter estimates.

A. Steady-State Analysis—Stator Resistance Adaptation

At low speeds, the stator resistance estimate plays an important role in the speed and position estimation, particularly in loaded conditions. To extract the sensitivity of the d-axis current estimation error $\tilde{i}_d$ and the speed correction term ω_ε to the parameter errors, the combined observer is investigated in steady state. The position estimation error is assumed zero because the HF signal injection method is in use.

In steady state, the equations of the PMSM and the adaptive observer can be written as

$$\mathbf{u}_s = R_s \mathbf{i}_s + \omega_m \mathbf{J}(\mathbf{L}\mathbf{i}_s + \boldsymbol{\psi}_{pm}) \tag{12}$$

$$\mathbf{u}_s = \hat{R}_s \hat{\mathbf{i}}_s + (\omega_m - \omega_\varepsilon)\mathbf{J}(\mathbf{L}\hat{\mathbf{i}}_s + \hat{\boldsymbol{\psi}}_{pm}) - \boldsymbol{\lambda}\tilde{\mathbf{i}}_s \tag{13}$$

respectively. The estimated quantities are expressed in terms of their actual values and estimation errors, i.e. $\hat{x} = x - \tilde{x}$. The stator voltage is eliminated by combining (12) and (13), yielding

$$R_s \mathbf{i}_s + \omega_m \mathbf{J}(\mathbf{L}\mathbf{i}_s + \boldsymbol{\psi}_{pm}) \tag{14}$$
$$= (R_s - \tilde{R}_s)(\mathbf{i}_s - \tilde{\mathbf{i}}_s)$$
$$+ (\omega_m - \omega_\varepsilon)\mathbf{J}[\mathbf{L}(\mathbf{i}_s - \tilde{\mathbf{i}}_s) + (\boldsymbol{\psi}_{pm} - \tilde{\boldsymbol{\psi}}_{pm})] - \boldsymbol{\lambda}\tilde{\mathbf{i}}_s$$

When it is assumed that the estimation errors and the speed correction term ω_ε are small, their products can be omitted and the equation reduces to

$$(R_s \mathbf{I} + \omega_m \mathbf{J}\mathbf{L} + \boldsymbol{\lambda})\tilde{\mathbf{i}}_s \tag{15}$$
$$= -\tilde{R}_s \mathbf{i}_s - \omega_m \mathbf{J}\tilde{\boldsymbol{\psi}}_{pm} - \omega_\varepsilon \mathbf{J}(\mathbf{L}\mathbf{i}_s + \boldsymbol{\psi}_{pm})$$

The q component of the current estimation error is zero in steady state because it is driven to zero by the adaptation mechanism (8). The observer gain $\boldsymbol{\lambda}$ is proportional to the rotor speed and is also omitted since low speeds are investigated. In addition, the d component of the stator current is ignored

TABLE I
MOTOR DATA

Nominal voltage U_N	370 V
Nominal current I_N	4.3 A
Nominal frequency f_N	75 Hz
Nominal torque T_N	14.0 Nm
Stator resistance R_s	3.59 Ω
Direct-axis inductance L_d	36.0 mH
Quadrature-axis inductance L_q	51.0 mH
PM flux ψ_{pm}	0.545 Vs
Total moment of inertia	0.015 kgm^2

because it is controlled to a considerably smaller value than the q component. In the component form, the result is

$$\begin{bmatrix} R_s & -L_q i_q \\ \omega_m L_d & \psi_{pm} \end{bmatrix} \begin{bmatrix} \tilde{i}_d \\ \omega_\varepsilon \end{bmatrix} = \begin{bmatrix} 0 & 0 \\ -i_q & -\omega_m \end{bmatrix} \begin{bmatrix} \tilde{R}_s \\ \tilde{\psi}_{pm} \end{bmatrix} \tag{16}$$

For investigating the sensitivity of $\tilde{i}_d$ and ω_ε to the parameter errors, the variables $\tilde{i}_d$ and ω_ε are solved from (16). Assuming $R_s \psi_{pm} \gg \omega_m L_d L_q i_q$, the result is

$$\begin{bmatrix} \tilde{i}_d \\ \omega_\varepsilon \end{bmatrix} = -\frac{1}{\psi_{pm}} \begin{bmatrix} \frac{L_q}{R_s} i_q^2 & \frac{L_q}{R_s} \omega_m i_q \\ i_q & \omega_m \end{bmatrix} \begin{bmatrix} \tilde{R}_s \\ \tilde{\psi}_{pm} \end{bmatrix} \tag{17}$$

It can be seen that both variables on the left-hand side depend both on $\tilde{R}_s$ and $\tilde{\psi}_{pm}$. The correction term ω_ε is selected for parameter adaptation because i_q has more effect on $\tilde{i}_d$ than on ω_ε. When using relative parameter errors, the speed correction term is

$$\omega_\varepsilon = -\frac{1}{\psi_{pm}} \left(R_s i_q \frac{\tilde{R}_s}{R_s} + \psi_{pm} \omega_m \frac{\tilde{\psi}_{pm}}{\psi_{pm}} \right) \tag{18}$$

The dependence between the stator resistance error and ω_ε is proportional to the resistive voltage drop, whereas the dependence between the PM flux error and ω_ε is proportional to the back-emf.

The effect of the parameter errors on ω_ε was calculated numerically, using the parameters given in Table I. The effect of the stator resistance error $\tilde{R}_s / R_s = -0.1$ is illustrated in Fig. 2(a), and the effect of the PM flux error $\tilde{\psi}_{pm} / \psi_{pm} = -0.1$ in Fig. 2(b). The results obtained from the approximate equation (18) are shown. In addition, the figures show the results obtained using (15) and the results of steady-state simulations. The speed values used for the calculation and simulation in Fig. 2(a) are -0.067 p.u., 0 p.u., and 0.067 p.u, and the torque values used in Fig. 2(b) are $-T_N$, 0, and T_N. According to Fig. 2, the assumptions used in (18) result in only a small error. At low speeds and large values of load torque, the resistive voltage drop dominates, and the stator resistance is thus a reasonable selection for the adaptation.

For the stator resistance adaptation, the closed-loop system shown in Fig. 3 is investigated. The gain

$$F_R = -i_q / \psi_{pm} \tag{19}$$

corresponds to (17) with $\tilde{\psi}_{pm} = 0$. The stator resistance is estimated by integration from ω_ε, the transfer function of the adaptation mechanism thus being

$$G_R(s) = -k_R / s \tag{20}$$

408

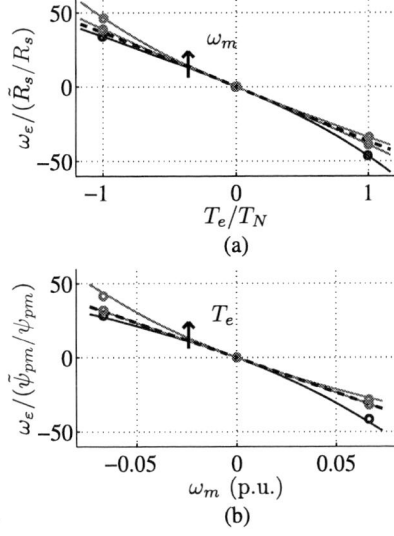

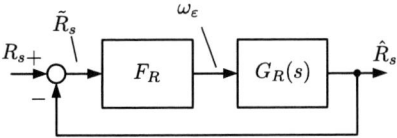

Fig. 2. Effects of parameter estimation errors on speed correction term ω_ε: (a) effect of resistance error $\tilde{R}_s/R_s = -0.1$ as a function of torque for various rotor speed values and (b) effect of PM flux error $\tilde{\psi}_{pm}/\psi_{pm} = -0.1$ as a function of rotor speed for various electromagnetic torque values. Solid lines show the calculated values from (15) without additional assumptions, dashed line shows calculated values from (18), and the circles are simulated values.

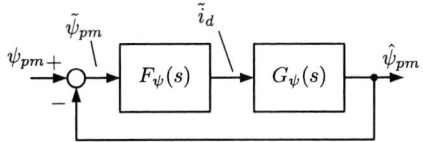

Fig. 3. Closed-loop system of the stator resistance adaptation.

where k_R is the stator resistance adaptation gain. The closed-loop system has a bandwidth

$$\alpha_R = i_q k_R / \psi_{pm} \tag{21}$$

which can be affected by properly selecting the gain k_R.

B. Small-Signal Analysis—PM Flux Adaptation

At medium and high speeds, the speed correction term ω_ε is not available since the HF signal injection is not used. The parameter adaptation can be based on $\tilde{i}_d$, which is obtained from the adaptive observer. Small-signal analysis is used for obtaining the sensitivity of $\tilde{i}_d$ to the parameter errors.

For the analysis, the rotor speed is assumed constant, and the observer is written assuming zero orientation error, yielding

$$\dot{\hat{\psi}}_s = \mathbf{u}_s - \hat{R}_s \mathbf{i}_s - \omega_m \mathbf{J}\hat{\psi}_s + \lambda\tilde{\mathbf{i}}_s \tag{22a}$$

$$\hat{\mathbf{i}}_s = \mathbf{L}^{-1}(\hat{\psi}_s - \hat{\psi}_{pm}) \tag{22b}$$

where only the stator resistance and the PM flux errors are taken into account. The stator flux error $\tilde{\psi}_s$ and the stator current error $\tilde{\mathbf{i}}_s$ are solved by subtracting the estimates in (22) from the actual variables solved from (1) and (2). The result is

$$\dot{\tilde{\psi}}_s = -R_s \mathbf{i}_s + \hat{R}_s \hat{\mathbf{i}}_s - \omega_m \mathbf{J}\tilde{\psi}_s - \lambda\tilde{\mathbf{i}}_s \tag{23a}$$

$$\tilde{\mathbf{i}}_s = \mathbf{L}^{-1}(\tilde{\psi}_s - \tilde{\psi}_{pm}) \tag{23b}$$

Fig. 4. Closed-loop system of the PM flux adaptation.

The estimates are replaced with the actual quantities and their errors, i.e. $\hat{x} = x - \tilde{x}$, and the estimation errors are assumed small so that their products can be omitted. In addition, (23b) is substituted for the current error in (23a), the result being

$$\dot{\tilde{\psi}}_s = \underbrace{\left(-R_s \mathbf{L}^{-1} - \omega_m \mathbf{J} - \lambda\mathbf{L}^{-1}\right)}_{\mathbf{A}} \tilde{\psi}_s$$
$$+ \underbrace{\left(R_s \mathbf{L}^{-1} + \lambda\mathbf{L}^{-1}\right)}_{\mathbf{B}} \tilde{\psi}_{pm} - \tilde{R}_s \mathbf{i}_s \tag{24a}$$

$$\tilde{\mathbf{i}}_s = \underbrace{\mathbf{L}^{-1}}_{\mathbf{C}} \tilde{\psi}_s \underbrace{-\mathbf{L}^{-1}}_{\mathbf{D}} \tilde{\psi}_{pm} \tag{24b}$$

The stator resistance error can be detected from $\tilde{\mathbf{i}}_s$ only when the motor is loaded, and when the stator current $\mathbf{i}_s$ varies, the gain from $\tilde{R}_s$ to $\tilde{\mathbf{i}}_s$ changes. Because the stator resistance is estimated at low speeds using the HF signal injection, only the PM flux is selected for the adaptation at medium and high speeds. It is also to be noted that only two quantities, the current error components $\tilde{i}_d$ and $\tilde{i}_q$, can be used for the adaptation and only two quantities can thus be estimated. Since the rotor speed is estimated using the q component of the current estimation error, the d component is selected for the PM flux adaptation.

For investigating the behavior of the PM flux adaptation, the closed-loop system of Fig. 4 is studied. The transfer function from the scalar-valued flux error $\tilde{\psi}_{pm}$ to the current error $\tilde{i}_d$ is obtained from

$$F_\psi(s) = [1 \ 0]\left\{\mathbf{C}(s\mathbf{I} - \mathbf{A})^{-1}\mathbf{B} + \mathbf{D}\right\}[1 \ 0]^T \tag{25}$$

Integration is used for the adaptation of $\hat{\psi}_{pm}$, and the transfer function corresponding to the adaptation is

$$G_\psi(s) = -k_\psi \hat{L}_d / s \tag{26}$$

where k_ψ is the adaptation gain. The closed-loop transfer function

$$G_{c\psi}(s) = \frac{F_\psi(s)G_\psi(s)}{1 + F_\psi(s)G_\psi(s)} \tag{27}$$

from ψ_{pm} to $\hat{\psi}_{pm}$ can be evaluated in any operating point.

The closed-loop transfer function (27) was analyzed numerically in different operating points. The parameter values given in Table I were used for the calculations. Fig. 5 shows poles and Fig. 6 unit step responses obtained with a constant adaptation gain $k_\psi = 0.2\omega_B$, where ω_B is the base angular speed. The speed varies from 0 to 1 p.u. in Fig. 5. The stator current and the rotational direction have no effect on the results. Due to symmetry, only the upper half-plane is shown in the pole plot. The step responses in Fig. 6 were obtained at speeds $\omega_m = 0.25$ p.u., 0.5 p.u., 0.75 p.u., and 1.0 p.u.

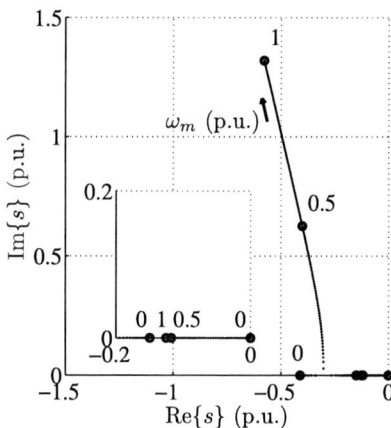

Fig. 5. Pole plot of the closed-loop system of PM flux adaptation. The vicinity of the origin is also shown as a magnification.

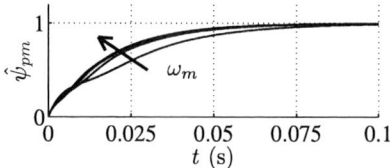

Fig. 6. Step response of the closed-loop system of PM flux adaptation for various rotor speeds.

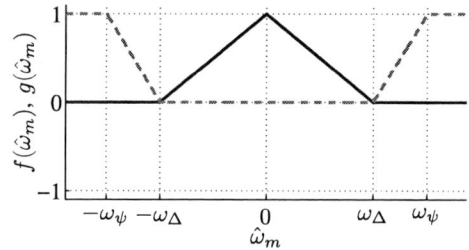

Fig. 7. Functions $f(\hat{\omega}_m)$ (solid) and $g(\hat{\omega}_m)$ (dashed) as functions of estimated rotor speed.

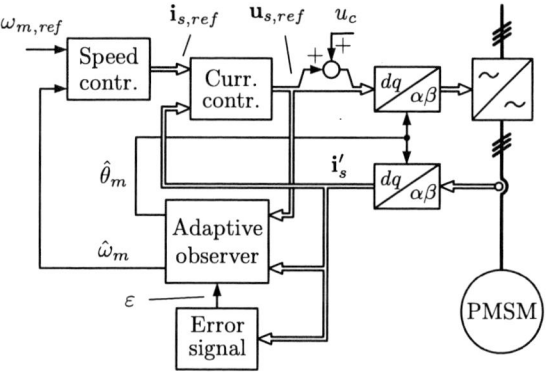

Fig. 8. Block diagram of the control system.

According to the results, the small constant gain gives a relatively slow response. However, the actual PM flux changes slowly since it depends on the temperature. Although there are complex poles, the dominant pole is real-valued and it is almost constant as the speed increases above 0.5 p.u. Hence, the response time to a changing PM flux does not vary significantly at higher speeds, which can also be seen from the step responses in Fig. 6.

C. Gain Scheduling

The adaptation of the stator resistance is in use only at low speeds in which the HF signal injection method is used. The PM flux adaptation is not used simultaneously with the stator resistance adaptation, and is enabled when the rotor speed is higher than ω_Δ. The adaptation gains k_R and k_ψ are changed with the varying rotor speed and stator current.

In order to have a nonnegative bandwidth (21) for the stator resistance adaptation, the gain k_R and the current i_q must have the same sign. A constant bandwidth α_R is infeasible, since $i_q = 0$ would imply infinite adaptation gain. A signum function in the gain k_R could cause chattering near zero i_q. Therefore, the gain k_R is changed proportionally to i_q, i.e.

$$k_R = \alpha_{R0} f(\hat{\omega}_m) \hat{\psi}_{pm} \frac{i_q}{I_B^2} \qquad (28)$$

which results in the adaptation bandwidth

$$\alpha_R = \alpha_{R0} f(\hat{\omega}_m) \frac{i_q^2}{I_B^2} \qquad (29)$$

The parameter α_{R0} is a constant corresponding to the adaptation bandwidth at zero speed and at approximately nominal

load, I_B is the base value of the current, and $f(\hat{\omega}_m)$ is a speed-dependent function shown in Fig. 7.

The gain k_ψ for the PM flux adaptation is varied according to

$$k_\psi = k_{\psi 0} \, g(\hat{\omega}_m) \qquad (30)$$

where $k_{\psi 0}$ is a positive constant and $g(\hat{\omega}_m)$ is a speed-dependent function shown in Fig. 7. At speeds higher than ω_ψ, the gain k_ψ is thus kept constant.

V. SIMULATION RESULTS

The proposed method was investigated by means of simulations and laboratory experiments. Fig. 8 shows the block diagram of the control system comprising cascaded speed and current control loops. PI-type speed control with active damping is used. The data of the six-pole interior-magnet PMSM (2.2 kW, 1500 rpm) are given in Table I. The base values for voltage, current, and angular speed are defined as $U_B = \sqrt{2/3}U_N$, $I_B = \sqrt{2}I_N$, and $\omega_B = 2\pi f_N$, respectively. The electromagnetic torque is limited to 22 Nm, which is 1.57 times the nominal torque T_N. The high-frequency carrier excitation signal has a frequency of 833 Hz and an amplitude of 40 V. The threshold speeds $\omega_\Delta = 0.13$ p.u. and $\omega_\psi = 0.2$ p.u., the resistance adaptation bandwidth $\alpha_{R0} = 0.01$ p.u., and the constant $k_{\psi 0} = 0.2\omega_B$. The current and speed control bandwidths are 5.33 p.u. and 0.067 p.u., respectively, and the bandwidth $\alpha_i = 0.067$ p.u.

The MATLAB/Simulink environment was used for the simulations. Fig. 9 shows results obtained at zero speed reference. Except the stator resistance, the parameter values used in the controller were equal to those of the motor model. In the

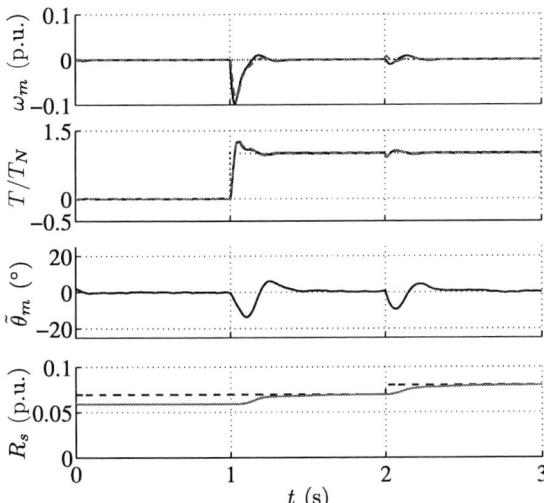

Fig. 9. Simulation results showing stator resistance adaptation. First subplot shows electrical angular speed of the rotor (solid), its estimate (dashed), and its reference (dotted). Second subplot shows the load torque reference (dotted), the electromagnetic torque (solid), and its estimate (dashed). Third subplot shows the estimation error of the rotor position. Last sublot shows the stator resistance (dashed) and its estimate (solid).

beginning of the simulation, the stator resistance estimate is 15 % smaller than the actual stator resistance. A nominal load torque step is applied at $t = 1$ s, and the stator resistance estimate starts converging to the actual resistance immediately. At $t = 2$ s, a 15 % step increase occurs in the stator resistance, and its estimate again follows the actual stator resistance. The stator resistance estimate converges close to the actual resistance in less than 1 s.

Results showing the behavior of the estimated PM flux are depicted in Fig. 10. The estimated flux is 15 % bigger than its actual value in the beginning of the simulations, and other parameter estimates are equal to the actual values in the motor model. The speed reference is changed from zero to 0.5 p.u. at $t = 0.5$ s, and a nominal load torque step is applied at $t = 1$ s. In Fig. 10(a), the parameter adaptation is not in use, whereas in Fig. 10(b), the adaptation is used. Fig. 10(a) shows that the erroneous PM flux results in an error in the rotor position estimate both at no load and when a load is applied. In addition, the electromagnetic torque is smaller than the estimated torque. According to Fig. 10(b), the adaptation practically removes the PM flux error in less than 0.2 s after the motor is started, and the errors in the rotor position and the torque are avoided.

VI. EXPERIMENTAL RESULTS

The experimental setup is illustrated in Fig. 11. The PMSM is fed by a frequency converter that is controlled by a dSPACE DS1103 PPC/DSP board. Mechanical load is provided by a PMSM servo drive. An incremental encoder is used for monitoring the actual rotor speed and position. The nominal DC-link voltage is 540 V, and the switching frequency and the sampling frequency are both 5 kHz. The dc-link voltage of the converter is measured, and a simple current feedforward

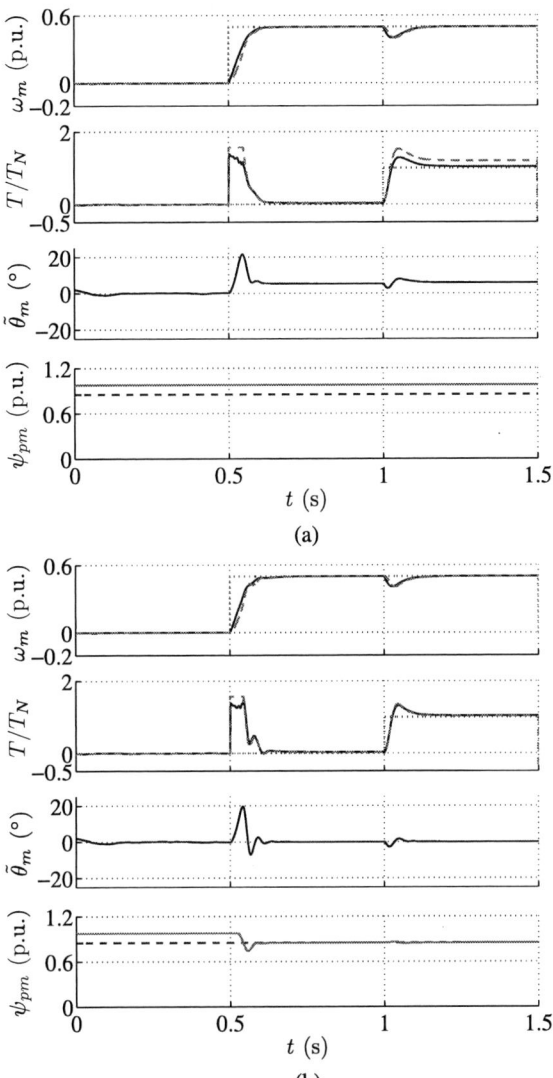

Fig. 10. Simulation results showing PM flux adaptation: (a) without parameter adaptation (b) with parameter adaptation. Explanations of the curves are as in Fig. 9 except that the last subplot shows the PM flux (dashed) and its estimate (solid).

compensation for dead times and power device voltage drops is applied.

Results obtained in low-speed operation are depicted in Figs. 12 and 13, showing the behavior of the stator resistance adaptation. Additional 1-Ω resistors were added between the frequency converter and the PMSM as shown in Fig. 11. The resistance was changed stepwise by opening or closing a manually-operated three-phase switch connected in parallel with the resistors. The experiment in Fig. 12 corresponds to the simulation in Fig. 9. The error in the stator resistance is decreased after the load torque step at $t = 1$ s. The estimated stator resistance follows well the actual resistance after the stepwise increase at $t = 2$ s.

In the experiment of Fig. 13, the drive was operating at very low speed ($\omega_m = -0.05$ p.u.) in the regenerating mode. The load torque was at the positive nominal value. The stator

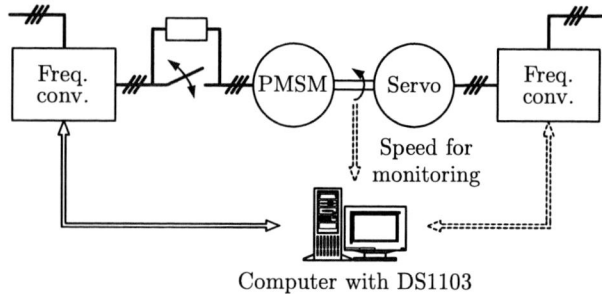

Fig. 11. Experimental setup. Mechanical load is provided by a servo drive.

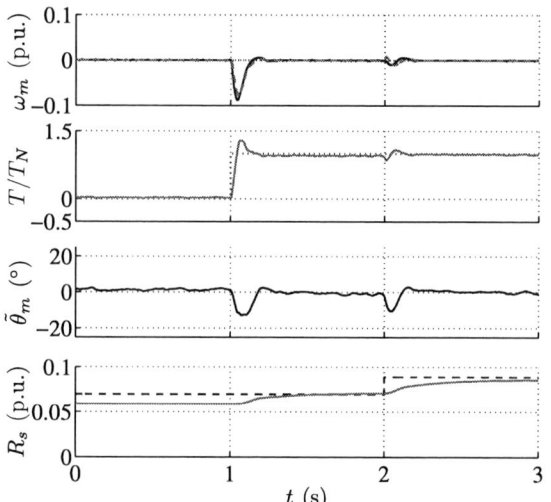

Fig. 12. Experimental results showing stator resistance adaptation. First subplot shows the electrical angular speed of the rotor (solid), its estimate (dashed), and its reference (dotted). Second subplot shows the load torque reference (dashed) and the electromagnetic torque estimate (solid). Third subplot shows the estimation error of the rotor position. Last sublot shows the stator resistance (dashed) and its estimate (solid).

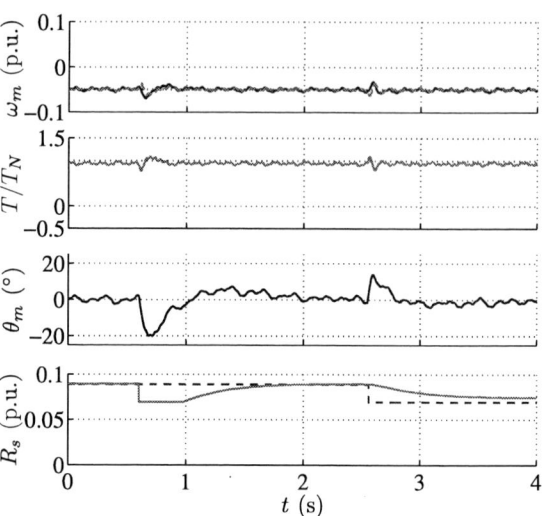

Fig. 13. Experimental results showing stator resistance adaptation. Explanations of the curves are as in Fig. 12.

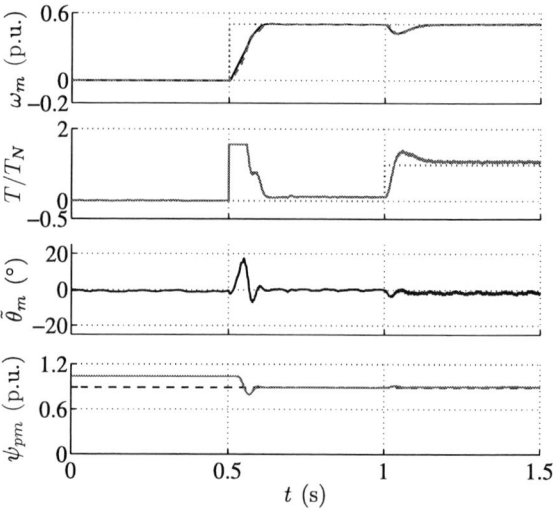

Fig. 14. Experimental results showing PM flux adaptation. Explanations of the curves are as in Fig. 12 except that the last subplot shows the PM flux (dashed) and its estimate (solid).

resistance estimate was forced to an incorrect value at $t \approx 0.6$ s. When the resistance adaptation is activated again at $t \approx 1$ s, the estimated resistance returns close to the actual resistance in about 1 second. After the stepwise decrease in the resistance at $t \approx 2.5$ s, the estimated resistance settles close to the new value. It is to be noted that in the experiments in Figs. 12 and 13, the inverter unidealities contribute to the resistance seen by the controller. Therefore, the estimated resistance is not precisely the stator resistance shown in the figures.

Figs. 14 and 15 show results in the medium-speed operation, where the PM flux adaptation is in use. The experiment in Fig. 14 corresponds to the simulation in Fig. 10. The results are comparable to that of the simulation, and the initial 15 % PM flux error is rapidly removed after the acceleration at $t = 0.5$ s. The PM flux estimate also stays close to the actual flux when the load torque step is applied. Fig. 15 shows results in constant-speed operation, the rotor speed being 0.5 p.u. In Fig. 15(a), the load torque was at the positive nominal value, while in Fig. 15(b) the load torque was at the negative nominal value. The drive operated thus in the motoring and in the regenerating modes in Figs. 15(a) and

15(b), respectively. The PM flux estimate was forced to an erroneous value at $t \approx 0.5$ s, and the adaptation was enabled again at $t \approx 1$ s. The erroneous PM flux estimate causes an error in the electromagnetic torque estimate, and the position estimation error also impairs the performance of the drive. After $t \approx 1$ s, the estimated PM flux quickly converges close to its actual value, leading to a reduced position estimation error and improved torque estimation accuracy.

VII. CONCLUSIONS

This paper proposed a method for the estimation of the stator resistance and the PM flux in a sensorless PMSM drive. The adaptive observer augmented with an HF signal injection technique at low speeds was used for the adaptation of the parameters in addition to the speed and position estimation. The system was investigated by means of steady-state and

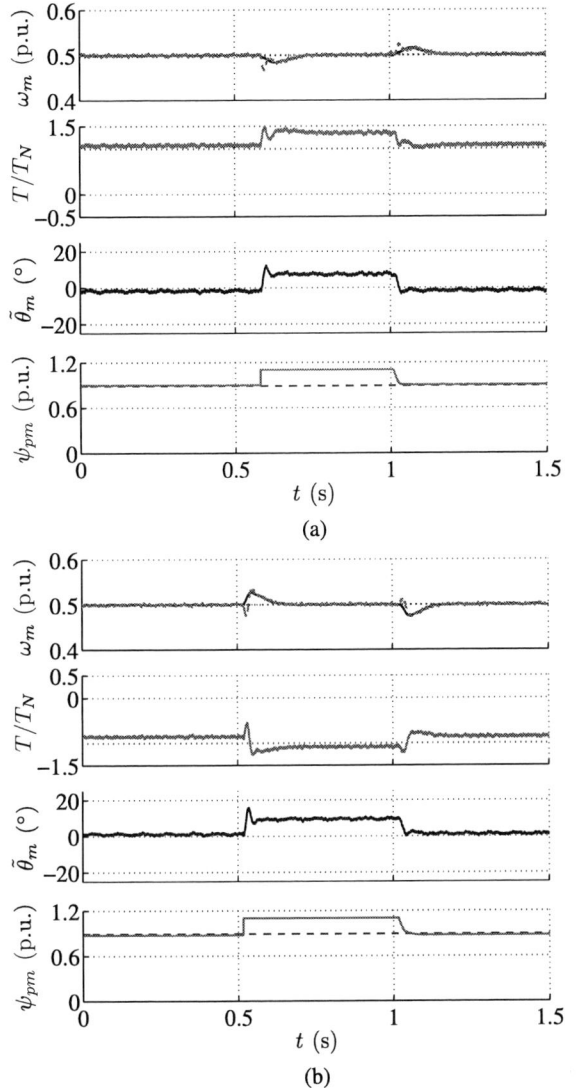

Fig. 15. Experimental results showing PM flux adaptation. Explanations of the curves are as in Fig. 14.

small-signal analyses in order to develop reasonable adaptation algorithms. The simulation and experimental results show that the stator resistance adaptation reduces the resistance error significantly. Even though the high-frequency signal injection removes the position estimation error in steady state even without any resistance adaptation, the good accuracy of the resistance estimate reduces estimation errors when the signal injection is not in use. The decreased speed correction term due to the resistance adaptation also improves the dynamic performance of the combined observer. The PM flux adaptation reduces the position estimation error at medium and high speeds and improves the electromagnetic torque estimation accuracy. The parameter estimates converge rapidly close to the actual parameters, and the the sensitivity to the parameter variations is reduced.

ACKNOWLEDGEMENT

The authors gratefully acknowledge the financial support given by ABB Oy, Walter Ahlström foundation, and KAUTE foundation.

REFERENCES

[1] R. Wu and G. R. Slemon, "A permanent magnet motor drive without a shaft sensor," *IEEE Trans. Ind. Applicat.*, vol. 27, no. 5, pp. 1005–1011, Sept./Oct. 1991.

[2] R. B. Sepe and J. H. Lang, "Real-time observer-based (adaptive) control of a permanent-magnet synchronous motor without mechanical sensors," *IEEE Trans. Ind. Applicat.*, vol. 28, no. 6, pp. 1345–1352, Nov./Dec 1992.

[3] M. Schroedl, "Sensorless control of AC machines at low speed and standstill based on the INFORM method," in *Conf. Rec. IEEE-IAS Annu. Meeting*, vol. 1, San Diego, CA, Oct. 1996, pp. 270–277.

[4] P. L. Jansen and R. D. Lorenz, "Transducerless position and velocity estimation in induction and salient AC machines," *IEEE Trans. Ind. Applicat.*, vol. 31, no. 2, pp. 240–247, March/April 1995.

[5] M. Tursini, R. Petrella, and F. Parasiliti, "Sensorless control of an IPM synchronous motor for city-scooter applications," in *Conf. Rec. IEEE-IAS Annu. Meeting*, vol. 3, Salt Lake City, UT, Oct. 2003, pp. 1472–1479.

[6] A. Piippo, M. Hinkkanen, and J. Luomi, "Sensorless control of PMSM drives using a combination of voltage model and HF signal injection," in *Conf. Rec. IEEE-IAS Annu. Meeting*, vol. 2, Seattle, WA, Oct. 2004, pp. 964–970.

[7] K.-H. Kim, S.-K. Chung, G.-W. Moon, I.-C. Baik, and M.-J. Youn, "Parameter estimation and control for permanent magnet synchronous motor drive using model reference adaptive technique," in *Proc. IEEE IECON'95*, vol. 1, Orlando, FL, Nov. 1995, pp. 387–392.

[8] T. Sebastian, "Temperature effects on torque production and efficiency of PM motors using NdFeB magnets," *IEEE Trans. Ind. Applicat.*, vol. 31, no. 2, pp. 353–357, Mar./Apr. 1995.

[9] R. Krishnan and P. Vijayraghavan, "Fast estimation and compensation of rotor flux linkage in permanent magnet synchronous machines," in *Proc. IEEE ISIE'99*, vol. 2, Bled, Slovenia, July 1999, pp. 661–666.

[10] M. Elbuluk, L. Tong, and I. Husain, "Neural-network-based model reference adaptive systems for high-performance motor drives and motion controls," *IEEE Trans. Ind. Applicat.*, vol. 38, no. 3, pp. 879–886, May/June 2002.

[11] P. Niazi and H. Toliyat, "On-line parameter estimation of permanent magnet assisted synchronous reluctance motor drives," in *Proc. IEEE IEMDC'05*, San Antonio, TX, May 2005, pp. 1031–1036.

[12] S. Wilson, G. Jewell, and P. Stewart, "Resistance estimation for temperature determination in PMSMs through signal injection," in *Proc. IEEE IEMDC'05*, San Antonio, TX, May 2005, pp. 735–740.

[13] X. Xi, Z. Meng, L. Yongdong, and L. Min, "On-line estimation of permanent magnet flux linkage ripple for PMSM based on a kalman filter," in *Proc. IEEE IECON'06*, Paris, France, Nov. 2006, pp. 1171–1175.

[14] M. Eskola and H. Tuusa, "Comparison of MRAS and novel simple method for position estimation in PMSM drives," in *Proc. IEEE PESC'03*, vol. 2, Acapulco, Mexico, June 2003, pp. 550– 555.

[15] K.-W. Lee, D.-H. Jung, and I.-J. Ha, "An online identification method for both stator resistance and back-emf coefficient of PMSMs without rotational transducers," *IEEE Trans. Ind. Electron.*, vol. 51, no. 2, pp. 507–510, Apr. 2004.

[16] S. Ichikawa, M. Tomita, S. Doki, and S. Okuma, "Sensorless control of permanent-magnet synchronous motors using online parameter identification based on system identification theory," *IEEE Trans. Ind. Electron.*, vol. 53, no. 2, pp. 363–372, Apr. 2006.

[17] A. Piippo, M. Hinkkanen, and J. Luomi, "Analysis of an adaptive observer for sensorless control of PMSM drives," in *Proc. IEEE IECON'05*, Raleigh, NC, Nov. 2005, pp. 1474–1479.

[18] M. Corley and R. D. Lorenz, "Rotor position and velocity estimation for a salient-pole permanent magnet synchronous machine at standstill and high speeds," *IEEE Trans. Ind. Applicat.*, vol. 43, no. 4, pp. 784–789, July/Aug. 1998.

413

Development of 150000 r/min, 1.5 kW Permanent-Magnet Motor for Automotive Supercharger

Toshihiko Noguchi *, *IEEE Senior Member*, and Masaru Kano *

* Nagaoka University of Technology
Address: 1603-1 Kamitomioka, Nagaoka, Niigata 940-2188, Japan
Phone: +81-258-47-9510, Fax: +81-258-47-9500
e-mail: tnoguchi@vos.nagaokaut.ac.jp

Abstract—**This paper discusses an optimum design of an ultra high-speed permanent-magnet synchronous motor (PMSM), which is applied to a supercharger of an automotive engine. Although the motor is driven by an inverter with a 12-V DC bus voltage due to an automotive power source, it achieves the maximum rotating speed of 150000 r/min and the rated output of 1.5 kW. Since the power source strictly restricts the motor terminal voltages and the fundamental operating frequency is as high as 2500 Hz, it is significant to pursue further reduction of the synchronous impedance in the motor, paying attention to its permeance coefficient. In the paper, a FEM-based electromagnetic field analysis is conducted, followed by a theoretical discussion on the optimum machine design. In addition, the mechanical structure is discussed to produce a real machine. The developed prototype has a variety of unique features from electrical and mechanical viewpoints, and some experimental test results are presented to demonstrate its potential.**

Index Terms— **ultra high-speed, permanent-magnet synchronous motor, supercharger, electromagnetic field analysis, and permeance coefficient.**

I. INTRODUCTION

Superchargers are often used on automotive combustion engines to enhance the engine output power and to reduce the physical engine size at the same time. Not only the performance improvement in a power-weight ratio of the engines but also various advantages over common power plants can be expected, such as a quality improvement of the exhaust gas, a fuel-efficiency improvement with respect to the output power, and a quick response of the engine torque. A conventional supercharger has a mechanical linkage with the engine, which uses a timing belt from the crankshaft via overdrive devices, and it compresses inlet air to the engine cylinders by means of the mechanical power provided by the engine. Fig. 1 illustrates a mechanical configuration of the traditional supercharging system. As shown here, many of such conventional superchargers employ a positive displacement compressor rather than a centrifugal compressor because of limitation of the operating speed due to the low engine-rotation. Low-efficiency and low boost-pressure are, however, major drawbacks of the positive displacement compressor, which prevents further performance improvement of the supercharged engines.

On the other hand, an electrical drive of the supercharger is a very promising approach as a next-generation axulially machine in future automobiles. Fig. 2 depicts an outline of the investigated supercharging system for the automotive combustion engines. Since the electrically driven supercharger allows to employ the centrifugal compressor instead of the positive displacement compressor, highly efficient operation can be achieved, e.g., higher operating speed over 100000 r/min, higher boost-pressure, smaller mechanical dimensions, faster response of the inlet air compression, etc. Furthermore, there is another advantage of the electric supercharger in a mechanical-linkage-free structure, which reduces overall mechanical losses and eliminates complicated link mechanism.

This paper discusses an optimum design of an ultra high-speed permanent-magnet synchronous motor (PMSM), which is specifically applied to the electric supercharger of the automotive engines. Although the PMSM is operated by an inverter with a 12-V DC bus voltage for an adjustable-speed drive, it must attain the maximum rotating speed of 150000 r/min and the rated output power of 1.5 kW for electrification of the centrifugal-compressor-based supercharger. There are various difficult technical problems in the machine

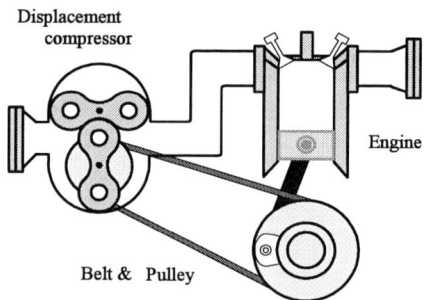

Fig. 1. System configuration of conventional supercharging system.

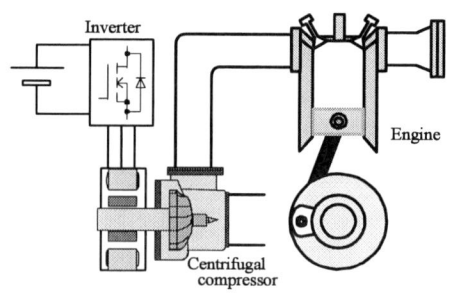

Fig. 2. System configuration of electric supercharging system.

TABLE I. Target specifications of ultra high-speed motor.

Assumed engine	1.5 L class
Rated output power	1.5 kW
Rated speed	150000 r/min
Rated torque	0.0955 Nm
Overload capacity and duration	200 %, 1 s

TABLE II. Electrical and mechanical design parameters of ultra high-speed motor.

Rated voltage	2.96 V/phase
Rated current	195 A
Number of phases	3 phase
Number of poles	2 poles
Stator configuration	Concentrated winding structure
Winding configuration	1 turn, 2 parallels per phase
Number of stator slots	6 slots
Stator outer diameter	92 mm
Stator inner diameter	28 mm
Stator stack length	30 mm
Stator tooth width	10 mm
Electromagnetic steel plates	10JNEX900 (0.1-mm thick, 6.5-% silicone, μ_s=23, B_{max}=1.8 T)
Rotor shaft diameter	12 mm
Permanent magnet	N-42SH Nd-Fe-B (Br=1.28 T, bHc=955 kA/m, BH_{max}=310 kJ/m^3)
Bearings	Angular ceramic-ball bearing with grease lubrication

design, e.g., drastic reduction of the synchronous impedance, minimization of the iron and the copper losses, further improvement of the power-weight ratio, etc., which should simultaneously be solved with compromise to some extent. In addition, these electrical design requirements must be satisfied all together with a compact and robust mechanical structure design. In the design process, a finite-element-method (FEM)-based electromagnetic field analysis is conducted to make the fine adjustment of the detailed machine shape and to seek the best design parameter set, focusing on a permeance coefficient of the PMSM. The mechanical structure is also discussed to create a real prototype machine, which is experimentally examined to confirm the basic performance as a first step of the system development.

II. REQUIREMENTS AND DESIGN SPECIFICATIONS FOR ULTRA HIGH-SPEED MOTOR

A. Requirements for Ultra High-Speed Motor

Assuming that the centrifugal compressor is employed for a supercharger mounted on a 1.5-litter class gasoline engine, the ultra high-speed motor is required to output the mechanical power from 1.5 to 2 kW at 150000 r/min or higher. Particularly, when boosting the inlet air compression, extremely fast response is indispensable to accelerate the compressor from several thousand r/min to the rated operating speed of 150000 r/min within approximately 0.5 s, which is almost comparable with a response time of the conventional supercharger. In order to meet with this requirement, the ultra high-speed motor

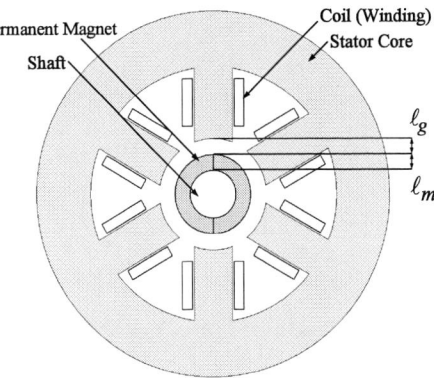

Fig. 3. Cross section diagram of ultra high-speed SPMSM.

TABLE III. Design conditions of five different permeance coefficients.

Design type	#1	#2	#3	#4	#5
Air gap length ℓ_g (mm)	6	5	4	3	2
Permanent magnet thickness ℓ_m (mm)	2	3	4	5	6
Permeance coefficient	0.33	0.6	1	1.67	3

must have an overload capacity double of the rated output power at least for one second. Taking these requirements for the motor into account, the target specifications are determined as listed in TABLE I.

B. Basic Design Concept of Ultra High-Speed Motor

On the basis of the target specifications, detailed electrical and mechanical design parameters must be investigated prior to the numerical analysis for the best solution. In order to achieve the highest efficiency among various electric machines, a two-pole three-phase surface permanent-magnet synchronous motor (SPMSM) is selected as the best rotating machine for the ultra high-speed motor drive because of no magnetizing current required unlike an induction motor or a reluctance motor. A strong Nd-Fe-B permanent magnet of the rotor allows not only improvement of the motor efficiency but also drastic size reduction of the rotor to restrict its circumference speed and to mitigate the centrifugal force effect, which is of vital importance in the ultra high-speed motor. Using such powerful Nd-Fe-B permanent magnet that has 310-kJ/m^3 BH_{max} makes a wide-air-gap design possible to reduce the synchronous reactance and to obtain a sinusoidal electromotive force (e.m.f.) regardless of the concentrated stator winding structure.

On the other hand, the stator has a six-teeth and six-slot structure and the concentrated windings because the motor is required to reduce the leakage inductance thoroughly as well as the synchronous reactance. It should be noted that each phase has a pair of single-turn windings in parallel. The stator core must be designed to have as low iron-losses as possible even at 150000 r/min. Therefore, high-performance 6.5-% silicone electromagnetic steel plates of which thickness is only 0.1 mm are employed to compose the laminated stator iron core. TABLE II is a summary of the basic conceptual design for the ultra high-speed SPMSM to be investigated in the paper.

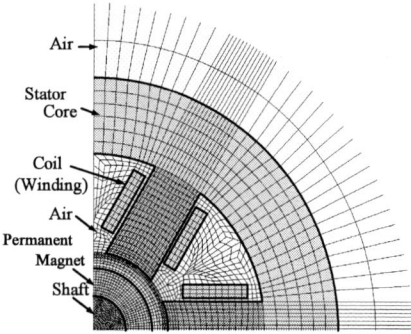

Fig. 4. Generated mesh for FEM analysis only in quarter portion.

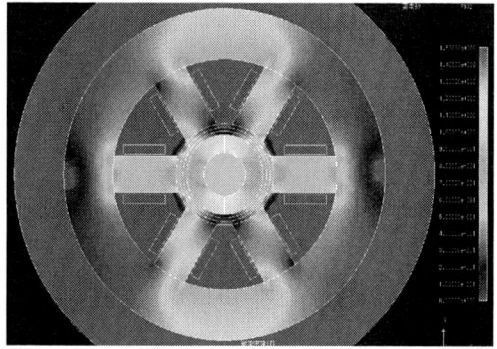

Fig. 5. Example of flux density distribution at rated operation.

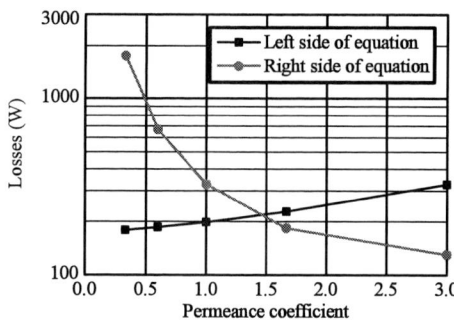

Fig. 6. Loss characteristics with respect to permeance coefficient.

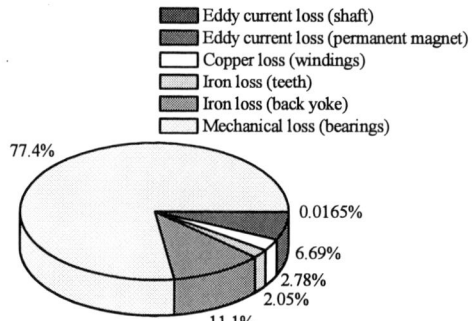

Fig. 7. Numerically calculated losses of design type #4.

III. THEORETICAL DISCUSSION AND ELECTROMAGNETIC FIELD ANALYSIS FOR OPTIMUM MOTOR DESIGN

A. Theoretical Discussion for Optimum Motor Design

It is well known that the operation characteristics of the SPMSM dominantly depend on the permeance coefficient because it determines an operating point of the permanent magnet on its B-H curve. Assuming that the investigated SPMSM has uniform permeance distribution along the air gap, the permeance coefficient p_u of the SPMSM is expressed by the following equation:

$$p_u = \frac{\ell_m}{a_m} \frac{a_g}{K_C \ell_g} = \frac{\ell_m}{D_m - \ell_m} \frac{D_m + \ell_g}{K_C \ell_g}, \quad (1)$$

where ℓ_m is a permanent-magnet thickness, a_m is an averaged cross sectional area of the permanent magnet, a_g is an averaged cross sectional area of the air gap between the rotor and the stator, ℓ_g is an air gap length, D_m is an outer diameter of the permanent magnet, and K_C is a Carter's coefficient. Since K_C normally takes a value of approximately 1.2 to 1.5, a_g can be regarded as almost same as $a_m K_C$; thus, the following approximated expression is obtained:

$$p_u \approx \frac{\ell_m}{\ell_g}. \quad (2)$$

This equation indicates that the permeance coefficient is determined by the ratio between ℓ_m and ℓ_g as illustrated in Fig. 3. The permeance coefficient p_u is basically proportional to the e.m.f. if other physical dimensions of the investigated motor are not changed.

On the other hand, in order to improve the total efficiency of the motor drive system including the inverter, the inverter loss must be taken into account. The maximum efficiency condition of the motor drive system can be expressed as

$$W_i^{st} + W_e^{mag} + W_m = W_c + W_{Ron}, \quad (3)$$

where W_i^{st} is an iron loss of the stator core, W_e^{mag} is an eddy-current loss of the permanent magnet, W_m is a mechanical loss (bearing friction loss), W_c is a copper loss of the stator windings, and W_{Ron} is a conduction loss of the inverter. The left-hand side terms of (3) are not contingent upon the line currents of the motor, while the right-hand side terms are almost proportional to square of the currents. There is a switching loss as well as the conduction loss in the inverter, but the switching loss does not affect the maximum efficiency condition indicated by (3) because it is proportional to the line currents.

B. FEM-Based Electromagnetic Analysis and Results

TABLE III shows analytical conditions of the ultra high-speed SPMSM, where five combinations of the permeance coefficient, i.e., the permanent-magnet thickness ℓ_m and the air gap length ℓ_g, are investigated. Figs. 4 and 5 are a quarter portion of the generated mesh (7692 elements and 7909 nodes in the whole mesh) and an example of the resultant magnetic-flux density distribution, respectively. As shown in this example, the mean value of the flux density is approximately 0.45 T in the air gap, while significant magnetic saturation is hardly observed in the iron core. Fig. 6 shows the loss analysis result, where the left-hand side terms and the right-hand side terms in (3) are separately drawn with respect to p_u. It can be found from this figure that (3) is satisfied at $p_u = 1.5$, i.e., the total loss is minimized at this

(a) Stator core made of 0.1-mm thick, 6.5-% silicone electromagnetic steel plates and single-turn stator windings.

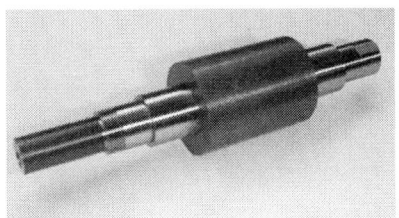

(b) Nd-Fe-B permanent-magnet rotor before mechanical reinforcement using carbon fiber.
Fig. 8. Photographs of stator and rotor.

Fig. 9. Three-dimensional computer graphic of motor assembly.

condition. Therefore, the design type of #4 (ℓ_m = 5 mm and ℓ_g = 3 mm), which has the nearest permeance coefficient to p_u = 1.5, is introduced to the prototype machine production. Fig. 7 represents numerically calculated losses of the design type #4. As can be seen in the figure, the mechanical friction loss is dominant, which accounts for 77.4 %, while the stator iron loss and the eddy-current loss on the permanent magnet are merely 19.8 % of the total loss. The electrical efficiency except for the mechanical loss is approximately 95 %.

IV. ELECTRICAL AND MECHANICAL STRUCTURE OF PROTOTYPE MACHINE

As described in the basic design concept of the ultra high-speed motor, the prototype has a special electrical and mechanical structure. Fig. 8 shows photographs of the stator and the rotor. The laminated stator core consists of approximately 300 sheets of 6.5-% silicone electromagnetic steel plates, of which outer diameter is 92 mm, inner diameter is 28 mm, and axially stack length

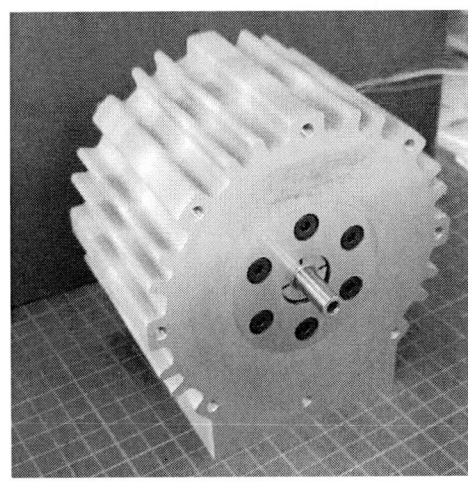

(a) Front view (load side) of prototype machine.

(b) Rear view (opposite side of load) of prototype machine.
Fig. 10. Exterior photographs of prototype machine.

is 30 mm. An alphabetically "b"-shaped stator-winding bar is inserted in each stator tooth with keeping electrical insulation from the stator core by polyimide taping, and is connected to a neutral-point end ring together with the other stator windings. Each of the stator windings has cross sectional area of 31.2 mm², resulting in current density of approximately 6.25 A/mm² at rated load. Every clearance between the tooth and the stator winding is less than 0.5 mm, which effectively improves the magnetic coupling by reducing the leakage inductance. On the other hand, the rotor is simply assembled with a steel shaft and the ring-shaped Nd-Fe-B permanent magnet, and is magnetized so that the flux distribution becomes sinusoidal. The photograph of the rotor shows exterior of the permanent magnet before mechanical reinforcement with carbon fiber. After assembling the rotor, 1.5-mm thick layer of the carbon fiber is formed with special epoxy resin on the permanent magnet surface against large centrifugal force. Fig. 9 illustrates a three-dimensional computer graphic (a bird's-eye view) of the prototype motor assembly. All of the metal components are made with highly precise NC machining tools. Particularly, bearings are the most important parts to realize the ultra high-speed operation. In the real machine, a pair of super-precise angular ceramic ball bearings is used because the prototype is just

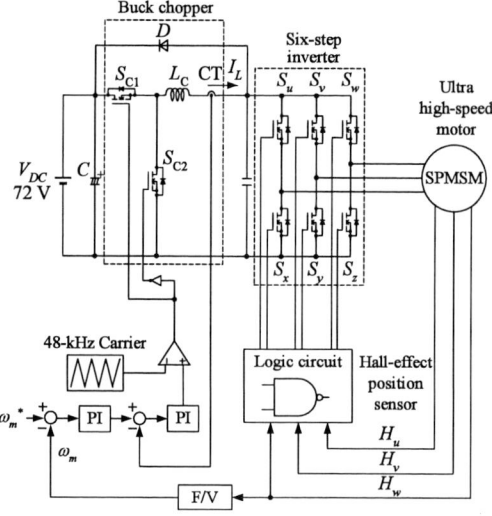

Fig. 11. Schematic diagram of pseudo current-source inverter drive.

TABLE IV. Measurement result of motor parameters.

Motor parameters	Designed value	Measured value
E.m.f constant (10^{-5} V/r/min)	1.96	1.89
R_a (mΩ)	0.054	0.072
L_a (μH)	0.056	0.07

experimentally examined and requires neither high reliability nor mechanical endurance in the laboratory tests. However, the prototype is designed and created to have extremely high accuracy of μm-order, especially in the bearing and shaft part. Fig. 10 shows a front view and a rear view of the assembled prototype motor.

V. EXPERIMENTAL SYSTEM AND TEST RESULTS

A prototype motor was tested to confirm several basic operation characteristics with an experimental setup illustrated in Fig. 11. Since the fundamental operating frequency of the test motor is over 2 kHz, a pseudo current-source inverter was employed to drive the motor instead of a common voltage-source PWM inverter. The pseudo current-source inverter features a current-controlled buck chopper part and a six-step inverter part. The former has a DC bus current feedback loop and controls the DC bus current in accordance with a torque command coming from a speed control loop. Therefore, it can be regarded as a regulated current-source, which achieves PAM of the DC bus current. This approach is superior to ordinary thyristor-based power converters because it operates at the switching frequency of 48 kHz, resulting in drastic reduction of the DC bus inductance and in millisecond-order fast response of the DC bus current control. On the other hand, the six-step inverter commutates the DC bus current, and generates 120-deg conduction patterns of the line currents. Every time the current is commutated, the synchronous inductance of the motor and the line inductance generate surge voltages across the inverter terminals. In the case of the pseudo current source inverter, however, the DC bus voltage clamps the surge voltages via body-diodes in the inverter MOSFETs and a bypass diode of the buck chopper. The

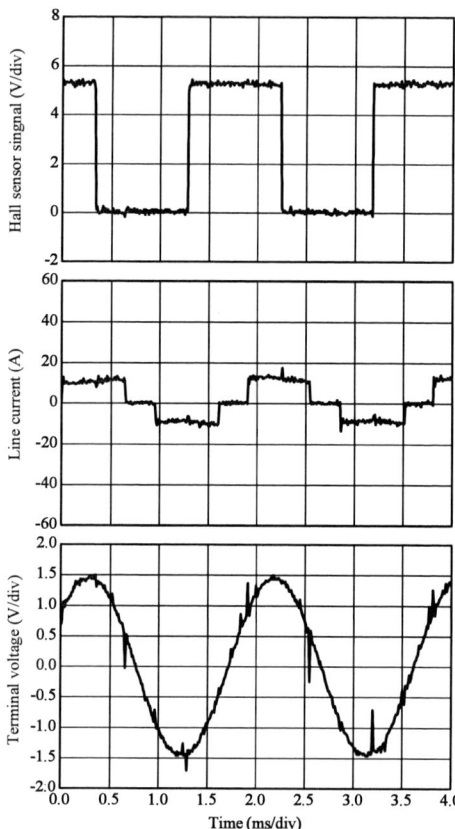

Fig. 12. Operating waveforms at 33000 r/min under no-load condition.

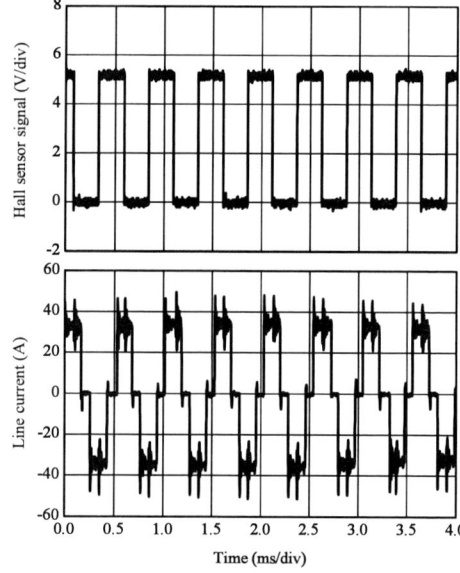

Fig. 13. Operating waveforms at 120000 r/min under no-load condition.

current commutation is controlled by a logic circuit where three-phase Hall-effect sensor signals generate the 120-deg conduction patterns.

TABLE IV indicates comparison between the designed and the experimentally measured motor parameters. As listed here, both of the winding resistance and the inductance of the real machine are slightly higher than the designed values. It is inferred that the higher resistance is due to contact resistance between the winding bars and

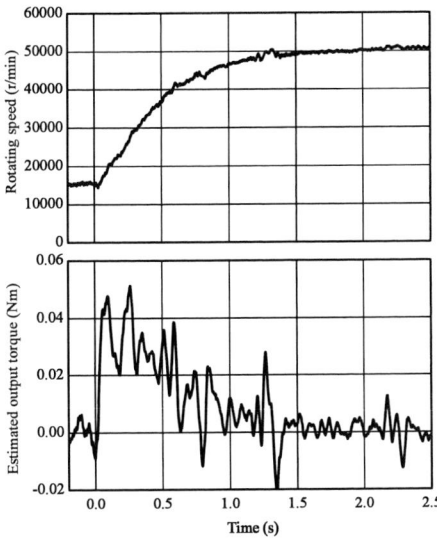

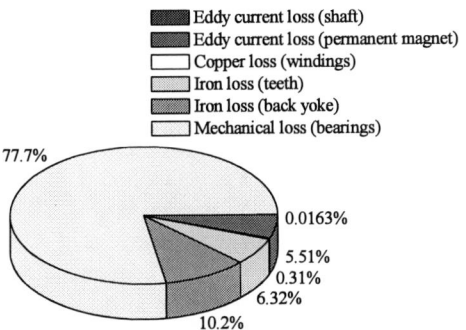

Fig. 14. Speed step response from 15000 to 50000 r/min and experimentally estimated output torque.

Eddy current loss (shaft)
Eddy current loss (permanent magnet)
Copper loss (windings)
Iron loss (teeth)
Iron loss (back yoke)
Mechanical loss (bearings)

77.7%

0.0163%
5.51%
0.31%
6.32%
10.2%

Fig. 15. Experimentally estimated loss-analysis result.

the neutral-point end ring and that the higher inductance includes line inductance and the leakage inductance.

Fig. 12 shows waveforms of the Hall-effect position sensor signal, a line current and a motor terminal voltage, which were measured at approximately 33000 r/min in a test run. The current has low amplitude of approximately 10 A because of no-load operation, and its 120-deg conduction pattern is generated synchronously with the output signal of the Hall-effect position sensor. Since the motor terminal voltage is sinusoidal without conspicuous harmonic distortion, which means the e.m.f. is properly generated by the permanent magnet on the rotor in spite of the concentrated winding structure of the stator. Fig. 13 demonstrates operating waveforms of the prototype at 120000 r/min under no-load condition. As can be seen in the figure, an excellent current waveform with 120-deg conduction pattern is properly obtained, which is in phase with the Hall-effect position sensor signal.

In addition, an acceleration test was conducted to check the output-torque controllability. In general, it is rather difficult to measure the mechanical output at such an ultra high-speed as 150000 r/min, so the output torque of the test motor was estimated by acceleration in the speed step response waveform and a design value of the rotor inertia. From Fig. 14, it is inferred that the maximum output torque in the acceleration from 15000 to

50000 r/min is 0.05 Nm, which is half of the rated value. Although the output torque waveform is quite choppy as indicated in Fig. 14, it can be found from the waveform envelope that the PI regulator in the speed loop linearly controls the torque.

Fig. 15 shows an experimentally estimated loss analysis result. There can be seen a good agreement with the numerically analyzed result shown in Fig. 7 in the overall loss percentage, where the mechanical friction loss caused by the bearings is dominant and the stator iron losses and the eddy-current loss of the permanent magnet account for 22.0 % of the total loss.

VI. CONCLUSION

This paper described development of an ultra high-speed SPMSM, of which ratings are 150000 r/min and 1.5 kW. The prototype has various unique features from electrical and mechanical viewpoints to realize the ultra high-speed drive fed by a 12-V DC power source like batteries. A 120000-r/min operation under no-load condition and speed step response to 50000 r/min were examined, and proper operation characteristics were confirmed through some experimental tests. Furthermore, experimentally estimated loss analysis result agreed very well with the FEM-based electromagnetic field analysis result. Consequently, the dominant loss factor was found to be a mechanical loss dissipated by the bearings.

REFERENCES

[1] M. Okawa, "Design Manual of Magnetic Circuit and PM Motor," *Sogo Research*, 1989 (in Japanese).

[2] T. Koganezawa, I. Takahashi, and K. Ohyama, "Sensorless Speed Control of a PM Motor by a Quasi-Current Source Inverter," *IEE-Japan Proc. of Ind. App. Soc. Ann. Conf.*, p. 175, 1992 (in Japanese).

[3] I. Takahashi, T. Koganezawa, T. Su G., and K. Ohyama, "A Super High Speed PM Motor Drive System by a Quasi-Current Source Inverter." *IEEE Trans. on Ind. App.*, vol. 30, no. 3, p.p. 683-690, 1994.

[4] B. –H. Bae, and S. –K. Sul, "A Compensation Method for Time Delay of Full-Digital Synchronous Frame Current Regulator of PWM AC Drives," *IEEE Trans. on Ind. App.*, vol. 39, no. 3, p.p. 802-810, 2003.

[5] B. –H. Bae, S. –K. Sul, J. –H. Kwon, and J. –S. Byeon, "Implementation of Sensorless Vector Control for Super-High-Speed PMSM of Turbo-Compressor," *IEEE Trans. on Ind. App.*, vol. 39, no. 3, p.p. 811-818, 2003.

[6] T. Noguchi, Y. Takata, Y. Yamashita, Y. Komatsu, and S. Ibaraki, "220000r/min, 2-kW Permanent Magnet Motor Drive for Turbocharger", *IEE-Japan Int. Power Elec. Conf. (IPEC) -Niigata*, p.p. 2280-2285, 2005.

[7] T. Noguchi, Y. Takata, Y. Yamashita, Y. Komatsu, and S. Ibaraki, "220000r/min, 2-kW PM Motor Drive for Turbocharger", *IEE-Japan Trans. on Ind. App.*, vol. 125, no. 9, p.p. 854-861, 2005 (in Japanese).

[8] T. Noguchi, Y. Takata, Y. Yamashita, and S. Ibaraki, "160,000-r/min 2.7-kW Electric Drive of Supercharger for Automobiles." *The Sixth Int. Conf. on Power Elec. and Drive Sys. (PEDS) –Kuala Lumpur*, p.p. 1380-1385, 2005.

[9] C. Zwyssig, M. Duerr, D. Hassler, and J. W. Kolar, "An Ultra-High-Speed, 500000 rpm, 1 kW Electrical Drive System," *The Fourth Power Conv. Conf. (PCC) –Nagoya*, CDROM, 2007.

Analysis and Performance Evaluation of Radial Flux Air-Cored Permanent Magnet Machines with Concentrated Coils

P.J. Randewijk, M.J. Kamper and R-J. Wang

Department of Electrical & Electronic Engineering, Stellenbosch University, Stellenbosch, South Africa

Abstract—In this paper two different concentrated coil configurations for Radial Flux Air-cored Permanent Magnet (RFAPM) machines are analysed and evaluated against a RFAPM machine that utilises overlapping coils. The comparative analysis of the different machines are done analytically. The analytical results are compared with finite element analysis results of simplified linearised equivalent models for the different coil configurations. The finite element analysis results correlate well with the analytical results in predicting a higher torque factor for small diameter, short stack RFAPM machines with concentrated coil configurations than for machines with overlapping coil configurations, with the same coil side-width values.

Index Terms—air-cored, concentrated coil, permanent magnet, radial fluxtheoretically

LIST OF SYMBOLS

Roman Symbols

a	the number of parallel circuits per phase
B_{p_1}	peak fluxdensity of the first harmonic
C_M	machine constant
E_p	peak sinusoidal phase voltage
h	height/thickness of the stator coils
h_m	magnet height/thickness
h_y	yoke height/thickness
I_p	peak sinusoidal phase current
k_f	fill factor
k_λ	flux-linkage factor
k_q	coil per phase to pole ratio
K_T	torque factor
ℓ	active copper length of the stator conductors
ℓ_e	end turn length of the stator conductors
ℓ_g	air gap length
N	number of turns per coil
p	number of poles
P_{cu}	copper losses
P_e	electrical power
P_m	mechanical power
q	number of coils per phase
Q	total number of coils ($Q = 3q$)
r_n	nominal stator radius
R_{cu}	copper resistance
R_{ph}	total copper resistance per phase
V_{cu}	total copper volume of the stator
w	coil side-width of the stator coils

Greek Symbols

α	relative angle between the rotor and the stator
δ	the angle measured from the centre of the coil side
Δ	$\frac{1}{2}$ coil side-width angle of the stator coils
λ_1	flux-linkage of a single turn
λ	flux-linkage
ρ_{cu}	copper conductance
θ_p	pole pitch angle
θ_q	coil pitch angle
ζ	end turn to active copper length ratio

I. INTRODUCTION

In this paper two different concentrated coil configurations for Radial Flux Air-cored Permanent Magnet (RFAPM) machines are analysed and evaluated against a RFAPM machine that utilises overlapping coils. The construction of RFAPM machines, as shown in Fig. 1, is similar to that of dual rotor radial flux toroidal-wound permanent magnet machines, [1]. The difference are that the stator of the RFAPM consists of air-cored coils, instead of iron-cored toroidal coil, and that the rotor uses a north-south magnet polarity configuration, instead of a north-north configuration. Concentrated coils are also sometimes referred to in the literature as concentrated windings [2].

The main reasons for using concentrated coils above overlapping coils is to reducing the manufacturing costs of the machine and still be able to produce the same amount of torque. Using concentrated coils allow for:

- a simpler coil construction, [3] which could ultimately lead to automated manufacturing of the stator and
- smaller end-turn lengths of the coils implying less copper being used, [2].

Concentrated coil RFAPM machines are intended for low cost, low power, direct drive wind-generator applications.

II. THEORETICAL ANALYSIS

The comparative analysis of the machines under consideration will start with the calculation of the flux-linkage.

It was found that with the carefull sizing and spacing of the pole magnets, the airgap flux density in an axial flux air-cored permanent magnet machine can be made quasi-sinusoidal, [4] and [5]. For the theoretical analysis, it will therefore be assumed that the air gap flux density of the RFAPM machine

978-1-4244-0644-9/07/$25.00 ©2007 IEEE

Fig. 1. A 3D view of a 16 pole RFAPM machine with concentrated coils.

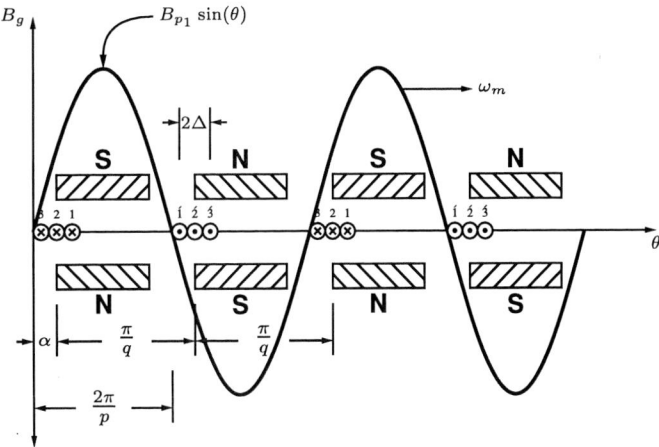

Fig. 3. A 2D linearised cross-sectional view of one phase of a RFAPM machine with overlapping stator coils in relation to the pole positioning and relative flux distribution.

is also sinusoidal and that the peak flux density, B_{p_1}, is the same for all three types of machines under consideration.

A. RFAPM machine with overlapping coils

A three-dimensional view of the typical coil configuration of a RFAPM machine with overlapping stator coils are shown in Fig. 2. A two-dimensional linearised cross-sectional view along the nominal stator radius of only one phase of the overlapping stator coil configuration, with a sinusoidal radial flux density, a coil pitch, θ_q, equal to the pole pitch, $\theta_p = \frac{2\pi}{p}$, a coil position α with respect to the flux density wave and a coil side with of 2Δ can be represented as shown in Fig. 3.

For the analysis we assume that the stator thickness is much smaller than the nominal stator radius, i.e. $h \ll r_N$ allowing us to consider all the turns to be situated on the nominal stator radius. To begin the analysis, we start by looking at a single turn, say 1 and 1' of Fig. 3, as shown in Fig. 4.

The flux-linkage for this turn at position δ inside the coil,

can be calculated as

$$
\begin{aligned}
\lambda_1 &= \int_0^\ell \int_{\alpha+\delta}^{\alpha+\frac{2\pi}{p}-\delta} B_{p_1} \sin\left(\frac{\theta p}{2}\right) r_n d\theta dz \\
&= \frac{4}{p} B_{p_1} \cos\left(\alpha \frac{p}{2}\right) \cos\left(\delta \frac{p}{2}\right) r_n \ell .
\end{aligned}
\tag{1}
$$

The total flux-linkage for N number of turns, can be calculated by integrating with respect to δ across the entire coil side-width (i.e. between $-\Delta$ and Δ), dividing by the coil side-width (i.e. 2Δ) to get the average flux-linkage and multiplying the result by N.

The total flux-linkage for a typical coil with a wide coil side-width can thus be calculated as follows,

$$
\begin{aligned}
\lambda_N &= \frac{N}{2\Delta} \int_{-\Delta}^{\Delta} \frac{4}{p} B_{p_1} \cos\left(\alpha \frac{p}{2}\right) \cos\left(\delta \frac{p}{2}\right) r_n \ell d\delta \\
&= \frac{4}{p} B_{p_1} N \cos\left(\alpha \frac{p}{2}\right) r_n \ell k_\lambda
\end{aligned}
\tag{2}
$$

with k_λ the flux-linkage factor given by

$$
k_\lambda = \frac{\sin\left(\Delta \frac{p}{2}\right)}{\Delta \frac{p}{2}} .
\tag{3}
$$

Fig. 2. A 3D view of the typical coil configuration of a RFAPM machine with overlapping stator coils.

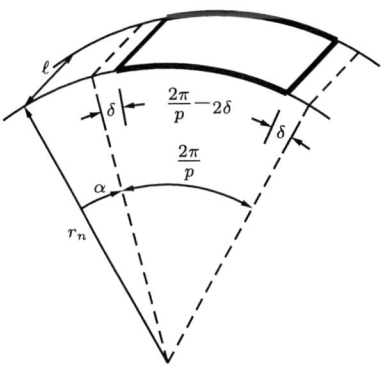

Fig. 4. A single turn of an overlapping stator coil of a RFAPM machine.

The maximum coil side-width will be equal to $\frac{\pi}{Q}$ *mechanical* degrees, i.e.

$$\Delta_{|max} = \frac{\pi}{2Q} \ . \tag{4}$$

The induced voltages per coil can now be calculated as

$$
\begin{aligned}
e_{coil} &= \frac{d}{dt}\lambda_N \\
&= \frac{d}{dt}\tfrac{4}{p}NB_{p_1}\cos\left(\omega_m\tfrac{p}{2}t\right)k_\lambda r_n\ell \\
&= -2\omega_m NB_{p_1}k_\lambda r_n\ell\sin\left(\omega_m\tfrac{p}{2}t\right) \ ,
\end{aligned}
\tag{5}
$$

with $\alpha = \omega_m t$, where ω_m is the mechanical speed at which the rotor is turning and therefore the peak sinusoidal phase voltage can be written as

$$E_p = 2\tfrac{q}{a}\omega_m NB_{p_1}k_\lambda r_n\ell \ . \tag{6}$$

The torque of a rotating electrical machine is given by

$$T_m = \frac{P_m}{\omega_m} \ . \tag{7}$$

By ignoring the mechanical (i.e. windage and friction) losses in the machine, the mechanical power input will be equal to the electrical power output. To aid with the comparative analysis of the different machines under consideration, the phase voltage and the phase current are deemed to be in phase and sinusoidal. The mechanical power of the machine can thus be written as

$$P_m = P_e = \tfrac{3}{2}E_p I_p \ . \tag{8}$$

From the simplified equivalent circuit of an air-cored, permanent magnet generator shown in Fig. 5, the power delivered to the electrical load, P_s, can be calculated as

$$P_s = P_e - P_{cu}$$

with

$$P_{cu} = 3\frac{I_p^{\,2}}{2}R_{ph} \ . \tag{9}$$

The resistance of copper wire, is given by:

$$R_{cu} = \frac{\rho_{cu}\left(2\ell + \ell_e\right)}{A_{cu}} \tag{10}$$

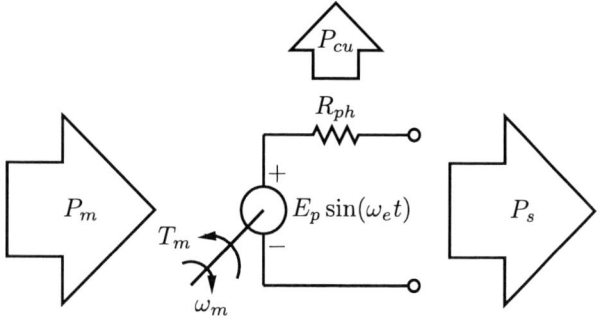

Fig. 5. Power flow and equivalent circuit of an air-cored permanent magnet machine.

For overlapping coils, [3] the end-turn length are given by:

$$\ell_e = 2\left(\frac{\pi}{Q}\right)r_n + 4h \tag{11}$$

The area of the copper wire could be approximated by:

$$A_{cu} = \frac{(h \times w)k_f}{N} \tag{12}$$

Substituting (12) into (10), the total resistance of a coil with N windings, can thus be calculated as

$$R_{cu} = \frac{N^2\rho_{cu}(2\ell + \ell_e)}{k_f hw} \ , \tag{13}$$

or in terms of Δ as

$$R_{cu} = \frac{N^2\rho_{cu}(2\ell + \ell_e)}{k_f h 2 r_n \Delta} \ , \tag{14}$$

using the approximation that

$$\Delta \approx \frac{w}{2r_n} \ . \tag{15}$$

The resistance per phase, with q number of coils per phase and a number of parallel circuits, can thus be calculated as

$$
\begin{aligned}
R_{ph} &= \frac{\frac{q}{a}R_{cu}}{a} \\
&= \frac{N^2 q \rho_{cu}(2\ell + \ell_e)}{a^2 k_f h 2 r_n \Delta} \ .
\end{aligned}
\tag{16}
$$

From (9) and (16):

$$
\begin{aligned}
I_p &= \sqrt{\frac{2P_{cu}}{3R_{ph}}} \\
&= \sqrt{\frac{4P_{cu}a^2 k_f h r_n \Delta}{3N^2 q \rho_{cu}(2\ell + \ell_e)}}
\end{aligned}
\tag{17}
$$

The developed torque in terms of the machine parameters, can be calculated from (7), by substituting (6) and (17) into (8) resulting in

$$
\begin{aligned}
T_m &= \frac{\tfrac{3}{2}E_p I_p}{\omega_m} \\
&= 3\tfrac{q}{a}NB_{p_1}k_\lambda r_n\ell\sqrt{\frac{4P_{cu}a^2 k_f h r_n \Delta}{3N^2 q \rho_{cu}(2\ell + \ell_e)}} \ .
\end{aligned}
\tag{18}
$$

B. RFAPM machine with concentrated Type I coils

The first type of concentrated (or non-overlapping) coil configuration investigated, is where the centres of the coil sides are symmetrically spaced around the circumference of the stator. Three adjacent coils for this (say) Type I concentrated coil configuration are graphically depicted in Fig. 6.

In order to construct a three-phase machine using concentrated coils, three coils, one for each phase, need to be equally spaced over 2, 4, 8, etc. number of poles. In order to obtain the maximum flux-linkage, the average coil width should strive to $\frac{2\pi}{p}$ *mechanical* degrees or π *electrical* degrees. Intuitively, the best results will be obtained with three coils per four poles, as shown in Fig. 7.

Fig. 6. A 3D view of the concentrated Type I stator coil configuration of RFAPM machine.

Analytically, the flux-linkage can be calculated as

$$\lambda_1 = \int_0^\ell \int_{\alpha+\delta}^{\alpha+\frac{\pi}{Q}-\delta} B_{p_1} \sin\left(\theta\frac{p}{2}\right) r_n d\theta dz \qquad (19)$$

and thus for N conductors as

$$\lambda_N = \frac{N}{2\Delta} \int_{\Delta}^{-\Delta} \lambda_1 d\delta . \qquad (20)$$

As the rotor and the stator move relative to one another, the angle α, with $\alpha = \omega_m t$, will change. The flux-linkage and thus the induced voltage will vary sinusoidally for a constant mechanical speed, ω_m. The maximum flux-linkage for the Type I configuration, will occur at

$$\alpha = \frac{1}{2}\left(\frac{2\pi}{p} - \frac{\pi}{Q}\right) . \qquad (21)$$

By substituting (21) into (19), equation (19) simplifies to

$$\lambda_1 = \frac{4}{p} B_{p_1} r_n \ell \sin(\gamma) , \qquad (22)$$

with

$$\gamma = \frac{\pi}{2Q}\frac{p}{2} - \delta\frac{p}{2} . \qquad (23)$$

By substituting (23) into (22) and (22) into (20), equation (20) reduces to

$$\lambda_N = \frac{4}{p} N B_{p1} r_n \ell k_\lambda , \qquad (24)$$

which is similar to equation (2), with only a different the value of k_λ, namely

$$k_\lambda = \frac{\cos\left(\frac{\pi}{2Q}\frac{p}{2} - \Delta\frac{p}{2}\right) - \cos\left(\frac{\pi}{2Q}\frac{p}{2} + \Delta\frac{p}{2}\right)}{2\Delta\frac{p}{2}} . \qquad (25)$$

The analysis will furthermore be exactly the same as for the overlapping coils, except for the end-turn length, which for this Type I configuration, can be calculated as

$$\ell_e = 2\left(\frac{\pi}{Q}\right) r_n . \qquad (26)$$

C. RFAPM machine with concentrated Type II coils

In order to maximising the flux-linkage further, i.e. get the coil widths closer to π *electrical* degrees, the concentrated coils can be spaced such that the coil sides of the adjacent coils touch as shown in Fig. 8 and what will be called a Type II concentrated coil configuration. In Fig. 9 the layout of this Type II configuration with respect to the pole placement is shown for three coils per four poles.

For the Type II configuration, the flux-linkage can be calculated as

$$\lambda_1 = \int_0^\ell \int_{\alpha+\delta}^{\alpha+\frac{2\pi}{Q}-2\Delta-\delta} B_{p_1} \sin\left(\theta\frac{p}{2}\right) r_n d\theta dz , \qquad (27)$$

with the maximum flux-linkage occurring at

$$\alpha = \frac{1}{2}\left[\frac{2\pi}{p} - \left(\frac{2\pi}{Q} - 2\Delta\right)\right] . \qquad (28)$$

By substituting (28) into (27), equation (27) simplifies to

$$\lambda_1 = \frac{4}{p} B_{p_1} r_n \ell \sin(\gamma) , \qquad (29)$$

with

$$\gamma = \frac{\pi}{Q}\frac{p}{2} - \Delta\frac{p}{2} - \delta\frac{p}{2} . \qquad (30)$$

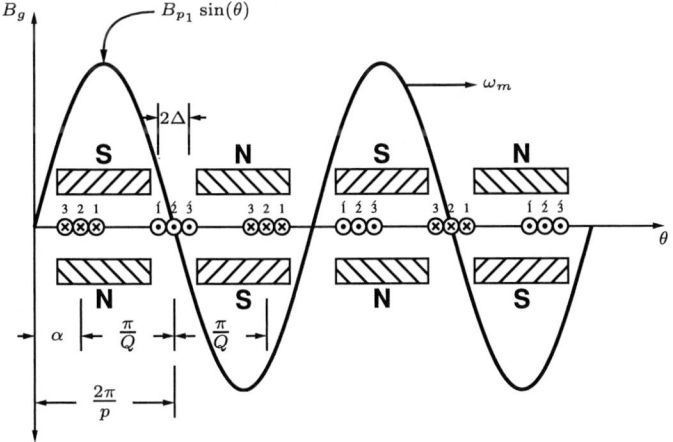

Fig. 7. A 2D linearised cross-sectional view of a RFAPM machine with concentrated Type I stator coils showing all three three-phase coils (with 3 coils per 4 poles).

Fig. 8. A 3D view of the concentrated Type II stator coil configuration of RFAPM machine.

423

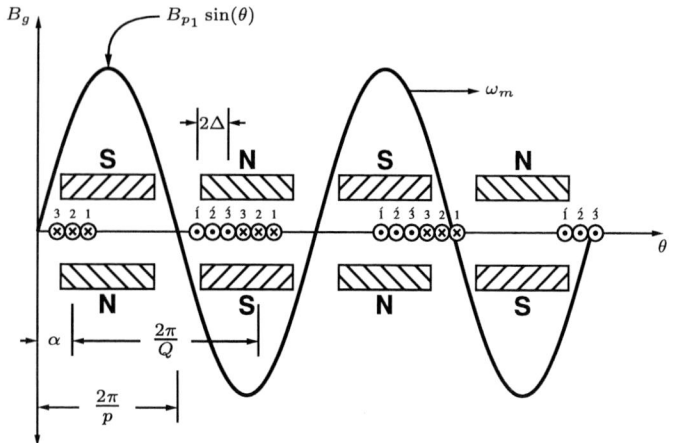

Fig. 9. A 2D linearised cross-sectional view of a RFAPM machine with concentrated Type II stator coils showing all three three-phase coils (with 3 coils per 4 poles).

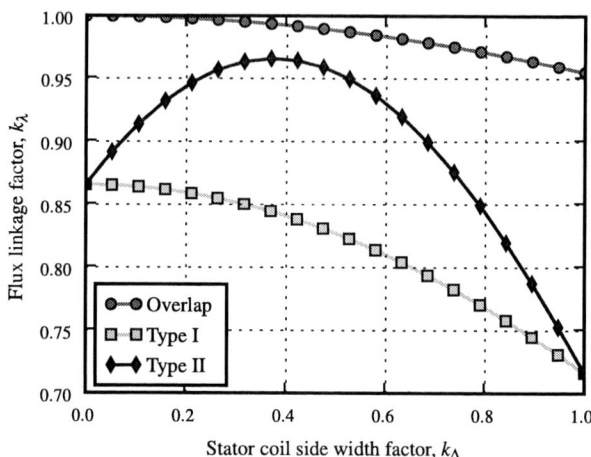

Fig. 10. Flux linkage vs. stator coil side-width

For N conductors, the flux-linkage for the Type II configuration can also be calculated using (20). Thus by substituting (30) into (29) and (29) into (20), equation (20) reduces to

$$\lambda_N = \tfrac{4}{p} N B_{p1} r_n \ell k_\lambda \ , \tag{31}$$

which is once again similar to equation (2) with again a different k_λ value, namely

$$k_\lambda = \frac{\cos\left(\frac{\pi}{Q}\frac{p}{2} - 2\Delta\frac{p}{2}\right) - \cos\left(\frac{\pi}{Q}\frac{p}{2}\right)}{2\Delta\frac{p}{2}} \ . \tag{32}$$

The analysis will furthermore be exactly the same as for the overlapping coils, with the only difference the value of Δ and the end-turn length, which for the Type II configuration can be calculated as

$$\ell_e = 2\left(\frac{2\pi}{Q} - 2\Delta\right) r_n \ . \tag{33}$$

D. Comparative analysis between the different machines

From the previous sections it is clear that the only difference in the analysis between the overlapping, Type I and Type II stator coil configurations are with respect to the different number of coils, Q, the variation in flux-linkage, as given by k_λ, the possibility to vary the coil side-widths, expressed in term of the angle Δ, and the coil's end-turn lengths, ℓ_e. The torque can therefore be expressed from (18) as a common factor, C_M, multiplied by a machine specific torque factor, K_T, as

$$T_m = C_M K_T \ , \tag{34}$$

with

$$C_M = B_{p1}\sqrt{\frac{12 P_{cu} k_f h r_n{}^3 \ell}{\rho_{cu}}} \ , \tag{35}$$

$$K_T = k_\lambda \sqrt{\frac{q\Delta}{(2+\zeta)}} \tag{36}$$

and

$$\zeta = \frac{\ell_e}{\ell} \ , \tag{37}$$

the ratio between the end-turn length and the active copper length.

We will begin our comparative analysis of the different stator coil configurations by comparing the difference in the flux-linkage factor, as express by (3), (25) and (32), for a common 16 pole rotor configuration. In Fig. 10 the different flux-linkage factors are plotted against the stator coil side-width factor, k_Δ, with

$$k_\Delta = \frac{\Delta}{\Delta_{|max}} \ . \tag{38}$$

From this graph it can be seen that the maximum flux-linkage of the Type II concentrated coil configuration occurs with the coil side-width at 37 % of its maximum possible value, with $k_\lambda = 0.966$. For the overlapping coil configuration, the coil side-width is usually very close to its maximum possible value, as was shown in Fig. 2, resulting in $k_\Delta \approx 1.0$ and therefore $k_\lambda \approx 0.955$.

The flux-linkage is only affected by the number of coils and the coil side-width. On the other hand, from (36), the torque factor is also affected by ζ. Thus, the smaller the end-turn length in comparison with the active stack length, the higher the torque factor would be.

The end-turn lengths of all the coil configurations depend on the nominal radius and the number of stator coils, as given by (11), (26) and (33). The overlapping coil configuration's end-turn length is also dependant on the the coil height, while the Type II concentrated coil configuration's end-turn length is again also dependant on the coil side-width angle, Δ. This implies that for a short stack machine[1] with a fairly large coil height in comparison to the nominal stator radius, a concentrated coil configuration should have a higher torque

---

[1]A short stack machine is a machine with a small active copper length, ℓ, in comparison to its nominal stator radius, r_n.

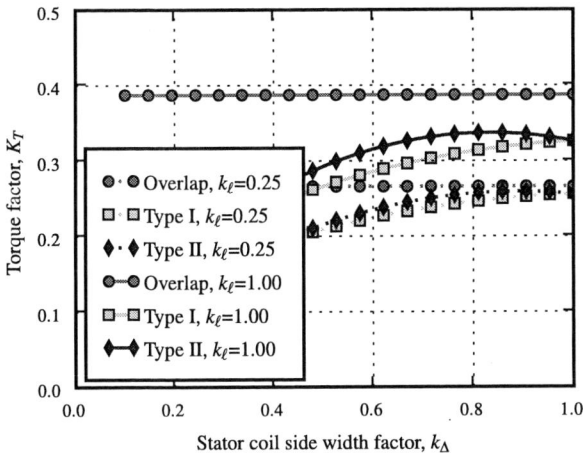

Fig. 11. Torque factor vs. stator coil side-width, with $k_h = 0.1$

factor than an overlapping coil configuration. This is confirmed in Fig. 11 and Fig. 12 for $k_h = 0.10$ and $k_h = 0.02$ respectively, with k_ℓ the ratio of the active stack length to nominal stator radius, defined as

$$k_\ell = \frac{\ell}{r_n} \qquad (39)$$

and k_h, the ratio of the coil height to nominal stator radius, defined as

$$k_h = \frac{h}{r_n} . \qquad (40)$$

For Fig. 11 and Fig. 12, the coil side-width for the overlapping coil configuration is kept constant at 100 % of the maximum coil side-width of $\frac{\pi}{Q}$ *mechanical* degrees.

From Fig. 11 it is evident that a machine with a Type II concentrated coil configuration with $k_h = 0.1$ and $k_\ell = 0.25$ will have almost the same torque factor as a machine with an

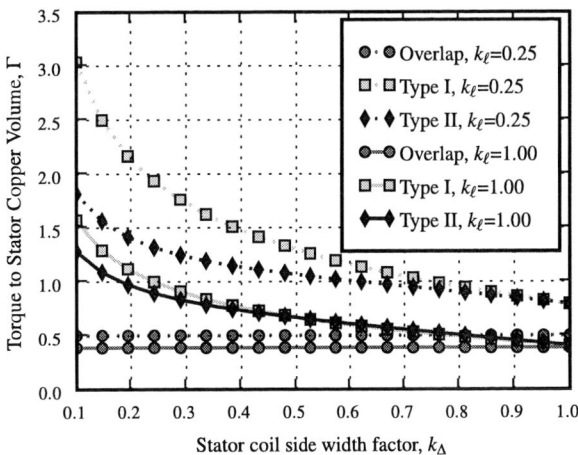

Fig. 13. Torque to Stator Copper Volume Ratio vs. stator coil side-width, with $k_h = 0.10$

overlapping coil configuration, with $k_\Delta = 86\,\%.$[2]

Another important consideration is the copper volume of the stator, calculated as

$$V_{cu} = 2\Delta r_n h k_f (2\ell + \ell_e)Q . \qquad (41)$$

More important than the actual copper volume is maybe the ratio between the torque and the copper volume, defined for comparative analysis as

$$\Gamma = \frac{K_T}{V_{cu}} . \qquad (42)$$

In Fig. 13 and Fig. 14 Γ versus the coil side-width factor are shown for $k_h = 0.10$ and $k_h = 0.02$ respectively. These graphs clearly show that even for $k_\ell = 1.0$ and $k_h = 0.02$, both concentrated coil configuration type machines can in

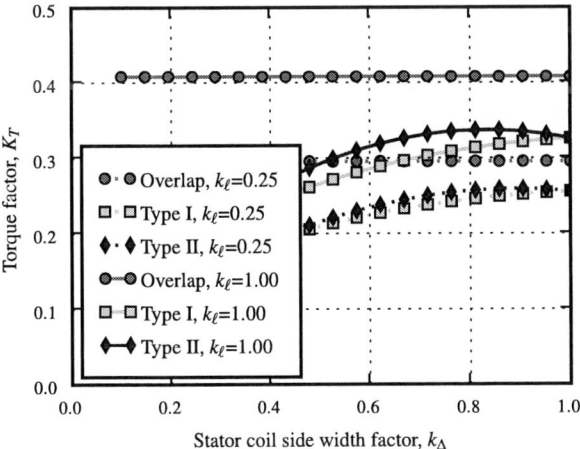

Fig. 12. Torque factor vs. stator coil side-width, with $k_h = 0.02$

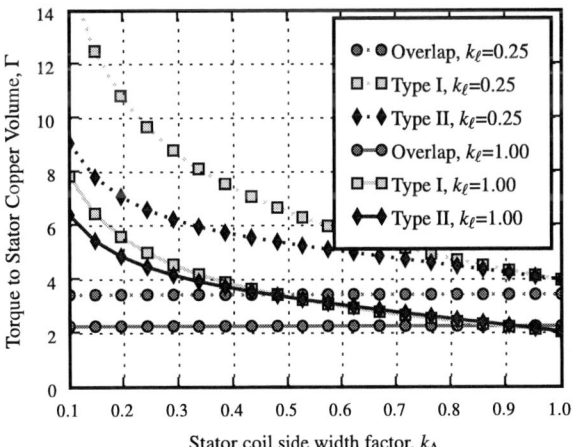

Fig. 14. Torque to Stator Copper Volume Ratio vs. stator coil side-width, with $k_h = 0.02$

[2]With $r_n = 100\,\text{mm}$, this translates to a coil height, $h = 10\,\text{mm}$ and a active stack length, $\ell = 25\,\text{mm}$.

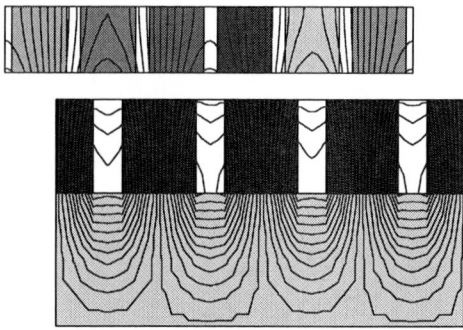

Fig. 15. Simplified equivalent linear model for the RFAPM machine with concentrated Type I coils.

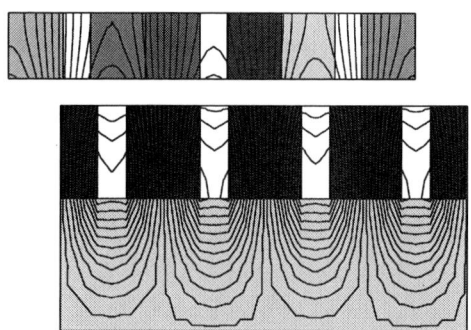

Fig. 16. Simplified equivalent linear model for the RFAPM machine with concentrated Type II coils.

TABLE I
COMPARISON BETWEEN THE FEA AND THE THEORETICAL ANALYSIS FOR
THE TYPE I MACHINE.

$p = 32$, $r_n = 1.0$, $k_\ell = 0.25$, $k_h = 0.1$				
k_Δ	$T_{m|FEA}$	$T_{m|Theory}$	$\lambda_{N|FEA}$	$\lambda_{N|Theory}$
0.2	1.067	1.120	0.443	0.430
0.4	1.048	1.100	0.438	0.420
0.6	1.022	1.070	0.426	0.410
0.8	0.981	1.030	0.411	0.400
0.9	0.957	1.010	0.400	0.390

TABLE II
COMPARISON BETWEEN THE FEA AND THE THEORETICAL ANALYSIS FOR
THE TYPE II MACHINE.

$p = 32$, $r_n = 1.0$, $k_\ell = 0.25$, $k_h = 0.1$				
k_Δ	$T_{m|FEA}$	$T_{m|Theory}$	$\lambda_{N|FEA}$	$\lambda_{N|Theory}$
0.2	1.098	1.090	0.494	0.470
0.4	1.135	1.150	0.508	0.490
0.6	1.124	1.160	0.492	0.470
0.8	1.055	1.100	0.451	0.430
0.9	0.998	1.040	0.424	0.410

general, theoretically deliver more torque per copper volume than for the overlapping coil configuration type machine. For Fig. 13 and Fig. 14, the coil side-width for the overlapping coil configuration is again kept constant at 100 % of the maximum coil side-width of $\frac{\pi}{Q}$ *mechanical* degrees.

III. FINITE ELEMENT ANALYSIS

In this section the analytical (theoretical) results in section II are compared with results obtained from Finite Element Analysis (FEA). The FEA was done using simplified equivalent linear machine models of the different coil configurations as shown in Fig. 15 and Fig. 16 for the concentrated Type I and Type II configuration respectively. Table I shows the comparison between the FEA and the theoretical analysis of the Type I concentrated coil configuration, while Tables II shows the comparison between the FEA and the theoretical analysis of the Type II concentrated coil configuration. The FEA and theoretical results in Table I and Table II were normalised with respect to the results obtained for the overlapping coil configuration.

IV. CONCLUSION

Although the torque factor of the Type II concentrated coil configuration, for the best cast scenario as shown in Fig. 11, was calculated to be 16 % less than that of the overlapping coil configuration, the torque to stator copper volume at the same coil side-width value of 81 % was calculated to be 45 % more, as shown in Fig. 13. The FEA results that was done for similar small diameter, short stack RFAPM machines correlate well with the analytical results, in predicting that for the Type II concentrated coil configuration, a higher torque factor than for the overlapping coil configuration, with the the same coil-side width values, can be obtained.

REFERENCES

[1] R. Qu and T. A. Lipo, "Dual-rotor, radial-flux, toroidally wound, permanent-magnet machines," *IEEE Trans. Ind. Appl.*, vol. 39, no. 6, pp. 1665–1673, 2003.

[2] J. Cros and P. Viarouge, "Synthesis of high performance pm motors with concentrated windings," *IEEE Trans. Energy Convers.*, vol. 17, no. 2, pp. 248–253, Jun. 2002.

[3] M. Kamper, F. Rossouw, and R.-J. Wang, "Analysis and performance evaluation of axial flux air-cored stator permanent magnet machine with concentrated coils," in *IEEE International Electric Machines and Drives Conference (IEMDC)*, May 2007.

[4] D. N. Mbidi, "Design and evaluation of a 300 kw double stage axial-flux permanent magnet generator." Master's thesis, Stellenbosch University, 2001.

[5] R.-J. Wang, "Design aspects and optimisation of an axial field permanent magnet machine with an ironless stator," Ph.D. dissertation, Stellenbosch University, Stellenbosch, South Africa, 2003.

Analysis and Experimental Investigation for Field-Control Capability of a Novel Hybrid Excitation Claw-Pole Synchronous Machine

Yang Chengfeng, Lin Heyun, Liu Xiping, Fang Shuhua, and Guo Jian

School of Electrical Engineering, Southeast University, 2 Si-Pai-Lou, Nanjing, 210096, P.R. China

Abstract: **The structure and operation principle of a novel hybrid excitation claw-pole synchronous machine (HECSM) are introduced. The field-control capability based on the equivalent magnetic circuit method is analyzed. A 3D finite element analysis (FEA) is used to analyze no-load magnetic field distributions, air-gap flux, phase flux linkage and back EMF variation for different field currents. Furthermore, torque characteristics of the HECSM for different field currents are explored. A 2kW prototype is manufactured, and the experiment for field-control capability is carried out. Experimental results coincide well with those of FEA. The experimental and FEA results indicate that the flux of the HECSM can be adjusted in a wide range with a relatively low field current.**

Index terms: **claw-pole, field-control, field weakening, finite element analysis, hybrid excitation, synchronous machine.**

I. INTRODUCTION

Recently, there is an increasing tendency to study hybrid excitation machines for flux weakening operation. This kind of electrical machine combines the advantages of permanent magnet (PM) machines with the possibility of controllable flux by field windings and changing air-gap magnetic flux of electric machines easily [1]-[3].

A novel hybrid excitation claw-pole synchronous machine (HECSM) with asymmetrically stagger PM is presented and developed [2]-[3]. This machine combines the fixed excitation of PM with the variable flux given by a circumferential field winding located on the inner stator. In this manner, air-gap flux can be controlled over a wide range with a minimum of conduction losses and without demagnetization risk for the PM.

Fig. 1 shows the structure of the HECSM, which consists of a rotor, an outer stator and an inner stator. The surface of claw-pole on rotor is divided into two sections, one is PM pole, and the other is a solid iron pole.

PMs create a nearly constant flux, and the field winding current generates a variable flux. These two flux components converge in the air gap, and the flux generated by field windings goes to add or subtract to PM flux according to the direction of field current. Fig.2 outlines the per pole flux path under different field currents.

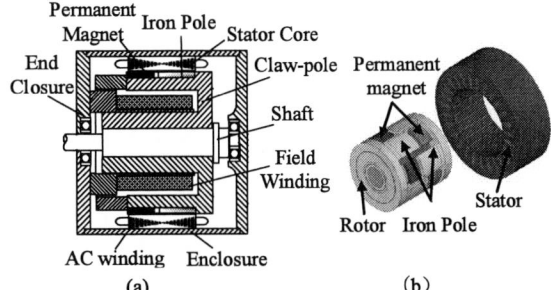

Fig.1　Structure of the HECSM. (a) Magnetic Structure. (b) Stator and rotor model.

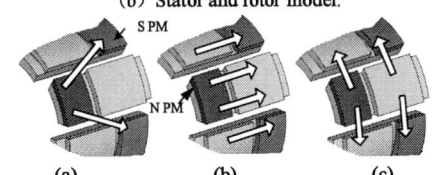

Fig.2　Operation principle of the HECSM. (a) PM flux. (b) Demagnetizing effect of the field flux. (c) Magnetizing effect of the field flux.

II. ANALYSIS FOR FIELD-CONTROL CAPABILITY

A. Equivalent Magnetic Circuit

By neglecting saturation and the soft iron reluctance, the equivalent magnetic circuit for one pole pair can be obtained based on the HECSM structure, see Fig.3. Most of the flux created by the PM crosses the air-gap and flows into the stator, closing the loop through the laminated stator core and the next consecutive PM. Flux created by the field winding follows a low reluctance path crossing the loop along laminated stator core and the adjoining iron pole. The equivalent magnetic circuit as shown in Fig.3 considers both types of machine excitation of PMs and field current, the spatial flux distribution in the machine as well as flux leakages [4].

A careful analysis of this circuit allows separation of the circuit for both PM and field flux. Flux produced by the field current, marked by the large dashed line, flows through stator and rotor core, and iron poles. On the other hand, PM flux, marked by the smaller dashed line, flows primarily in the stator and rotor core and in the PMs themselves.

This work is supported by National Science Foundation of China (50337030).

978-1-4244-0644-9/07/$25.00 ©2007 IEEE

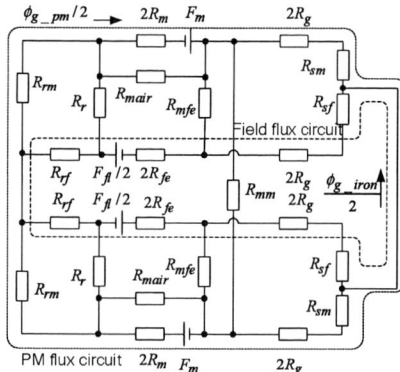

Fig.3 Equivalent magnetic circuit of HECSM

B. Air-gap Flux and Field-control Ratio

By dividing the air-gap surface into two sections, one corresponding to the PM and the other one to the iron pole areas, the total air-gap flux can be divided into two components associated with each of these sections: the iron pole air-gap flux ϕ_{g_iron} and the PM air-gap flux ϕ_{g_pm}, see Fig.4. When current is injected into the field winding, in either positive or negative directions, flux of the iron pole section changes linearly if the iron saturation is neglected.

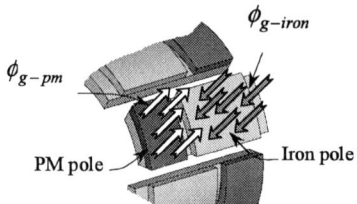

Fig.4 Air-gap flux components

On the other hand, flux associated with the permanent magnet is invariant. The combination of these two fluxes results in the total air-gap flux. This resultant flux can be either the summation or subtraction of each component, depending upon the direction of the field current. Therefore, the total air-gap flux can be expressed as

$$\phi_g = \phi_{g\_iron} + \phi_{g\_pm} \tag{1}$$

where

$$\phi_{g\_iron} = \frac{N_{fl}I_{fl}}{2R_g + R_{g\_in} + R_{g\_out}} \tag{2}$$

where $N_{fl}I_{fl}$ are the field ATs, R_g, R_{g_in} and R_{g_out} are the reluctance of air-gap, inner part and outer part of the additional air-gap, respectively. The field current I_{fl} is positive for magnetizing flux, negative for demagnetizing.

The second term of (1) is

$$\phi_{g\_pm} = \frac{R_{mm}R_{lg}\phi_r}{(R_{lg} + R_{g\_in} + R_{g\_out})(4R_g + R_{mm}) + 2R_gR_{mm}} \tag{3}$$

where R_{mm} are the leakage of one PM pole to its adjacent PM pole, ϕ_r is the remanent flux due to the PM remanence, and

$$R_{lg} = R_{mair} \, // \, R_{mfe} \, // \, 2R_m \tag{4}$$

where R_{mair}, R_{mfe} are the leakage of one PM pole to the air and to the iron pole on the same claw-pole surface, respectively. R_m is the reluctance of a PM pole.

Equations (2) and (3) show the decoupled effect between each air-gap flux components. The ϕ_{g_iron} is primarily dependent on the magnitude and direction of the field current. Conversely, the ϕ_{g_pm} is mainly a function of the PM magnetic parameters. Under no saturation, both components cross the air-gap by independent path: iron and PM poles, respectively. As a result, low field ATs is required to manipulate the iron flux component, and there is no demagnetizing action over the PM.

Then, the field-control ratio can be obtained

$$K = \phi_{g\_iron} / \phi_{g\_pm} \tag{5}$$

This ratio is a very important parameter to evaluate the field-control capability for a hybrid excitation machine.

III. 3D FINITE ELEMENT ANALYSIS OF THE HECSM

A. 3D Finite Element Analysis of the HECSM

A 2-kW prototype is manufactured as shown in Fig.5. Main data of the prototype is as follows: outer diameter, inner diameter and stack length of stator is 168mm, 102mm and 60mm, respectively, PM thickness is 5mm, axial length of PM pole and iron pole is 20mm and 35mm, respectively, and the number of stator slots, claw-pole poles and field winding turns is 27, 8 and 1500, respectively.

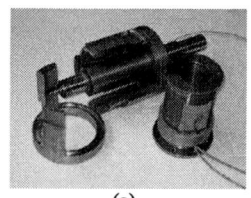

(a)   (b)

Fig.5 HECSM 2-kW prototype. (a) Rotor assembly. (b) Stator.

A 3D finite element analysis (FEA) is employed to determine no-load flux distribution and the field-control capability of the HECSM [5]. The mesh of the HECSM is shown in Fig.6. The elements number of stator and rotor are 38743 and 40406, respectively.

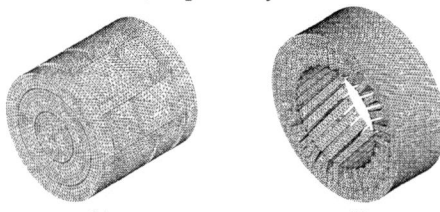

(a)   (b)

Fig.6 Mesh for HECSM. (a) Rotor. (b) Stator.

To evaluate the field-control capacity, FE analysis of 2-pole section is studied under field current variation. Fig.7 shows the portion under analysis.

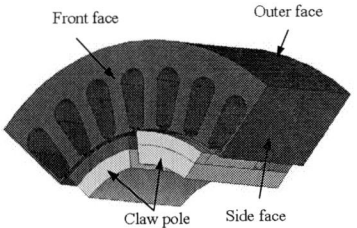

Fig.7 A pair claw-pole and stator model

Vector variation of the flux density across the air-gap and stator, for the no-load condition, are depicted in Fig. 8, 9 10 and 11. Magnitude and direction change according to the magnitude and direction of the field current. For no-field excitation, the only flux present into the machine is that imposed by the PM, and the direction in the stator core corresponds to that shown in Fig.9(a), 10(a) and 11(a). Oblique flux circulation, with respect to the shaft, is found into the stator yoke. At the air-gap, flux crosses mostly by the PM poles; only a few portion cross in front of the iron poles which constitutes leakage as shown in Fig.8(a).

When the field flux acts in such a way that it subtracts from the PM flux in the air-gap, which makes demagnetizing effect, stator core is crossed axially (parallel to the shaft), as is shown in Fig 9(b), 10(b) and 11(b). Tangential (perpendicular to the shaft) flux density distribution at the stator core is encountered when field flux adds to the PM flux at the air-gap, as is depicted in Fig. 9(c), 10(c) and 11(c).

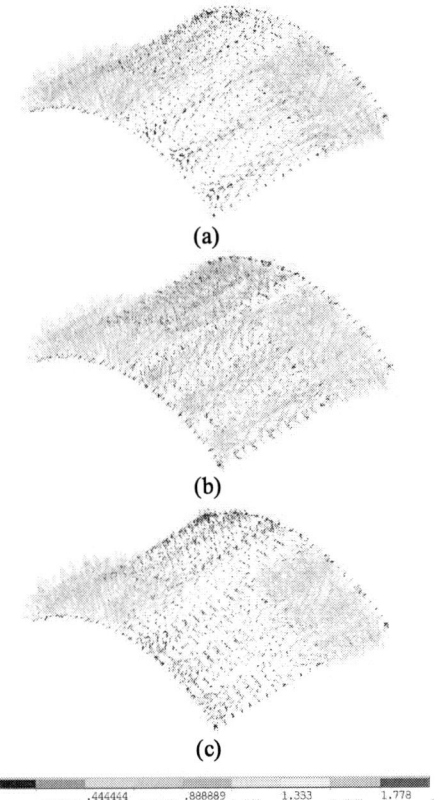

Fig.8 Air-gap flux density distribution over 2 poles for different field current. (a) 0A. (b) -0.5A with field-weakening flux. (c) 0.5A with field-boosting flux.

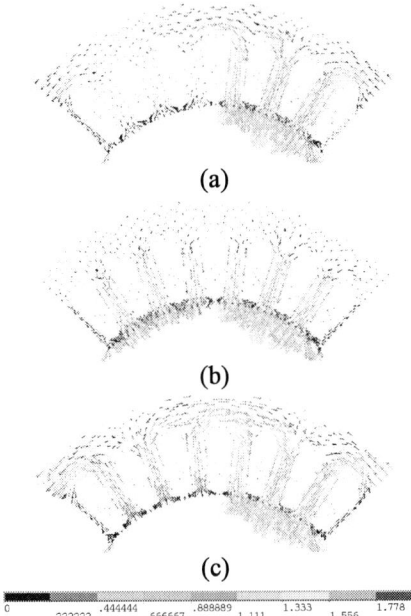

Fig.9 Flux density distribution and directions over the front face of stator for different field currents. (a) 0A with stagger flux. (b) -1A with axial field-weakening flux. (c) 1A with tangential field-boosting flux.

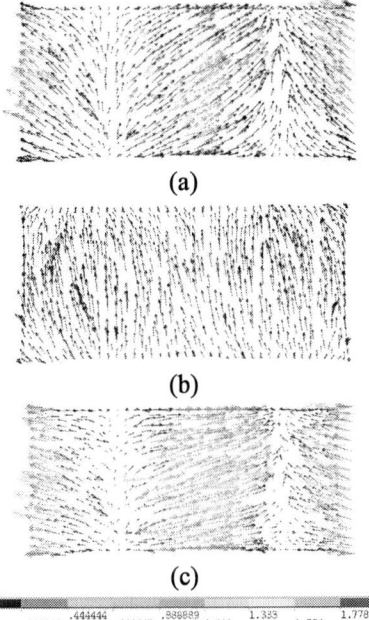

Fig.10 Flux density distribution and directions over the outer face of stator for different field currents. (a) 0A with stagger flux. (b) -1A with axial field-weakening flux. (c) 1A with tangential field-boosting flux.

Fig.12 shows the air-gap flux for three different conditions of the field current. From this result, it can be seen that the flux over the PM is unidirectional and its magnitude is almost constant, however the direction of the flux over the iron pole surface changes according to the magnitude and direction of the field current. These two components make the total air-gap flux vary, either weakening or boosting the air-gap flux from the non-field current condition. Analyzing both fluxes separately as shown in Fig.13, it can be noticed that the PM does not maintain a constant value of flux due to the saturation of the iron. With a variation of ±1A the flux can be modified in a range from 0.01 to 0.58 mWb in Fig.13.

429

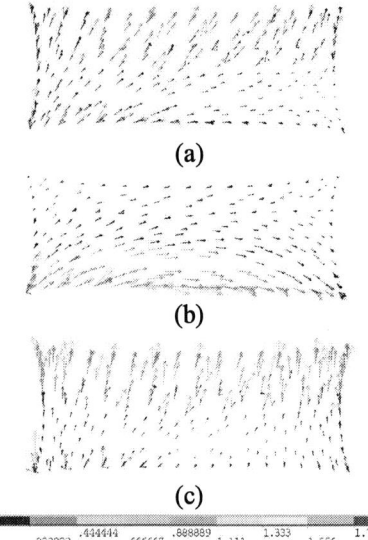

(a)

(b)

(c)

Fig.11 Flux density distribution and directions over the side face of stator for different field currents. (a) 0A with stagger flux. (b) -1A with axial field-weakening flux. (c) 1A with tangential field-boosting flux.

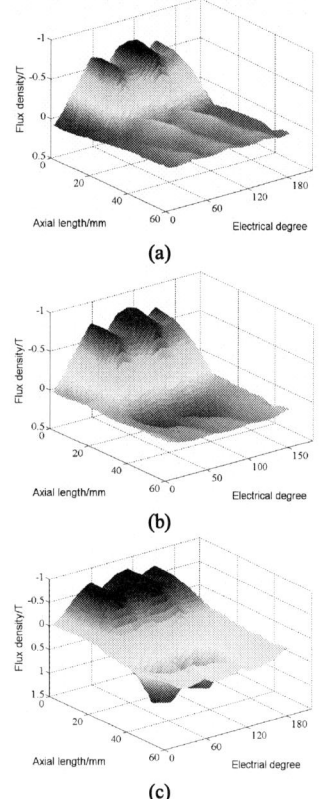

(a)

(b)

(c)

Fig.12 Air-gap flux distribution for different field currents. (a) 1A with magnetizing effect. (b) No field current. (c) -1A with demagnetizing effect.

Fig.14 shows the variation of the flux that crosses the stator in axial direction. When fluxes of each portion of the air-gap are in the opposite direction, this flux became larger.

FEA shows that the air-gap flux can be adjusted in a wide range with relatively low field currents as shown in Fig.15. Numerically, FEA predicts that the air-gap flux changes over a range ±60% with respect to the no-field current with a variation of ±1A. Experimental fluxes per pole obtained from the prototype are compared with

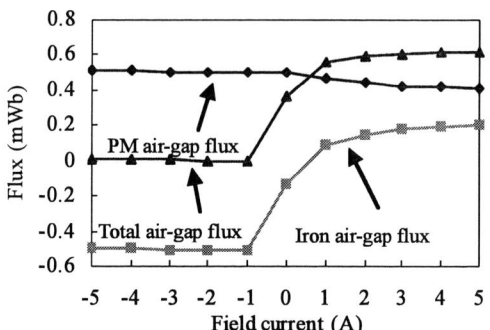

Fig.13 Flux distribution in the air-gap

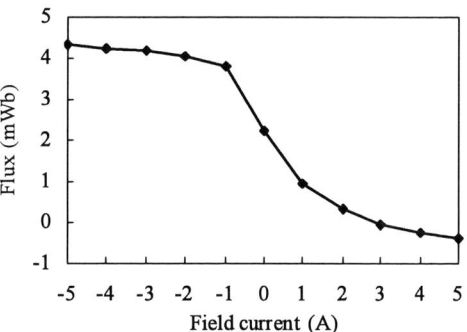

Fig.14 Flux crossing stator in axial direction

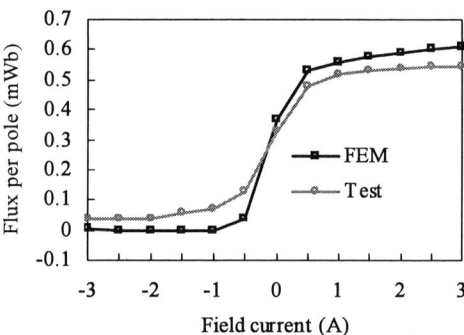

Fig.15 Experimental results comparison for the air-gap flux

FEA results in Fig.15. It can be seen that a wide range of air-gap flux control can be achieved with a relatively small field current. With a variation of ±1A, the flux per pole varies in a range of 0.10-0.55 mWb, with 0.35 mWb under no-field current.

B. Flux linkage and Back EMF Analysis of the HECSM

Time performance of back EMF can be evaluated by the space derivative of the flux linked with windings and can be expressed as

$$e(t) = -\frac{d\Psi(\theta)}{d\theta}\frac{d\theta}{dt} = -\frac{d\Psi}{d\theta}\omega_r \qquad (6)$$

where e is the back EMF per phase, Ψ is the air-gap flux linkage per phase, θ is the rotor position and ω_r is the angular speed of the rotor. The back EMF is obtained by using numerical derivation of the linked flux linkage. The flux linkage linked for a given position of the rotor can be expressed as

$$\Psi(\theta) = \frac{1}{S} \int A\, dS \qquad (7)$$

where S is the cross sectional area of the winding. By changing the rotor position in small steps different sets of rotor position-linked flux linkage values are obtained. Using numerical interpolation and numerical derivation, the behavior of the back EMF can be found easily [6].

Rotor position-linked flux linkage values have been obtained using 3D FEA for different positive and negative field current values and shown in Fig.16 (a). Taking the numerical derivative of the flux linkage waveform will result in back EMF variation displayed in Fig.16 (b) for various values of the field current. It is obvious that the machine back EMF can be easily increased or reduced by applying positive or negative current to the field winding.

Fig.16(c) shows the experimental field-control capability of the HECSM. The measured no-load output voltage is presented for a machine speed of 900 rpm. Increases and decreases of the output voltage with respect to the zero field current condition are obtained. With a variation of ±1A the flux is modulated over a range ±60% with respect to no field current. The experimental results coincide well with those of the FEA.

C. Torque Analysis of the HECSM

Analyses of the HECSM torque examining the torque for various field currents, peak torque at rated stator current conditions is also accomplished. Fig.17 shows the static torque characteristic of the machine. The plot is obtained for different field currents cases and implies that the machine torque range can be increased by nearly 55% with respect to the no field current case.

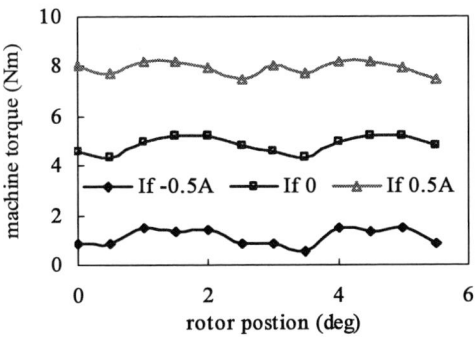

Fig.17 Static torque characteristic of the HECSM for different field currents values at rated conditions.

Fig.18 shows the torque vs. armature current for motor operation. As one sees from these characteristics, machine torque increases as the armature current is increased. Torque variation against armature current for a given DC field current is close to linear for small values of field current.

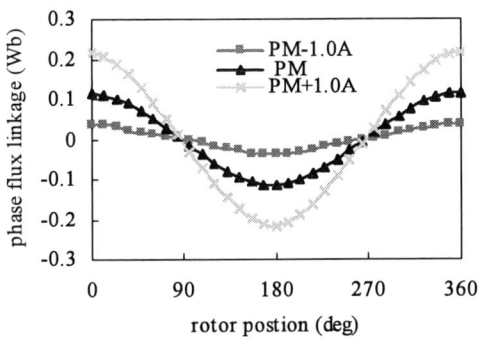

(a)

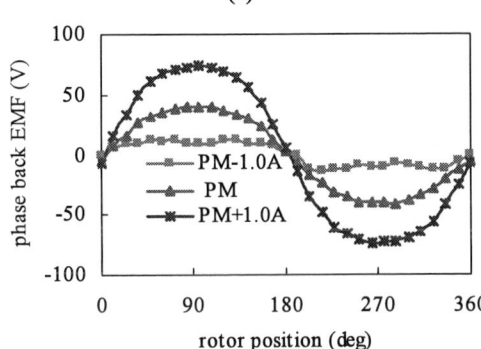

(b)

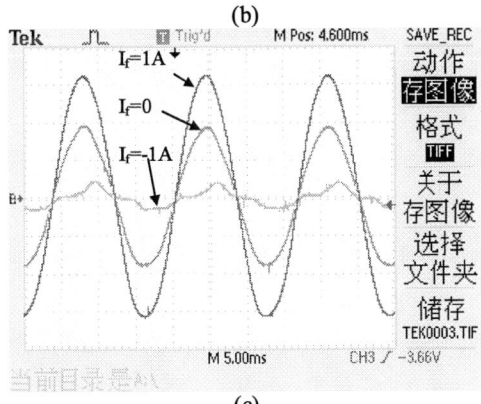

(c)

Fig.16 Phase flux linkage and back EMF variation for different field currents. (a) phase flux linkage. (b) phase back EMF under speed of 900 rpm. (c) phase back EMF experimental waveform.

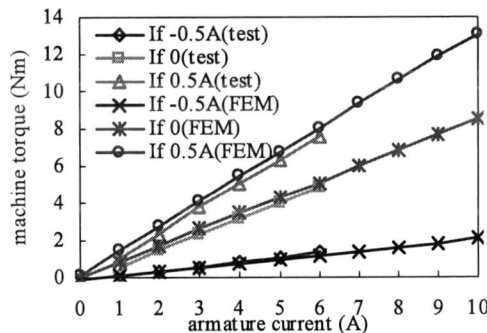

Fig.18 Machine torque for various values of armature and field currents

IV. CONCLUSION

This paper analyzes field-control capability based upon the equivalent magnetic circuit, 3D finite element analysis and experiments for prototype of the HECSM. The magnetic structure allows for easy control of the air-gap flux over a wide range, using only a relatively small field current. The flux in the air-gap is determined by the direction and magnitude of the field current. There is no risk of demagnetization for the PM because the control action is exercised over the iron poles.

The field-control capability depends upon the ratio of iron pole air-gap flux to PM pole air-gap flux. Magnitude of the air-gap flux density changes according to the magnitude and direction of the field current. The Experimental and FEA results agree that a nearly 60% control range can be obtained at no load and 55% control range can be obtained at full load with a relatively modest requirement for field current.

REFERENCES

[1] E. Spooner, S A W Khatab, N G Nicolaou, "Hybrid excitation of AC and DC machine," *Journal of University of Manchester Institute of Science and Technology*, 1989, No.3, pp. 48-52.

[2] Yang Chengfeng, Lin Heyun, "Research of the magnetic circuit configuration and its field-control capabilities of hybrid excitation claw-pole synchronous machine", *Small & Special Electrical Machines*, 2007, No.7, pp. 12-15.

[3] Yang Chengfeng, Lin Heyun, "3D-FEA for field-control performance of hybrid excitation claw-pole synchronous machine with asymmetrically stagger PM. *Journal of Southeast University*, 2007, No.4, pp.614-617.

[4] Tapia J A, Leonardi F, Lipo T A, "Consequent pole permanent magnet machine with field weakening capability", *IEEE International Electric Machines and Drives Conference*, Cambridge, MA, USA, 2001, pp. 126-131.

[5] Yang Chengfeng, Lin Heyun, "Magnetic field analysis of hybrid excitation brushless claw-pole motor with three-dimensional finite element method", *Proc. of 8th Int. Conf. on Electrical Machines and Systems (ICEMS 2005)*, Nanjing (China), Sep. 2005, pp.664-666.

[6] M. Aydin, S. Huang and T. A. Lipo, "A new axial flux surface mounted permanent magnet machine capable of field control", *IEEE Industry Applications Annual Meeting*, Oct 2002, pp. 1250-1257.

[7] M. Aydin, "Axial flux surface mounted PM machines for traction drive applications", *PhD Preliminary Report*, U niversity of Wisconsin-Madison, 2002.

A single-Capacitor Turn-off Snubber for Interleaved Boost Converter with Coupled Inductor

S. -Y. Tseng, [#]J. -Z. Shiang and Y. -H. Su

Digital Application and Renewable Energy Lab.
(GPEARL)
Department of Electrical Engineering
Chang-Gung University
Kwei-Shan Tao-Yuan , Taiwan, R.O.C
E-mail: sytseng@alumni.ccu.edu.tw
TEL : +886-3-2118800
FAX :+886-3-2118026

[#]Department of Electrical Engineering
Chien Kuo Technology University
Changhua, Changhua City, Taiwan, R.O.C

Abstract--**This paper proposes an interleaved boost converter associated with a coupled inductor with a single-capacitor turn-off snubber to reduce the switching loss at turn-off transition. The proposed turn-off snubber is also used to limit rising rate of voltages across switches for reducing spike voltage. In this circuit structure, the proposed converter is operated in an interleaved fashion to reduce output current ripple and with a coupled inductor to increase step-up voltage ratio. As compared with the counterparts of conventional converter topologies, the proposed converter has the merits of less component count, higher efficiency, smaller size, and they are easier to implement. Hardware measurement obtained from experimental prototype have been verified these merits.**

Keywords: **Interleaved boost converter, single-capacitor turn-off snubber, coupled inductor.**

I. INTRODUCTION

Using the photovoltaic (PV) power system as an alternative energy resource has been widely discussed during the last decade due to the rapid growth of power electronics techniques [1]–[2]. It is also the most effective, less expensive, harmless and less environmental pollution effect of renewable energy source. This energy can be converted into electrical energy through implementation of PV array. Recently, PV system applications are well recognized and widely used in electric power technologies, such as solar power generation for grid connection, solar vehicle constriction, battery charging, water pumping, satellite power system, and so on.

Using a single stage as dc/dc converters has been widely used in a PV power system. To apply to grid connection, the PV power system usually needs a higher input to output voltage transfer ratio. For achieving a high step-up voltage ratio, a push-pull, half-bridge or full-bridge converter associated with a transformation and high turns ratio can be adopted [3]–[5]. Due to high turns ratio of transformer, there exists a large amount of leakage inductance in the converters, resulting in high spike voltage across switch and low conversion efficiency. As a result, they need a large capacity of

soft-switching circuit or snubber to solve the problem, leading to high cost. To trade off two problems for achieving high step-up voltage ratio and for solving leakage inductance, a coupled-inductor boost converter is used in this application.

When the coupled-inductor boost converter is used as a high step-up voltage ratio converter, it can use an active-clamp circuit to recover the energy trapped in leakage inductor and to reduce switching loss at turn-on transition. However, its switching loss at turn-off transition does not reduce. To solve this problem, a passive turn-off snubber can be used, as shown in Fig. 1. In this paper, we focus on solving the switching loss at turn-off transition. Additionally, to reduce ripple current and increase powering capability, two converters operated with an interleaved fashion, as shown in Fig. 2(a), are usually adopted [6]–[7], which also can achieve fast dynamic response and small ripple voltage. Although the two boost converters can achieve a soft-switching feature, their component count and cost are increased significantly. We propose soft-switching interleaved boost converter associated with a coupled inductor with a single-capacitor snubber, as shown in Fig. 2(f), to release the above discussed drawbacks. The converter requires only a resonant capacitor C_S which is associated with couple-inductors L_{12} and L_{22} to function as a lossless turn-off snubber, reducing switching loss and component counts significantly. In the full paper, design of the proposed single-capacitor snubber will be presented in detail.

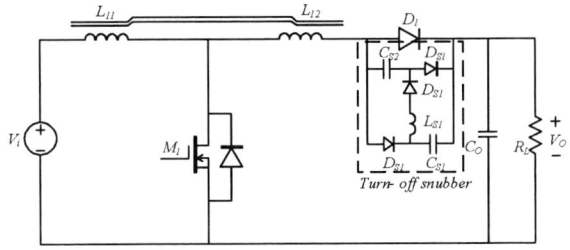

Fig. 1. Schematic diagram of convention coupled-inductor boost converter with a lossless turn-off snubber.

978-1-4244-0644-9/07/$25.00 ©2007 IEEE

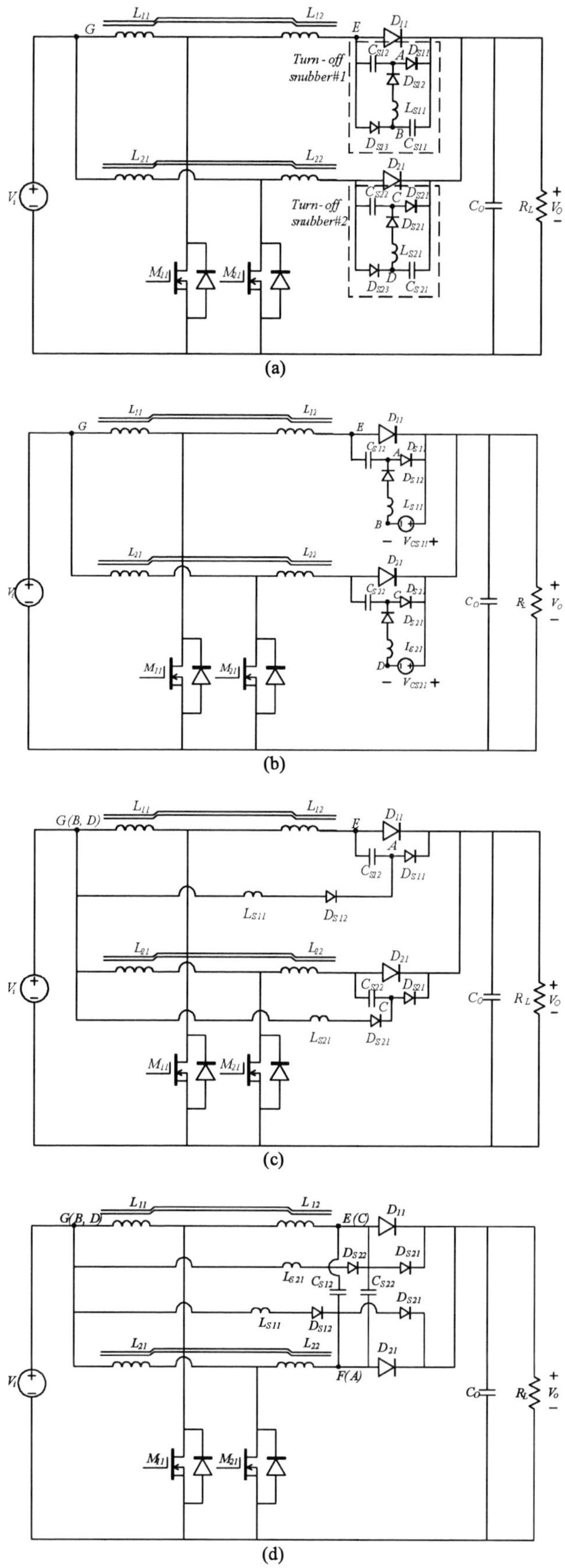

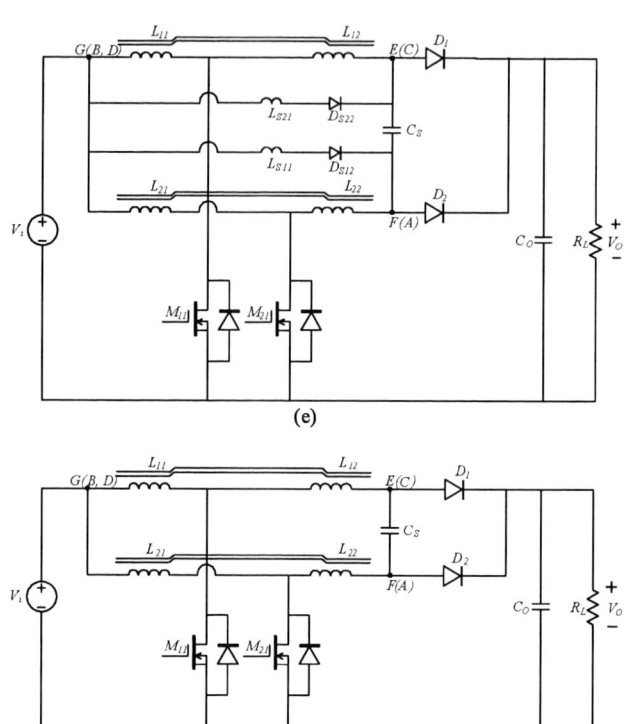

Fig. 2. Illustration of the proposed interleaved coupled-inductor boost converter derived from two turn-off snubbers.

II. PRIVATION OF THE PROPOSED CONVERTER

To reduce witching loss, a lossless turn-off snubber is inserted in a conventional PWM power converter, as shown in Fig. 1. When switch M_1 is turned on, capacitors C_{S1} and C_{S2} are charged through inductor L_{S1} and diode D_{S2} in a resonant manner. At the end of the resonant interval, capacitors C_{S1} and C_{S2} are charged $(NV_i + V_O)$, and are clamped at $(NV_i + V_O)$ until switch M_1 is turned off. When switch M_1 is turned off, the charges stored in capacitors C_{S1} and C_{S2} are discharged to the output load through diodes D_{S1} and D_{S3}, respectively. Thus, switch M_1 is turned off with zero-voltage transition (ZVT). As mentioned above, although it can achieve soft-switching feature, its output current ripple is relatively large for high current and low output power applications. Therefore, to reduce output current ripple, an interleaved scheme is usually adopted. In the following, the proposed interleaved coupled-inductor boost converter with a single-capacitor snubber is derived.

Two lossless turn-off sunbbers are used in an interleaved boost converter with coupled inductor to reduce switching losses, as shown in Fig. 2(a). In Fig. 2(a), voltage across capacitors C_{S11} and C_{S21} can be respectively replaced witch two dc voltage V_{CS11} and V_{CS21} to simplify the circuit structure of the boost converter. When voltages across capacitors C_{S11} and C_{S21} are replaced with dc voltages, the energies stored in capacitors C_{S11} and C_{S21} do not need to discharge their charges. Thus, diode D_{S13} and D_{S23} can be removed, as shown in Fig. 2(b). If voltage V_{CS11} or V_{CS21} is equal to $(V_D - V_i)$, nodes B, D and G have the same potential. Thus, they can be merged as the same node G, as shown in Fig. 2(c). Based on the operational principle of an interleaved coupled-inductor boost converter and turn-off snubber, operational states of diode D_{11} (or D_{21}) are the same as diode D_{S21} (or D_{S11}) except that the operational duration of turn-off snubber is operated within resonant mode. Since the duration of resonant mode is much shorter than a period of the proposed converter, nodes F and A (or E and C) can be combined as the same node F (or E), as shown in Fig. 2(d). It will not affect its original operational principle. In Fig. 2(d), because capacitors C_{S12} and C_{S22} are connected in parallel, they can be respectively merged and replaced with capacitor C_S. Similarly, since two pairs of diodes (D_{11}, D_{S21}) and (D_{S11}, D_{21}) are respectively connected in parallel, they can be respectively replaced by diodes D_1 and D_2, as shown in Fig. 2(e). From Fig. 2(e), it can be seen that the branches of coupled inductors L_{11} and L_{12}, and inductor L_{S21} and diode D_{S22} form a closed path. Additionally, their current directions are uni-directional in the derived converter. According to operational principle of circuit, they can be combined and replaced by the coupled inductor L_{11} and L_{12}. Similarly, the coupled inductors L_{21} and L_{S22}, and inductor L_{S11} connected with diode D_{S12} in series can be combined and replaced by the coupled inductors L_{21} and L_{22}, as shown in Fig. 2(f). Form Fig. 2(f), it can be observed that the derived boost converter requires only a resonant capacitor C_S, which is associated with coupled inductors L_{11} (or L_{12}) and L_{21} (or L_{22}) to function as a lossless turn-off snubber, reducing switching loss and component counts significantly.

III. OPERATIONAL PRINCIPLE OF THE PROPOSED CONVERTER

The proposed converter consists of an interleaved coupled-inductor boost converter and a single-capacitor turn-off snubber, as shown in Fig. 2(f). In this circuit structure, the coupled inductor is used to increase step-up ratio and the single-capacitor turn-off snubber is adopted to reducing switching loss. According to operational principle of circuit, operation of the proposed converter is divided into 12 modes, as shown in Fig. 3, and their key waveforms illustrated in Fig. 4. Since operation modes between $t_0 \sim t_6$ are similar to those between $t_6 \sim t_{12}$ except that the operation of switch changes form M_1 to M_2. Thus, each operational mode within half one switching cycle is briefly described as follows.

Mode 1 [Fig. 3(a); $t_0 \leq t < t_1$]: Before t_0, diodes D_1 and D_2 are in freewheeling through inductor $L_{11} \sim L_{22}$. At the moment, current $I_{i11}(= I_{i12})$ is equal to diode current I_{D1} and current $I_{i21}(= I_{i22})$ is similarly equal to diode current I_{D2}. When $t = t_0$, switch M_1 is turned on. Switch current I_{DS1} is the difference between inductor current I_{i11} and I_{i12} ($= I_{D1}$). Within this time interval, switch current I_{DS1} abruptly increases, while diode current I_{D1} quickly decreases.

Mode 2 [Fig. 3(b); $t_1 \leq t < t_2$]: At t_1, switch current I_{DS1} is equal to inductor current I_{i11}. As a result, diode D_1 is reversely biased. During this time interval, capacitor current I_{CS} varies from 0 to a negative value. Additionally,

diode current I_{D2} is the sum of inductor current I_{i22} and capacitor current I_{CS}.

Mode 3 [Fig. 3(c); $t_2 \leq t < t_3$]: When $t = t_2$, the negative capacitor current I_{CS} equal inductor current I_{i22}. At the moment, diode D_2 is in reversely bias. During this time interval, snubber capacitor C_S and inductors L_{21} and L_{22} form a resonant network and they start to resonate. Additionally, the energy stored in output capacitor C_O is released to load.

Mode 4 [Fig. 3(d); $t_3 \leq t < t_4$]: At t_3, voltage V_{CS} across snubber capacitor C_S is charged to $-(V_O + NV_i)$. At the same time, diode D_2 is forwardly biased. As a result, snubber capacitor C_S stops charging. Within this time interval, the energy stored in inductor L_{m12} is transferred to primary winding N_{11} through coupled inductors L_{11} and L_{12}. Switch current I_{DS1} increases linearly. In addition, the energies stored in inductors L_{m21} and L_{m22} is released to load and inductor currents I_{L21} and I_{L22} decrease linearly.

Mode 5 [Fig. 3(e); $t_4 \leq t < t_5$]: At t_4, switch M_1 is turned off. At the same time, the energies stored in inductors L_{m11} and L_{m12} is transferred to load through sunber capacitor C_S. The voltage V_{CS} across snubber capacitor C_S is discharged from $-(V_O + NV_i)$ to 0. Therefore, switch M_1 is operated with zero-voltage transition (ZVT) at turn-off transition.

Mode 6 [Fig. 3(f); $t_5 \leq t < t_6$]: when $t = t_5$, the energy stored in snubber capacitor C_S is released to 0. At the same time, diode D_1 is in freewheeling through inductors L_{11} and L_{12}. During this time interval, diode D_2 is also in freewheeling through inductors L_{21} and L_{22}. When switch M_2 is turned on at end of mode 6, the other half one switching cycle will start.

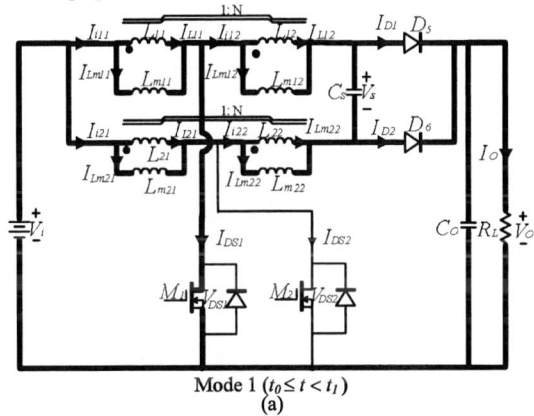

Mode 1 ($t_0 \leq t < t_1$)
(a)

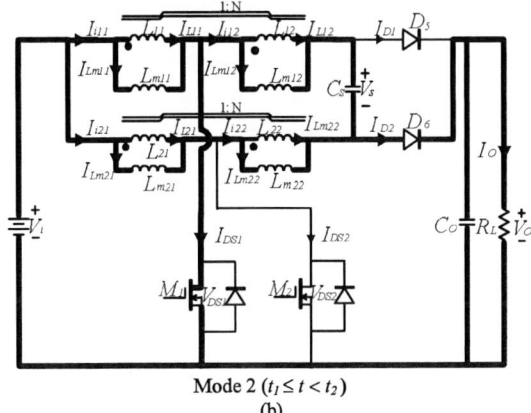

Mode 2 ($t_1 \leq t < t_2$)
(b)

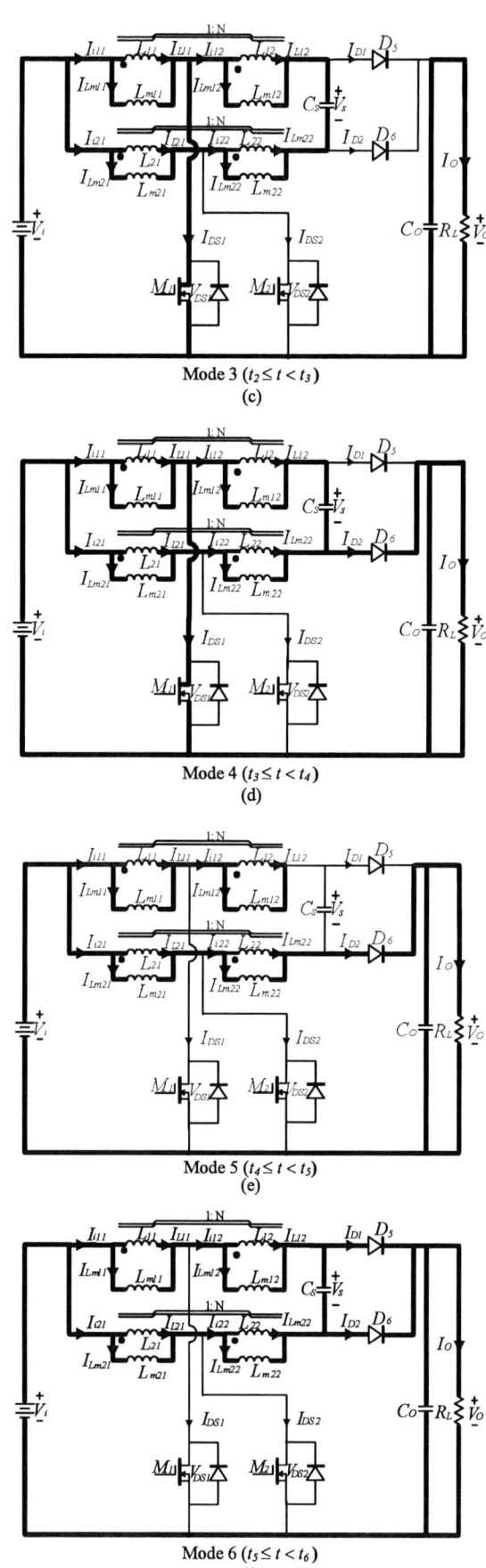

Mode 3 ($t_2 \le t < t_3$)
(c)

Mode 4 ($t_3 \le t < t_4$)
(d)

Mode 5 ($t_4 \le t < t_5$)
(e)

Mode 6 ($t_5 \le t < t_6$)
(f)

Fig. 3. Operational modes of the proposed converter during half one switch cycle.

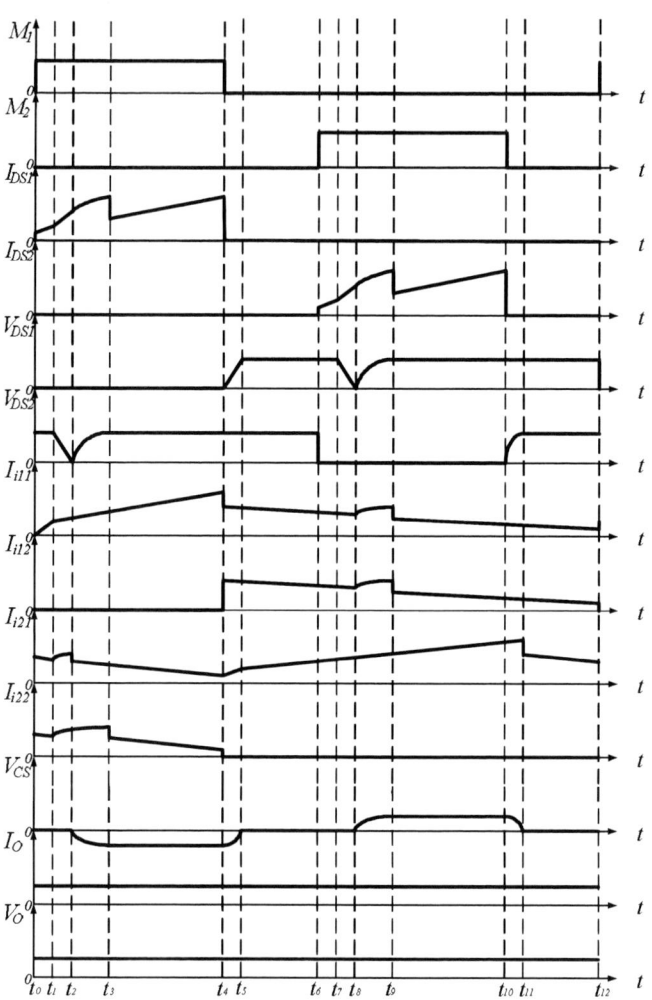

Fig. 4. Key waveforms of the proposed converter operating over one switching cycle.

IV. DESIGN OF THE PROPOSED CONVERTER

To realize the proposed soft-switching converters systematically, design of inductors $L_{m11} \sim L_{m22,}$ and the snubber capacitor C_S are presented as follows:

A. Transfer function M

Since the proposed converter is operated in CCM, the relationship between V_i and V_O can be attained with volt-second balance principle. It can be expressed by

$$V_i D T_S - (\frac{V_O - V_i}{N+1})(1-D)T_S = 0, \qquad (1)$$

where D is the duty ratio, T_S is the period, V_i represents input voltage and V_O is output voltage. From (1), it can be seen that transfer function M of the proposed converter can be determined as

$$M = \frac{V_O}{V_i} = \frac{(1+ND)}{1-D}, \qquad (2)$$

where N is turns ratio of coupled inductor.

B. Coupled inductors $L_{m11} \sim L_{m22}$

Due to the proposed converter operated in CCM, the coupled inductors $L_{m11} \sim L_{m22}$ must be respectively greater than the boundary inductors $L_{m11B} \sim L_{m22B}$ which are the proposed converter operated at the boundary of CCM and DCM. When the proposed converter is operated at the boundary condition, its current waveforms is shown in Fig. 5. Form Fig. 5, it can be observed that the peak value of inductor current $I_{i11(1)}$ can expressed as

$$I_{i11(1)} = \frac{V_i}{L_{m11B}} DT_S. \tag{3}$$

While, inductor current $I_{i11(2)}$ can also be determined by

$$I_{i11(2)} = I_{i12(1)} = \frac{I_{i1(1)}}{N+1} = \frac{V_i}{(N+1)L_{m11B}} DT_S. \tag{4}$$

In (4), it can be derived the average current $I_{i12(av)}$ as follows:

$$I_{i12(av)} = \frac{(1-D)V_i}{2(N+1)L_{m11B}} DT_S. \tag{5}$$

Since the proposed converter is operated in an interleaved fashion, its output current I_O is equal to twice of inductor current $I_{i12(av)}$. Therefore, the boundary inductor $L_{m11B} (= L_{m21B})$ can be determined by

$$L_{m11B} = \frac{D(1-D)V_i}{4(N+1)I_O} T_S. \tag{6}$$

Additionally, the boundary inductor $L_{m12B} (= L_{m22B})$ can be also expressed as

$$L_{m12B} = \frac{D(1-D)N^2 V_i}{4(N+1)I_O} T_S. \tag{7}$$

From (6) and (7), it can be seen that the inductances L_{m11} and L_{m12} must be respectively greater than the boundary inductor L_{m11B} and L_{m12B}. Similarly, inductances L_{m21} and L_{m22} are also greater than the boundary inductances L_{m21B} and L_{m22B}.

C. Snubber Capacitor C_S

In the proposed converter, snubber capacitor C_S resonates with inductors L_{m11} and L_{m12}, or L_{m21} and L_{m22} to smooth out switch voltage at turn-off transition. The energy stored in C_S can be expressed by

$$W_{CS} = \frac{1}{2} C_S (V_O + N V_i)^2. \tag{8}$$

To completely eliminate turn-off loss of switch, the energy stored in capacitor C_S must be at least equal to the turn-off loss W_{Soff}. According to analysis principle of switching loss, turn-off loss W_{Soff} can be determined as

$$W_{Soff} = \frac{t_{Soff}}{2} \left(\frac{N V_i - V_O}{N+1} \right) I_{DP}, \tag{9}$$

where t_{off} is the time interval of switch M_1 or M_2 at turn-off transition and I_{DP} is its peak current value. From (8) and (9), it can be found that snubber capacitor C_S can be expressed as

$$C_S \geq \frac{t_{Soff}(N V_i - V_O)}{(N+1)(V_O + N V_i)^2} I_{DP}. \tag{10}$$

In (10), t_{Soff} is chosen with 500 ns in practical design considerations.

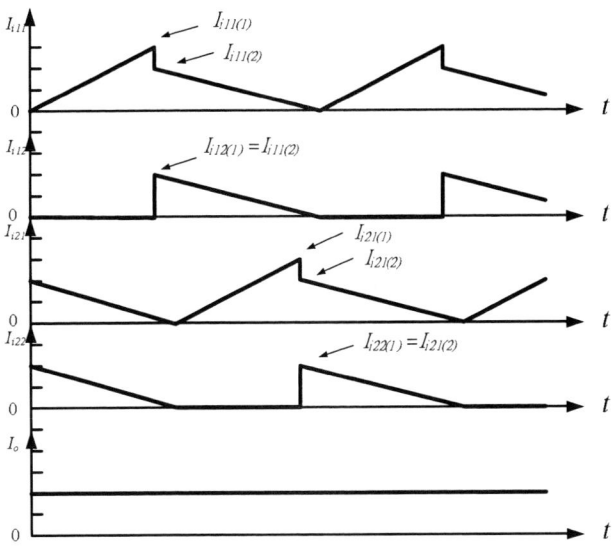

Fig. 5. Conceptual waveforms of inductor current of the proposed converter with coupled inductor operated in the boundary of CCM and DCM.

V. MEASURED RESULTS

To verify the performance of the proposed soft-switching converter, a prototype for supplying power to inverter of grid-connection system and using a PV array as a source with the following specifications was implemented.

☐ input voltage V_i: DC 48 V (PV arrays),
☐ switching frequency f_S: 50 kHz,
☐ output voltage V_O: DC 350 V, and
☐ maximum output current $I_{O(max)}$: 2 A,
☐ maximum output power $P_{O(max)}$: 700 W.

According to (10), values of snubber capacitor C_S can be calculated as 0.6 nF. In our design sample, a capacitor with 1nF is adopted. The components of power stage in the proposed boost converter were determined as follows:

☐ M_1, M_2: IRFP250 × 2,
☐ L_{m11}, L_{m21}: 20 μH,
☐ L_{m12}, L_{m22}: 2 mH,
☐ D_1, D_2: UF3031,
☐ C_O: 470 uF/400 V, and
☐ inductor core (L_{m11} and L_{m21}, or L_{m21} and L_{m22}): EE-55.

Measured switches voltage and current waveforms of the proposed converter are shown in Fig. 6. Fig. 6(a) shows those of switch M_1, while Fig. 6(b) shows those of switch M_2. From Fig. 6, it can be seen that switches M_1 and M_2 can be operated with ZVT at turn-off transition. Fig. 7 shows measured output inductor current I_L and output current I_O waveforms of the conventional single coupled-inductor boost converter. While, Fig. 8 shows measured those waveforms of the proposed interleaved boost converter. From Figs. 7 and 8, it can be observed that the proposed interleaved boost converter can reduce output current ripple.

To make a fair comparison, the hardware components of the proposed converter, hard-switching boost converter and those with two turn-off snubber are kept as the same as possible. Fig. 9 shows the plots of output voltage and current waveforms of the proposed converter under step-load changes between 20 % and 100 % with respectively rate of 1 kHz and a duty ratio of 50 %. From Fig. 9, it can be observed that although the proposed converter is using less component counts, it can yield a good dynamic performance. Comparison between the efficiencies of the proposed converter and the discussed converter with hard switching is illustrated in Fig. 10. In Fig. 10, the proposed converter can achieve efficiency of 87 % under full load. It can be observed that the proposed converter cannot always yield higher efficiencies than the other one under various operating condition. It has a trend that, at lower output load, the proposed converter can yield higher efficiency than the other one. While, at heavy output load, the proposed converter and the one with hard switching can yield almost the same conversion efficiency. These reasons behind are that given a fixed power level, a higher output load will result in higher energy losses at turn-on transition. The energy losses are nearly switch loss W_{Soff}.

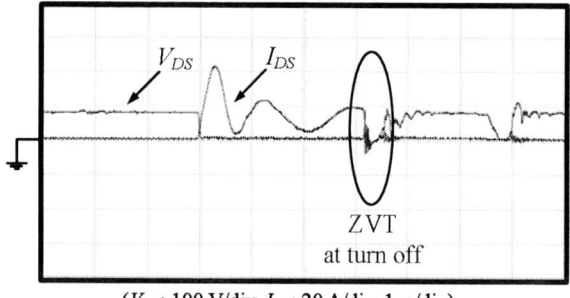

(V_{DS}: 100 V/div, I_{DS}: 20 A/div, 1μs/div)

(a)

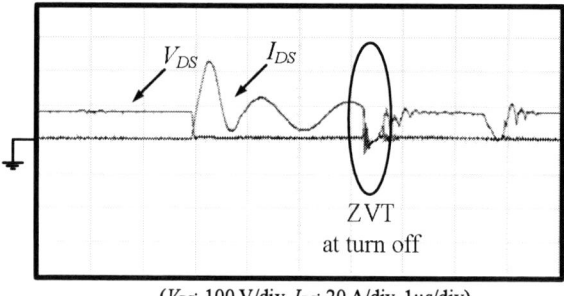

(V_{DS}: 100 V/div, I_{DS}: 20 A/div, 1μs/div)

(b)

Fig. 6. Measured voltage V_{DS} and current I_{DS} waveforms of (a) switch M_1, and (b) switch M_2 under 50 % of full load.

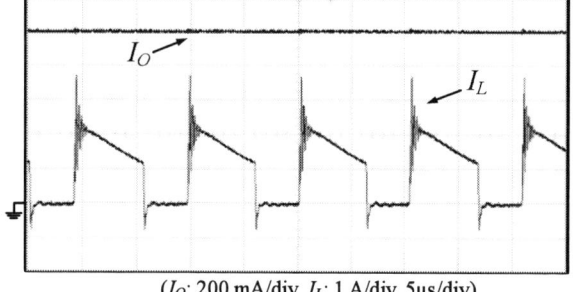

(I_O: 200 mA/div, I_L: 1 A/div, 5μs/div)

Fig. 7. Measured inductor current I_L and output current I_O waveforms of single boost converter with coupled inductor.

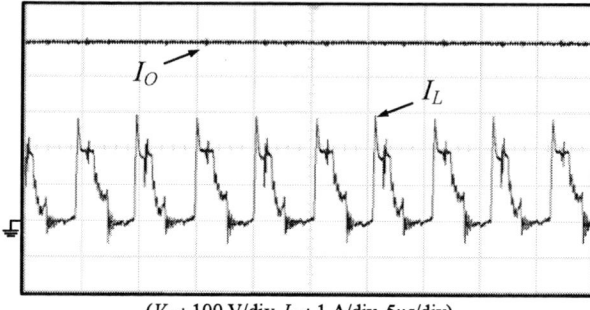

(V_{DS}: 100 V/div, I_{DS}: 1 A/div, 5μs/div)

Fig. 8. Measured inductor current I_L and output current I_O waveforms of the proposed interleaved boost converter with coupled inductor.

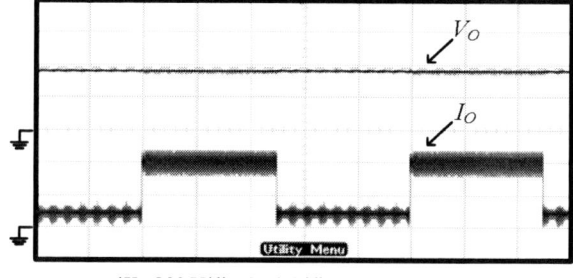

(V_O: 200 V/div, I_O: 1 A/div, 500 ms/div)

Fig. 9. Output voltage V_O and current I_O under step-load changes between 20 % and 100 % of full load of the proposed interleaved boost converter.

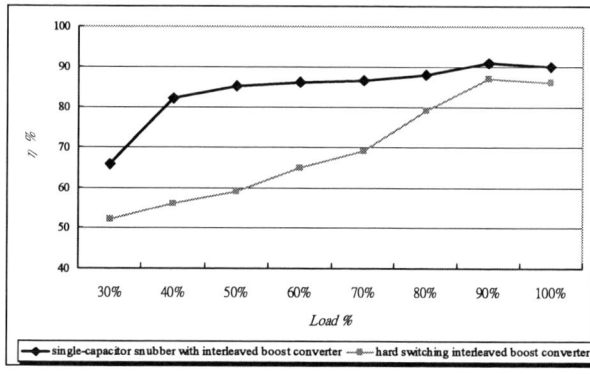

Fig. 10. Comparison between efficiencies of the discussed interleaved coupled-inductor boost converter with hard switching and with the proposed single-capacitor snubber from light load to heavy load.

VI. CONCLUSION

In this paper, the interleaved coupled-inductor boost converter with a single-capacitor turn-off snubber to smooth out turn-off transition has been proposed. Their operational principle, stead-state analysis and design have been described in detail. Additionally, it has been found that the proposed soft-switching converter is using less number of component counts to achieve almost the same dynamic performance as their counterparts. In particular, from the efficiency comparison, it has been shown that the proposed converter can yield higher efficiency than their counterparts at light load condition and almost the same conversion efficiency of the one with hard switching at heavy load condition. An experimental prototype for a grid-connection application

(700 W, 350 V/2 A) has been built and evaluated, achieving the efficiency of 87 % under fuel loud condition and verifying feasibility of the proposed snubber.

REFERENCES

[1] J. L. Duarte, J. A. A. Wijntjens, and J. Rozenboom, "Designing light sources for solar-powered systems," *in Proc. 5th European Conference on Power Eletronics and Applications*, vol. 8, 1993, pp. 87–82.

[2] U. Germann and H. G. Langer, "Low cost DC to AC converter for photovoltaic power conversion in residential applications," *Proc. IEEE PESC'93 Conference*, 1993, pp. 588–594.

[3] J.-C. Hung, *et al.*, "An Active-Clamp Push-Pull Converter for Battery Sourcing Applications," *Proceedings of the APEC*, Vol. 2, 2005, pp. 1186–1192.

[4] H. Sakamoto, *et al.*, "A Self Oscillated Half Bridge Converter Using Impulse Resonant Soft-Switching," *Proceedings of the INTELEC*, 2002, pp. 227–231.

[5] T.-F. Wu, *et al.*," Applications of Soft-Switching Full-Bridge Converter and Rotational Electric Field to Transdermal Drug Delivery," *Proceedings of the APEC*, Vol. 2, 2004, pp. 919–925.

[6] P.-L. Wong, *et al.*," Performance improvements of interleaving VRMS with coupled inductors," *Proceedings of the Applied Power Electronics Conference and Exposition*, 2000, pp. 973–978.

[7] C. Chang, *et al.*," Current ripple bounds of interleaved power converters," *IEEE Transactions on Aerospace and Electronic Systems*, 1996, pp. 1505–1508.

Buck-Boost Converter Associated with Active Clamp Forward Converter for PV Power System

S. -Y. Tseng, W. –C. Chen, Y. –J. Li and * J. -S. Kuo

Department of Electric Engineering
Chang Gung University
Kwei-Shan Tao-Yuan,Taiwan, R.O.C
Tel: 03-2118800
Email: sytseng@mail.cgu.edu.tw

*Department of Electric Engineering,
Chienkuo Technology University
Changhua, Changhua City, Taiwan, R.O.C

Abstract--This paper presents an LED lighting system with a PV source. The circuit structure of the proposed system adopts buck-boost converter associated with forward converter. In this research, buck-boost converter is used as a charger for charging battery, and forward converter with active clamp circuit is adopted to drive LED lighting, and to respectively recover and reset energies stored in leakage inductor and magnetizing inductor of the transformer. To simplify the circuit structure, switches of two converters are integrated with a synchronous switch technique. With this approach, the proposed system has several merits, which are a less component counts, lighter weight, smaller size and higher conversion efficiency. Compared with the conventional converter with hard-switching circuit, the proposed one can improve conversion efficiency of 6% and achieve efficiency about 82% under full load when the proposed system is working in the driving LED condition. Experimental results obtained from a prototype with the output voltage of 10 V and output power of 20 W have verified its feasibility.

Keywords: LED lighting system, buck-boost converter, forward converter, active clamp circuit, synchronous switch technique.

I. INTRODUCTION

In recent years, light emitting diodes (LEDs) are becoming more prevalent in a wide variety of applications. Due to advances over the past few decades have increased the efficiency of LEDs many times [1], their applications have grown to include automotive taillights, LCD back lights, traffic signals, and electronic signs [2]-[3]. Additionally, serious greenhouse effect and environmental pollution caused by overusing fossil fuels have disturbed the balance of global climate. To reduce emission of exhausted gases, zero-emission renewable energy sources have been rapidly developed. One of these sources is photovoltaic (PV) arrays, which is clean, quiet and an efficient method for generating electricity. As mentioned above, this paper proposes an LED driving system, which adopts the PV array

energy for traffic signal or electronic sign applications.

In traffic signal or electronic sign applications using PV energy, the power system will inevitably need batteries for storing energy during the day and for releasing energy during the night. Thus, it needs a charger and discharger, as shown in Fig. 1. Since the proposed power system belongs in the low power applications, buck, boost, buck-boost, flyback and forward converters are more suitable for the proposed power system applications [4]~[10].

In these topologies of the low power applications, according to the relationship among PV input voltage V_{PV}, battery voltage V_B and output voltage V_O, the proposed system can choose features of step-up and step-down voltage ratio simultaneously as the charger or discharger for a wide applications. For this reason, the charger of the proposed power system adopts buck-boost converter and the discharger of one uses forward converter. Since forward converter exits two problems, which are the energies trapped in leakage inductor and magnetizing inductor of transformer T_f, it needs a snubber circuit to recover these energies. Thus, forward converter can use an active clamp circuit to solve these problems. To simplify the proposed circuit structure and increase conversion efficiency, a bidirectional buck-boost converter and active clamp forward converter are used, as shown in Fig. 2. Since the charger and discharger of the proposed power system are operated in complementary and with switch S_1 to control their operational states, the inductor L_1 of buck-boost converter and primary winding inductor L_m of transformer T_f can be merged. Thus, switches between the bidirectional buck-boost converter and active clamp forward converter are integrated with the synchronous switch technique [11] to reduce their component counts, as shown in Fig. 3(c). With this circuit structure, the proposed one can yield high efficiency, reduce weigh, size and volume, and increase the discharging time of battery significantly.

978-1-4244-0644-9/07/$25.00 ©2007 IEEE

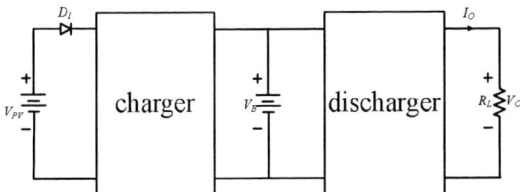

Fig. 1. Block diagram of PV power system for LED lighting.

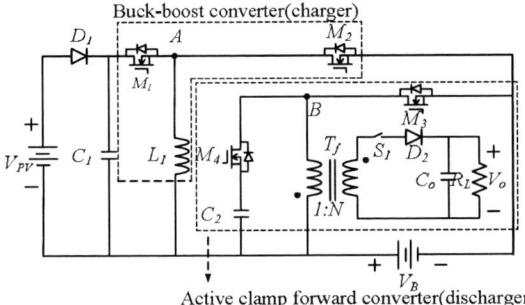

Fig. 2. Schematic diagram of the conventional PV power system.

II. DERIVATION OF THE PROPOSED CONVERTER

In general, an LED lighting system with PV source consists of a charger and discharger, as shown in Fig. 1. Although a charger and dischger can charge battery during the day and drive LED lighting during the night, they need more component counts and driving circuit, resulting in higher cost, and larger volume and size. Additionally, to increase conversion efficiency of the LED lighting, a soft-switching circuit is inserted into a charger or discharger. As mentioned above, a birdirectional buck-boost converter as a charger and an active clamp forward converter as a discharger with synchronous switch technique [4] to integrate switches of two converters for solving above problems are proposed. In the following, the proposed LED lighting system with synchronous switch technique is derived.

Buck-boost converter and active clamp forward converter are integrated in an LED lighting system, as shown in Fig. 2. Since the battery charger and discharger are operated in complementary and use switch S_1 to control their operational states, nodes A and B are combined as a node C, as shown in Fig. 3(a). It will not affect the original operational of converters. Additionally, switches M_1 and M_4 share a common node and they can be operated in synchronous, switch pair (M_1, M_4) can be integrated with switch integration. Similarly, switch pair (M_2, M_3) can also combined, as shown in Fig. 3(b). In Fig. 3(a), inductor L_1 and magnetizing inductor L_m of transformer T_f are connected in parallel. If magnetizing inductor L_m is used as a choke inductor of buck-boost converter, it can make the one operated in continuous conduction mode (CCM) or discontinuous conduction mode (DCM) which avoids

a complete saturation of transformer core between light load and heavy load. Inductor L_1 and magnetizing inductor L_m can be merged as the inductor L_{1m}.

Due to two converters operated in complementary, voltage across switches M_1 and M_4 (or M_2 and M_3) are the same value. Therefore, diodes D_{B141}, D_{B142}, D_{B231}, and D_{B232}, can be shorted, while diodes D_{F141}, D_{F142}, D_{F231} and D_{F232} can be removed. Capacitors C_1 and C_2 are connected in parallel. As a result, capacitors C_1 and C_2 can be combined with a single capacitor C_c, as shown in Fig. 3(c). Note that switch S_1 can be operated by manual or automatic method to control the operational states of the proposed system. From Fig. 3(c), it can be seen that the proposed circuit structure can use a less component counts to achieve the same femctions. Thus, the proposed system can reduce its cost, weight and size and can increase its conversion efficiency significantly.

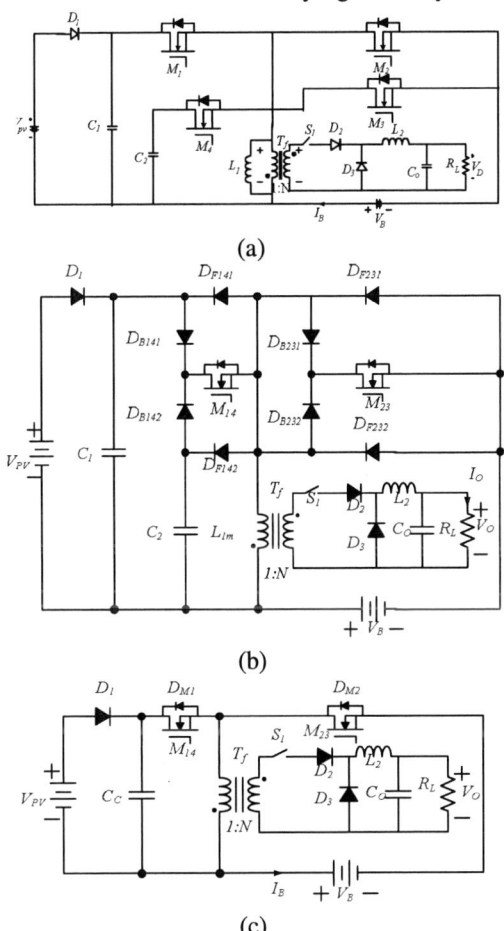

Fig. 3. Derivation of the proposed PV power system for LED lighting.

III. OPERATIONAL PRINCIPLE OF THE PROPOSED CONVERTER

The proposed converter, as shown in Fig. 3(c), consists of buck-boost converter as a charger and active clamp forward converter as a discharger. Since changer and discharger are operated in complementary, their operational principle can be described respectively as follows:

A. Charging state for battery

In charging state for battery, buck-boost converter is adopted. Therefore, switch M_1 is controlled to regulate the charging current of battery, and switch S_1 is kept in off state. According to operational mode of buck-boost converter, when the one is operated in CCM, its operational modes is divided into 5 modes, as shown in Fig.4. At t_0, switch M_1 is turned on. Inductor current I_{Lm} increases linearly. During this time interval, buck-boost converter is in storing energy state. At t_1, switch M_1 is turned off and M_2 is kept in off state. Since inductor current I_{Lm} must be in smooth transition, energy stored in inductor L_m is transferred to junction capacitors of switches M_1 and M_2. When $t = t_2$, voltage V_{DS1} across switch M_1 reaches the battery voltage V_B, body diode of switch M_2 is forwardly biased. At t_3, switch M_2 is turned on. Since body diode of switch M_2 is forwardly biased, switch M_2 can be operated with zero voltage switching (ZVS) at turn-on transition. At t_4, switch M_2 is turned off and switch M_1 is kept in off state, energy stored in inductor L_m is released to junction capacitors of switches M_1 and M_2. When $t = t_5$, a switching cycle is complete. Detailed description of the operational principle can be found from [12].

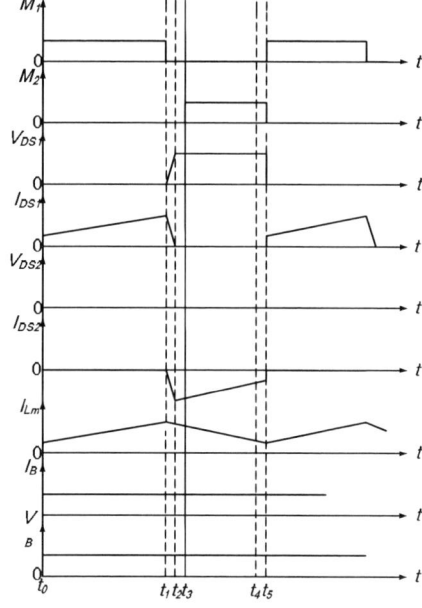

Fig. 4. Concept waveforms of the key components in buck-boost converter.

B. discharging state for LED lighting

The proposed lighting system with LED is applied to a traffic sign. Since source of lighting system is supplied by battery, its discharging efficiency is very important for increasing lighting time under the same energy stored in battery. To achieve higher conversion efficiency, an active clamp forward converter is adopted. The active-clamp circuit can recover the energy trapped in magnetizing and leakage inductors of transformer T_f, and to achieve ZVS at turn-on transition. Its operation can be divided into 8 modes and their equivalent circuit and key waveforms are illustrated in Figs. 5 and 6, respectively. In the following, each operational mode is described brielfly.

Mode 1 [Fig. 5(a); $t_0 \le t < t_1$]: Before t_o, switches M_1 and M_2 are in the off state and body diode D_{M2}, diodes D_2 and D_3 are in forwardly bias. Since body diode D_{M2} of switch M_2 is forwardly biased, voltage across leakage inductor L_K is nearly equal to V_B and voltage across switch M_2 is equal to 0. When $t = t_0$, switch M_2 is turned on. At the same time, switch M_1 is operated with ZVS at turn-on transition. Within this time interval, inductor current I_{LK} abruptly increases. Additionally, energy stored in inductor L_2 is released to load and its current I_{L2} linearly decreases.

Mode 2 [Fig. 5(b); $t_1 \le t < t_2$]: When $t = t_1$, switch M_2 still stays in the on state, while switch M_1 is kept in the off state. Inductor current I_{LK} is equal to the sum of I_{Lm} and I_{N1}. At the same time, current I_{N1} is equal to NI_{L2} ($=NI_{N2}$) where I_{L2} is the initial value of inductor L_2 operated in CCM. Diode D_3 is reversely biased, while diode D_2 still stays in forwardly bias. During this time interval, switch current I_{DS1} is equal to I_{LK} and its value increases linearly. Additionally, inductor current I_{Lm} is increased from a negative value to a positive value. Current I_{L2} is also increased linearly.

Mode 3 [Fig. 5(c); $t_2 \le t < t_3$]: At t_2, switch M_2 is turned off and M_1 is kept in the off state. Within this time interval, current I_{LK} abruptly decreases and is clamped to inductor current I_{Lm}. The energies stored in leakage and magnetizing inductors are released to capacitors C_{M1} and C_{M2}. Thus, voltage across capacitor C_{M1} changes from ($V_B + V_{CC}$) to 0. While, voltage across capacitor C_{M2} varies form 0 to ($V_B + V_{CC}$). Inductor current I_{L2} is decreased linearly and the energy stored in one is released to load.

Mode 4 [Fig. 5(d); $t_3 \le t < t_4$]: At t_3, body diode D_{M1} of switch M_1 is in forwardly bias and inductor current I_{Lm} is equal to I_{Lm}. At the moment, inductors, which include leakage and magnetizing inductors, resonate with capacitor C_C. Within this time interval, diode D_2 is in freewheeling through inductor L_2 and inductor current I_{L2} decreases linearly.

Mode 5 [Fig. 5(e); $t_4 \leq t < t_5$]: At t_4, switch M_1 is turned on. Since body diode D_{M1} of switch is forwardly biased before t_4, switch M_1 is operated with ZVS at turn-on transition. During this time interval, inductors L_K and L_m, and capacitor C_C stay in resonant manner. Inductor current L_2 still decreases and it supplies energy to load.

Mode 6 [Fig. 5(f); $t_5 \leq t < t_6$]: When $t = t_5$, the resonant current I_{Lk} ($=I_{DS1}$) is equal to 0. Current I_{Lk} increases from 0 to a positive value through switch M_1. Therefore, body diode D_{M1} is reversely biased. During this time interval, current I_{L2} still decreases linearly.

Mode 7 [Fig. 5(g); $t_6 \leq t < t_7$]: At t_6, switch M_1 is turned off. Since the resonant current I_{Lk} is a positive value, it will charge capacitor C_{M1} and discharge capacitor C_{M2}. Within this time interval, energy stored in inductor L_2 is released to load and inductor current I_{Lk} decreases linearly.

Mode 8 [Fig. 5(h); $t_7 \leq t < t_8$]: At t_7, energy stored in capacitor C_{M2} is completely discharged. Therefore, body diode D_{M2} is forwardly biased. At the same time, voltage across switch M_2 is equal to 0. During this time interval, energy stored in inductor L_2 is still released to load. When switch M_2 is turned on at the end of mode 8, a new switching cycle will start.

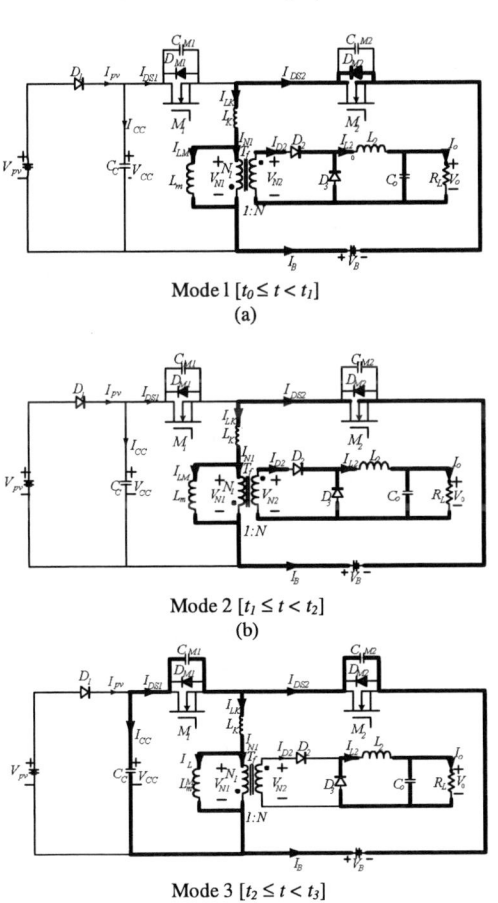

Mode 1 [$t_0 \leq t < t_1$]
(a)

Mode 2 [$t_1 \leq t < t_2$]
(b)

Mode 3 [$t_2 \leq t < t_3$]
(c)

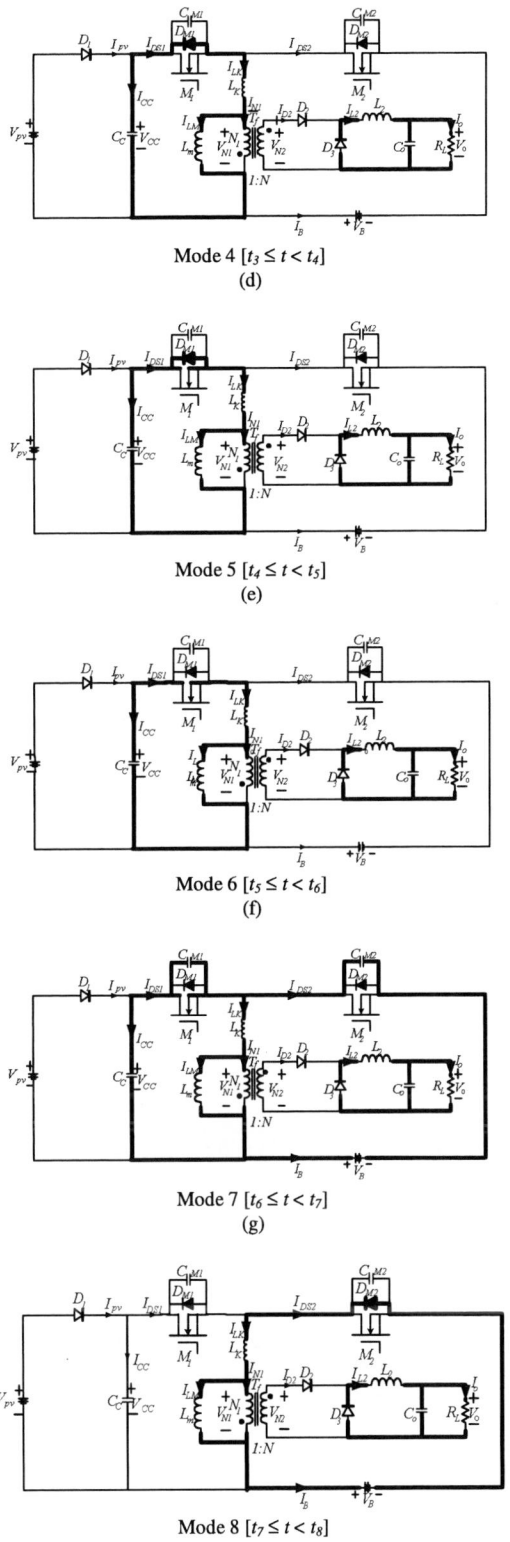

Mode 4 [$t_3 \leq t < t_4$]
(d)

Mode 5 [$t_4 \leq t < t_5$]
(e)

Mode 6 [$t_5 \leq t < t_6$]
(f)

Mode 7 [$t_6 \leq t < t_7$]
(g)

Mode 8 [$t_7 \leq t < t_8$]
(h)

Fig. 5. Equivalent circuit modes of the proposed active clamp forward converter over a switching cycle.

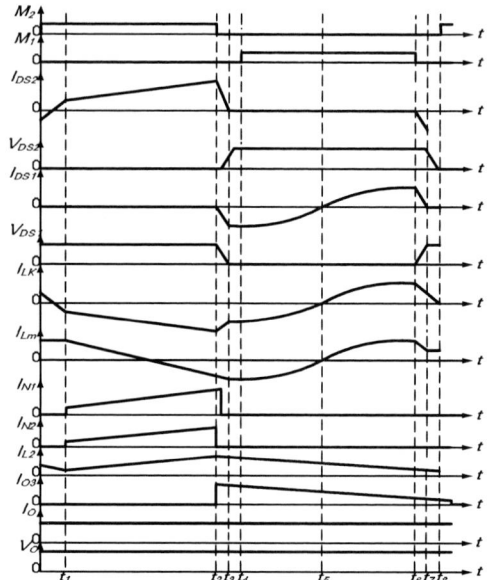

Fig. 6. Conceptual voltage and current waveforms of the key components of active clamp forward converter.

IV. DESIGN OF THE PROPOSED LIGHTING SYSTEM

The proposed system is composed of buck-boost converter and active clamp forward converter. Since switches and inductors in the buck-boost converter and the active clamp forward converter are integrated with the synchronous switch technique, design of the proposed lighting system must satisfy requirements of each circuit. In the following, buck-boost converter and active clamp forward converter are analyzed briefly.

A. buck-boost converter

Since buck-boost converter is operated with varying duty ratio manner to control the charging current of battery, its design only avoid a completely saturation of inductor. Therefore, duty ratio D_1 and inductor L_m are analyzed in the following.

A. 1. Duty ratio D_1

To determine duty ratio, we must first attain input to output voltage transfer ratio M_1. According to volt-second balance principle, the following equation can be obtained:

$$V_{PV}D_1T_S + (-V_B)(1-D_1)T_S = 0, \qquad (1)$$

where V_{PV} is output voltage of PV, V_B is voltage across battery and T_s represents the period of the proposed converter. From (1), it can be found that transfer ratio M_1 can be expressed as

$$M_1 = \frac{V_B}{V_{PV}} = \frac{D}{1-D}, \qquad (2)$$

Based on the energy conservation, the charging current I_B of battery can be indicated by

$$I_{B(av)} = \frac{1-D_1}{D_1}I_{PV(av)}, \qquad (3)$$

where $I_{PV(av)}$ is an average output current of PV. When current $I_{PV(av)}$ is fixed, current I_B is proportional to duty ratio D_1.

A. 2. Inductor L_m

By applying the Faraday's law, N_1 of the transformer T_f can be given as

$$N_1 = \frac{DV_{PV}T_S}{A_C \Delta B} = \frac{D_1V_{PV}T_S}{\Delta \phi}, \qquad (4)$$

where A_C is the effective cross-section area of the transformer core, $\triangle B$ is the working flux density and $\triangle \phi$ represents the working flux. According to analysis principle of magnetic circuit, the working flux $\triangle \phi$ can expressed as

$$\Delta\phi = \frac{F}{R} = \frac{N_1 \Delta I_{Lm}}{R} \qquad (5)$$

where F is a magnetic force, $\triangle I_{Lm}$ is a variation of inductor current I_{Lm} when switch M_1 is turned on and R represents a magnetic resistance of transformer core. To determine the relationship between the average output current $I_{PV(av)}$ and $\triangle I_{Lm}$, concept waveforms of inductor current I_{Lm} and $I_{PV(av)}$ is shown in Fig. 7. From Fig. 7, it can be seen that inductor current $I_{Lm(av)}$ can be derived as follows:

$$I_{Lm(av)} = \frac{[I_{Lm(0)} + I_{Lm(1)}]D_1}{2}, \qquad (6)$$

where $I_{Lm(0)}$ is the initial value of inductor current I_{Lm} operated in CCM and $I_{Lm(1)}$ is the maximum value of I_{Lm}. Current $I_{Lm(1)}$ can be also expressed by

$$I_{Lm(1)} = I_{Lm(0)} + \frac{V_{PV}}{L_m}D_1T_S. \qquad (7)$$

According to (6) and (7), $I_{Lm(av)}$ can be determined and expressed by

$$I_{Lm(av)} = D_1 I_{Lm(0)} + \frac{V_{PV}}{2L_m}D_1{}^2T_S. \qquad (8)$$

Since the average inductor current $I_{Lm(av)}$ is equal to $I_{PV(av)}$, current $I_{B(av)}$ can be expressed by

$$I_{B(av)} = (1-D_1)I_{Lm(0)} + \frac{V_{PV}}{2L_m}D_1(1-D_1)T_S. (9)$$

When $I_{Lm(0)}$, V_{PV}, L_m and T_S are specified, the relationship between $I_{B(av)}$ and duty ratio D_1 is listed in (9). From (4) and (5), it can be seen that when V_{PV}, D_1, T_S and N_1 are specified, the operational range of flux $\triangle \phi$ is limited. Since $\triangle \phi$ is proportional to $\triangle I_{Lm}$, $\triangle I_{Lm}(=\triangle I_{Lm(1)} - \triangle I_{Lm(0)})$ is also limited.

B. Active Clamp Forward converter

In design of the active clamp forward converter, determination of duty ratio D_2 transformer T_f, active clamp capacitor C_C and analyzed briefly.

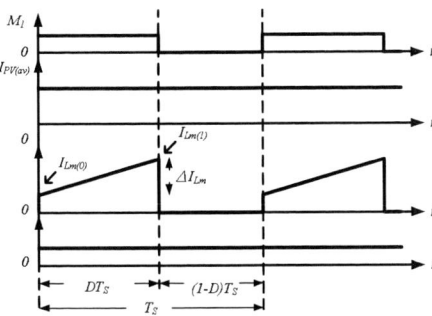

Fig. 7. Concept current waveforms of key components in buck-boost converter.

B.1. Duty ratio D_2

To determine duty ratio, we must first attain input to output voltage transfer ratio M. Since the active clamp circuit only helps switch M_2 to achieve soft-switching feature, it does not affect transfer ratio M_2 of the proposed forward converter. That is, transfer ratio M_2 will be the same as the conventional one. According to volt-second balance principle, the following equation can be obtained:

$$(NV_B - V_O)D_2T_S + (-V_O)(1-D_2)T_S, \quad (10)$$

where $N(=N_2/N_1)$ is the turns ratio of transformer T_f. From (10), it can be found that transfer ratio M_2 can be expressed as

$$M_2 = \frac{V_O V_B}{ND_2} \quad (11)$$

According to (11), a large duty ratio D_2 corresponds to a smaller transformer turns ratio N, which results in a lower current stress imposed on switches M_1 and M_2, as well as voltage stress on freewheeling diode D_2. However, in order to accommodate variations of load, line voltage and component value, it is better to select an operating range as $D = 0.35 \sim 0.4$.

B.2. Transformer T_f

Once the duty is select, the turns ratio of the transformer T_f can be determined using (11), which yields

$$N = \frac{VD}{D_2 V_B} \quad (12)$$

By applying the Faraday's low, N_1 of the transformer T_f can be given as

$$N_1 = \frac{D_2 V_B T_S}{A_C \Delta B} \quad (13)$$

To design the proposed lighting system, design of N_1 must be satisfied with (4) and (13) simultaneously. In the proposed lighting system, V_{pv} is greater than V_B. Thererfore, design of N_1 only meets (4). According to (4) and (12), N_2 can be determined.

To achieve a ZVS feature, the energy stored in inductor L_m must satisfy the following inequality:

$$\frac{L_m(I_{Lm(tv6)} - I_{Lm(tv7)})^2}{2} \geq \frac{(C_{M1} + C_{M2})V_{DS}^2{}_{(max)}}{2} \quad (14)$$

where $I_{Lm(tv6)}$ is the magnetizing inductor current at time t_6, $I_{Lm(tv7)}$ is that at time t_7, C_{M1} and C_{M2} are respectively the junction capacitors of switches M_1 and M_2, and $V_{DS(max)}$ is the voltage across switch M_1 (or M_2) and its value is equal to $(V_B + V_{CC})$. According to volt-second balance principle, the voltage V_{CC} can be expressed by

$$V_{CC} = \frac{D_2 V_B}{(1-D_2)} \quad (15)$$

Once C_{M1}, C_{M2}, and $I_{Lm(tv6)}$ and $I_{Lm(tv7)}$ are specified, the inequality of the magnetizing inductor for L_m can be expressed as

$$L_m \geq \frac{(C_{M1} + C_{M2})V_B^2}{(1-D)^2(I_{Ln(tv6)} - I_{Ln(tv7)})^2} \quad (16)$$

To design the magnetizing inductor L_m, it must satisfy (4) and (16) simultaneously.

B.3 Output Filter L_2 and C_O

Since the proposed forward converter is operated in *CCM*, output filter inductor L_2 must be large enough to maintain *CCM* operation. The inductance of L_2 must satisfy the following inequality:

$$L_2 \geq \frac{V_O(1-D)T_S}{\Delta I_{L2(max)}} \quad (17)$$

where $\Delta I_{L2(max)}$ is the maximum ripple of output inductor current I_{L2}. When the maximum current ripple is specified, the minimum magnetizing inductance can be determined.

The output capacitor C_O is primarily designed for reducing ripple voltage. The ripple voltage across output capacitor C_O is determined as follows:

$$\Delta V_{rco} = \Delta Q_{co}/C_o + \Delta I_{L2(max)} * ESR$$
$$= 1/L_o \times 1/2 \times \Delta I_{L2(max)}/2 \times T_S/2 + \Delta I_{L2(max)} * ESR$$
$$= \Delta I_{L2(max)}/L_O(1/8f_s + C_O * ESR) \quad (18)$$

where *ESR* is the equivalent series resistance. For alumimum electrolytic capacitors, the product of $C_O * ESR$ is much less than $1/8f_s$ and it can be neglected. Thus, capacitor C_O is selected as

$$C_O = \frac{\Delta I_{L2(max)}}{8 f_s \Delta V_{rco}} \quad (19)$$

V. MEASURED RESULTS

To verify the performances of the proposed lighting system, a prototype which is a charger and discharger of an *LED* lighting system, and uses a PV arrays as a source with the following specifications was implemented.

A. Buck-boost Converter (charger)

- Input Voltage V_{PV}: DC 17~20 V (PV arrays),
- Switching frequency f_{S1} : 250 kHz,
- Output Voltage V_B: DC 6 V (battery : 6 v/2.3 Ah), and
- Maximum output current $I_{B(max)}$: 2.3 A

B. Active clamp forward converter (discharger)

- Input voltage V_B : DC 6 V_{dc},
- Switching frequency f_{s2} : 250 kHz,
- Output voltage Vo : 10 V_{dc}, and
- Output current Io : 2 A.

Since the charging current I_B of battery can be changed by duty ratio of switch M_1 in buck-boost converter, it depends on duty ratio. Figs. 8 (a) and (b) respectively show the measured voltage V_{DS} waveform of switch M_1 and current I_B waveform under duty ratios of 0.2 and 0.3, from which it can be seen that the charging current of battery is proportional to duty ratio. When the propose lighting system is operated in the discharging state, active clamp forward is in working. Measured voltage V_{DS} and current I_{DS} waveform of switches M_1 and M_2 are respectively shown in Figs. 9 and 10. Figs. 9 (a) and (b) Shows those under 30 % of full load. While, Fig. 10 shows those under full load. From Figs. 9 and 10, it can be seen that switches M_1 and M_2 are operated with ZVS at turn-on transition. Comparison of conversion efficiency between forward converter with hard switching and one with active clamp circuit from light load to heavy load is shown in Fig. 11, illustrating that the efficiency of the proposed converter is higher than the hard-switching one. Its efficiency is 82 % under full load. Fig. 12, shows step-load charge between 30 % of the full load and the full load, from which it can be observed the voltage regulation of output voltage V_O has been limited within ±2 %.

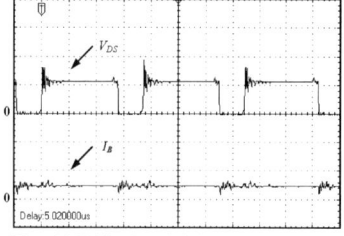

(V_{DS}: 20 V/div, I_{DS}: 0.2 A/div, 1 μs/div)

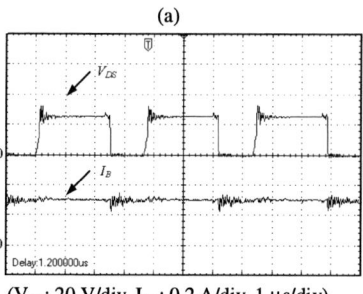

(V_{DS}: 20 V/div, I_{DS}: 0.2 A/div, 1 μs/div)

(b)

Fig. 8. Measured voltage V_{DS} of switch M_1 and the charged current I_B waveform operated in duty ratio of (a) 0.2 and (b) 0.3 for working in the charged state.

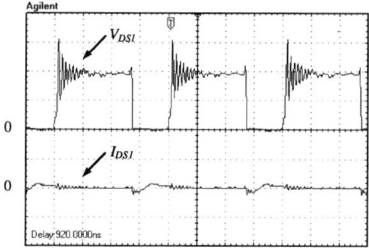

(10 V/div , 5 A/div , 1 μs/div)

(a)

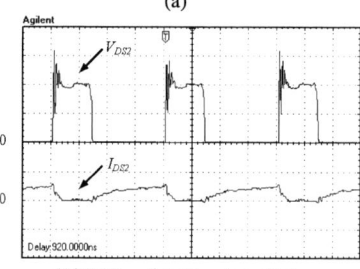

(10 V/div , 5 A/div , 1 μs/div)

(b)

Fig. 9. Measured voltage V_{DS} and current I_{DS} waveform of (a) switch M_1 and (b) switch M_2 for working in the discharged state under 30% of full load.

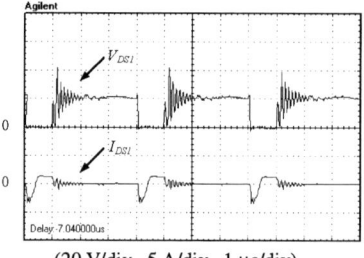

(20 V/div , 5 A/div , 1 μs/div)

(a)

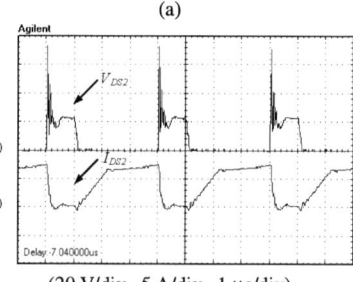

(20 V/div , 5 A/div , 1 μs/div)

(b)

Fig. 10. Measured voltage V_{DS} and current I_{DS} waveform of (a) switch M_1 and (b) switch M_2 for working in the discharged state under full load.

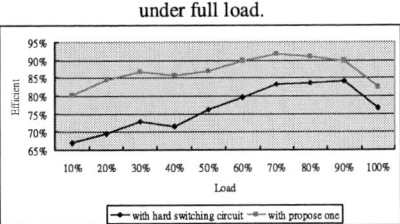

Fig. 11. Comparison conversion efficiency between the conventional hard-switching flyback converter and the proposed one from light load to heavy load for working in the discharged state.

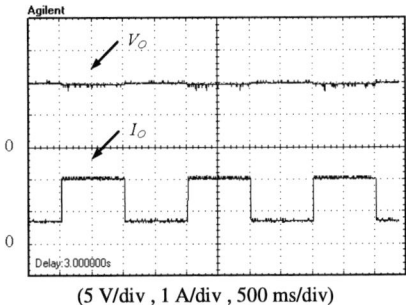

(5 V/div , 1 A/div , 500 ms/div)

Fig. 12. Output voltage V_O and output current I_O under step-load changes between 30% and 100% of full load of the proposed forward converter operated in the discharged state.

VI. CONCLUSION

In this paper, a buck-boost converter associated with active clamp forward converter has been successfully integrated with synchronous switch technique for an LED lighting system. Operational principle, steady-state analysis and design of the proposed lighting system have been analyzed. Additionally, a prototype in which output voltage and current of charger are 6V/2.5A and those of discharger are 10V/2A was implemented. The proposed active clamp forward converter cam achieve the efficiency around 82 % under full load. From experimental results, it can be seen that the proposed lighting system has been verified its feasibility.

REFERENCES

[1]. S. Kohraku and K. Kurokawa, "New Methods For Solar Cells Measurement By Led Solar Simulator," *Proceedings of Photovoltaic Energy Conversion*, Vol. 2, pp. 1977-1980, 2003

[2]. N. Narendran and Y. Gu, "Life of LED-based White Light Sources, " *IEEE Trans. on Display Technology*, Vol. 1, pp. 167-171, 2005

[3]. C. –C. Chen, C. –Y. Wu, T. –F. Wu, "LED back-light driving system for LCD Panels," *Proceedings of Applied Power Electronics Conference and Exposition*, pp. 19-23, 2006

[4]. Tsai-Fu. Wu, T. –H. Yu, "Unified Approach to Developing Single Stage Power Converters, "*IEEE Trans. On Aerospace and Electronic Systems* Vol. 34, pp. 211-223, 1998

[5]. M. Karppanen ., T. Suntio., M. Sippola, "Dynamical Characterization of Input-Voltage-Feedforward-Controlled Buck Converter," *IEEE Trans. On Industrial Electronics* Vol. 54, pp. 1005 – 1013, 2007

[6]. L. –H. Barreto, M. –C. Sebastiao, L. –C. Freitas, E. –A. Coelho, V. –J. Farias, J. –B. Vieira, "Analysis of a soft-switched PFC boost converter using analog and digital control circuits," *IEEE Trans. On Industrial Electronics* Vol. 52, pp. 221 – 227, 2005

[7]. K. –C. Wu, "A comprehensive analysis of current-mode control for DCM buck-boost converters," *IEEE Trans. On* Vol. 51, pp. 733 – 735, 2004

[8]. B. –R. Lin and F. –Y. Hsieh, "Soft-Switching Zeta–Flyback Converter With a Buck–Boost Type of Active Clamp," *IEEE Trans. On Industrial Electronics* Vol. 54, pp. 2813 – 2822, 2007

[9]. Y. –K. Lo, T. –S Kao, J. –Y. Lin, "Analysis and Design of an Interleaved Active-Clamping Forward Converter," *IEEE Trans. On Industrial Electronics* Vol. 54, pp. 2323 – 2332, 2007

[10]. S. –S. Lee, S. –W. Choi, G. –W. Moon, "High-Efficiency Active-Clamp Forward Converter With Transient Current Build-Up (TCB) ZVS Technique," *IEEE Trans. On Industrial Electronics* Vol. 54, pp. 310 – 318, 2007

[11]. L. -P. Wong, D. -K. Chang, M. -H. Chow and Y. -S. Lee., "Interleaved three-phase forward converter using integrated transformer," *IEEE Trans. On Industrial Electronics* Vol. 52, pp. 1246 – 1260, 2005

[12]. M. K. Kazimierczuk. and R. I. Cravens., "Input impedance of closed-loop PWM buck-boost DC-DC converter for CCM," *IEEE International Symposium on Circuits and Systems* Vol. 3, pp. 2047-2050, 1995

Comparison of Three-Phase DC-DC Converters vs. Single-Phase DC-DC Converters

Christian P. Dick, Andreas König, Rik W. De Doncker
Institute for Power Electronics and Electrical Drives
RWTH Aachen University
Jaegerstr. 17/19, D - 52066 Aachen, Germany
Phone: +49 241 80-96936, Fax: +49 241 80-92203
Email: di@isea.rwth-aachen.de

Abstract—In this paper a comparison of common single-phase and three-phase DC-DC resonant and non-resonant converter bridges on application level is carried out. Device stresses and efficiencies are qualified at constant transferred power and constant terminal voltages. As a result, this paper presents the predominance of three-phase topologies when talking about switching losses in silicon devices of non-resonant solutions. Subsequently transformer efficiency is discussed in detail for specific core geometries. Finally the stress on the input capacitors is qualified. Concerning losses and charge to be buffered the predominance of the three-phase topology is shown.

I. Nomenclature

a	shaft length
A_w	copper window
A_c	core cross sectional area
B_pp	maximum flux density excitation
f, $\omega = 2\pi f$	operating frequency
I_AC	current in the AC-link
I_RMS	RMS value of I_AC
I_AVG	average value of I_AC
N	number of turns on transformer primary side
$\overline{P}$	average power being transferred
P	transient power being transferred
P_c	core losses
P_w	winding losses
$P_\mathrm{sw,0}$	constant
V_c	core volume
V_w	winding volume
Q	charge to be buffered
$r_\mathrm{T,eff}$	transformer winding ratio
$1,2,3$	indices 1,2,3-leg-bridge

II. Introduction

DC-DC converters are widely used in a variety of applications. At the same time numerous topologies can be found in literature. In this paper one majority of converters is discussed and compared to each other, being topologies with an AC-link. Fig. 1 gives an overview of the topologies. The two or three-phase active input bridge is followed by an AC-link, which consists of passives. The rectifier bridge can optionally contain active devices. The comparison is carried out on application level which means, that device stresses or costs are regarded at constant given voltages U_in and U_out, and at a constant given transferred power $\overline{P}$.

All topologies discussed in this paper consist of one, two or three half bridges at the input side, consisting of two switching devices each. Hence, topologies with one or two inverter legs belong to the single-phase topologies,

Fig. 1. Block Diagram of the Compared DC-DC Converters

circuits with three inverter legs are three-phase topologies; an overview is given in Table I. First comparisons are given in [1], [2] and [3]. Introductions to resonant converters and their waveforms are presented in [4] and [5]. Investigations of hard switched DC-DC converters connected to photovoltaic modules can be found in [6].

A. Topologies: Fundamentals and Assumptions

Each inverter leg of the discussed topologies (compare Table I) is operated at a duty cycle of 50%. The three-phase bridge operates at a constant phase shift between the inverter legs of 120°. The single-phase 2-leg-bridge offers an additional degree of freedom, a variable phase shift between the inverter legs, indicated with the angle β. With $\beta = 0$ there is no voltage in the AC-link ($U_\mathrm{AC} = 0$), the transferred power is zero. At $\beta = \pi$ the effective (RMS-) AC-voltage of U_AC is maximum which results in a maximum transferred power. In this paper the general equation for the transferred power depending on β is given. However, all the results are calculated for conditions at nominal power, when also the device stresses are maximum. It is assumed that due to the design β equals π at nominal power.

The AC-link also offers several degrees of freedom. Besides purely inductive AC-links, resonant circuits, such as the series and the series-parallel resonant circuit are regarded. All discussed topologies contain a transformer providing galvanic isolation. Fig. 2 shows an exemplary single-phase equivalent circuit of a resonant AC-link. With the pictured devices all preconditions are given for the following calculations of the AC-link.

The considered resonant bridges have a series resonant capacitor C_res, the non-resonant bridges only have the transformer in the AC-link. There is one assumption in the calculations, i.e. the transformer is only described by one passive element being the total leakage inductances $L_\sigma = L_{\sigma,1} + L'_{\sigma,2}$ in its single-phase equivalent circuit (Fig. 2).

A diode rectifier forces the currents to the waveforms depicted in Table I. It should be noted that when com-

978-1-4244-0644-9/07/$25.00 ©2007 IEEE   448

	Active Input Bridge	Voltage Waveform	Current Waveform (Resonant)	Current Waveform (Non-Resonant)
1-Leg-Bridge				
2-Leg-Bridge				
3-Leg-Bridge				

TABLE I

ACTIVE INPUT BRIDGES OF DC-DC CONVERTERS

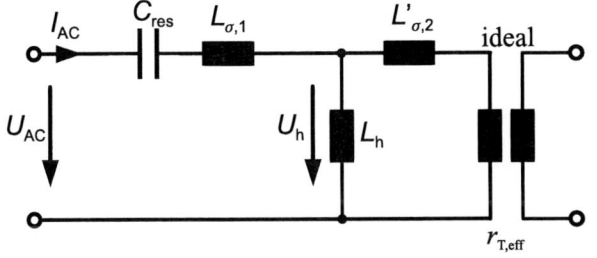

Fig. 2. Single-Phase Equivalent Circuit of an Exemplary AC-Link

it is assumed that the operation takes place in the load independent point with the series resonance being the operating frequency such that there is no dependence on the output voltage.

B. Comparison: Principles and Quantities

Based on the resulting current and voltage waveforms in Table I, different calculations are carried out. In contrast to literature this paper gives a detailed comparison on the topologies by calculating device stresses of active and passive components.

First silicon losses of the input bridge are considered. At constant silicon area, corresponding to constant costs of the bridges, conduction and switching losses are qualified. The diode rectifier is not considered; a similar behavior compared to the active input bridge is assumed.

The transformer is discussed concerning efficiency. With the assumption of equal flux densities and current densities the problem is traced back to a geometric consideration. For a specific core geometry conclusions were found, but the geometries do not avoid an outer leakage field. There are more general conclusions given, regarding the outer leakage field.

The DC-input capacity is analyzed concerning the RMS current corresponding to losses and the voltage ripple which should be avoided. The output capacity is assumed to be stressed in a similar way.

paring the two single-phase bridges this diode rectifier is identical. The only difference within these two converters is, besides the input bridge, the transformer's winding ratio. It is assumed that the single-phase transformers have an identical core shape but different numbers of windings at the primary side, i.e. $N_1 = 0.5 N_2$, $r_{\mathrm{T,eff},1} = 0.5\, r_{\mathrm{T,eff},2}$.

In the non-resonant bridges the quantity U'_{out} is used and is equal to the output voltage transferred via the transformer to the input voltage ($U'_{\mathrm{out}} = r_{\mathrm{T,eff}} U_{\mathrm{out}}$). Consequently, the output voltage transferred to the primary side differs by a factor of two for the single-phase bridges ($U'_{\mathrm{out},2} = 2\, U'_{\mathrm{out},1}$).

The interrelation between input U_{in} and output voltages U_{out}, and the average transferred power $\overline{P}$ depends on the topology and the modulation scheme. It is indicated when necessary in the tables below. In the resonant bridges

	$I_{AC}(t)$	$P(t)$	I^2_{RMS}	R_{ds} p. Switch	I_{AVG}	$P_{loss} = P_{cond}$
1-Leg-Bridge	$\frac{\overline{P}\cdot\pi}{U_{in}}\sin(\omega t)$	$\overline{P}\cdot\frac{\pi}{2}\lvert\sin(\omega t)\rvert$	$\frac{\overline{P}^2\cdot\pi^2}{2U^2_{in}}$	R_{ref}	$2\frac{\overline{P}}{U_{in}}$	$\frac{\overline{P}^2\cdot\pi^2}{2U^2_{in}}R_{ref} + 2\overline{P}\frac{U_{fwd}}{U_{in}}$
2-Leg-Bridge	$\frac{\overline{P}\cdot\pi}{2U_{in}}\sin(\omega t)$	$\overline{P}\cdot\frac{\pi}{2}\lvert\sin(\omega t)\rvert$	$\frac{\overline{P}^2\cdot\pi^2}{8U^2_{in}}$	$2R_{ref}$	$\frac{\overline{P}}{U_{in}}$	$\frac{\overline{P}^2\cdot\pi^2}{2U^2_{in}}R_{ref} + 2\overline{P}\frac{U_{fwd}}{U_{in}}$
3-Leg-Bridge	$\frac{\overline{P}\cdot\pi}{3U_{in}}\sin(\omega t)$	$\overline{P}\cdot\frac{\pi}{3}\sin(\omega t)$ for $\pi/3 \le \omega t \le 2\pi/3$ periodical in $\pi/3$ or $6f$	$\frac{\overline{P}^2\cdot\pi^2}{18U^2_{in}}$	$3R_{ref}$	$\frac{2}{3}\frac{\overline{P}}{U_{in}}$	$\frac{\overline{P}^2\cdot\pi^2}{2U^2_{in}}R_{ref} + 2\overline{P}\frac{U_{fwd}}{U_{in}}$

TABLE II

SILICON LOSSES OF RESONANT BRIDGES

	$\overline{P}$	I_{RMS}	R_{ds} p. Switch	I_{AVG}	I_{off}	P_{cond}	P_{sw}	Remarks
1-Leg-Bridge	$\frac{U'_{out}\cdot\left(U^2_{in}-(2U'_{out})^2\right)}{16\cdot U_{in}f L_\sigma}$ @ $N_1 = 0.5 N_2$	$2I_2$	R_{ref}	$2I_{a2}$	$2I_{o2}$	$4I^2_2 R_{ref} + 2I_{a2}U_{fwd}$	$4\frac{P_{sw,0}I_{o2}}{1\,A}$	$r_{T,1} = 0.5\,r_{T,2}$
2-Leg-Bridge	$\frac{U_{in}U'_{out}}{8f L_\sigma}\left(2\frac{\beta}{\pi} - \frac{U'^2_{out}}{U^2_{in}} - \frac{\beta^2}{\pi^2}\right)$	I_2	$2R_{ref}$	I_{a2}	I_{o2}	$4I^2_2 R_{ref} + 2I_{a2}U_{fwd}$	$4\frac{P_{sw,0}I_{o2}}{1\,A}$	Results @ $\beta = \pi$
3-Leg-Bridge	$\frac{U'_{out}\cdot\left(U^2_{in}-U'^2_{out}\right)}{9\cdot U_{in}f L_\sigma}$	$\approx \frac{2}{3}I_2$	$3R_{ref}$	$\frac{2}{3}I_{a2}$	$\frac{1}{2}I_{o2}$	$4I^2_2 R_{ref} + 2I_{a2}U_{fwd}$	$3\frac{P_{sw,0}I_{o2}}{1\,A}$	for $U'_{out} > 0.5\,U_{in}$

TABLE III

SILICON LOSSES OF NON-RESONANT BRIDGES

III. SILICON LOSSES OF THE INPUT BRIDGE

A. Conduction Losses P_{cond}

The on-state losses are determined with the on-state resistance R_{ds} and the forward voltage drop U_{fwd} by eq. (1),

$$P_{cond} = \sum \left(R_{ds}I^2_{RMS} + U_{fwd}I_{AVG} \right) \tag{1}$$

with the rms-value I_{RMS} and the average value I_{AVG} of the current I_{AC}. I_{AC} always flows either through the upper or the lower switch such that at all points in time R_{ds} and U_{fwd} of one device per half bridge cause losses. Thus, loss calculation is carried out with I_{RMS} and I_{AVG} of one device per half bridge. There is a constraint for the non-resonant bridges: Due to the negative current in the devices the calculation with U_{fwd} is valid only if Mosfets are used and if the channel is conducting - not the intrinsic diode. In case of resonant bridges the calculation is also valid for IGBTs, due to the always positive current through the devices.

The comparison is carried out on application level for constant transferred power $\overline{P}$. Thus, I_{RMS} and I_{AVG} are derived from the current waveforms in Table I, the results are depicted in Table II and III. Furthermore, the total silicon area of each input bridge, thus costs, is kept constant. At a constant input voltage the thickness of silicon is fixed, such that the resulting R_{ds} is simply increased by the number of inverter legs. For the 1-leg-bridge the reference value R_{ref} is introduced.

As a result it can be seen that the single-phase bridges as well as the three-phase bridges show identical conduction losses at constant silicon area (compare Tables II and III). From this point of view, there is no topology to be preferred.

B. Switching Losses P_{sw}

In resonant converters, the silicon devices are switched at zero current which results in soft-switching behavior. For high efficiency applications the resonant tank should be operated slightly above resonant frequency, resulting in a turn-off instance at near zero current (near-ZCS). Then the remaining current is used to achieve zero-voltage switching at turn on (ZVS). Due to both ZVS and near-ZCS the switching losses are neglected for resonant bridges in this comparison.

In non-resonant DC-DC-converter topologies, the switching losses are significant and have therefore to be considered in this comparison. Although ZVS is assumed to be achieved at turn on, the switches are turned off hard. Thus, besides the RMS-value of the current I_{AC} the turn-off current I_{off} is important for the comparison of the converters. The switching losses can be derived via:

$$P_{sw} = f \cdot \sum \left(\frac{1}{2}U_{in}I_{off}t_{sw} \right) = \frac{P_{sw,0}}{1\,A} \cdot \sum I_{off} \tag{2}$$

For a better readability of Table III some quantities are referred to the corresponding quantities of the 2-leg-bridge as follows:

$$I_{RMS,2} = I_2 = \frac{\sqrt{3}}{96}\frac{U_{in} - U'_{out}}{U_{in}f L_\sigma}\sqrt{\frac{(U_{in} + U'_{out})^3}{U_{in}}}$$

$$I_{\text{off},2} = I_{o2} = \frac{U_{\text{in}}^2 - U_{\text{out}}'^2}{16 \cdot U_{\text{in}} f \, L_\sigma} \tag{3}$$

$$I_{\text{AVG},2} = I_{a2} = \frac{1}{128} \frac{U_{\text{in}}^3 + U_{\text{in}}^2 U_{\text{out}}' - U_{\text{in}} U_{\text{out}}'^2 - U_{\text{out}}'^3}{U_{\text{in}}^2 f \, L_\sigma}$$

In the non-resonant 1-leg and 2-leg-bridge topologies the maximum current is always switched off, compare the waveforms in Table I. In the 3-leg-bridge however the current I_{off} is smaller than the maximum current, resulting in lower overall switching losses as indicated in Table III.

IV. Transformer Comparison

This section on comparing single-phase with three-phase transformers is divided in five parts. First, it will be shown, that at constant peak-to-peak flux density B_{pp}, the specific losses are constant, resulting in a relation between the number of turns on the primary side N and the core's cross sectional area A_c. Thus, core losses only depend on core volume. In the second step the calculation of winding losses is traced back to a geometric problem. Subsequently, all geometric relations are considered with a given three-phase core. The argumentation to lower losses of single-phase transformers is presented for the given geometries. Finally, optimized core geometries avoiding an outer leakage flux are discussed such that there are some general conclusions on the advantages and disadvantages of three-phase transformers vs. single-phase transformers.

The single-phase 1-leg-bridge is not regarded, since with $N_1 = 0.5 \, N_2$, $A_{w1} = A_{w2}$ and identical core geometry both single-phase transformers show a similar performance. Thus, only the indices 2 and 3 are found in this section.

A. Core Losses P_c

The voltage U_h at the mutual inductance of the transformer can be approximated with the voltage waveforms $U_{\text{AC}}(t)$ in Table I for the non-resonant converters and also for the resonant converters in case of no-load operation. Hence, the peak to peak flux density B_{pp} can be calculated with $U_h(t) = -N \cdot A_c \cdot \frac{dB}{dt}$ via integration which yields in eq. (4).

$$\binom{B_{\text{pp2}}}{B_{\text{pp3}}} = \frac{U_{\text{in}}}{f} \binom{\frac{1}{2N_2 A_{c2}}}{\frac{2}{9N_3 A_{c3}}} \tag{4}$$

With the above introduced assumption that the excitation of the cores should be equivalent ($B_{\text{pp2}} = B_{\text{pp3}}$), the boundary condition in eq. (5) is fixed for the following steps.

$$N_3 A_{c3} = \frac{4}{9} N_2 A_{c2} \tag{5}$$

Specific (local) core losses $P_{c,\text{sp}}$ for ferrites are modeled in [7] and can be estimated by eq. (6),

$$P_{c,\text{sp}} = k_i \frac{1}{T} \left(B_{\text{pp}}\right)^{\beta - \alpha} \int_0^T \left|\frac{dB}{dt}\right|^\alpha dt \tag{6}$$

with α, β Steinmetz parameters, and k_i a constant derived in [7]. The concrete specific losses of the single-

phase and the three-phase core and their relation, taking eq. (5) into account, turn out to be:

$$\frac{P_{c,\text{sp},2}}{P_{c,\text{sp},3}} = \frac{k_i \cdot (B_{\text{pp2}})^{\beta - \alpha} \cdot \left(\frac{U_{\text{in}}}{N_2 A_2}\right)^\alpha}{\frac{k_i}{3}(B_{\text{pp3}})^{\beta - \alpha} \left[2\left(\frac{U_{\text{in}}}{3N_3 A_3}\right)^\alpha + \left(\frac{2U_{\text{in}}}{3N_3 A_3}\right)^\alpha\right]}$$

$$\overset{(5)}{=} \frac{2}{3}\left(\frac{3}{4}\right)^\alpha + \frac{1}{3}\left(\frac{3}{2}\right)^\alpha \tag{7}$$

For a Steinmetz parameter of $\alpha = 1.3$, which is a realistic value, the fraction of eq. (7) turns out to be approximately one (1.02). Thus, for the calculation of core losses only the core volume has to be determined; the relation between the core losses of the single or three-phase transformer is determined by eq. (8):

$$\frac{P_{c2}}{P_{c3}} = \frac{V_{c2}}{V_{c3}} \tag{8}$$

B. Winding Losses P_w

For the winding losses only the DC-losses are considered. The Skin- and proximity effect can be minimized in both cases single- and three-phase transformers by equal methods, e.g. using litz-wire. The losses can be derived conventionally by eq. (9).

$$P_w = 2\rho \frac{l \cdot N}{\frac{A_w}{2N}} \cdot I_{\text{RMS}}^2 = \rho J_{\text{RMS}}^2 \cdot V_w \quad , \tag{9}$$

with ρ specific resistance, N, number of turns on primary side, l medium turn length and $\frac{A_w}{2N}$ cross section of one turn. The factor of two is reasoned by calculating twice the losses of the primary winding. J_{RMS} is the effective current density, regarding a constant copper fill factor for single- and three-phase transformer, defined with eq. (10):

$$I_{\text{RMS}} = J_{\text{RMS}} \frac{A_w}{2N} \tag{10}$$

The relation between the copper losses of the single and the three-phase transformer is determined by eq. (11):

$$\frac{P_{w2}}{P_{w3}} = \frac{V_{w2}}{V_{w3}} \frac{J_{\text{RMS2}}^2}{J_{\text{RMS3}}^2} \tag{11}$$

With the additional assumption of constant current densities $J_{\text{RMS2}} = J_{\text{RMS3}}$, which corresponds to constant flux densities B_{pp} in the core, the relation between the winding losses only depends on geometry.

$$\frac{P_{w2}}{P_{w3}} = \frac{V_{w2}}{V_{w3}} \tag{12}$$

C. Geometric Considerations

Fig. 3 and Fig. 4 show the single-phase and the three-phase core being discussed. It is assumed that all coils are wound onto the round shafts. For the single-phase transformer there are two windings on two shafts, for the three phase version there are six windings on three shafts. There are no further constraints on winding configurations.

For solving eqs. (8) and (12), the following set of equations is derived from Fig. 3 and 4:

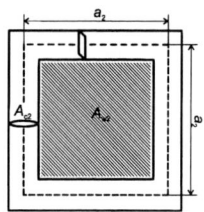

Fig. 3. Geometry of the Single-Phase Core

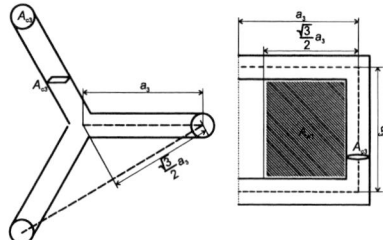

Fig. 4. Geometry of the Three-Phase Core

$$A_{\text{w2}} = \left(a_2 - \sqrt{A_{\text{c2}}}\right)\left(a_2 - \sqrt{\frac{4}{\pi}A_{\text{c2}}}\right)$$

$$V_{\text{c2}} = 4A_{\text{c2}}a_2$$

$$V_{\text{w2}} = 2\pi\left(a_2 - \sqrt{A_{\text{c2}}}\right)\left(\frac{1}{4}a_2^2 - \frac{A_{\text{c2}}}{\pi}\right)$$

$$A_{\text{w3}} = \left(a_3 - \sqrt{A_{\text{c3}}}\right)\left(\frac{\sqrt{3}}{2}a_3 - \sqrt{\frac{A_{\text{c3}}}{\pi}}\right) \quad (13)$$

$$V_{\text{c3}} = 9A_{\text{c3}}a_3$$

$$V_{\text{w3}} = 3\pi\left(a_3 - \sqrt{A_{\text{c3}}}\right)\left(\frac{3}{4}a_3^2 - \frac{A_{\text{c3}}}{\pi}\right)$$

D. Derivation of Losses

For the comparison and the calculation of the relative losses in eqs. (8) and (12) of the single and three-phase transformers all necessary equations are given with eqs. (5), (10) and (13). As input parameters the design parameter $\frac{N_2}{N_3}$ was to be chosen, whereas the current relation is fixed from Table II to $\frac{I_{\text{RMS2}}}{I_{\text{RMS3}}} = \frac{3}{2}$. The set of equations is nonlinear and is not solved analytically but numerically. Thus, a three-phase reference core, compare Fig. 5, is given with $a_3 = 3\,\text{cm}$, $A_{\text{c3}} = 0.8\,\text{cm}^2$, $A_{\text{w3}} = 4.4\,\text{cm}^2$, $V_{\text{c3}} = 21.6\,\text{cm}^3$ and $V_{\text{w3}} = 128.9\,\text{cm}^3$.

Fig. 5. Three-Phase Reference Core on Socket

Figure 6 shows a plot of the relative core and copper losses, with the design parameter $\frac{N_2}{N_3}$ being a variable. With an increasing number of turns on the single-phase core related to the number of turns on the three-phase core, $\frac{P_{\text{w2}}}{P_{\text{w3}}}$ increases and $\frac{P_{\text{c2}}}{P_{\text{c3}}}$ decreases. Within the calculations geometric parameters a_2, A_{c2}, A_{w2}, V_{c2} and V_{w2} are adapted correspondingly.

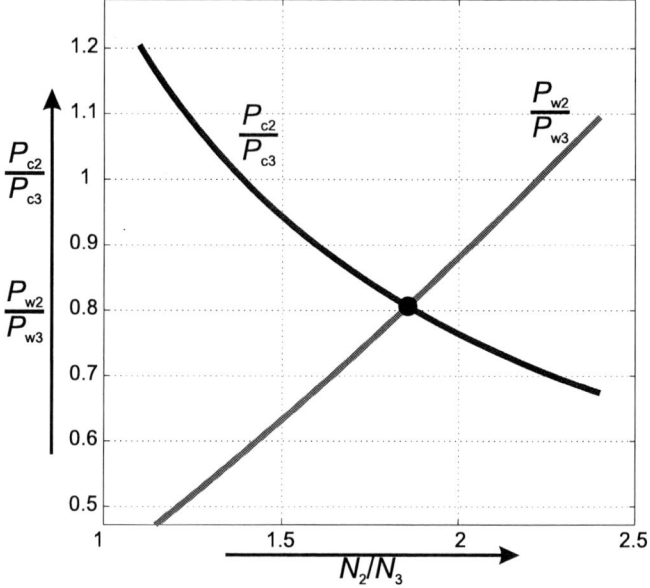

Fig. 6. Relative Core and Copper Losses

This figure does not have a direct conclusion on the absolute total losses. However, in case one transformer (single- or three-phase) is designed to have a constant relation between core and copper losses $\frac{P_{\text{c}}}{P_{\text{w}}} = k_{\text{const}}$, e.g. $k_{\text{const}} = 1$, and the other transformer (three- or single-phase) should show the same relation $\frac{P_{\text{c2}}}{P_{\text{w2}}} = \frac{P_{\text{c3}}}{P_{\text{w3}}} = k_{\text{const}}$, then these assumptions lead to eq. (14):

$$\frac{P_{\text{c2}}}{P_{\text{c3}}} \overset{!}{=} \frac{P_{\text{w2}}}{P_{\text{w3}}} \quad (14)$$

The point, for which eq. (14) is fulfilled, are indicated in Fig. 6 by a dot.

In case a relative winding ratio of $\frac{N_2}{N_3} = 1.85$ could be realized, the single-phase transformer would show nearly 20% less losses compared to the three-phase transformer for the given geometries.

In order to verify this number of 20%, another reference core was applied for the same calculation. The new geometry differs from Fig. 4, such that the heigth or the shaft length $a_{3,\text{shaft}}$ is increased related to the distance of the shaft to the geometry center $a_{3,\text{dist}}$. The second reference core shows the following geometric parameters: $a_{3,\text{shaft}} = 5.15\,\text{cm}$, $a_{3,\text{dist}} = 3.03\,\text{cm}$, $A_{\text{c3}} = 0.78\,\text{cm}^2$, $A_{\text{w3}} = 14.4\,\text{cm}^2$, $V_{\text{c3}} = 26.4\,\text{cm}^3$ and $V_{\text{w3}} = 133.4\,\text{cm}^3$.

The calculations with the second three-phase reference core were carried out analogous to the calculations above. Compared to Fig. 6 the calculations lead to one result that at $\frac{N_2}{N_3} = 0.93$, the three-phase transformer shows 40% less losses compared to the single phase.

Both divergent results show that future research has to be carried out in this field. The author still expects for both reference cores that efficiency could be increased when shifting winding losses to core losses.

E. Conclusions Regarding an Outer Leakage Field

The discussed geometries both show an outer leakage field, which could disturb the surrounding electronics and which could also be the reason for losses in other metal components. To avoid this for the single-phase transformer, one could for example mount both windings, the primary and secondary one, not on two shafts but on one. Thus, both windings are on one side surrounded by the core which forces the magnetic field at the outer side of the winding to zero. The leakage flux, necessary for L_σ and thus necessary for all discussed topologies, must be established inside the transformer.

The discussed three-phase transformer shows the same problem of an outer leakage field as the single-phase transformer. Although the leakage field is not that serious for the three-phase transformer, it should be avoided. To achieve this, additional vertical shafts, compare Fig. 5, should be installed, which do not carry any winding. These shafts could be much smaller in their cross sectional area, because the main flux is carried by the shafts surrounded by the windings.

These necessary geometry modifications influence the calculations above.

V. Capacitor Comparison

The comparison of the input capacitors C indicated in the topologies in Table I is carried out concerning two items, resulting voltage ripple and losses. From the current waveform in the AC-link, from the switching pattern and from a constant input current $\overline{P}/U_{in}$ the current $I_{cap}(t)$ through the capacitor is derived. The calculation of the RMS value yields in a conclusion on losses. Furthermore the charge to be buffered is calculated by

$$Q_{cap} = \int_{t_1}^{t_2} I_{cap}(t)dt \quad , \tag{15}$$

with t_1 and t_2 being two consecutive time instances of zero crossings of the capacitors current. The calculated charge leads to a conclusion on the resulting voltage ripple.

Furthermore, in case of a resonant bridge, the voltage stress of the series-resonant capacitor C_{res} (see Fig. 2) is qualified.

A. Resonant Bridges

1) Input Capacity C: The calculations, summarized in Table V, show that the stresses can be determined analytically. For a comparison of the topologies the relative stress is indicated related to the 2-leg-bridge. It can be seen, that the 1-leg-bridge needs a capacity 5.2 times larger, and the 3-leg-topology needs a capacity 35 times smaller than the 2-leg-bridge in order to gain the same voltage ripple. This corresponds to the results, not deduced analytically, in [4]. Furthermore, at a constant equivalent series resistance the

	Peak Voltage across C_{res}	Remarks
1-Leg-Bridge	$\dfrac{\overline{P}}{2U_{in}} \cdot \dfrac{1}{fC_{res}}$	1 series capacitor is necessary
2-Leg-Bridge	$\dfrac{\overline{P}}{4U_{in}} \cdot \dfrac{1}{fC_{res}}$	1 series capacitor is necessary
3-Leg-Bridge	$\dfrac{\overline{P}}{6U_{in}} \cdot \dfrac{1}{fC_{res}}$	3 series capacitors are necessary

TABLE IV

Stresses on the Resonant Capacitor

1-leg-bridge shows losses 16.8 times larger, and the 3-leg-topology shows losses being 4.8 times smaller than the losses at the 2-leg-bridge.

2) Series Resonant Capacity C_{res}: The voltage stress on a series resonant capacity C_{res} scales with the AC current I_{AC} and thus, it is load dependent. In a resonant converter this stress is sinusoidal and classified with the peak voltage across the capacitor C_{res} in Table IV.

B. Non-Resonant Bridges

The calculations, summarized in Table VI, show that the stresses can be determined depending on the output voltage referred to the primary side. For a comparison of the topologies, the relative stress is indicated related to the 2-leg-bridge. For a realistic comparison it is assumed, that the output voltage referred to the primary side is 95% of the input voltage, and consequantly half of that value for the 1-leg-bridge. It can be seen that for a constant voltage ripple the 1-leg-bridge needs a capacity 4.2 times larger, and the 3-leg-topology needs a capacity 16 times smaller than the 2-leg-bridge. Furthermore, at constant equivalent series resistance the 1-leg-bridge shows losses 8.2 times larger, and the 3-leg-topology shows losses being 31 times smaller than the losses of the 2-leg-bridge.

VI. Conclusions

This paper presents a comparison of DC-DC converter topologies on application level. At constant transferred power and at constant terminal voltages the most common single and three-phase topologies, resonant and non-resonant bridges, are compared mostly by means of analytical equations. It is shown that the silicon conduction losses are constant and independent from the topology at constant silicon area. The non-resonant three-phase bridges show the advantage of 25% reduced switching losses compared to the single-phase non-resonant bridges.

A method to compare different transformers and transformer geometries to each other, is presented, using only general assumptions being a constant maximum core excitation B_{pp}, a constant current density J_{RMS} and a constant ratio between core and copper losses. It is shown that with the two given three-phase reference cores no conclusion can be made on the predominance of either the single-phase transformer or the three-phase one. Future research on transformer core geometries, especially regarding the outer leakage field, has to be carried out.

	$I_{cap}(t)$	$I_{cap,RMS}$	rel. losses $\propto I^2_{cap,RMS}$	Q_{cap} to be buffered	rel. voltage ripple $\propto Q_{cap}$
1-Leg-Bridge	$\dfrac{\overline{P}}{U_{in}}\left\{\begin{array}{l}1-\pi\sin(\omega t)\ \text{for}\ 0\leq\omega t\leq\pi \\ 1 \qquad\qquad \text{for}\ \pi\leq\omega t\leq 2\pi\end{array}\right.$ periodical in 2π or f	$\dfrac{\overline{P}}{U_{in}}\underbrace{\sqrt{\dfrac{\pi^2}{2}-1}}_{=1.9836}$	16.8	$\dfrac{\overline{P}}{\omega U_{in}}\underbrace{\left[2\arcsin\!\left(\dfrac{1}{\pi}\right)+2\pi\cos\left(\arcsin\!\left(\dfrac{1}{\pi}\right)\right)-\pi\right]}_{=3.4627}$	5.236
2-Leg-Bridge	$\dfrac{\overline{P}}{U_{in}}\left(1-\dfrac{\pi}{2}|\sin(\omega t)|\right)$	$\dfrac{\overline{P}}{U_{in}}\underbrace{\sqrt{\dfrac{\pi^2}{8}-1}}_{=0.4834}$	1	$\dfrac{\overline{P}}{\omega U_{in}}\underbrace{\left[2\arcsin\!\left(\dfrac{2}{\pi}\right)+\pi\cos\left(\arcsin\!\left(\dfrac{2}{\pi}\right)\right)-\pi\right]}_{=0.6613}$	1
3-Leg-Bridge	$\dfrac{\overline{P}}{U_{in}}\left(1-\dfrac{\pi}{3}\sin(\omega t)\right)$ for $\pi/3\leq\omega t\leq 2\pi/3$ periodical in $\pi/3$ or $6f$	$\dfrac{\overline{P}}{U_{in}}\underbrace{\sqrt{\dfrac{1}{2}\left(\dfrac{\pi^2}{9}-1\right)}}_{=0.2198}$	0.21	$\dfrac{\overline{P}}{\omega U_{in}}\underbrace{\left[2\arcsin\!\left(\dfrac{3}{\pi}\right)+\dfrac{2\pi}{3}\cos\left(\arcsin\!\left(\dfrac{3}{\pi}\right)\right)-\pi\right]}_{=0.0189}$	0.028

TABLE V

STRESSES ON INPUT CAPACITY OF RESONANT BRIDGES

	$I_{cap}(t)$	$I_{cap,RMS}$	rel. losses $\propto I^2_{cap,RMS}$	Q_{cap} to be buffered	rel. voltage ripple $\propto Q_{cap}$
1-Leg-Bridge	$\dfrac{\overline{P}}{U_{in}}\cdot\left\{\begin{array}{ll}1+2\dfrac{U_{in}}{U'_{out}}\left(1-\dfrac{t}{t_1}\right) & 0\leq t\leq t_1 \\ 1 & t_1\leq t\leq\frac{1}{2f}\end{array}\right.$ for $\frac{1}{2f}\leq t\leq\frac{1}{f}$, periodical in f, $t_1=\dfrac{U_{in}-2U'_{out}}{4fU'_{out}}$	$\dfrac{\overline{P}}{U_{in}}\underbrace{\sqrt{4\dfrac{U_{in}^2}{3U_{out}'^2}-2}}_{=1.9772}$ for $U'_{out}\to 0.95\frac{U_{in}}{2}$	8.19	$\dfrac{\overline{P}}{16fU_{in}}\underbrace{\dfrac{4U_{in}'^3+2U_{out}'^3-7U_{in}U_{out}'^2+4U_{in}^2U'_{out}}{U_{in}^2U'_{out}}}_{=9.5473}$ for $U'_{out}\to0.95\frac{U_{in}}{2}$	4.219
2-Leg-Bridge	$\dfrac{\overline{P}}{U_{in}}\cdot\left\{\begin{array}{ll}1+2\dfrac{U_{in}}{U'_{out}}\left(1-\dfrac{t}{t_1}\right) & 0\leq t\leq t_1 \\ 1-2\dfrac{U_{in}}{U'_{out}}\dfrac{t-t_1}{\frac{1}{2f}-t_1} & t_1\leq t\leq\frac{1}{2f}\end{array}\right.$, periodical in $2f$, $t_1=\dfrac{U_{in}-U'_{out}}{4fU'_{out}}$	$\dfrac{\overline{P}}{U_{in}}\underbrace{\sqrt{4\dfrac{U_{in}^2}{3U_{out}'^2}-1}}_{=0.6909}$ for $U'_{out}\to0.95U_{in}$	1	$\dfrac{\overline{P}}{16fU_{in}}\underbrace{\dfrac{4U_{in}'^3+U_{out}'^3-3U_{in}U_{out}'^2}{U_{in}^2U'_{out}}}_{=2.2630}$ for $U'_{out}\to0.95U_{in}$	1
3-Leg-Bridge	$\dfrac{\overline{P}}{U_{in}}\cdot\dfrac{1}{2U'_{out}U_{out}'^2}\cdot\dfrac{2U_{out}'^3-U_{in}^3}{2U_{out}'^3-U_{in}^3}\cdot$ $\left\{\begin{array}{l}+U_{in}^2U'_{out}-2U_{in}U_{out}'^2+6fU_{out}'^2(2U_{in}-U'_{out})t \\ +3U_{in}^2U'_{out}-4U_{in}U'_{out}+12fU_{out}'^2(U_{in}-U'_{out})t\end{array}\right.$ for $0\leq t\leq\frac{U_{in}-U'_{out}}{3fU_{in}}$, for $\frac{U_{in}-U'_{out}}{3fU_{in}}\leq t\leq\frac{1}{6f}$ periodical in $6f$	$\dfrac{\overline{P}}{U_{in}}\underbrace{\dfrac{\sqrt{U_{in}'^4-4U_{in}'^3U'_{out}+16U_{in}'^2U_{out}'^2-12U_{out}'^4}}{6U'_{out}(U_{in}+U'_{out})}}_{=0.1229}$ for $U'_{out}\to0.95U_{in}$	0.03	$\dfrac{\overline{P}}{16fU_{in}}\underbrace{\left[\dfrac{2}{3}\dfrac{(U_{in}'^2U'_{out}+2U_{out}'^3-U_{in}^3-2U_{in}U_{out}'^2)^2}{U_{in}'^2U'_{out}(2U_{in}-U'_{out})(U_{in}^2-U_{out}'^2)}\right]}_{=0.1348}$ for $U'_{out}\to0.95U_{in}$	0.060

TABLE VI

STRESSES ON INPUT CAPACITY OF NON-RESONANT BRIDGES

The input capacity of the three-phase bridges needs to buffer more than an order of magnitude less charge compared to the single-phase 2-leg-bridge. The single-phase bridge has to buffer about five times the charge of the 2-leg-bridge. With a given equivalent series resistance losses are also reduced drastically when using three-phase topologies.

REFERENCES

[1] R. STEIGERWALD: *A Comparison of Half-bridge DC-DC Soft-Switched Converter Topologies*, IEEE Transactions on Power Electronics, Vol. 3, No. 2, p. 174-182, 1988

[2] R. STEIGERWALD, R. DE DONCKER, M.H. KHERALUWALA: *A Comparison of High-Power DC-DC Soft-Switched Converter Topologies*, IEEE Transactions on Power Electronics, Vol. 32, No. 5, p. 1139-1145, September/October 1996

[3] R. DE DONCKER, D.M. DIVAN, M.H. KHERALUWALA: *A Three-Phase Soft-Switched High-Power-Density dc/dc Converter for High-Power Applications*, IEEE Transactions on Industry Applications, Vol. 27, No. 1, p. 63-73, January/Febuary 1991

[4] J. JACOBS, A. AVERBERG, R. DE DONCKER: *Multi-Phase Series Resonant DC-to-DC Converters: Stationary Investigations*, PESC 2005 Conference in Recife, Brasil

[5] J. JACOBS: *Multi-Phase Series Resonant DC-to-DC Converters*, Dissertation, RWTH Aachen, Aachener Beiträge des ISEA, Shaker Verlag, 2005,

[6] C.P. DICK, K. RIGBERS, H. RADERMACHER, R. DE DONCKER: *Investigations on the Controllability and the Design of DC/DC Converters connected to PV - Generators*, Electrical Power Quality and Utilization, Journal, ISSN 1234-6799, Vol.12, No.2, 2006

[7] K. VENKATACHALAM, C.R. SULLIVAN, T. ABDALLAH, H. TACCA: *Accurate prediction of ferrite core loss with nonsinusoidal waveforms using only Steinmetz parameters*, 2002. Proceedings. 2002 IEEE Workshop on Computers in Power Electronics p. 36-41, June 2002

Applying Modified One-Comparator Counter-Based PWM Control Strategy to Flyback Converter

K. I. Hwu, and Y. H. Chen
Center for Power Electronics Technology
National Taipei University of Technology, Taiwan
eaglehwu@ntut.edu.tw

Abstract- **In this paper, the field programmable gate arrays (FPGA) technique is used to control the flyback converter. The proposed control loop design, together with the one-comparator counter-based PWM control strategy without any analog-to-digital converter (ADC) used is presented and applied to controlling the flyback converter under the discontinuous current mode (DCM). The detailed illustration of the proposed control scheme is provided, together with some experimental results to verify its effectiveness.**

I. INTRODUCTION

As generally acknowledged, there are many disadvantages existing in the analog controller, such as large system delay, noise interference, parameter variations due to thermal variations, and so on, thereby reducing the performance of the DC-DC converter. Therefore, many researches have presented digital control methods [1-16] to overcome these problems. And, among them, the microcomputer (uP) [3-5] or the digital signal processor (DSP) [6-12] has been utilized, whose switching frequency for the DC-DC converter is limited to some extent. Consequently, to overcome such this problem, a field programmable gate arrays (FPGA) based technique [13-16] is applied to controlling the forward the DC-DC converter. Unlike uP and DSP, FPGA has no problem in timing sequence, thereby allowing many processes to go at the same time with the system delay reduced as minimum as possible.

Recently, DC-DC converter, isolated or nonisolated, are getting more and more attractive in industrial applications. Among them, the flyback converter is getting more and more popular due to its inherent advantages, such as simple structure, isolation, multi-output, low cost, etc. As generally acknowledged, the flyback converter is widely operated in the discontinuous current mode (DCM) to get high stability and fast dynamic due to its bilinear characteristics, similar to the boost behavior response. And hence the flyback converter offers the output power up to about 100W. If the flyback converter provides larger output power then new technologies are applied to the flyback converter and this is not considered herein. The aim of this paper focuses on how to digitalize the controller of the flyback converter based on the one-comparator counter-based PWM control strategy [17] without any analog-to-digital converter (ADC) used. As known, the one-comparator counter-based PWM control strategy is based on the triangular-like ripple of the output voltage. However, the ripple of the output voltage of the flyback converter is nontriangular and hence the one-comparator counter-based PWM control strategy can not be

utilized herein. To overcome this problem, the proposed control technique is presented to make this one-comparator counter-based PWM control strategy more flexible in use. The following are operating principles illustrated and some experimental results, to be provided to verify the effectiveness of the proposed control scheme.

II. PROPOSED OVERALL SYSTEM CONFIGURATION

Fig. 1 displays the overall system with the proposed control strategy. The main power stage is a flyback converter. A DCR snubber to protect the main switch S_w contains one resistor R_{is}, one capacitor C_{is}, and one diode D_{is}. The output voltage information, sent to FPGA, is obtained through one high-speed comparator via one high-speed photo-coupler after the voltage divider composed of R_{f1} and R_{f2}. Inside FPGA, there are two modules given. One is a main controller that is used to be in charge of the system timing as well as to tackle the switching timing. The other is a proportional-integral (PI) controller that is used to produce the suitable control effort due to the output voltage error. Also, since the current sourcing from FPGA is not sufficient to drive the MOSFET switches, the gate drivers are added to upgrade the FPGA capability of current sourcing. And, since the primary side is isolated from the secondary side, photo-couplers are used to transfer signals between two sides.

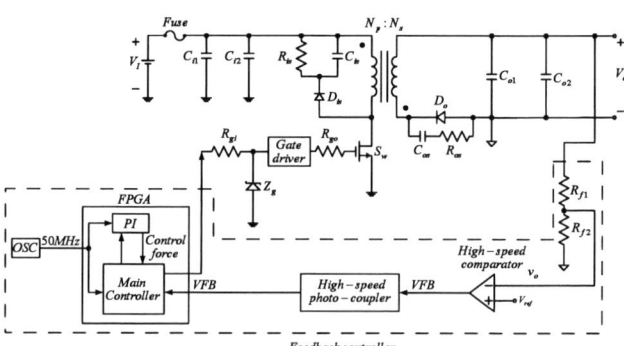

Fig. 1. Proposed system configuration.

III. OVERVIEW OF ONE-COMPARATOR COUNTER-BASED PWM CONTROL AND ITS CONTROL LOOP

Basic operating principles of the one-comparator counter-based control [17] are based on the fact that the sensed output voltage ripple is triangular-like, to be shown in Fig. 2. In Fig. 2, there are two counters created in FPGA whose system clock is set to 100MHz, i.e. the corresponding period is 10ns. One is *PWM_COUNT* employed to count the period T_s of the 9-bit digital PWM signal and the other is *COUNT* utilized to obtain the output voltage information. And, T_s corresponds to 512CLK, i.e. 5.12μs . The moment *PWM_COUNT* becomes zero, *COUNT* is set to zero and the main switch S_1 gets turned on. Besides, the high-speed comparator shown in Fig. 1 is utilized to determine the relationship between the sensed output voltage v_o sent to the negative terminal of the high-speed comparator and the output voltage reference V_{ref} sent to the positive terminal of the high-speed comparator.

As shown in Fig. 2, the output voltage ripple is approximately triangular. As S_1 is turned on during the duty cycle, v_o is to increase. As soon as v_o reaches V_{ref} , the comparator output signal *VFB* changes its status from the high level to the low level, i.e. *VFB* = 0, thereby creating a negative-edged signal that is sent to FPGA. At this moment, *COUNT* starts counting from zero. As soon as *VFB* changes its status from the low level to the high level, *VFB* = 1, *COUNT* stops counting and the resulting value of *COUNT* is saved as *REG*, which is utilized to represent the feedback output voltage information. In this case, the center point of the ripple of v_o locates at V_{ref} , implying that the average or DC value of v_o is equal to V_{ref} . At this moment, *REG* is 256CLK, i.e. the corresponding elapsed time of *COUNT* is 2.56μs that is half of the switching period T_s. On the other hand, if the center point of the ripple of v_o is below the level of V_{ref} , implying that the average or DC value of v_o is smaller than V_{ref} , then the resulting value of *REG* is smaller than 256, i.e. the corresponding sensed output voltage error, obtained by subtracting the value of *REG* from 256, is positive. And, if the center point of the ripple of v_o is beyond the level of V_{ref} , implying that the average or DC value of v_o is larger than V_{ref} , then the resulting value of *REG* is larger than 256, i.e. the corresponding sensed output voltage error is negative. Therefore, the larger the error in the sensed output voltage, the larger the control effort that determines the next duty cycle.

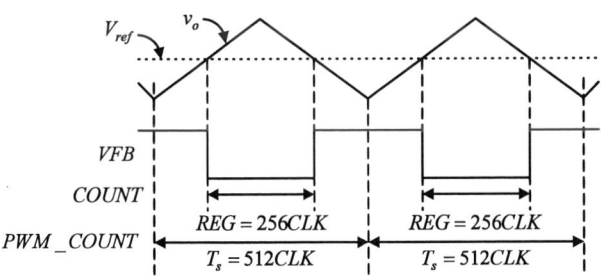

Fig. 2. Center point of the ripple of v_o equal to V_{ref} .

IV. PROPOSED CONTROL STRATEGY

A. *Basic operating principles*

Based on the mention above, the output voltage information is obtained by the triangular-like ripple of the output voltage. However, for the flyback converter to be considered, the ripple of the output voltage is nontriangular but pulsating. Thus, the one-comparator counter-based PWM

control is not suitable herein. That is to say, there are some modifications required herein.

The design concept begins from the flyback converter operating in the boundary current mode (BCM). Based on (1) and (2) for the input power P_i and the output power P_o respectively, if P_i is equal to P_o, then the peak-to-peak current at the secondary is proportional to the square of the output current I_o, and hence the output voltage ripple is proportional to I_o if the output voltage ripple is dominated by the equivalent series resistance (ESR) of the output capacitor.

$$P_i = \frac{1}{2} L_s I_{sp}^2 f_s \qquad (1)$$

$$P_o = V_o I_o \qquad (2)$$

where L_s is the secondary-side self-inductance current, I_{sp} is the secondary-side peak-to-peak current and f_s is the switching frequency.

Figs. 3 and 4 show the output voltage information under the boundary current mode (BCM) and the discontinuous current mode (DCM) in the steady state, respectively. *COUNT* starts counting as *VFB* is changed from the high level to the low level, whereas *COUNT* stops counting as *VFB* is changed from the low level to the high level. Therefore, the low-level duration x_1 in BCM is longer than that x_2 in DCM. And hence, the output voltage error in BCM is smaller than that in DCM, implying that the DC value of output voltage in BCM is smaller than that in DCM. That is to say, the less the load is, the higher the output voltage is.

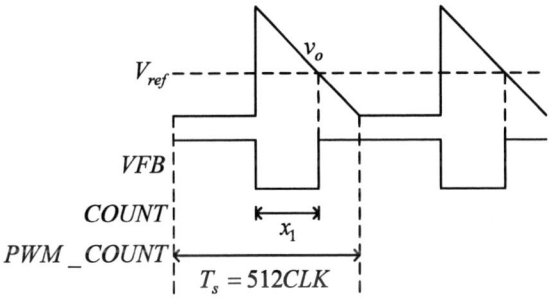

Fig. 3. Output voltage information in BCM.

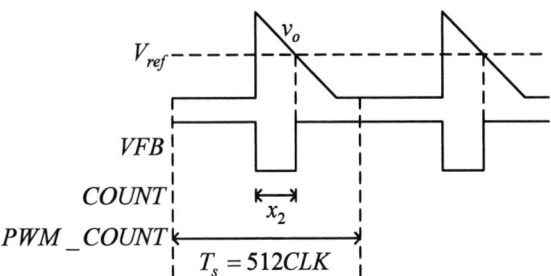

Fig. 4. Output voltage information in DCM.

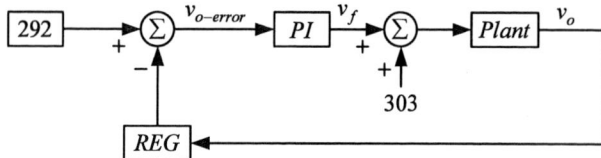

Fig. 5. Proposed control loop.

B. *Parameter tuning philosophy*

Fig. 5 shows the proposed control loop design. The value of *REG* is subtracted from the digital reference 292 to get the output voltage error $v_{o\text{-}error}$ which is sent to the PI controller to get a proper control effort v_f added to 303 so as to obtain an appropriate PWM control signal to drive the main switch S_w in the plant. Under BCM in the steady state, the duty cycle is 43%, which is based on specifications to be mentioned later. Also, the system clock of FPGA is set 100MHz and 9-bit DPWM is taken and hence the switching frequency is 195kHz, digital reference is set to 292 over all the output current range in DCM. As for tuning the parameters of the PI controller, under the rated output load, let the integral factor k_i to be zero and change the proportional factor k_p so as to obtain the output voltage which is about 80% of the desired output voltage, and after this, increase the value of k_i to some value so as to get the desired output voltage within the specifications and with oscillation as minimum as possible. From the time onward, apply such a set of controller parameters to over all the output current range, with some fine tuning added.

V. EXPERIMENTAL RESULTS

Before entering into this topic, there are some specifications given as follows: (i) rate DC input voltage is 96V; (ii) rated DC output voltage is 48V; (iii) rated DC output power is 72W; (iv) switching frequency is set to 195kHz; (v) turns ratio of the main transformer is 10:15; and (vi) parameters of the PI controller are $k_p = 0.25$ and $k_i = 0.125$.

Figs. 6 to 8 show the waveforms of the gate driving signal v_{gs} of the switch S_w, the voltage v_{ds} across S_w, the secondary-side current i_s and the output voltage v_o, for no load, half load and rated load respectively. From the experimental results, it is obvious the proposed control scheme can stabilize the flyback converter and this converter operates in BCM at rated load and in DCM below rated load. Besides, the values of the output voltages shown in Figs. 6 to 8 are close to 50V measured by one isolated probe with an attenuation factor of 50. Besides, using one voltage meter, the measured values of the output voltage versus load current are shown in Table I, with all the values of the output voltage close to 48V. According to the mention above, the effectiveness of the proposed control topology is verified.

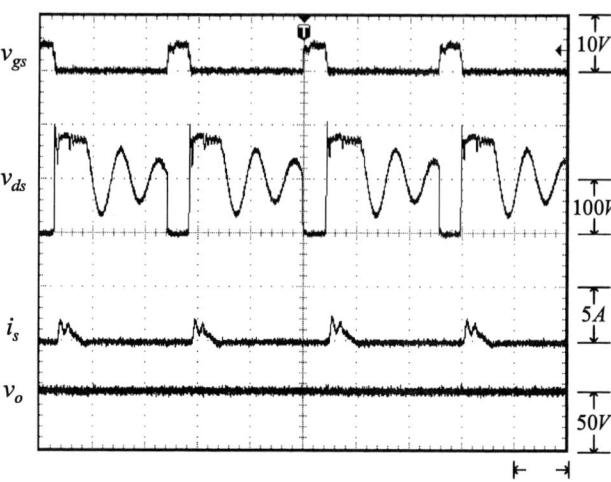

Fig. 6. Waveforms relative to no load from top to bottom: (a) v_{gs} of S_w; (b) v_{ds} across S_w; (c) i_s; (d) v_o.

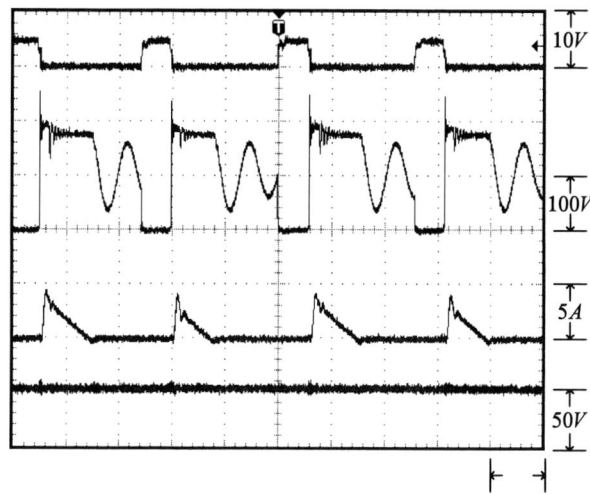

Fig. 7. Waveforms relative to half load from top to bottom: (a) v_{gs} of S_w; (b) v_{ds} across S_w; (c) i_s; (d) v_o.

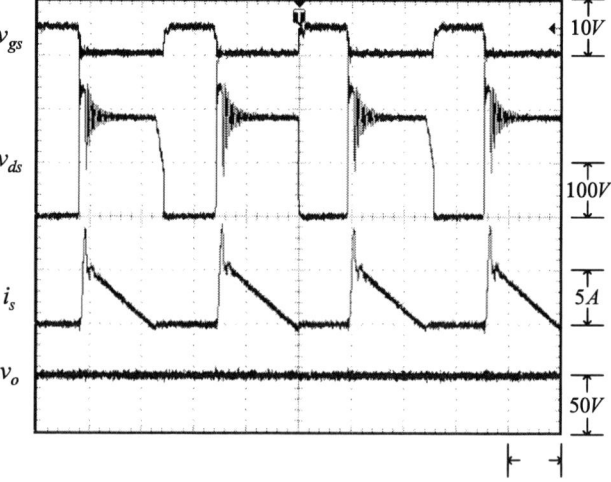

Fig. 8. Waveforms relative to rated load from top to bottom: (a) v_{gs} of S_w; (b) v_{ds} across S_w; (c) i_s; (d) v_o.

TABLE I.
OUTPUT VOLTAGE AS A FUNCTION OF LOAD CURRENT

Load current	Output voltage
0.15 A	48.18 V
0.30 A	48.13 V
0.45 A	48.10 V
0.60 A	48.09 V
0.75 A	48.09 V
0.90 A	48.08 V
1.05 A	48.07 V
1.20 A	48.07 V
1.35 A	48.06 V
1.50 A	48.00 V

VI. CONCLUSION

In this paper, based on FPGA, the proposed control loop design, together with the one-comparator counter-based PWM control strategy, is used to control the flyback converter without any ADC used. By doing so, the capability of the one-comparator counter-based PWM control strategy is enhanced and is not only applied to the flyback converter operating in DCM below rated load, but also to any type of DC-DC converter operating in DCM.

REFERENCES

[1] B. K. Bose, "Microcomputer Control of Power Electronics and Drives," *in IEEE Press* (New York: 1996, pp. 640).

[2] L. Brush, "Trends in digital power management: power converter and system demand characteristics," *IEEE APEC'05*, pp. 161-166, 2005.

[3] D. He and R. M. Nelms, "Current-mode control of a DC-DC converter using a microcontroller: implementation issues," *IEEE IPEMC'04*, pp. 538-543, 2004

[4] D. He, W. Dilliard and R. M. Nelms, "Microcontroller implementation of current-mode control for a discontinuous mode boost converter," *IEEE IECEC'04*, pp. 255-260, 2004.

[5] D. He and R. M. Nelms, "Peak current-mode control for a boost converter using an 8-bit microcontroller," *IEEE PESC'03*, pp. 938-943, 2003.

[6] P. T. Tang and C. K. Tse, "Design of DSP-based controller for switching power converters," *IEEE TENCON'96*, pp. 889-894, 1996.

[7] A. Prodic, D. Maksimovic and R. W. Erickson, "Design and implementation of a digital PWM controller for a high-frequency switching DC-DC power converter," *IEEE IECON'01*, pp. 893-898, 2001.

[8] C. H. Chan and M. H. Pong, "DSP controlled power converter," *IEEE PEDS'95*, pp. 364-369, 1995.

[9] Guo Liping, J. Y. Hung and R. M. Nelms, "ID controller modifications to improve steady-state performance of digital controllers for buck and boost converters," *IEEE APEC'02*, pp. 381-388, 2002.

[10] S. Chattopadhyay and S. Das, "A digital current mode control technique for DC-DC converters," *IEEE APEC'05*, pp. 885-891, 2005.

[11] Z. H. Jiang, X. D. Sun and L. P. Huang, "The controller of high frequency and high dynamic performance dual-boost PFC module based on DSP," *IEEE IECON'03*, pp. 249-254, 2003.

[12] L. Guo, J. Y. Hung and R. M. Nelms, "Digital controller design for buck and boost converters using root locus techniques," *IEEE IECON'03*, pp. 1864-1869, 2003.

[13] M. M. Islam, D. R. Allee, S. Konasani and A. A. Rodriguez, "A low-cost digital controller for a switching DC converter with improved voltage regulation," *IEEE Power Electronics Letters*, vol. 2, no. 4, pp. 121-124, 2004.

[14] S. Saggini, G. Garcea, M. Ghioni and P. Mattavelli, "Analysis of high-performance synchronous-asynchronous digital control for dc-dc boost converters," *IEEE APEC'05*, pp. 892-898, 2005.

[15] K. Wang, N. Rahman, Z. Lukic and A. Prodic, "All-digital DPWM/DPFM controller for low-power DC-DC converters," *IEEE APEC'06*, pp. 719-723, 2006.

[16] Y. T. Yau, K. I. Hwu and Yung-Shan Chou, "Forward converters using a CPLD-based control technique to obtain a fast transient load response", *IEEE PEDS'03*, pp. 359-364, 2003.

[17] K. I. Hwu and Y. T. Yau, "Applying a counter-based PWM control scheme to an FPGA-based SR forward converter," *IEEE APEC'06*, pp. 1396-1400, 2006.

Analysis of Conducted EMI Reduction on a Boost Converter Using Progressive Inductor Winding Technique

Kritsada Saritsiri, Werachet Khan-ngern

Faculty of Engineering,
Research Center for Communications and Information Technology,
King Mongkut's Institute of Technology Ladkrabang,
Thailand, E-mail: kritsada_mr@yahoo.com, kkveerac@kmitl.ac.th

Abstract

This paper presents the analysis of conducted electromagnetic interference (EMI) reduction on a boost converter using of progressive winding on the inductor. Two layer of inductor winding is modeled. The EMI of a boost converter can be caused by switching dv/dt rates and interacting with parasitic capacitance in the inductor. This paper focuses on the modeling and analysis on the parasitic capacitance of the progressive inductor winding. The EMI sources of the inductor normal winding are identified and measured on both of discontinuous and continuous current modes. Finally, the experimental EMI spectrum, in associated with the voltage and current waveform of MOSFET, for both of normal and progressive winding is compared and analyzed to confirm the advantage of a progressive winding technique.

Keywords : conducted electromagnetic interference, progressive inductor winding techniques

Contents:

1. Introduction

Electromagnetic interference (EMI) is an important consideration for boost converter [1], because the conducted or radiated EMI noise can be generated by electronic component operation. The parasitic capacitance [2-3] of the inductor of a boost converter can cause the EMI noise. The higher parasitic capacitance in the inductor results the higher EMI emission. Since progressive winding techniques can significantly reduce parasitic capacitance in the inductor. So, the EMI noise generated by a normal winding could be reduced by progressive winding techniques in some frequency ranges [4]. This paper presents the parasitic modeling in the proposed technique. Some experimental results show the comparison between the conducted EMI frequency response of the conventional winding and the progressive winding. The boost converter is operated in two different modes: continuous and discontinuous inductor current.

2. Proposed technique

The characteristic of the inductor is depended on the effect of the wiring resistance and the parasitic capacitance. The equivalent circuit and the frequency response are shown in Figures 1(a) and 1(b).

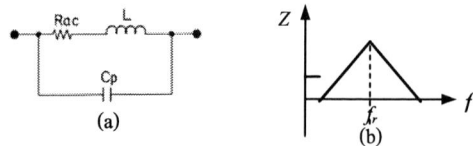

(a)   (b)

Fig. 1 (a) Equivalent circuit of the inductor at high frequency. (b) Impedance response

Self resonant frequency (SRF, f_r) shown in Fig.1 (b) results the value of parasitic capacitance.

Fig.2 Normal winding of inductor for boost converter.

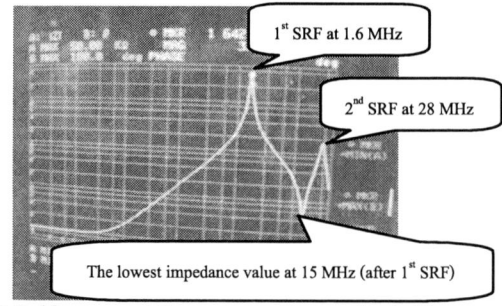

Fig.3 Self-resonant frequency of normal winding inductor

Normal or conventional winding is continuous winding. The starting and ending windings and its modeling show in Figure 2. The winding model presents the modeling of the parasitic capacitance of the normal winding. The self-resonant frequency of normal winding is shown in Figure 3. For the normal winding, the SRF occurs 2 times. The SRF is occurred at 1.6 MHz and at 28 MHz.

Figure 4 presents winding layout and the modeling of the progressive. The frequency response shown in Figure 5 results the advantage of increasing the SRF up to 4 MHz.

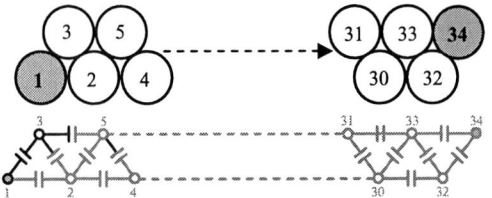

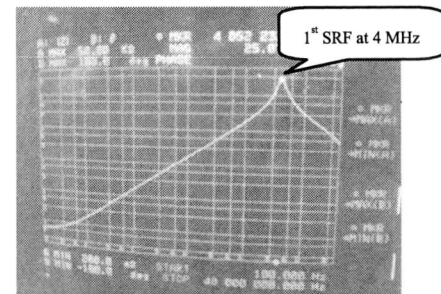

Fig. 4 Progressive winding and the modeling

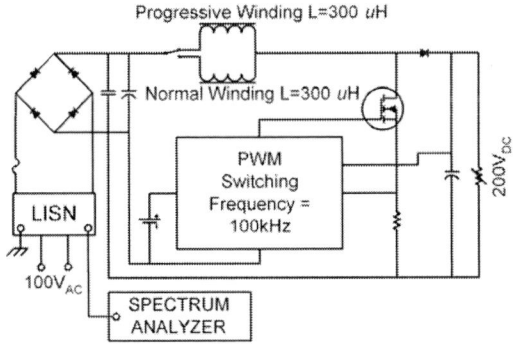

Fig. 5 Self-resonant frequency of the progressive inductor winding

3. Some experimental results

Figure 6 shows the EMI noise measurement test setup. The input voltage is 100 V_{AC} and the output voltage is 200 V_{DC}. The switching frequency is 100 kHz. The inductance value and number of turn for both winding techniques are 300 µH and 34 turns, respectively. Figure 7 shows the layout of the boost converter. The testing power is 70 W and 140 W operated in discontinuous and continuous modes respectively.

For measurement the conducted EMI noise, the ADVANTEST R3132 spectrum analyzer is used to measure the conducted EMI emission of the boost converter. The peak mode is used to present the worst case condition.

Fig.7 The layout of the boost converter

3.1 The EMI level of Normal and progressing winding operated in continuous and discontinuous mode

Normal winding

Some of conducted EMI noise spectra for normal winding inductor when the boost converter operated in discontinuous and continuous mode are shown in Figs.8 and 9 respectively. Figure 10 shows the comparison of EMI level of those continuous and discontinuous operating modes. It is shown that the EMI level of that continuous mode higher than of that discontinuous about 5 dB at beyond 20 MHz.

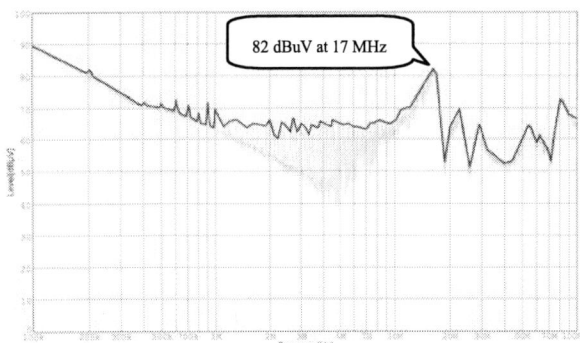

Fig.8 EMI level for normal winding in discontinuous mode

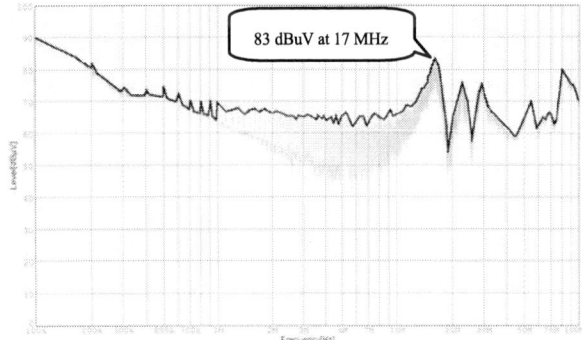

Fig.9 EMI level for normal winding in continuous mode

461

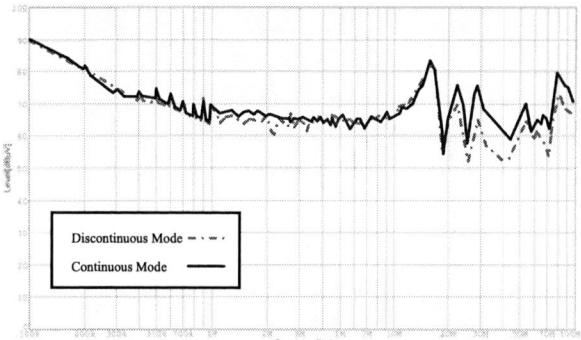

Fig.10 The comparison of EMI level between discontinuous and continuous mode for normal winding

Progressive winding

Fig.11 shows the comparison of EMI level between discontinuous and continuous mode of the progressive winding. It is quite similar to the normal winding. It is shown that the EMI level of that continuous mode is higher than of that discontinuous about 5 dB at beyond 10 MHz.

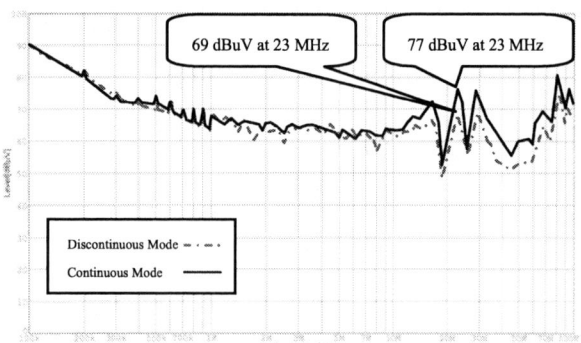

Fig.11 The comparison of the EMI level between discontinuous and continuous mode for progressive winding inductor.

3.2 The EMI comparison between normal and progressing winding operated in continuous mode

The comparison of EMI between normal and progressive winding when the boost converter operated in discontinuous and continuous mode are shown in Figs.12 and 13 respectively. At the discontinuous mode, the EMI from both winding techniques are peaked nearly in a frequency range from 100 kHz to 10 MHz. While in the frequency range from 10 MHz to 20 MHz, the EMI of progressive winding is less than of that normal winding technique up to 15 dB.

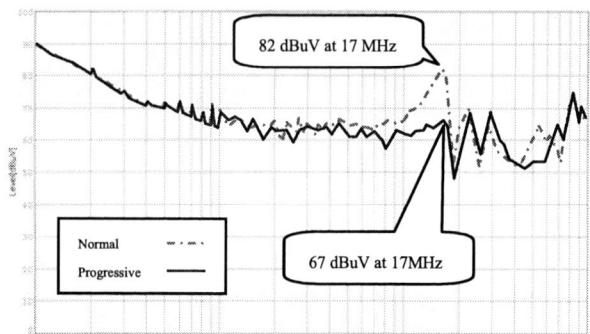

Fig.12 The comparison of the EMI level between normal and progressive windings in the discontinuous mode

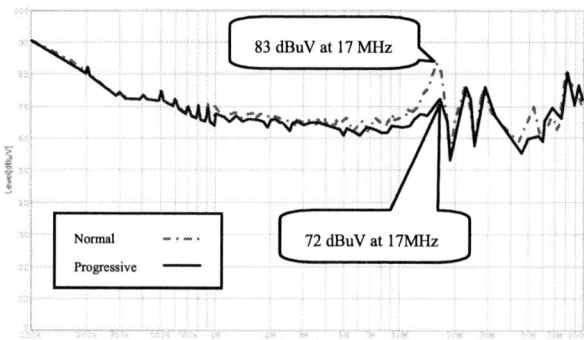

Fig.13 The comparison of the EMI level between normal and progressive winding in the continuous mode

In continuous mode operation, the EMI level result is quite similar to of that discontinuous mode.

4. Voltage and current analysis

This section shows some voltage and current waveform to analyze the EMI phenomena. In the final paper will present all condition of each winding type as the following:

1. Voltage and current waveform of the main switch that operated in discontinuous mode
2. Voltage and current waveform of the main switch that operated in continuous mode
3. Voltage and current waveform of the inductor that operated in discontinuous mode
4. Voltage and current waveform of the inductor that operated in continuous mode

Some experimental waveforms

The expanded the waveform of the main switch (case 1) in Figure 14 shows the I_D waveform of normal inductor winding resulting a high ringing. The frequency of this ringing is 16.7 MHz which is corresponding to the EMI peak of that normal winding at 17 MHz as shown in Figure 12.

In case of the progressive inductor winding, the peak of EMI occurs a little of I_D waveform ringing shown in Figure 15. The frequency of this ringing is 33.8 MHz which is corresponding to the EMI peak of that progressive winding at 33.8 MHz as shown in Figure 13.

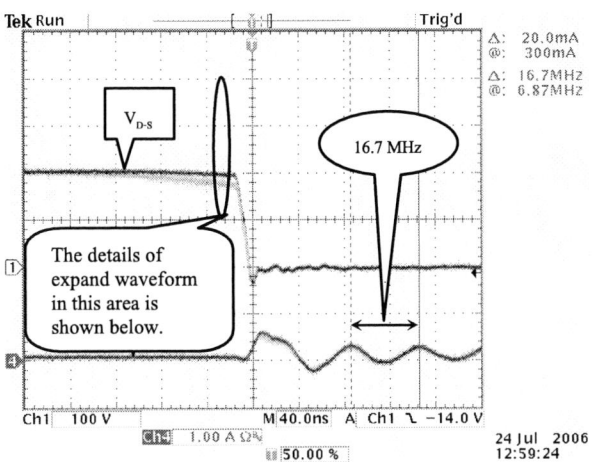

Fig.14 Voltage and current waveforms of the main switch using the normal inductor winding in discontinuous mode

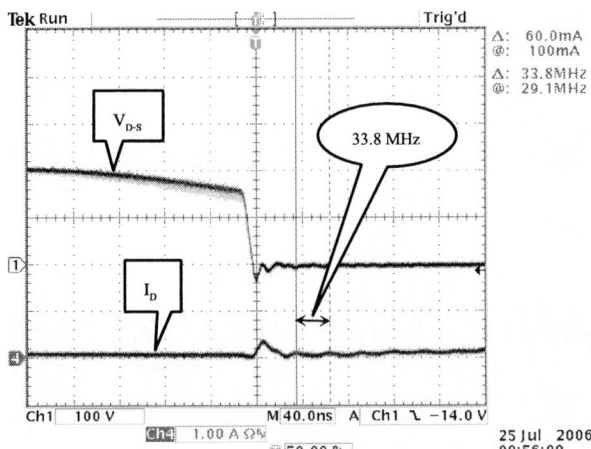

Fig.15 Voltage and current waveforms of the main switch using the progressive inductor winding in discontinuous mode.

The analysis will focused on voltage and current waveforms of the inductor. Figure 16 shows the ringing frequency of the normal winding case that operated in continuous mode. The EMI frequency at 17 MHz shown in Figure 12 is confirmed by the ringing of the inductance current shown in Figure 16.

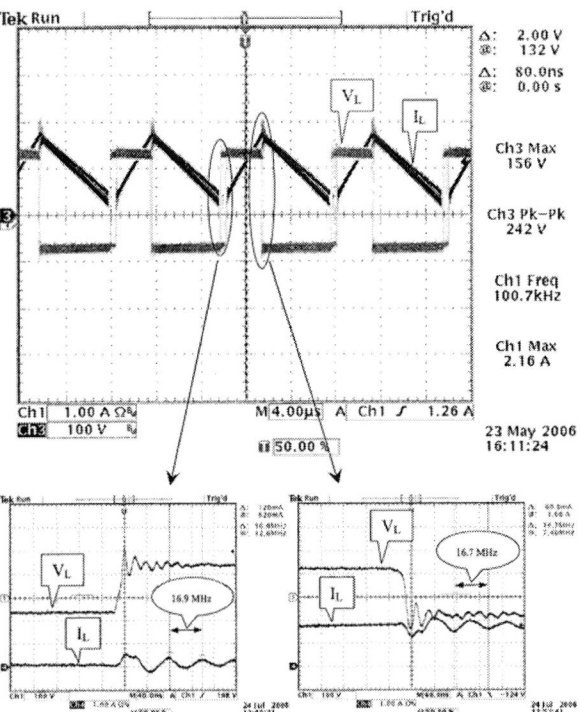

Fig.16 Voltage and current waveform of the inductor using the normal inductor winding in continuous mode

In case of progressive winding, the EMI peak at 33.8 MHz with lower EMI than that of normal winding case can be confirmed with good agreement by the ringing frequency in Figure 17.

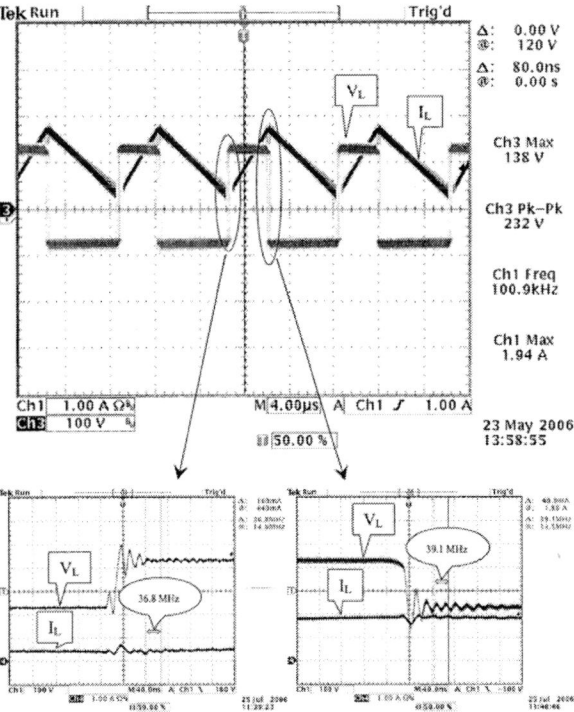

Fig.17 Voltage and current waveform of the inductor using the progressive inductor winding in continuous mode

5. Conclusion

The progressive inductor winding technique can reduce the parasitic capacitance in the inductor. It results the reduction of EMI noise in a boost converter. The frequency range between 10 MHz to 20 MHz is the frequency range of the significant EMI reduction using the improvement of parasitic capacitance of the progressive winding technique. The EMI achievement can be confirmed by the experiment results in both discontinuous and continuous modes.

The details of the winding modeling and the analysis of EMI phenomena of both operating modes will be completed in the final paper.

References

[1] M.N. Gitau; "Modeling Conducted EMI Noise Generation and Propagation in Boost Converters"; IEEE International Symposium on Volume 2, 4-8 December 2000, pp.353-358.

[2] Massarini, A.; Kazimierczuk, M.K.; Grandi, G.; "Lumped parameter models for single- and multiple-layer inductors"; PESC '96., 27th Annual IEEE, 23-27 June 1996, pp.295-301.

[3] Massarini, A.; Kazimierczuk, M.K.; "Self-capacitance of inductors"; Power Electronics, IEEE Transactions onVolume 12, Issue 4, July 1997, pp. 671 – 676.

[4] John C.Fluke.; "Controlling Conducted Emission by Design"; New York, VNR, 1991.

Practical Issues Concerned with Zero sequence component and Harmonic Compensation in Four-Wire systems

E. Pashajavid*, K. Kanzi*, and M. Tavakoli Bina*

* Faculty of Electrical Engineering, K. N. Toosi University of Technology, P. O. Box 16315–1355, Tehran 16314, Iran,
E-mail: tavakoli@ieee.org

Abstract-- A minimization problem for the instantaneous active and reactive currents give the optimal solution for the three-wire three-phase systems. This basic solution is simple to implement, but posing practical issues when it is applied to a four-wire three-phase system. The generalized theory of instantaneous powers has been proposed as an alternative to the optimal solution for a general three-phase system. Various simulations performed using this method, showing similar issues like those of the optimal solution for the four-wire systems when the power system is unbalanced or distorted. This paper mathematically shows that the optimal solution can be presented like the generalized theory, concluding no difference between the two methods but their appearances. Further, complementary analytical discussions are suggested to overcome the stated problem. Finally, the improved solution is simulated for an unbalanced distorted four-wire system. Comparing the complementary solutions with those of previously emerged generalized definitions confirms that the proposed solution provides almost a complete cancellation of source zero sequence current under unbalanced three-phase load voltages.

Index Terms-- Four-wire systems, generalized method, harmonic distortion, power definitions, zero sequence component.

I. INTRODUCTION

Non-linear loads are used extensively in power systems, making the three-phase load terminal voltages unbalanced and distorted. There are several ways to achieve non-linear current compensation; the active power filter (APF) is the most common choice. Among various suggested compensation strategies for the three-phase APF, the generalized theory of instantaneous powers provides interesting definitions that are usefully related and formulated [1-2]. This method potentially can produce reference currents for the APF to supply reactive power and current harmonics of the load. This method functions perfectly as long as the power system voltage consists of no zero or negative sequence components. A complementary strategy is introduced in [3], which introduces a solution for three-wire systems based on modeling an optimization problem. Optimal solutions are simple, and can be fully implemented for isolated power systems. Further, the solutions are modified to be applied to unbalanced four-wire systems in which the load bus contains zero sequence voltage. This paper analyzes an optimal solution in order to obtain different suggested power definitions; in particular, the generalized theory of instantaneous power definition (*GTIP*). It is then followed by simulation of issues concerned with zero sequence components, and some suggestions are proposed to get desirable compensation rule. Effectiveness of the proposed suggestions is verified by MATLAB/SIMULINK simulation results.

II. HOW POWER DEFINITIONS ARE DERIVED?

There are currently different definitions available in three-phase power systems. These are based on instantaneous strategies that are used for compensating load power excluding its average active power. Definitions look dissimilar, but some of them are basically derived by the same root. In principle, they introduce similar compensation issues for three-phase four-wire systems, which here this fact is described.

A. Basic optimized solution

Consider a power system load bus voltage of $\mathbf{v}(t) = [v_a(t) \quad v_b(t) \quad v_c(t)]^t$ as shown by Fig. 1(a), including the source, the load and the compensator. Also, let us assume the three-phase load instantaneous current ($\mathbf{i}_l(t) = [i_{la}(t) \quad i_{lb}(t) \quad i_{lc}(t)]^t$) is made up of two parts; the active current ($\mathbf{i}_{lp}(t) = [i_{lap}(t) \quad i_{lbp}(t) \quad i_{cp}(t)]^t$) and remaining part that do not contribute in active power (here it is called inactive current $\mathbf{i}_{lq}(t) = [i_{laq}(t) \quad i_{lbq}(t) \quad i_{lcq}(t)]^t$). It is now possible to construct an optimization problem, aiming at minimizing the sum of squares of active parts in the form of active power [4]. One constraint is added to emphasize that inactive part has no share in active power as follows:

$$\text{minimize} \sum_{k=a,b,c} (i_{lk}(t) - i_{lkq}(t))^2$$
$$\text{subject to}:$$
$$i_{laq}(t)v_a(t) + i_{lbq}(t)v_b(t) + i_{lcq}(t)v_c(t) = 0 \tag{1}$$

Solving (1) using the well-known Lagrange multiplier method leads to the following optimal solution:

This work was performed in the Research Laboratory of K. N. Toosi University of Technology.

978-1-4244-0644-9/07/$25.00 ©2007 IEEE

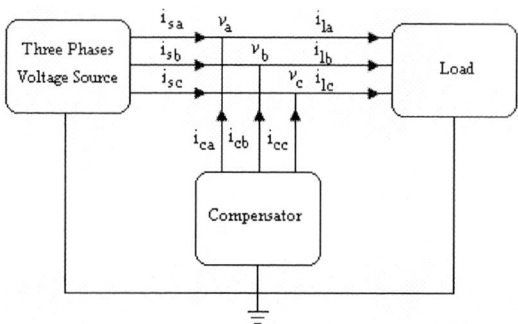

(a)

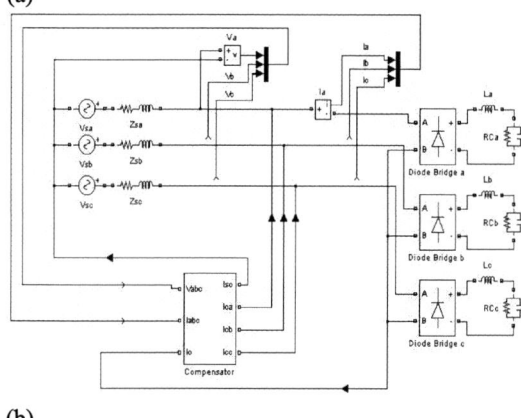

(b)

Figure 1: (a) A typical interconnection of the source, the load and the compensator across a three phase four-wire power system , (b) Non linear and unbalanced power system

$$\begin{bmatrix} i_{lap}(t) \\ i_{lbp}(t) \\ i_{lcp}(t) \end{bmatrix} = \frac{i_{la}(t)v_a(t)+i_{lb}(t)v_b(t)+i_{lc}(t)v_c(t)}{v_a(t)^2+v_b(t)^2+v_c(t)^2} \begin{bmatrix} v_a(t) \\ v_b(t) \\ v_c(t) \end{bmatrix},$$

$$\begin{bmatrix} i_{laq}(t) \\ i_{lbq}(t) \\ i_{lcq}(t) \end{bmatrix} = \begin{bmatrix} i_{la}(t) \\ i_{lb}(t) \\ i_{lc}(t) \end{bmatrix} - \begin{bmatrix} i_{lap}(t) \\ i_{lbp}(t) \\ i_{lcp}(t) \end{bmatrix}$$

(2)

These results have already appeared in literature, where replacing the eminent relation $i_{la}(t)v_a(t)+i_{lb}(t)v_b(t)+i_{lc}(t)v_c(t)$ with the load active power definition ($p_l(t)$) simplifies (2) in the form of

$$i_{lp}(t)=\frac{p_l(t)}{v_a(t)^2+v_b(t)^2+v_c(t)^2}\mathbf{v}(t)=\frac{p_l(t)}{\mathbf{v}(t).\mathbf{v}(t)}\mathbf{v}(t),$$

$$i_{lq}(t)=\mathbf{i}_l(t)-\mathbf{i}_{lp}(t)$$

(3)

Let us call (3) as the optimal solution (OS) throughout this paper, which provides reference currents for the compensator to follow.

B. Derivation of the generalized theory of instantaneous powers

Interesting definitions are introduced in [1-2] as the generalized theory of instantaneous powers (GTIP) as

below:

$$p_l(t)=\mathbf{v}(t).\mathbf{i}_l(t) \ , \ \mathbf{q}_l(t)=\mathbf{v}(t)\times\mathbf{i}_l(t),$$

$$\mathbf{i}_{lp}(t)=\frac{p_l(t)}{\mathbf{v}(t).\mathbf{v}(t)}\mathbf{v}(t), \ \mathbf{i}_{lq}(t)=\frac{\mathbf{q}_l(t)\times\mathbf{v}(t)}{\mathbf{v}(t).\mathbf{v}(t)}$$

(4)

We have been trying to implement the GTIP of (4) as a compensation rule for an active filter to be employed for a three-phase four-wire distribution system. Taking a closer look into the GTIP definitions indicates that they can be derived from the OS formulation (3). In fact, the active current of the load ($\mathbf{i}_{lp}(t)$) defined by the OS in (3) and the GTIP in (4) are identical. From the OS, inactive current $\mathbf{i}_{lq}(t)$ in (3) can be expanded as follows:

$$\mathbf{i}_{lq}(t)=\mathbf{i}_l(t)-\mathbf{i}_{lp}(t)=\mathbf{i}_l(t)-\frac{p_l(t)}{\mathbf{v}(t).\mathbf{v}(t)}\mathbf{v}(t)=$$

$$\frac{\mathbf{i}_l(t)[\mathbf{v}(t).\mathbf{v}(t)]-[\mathbf{v}(t).\mathbf{i}_l(t)]\mathbf{v}(t)}{\mathbf{v}(t).\mathbf{v}(t)}=\frac{[\mathbf{v}(t)\times\mathbf{i}_l(t)]\times\mathbf{v}(t)}{\mathbf{v}(t).\mathbf{v}(t)}$$

(5)

Comparing (5) with (4) verifies that by defining $\mathbf{q}_l(t)=\mathbf{v}(t)\times\mathbf{i}_l(t)$ the GTIP gives exactly identical solutions as those of the OS. Therefore, it is expected similar issues for the GTIP like those of the OS.

III. ANALYZING THE EFFECTS OF LOAD VOLTAGE UNBALANCING AND DISTORTION ON *GTIP* PERFORMANCE

As aforementioned, the GTIP gives exactly identical solutions like those of the OS. Hence the OS is used to analyze the issues concerned with the GTIP. According to (3), compensation reference currents can be obtained by replacing $p_l(t)$ with its average part $\bar{p}_l(t)$. Let us assume that $\mathbf{v}$ includes all sequences ($\mathbf{v}(t)=\mathbf{v}(t)^+ + \mathbf{v}(t)^- + \mathbf{v}(t)^o$, where $\mathbf{v}(t)^+$, $\mathbf{v}(t)^-$ and $\mathbf{v}(t)^o$ denote positive, negative and zero sequences of $\mathbf{v}(t)$ respectively) [5]. So the source current after compensation can be derived as follows:

$$\mathbf{i}_s(t)=(\lambda)\mathbf{v}(t)^+ + (\lambda)\mathbf{v}(t)^- + (\lambda)\mathbf{v}(t)^o, \lambda=\frac{\bar{p}_l(t)}{\mathbf{v}(t).\mathbf{v}(t)}$$

(6)

It can be seen from (6) that each voltage component creates its corresponding sequence current. The load zero sequence voltage $\mathbf{v}(t)^o$ produces source-end zero sequence current, producing unbalanced and distorted source currents as well as the source-end neutral-wire current. Also the load negative sequence voltage $\mathbf{v}(t)^-$ produces source-end negative sequence current, which leads to source-end current distortion along with power factor reduction. Another issue producing some concerns is $\mathbf{v}(t).\mathbf{v}(t)$ in the denominator of λ . Because of the available voltage sequences, $\mathbf{v}(t).\mathbf{v}(t)$ has oscillating component in addition to the average value that makes the source-end currents distorted. To overcome these issues, we suggest replacing $\mathbf{v}(t)$ with $\mathbf{v}(t)^+$ in (3). Hence the new source-end current can be obtained as follows:

$$\mathbf{i}_s(t) = \frac{\overline{p}_l(t)}{\mathbf{v}(t)^+ . \mathbf{v}(t)^+} \mathbf{v}(t)^+ \qquad (7)$$

Using (7), both negative and zero sequence currents can be removed from the source-end currents; but, this situation functions perfectly as long as the load positive sequence voltage is no distorted. Otherwise, the source-end current will be distorted because $\mathbf{v}(t)^+ . \mathbf{v}(t)^+$ will have oscillating part on top of an average value. Moreover, since $\mathbf{v}(t)_1^+ \times \mathbf{i}_s(t) \neq 0$, source-end power factor of the fundamental components differs from unity. It can summarize that the source-end currents after compensation should be as follows:

$$\mathbf{i}_s(t) = \frac{\overline{p}_l(t)}{\mathbf{v}(t)_1^+ . \mathbf{v}(t)_1^+} \mathbf{v}(t)_1^+ \qquad (8)$$

In this situation, $\mathbf{v}(t)_1^+ . \mathbf{v}(t)_1^+$ includes only an average value, i.e. $\mathbf{v}(t)_1^+ \times \mathbf{i}_s(t) = 0$. So the source-end currents are balanced and sinusoidal. Consequently, the source-end neutral-current is theoretically zero. Finally, the power factor is unity. Using (5) and (8), the improved *GTIP* compensation reference currents can be derived as follows:

$$
\begin{aligned}
\mathbf{i}_c(t) &= \mathbf{i}_l(t) - \frac{\overline{p}_l(t)}{\mathbf{v}(t)_1^+ . \mathbf{v}(t)_1^+} \mathbf{v}(t)_1^+ = \\[2mm]
&\frac{(\mathbf{v}(t)_1^+ . \mathbf{i}_l(t) - \overline{p}_l(t))}{\mathbf{v}(t)_1^+ . \mathbf{v}(t)_1^+} \mathbf{v}(t)_1^+ + \frac{(\mathbf{v}(t)_1^+ . \mathbf{v}(t)_1^+) \mathbf{i}_l(t) - (\mathbf{i}_l(t) . \mathbf{v}(t)_1^+) \mathbf{v}(t)_1^+}{\mathbf{v}(t)_1^+ . \mathbf{v}(t)_1^+} = \\[2mm]
&\frac{(p_{l1}^+(t) - \overline{p}_l(t))}{\mathbf{v}(t)_1^+ . \mathbf{v}(t)_1^+} \mathbf{v}(t)_1^+ + \frac{(\mathbf{v}(t)_1^+ \times \mathbf{i}_l(t)) \times \mathbf{v}(t)_1^+}{\mathbf{v}(t)_1^+ \cdot \mathbf{v}(t)_1^+} = \\[2mm]
&\frac{(p_{l1}^+(t) - \overline{p}_l(t))}{\mathbf{v}(t)_1^+ . \mathbf{v}(t)_1^+} \mathbf{v}(t)_1^+ + \frac{\mathbf{q}_{l1}^+(t) \times \mathbf{v}(t)_1^+}{\mathbf{v}(t)_1^+ \cdot \mathbf{v}(t)_1^+}
\end{aligned}
\qquad (9)
$$

IV. ANALYSIS ANS SIMULATIONS

To distinguish the advantages of the proposed improved method over the conventional ones, various simulations are arranged and analyzed. Here it is simulated the suggested improvement of (9), where the outcomes are compared with *GTIP* to examine their advantages as well as disadvantages. Figure 1(a) is considered and simulated with SIMULINK, where three single-phase rectifiers are introduced as the load like it is shown by Fig. 1(b). Two cases are examined to describe the differences between the performance of the discussed methods; first, balanced sinusoidal load voltages and second, distorted and unbalanced load terminal voltages. Three voltages of the first case are sinusoidal 50 Hz with the RMS value of 220 V. The following distorted unbalanced voltages are applied to the load terminal for the second case:

$$
\begin{aligned}
v_a(t) &= 311\sin(\omega t) + 40\sin(3\omega t - 18°) \\
v_b(t) &= 280\sin(\omega t - 72°) + 35\sin(3\omega t - 90°) \\
v_c(t) &= 280\sin(\omega t + 90) + 20\sin(3\omega t - 51°)
\end{aligned}
\qquad (10)
$$

Also, the single-phase non-linear load parameters are as below:

$$
\begin{aligned}
R_a &= 2\,\Omega, \quad L_a = 1\,mH, \quad C_a = 1\,mF \\
R_b &= 3.87\,\Omega, \quad L_b = 1\,mH, \quad C_b = 1\,mF \\
R_c &= 4.84\,\Omega, \quad L_c = 1\,mH, \quad C_c = 1\,mF
\end{aligned}
\qquad (11)
$$

A. Case 1: Balanced sinusoidal load terminal voltage

Considering Figs. 1(a) and 1(b), simulation of the first case is arranged with unsymmetrical loads of (11), while load terminal voltages are *balanced*. Since the source is inherently balanced, this situation causes balanced source-end currents flowing to the load terminal despite unsymmetrical load assumption. In practice, this cannot be implemented, while nearly all researches use these fictitious assumptions in their compensating cases.

Figure 2 shows the simulation results for the first case. In this picture, Figs. 2(a)–(b) depict the load terminal balanced voltages and asymmetrical distorted load currents, respectively. It can be seen that the load currents are considerably distorted, showing significant load zero sequence current on the fourth-wire as it is illustrated by Fig. 2(e). As can be seen from Fig. 2, *GTIP* and the proposed method completely compensate source current and source neutral current as it is expected by considering balanced terminal voltages.

B. Case 2: Unbalanced and distorted load terminal voltages

Load terminal voltages are distorted and unbalanced like (9) for simulations of the second case. This situation is normally true for power systems, especially for distribution systems. Meanwhile, the distortion might be somehow exaggerated to distinguish performance of the discussed compensation methods in cancellation of zero sequence current of the fourth-wire and harmonic compensation. Simulation results for the second case are given by Figure 4, where Figs. 3(a), (b) and (g) depict the load terminal distorted unbalanced voltages, asymmetrical distorted load currents and load side neutral current respectively. Figs. 3(c) – (e) depict effects of voltage sequences on source currents and Figs. 3(h) – (j)

467

depict effects of voltage sequences on source neutral current. As it can be seen from Figs. 3(f) and (k), the proposed method completely compensates source current and source neutral current.

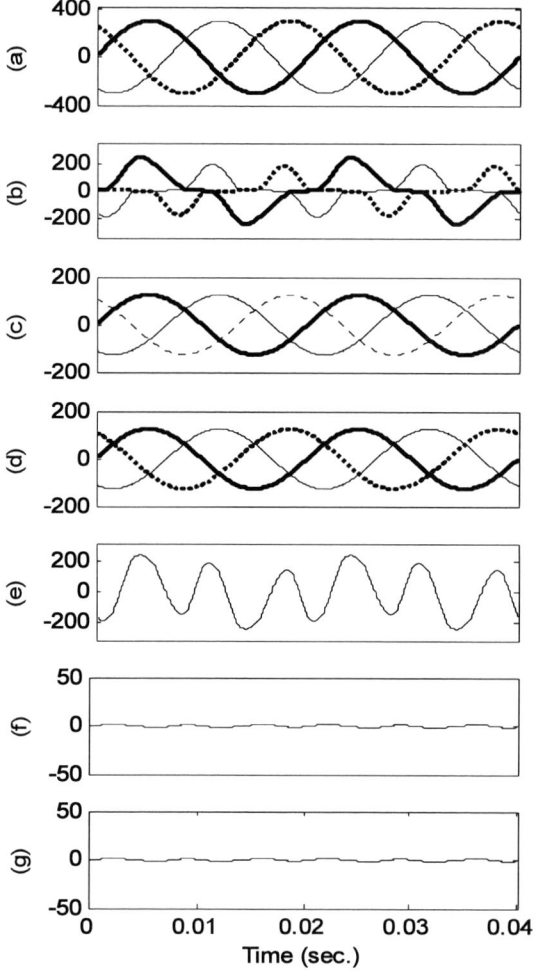

Figure 2: Simulation results for the case 1: (a) three phase balanced load voltages (V), (b) asymmetrical distorted load currents (A), (c)–(d) three phase source currents (A) after compensation using the *GTIP* method and the proposed method, (e) the load neutral current (A), (f)–(g) the source neutral current (A) after compensation using the *GTIP* and the proposed methods.

Table I summarizes the THD of the source currents after compensation that are gathered from simulation results for all the discussed compensating methods. Simulations show that while the uncompensated case is significantly undesirable, the conventional GTIP is ineffective under the real condition of the case 2. When the zero sequence voltage is subtracted from the main voltage, still the compensation produces undesirably high THD%. Nevertheless, the neutral-wire current is lowered.

Additionally, inclusion of the negative sequence voltage in the compensation rule has a more suitable THD% compared to that of the inclusion of zero sequence voltage. Meanwhile, the proposed method lowers the THD% for both theoretical case 1 and the real case 2 down to acceptable values. Practically, the compensating device operates effectively to supply part of the load currents.

Also, Table II compares the RMS values of the source fourth wire current after compensation. It can clearly be observed the advantages of the proposed method in Section III over the *GTIP*.

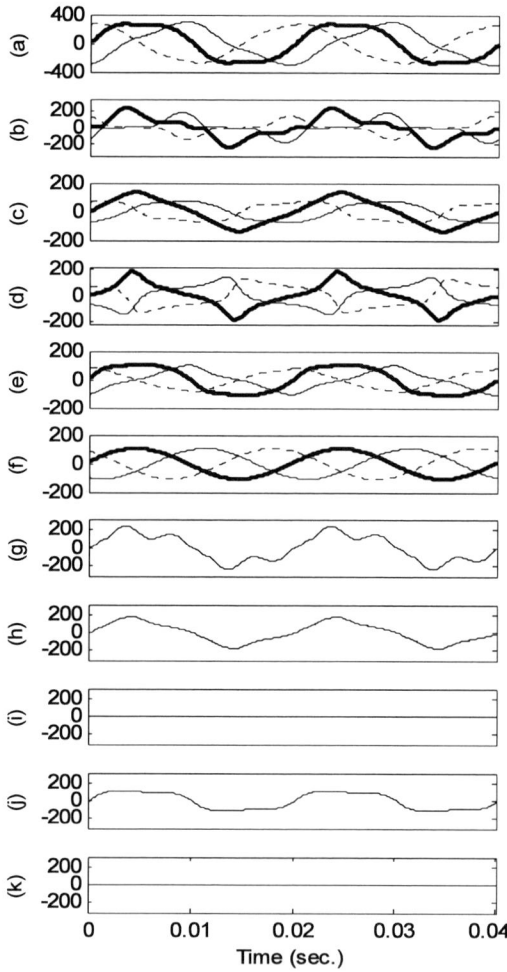

Figure 3: Simulation results for the case 2: (a) three phase balanced load voltages (V), (b) asymmetrical distorted load currents (A), three phase source currents (A) after compensation using (c)–(f) the *GTIP* method, the *GTIP* without zero seqence component, the *GTIP* without negative sequence component, and the proposed method, (g) the load neutral-wire current (A), the source neutral current (A) after compensation using (h)–(k) the *GTIP* method, the *GTIP* method without zero seqence component, the *GTIP* without negative seqence component, and the proposed method.

TABLE I
THE THD% OF THE SOURCE-END CURRENTS SIMULATED UNDER VARIOUS COMPENSATION METHODS.

Phases	Case 1			Case 2		
	A	B	C	A	B	C
Uncompensated	36.12	63.36	69.5	39.7	62.83	60.3
GTIP $\mathbf{v}$	0.7	0.9	0.9	12	12	23.5
GTIP $\mathbf{v} - \mathbf{v}^o$	0.7	0.9	0.9	34	38.2	35.8
GTIP $\mathbf{v} - \mathbf{v}^-$	0.7	0.9	0.9	12.1	20.1	8.6
Proposed method	0.7	0.9	0.9	0.8	1	1

TABLE II
THE RMS VALUE OF THE SOURCE-END NEUTRAL-WIRE CURRENTS (A)

	Case 1	Case 2
Uncompensated	150.5	142
GTIP $\mathbf{v}$	2.05	111
GTIP $\mathbf{v} - \mathbf{v}^o$	2.05	1.07
GTIP $\mathbf{v} - \mathbf{v}^-$	2.05	90
Proposed method	2.05	1.05

V. CONCLUSION

This paper reviews conventional theory of instantaneous powers, and describes their advantages for three-phase three-wire systems. Nevertheless, these methods are unable to effectively operate in three-phase four-wire systems. The fourth wire at the source-end carries zero sequence currents despite the application of conventional GTIP compensation rule. This is shown by MATLAB SIMULINK simulations. Then, some suggestions are presented to improve the GTIP. Suggested improvements are also simulated with SIMULINK. Simulation results confirm that the suggested methods lowers the source-end zero sequence current (and the neutral-wire current) down to acceptable levels. Moreover, the distortion and unbalanced cases are considered to show that the modified theory of the GTIP provides useful solutions.

ACKNOWLEDGEMENT

The authors would like to thank the support of the research Laboratory of power quality and reactive power control in K. N. Toosi University of Technology.

REFERENCES

[1] F. Z. Peng and J. S. Lai, "Generalized instantaneous reactive power theory for three-phase power systems," IEEE Transactions Instrum. Meas., vol. 45, No. 1, pp. 293–297, 1996.

[2] F. Z. Peng, George W. Ott, Jr. and Donald J. Adams, "Generalized instantaneous reactive power theory for three-phase power systems", IEEE Transactions on Power Electronics, Vol. 13, NO. 6, November 1998.

[3] M. Tavakoli Bina, "A New Complementary Approach to Inactive Power Compensation", IEEE PESC, 2003, Electronic Record.

[4] S. Fryze, "Effective wattles and apparent power in electrical circuits for the case of non sinusoidal waveform of current and voltage", in elektrotechnische zeitschr., vol. 53, pp. 596-599, 1932.

[5] Gerardus C. Paap, "Symmetrical Components in the Time Domain and Their Application to Power Network Calculations", IEEE Trans. On power systems, Vol. 15, NO. 2, pp 522-528, May 2000.

Automated Design and Implementation of Resonant Controllers for Current Control of Shunt Active Filters

W. Lenwari*, M. Sumner**, and P. Zanchetta**

* Department of Control System and Instrumentation Engineering, King Mongkut's University of Technology Thonburi,
126, Prachauthit Rd, Bangmod, Tungkru, Bangkok 10140, Thailand
** School of Electrical and Electronic Engineering, University of Nottingham,
University Park, Nottingham NG7 2RD, United Kingdom

Abstract–This paper presents the automated design of resonant controllers for current control of three phase shunt active filters. The design methods employs a Genetic Algorithm (GA) based routine. The resonant control compensator is introduced in order to improve the control accuracy and performance for ac input signals. An accurate harmonic current control for shunt active filters can be achieved using resonant compensators. In this paper, the current control system is designed to operate with a single dq reference frame, synchronous with the 50 Hz supply voltage, for the compensation of all main harmonic currents, 5th, 7th, 11th and 13th. The design procedure and the principle of the proposed control method are presented. Experimental results confirm the accuracy and effectiveness of the GA-designed compensators, showing excellent harmonic current tracking performance.

Index Terms—Power Quality, Harmonic, Current Control, Shunt Active Filter, Resonant Controller

I. INTRODUCTION

The harmonic problems in electrical power systems have sparked the research that has led to the extensive understanding of power quality issues. The rapid increase of non linear loads, particularly power electronic systems in both industrial and domestic power distribution is leading to a degradation of the quality of the transmitted power [1], [2]. Power electronic converters act as sources of voltage or current harmonics, and if these are of a sufficient size, system voltage distortion and even grid stability problems can occur [3]. The Shunt Active Filter (SAF) is one such mitigation device which is designed to inject current harmonics into the distribution grid, which should exactly cancel the harmonic currents caused by disturbing non-linear loads [2], [4]. The performance of SAF is normally determined by:

a) The precise current control to match the reference signal
b) The ability to identify the accurate reference signals for the control system

A variety of current control methods have been published, with the objective of tracking the reference currents with the lowest possible error [5]-[12], [18]-[19]. Theses approaches range from basic hysteresis control [5] to multi-reference axis controllers [6], fuzzy logic control [7], sliding mode control [8], nonlinear control [9],

predictive controllers [10] and Proportional plus Integral (PI) plus zero pole controllers designed using Genetic Algorithms [11]. Processing and sampling delays, when implementing the control using a microprocessor, cause difficulty in control design for compensating phase errors of harmonic components. When including these delays, it becomes even more difficult to design a sufficient high bandwidth controller to reduce the steady state error in the harmonic control. In a dq rotating reference frame synchronous to the 50Hz fundamental supply voltage, PI or PI + Derivative (PID) controllers can control fundamental values accurately. However, the system harmonics appear as ac signals in this reference frame and therefore PI and PID controllers have limited effect on these signals. The bandwidth of PI or PID controllers is usually kept below 200 Hz [12], and the SAF using this form of control may actually exacerbate the grid harmonic problem.

In [18]-[19], the authors have proposed the current control for shunt active filters based on resonant compensators and have obtained the excellent steady state control of harmonic currents. However, the design procedure was complex due to the transfer function of resonant compensators, with a high numbers of parameters to be designed [19]. The purpose of this work is to investigate the use of GA applied to the design of resonant compensators as proposed in [19] for the compensation of 5th, 7th, 11th and 13th harmonics. The design procedure using GA and the principle of the control method are discussed. Experimental results are presented to confirm the effectiveness and accuracy of the GA-designed controller.

II. CONTROL OF SHUNT ACTIVE FILTERS IN THE SYNCHRONOUS ROTATING FRAME

The overall current control scheme of the SAF is shown in Fig. 1. It employs a single dq frame of reference which is fixed to the measured supply voltage vector. The three phase supply voltages are measured and a phase locked loop (PLL) [17] is used to derive the voltage angular position θ with respect to the "a" phase winding. The d and q axis voltage are obtained, and the PLL operates to drive the q axis voltage to zero. The measured phase currents are transformed to the synchronous dq reference frame using θ and appear as dc components. There are three control loops as shown in the

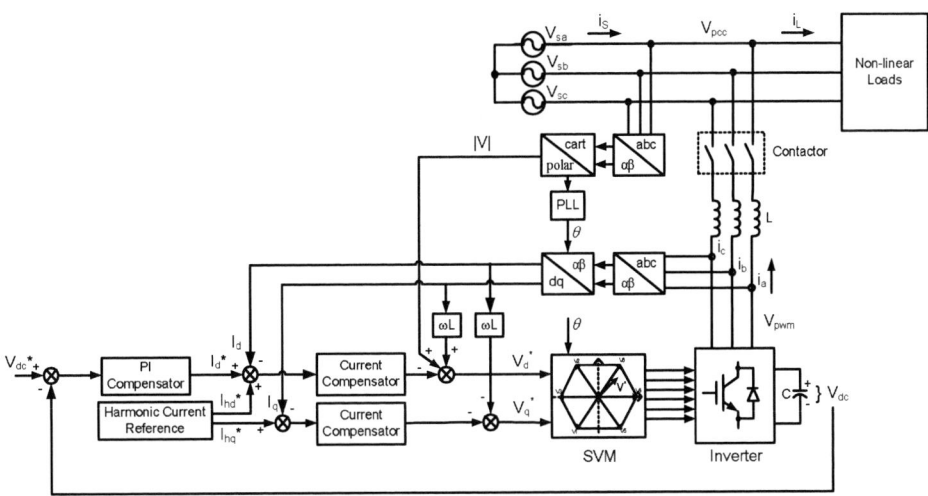

Fig. 1. Shunt active filter control structure

configuration of Fig. 1: the dc link voltage control, and the d and q axis current controllers. For the dc link voltage control, V_{dc} is kept constant at its reference value $V_{dc}*$ by a standard PI compensator. The PI compensator output is the d-axis current reference for the current control loop. The bandwidth of the dc link voltage controller is low (10 Hz) and can be designed separately from the current control loop.

A. Proportional+Resonant Compensator

Proportional plus resonant type compensators have been proposed in the literature for current control at fundamental frequency [13]-[14] and have demonstrated excellent control properties. When the resonant controller is applied to signals tuned to its resonant frequency, its performance is similar to the Proportional + Integral (PI) compensator for a dc signal [14]. It therefore offers significant improvement to the PI or PID approach, and can potentially be exploited as a high performance harmonic current controller for a shunt active filter. It can be derived mathematically by transforming an ideal PI compensator employed in a synchronous rotating frame rotating at the frequency of interest, to the stationary reference frame [14]. In [19], the current control based on resonant compensators was proposed to track four main harmonic currents and an excellent control performance was obtained. The current compensator proposed in [19] is given in (1) for the continuous case

$$C(s) = \left[K_p + \frac{K_{r1}\omega_1 s - K_1 \omega_1^2}{s^2 + (\omega_1/Q_1)s + \omega_1^2} + \frac{K_{r2}\omega_2 s - K_2 \omega_2^2}{s^2 + (\omega_2/Q_2)s + \omega_2^2} \right] \left[\frac{(s-z_1)(s-z_2)}{(s-p_1)(s-p_2)} \right] \tag{1}$$

where K_p is the proportional gain. K_{r1} and K_{r2} are the gain of resonant terms. K_1 and K_2 are used to precisely position the zeros of resonant terms. Q is introduced in (1) to represent the ratio of the centre or resonant frequency, f_0, to the -3dB bandwidth which is f_H-f_L. The upper and lower frequencies, f_H and f_L, are defined as the frequencies where the gain has dropped to 0.707 of the gain at centre frequency. It should be noted that two lead

compensators are added to the transfer function in (1) to stabilize the control system. Equation (1) comprises of one proportional term and two resonant terms having resonant frequencies ω_1 at 300Hz and ω_2 at 600Hz. These two resonant frequencies correspond to the appearance of harmonic currents in a dq frame of reference. 5th and 7th harmonics are seen as ac 300Hz and 11th and 13th harmonics are seen as ac 600Hz as given in (2)

$$\begin{aligned} i_{d5} &= K_5 \cos 6\theta, & i_{d7} &= K_7 \cos 6\theta \\ i_{q5} &= -K_5 \sin 6\theta, & i_{q7} &= K_7 \sin 6\theta \\ i_{d11} &= K_{11} \cos 12\theta, & i_{d13} &= K_{13} \cos 12\theta \\ i_{q11} &= -K_{11} \sin 12\theta, & i_{q13} &= K_{13} \sin 12\theta \end{aligned} \tag{2}$$

where K_5, K_7, K_{11} and K_{13} are the amplitude of each harmonics. Each of the two pairs of harmonics appears at the same frequency, with a difference in the sign of the q-axis current. Therefore, one resonant compensator is able to compensate two harmonic currents. This is a benefit of designing the control in the dq reference frame rather than in a stationary reference frame.

Since the control strategy was digitally implemented in this work, the simplified current control loop for the SAF in z-domain is shown in Fig. 2. The supply impedance is usually ignored in the design as it is unknown, and assumed to be much smaller than the SAF impedance. The PWM inverter can be assumed to be a zero-order hold circuit. The continuous-time compensator (1) is discretized by the forward rectangular rule approximation given in (3) to give (4) where T_s is the sampling time.

$$s = \frac{z-1}{T_s} \tag{3}$$

$$D(z) = \left(K_p + \frac{Az+B}{z^2 + Cz + D} + \frac{Ez+F}{z^2 + Gz + H} \right) \frac{(z-a)(z-c)}{(z-b)(z-d)} \tag{4}$$

471

Where:-

$$A = K_{r1}\omega_1 T_s \quad , \qquad B = -K_{r1}\omega_1 T_s - K_1\omega_1^2 T_s^2 \quad ,$$

$$C = (\omega_1/Q_1)T_s - 2 \quad , \quad D = 1 + \omega_1^2 T_s^2 - (\omega_1/Q_1)T_s \quad ,$$

$$E = K_{r2}\omega_2 T_s \quad , \qquad F = -K_{r2}\omega_2 T_s - K_2\omega_2^2 T_s^2 \quad ,$$

$$G = (\omega_2/Q_2)T_s - 2 \, , \quad H = 1 + \omega_2^2 T_s^2 - (\omega_2/Q_2)T_s \, .$$

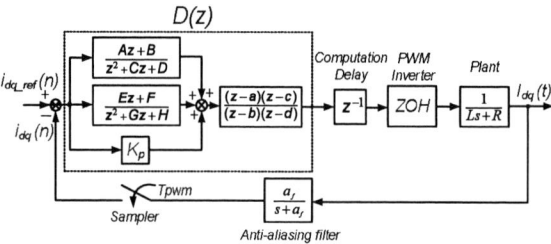

Fig. 2. dq reference frame current control loop in discrete domain

III. GENETIC ALGORITHM APPLIED TO THE DESIGN OF THE CURRENT CONTROLLER

Genetic Algorithm is a stochastic global search method that is inspired by the theories of evolution and natural selection [15]. It operates on a population of potential solutions, termed individuals, applying the principle of evolution, simulated by means of mathematical operations that mimic the process of selection, crossover and mutation. The basis of GAs is that a "population" of problem solutions is composed of "chromosomes" which are strings encoding problem solutions. These chromosomes are formed from the combinations of optimization variables called "genes". A GA-based routine starts with a population of individuals generated randomly, and during each iteration generates successive populations of strings through three operators: reproduction, crossover and mutation.

Reproduction is a process in which individuals in the current population are evaluated using a measure of their objective function values, called "fitness function". A fitness function measures the fitness of an individual to survive in a population of individuals. The genetic algorithm will seek the solution that maximizes or minimizes the fitness function. Once an individual has been selected for reproducing, an exact replica of the string is made. After reproduction, simple *crossover* is performed on the newly reproduced chromosomes, regulated by the crossover probability. Crossover is a genetic operator used to vary a chromosome or chromosomes from one generation to the next. It is an analogy to reproduction and biological crossover, upon which genetic algorithms are based. Chromosomes are mated at random and each pair of these individuals undergoes crossing over. All data beyond that crossover point in the organism individual is swapped between the two parent organisms.

The mechanics of reproduction and crossover are simple, involving random number generation, string copies and some partial string exchanges. *Mutation* is the random alteration of the value of a gene. This operator plays a secondary role in the simple GA and to obtain good results it is generally regulated by a small mutation probability, which is of the order of one mutation per thousand genes. GA will generate at each step a new generation of solutions using the operations of mutation and crossover and selecting the best individuals for the population at the following step. The results of this evolutionary manipulation are new solutions, termed offspring. The new sets of solutions produced, representing a new generation, are then interbred by the same means as the parent solutions. The process stops when a termination criterion e.g. the maximum number of generations has been reached. The output of the GA optimisation should, in theory, converge at the best possible solution. Unlike other optimisation algorithms, because the GA performs a search of the solution space in parallel and not by point to point, it is far less likely to converge at local optimum but more likely to reach the global optimum solution.

With reference to the GA structure, the evaluation of the fitness function consists of a routine test which can be performed either on-line on the experimental rig or off-line using the computer simulations. An on-line approach is not recommended for the active filtering optimisation. Since the shunt active filters control the current through very low impedance represented by the filter inductor, the system can very easily trip particularly due to the over-current protection. In addition, the active filter system needs to be fully functional and stable on start up. In this work, the off-line approach is therefore chosen. The simulation is iteratively used to perform the optimisation. Specific original software has been developed within the MATLAB environment, which uses Genetic Algorithms to automatically tune regulators in the discrete z-domain also for any other kind of applications [16]. It can automatically select the best parameters for the active filter current controllers.

In this paper, the program recursively runs a MATLAB/Simulink simulation reproducing the whole active filter system, testing each of the individuals in the current population. The model of the shunt active filter system includes all components, saturation limits of the controller and a distorted supply voltage. A GA optimisation is applied to find the controller parameters of the discrete-time resonant controller having the equation as given in (4) where there are eleven parameters (K_p, K_{r1}, K_1, Q_1, K_{r2}, K_2, Q_2, z_1, z_2, p_1, and p_2) that need to be optimised since the two resonant frequencies, ω_1 and ω_2, are fixed to 300 and 600Hz respectively. The fitness function can be defined based on the target specification such as transient overshoot, rise time, steady state error, steady state ripple, bandwidth of the controller and so on. The evaluation of the fitness function used in this work is made on the basis of the Integral of Absolute Error (IAE). The current error is the different between the phase reference current (i_a*) and the actual current (i_a). The output of the fitness function is called "fitness value". The fitness function is calculated by the algorithm at each optimization step, assesses the performance of several possible control solutions, and selects which values will evolve to the next generation. In this project, individuals with a lower value of fitness value have a higher probability of contributing one or more offspring in the next generation.

IV. EXPERIMENTAL RESULTS

The experimental rig comprised a 10kVA SAF controlled by a TMS320C6711 DSP. A sampling frequency of 5 kHz is chosen for the implementation to side the prototype with performance of commercial high power SAF. Data acquisition and pulse generation are coordinated by an FPGA. Space Vector Modulation is used for the modulation strategy. The shunt active filter inductance and resistance are 7.5 mH and 0.4 Ω respectively. The tests are performed with a balanced real main supply. For the following tests, the harmonic references were derived from a sinusoidal reference generator in the DSP controller rather than from detecting the harmonics in a non-linear load. This allows the performance of the current controller itself to be evaluated, independent of the dynamics of the harmonic identifier.

Experimental results are presented showing the current controller behaviors when a GA-designed controller is employed. Figs. 3-6 show steady state control of a 5th, 7th, 11th, and 13th harmonic respectively. Controller reference currents, i_a^*, are made available to the data acquisition system (DAQ) via digital to analogue converters. One can observe that the real current rapidly tracks the reference where a small error in amplitude and phase can be noticed. This shows the effectiveness of the GA-designed resonant controller operating in a synchronously rotating dq reference frame for accurate steady state control of the harmonic currents generated by a SAF.

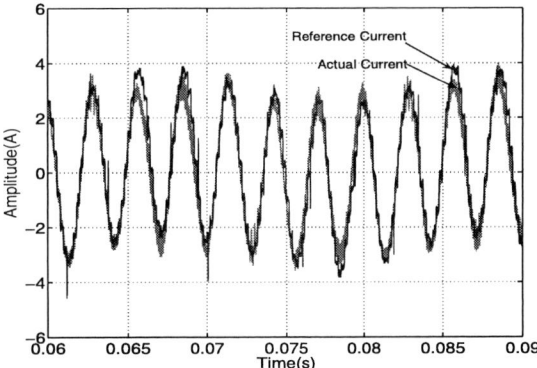

Fig.4. 7th harmonic current control

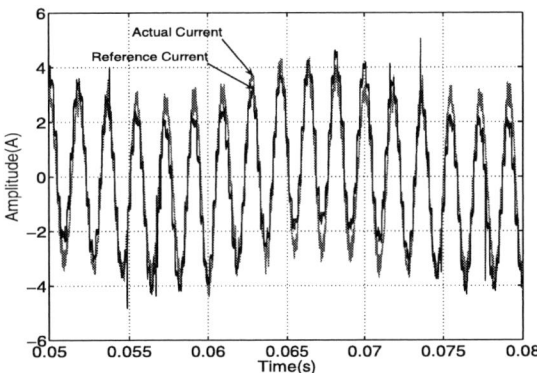

Fig.5. 11th harmonic current control

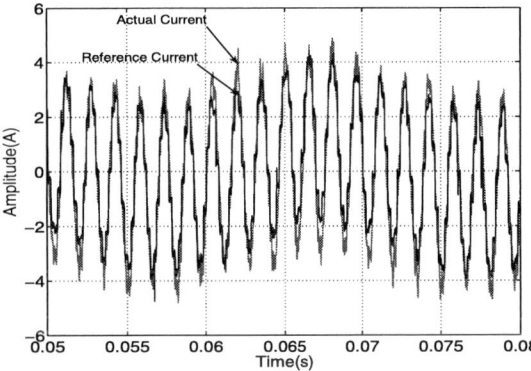

Fig.6. 13th harmonic current control

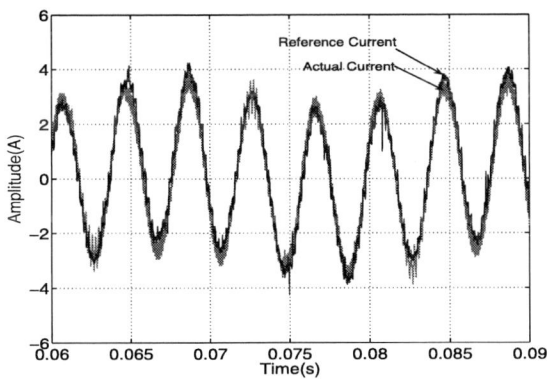

Fig.3. 5th harmonic current control

V. CONCLUSIONS

This paper has described the automated design of the resonant controller for the shunt active filter using a GA. The benefit of this design technique is that it can certainly reduce the commissioning time, spent in the design procedure. Without the use of GA, it is noticed that a high number of controller parameters require a hand-tuned design certainly causing the difficulty in the control design. A high performance current control based on resonant compensators has been designed and successfully implemented in this paper. The GA-designed controller can precisely control 5th, 7th, 11th and 13th harmonic currents with small magnitude and phase errors. A resonant controller is designed to be used in a dq reference frame rotating at the frequency of the fundamental supply voltage. The advantage of designing the control in the dq frame of reference is that four main harmonics can be controlled with a single compensator having only two resonant terms. The experimental results show excellent harmonic current tracking performance of the GA-designed controller.

REFERENCES

[1] R. C. Dugan, M.F. McGranaghan, and H. W. Beaty, *Electrical Power System Quality*, McGraw-Hill, 1996.

[2] H. Akagi, "New trends in active filters for power conditioning," *IEEE Trans. Ind. Applicat.*, vol. 32, pp. 1312-1322, Nov./Dec. 1996.

[3] R. Yacamini, "Power system harmonics: Part 3. Problems caused by distorted supplies," *IEE Power Engineering Journal*, vol. 9, no. 5, pp. 233 – 238, Oct. 1995.

[4] F. Z. Peng, "Application issues of active power filters," *IEEE Industry Applications Magazine*, vol. 4, no. 5, pp. 21-30, Sept./Oct. 1998.

[5] L. Malesani, P. Mattavelli, and P. Tomasin, "High-performance hysteresis modulation technique for active filters," *IEEE Trans. Power Electron.*, vol. 12, no. 5, pp.876-884, Sept. 1997.

[6] M. Sumner, B. Palethorpe, and D.W.P. Thomas, "Impedance measurement for improved power quality - Part 2: a new technique for stand alone active shunt filter control," *IEEE Trans. Power Delivery*, vol. 19, no. 3, pp. 1457-1463, Jul. 2004.

[7] S.K. Jain, P. Agrawal, and H.O. Gupta, "Fuzzy logic controlled shunt active power filter for power quality improvement," *IEE Proceedings on Electric Power Applications*, vol. 149, Issue: 5, pp. 317 – 328, Sept. 2002.

[8] N. Mendalek, K. Al-Haddad, F. Fnaiech, and L.A. Dessaint, "Sliding mode control of 3-phase shunt active filter in the d-q frame," *in Proc. of Power Electronics Specialists Conference, PESC 02*, vol.1, Jun 2002, pp. 369-375.

[9] N. Mendalek, and K. Al-Haddad, "Modeling and nonlinear control of shunt active power filter in the synchronous reference frame," *in Proc. of Ninth International Conference on Harmonics and Quality of Power*, vol. 1, Oct. 2000, pp. 30 – 35.

[10] S. G. Jeong, and M. H. Woo, "DSP-based active power filter with predictive current control," *IEEE Trans. Ind. Electron.*, vol. 44, no. 4, pp. 329-336, Jun. 1997.

[11] V. Diana, M. Sumner, P. Zanchetta, and M. Marinelli, "The use of genetic algorithms for the design of current controllers for active shunt filters," *in Proc. of 29th Annual Conference of the IEEE Industrial Electronics Society, IECON'03*, Virginia, USA, Nov. 2003, vol. 3, pp.2005-2010.

[12] D. Butt, "An investigation of harmonic correction techniques using active filtering," PhD thesis, The University of Nottingham, 1999.

[13] Y. Sato, T. Ishizuka, K. Nezu, and T. Kataoka, "A new control strategy for voltage-type PWM rectifiers to realize zero steady-state control error in input current," *IEEE Trans. Ind. Applicat.*, vol. 34, no. 3, pp. 480-486, May/Jun. 1998.

[14] D.N. Zmood, D.G. Holmes, and G. Bode, "Frequency domain analysis of three phase linear current regulators," *IEEE Trans. Ind. Applicat.*, vol. 37, no. 2, pp. 601-610, Mar./Apr. 2001.

[15] Z. Michalewicz, *Genetic algorithms + Data structures = Evolution programs*, Springer-Verlag, 1996.

[16] P. Zanchetta, M. Sumner, F. Cupertino, M. Marinelli, and E. Mininno, "Online and off-line control design in power electronics and drives using genetic algorithms," *in Proc. of IEEE 29th Industry Applications Society Annual Meeting, IAS 94*, vol. 2, pp. 864 – 871, Oct. 1994.

[17] V. Kaura and V. Blasko, "Operation of a phase locked loop system under distorted utility conditions," *IEEE Trans. Ind. Applicat.*, vol. 33, no. 1, pp. 58-63, Jan/Feb. 1997.

[18] W. Lenwari, M. Sumner, and P. Zanchetta, "Design and analysis of high performance current control for shunt active filters," *in Proc. of the 3rd IET International Conference on Power Electronics, Machines and Drives, PEMD 2006*, pp.90–95, Apr. 2006.

[19] W. Lenwari, M. Sumner, P. Zanchetta, and M. Culea, "A high performance harmonic current control for shunt active filters based on resonant compensators," *in Proc. of 32nd Annual Conference of the IEEE Industrial Electronics Society, IECON'06*, pp.2109–2114, Nov. 2006.

A Modular Structured Multilevel Inverter Active Power Filter with Unified Constant-Frequency Integration Control for Nonlinear AC Loads

P. Y. Lim
School of Engineering and Information Technology
Universiti Malaysia Sabah
Sabah, Malaysia
lpy@ums.edu.my

N. A. Azli
Department of Energy Conversion
Faculty of Electrical Engineering
Universiti Teknologi Malaysia
Johor, Malaysia
naziha@ieee.org

Abstract— **Active power filter (APF) has a vital role for compensating the reactive power and harmonics from a group of nonlinear loads connected at the common point of coupling to ensure that the resulting total current drawn from the main incoming supply is sinusoidal. This paper is to present the performance of a Modular Structured Multilevel Inverter (MSMI) APF with Unified Constant Frequency Integration (UCI) control technique for compensating the distorted current caused by various nonlinear ac loads. MSMI is well known for its capability in reducing the voltage and current stresses on the semiconductors switches and providing the convenience for future expansion in order to achieve higher power ratings for the APF system. Whereas, the UCI control technique is adopted due to its constant switching frequency operation and simple analog circuitry. To demonstrate the capability of the single phase MSMI APF with UCI control technique for compensation of harmonics caused by various AC nonlinear loads, simulation results are presented and discussed.**

Index Terms--**Active power filter, current harmonics compensation, modular structured multilevel inverter, unified constant frequency integration control, nonlinear loads.**

I. INTRODUCTION

Active power filter (APF) has been widely used for harmonics and reactive power compensation and it now becomes a mature technology for improving the power quality. Recently, semiconductor constraints and the voltage stress on the semiconductor switches have drawn the attraction of the researchers. In order to achieve higher voltage level of the filtering operation, inverter configuration such as Hybrid Asymmetric Multilevel Inverter [1] was proposed.

The performance of the APF inverter depends very much on the control scheme. Many technical papers related with active power filters and their control methods were presented. Some researchers spent their effort in developing the APF system with Unified Constant-Frequency Integration (UCI) controller [2][3] recently. APF controlled by UCI Controller has the attractive advantages comparing to the other control method such as the PQ Theory [4], Linear Current Control, Digital Deadbeat Control and Hysterisis Control [5]. Those proposed control approaches need to sense the input voltage, load current and then calculate the harmonic reactive

component in the load in order to generate the reference current for controlling the inverter to produce a current that is equal to the amplitude and opposite in direction of the reactive current of nonlinear loads. Those control methods require fast and real-time calculation which require several high precision analogue multipliers or high speed DSP chip with fast A/D converter that yields high cost, complexity and low stability.[3]

As for the UCI controller, it is based on the theory of one cycle control [6], which employs an integrator with reset port as its core component to control the duty ratio of the APF inverter switches. This method ensures that the current drawn from the AC main supply will be sinusoidal waveform with different non-linear AC loads connected to the supply. No calculation of reference current for the controller is required which greatly simplifies the control circuit.

This paper is to realize the implementation of UCI Controller to Modular Structured Multilevel Inverter (MSMI) APF which will generate harmonics current that cancels the harmonics current from a nonlinear load in order to form better sinusoidal incoming supply. MSMI has an advantage of reducing the voltage stress of switching devices. Furthermore, the modularized circuit layout of MSMI allows expansion of the APF structure to be done easily.[7] This will be convenient if expansion of APF is required to meet higher power level demand, such as in power transmission line. The performance of a single phase simulation of MSMI APF which is connected to different AC nonlinear loads will be presented for the purpose of demonstration.

II. OPERATION OF THE MSMI WITH UCI CONTROLLER

MSMI is an inverter structure, which has cascaded inverters with Separate DC Sources (SDCs). This topology allows expansion of the number of levels, which provides flexibility in increasing of higher voltage level. This can be done easily as the inverter has modularized circuit layout. Furthermore, the cascaded inverter structure help to reduce voltage stress on switches, as lower voltage will be imposed by the DC side capacitor voltage to the switches or in other words, switches only have to bear on smaller value of voltage.

978-1-4244-0644-9/07/$25.00 ©2007 IEEE

Fig.1 shows the structure of a 5-level MSMI APF.[8] When this multilevel inverter is implemented as an APF, the phase voltage of this particular APF is listed as in Table I based on Kirchoff's voltage law, considering the DC capacitor voltages as V_{o1} and V_{o2} and they are equal to V_{dc} respectively.

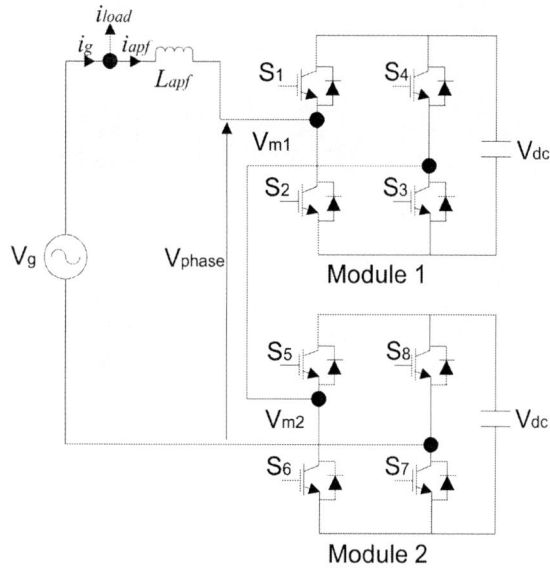

Fig. 1 Modular Structured Multilevel Inverter

Table 1 Switching Function Table

S1	S2	S3	S4	S5	S6	S7	S8	V_{phase}
				1	0	1	0	$2V_{dc}$
1	0	1	0	1	0	0	1	V_{dc}
				0	1	1	0	V_{dc}
				0	1	0	1	0
				1	0	1	0	V_{dc}
1	0	0	1	1	0	0	1	0
				0	1	1	0	0
				0	1	0	1	$-V_{dc}$
				0	1	0	1	$-2V_{dc}$
0	1	0	1	0	1	1	0	$-V_{dc}$
				1	0	0	1	$-V_{dc}$
				1	0	1	0	0
				0	1	0	1	$-V_{dc}$
0	1	1	0	0	1	1	0	0
				1	0	0	1	0
				1	0	1	0	V_{dc}

The proposed configuration is as shown in Fig. 2.[8] Through the current transformer, the mains current will be compared with the integrated voltage signal from each module of inverter to generate pulses for R-S flip-flops for triggering the inverter switches.

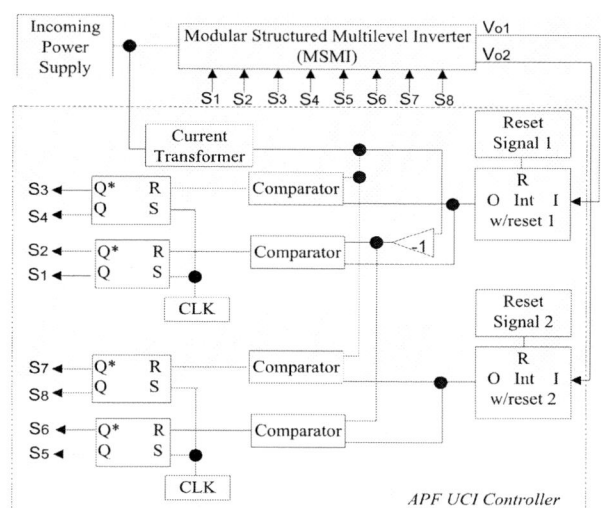

Fig. 2 Conceptual diagram of the proposed structure of single phase MSMI APF with UCI Controller.

III. SIMULATION RESULTS

Computer simulations using MATLAB / SIMULINK were used to evaluate the performance of the MSMI APF with UCI controller under different AC load conditions. The developed system is shown in Fig. 3.

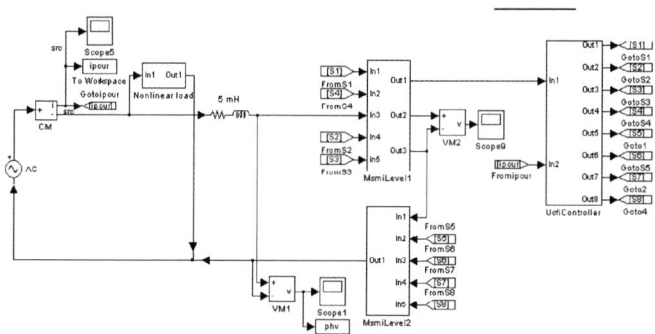

Fig.3 MSMI APF system blocks

In MATLAB / SIMULINK simulations, the nonlinear RC and RL load will be connected to the AC supply and the APF performance will be analysed. First, a diode rectifier with RC load was connected to the AC main inlet as shown in Fig. 4.

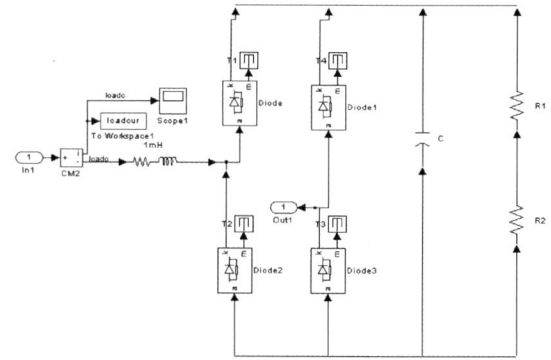

Fig.4 MATLAB / SIMULINK subsystem blocks of nonlinear diode rectifier with RC loads.

Fig.5 shows that without the compensation from the APF, the AC current waveform was highly distorted by the nonlinear RC load. The indicated Total Harmonic Distortion in Fig. 6 is 146.96% based on the first hundred harmonics components.

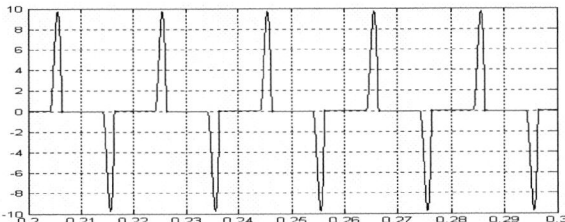

Fig.5 Distorted source current caused by diode rectifier with RC load.

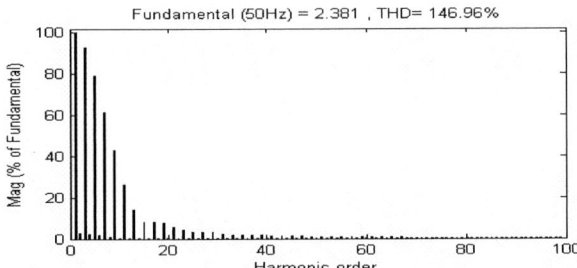

Fig.6 Harmonic current components of current drawn by RC load without APF compensation.

By connecting the shunt MSMI APF in between the incoming supply and the nonlinear RC load, the harmonic problem can be mitigated. Compensation current from APF in Fig.7 will compensate the distorted source current waveform.

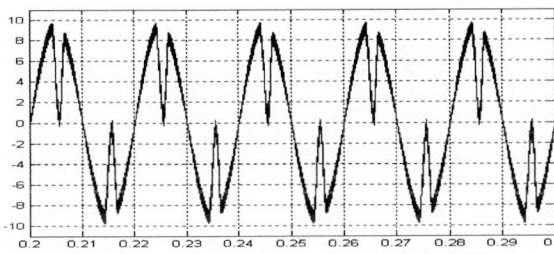

Fig.7 Compensation current produced by MSMI APF.

After compensated by the proposed configuration of MSMI APF, the current waveform drawn from the AC incoming supply is now nearly similar to sinusoidal wave shape as shown in Fig.8. The THD of AC supply current after compensation is greatly reduced from 146.96% to 9.84% as in Fig.9.

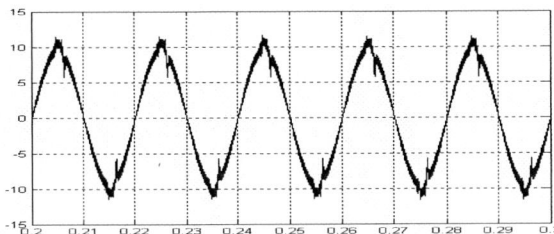

Fig.8 Source current waveforms after compensation.

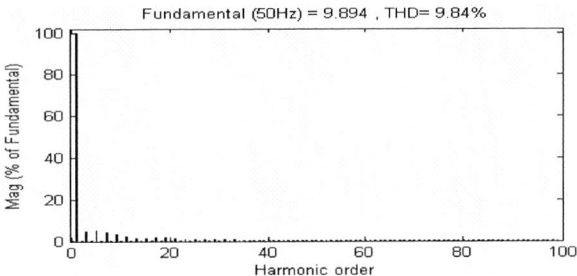

Fig.9 Harmonic spectrum of source current after compensation.

A load disturbance has been created to test the ability and flexibility of the proposed APF configuration. At 0.3 second of the simulation time, the load suddenly changes from 250 Ω to 200 Ω. Fig.10 shows the increment of current peak due to the reduction of load.

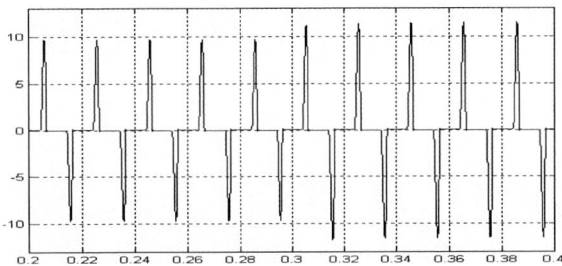

Fig. 10 Load changes from 250 Ω to 200 Ω at 0.3s.

The UCI controller able to sense the changes of load at 0.3 second and compensation was done immediately within a cycle of waveform as in Fig.11. Fig.12 shows the compensated source current from the AC input which remains sinusoidal under load changing condition.

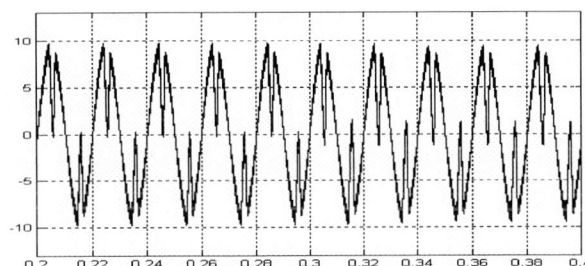

Fig. 11 Compensation current to the changing load at 0.3s

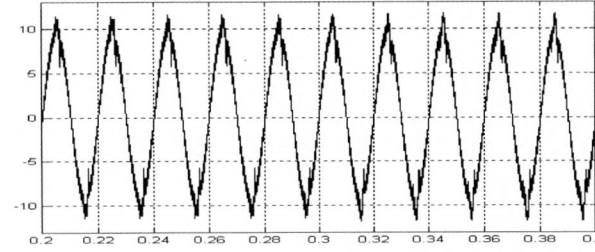

Fig 12 : Source current drawn from the AC input with change in load at 0.3s

Fig.13 is the phase voltage of this particular MSMI APF, which indicates 5 levels of voltage including the reference level. The waveform is closer to sinusoidal compared to the conventional single stage APF that produces square wave voltages.

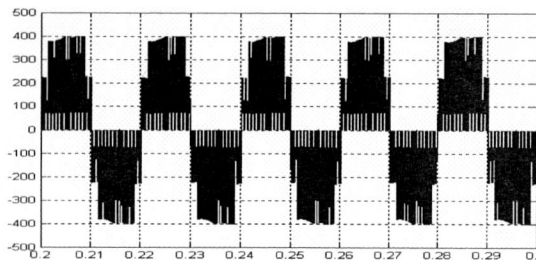

Fig.13 Phase voltage of APF in 5-level.

The proposed configuration of APF also tested for compensating the current drawn by diode rectifier with RL load. Fig.14 is the subsystem block of nonlinear diode rectifier with RL load connected to the AC main supply in MATLAB / SIMULINK.

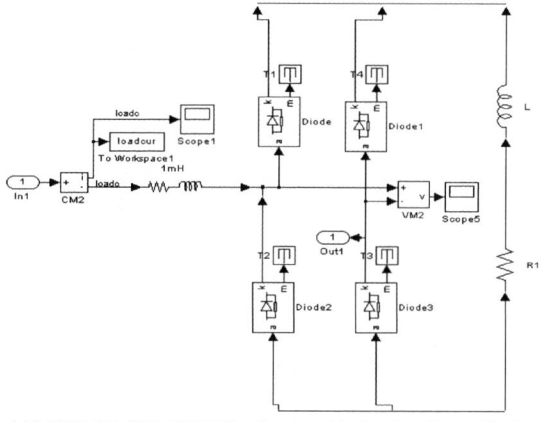

Fig.14 MATLAB / SIMULINK subsystem block of nonlinear diode rectifier with RL loads.

The current drawn by the RL load is shown in Fig.15 and the Total Harmonic Distortion percentage obtained using FFT analysis is 42.96%.

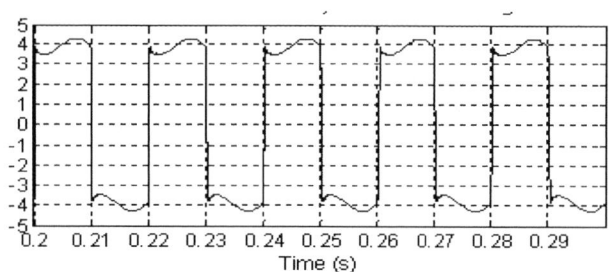

Fig 15. Distorted source current caused by diode rectifier with RL load.

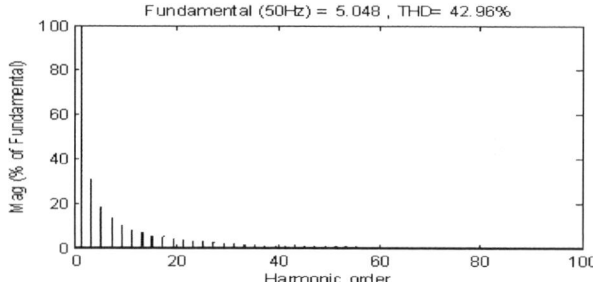

Fig 16. Harmonic current components of current drawn by RL load without APF compensation.

Fig. 17 depicts the compensation current from the APF to compensate the distorted current caused by nonlinear RL load.

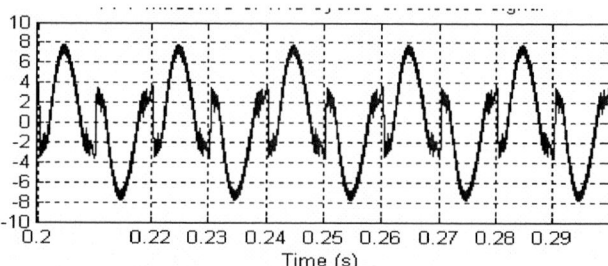

Fig. 17 Compensation current produced by MSMI APF.

Fig.18 and Fig.19 show the compensated sinusoidal AC current where the Total Harmonic Distortion is reduced to 22.42%.

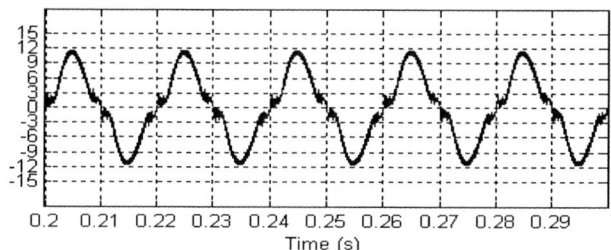

Fig.18 Source current waveforms after compensation.

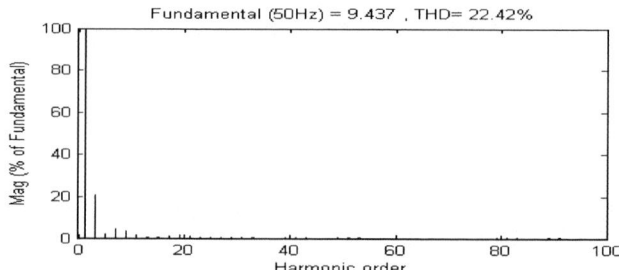

Fig. 19 Harmonic spectrum of source current after compensation.

Referring to Fig.20, it shows that the proposed APF configuration is connected to compensate the AC incoming current. Then, it was disconnected from the system at 0.22 seconds and reconnected at 0.32 seconds. There is an increase in current during the reconnection of APF to the system. However, it recovered immediately within half cycle. It is obvious that the proposed configuration of APF with UCI Controller has fast response in operation.

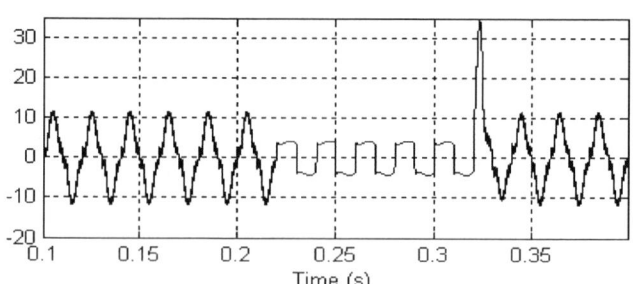

Fig. 20 Simulated source current waveforms with the APF disconnected at t = 0.22ms and reconnected at t = 0.32s.

IV. CONCLUSION

It has been shown in this paper that the MSMI APF with UCI controller possesses high capability in compensating the distorted current caused by different load. Fast response of the controller was observed when the APF was switched on and off from the system. This is due to the reason that the UCI controller performs current compensation in the manner of cycle by cycle. The capability of the MSMI APF with UCI controller is undoubtedly providing a flexible solution for power quality control. Due to the simplicity and flexibility of the circuitry, this configuration of APF is applicable and is very suitable for industrial power filtering purposes.

V. REFERENCES

[1] Lopez, M. G., Moran, L. T, Espinoza, J. C.; Dixon, J. R. "Performance Analysis of a Hybrid Asymmetric Multilevel Inverter for High Voltage Active Power Filter Applications", *Industrial Electronics Society, 2003. IECON '03. The 29th Annual Conference of the IEEE*, Vol: 2, 2-6 Nov 2003, Pg 1050-1055

[2] L. Zhou, and K. Smedley. "Unified Constant-frequency Integration Control of Active Power Filters." *APEC2000*, p.406-412.

[3] C. M. Qiao, K. M. Smedley, and F. Maddaleno, "A Comprehensive Analysis and Design of a Single Phase Active Power Filter with Unified Constant-frequency Integration Control." presented at the *2001 IEEE 32nd Annual Power Electronics Specialists Conference, PESC*, vol. 3, pp.1619-1625, June 2001.

[4] Haque, M. T. "Single Phase PQ Theory for Active Filters". *2002 IEEE Region Conference on Computers, Communications, Control and Power Engineering*. Oct 28-31. TENCON '02 Proceedings. 2002. vol. 3: 1941-1944.

[5] Simone Buso, Luigi Malesani, "Comparison of Current Control Techniques for Active Filter Applications" *IEEE Trans on Industrial Electronics*. Vol. 45. No. 5 October 1998.

[6] Cuk, S. and Smedley,K. "One-cycle Control of Switching Converter", *PESC 1991*, p.888-96.

[7] Azli, N.A.; Yatim, A.H.M, " Modular Structured Multilevel Inverter For High Power AC Power Supply Applications" *Industrial Electronics, 2001. Proceedings. ISIE 2001. IEEE International Symposium* on Volume: 2 , 12-16 June 2001 Pages:728 - 733 vol.2.

[8] Azli, N.A.: Lim, P.Y., "Modular Structured Multilevel Inverter with Unified Constant-Frequency Integration Control for Active Power Filters" *Power Electronics and Drives Systems, 2005, PEDS 2005 International Conference* on Volume 2, 28-01 Nov.2005 Pages 1312 -1316.

This page intentionally left blank.

Author Index

A

Abdi, Ehsan 1096
Abe, Seiya..1388
Abjadi, N. R. 1442
Achara, P. 394
Adélaide, L. 569
Adya, A. 731
Afjei, E..722
Agarwal, Pramod 1810
Agarwal, Vineeta 1891
Ahmadian, H. Molla 1147
Ahn, Jin-Woo...1857
Almardy, M. 677
Alonge, F. 959
Amaral, Acácio M. R. 587, 643
Amirudin, Dessy 534
Amrane, F. 569
An, Young-Joo 1527
Andersen, P. Scavenius.............................886
Ang, Y. 382
Ang, Yong-Ann 376
Aodsup, K. 937
Arab, G.R. 1449
Aree, P. 703
Arvindan, A. N. 480
Ashaibi, Ahmed Ali 1242
Ataei, S. 722
Athab, Hussain S. 869, 874
Attaviriyanupap, Pathom 1102
Auger, Francois..1368
Ayob, S. M.........................1274, 1363, 1682
Azli, N. A. 475, 1041, 1274, 1363, 1682

B

Bac, Nguyen Xuan....................................1501
Baharom, R. 1626
Baiju, M.R. 1047
Banerjee, Subrata.....................................812
Bartholet, M.T. 257
Batzies, Ekkehard 1316
Bhat, A.K.S. 677
Bhuvaneswari, G. 310
Bi, C. 1082
Bina, M. Tavakoli 465, 1060, 1065, 1799
Binder, A. 249
Bingham, C. M. 382
Bingham, Chris 376
Binh, Tran Cong 1195
Biswas, S. K. 1352, 1605
Blaabjerg, Frede................226, 541, 1247, 1376
Boonchiam, P. N. 937, 1851
Boonyaroonate, Itsda 1078, 1383
Bosing, M. 1160
Branco, P.J. Costa.....................................917
Brauer, Helge J. 716
Buatti, Gustavo M......................................643

Bunlaksananusorn, C. 977, 1575

C

Cangemi, T. 959
Cardoso, A. J. Marques 587, 643
Carstensen, Christian 912
Chalermyanont, Kusumal 295
Champa, P. 703
Chan, K. W. 1394, 1727, 1804
Chan, Shun-Yu 305
Chang, David..305
Chang, Tsin-Yuan 581
Chang, Y. D. 1321
Chao, Ma Xian 665
Chatratana, S. 1495
Chaudhari, M. A. 1708
Chen, Jiaxin 510
Chen, L. 1538
Chen, Sufen 1262
Chen, W. C. 440
Chen, Wei 1636
Chen, Y. H. 456, 1278
Chen, Y. M. 1321
Chenfeng, Yang...745
Cheng, Chien-Lung 749, 1703
Cheng, K. W. Eric 1691, 1697
Cheng, K.W.E. 1727
Cheng, Qiang..1330
Chengfeng, Yang...............................270, 427
Chereau, Vinciane 1368
Chern, Shyi-Ching 749, 1703
Cheung, N.C. 1727
Cheung, Norbert C 1691, 1697
Chi, Chien-An 388
Chiang, Wen-Jung 824
Chien, F.T. 660
Chin, Li-Yuan 305
Chivite-Zabalza, F. Javier..............788, 796, 804
Cho, B. H. 401
Cho, Kyu Min 665, 1142
Choi, S. J. 401
Choudhary, Sonika 1757
Chrin, P. 1575
Chudamani, R. 1827
Chudoung, Nakharet 1213
Chun, Tae-Won 1857
Chunkag, V. 863
Ciobotaru, M. 226
Colak, Baris 1507
Corradini, L. 600
Cosic, A. 1301
Cruden, A. 1182

D

Dahlan, N.Y..527
Dahono, P. A. 1267
Dahono, Pekik Argo 534

Author Index

Dai, Z. 1885
Dananjayan, P. 626
Davat, Bernard..1
Deb, N. K. 1352, 1605
Dehbonei, H. 1657
Deleroi, W. 1495
Deng-Em, S. 1851
Deni, ...534, 1267
Densei-Lambda, K.K. 280
Desai, Hardik P. 829
Dhomane, G. A. 1590
Dick, Christian P. 448
D'ippolito, F. 959
Ditmanson, C. 556
Doki, Shinji..999
Doncker, R. W. De 907, 912, 1160
Doncker, Rik W. De 213, 327, 333, 448, 710, 716
Dong, Lei 1691
Dong, Ming-Chui.................................607, 614
Dong, Yang 1340
Dorkmai, Pramoch.....................................697
Dorrell, David G 886, 922, 1167, 1174
Duan, S.X. 551
Duan, Shanxu....................................836, 842
Dwivedi, Avneesh.....................................1757
Dzung, Phan Quoc 1195, 1202, 1501

E

Ekkaravarodome, Chainarin 1383
Ertan, H. Bulent 1507
Eskandari, B..................................1060, 1065

F

Fang, D. Z. 1394, 1804
Fang, Kuo-Lun.................................1610, 1712
Fang, Tzu-Hsuan......................................1717
Fei, Wanmin....................................350, 354, 1672
Ferraz, Antonio 817
Ferreira, O.C. 1017
Fidler, Peter 327
Fingerhuth, S. 1160
Finney, S.J. 299, 1242
Finney, Steve J......................................1255
Foroosh, S. Chini 1465
Forsyth, Andrew J. 788, 796, 804
Foster, M. P. 382
Foster, Martin 376
Fuengwarodsakul, Nisai H. 710
Fukuda, Shoji 1070
Fukushima, K. 1885

G

Gairola, Sanjay 738, 899
Gao, F. 1247, 1376
Garg, Vipin 310
Geethalakshmi, B.....................................626

Goel, P.K. 941
Gonthier, L. 322
Gopinath, Anish 1047
Goyal, Devendra......................................1520
Grant, D M 368
Grantham, Colin 1284
Gruber , W. 574
Guan, Xiaohan..................................994, 1752
Gueldner, H. 1006
Guldner, H. 556
Guo, Youguang 275, 510, 1662
Gupta, H.O. 1810
Gupta, J.R.P 731

H

Hai, Quach Thanh....................................1033
Hajian, M. 1449
Hamzah, M.K. 527, 1626
Hamzah, N.R. 527, 1626
Han, Ying-Duo 607, 614
Hansen, P. E. 886
Haque, M. Tarafdar 620
Harada, Y...1885
Hasegawa, Masaru 1543
Hellinger, R. 1006
Hennen, Martin D. 716, 907
Hew, W.P. 1514
Heyun, Lin 270, 427, 745
Higuchi, Kohji 280
Hinkkanen, Marko 406
Hirokawa, Masahiko 1388
Hirota, Atsushi 1740
Ho, Shine-Tzong 498, 1788
Hoang, Nguyen Minh 1195, 1501
Hofmann, W. 781, 1538
Hotait, Hadi A 299, 1255
Hothongkham, Prasopchok.............................1236
Hsieh, C. T. 1567
Hsu, Chih-Jen 286
Huang, P. L. 1321
Hung, Tsung-You 1762, 1767
Hwu, K. I. 338, 456, 692, 1278

I

Idris, Z. 527
Iov, F. 226
Ishitobi, Manabu 504
Islam, S. 1834
Iso, Osamu 1102

J

Jang, B.H. 1657
Jangjaempradit, Saksit 1641
Jangwanitlert, A. 989, 1412
Janjornmanit, Suchart 1327
Jayashree, E. 1555

Author Index

Jeevananthan, S. 1221
Jegathesan, V. 1677
Jerome, Jovitha 1677
Jeung, Giwoo 1092
Jian, Guo 270, 427, 745
Jou, Hurng-Liahng 493, 824
Jovanovic, Milutin G 922
Junge, Christian 1533
Jwo, W. S. 1560

K

Kadir, M. N. Abdul 1514
Kaewsingha, Aswin 1327
Kamnarn, U. 863
Kamper, M.J. 420, 1017, 1295
Kando, M. 488
Kang, Yong 551, 836
Kano, Masaru 414
Kanthaphayao, Y. 863
Kanzi, K. 465
Karunakar, K 1620
Karutz, P. 574
Kasal, Gaurav Kumar 357
Kasper, K. A. 1160
Kavitha, A. 595
Kazimierczuk, Marian K. 1136
Kennel, R.M. 1017
Kerz, O. 363
Khaehintung, Noppadol 847, 1429
Khajeh, A. 1455
Khalil, Ahmed G. Abo- 1471
Khan, P. K. Shadhu 869, 874
Khan-Ngern, Werachet 460, 1335
Khomfoi, Surin 1055, 1228
Khun, C. 488
Kim, Dong-Hun 1092
Kim, H. S. 1846
Kim, Hee Jun 665, 1142
Kim, Heung-Geun 1092
Kim, I. C. 1846
Kim, In Dong 1092
Kinnaraes, V. 1483
Kinnares, V. 1356, 1489
Kinnares, Vijit 1236
Kittiratsatcha, S. 977
Ko, S. H. 1657, 1846, 1846
Ko, T.K. 1657
Ko, Yi-Pin 388
Kobayashi, Takayuki 265
Kock, H.W. De 1017
Koenig, Andreas 327
Kohama, Teruhiko 1417
Kok, W. Sae- 368
Kolar, J.W. 257, 574
Kongsuk, P. 937
Kongthawornwattana, P. 977
Konig, Andreas 448

Krein, Philip T. 221
Krismadinata, ...1290
Kubota, Hisao 265
Kulvitit, Youthana 342, 697
Kumar, S. Ganesh 1632
Kumar, S. Krishna 1632
Kumchaiyo, Ruthapong 1078
Kunakorn, Anantawat 1429
Kuo, J. S. 440
Kuo, Jian-Long 1717, 1722
Kurokawa, F. 968, 1398
Kusuhara, Yoshito 954
Kwok, K. W. 1727
Kwok, Y. L. 1727
Kwon, Soon Kurl 504

L

Laczynski, T. 1645
Lafzi, A. 620
Lai, Y. M. 1262
Lai, Yen-Shin 1586
Lakhdari, Z. 569
Lan, Yi-Hung 749
Lee, Chien-Min 1586
Lee, Dong-Choon 1471
Lee, Dong-Hee 1527, 1857
Lee, Hong Hee 1027, 1033
Lee, S. R. 1657, 1846
Lee, S. W. 1657, 1846
Lee, Yuang-Shung 286, 388
Lei, Dong 949, 1340, 1697
Lei, Yuzhou 291, 1777
Leibfried, T. 363, 726
Lenke, Robert U. 213
Lenwari, W. 470
Leou, Rong Ceng 546
Lerdudomsak, Smith 999
Li, X. 551
Li, Y. J. 440
Liang, C. 1376
Liao, C.N. 660
Liao, Xiaozhong 1691
Lijie, Wang 949
Lim, P. Y. 475, 1041
Lim, S. H. 1846
Lim, T.C. 299
Lin, Chang-Hua 1610, 1712, 1762, 1767
Lin, Chih-Hong 1549
Lin, H. C. 1567
Lin, Hung-Chih 581
Lin, Min 1423
Lipo, Thomas A. 1308
Liu, B.Y. 551
Liu, Bangyin 836, 842, 1636
Liu, Dikai 275
Liu, Fei 842
Liu, Maw-Yang 1610, 1712

A-3

Author Index

Liu, Xian-Lin 1722
Liu, Yi-Hwa..546
Liu, Yuanchao...................291, 1615, 1752, 1777
Liu, Z. 551
Loh, P. C. 1247, 1376, 1620
Loh, Poh Chiang 541
Loron, Luc 1368
Lu, Haiyan 275
Lu, Y. 1727
Lu, Zhengyu.................................354, 1088
Luomi, Jorma..406

M

Ma, Yu....................................632, 1088, 1601
Macheiner, P. 880
Madawala, U. K. 648, 654
Makany, Ph. 569
Makino, Tomoaki 1740
Manmek, Thip 773
Mao, Peng 1615
Markadeh, Gh. R. Arab 1442
Marques, Gil D. 636
Martin, F. 363
Martins, J. F. 894, 1875
Masoum, Amir S. 767
Masoum, M.A.S. 767, 1834
Massoud, A.M. 299
Massoud, Ahmed M.....................................1255
Massoud, Ahmed 1242
Matsui, Keiju 1543
Matsui, N. 1398
Matsui, Y. 1109
Matsui, Yasuaki 1102
Matsuo, K. 1686
Matsuse, K. 394, 685
Matsuse, Kouki 521, 1460
Mattavelli, P..600
Mattavelli, Paolo 760
Mcmahon, Richard 1096
Medagam, Peda V 1477
Mekhilef, S. 1514
Meng, Peipei 632
Mertens, A. 1645
Meyer, Christoph 213
Milani, A. Roshan 620
Miri, A. M. 726
Mirmousa, H. 1404
Mishima, Tomokazu 563
Mithulananthan, N. 937
Mittal, A.P. 731
Mittal, Raghu K. 1757
Miura, T. 1686
Miyamoto, Hiroyuki 1773
Moallem, Ali 983, 1147
Modak, J. P. 1708
Moghani, J. S. 1455
Mondal, N. 1352, 1605

Moon, Y.H. 1657
Morimoto, Masayuki 1641, 1773
Morita, Katsuaki 1773
Moses, Paul S. 767
Mossner, K. 363
Mudannayake, Chathura P............................773
Mun, Sang Pil 504, 563
Mura, Florian 213
Muraoka, Hidekazu 563
Murthy, S.S......................................941, 1123, 1757

N

Nabeshima, Takashi 858, 1423, 1734
Naetiladdanon, Sumate 755
Nagai, Satoshi 1740
Nakagawa, Shin 954
Nakanishi, Hirotaka 858, 1734
Nakano, Kazushi 280
Nakano, Tadao...................................858, 1734
Nakaoka, Mutsuo 504, 563
Nakayama, Asahi 954
Nandhakumar, R. 1221
Nathakaranakule, Adisak.............................1383
Navi, K. 722
Nazarzadeh, Jalal 1782
Neammanee, B. 1129
Neuhaus, Christoph R..................................710
Ngern, W. Khan- 488, 515, 1667, 1794
Ngoc, Ha Pham 1102
Nguyen, Binhminh 1434
Nho, Eui-Chel 1527
Nho, Nguyen Van 1027, 1033
Nia, S.Hosein 1449
Ninomiya, T. 1885
Ninomiya, Tamotsu 954, 1388, 1417
Nishijima, Kimihiro 858, 1423, 1734
Nishimura, Jun 1460
Noguchi, Toshihiko 414, 1595, 1651
Noor, S.Z. Mohammad 1626
Norigoe, I. 1885
Nussbaumer, T. 257, 574

O

Obata, S. 671
Ogura, K. 1109
Oh, Won Seok 1142
Oka, Kazuo 521, 1460
Okuma, Shigeru 999
Omori, Hideki 563
Opanuruk, Puckapon 342
Oranpiroj, Kosol 318
Owatchaiphong, Satit 912
Ozdemir, Engin 1055
Ozdemir, Sule 1055

Author Index

P

Pai, Kai-Jun 1762, 1767
Pal, Jayanta 812
Palandurkar, M.V. 1708
Panda, Sanjib K. 852
Park, Hong-Geuk 1471
Park, J. H. 401
Pashajavid, E. 465
Passal, A. 322
Patel, H. K. 829
Pavitra, G. 1757
Peng, S.T. 930
Phuong, Le Minh 1195, 1202, 1501
Piboonwattanakit, K. 1667
Piippo, Antti 406
Pinto, A.J.P. 1123
Pires, A. J. 894, 917
Pires, V. Fernao 894, 1875
Plum, Thomas 327, 333
Pothana, Aravind 1152
Pothi, N. 1208
Pourboghrat, Farzad 1477
Prasad, Dinkar 812
Prasertsit, Anuwat 295
Premrudeepreechacharn, Suttichai 318, 1208
Pusorn, W. 1851

Q

Qian, Zhaoming 632, 1088, 1601
Qu, Yilong 1822, 1880

R

Rafael, Silviano 917
Rahim, Nasrudin Abd 1290
Rahimzadeh, S. 1799
Rahman, Muhammed Fazlur 1284
Rakpenthai, C. 1208
Ramalingam, C.S. 1827
Ramli, M. Z. 1274
Randewijk, P.J. 420, 1744
Rentzsch, M. 556
Ribeiro, Antonio C. 636
Ribeiro, Hugo 643
Ritchie, E. 1167
Rizqiawan, Arwindra 534
Rockhill, Andrew A. 1308
Rong, Runjie 541
Rossouw, F.G. 1295
Rost, J. 1006

S

Sadarangani, Chandur 1012, 1301
Saggini, S. 600
Saha, Bishwajit 504, 563
Saito, Y. 671

Sakulhirirak, D. 515
Salam, Z. 1274, 1363, 1682
Sanajit, N. 1412
Sangampai, Pairote 295
Sangwongwanich, Somboon 1213
Sankar, S. Siva 1632
Sano, Kohji 1595
Saparon, A. 527, 1626
Saritsiri, Kritsada 460
Sato, Akira 1651
Sato, S. 1109
Sato, Terukazu 858, 1423, 1734
Sawatpipat, P. 1840
Sawetsakulanond, B. 1356, 1483, 1489
Schmidt, I. 1115
Schneider, T. 249
Scholler, Tobias 1316
Schroder, D. 1187
Schuster, H. 1187
Sebastiao, Pedro J. 636
Sekine, T. 671
Sekiya, Hiroo 1136
Selvaraj, Jeyraj 1290
Senicar, Florian 1533
Sera, D. 226
Sezgin, Volkan 1507
Shah, Laxman 1182
Shahbazi, M. 1455
Shao, Shiyi 1096
Shariatmadar, S. Mohammad 1782
Sharma, Deepen 973
Sharma, V. K. 480
Shen, C.L. 930
Shi, Hu 949
Shiang, J. Z. 433, 1560
Shibano, Yusuke 265
Shisha, Samer 1012
Shuang, Gao 949, 1697
Shuhua, Fang 270, 427, 745
Silber, S. 257
Silva, J. Fernando 1875
Sim, J. M. 401
Sing, Bhim 941
Singer, A. 781
Singh, Bhim 58, 310, 357, 731, 738, 899, 1520, 1816
Singhal, Varun 1816
Sinha, S. 1352, 1605
Sirisuk, Phaophak 847, 1429
Sirisumrannukul, S. 1495
Skorokhod, Y.Y. 1868
Sode-Yome, A. 937
Soh, C.S. 1082
Soltani, J. 1442, 1449
Somsiri, P. 703
Son, Kwang-Myoung 1471
Songboonkaew, J. 989
Soter, Stefan 1533
Soulard, J. 1301

A-5

Author Index

Sousa, Duarte M. 636, 817
Srisongkram, W. 1851
Stone, D. A. 382
Stone, David 376
Stumberger, R.H. 1346
Su, Ching-Hung 1610, 1712
Su, Y.-H. 433
Subsingha, W. 1851
Sudhakar, S. Bala 1581
Sudmee, W. 1129
Suetsugu, Tadashi 1136
Sugawara, A. 1109
Sugimura, Hisayuki 504, 563
Sukita, S. 968
Sumner, M. 470
Sun, J. Q. 1394
Sun, Yu-Hua 493
Supriatna, E. G. 1267
Suryawanshi, H. M. 1590
Svechkarenko, D. 1301

T

Ta, Minh C. 1434
Tahami, F. 1147, 1465
Tai, Sio-Un 607, 614
Takeda, T. 1109
Takegami, Eiji 280
Tan, K. 1834
Tan, Siew-Chong 1262
Tan, Weipu 1822, 1880
Taniguchi, T. 1686
Tansatit, Tanvaa 342, 697
Tarateeraseth, V. 515
Tarnekar, S. G. 1708
Tayjasanant, T. 1840, 1862
Tedeschi, Elisabetta 760
Tenca, Pierluigi 1308
Teng, Jen-Hao 305
Teng, L. Y. 1041
Tenti, P. 600
Tenti, Paolo 760
Teo, K.K. 1082
Teodorescu, R. 226, 1247
Teshnizi, Hesameddin Mirzaee 983
Theinmontri, Surapon 295
Thrimawithana, D. J. 648, 654
Tiwari, S.K. 941
To, Huu-Phuc 1284
Tolbert, Leon M. 1055, 1228
Tomihisa, Yoshihiro 858, 1734
Tomioka, Satoshi 280
Tomita, H. 671
Trevisan, D. 600
Tsai, Y.T. 660
Tse, Chi K. 1262
Tseng, S. Y. 440, 433, 1321, 1560, 1567
Tsukakoshi, K. 1885

Tsunesada, Ryota...1417
Tungpimonrut, K. 703

U

Ueda, Shigeta...1070
Ulinuha, A. 1834
Uma, G. 595, 1555, 1632
Uyaisom, C. 1794

V

Vadirajacharya, K. 1810
Vaigundamoorthi, M. 1555
Vargas, Ismael Araujo-...........................788, 796
Vasudevan, Krishna..........................1152, 1827
Veerachary, M. 973, 1581
Veszpremi, K. 1115
Vilathgamuwa, D M 1247, 1620
Vinh, Pham Quang 1195, 1202, 1501
Viriya, P. 394, 685
Vishwakarma, Alok 1891
Vogelsberger, M.A. 1346
Volskiy, S.I. 1868
Vorlander, M. 1160

W

Walker, J. A. 1167
Wang, Chengzhi 1636
Wang, Chien-Ming 1610, 1712, 1762, 1767
Wang, Hua 1662
Wang, Peng 541
Wang, Qi 350
Wang, R-J. 420
Wang, Shoufang 1672
Wang, Shuhong 275
Wang, Shun-Chung 546
Wang, Xixi. 1691
Wangsathitwong, S. 1495
Watanabe, Kazushi 280
Watanabe, Takayuki 1136
Wegener, Ralf 1533
Weihrauch, N. C. 886
Welker, Volkmar 1316
Weller, A. 1006
Westermaier, C. 1187
Williams, Barry W............299, 1182, 1242, 1255
Wipasuramonton, P. 703
Wolbank, T.M. 880, 1346
Wong, Man-Chung 607, 614
Wu, Jinn-Chang 493, 824
Wu, Ming-Yi 1703
Wu, T. F 1321
Wu, Xinhui 852

X

Xiaozhong, Liao 949, 1340, 1697

Author Index

Xie, Xiaogao 1601
Xiping, Liu.................................270, 427, 745
Xu, Hai..1142
Xu, Jianxin..852
Xu, Pengwei...842
Xu, Yun 1636

Y

Yachiangkam, Samart.....................................1327
Yang, C. M. 1560
Yang, Yihan..1822, 1880
Yau, Y. T. 338, 692
Yeh, Jim-Chwen 749, 1703
Yeon, Jae Eul 665
Yingkayun, Krisda 318
Yongyuth, N. 685
Yoothanom, N. 515
Yoshida, Takatsugu 1070
Yoshimura, S. 671
Yoshioka, Satoshi 1543
Yossombut, K. 1840
Yun, S. T...401

Z

Zanchetta, P. 470
Zhan, Yuedong 1662
Zhang, Dongyan 994
Zhang, H.B. 299
Zhang, Junming 632
Zhang, Weiping 291, 994, 1330, 1615, 1752, 1777
Zhang, Xiaofeng 1088
Zhang, Xiaoqiang 291, 1777
Zhang, Yanli 350, 354, 1672
Zhao, Xusen..1752
Zhijun, E. 1804
Zhu, G.R. 551
Zhu, Jianguo 275, 510, 1662
Zirn, Oliver 1316
Zolghadri, M.R. 1404
Zolghadri, Mohammadreza...............................983
Zoller, T. 726
Zou, Yunping...1636

A-7